THE PHANEROZOIC GEOLOGY OF THE WORLD II

THE MESOZOIC, B

THE PHANEROZOIC GEOLOGY OF THE WORLD

Editors:
M. Moullade and A.E.M. Nairn

Centre de Recherches micropaléontologiques "Jean Cuvillier", Université de Nice, Parc Valrose, 06034 Nice Cédex (France)
Department of Geology, University of South Carolina, Columbia, S.C. 29208 (U.S.A.)

I. The Palaeozoic
II. The Mesozoic
III. The Cenozoic

THE PHANEROZOIC GEOLOGY OF THE WORLD II

THE MESOZOIC, B

Edited by

M. Moullade

Centre de Recherches micropaléontologiques "Jean Cuvillier"
Université de Nice
06034 Nice Cédex
France

and

A.E.M. Nairn

Department of Geology
University of South Carolina
Columbia, S.C. 29208
U.S.A.

ELSEVIER

Amsterdam — Oxford — New York — Tokyo   1983

ELSEVIER SCIENCE PUBLISHERS B.V.
Molenwerf 1
P.O. Box 211, 1000 AE Amsterdam, The Netherlands

Distributors for the United States and Canada:

ELSEVIER SCIENCE PUBLISHING COMPANY INC.
52, Vanderbilt Avenue
New York, NY 10017

Library of Congress Cataloging in Publication Data
(Revised for vol. 2)
Main entry under title:

The Mesozoic.

    (The Phanerozoic geology of the world ; 2)
    Includes bibliographies and indexes.
    1. Geology, Stratigraphic--Mesozoic. 2. Paleontology
--Mesozoic. I. Moullade, Michel. II. Nairn, A.E.M.
III. Series: Phanerozoic geology of the world ; 2.
QE651.P48 vol. 2 551.7s [551.7'6]   77-18920
ISBN 0-444-51671-4 (U.S. : v. A)
ISBN 0-444-41672-2 (U.S. : v. B)

ISBN 0-444-41672-2 (Vol. IIB)
ISBN 0-444-41584-X (Series)

Printed in The Netherlands

PREFACE

The best laid plans of mice and men
gang oft agly ...

R. BURNS

In this second volume, Mesozoic B, we present a further collection of contributions to the Mesozoic geology of the world. Yet the coverage is still incomplete, a measure of the amount of material which has to be surveyed. The articles presented emphasize the point made in the Introduction to volume A concerning the heterogeneity of style inevitable in the symposium approach. This heterogeneity reflects not only the cultural and scientific backgrounds of the contributing geologists, differences in their professional experiences, but also the nature of the geological evidence from the region being considered, the state of development of geological research in that area, etc... It is not always easy to find authors willing to undertake the onerous task of reviewing the Mesozoic of an appreciable region and this has resulted in several multi-authored contributions. In some cases the authors have been in close communication, in another case we, as editors, had the problem of co-ordination, a further source of heterogeneity.

The general philosophy of a paleogeographic approach and the separation of data from interpretation was outlined to the contributors but the actual form of presentation was the author's or authors' own; indeed it could not and should not be otherwise.

Our task has been one simply of trying to establish broad divisions within each text which are approximately the same from article to article. In this we have been very much helped by the ready co-operation of the contributors.

We have tried to present the articles in a regional context, and thus the first volume covered Asia in part, Africa and Australasia, while this volume covers the Americas in part, the Indian subcontinent s.l.

(India and Pakistan) extending the coverage of Asia, and Antarctica. As in the first volume there are gaps in the coverage, gaps which reflect the compromise between editors, publishers and contributors. We appreciate the consideration and tolerance of publishers and contributors, particularly those contributors whose articles were deferred because of inadequate regional coverage, while we attempted to assemble as comprehensive a group of articles as possible. Yet there are limits to the amount of delay which is tolerable and the present volume arises from this. The result is the absence of papers which have been delayed through difficulties in finding contributors, or illness or work assignments of contributors. We hope to present the missing articles in a third volume, which will also cover Europe, Maghreb, the Ocean Mesozoic floor and deal with broad geological issues affecting the Mesozoic.

Clearly from the foregoing, the success or failure of the Mesozoic of the World must be judged from the complete set rather than from individual volumes. The success it may have will be due to the contributors for their work and the publishers for their technical presentation. It remains for the editors to accept responsibility for the shortcomings. It remains for us to thank the contributors and the publishers for their co-operation and tolerance, and to hope that the reader will find within the followings a key to the appreciation of the fascinating problems posed by the events of the Mesozoic in different parts of the world.

M. MOULLADE
A.E.M. NAIRN
(July 1980)

CONTRIBUTORS

H.R. BALKWILL
Petro-Canada
Calgary, Alta. (Canada)

S.N. BHALLA
College of Science
Department of Geology
Salahdeen University
Arbel (Iraq)

C.M. BROWN
Bureau of Mineral Resources
Canberra, A.C.T. (Australia)

J. BUTTERLIN
Laboratoire de Géologie de l'Amérique latine
Ecole Normale Supérieure
F-9221 Saint-Cloud (France)

D.G. COOK
Geological Survey of Canada
Calgary, Alta. (Canada)

R.L. DETTERMAN
U.S. Geological Survey
Menlo Park, Calif. (U.S.A.)

A.F. EMBRY
Geological Survey of Canada
Calgary, Alta. (Canada)

E. HÅKANSSON
Institute of Historical Geology and Paleontology
Copenhagen (Denmark)

A.A. KURESHI
Staten Island College
Staten Island, N.Y. (U.S.A.)

N. MALUMIAN
Geological Survey of Argentina
Buenos Aires (Argentina)

J.C. MENDES
Institute of Geosciences
University of Sao Paulo
Sao Paulo (Brazil)

W.V. MARESCH
Institut für Mineralogie
Ruhr-Universität Bochum
D—4630 Bochum 1
(Bundesrepublik Deutschland)

A.D. MIALL
Institute of Sedimentary and Petroleum Geology
Calgary, Alta. (Canada)

F.E. NULLO
Geological Survey of Argentina
Buenos Aires (Argentina)

J.P. OWENS
U.S. Geological Survey
Reston, Va. (U.S.A.)

S. PETRI
Museu Paulista da USP, Ipiranga
01000 — Sao Paulo (Brazil)

C.J. PIGRAM
Bureau of Mineral Resources
Canberra, A.C.T. (Australia)

T.P. POULTON
Geological Survey of Canada
Calgary, Alta. (Canada)

V.A. RAMOS
Geological Survey of Argentina
Buenos Aires (Argentina)

A.C. RICCARDI
División Paleozoología Invertebrados
Museo de Ciencias Naturales
1900 La Plata (Argentina)

S.K. SKWARKO
Bureau of Mineral Resources
Canberra, A.C.T. (Australia)

P.N. STIPANICIC
Atomic Energy Organisation of Iran
Teheran (Iran)

M.R.A. THOMSON
British Antarctic Survey
Madingley Road
Cambridge CB3 0ET (Great Britain)

F.G. YOUNG
Home Oil Co. Ltd.,
Calgary, Alta. (Canada)

K. YOUNG
Department of Geology
University of Texas
Austin, Texas (U.S.A.)

CONTENTS

X

Chapter 1

ARCTIC NORTH AMERICA AND NORTHERN GREENLAND*[1]

H.R. BALKWILL, D.G. COOK, R.L. DETTERMAN, A.F. EMBRY, E. HÅKANSSON, A.D. MIALL, T.P. POULTON and F.G. YOUNG

INTRODUCTION

Mesozoic rocks occupy a broad belt across Arctic North America, from northern Alaska, through the Canadian Arctic Archipelago, to eastern North Greenland, and occur also to the south through the northern interior regions of Canada and along the margins of Baffin Bay (Fig. 1). They are found in a wide variety of physiographic settings, from glacier-covered mountains to featureless plains and submerged, ice-bound continental shelves. Thus, our knowledge of the Mesozoic is based on data that are highly variable in quality and quantity from place to place, derived from abundant excellent exposures in some areas, poor exposures in others, and limited well data and seismic profiles in still others. Along the Arctic continental shelf, seismic surveys and well control available in some places indicate that a relatively thick section of Mesozoic strata underlies much of the area. Further information on the Mesozoic will be slow in coming from the continental shelves, where drilling logistics are formidable due to the shifting Arctic ice pack, and Tertiary cover is as thick as 8 km.

In this paper, the Mesozoic history of Arctic North America is outlined from known land geology. None of the numerous regional tectonic models which have appeared in the literature to date is widely accepted by geologists familiar with the rocks, nor can any be adopted with confidence. The interpretations of the Mesozoic stratigraphy have evolved for the various areas mainly through the efforts of national geologic agencies (United States Geological Survey, Geological Survey of Canada and Geological Survey of Greenland). Macrofossils, especially ammonites, are the main basis for age determinations and correlations of the marine sedimentary rocks, although with the recent accumulation of well samples and interest in

nonmarine coal deposits, micropaleontology and palynology are beginning to supply much biostratigraphic data. For most of the region, biostratigraphic control is good and future work will result in only minor changes. In a few areas, where only reconnaissance work has been done, major changes and additions can be expected. Principal biostratigraphic and paleontologic information and references can be found in the geologic reports listed below, and in others by Imlay (1961, 1965, 1976), Tozer (1961, 1967, 1970a), Frebold (1964, 1970, 1975), Jeletzky (1964, 1966, 1970, 1971, 1980), Håkansson (1979), Imlay and Detterman (1973), Detterman et al. (1975), Brideaux and Fisher (1976), Brideaux (1977), Kemper (1977), Poulton (1978b), Davies and Norris (1980), Håkansson et al. (1981b), and Poulton et al. (1981).

In addition to references on the geology of individual areas cited in the text and correlation chart (Fig. 5), the following comprehensive regional syntheses, some of which include geophysical data, are recommended: Ziegler (1969), Brosgé and Tailleur (1970), Tailleur and Brosgé (1970), Detterman (1973), Grantz et al. (1975, 1979), Ahlbrandt et al. (1979), and Dutro (1981) for northern Alaska; Lerand (1973), Norris (1974), Yorath and Norris (1975), Young et al. (1976), Dixon (1981), Norris and Yorath (1981), and Norris (1981) for northern Yukon Territory and adjacent Mackenzie River delta region; Tassonyi (1969) and Yorath and Cook (1981) for the Great Bear Lake to Anderson Plain area; Ziegler (1969), Tozer (1970b), Plauchut (1971), Trettin et al. (1972), Miall (1975), Balkwill and Roy (1977), Balkwill (1978), Christie (1979), Trettin and Balkwill (1979), and Forsyth et

---

*[1]   Published with the permission of the Directors, Institute of Sedimentary and Petroleum Geology, United States Geological Survey, and Geological Survey of Greenland.

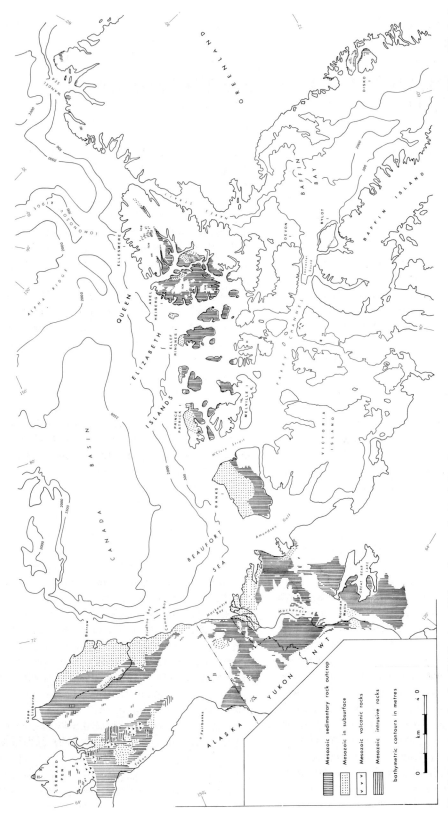

Fig. 1. Index map, showing distribution of Mesozoic strata in Arctic North America. Mesozoic dykes and sills in the essentially sedimentary succession of Queen Elizabeth Islands are not shown.

al. (1979) for the Canadian Arctic Archipelago; Dawes (1976), Håkansson (1979), and Håkansson et al. (1981b) for the Wandel Sea area of eastern North Greenland; and Rosenkrantz and Pulvertaft (1969), Henderson et al. (1976) and Keen and Hyndman (1979) for the Baffin Bay area.

Significant impetus for study of the Mesozoic of the Canadian and American Arctic was provided by the promise of hydrocarbon discovery in the sedimentary basins, which resulted, above all, in the discovery of a major oil field at Prudhoe Bay, northern Alaska, in 1967. Subsequent exploration resulted in important finds of gas in the Mackenzie River delta area and Melville Island, and oil and gas offshore Melville and Lougheed Islands.

RESPONSIBILITIES AND ACKNOWLEDGMENTS

Responsibilities for the content of this paper are approximately as follows: northern Alaska – Detterman; east-central part of Alaska and west-central part of Yukon – Detterman, Poulton; northern Yukon and Mackenzie River delta area – Young, Poulton, Embry; Great Bear Lake and Anderson Plain areas – Cook; Banks Island – Miall; Queen Elizabeth Islands – Balkwill, Embry; Bylot Island – Balkwill; central West Greenland – Balkwill, Håkansson, Poulton; northern Greenland – Håkansson; synthesis and co-ordination – Poulton, Embry, Balkwill. Colleagues of the Institute of Sedimentary and Petroleum Geology at Calgary and the Branch of Alaskan Geology at Menlo Park provided valuable information, advice, and ideas; D.K. Norris, Hillard Reiser and J.H. Wall critically read the manuscript. Drafting is by Dianna Campbell, Poulton and Balkwill.

GEOLOGIC SETTING

The Mesozoic history of the area is intimately related to the evolution of the Arctic Ocean and the northern Cordillera but at present there is little agreement among earth scientists regarding their mutual relations (see Vogt and Avery, 1974; Clark, 1975; Fujita, 1978). Lack of data from the critical areas of Canada Basin and the adjacent continental shelves makes all of the hypotheses to date speculative, and each has aspects which cannot be reconciled easily with known geology on land. Of particular significance

is the abundant evidence for an extensive pre-Cretaceous sediment source north of Alaska and possibly northwest of the Canadian Arctic Islands, in the area of the continental shelf and Canada Basin. The evidence and implications of this source terrane are briefly summarized in the discussion of the Mississippian to mid-Early Cretaceous tectono-stratigraphic regime.

Significant dislocation and juxtaposition of major geological terranes by strike-slip faulting, in large part during Mesozoic time, is becoming increasingly evident in rocks in the Cordilleran Orogen. Movements of this sort also must have played a major role in the Mesozoic geology of northern Alaska and Yukon (e.g. Richards, 1974; Norris and Yorath, 1981), and of North Greenland (Håkansson and Pedersen, 1981, and in prep.).

The Phanerozoic rocks of Arctic North America rest on Proterozoic sedimentary strata in some places and on older Precambrian crystalline rocks in others. They can be subdivided conveniently into three tectono-stratigraphic assemblages of approximately Cambrian to Missisippian, Mississippian to mid-Early Cretaceous and mid-Early Cretaceous to Tertiary ages, which correspond in concept with the Franklinian, Ellesmerian, and Brookian tectonic assemblages of Lerand (1973) and Norris and Yorath (1981). Each assemblage was dominated by a characteristic suite of tectonic and stratigraphic styles.

Cambrian to Mississippian regime

The early Phanerozoic regime in its early stage comprised a stable carbonate platform in the south and southeast, and a subsiding basin (Franklinian Geosyncline) receiving mixed carbonate and clastic sediment to the north and northwest (Churkin, 1969; Trettin et al., 1972; see Fig. 2). A tectonically active, in part volcanic, geanticline (Pearya) was present in the area of the northeastern Canadian Arctic Archipelago (Trettin, 1973). From Middle Devonian to Early Mississippian time from place to place, parts of the Franklinian Geosyncline were deformed and thrust southward, southeastward and southwestward, and thick successions of coarse clastic rocks (Brosgé and Dutro, 1973; Embry and Klovan, 1976) were deposited in front of the uplifted advancing orogens. Erosional denudation or subsidence of the deformed Franklinian rocks marks the end of this tectonostratigraphic regime.

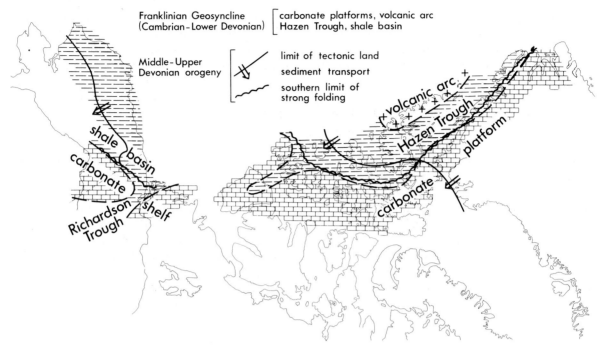

Fig. 2. Major tectonic provinces, Cambrian to Mississippian tectonic regime (after Churkin, 1969; H.P. Trettin, pers. comm., 1979; Embry and Klovan, 1976; Brosgé and Dutro, 1973). Franklinian provinces exemplified by Ordovician; Ellesmerian provinces by Frasnian and Fammenian (Upper Devonian) rocks.

Mississippian to mid-Early Cretaceous regime

In Early to Late Mississippian time from place to place, major depositional basins developed over areas underlain by deformed Franklinian strata: the Brooks-Mackenzie Basin of northern Alaska and adjacent mainland Canada, the Sverdrup Basin of the Canadian Arctic Archipelago, and the Wandel Sea Basin in eastern North Greenland (Fig. 3). The basins accumulated sediment as thick as 8000 m locally. Upper Paleozoic fill consisted of carbonate rocks, evaporites and clastic rocks whereas the lower Mesozoic fill was almost entirely clastic rocks, derived from relatively stable upland and cratonic source areas. The basins maintained their approximate positions throughout this regime and for the most part appear to have coincided in their extent with the limit of presently preserved sedimentary rocks.

The stable craton, with flat-lying to gently deformed Precambrian and lower Paleozoic strata and crystalline rocks was a major source of sediments during Mesozoic time.

The Brooks-Mackenzie Basin (new name coined to indicate the inferred integrity of the Alaskan "Brooks Geosyncline" and Canadian "Beaufort-Mac-

kenzie Basin" of previous authors, in late Paleozoic and early to middle Mesozoic times) contains a more or less concordant succession of Mississippian to Hauterivian strata, although onlap relations are common along the margins, and unconformities occur over intrabasinal arches. The interpretation of the essential unity of the basin ignores the probability that essentially similar rocks deposited in two entirely separate sedimentary basins may have been subsequently juxtaposed by large scale lateral motion on transcurrent faults. The basin margins against the North American craton to the southeast and east, and against an apparently extensive landmass to the north, are well documented in the stratigraphic record. The southern margin of the basin in pre-Late Jurassic time is enigmatic. A tectono-volcanic upland in central Alaska ("ancestral Brooks Geanticline"), which was a notable sediment source in Late Jurassic time, may have formed the southern margin of the basin earlier, although its position, and even its very existence during the Triassic, is speculative. Contrasting interpretations include those of Imlay and Detterman (1973) and Jeletzky (1975, 1977), who illustrated uplands in central Alaska and western parts of northern Yukon respectively, on the one hand, and Poulton et al.

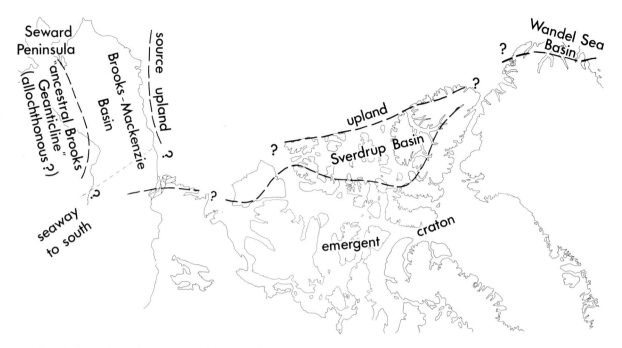

Fig. 3. Major tectonic provinces, Mississipian to mid -Early Cretaceous tectonic regime.

(1981) who emphasized the lack of evidence for such an upland prior to the Late Jurassic, on the other. The latter interpretation, which assumes some manner of emplacement of allochthonous terranes and no autochthonous central Alaska landmass prior to the Late Jurassic, is illustrated in this paper. Parts of this "upland" area, which was volcanically active in Jurassic time (Reiser et al., 1965) and has also yielded Neocomian radiometric (K-Ar) dates on volcanic and intrusive rocks (Detterman, 1973), may have drifted, as part of an Asian plate, into their present position and become juxtaposed against the remainder of northern Alaskan terrane during the Jurassic (Richards, 1974). Seward Peninsula, as part of a plate together with Chukotski Peninsula of Siberia, may be yet another allochthonous terrane that was emplaced in mid-Cretaceous time (Sainsbury and Bloomstein, 1977).

Connection of Brooks-Mackenzie Basin with the Mesozoic seas of western Canada via a narrow seaway through central or west-central Yukon has been speculated by many authors (e.g. Imlay and Detterman, 1973; Young et al., 1976; Jeletzky, 1977). However, others (e.g. Frebold, 1957) have repeatedly pointed out the lack of evidence for such a connection prior to Cretaceous time, and still others (e.g., Young et al., 1976; Poulton, in Poulton et al., 1981) showed that craton-connected shoreline clastic deposits extend southwestward nearly entirely across the northern Yukon. Thus whatever connections that existed between the Arctic and more southerly seas, must have been farther west, and presumably lay in Alaska where a landmass had been speculated to have been present by other authors. However, geologic relations in those areas are still speculative largely owing to poor exposure in some areas and to erosional removal of many of the pre-Barremian rocks in west-central Yukon and adjacent Alaska. Geologic understanding of this area, nevertheless, is of critical importance to any regional tectonic schemes that involve strike-slip faulting or rotation of Alaska (see below).

Sverdrup Basin is a northeast-elongated structural depression, 1300 km by 400 km, containing a concordant succession of upper Paleozoic and Mesozoic strata (Balkwill and Roy, 1979). It developed probably by graben-formation over deformed Proterozoic and lower Paleozoic strata in Late Mississippian time. Its formation may be related to post-orogenic thermal contraction and subsidence following the Franklinian orogeny (Sweeney, 1977). Mesozoic sedimentation in Sverdrup Basin was influenced by local evaporite diapirism and accompanied by basic sill and dyke intrusion that was widespread in the eastern part, and some basaltic volcanism (Balkwill, 1978; inset C of Fig. 8). Sverdrup Basin was bounded to the east and south by the stable platform and craton (see also Christie, 1979). The crustal structure has been interpreted by Forsyth et al. (1979) from various types of geophysical data.

The Wandel Sea Basin forms a NW–SE trending cover of the Franklinian–Caledonian junction in eastern North Greenland. Upper Paleozoic strata are widespread in the basin, whereas Mesozoic strata are present as local, structurally determined remnants, in part representing minor sub-basins. The Wandel Sea Basin was bounded to the southwest by the stable platform and craton and was probably juxtaposed to the Svalbard–Barents Sea complex through the Mesozoic prior to the opening of the North Atlantic Ocean (e.g., Irving, 1973; Smith and Briden, 1977; Christie, 1979).

The similarities of the marine faunas indicate long-lasting connections between Brooks-Mackenzie, Sverdrup, and Wandel Sea basins and with seas of Siberia, East Greenland and the North Sea although each of the North American basins is described here as a separate depositional basin during the early Mesozoic. A connection between Sverdrup Basin and Brooks-Mackenzie Basin from Late Jurassic time on is represented through Banks Island by as much as 400 m of pre-Barremian marine strata (Miall, 1975).

The source upland north of Alaska (Fig. 3) gave rise to Permian and Lower Triassic clastic wedges, comprising large volumes of mature quartz sandstone and siltstone that prograded southward in northern Alaska. It may be further indicated by eastward-overstepping relations of Triassic and Jurassic rocks in northwestern Yukon (Norris, 1974).

The existence of a structurally high rim bounding Sverdrup Basin on the northwest is indicated by unconformities, facies relations, and thickness trends in the Triassic to Hauterivian strata, but the rim was not a major sediment source (Meneley et al., 1975). Until stratigraphic information from the continental shelves and Canada Basin is obtained, the extent of this rim, and the age and origin of Canada Basin to its northwest will remain in question. The range of speculations is: Canada Basin may have been present and may not have been associated with significant lateral plate motion since Proterozoic (Meyerhoff, 1973), or Paleozoic (Churkin, 1973) time, and it may have been surrounded by an uplifted continental edge which subsequently submerged to form the continental shelves; Canada Basin may occupy the site of an earlier vast continental crustal block which subsided (Eardley, 1961), was rifted, and drifted away (Vogt and Ostenso,

1970; Herron et al., 1974; Yorath and Norris, 1975; Christie, 1979; Kerr, 1980), or fragmented by earth expansion (Carey, 1976); Alaska may have rotated counterclockwise away from northern Canada, the newly opened ocean being Canada Basin, and the highland (comprising Franklinian strata and possibly part of the North American craton) that was a sediment source area for northern Alaska having submerged to form the continental shelves (Carey, 1955; Hamilton, 1970; Tailleur, 1973; Newman et al., 1977; Grantz et al., 1979). Other more complex plate-tectonic schemes, partly involving combinations of the above hypotheses, have been proposed to account for the land area to the north (e.g., Fujita, 1978). In some plate-tectonic reconstructions (e.g., Irving, 1977; Smith and Briden, 1977), no cratonic landmasses lay immediately north or northwest of the North American Arctic throughout the Mesozoic.

Mid-Early Cretaceous to Late Cretaceous regime

Tectono-stratigraphic patterns characterizing the latest tectonic regime of the Mesozoic were clearly developed by Hauterivian to Barremian time (Fig. 4). They involved major deformation and uplift of Brooks Geanticline and the Cordilleran Orogen in northern Alaska and Yukon; subsidence of deep and narrow foredeep troughs north and east of those uplifts; disappearance of the significant northern landmass in the Canada Basin area and development of North Chukchi Basin off northwestern Alaska; transgression beyond the former basin margins far southward across the North American craton; sedimentation in the Baffin Bay area; accumulation of a thick pile of largely nonmarine sedimentary and volcanic rocks in Kobuk Basin of western Alaska, and development of Hope Basin west of it; and development of a distinct volcanic province in the western part of Wandel Sea Basin, and widespread deformation of this basin.

Throughout the Cretaceous, uplifts of the Cordilleran Orogen and Brooks Geanticline, and clastic wedges derived from them, migrated northward and northeastward in northern Alaska and Yukon. The name "Keele-Old Crow Landmass" has been used by previous authors (Jeletzky, 1975) for that part of the uplift in northwestern Yukon and northeastern Alaska. Brooks Geanticline is dominated by northerly-directed thrust faults and local mafic and felsic intrusions. The earliest sediments derived from these uplifts are Late Jurassic in age. Particularly significant

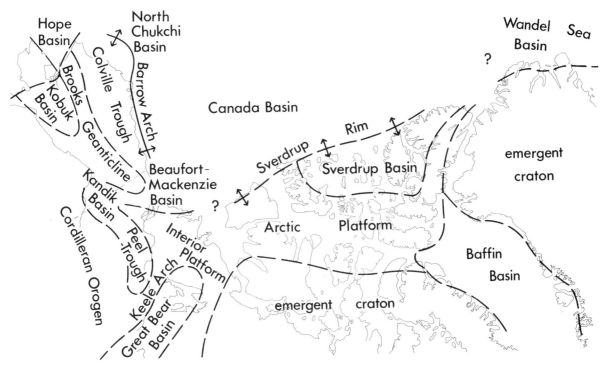

Fig. 4. Major tectonic provinces, mid-Early Cretaceous to Late Cretaceous tectonic regime. Subdivisions of Beaufort-Mackenzie Basin which are too small to appear on this map include, from west to east: Blow Trough, Cache Creek High, and Kugmallit Trough of Young et al., 1976 (see also Fig. 7).

uplift and eastward and northward expansion are indicated during the Barremian and Aptian by widespread clastic wedges that are particularly poorly sorted relative to underlying strata. Throughout the northern Cordilleran Orogen, the Mesozoic record is obscured by subsequent deformation, tectonic juxtaposition of dissimilar facies and erosion.

Kobuk Basin, an epi-orogenic depocentre south of the Brooks Geanticline in northwestern Alaska, received much volcanic fill as well as nonmarine clastic sediments from Aptian to Santonian time (see Patton, 1966, 1967). Radiometric (K-Ar) dates on intrusive rocks range from 82 to 100 m.y. (Detterman, 1973). Hope Basin is thought to have developed in Late Cretaceous time (Grantz et al., 1975).

A complex of strongly subsiding foredeeps in front of the geanticlines formed depocentres for flyschoid detritus derived from the tectonic highlands. They have various local names, including Colville Trough in northern Alaska, Blow Trough (western part of Beaufort-Mackenzie Basin) in northern Yukon, Kandik Basin in east-central Alaska, and Peel Trough farther southeast (see Fig. 4). The underlying rocks were disrupted by development of arches, horsts, and grabens,

not associated with volcanism, which established the relief over which Barremian and younger sediments were deposited.

Sverdrup Basin continued to be the main depocentre during this time interval in the Canadian Arctic Archipelago but, owing to transgression onto the Arctic Interior Platform, it was no longer a discrete sedimentary basin with strandline deposits. The Mesozoic rocks in the axial part of the basin were pierced by halokinetic diapirs.

The thick basinal successions contrast with thinner successions on a series of uplifts paralleling the edge of the continental shelf from Alaska (Barrow Arch) to the Canadian Arctic Islands (Sverdrup Rim) from Barremian at least into Cenomanian time (e.g., Fig. 6). Pre-Barremian Mesozoic rocks were largely eroded over Barrow Arch (Rickwood, 1970), and the younger Mesozoic succession is thicker and more complete to its south. The oceanward (north) side of the arch (i.e., North Chukchi Basin) is seen in seismic profiles off the northwest coast of Alaska (Grantz et al., 1975). The unstable craton margin on Banks Island and vicinity comprised small horsts that appear to be a part of this series of uplifts. Here also they were clearly

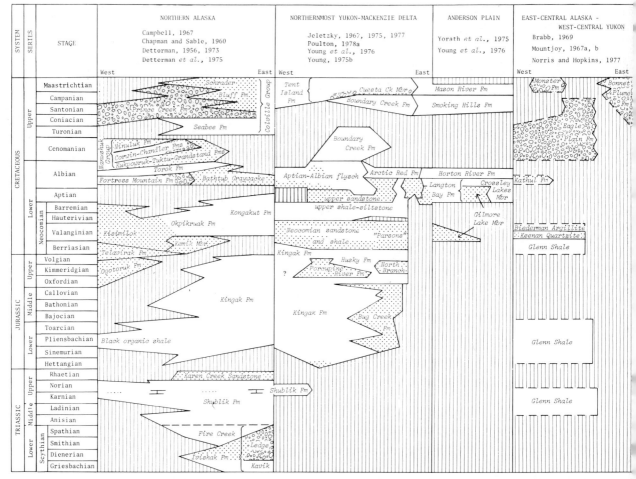

Fig. 5. Correlation chart, Mesozoic strata of Arctic North America and northern Greenland. Localities are shown on Fig. 1. Patterns indicate rock types as in Fig. 6, except pebbly pattern indicates various nonmarine clastic lithologies; vertical ruling-rocks absent or not identified.

two-sided, intermittently shedding sediments both oceanward and continentward during Barremian to Maastrichtian time (Miall, 1975, 1979).

Whatever extensive northern sediment source there was previously seems to have largely disappeared by mid-Early Cretaceous time, and all or a portion of the Canada Basin presumably was present. However, uplift which began in the Maastrichtian in northern Axel Heiberg and Ellesmere Islands contributed sediment to the south at this time (Fig. 17).

The emergent craton of North America continued to be a source area throughout Cretaceous time, although its sediment contribution was overshadowed in the west by that of the tectonic highlands. Those parts of the craton which experienced intermittent subsidence, transgression, and sedimentation from

Aptian to Maastrichtian time, are known as the Interior and Arctic Platforms (see Fig. 4). These shallow marine shelf areas were a complex of local, intermittent basins and arches each of which experienced its own depositional-erosional history (e.g. Miall, 1975; Yorath and Cook, 1981). Seaways extended southward into the western interior region of North America, and eventually to the Gulf of Mexico in Albian and Late Cretaceous time (Jeletzky, 1971; Williams and Stelck, 1975). The Interior Platform was uplifted in latest Cretaceous time, however.

Baffin Basin developed as part of the North Atlantic rift system during the Mesozoic. Although rifting of Baffin Bay—Labrador Sea area may have begun as long ago as Late Triassic time (see McMillan, 1973 and references there), the oldest known sedimentary

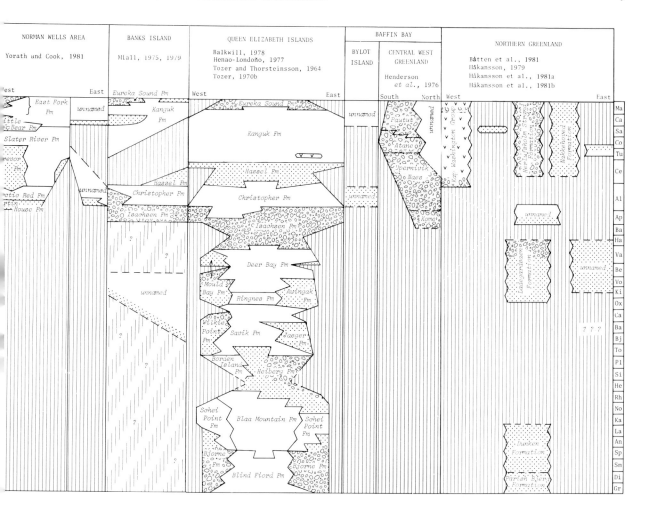

rocks (nonmarine) adjacent to Baffin Bay are perhaps Barremian (Pedersen, 1968), and the oldest marine sedimentary rocks are Cenomanian or Turonian (Birkelund, 1965). The basin is elongate in a northwest direction, parallel to the main trends in the underlying Precambrian crystalline rocks, and terminating where these trends diverge from a northwesterly direction. Almost the entire basin is offshore, and data are meager; Keen and Hyndman (1979) have discussed the geophysical data and interpretations most comprehensively.

By Late Cretaceous time a series of local marine and nonmarine basins were rapidly developed, infilled, and subsequently deformed in a NW—SE trending strike-slip mobile belt dissecting the center of the Wandel Sea Basin (Håkansson and Pedersen, 1981, and

in prep.). Deposition in two of the marine basins has been dated as Turonian–Coniacian and Santonian, respectively, whereas stratigraphic data are meager from the non-marine basins (Håkansson et al., 1981b). Basic sill and dyke intrusions of mainly Cretaceous age (Soper et al., 1981) have been restricted to the westernmost part of the basin, where also several kilometers of per-alkaline extrusives were accumulated in the Upper Cretaceous (Brown and Parsons, 1981; Batten et al., 1981).

Stratigraphic relationships

In summary of the foregoing, the Mesozoic spans two tectono-stratigraphic regimes, the late part of a Mississippian to mid-Early Cretaceous regime (Ellesmerian) and the early part of a subsequent mid-Early Cretaceous to Holocene regime (Brookian). Triassic to Hauterivian strata are confined to distinct sedimentary basins which developed in Mississippian time, in which the strata mostly are concordant on a thick succession of upper Paleozoic strata. In contrast, Barremian and younger Cretaceous strata are commonly widespread beyond the previous basin margins, where they are unconformable on a great variety of Paleozoic and Precambrian sedimentary and crystalline rocks. A similar stratigraphic pattern for the Mesozoic has been recognized worldwide (Kent, 1977).

A correlation chart and three regional cross-sections (Figs. 5–8) illustrate the most significant relations of the Mesozoic rocks in the sedimentary basins.

In northern Alaska (Fig. 6), the lower Mesozoic assemblage (below the Okpikruak and Kongakut formations approximately) is thin compared with the upper assemblage, totalling no more than 1200 m. It consists mainly of clastic rocks of northerly provenance and is truncated to the north, whereas the upper assemblage consists of clastic wedges, as thick as 7000 m, which were derived from tectonic highlands to the south, and prograded northward, thinning in that direction. The upper assemblage oversteps the lower to the north and extends onto the continental shelf area.

In northern Yukon and adjacent Mackenzie River delta area (Fig. 7), the lower assemblage (below Aptian and Albian flysch) displays gradual and irregular onlap to the southeast with the oldest Mesozoic strata ranging in age from Norian to Berriasian. The strata of the lower assemblage consists mainly of shallow water clastic rocks derived from the south and southeast. Normal faults in part controlled the southeastern basin margin during Jurassic and Neocomian time. Some of these are indicated by marked thickness changes in the "Parsons Sandstone". This informally named unit and adjacent ones in the Mackenzie delta area are being named formally and redescribed by Dixon (1981). The strata of the upper assemblage were derived mainly from tectonic uplifts to the southwest. They are thinner and finer grained away from these highlands. The line of the cross-section (Fig. 7) cuts the northeastern salient of this highland (Brooks Geanticline), and crosses an intra-basinal arch within Beaufort-Mackenzie Basin (Cache Creek High) over which the top portion of the lower assemblage is absent and the upper assemblage is thinner owing to onlap and the locally slower rate of subsidence. The

strata of the upper assemblage extend to the southeast far beyond the limits of the lower assemblages.

In Sverdrup Basin (Fig. 8), the lower assemblage (lower part of the Isachsen Formation and older) is markedly thicker than the upper assemblage, in contrast to what appears on the preceding cross-sections. Two alternating sedimentary patterns are evident: marginal sandstone with equivalent, thicker, basinal shale and siltstone (e.g., Bjorne and Blind Fiord formations); and basinwide sandstones which thicken toward the basin centre (e.g., Heiberg Formation). The upper assemblage consists of two major cycles (upper Isachsen, Christopher, and Hassel formations; Kanguk and Eureka Sound formations) which represent two basin fill episodes. This upper assemblage extends far beyond the limits of the Sverdrup Basin.

PALEOGEOGRAPHY AND GEOLOGIC HISTORY

The major sedimentary facies and tectonic elements of Arctic North America during the Mesozoic are illustrated by facies maps for representative time intervals (Figs. 9–17). None are restored palinspastically because of the absence to date of a widely accepted model for restoration; such a restoration probably would change the configurations in northern Alaska and Yukon significantly, and may well alter the relations of these areas with the Arctic Archipelago. Each of the timestratigraphic subdivisions into which we have divided the discussion of paleogeography and geologic history begins with a regional transgressive event recognized throughout Arctic North America.

Early Triassic rocks

The Mesozoic Era in Arctic North America began with a transgression in earliest Triassic time which followed a major latest Permian regression that left nearly the entire region emergent (Bamber and Waterhouse, 1971; Nassichuk et al., 1972; Detterman et al., 1975). The earliest sedimentation was confined to the basinal areas, where clearly defined marginal and basinal facies were established (Fig. 9). The marginal deposits consist mainly of sandstone with subordinate conglomerate and red shale, and the basinal deposits are interbedded shale and siltstone.

Lower Triassic strata of the Brooks-Mackenzie Basin are confined to the Ivishak Formation in the eastern and central parts of the Alaskan portion of the basin. There, Griesbachian shale and siltstone (Kavik

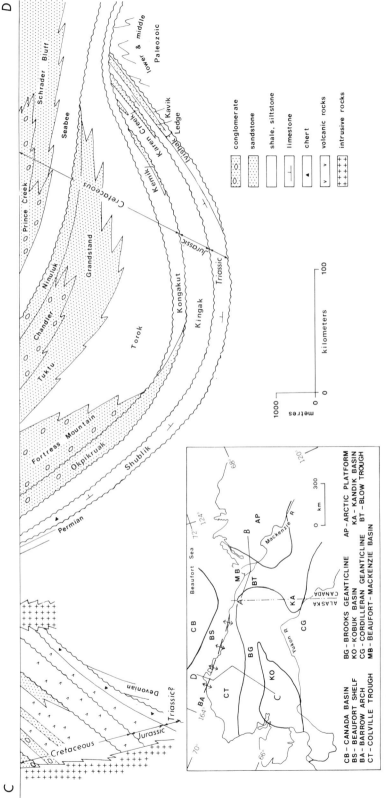

Fig. 6. Schematic stratigraphic cross-section (C—D) of Mesozoic strata across northern Alaska and locality map for this cross-section as well as for that of Fig. 7 (A—B).

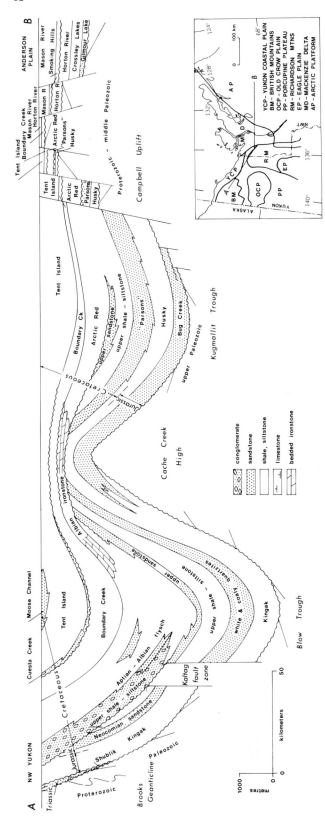

Fig. 7. Schematic stratigraphic cross-section of Mesozoic strata across northern Yukon and northwestern District of Mackenzie (Northwest Territories) of Canada. Location of cross-section shown as A–B on Fig. 6.

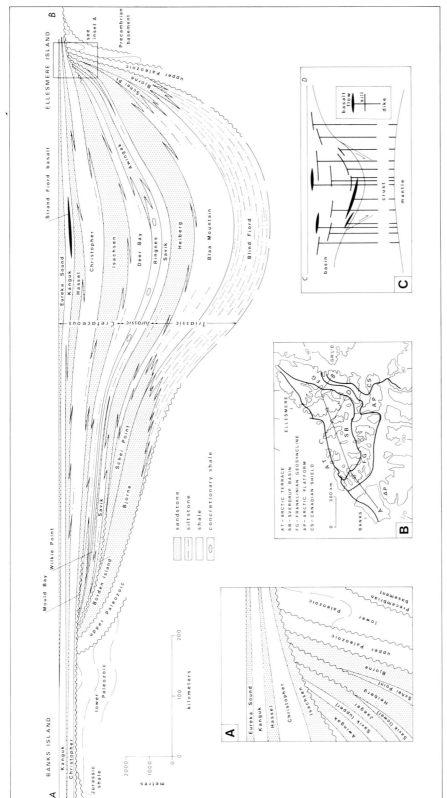

Fig. 8. Schematic stratigraphic cross-section (A–B) of Mesozoic strata across Canadian Arctic Archipelago. Inset A: detailed stratigraphic relationships on Ellesmere Island; inset B: locality map for cross-section A–B and for schematic crustal cross-section C–D; inset C: schematic crustal cross-section across Sverdrup Basin, showing attenuation below the basin and the abundant dykes and sills and flows within the upper Late Paleozoic through Mesozoic sedimentary rocks.

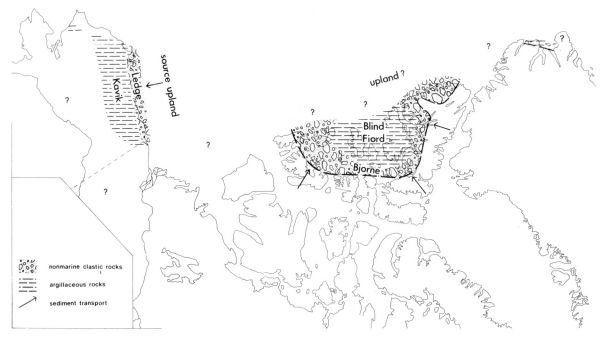

Fig. 9. Lower Triassic (Smithian) facies distribution map.

Member) are overlain by Dienerian to Spathian fluvial-deltaic sandstone and conglomerate (Ledge Sandstone Member) that grade southward from a northern land-mass into prodelta and deep marine, argillaceous strata, which partly resemble turbidites. Spathian deep marine shale and siltstone (Fire Creek Siltstone Member) overstepped the deltaic sandstone. The total Lower Triassic succession in the area is about 400 m thick.

A more substantial record is preserved in Sverdrup Basin where as much as 1500 m of Lower Triassic strata are present. The main source areas for this basin were the cratonic regions to the east and south. A thick succession of sandstone and conglomerate of fluvial and littoral origin rims the eastern and southern margins (Bjorne Formation). Toward the basin centre, this gives way to siltstone and shale (Blind Fiord Formation) (Fig. 9), which also extend as a basal shale to the northeastern basin margin. In the eastern part of the basin this facies change is very abrupt; it maintained a relatively fixed position throughout Early Triassic time. Also in that part of the basin the basinal strata have attributes of turbidites suggesting deep water conditions and the presence of a significant marine slope. They thin to the north, to less than 300 m at northern Axel Heiberg Island. In contrast, in western Sverdrup Basin, strandline and fluvial deposits (Bjorne Formation) prograded northward over argillaceous shallow shelf deposits (Blind Fiord Formation).

No evidence of deep water conditions during the Early Triassic has been noted in the west. The only indication of the existence of a northern source during this time interval is a gritty sandstone unit of Dienerian age at northern Axel Heiberg Island (Tozer, 1970).

In the Wandel Sea Basin the Permo-Triassic boundary probably lies somewhere in the top of the marine, virtually barren Parish Bjerg Formation, where the top sands have yielded both a few gastropods of Paleozoic affinity and a well preserved, fairly diverse Lower Triassic micro-flora (Håkansson, 1979). The boundary formation is overlain by approximately 100 m of marine shale and sandstone (lower part of Dunken Formation) containing Scythian (?Smithian) fossils (Kummel, 1953; Håkansson, 1979). The very restricted occurrence of Triassic sediments does not permit any paleogeographic reconstruction for this basin.

Middle and Late Triassic rocks

In the earliest Middle Triassic a major transgressive episode began throughout Arctic North America, reaching its maximum extent by Norian time. In Rhaetian time the sea retreated, resulting in a regional unconformity above Triassic strata.

On the northern flank of the Brooks-Mackenzie Basin the transgressive Middle and Upper Triassic strata are represented by interbedded shale, siltstone,

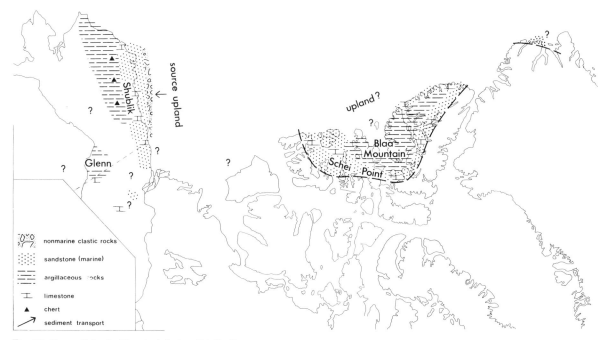

Fig. 10. Upper Triassic (Karnian) facies distribution map.

coquinoid limestone and sandstone (Shublik Formation) of marine shelf origin. The clastic sediments are of northern derivation; strandline deposits equivalent to these strata are presumed to occur farther north under the continental shelf (Fig. 10). In Alaska the Shublik Formation, 150 m thick, spans the entire Anisian to Norian interval whereas, to the east in northern Yukon, the Shublik Formation is thin, entirely Norian in age, of shallow marine character and rests unconformably on Precambrian through upper Paleozoic strata (Mountjoy, 1967a). Thus the area of sedimentation expanded both eastward and northward during Triassic time. Farther south in Alaska, Middle and Upper Triassic strata are mainly shale and chert (Shublik Formation, Glenn Shale) of deep water origin. The southern margin of the basin has not been defined but must have been considerably south of the present Brooks Range. Subsequent Upper Norian to Rhaetian regressive deposits along the northern rim of Brooks-Mackenzie Basin in northern Alaska are represented by a thin sheet sandstone (Karen Creek and Sag River Sandstones) of northerly derivation. Uppermost Triassic strata are unknown in the basin.

The Middle and Upper Triassic strata of Sverdrup Basin are thicker and more extensive than those of the Brooks-Mackenzie Basin. Along the southern and eastern margins of the basin, as much as 500 m of interbedded sandstone, siltstone, shale, and coquinoid limestone (Schei Point Formation) of marine shelf origin represent the Anisian to Norian Stages. In a few areas, littoral sandstones of Middle Triassic age occur near the basin edge, where they are overlain and overstepped by Karnian and Norian marine shale and limestone. In the basin, Anisian to Norian strata comprise as much as 2500 m of silty shale (Blaa Mountain Formation) of marine shelf and slope origin. Toward the northwestern margin of the basin, equivalent strata are thinner, and Karnian sandstones, apparently derived from a northern landmass, become common in the section. Uppermost Triassic beds are a regressive sequence of fine grained quartzose sandstone of Norian and Rhaetian age (Lower Heiberg Member). These strata are mainly of delta-front origin and were derived mainly from the east and southeast.

By Middle or Late Triassic time, a sufficient thickness of strata was present in the axial portion of Sverdrup Basin to mobilize underlying Carboniferous evaporites (Balkwill, 1978). The formation of evaporite-cored swells and complementary structural troughs had a marked effect on Middle and Late Triassic sedimentation, resulting in abrupt thickness variations within these strata.

An enormous amount of clastics reached Sverdrup Basin in latest Triassic time, essentially filling the basin, and the Triassic-Jurassic boundary in the basin centre lies in a nonmarine sandstone facies (Upper

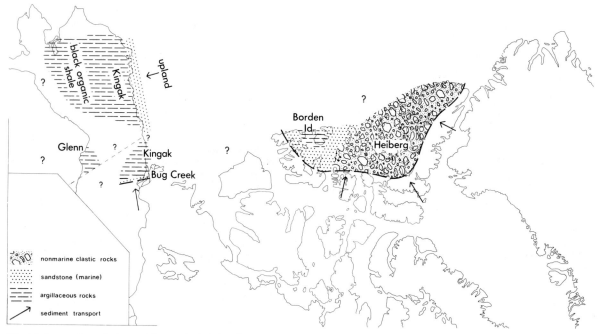

Fig. 11. Lower Jurassic (upper Sinemurian) facies distribution map.

Heiberg Member; Fig. 5).

Middle and possibly Upper Triassic strata comprise the upper 500 m of the Dunken Formation in the Wandel Sea Basin. The marine sandstone sequence with minor shale intercalations has yielded ammonites indicative of the lowermost Anisian zone from its lower part (Kummel, 1953; Håkansson and Heinberg, 1977).

Early to early Late Jurassic rocks

A complex history of transgressions and regressions with intermittent erosional episodes is represented by Hettangian through Lower Oxfordian basinal shale and marginal sandstone units. Some of the sandstones extended far across the basins (Fig. 5).

In northern Yukon and adjacent Mackenzie delta, strandline and marine shelf sandstones are preserved along the southeastern margin of Brooks-Mackenzie Basin (Bug Creek Formation; Jeletzky, 1967; Poulton et al., 1981; Dixon, 1981; Fig. 11). A complex history of shoreline migration and local tectonic control of sedimentation is recorded here. The basal transgression is dated in Brooks-Mackenzie Basin by Hettangian fossils at a few localities (Imlay and Detterman, 1973; Frebold and Poulton, 1977), but Sinemurian sandstone and shale appear to form the basal Jurassic deposits over much of the basin. The most extensive craton-ward transgression was in Bajocian to Bathonian

time; the most extensive regressive episodes indicated by basin-ward deposition of sandstone were in Pliensbachian and Early Oxfordian times (Poulton, 1978a). Northwestward, in northern Yukon and northern Alaska, the sandstones grade into shale and siltstone (Kingak Formation) which are as thick as 1200 m (Poulton, in Norris, 1981). Shale (Glenn Shale) and sandstone in east-central Alaska and west-central Yukon may represent marine connections with the Jurassic seas of the Cordillera to the south, but their geologic relations are not yet resolved.

A basal sandstone and glauconitic sandstone interbeds within the shale along the Arctic Coast area of Alaska were deposited near the northern margin of the basin. A few bentonite beds interlayered in the Kingak Formation of Alaska may be the product of the extensive Jurassic basic igneous activity that occurred contemporaneously in the landmass south of Brooks-Mackenzie Basin (Reiser et al., 1965).

In the Sverdrup Basin, Lower and Middle Jurassic strata are intertongued sandstone, siltstone and shale, as thick as 1200 m. No unequivocally Hettangian fossils have been recovered to date, although Hettangian strata are most likely present within the basin. Progressively younger strata form the basal transgressive deposits toward the basin margins, where Toarcian sandstones are the youngest basal strata preserved (Tozer and Thorsteinsson, 1964). From Hettangian to

Fig. 12. Middle Jurassic (middle Bajocian) facies distribution map.

Pliensbachian time, deltaic deposition occurred over much of the eastern and central Sverdrup Basin with thick quartzose sandstones deposited in delta-front and delta-plain environments (Upper Heiberg Member). To the west the deltaic sandstones grade into marine argillaceous sandstone, siltstone and shales (Borden Island Formation, lower part of Savik Formation), with fine grained sandstones of littoral origin occurring along the southwestern and northwestern margins (Borden Island Formation) (Fig. 11). Continued transgression eventually drowned the delta in Toarcian time. During the Toarcian and Bajocian, sand deposition in littoral and near-shore environments was restricted to the basin margins (Wilkie Point Formation, Jaeger Formation), while offshore shale and siltstone equivalents accumulated over much of the basin (lower part of Savik Formation) (Fig. 12). During a regressive period in Bathonian, Callovian and possibly Early Oxfordian time, glauconitic sands were deposited over a wide area and the basin margins underwent some erosion.

Previous reports of Bathonian fossils from the easternmost part of the Wandel Sea Basin (Greenarctic Consortium in Dawes, 1976) are questioned by Håkansson et al. (1981b), and the presence of Lower and Middle Jurassic strata have yet to be documented in this basin.

Late Jurassic to Early Cretaceous (Hauterivian) rocks

A major regional transgression in Oxfordian time advanced the shorelines over the former depositional limits. This pattern persisted through Late Jurassic to earliest Cretaceous time, and a major regressive episode followed in late Valanginian and Hauterivian time. Although later erosion has restricted Oxfordian through Valanginian strata almost exclusively to the basins, some occurrences of these strata on the craton to the south and east of the basin (e.g. Yorath et al., 1975; Miall, 1975) indicate their originally widespread character.

Over much of the Brooks-Mackenzie Basin, the preserved Upper Jurassic strata (Jeletzky, in Norris, 1981; Dixon, 1981) are shale and siltstone of offshore marine origin (upper part of Kingak Formation, Husky Formation, Glenn Shale) (Fig. 13). An Oxfordian to lower Volgian shallow marine and possibly deltaic sandstone unit in northern Yukon (Porcupine River Formation; Jeletzky, 1977), extended far into the basin and is as thick as 550 m. This unit probably had its source in cratonic regions to the south or southwest (Young, 1975a; Young et al., 1976; Fig. 13), although Jeletzky (1975, 1977) has argued for western sources. These sandstones grade to shale and siltstone toward the northeast and northwest (Husky and upper part of Kingak Formations). In northeastern Alaska

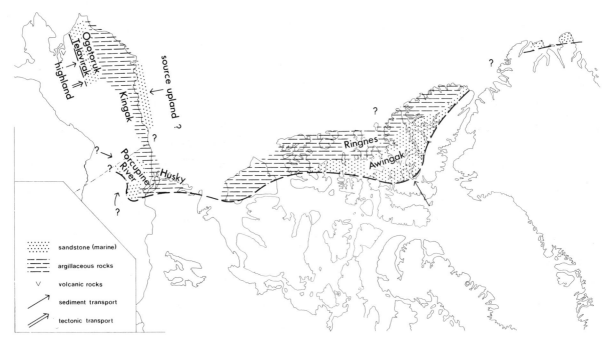

Fig. 13. Upper Jurassic (Upper Oxfordian) facies distribution map (including Kimmeridgian in North Greenland).

also, Upper Jurassic shales are siltier to the south (Detterman, 1973). Few other southerly derived sandstones, of strandline character, are preserved (e.g. North Branch Formation; Jeletzky, 1975; Fig. 5). As much as 3000 m of southerly derived, Upper Jurassic sandy turbidites were deposited in northwestern Alaska (Ogotoruk and Telavirak Formations; Cambell, 1967; Fig. 13). A northern landmass also may have contributed sediment as this time, although no Upper Jurassic littoral deposits of northern derivation are known.

In Berriasian time, much of the Alaskan part of the Brooks-Mackenzie Basin was a deep trough in front of the growing southern tectonic uplift. Turbidites derived from the south and pelagic shale were the predominant deposits during the Berriasian to Hauterivian interval (Kisimilok, Okpikruak and Kongakut formations). The basin shallowed to the north and east, where nearshore sandstone of northern derivation (e.g. Kemik Sandstone Member of the Kongakut Formation) occurs at the base of the shale succession (Fig. 14). The basin expanded eastward and northward throughout the Neocomian, and by Hauterivian time deep-water deposits occurred as far north as the north coast of Alaska, where pelagic shale (Kongakut Formation) onlaps older strata.

In contrast to the rocks of northern Alaska, the

Berriasian to Hauterivian strata of northern Yukon are mainly shallow marine and deltaic facies. Berriasian shallow marine sandstones are widespread in the southeast (lower sandstone division informally designated by Jeletzky, 1958; see also Jeletzky, in Norris, 1981) and give way to open marine shale and siltstone to the north (upper parts of Husky and Kingak formations). In Valanginian and Hauterivian time, sandy deltas prograded northward and westward away from the basin margins and the marginal areas underwent erosion (Myhr and Young, 1975). The quartzose deltaic sandstones ("Parsons Sandstone", Young et al., 1976; white and coaly quartzite informal divisions of Jeletzky, 1961; Keenan Quartzite, Brabb, 1969) apparently extended entirely across the northern Yukon part of the basin (see Fig. 14). In some areas normal faulting was contemporaneous with deltaic deposition, and marked thickness changes occur across faults (e.g. Dixon, 1981). Most significant uplifts of fault blocks and arches (e.g. Cache Creek High; see Fig. 7) occurred in approximately mid-Hauterivian time, and a regional unconformity is present above these strata over much of northern Yukon.

In the Sverdrup Basin, deltaic sedimentation followed the Oxfordian transgression. Oxfordian to Kimmeridgian deltaic and barrier-island sandstones (Awingak Formation) in the eastern and central parts of the

Fig. 14. Lower Cretaceous (Valanginian) facies distribution map.

basin (Fig. 13) were derived from the southeast and grade to the north and west into silty shale characterized by siltstone concretions as large as 5 m in diameter (Ringnes Formation, Balkwill et al., 1977). The delta was drowned in early Volgian time, and Volgian to lower Valanginian sandstone (Mould Bay Formation) is restricted mainly to the basin margins. A thick succession of shale and siltstone (Deer Bay Formation) was deposited in the basin centre at this time. Pebbly sandstone of probable Berriasian age within the Deer Bay shales along the northeastern basin margin may be evidence for the presence of a northern landmass. The deposition of these coarse clastic rocks may be related to a late Berriasian regression which has been documented in the eastern part of the basin (Kemper, 1977). A transgression in early Valanginian time was followed by a major regression in late Valanginian and Hauterivian time, when deltas prograded basinward and the basin margins were eroded (Fig. 14). Thick quartzose sandstones of delta plain and delta front facies (Isachsen Formation) extended into the basin centre as a result of this episode. The main source areas for these deltaic deposits appear to have been the craton to the east and south.

In the central part of the Wandel Sea Basin a mid-Oxfordian transgression was followed by deposition of approximately 250 m of clastic sediments recording a gradual development from marine fine grained sandstones through near coastal sands to sandy deltaic and limnic deposits (Ladegårdsåen Formation). The marine interval spans the Middle Oxfordian to the Lower Valanginian and displays an apparently regional delay in dinoflagellate appearance in the high Arctic relative to the standard boreal zonation based on ammonites and *Buchia* (Håkansson et al., 1981a). The Ladegårdsåen Formation overlies block faulted Triassic, upper Paleozoic, and older strata with angular unconformities. In the eastern part of the Wandel Sea Basin about 200 m of Kimmeridgian to Valanginian shale and sandstones were deposited in a muddy, barred coastal environment, which became more open marine in the Valanginian with the deposition of approximately 700 m of high energy coastal sands and shelf muds with sandy storm layers (Håkansson et al., 1981b). The two sub-basins appear to be unrelated, both in terms of depositional regimes and faunas.

Hauterivian to Cenomanian rocks

This time interval is characterized by particularly high subsidence rates at certain times in the well-developed foredeep troughs north of Brooks Geanticline and north and east of the Cordilleran Orogen (Fig. 4), and by the accumulation of great thicknesses of clastic strata throughout the entire region. It is also marked by widespread transgression of the craton which reached its maximum extent in about Middle Albian time, followed by regression in late Albian and Cenomanian time.

In northern Alaska, and adjacent northern Yukon and Mackenzie River delta area, the main depocentre was the sinuous foredeep which lay in front of the uplifted Brooks Geanticline and Cordilleran Orogen. A great variety of strata is present in the Alaskan part of the foredeep (Colville Trough) and, in general, it records the gradual filling of the basin. Barremian and Aptian strata have not been positively identified in northern Alaska but are possibly present in pelagic shales and turbidites of the upper part of the Okpikruak and Kongakut Formations in the basin centre. By Albian time, sediment supply from the Brooks Geanticline was overwhelming, taking the form of a northward-prograding delta-dominated shoreline in northwestern Alaska which persisted to Cenomanian time.

Along the southern margin of Colville Trough as much as 4000 m of Lower Albian turbidites are present (Fortress Mountain Formation and Bathtub Graywacke). To the north the sandstone and conglomerate turbidites of the Fortress Mountain Formation become thinner and finer, grading to pelagic shale of the Torok Formation. These strata are overlain by marine slope shales (Torok Formation) followed by shallow marine sandstone and shales (Tuktu and Kukpowruk formations) and topped by fluvial sandstones and conglomerates (Chandler and Corwin formations) to form a regressive basin fill package (Nanushuk Group) 4000 m thick (Figs. 5, 15; see Ahlbrandt et al., 1979). The main deltaic centres appear to have been in the western and central parts of Colville Trough. The fluvial strata grade to marine clastics to the north (Grandstand and Ninuluk formations) indicating that the foredeep was not entirely filled by the end of this time interval.

To the south of Brooks Geanticline, in Kobuk Basin, as much as 3000 m of largely nonmarine volcanic conglomerate and greywacke were deposited (Patton, 1966, 1967).

In northern Yukon and the adjacent Mackenzie River delta area, thick upper Hauterivian to Cenomanian strata, mainly of deep-water origin (informal upper shale-siltstone division of Jeletzky, 1958), occur in the western part of Beaufort-Mackenzie Basin (i.e., Blow Trough of Young et al., 1976; see Fig. 4). The littoral deposits were eroded during a regressive episode in early to middle Aptian time when sand deposition was widespread over the area (informal upper sandstone division of Jeletzky, 1958). The main source of the clastics appears to have been the Brooks Geanticline, although intrabasinal arches (e.g., Cache Creek High; see Fig. 7) may also have contributed. Clastics of cratonic source in Tuktoyaktuk Peninsula have been described by Dixon (1979, 1981). Major subsidence in Blow Trough and Kandik Basin took place in late Aptian to Albian time when as much as 4000 m of shale and sandy and conglomeratic turbidites were deposited (Young, 1977). These include the Aptian–Albian flysch division and the Kathul Graywacke (Fig. 5). The lower shale and siltstone unit of the Aptian–Albian flysch transgresses locally faulted Jurassic through Barremian strata. Away from Brooks Geanticline, the turbidite facies becomes thinner and changes to deep-water, silty shale (Arctic Red Formation) and bedded phosphatic ironstone (Fig. 15). Overlying Cenomanian shales (lower part of Boundary Creek Formation) are similar. Farther southeast along the series of foredeeps, in Peel Trough (Fig. 4), subsidence and deposition began in Aptian time with glauconitic sandstone and siltstone of western or southern derivation being laid down in the western part of the trough (Martin House and basal Arctic Red formations). These strata are overlain by and grade eastward into silty concretionary shales of the Arctic Red Formation (see Fig. 15). Middle Albian argillaceous strata onlap lower Paleozoic rocks to the east and north indicating basin expansion during that time. A basal transgressive Albian sandstone (Sans Sault Formation) is present locally. In the Late Albian to Cenomanian a similar pattern is indicated, with sandstones (lower part of Trevor Formation) to the west, and shales (lower part of Slater River Formation) to the east (Fig. 5). The southeastern limit of the foredeep trough was formed by Keele Arch which separated the foredeep from Great Bear Basin, the northern termination of a second, *en echelon* foredeep to the south (Fig. 4). The two troughs progressively encroached on Keele Arch to become linked in Late Albian time.

The northern Interior and Arctic Platforms (see Fig. 4) were transgressed during the late Aptian and Albian, when rapid subsidence and deposition characterized the foredeeps to the west. Albian shales and siltstones (Langton Bay, Horton River and Christopher Formations) (Fig. 15), in places underlain by an Aptian and Albian transgressive sandstone unit (Gilmore Lake Member), are present over a wide area. Local

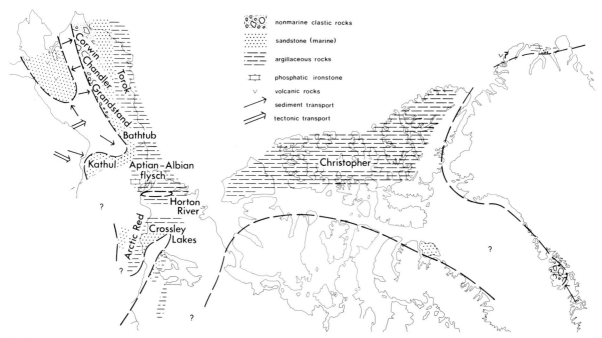

Fig. 15. Lower Cretaceous (middle Albian) facies distribution map (including Aptian in North Greenland).

arches and basins involved in the complex sedimentation history of much of the Interior Platform are described in detail by Yorath and Cook (1981). These strata, as thick as 500 m, represent deposits of an extensive marine shelf supplied both from the uplifted geanticlines to the west and the craton to the east. The eastern strandline deposits adjacent to the craton are not preserved. Upper Albian to Turonian strata are absent from most of the Interior Platform.

In the Sverdrup Basin, Barremian to middle Aptian strata consist of deltaic-fluvial sandstone (Isachsen Formation). These strata appear to grade to marine shales (Christopher Formation) to the northwest. Sverdrup Basin received mainly fine clastics from late Aptian to Middle Albian time (Christopher Formation) (Fig. 15). During the late Albian to Cenomanian regressive episode, deltaic-fluvial sandstones (Hassel Formation) were deposited in the basin. Prodelta equivalents of these strata probably occur to the northwest on the continental shelf.

Basic igneous activity, which occurred episodically throughout the history of the Sverdrup Basin (Fig. 8C), was common and widespread during the Barremian to Cenomanian time interval: there are basaltic lava flows in the Isachsen, Christopher and Hassel formations, and abundant gabbro dykes and sills with K-Ar ages of approximately 118 to 102 m.y. (Barremian to late

Albian). The rate of sediment accumulation in the Sverdrup Basin was fastest during the Albian (Balkwill, 1978). Crustal fracturing, mafic intrusion, and rapid basin subsidence characterize this time interval.

In the central part of the Wandel Sea Basin, two separate very small occurrences of Aptian to Lower Albian sands and mudstones have been interpreted as coastal and open marine remnants of a once complete cover (Håkansson et al., 1981b).

In the Baffin Bay region the oldest Mesozoic strata known are Barremian or Aptian nonmarine sandstones (Kome Formation) on the Nûgssuak Peninsula, central West Greenland. They are as thick as 200 m and rest on Precambrian granite (Henderson et al., 1976). Younger Cretaceous strata here (Upernivik Næs Formation) are also fluvial and deltaic sandstones. Along the western margin of Baffin Bay, at Bylot Island, Albian sandstones lie in paleotopographic hollows in crystalline basement.

Turonian to Maastrichtian rocks

A nearly complete record of this time-stratigraphic interval is preserved in the foredeep troughs of northern Alaska and Yukon, Sverdrup Basin and Baffin Bay, but significant gaps occur within the succession on the Interior Platform (Fig. 5). The marine sedimentary pattern in general is one of basin fill, with dark, organic-rich, tuffaceous marine shales predominant in the lower part of the interval and fluvial-deltaic sandstones in the upper part. The age of the change from marine shale to paralic sandstone varies over the area from Turonian in northern Alaska to Maastrichtian in Arctic Canada. By late Maastrichtian time a large part of the area was dominated by fluvial-deltaic deposition.

In northern Alaska, organic-rich black shale (Seabee Formation; 600 m max.) was deposited in Colville Trough unconformably over older Cretaceous strata during Turonian time. Abundant sediment supply from the Brooks Geanticline resulted in north-north-eastward deltaic progradation during Coniacian to Maastrichtian time (Fig. 16). Overlying the Seabee Formation are marine sandstone and shale (Schrader Bluff Formation) and fluvial sandstone and conglomerate (Prince Creek Formation). By Maastrichtian time, Colville Trough foredeep was essentially filled and most of the area was emergent (Fig. 17). Considerable sediment may have reached Canada Basin, and thick Coniacian to Maastrichtian marine and nonmarine strata probably underlie the continental shelf north of Alaska.

In the northern Yukon and adjacent continental shelf to the north and northeast of Brooks Geanticline a similar regressive succession of strata is present. Turonian to Campanian strata comprise 500 m of dark grey, partly bituminous shale of shallow to deep marine origin (Boundary Creek Formation; Fig. 16), which represents stable, largely stagnant conditions associated with inundation of the Beaufort-Mackenzie Basin and western Interior Platform areas. Abundant bentonite beds indicate frequent volcanic activity although their source has not been certainly identified. In Campanian time, eastward expansion of the Brooks Geanticline uplifted the western margin of Kandik Basin, progressively exposing the previously deposited Mesozoic strata. Late Campanian uplift and erosion of Cache Creek High, the intrabasinal arch in Beaufort-Mackenzie Basin, was followed by alluvial deposition (Cuesta Creek Member of Tent Island Formation) and marine transgression (lower Tent Island Formation). In Maastrichtian time, a delta complex with prodelta shale and siltstone (upper Tent Island Formation) followed by fluvial-deltaic sandstone in the Tertiary (Moose Channel Formation) prograded north and northeastward onto the present continental shelf area (Fig. 17; Young, 1975b, 1977).

Upper Cretaceous nonmarine sandstone, siltstone and shale of varying ages are also preserved here and there to the south of Brooks Geanticline in local basins (Eagle Plain, Bonnet Plume and Monster formations). The source areas for these sediments appear to have been the uplifted geanticlines to the north, west and south (Figs. 5, 16, 17).

In Peel Trough farther southeast along the foredeep, near Keele Arch (see Fig. 4), the pattern of deltaic progradation that was initiated in Late Albian time persisted into Turonian time. It is represented by deltaic sandstone (Trevor Formation) to the northwest and by black, organic-rich shale (Slater River Formation) to the southeast. The Slater River Formation is overlain by fluvial-deltaic sandstone of the Little Bear Formation. A second phase of deltaic progradation in this area is represented in the Santonian to Maastrichtian by westerly derived sandstones that overlie and grade to prodelta shale eastward (East Fork Formation; Yorath and Cook, 1981; Figs. 5, 16). Capping the succession in this area are Maastrichtian and Paleocene fluvial sandstone and conglomerate which filled the foredeep.

Over most of the northern Interior Platform no record exists for the Late Albian to Coniacian. How much of the record was destroyed during pre-Santonian erosion is not known. The foredeep troughs east of the Cordilleran Orogen continued to subside and to transgress northward, such that Turonian marine shales and siltstones overlie Paleozoic rocks in the northern part of Keele Arch (Yorath and Cook, 1981) suggesting that the Interior Platform was largely exposed from Late Albian to Coniacian times. The interior Platform was flooded again in late Santonian time and a basal transgressive sandstone is preserved here and there. From Santonian to Maastrichtian time the area was a shallow marine shelf receiving a low influx of fine sediment from source areas to the west and east and a considerable amount of volcanic dust. The resulting shales are thin, partly bituminous, bentonitic and pyritic (Smoking Hills and Mason River formations; Figs. 16, 17). The Interior and Arctic Platforms were emergent in late Maastrichtian time, except possibly on Banks Island where sedimentation may have been continuous into the Tertiary.

In Sverdrup Basin, a nearly continuous record of Turonian to Maastrichtian time is preserved. Turonian transgression drowned the Cenomanian deltas and the area became a shallow marine shelf receiving a low influx of fine terrigenous clastic and volcanic debris

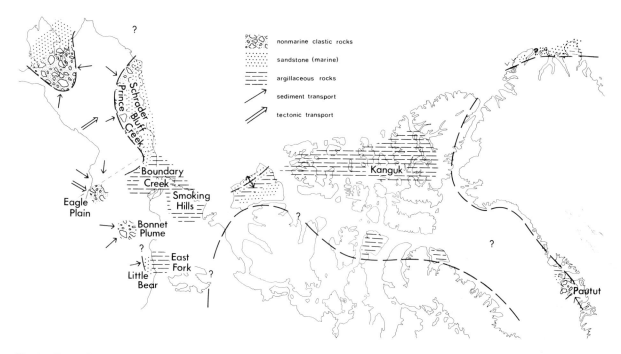

Fig. 16. Upper Cretaceous (lower Campanian) facies distribution map (including Turonian–Santonian in North Greenland).

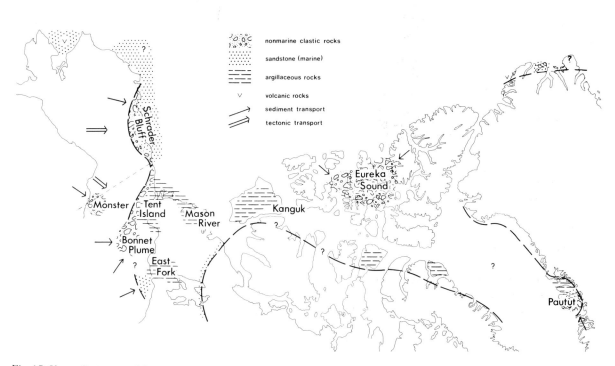

Fig. 17. Upper Cretaceous (Maastrichtian) facies distribution map.

from Turonian to Campanian time (Fig. 16). The Kanguk Formation, which encompasses these sediments, is thin (maximum 400 m), bituminous, bentonitic and pyritic. Basalt flows are present within the shales in the basin centre (Strand Fiord Formation) and shallow marine sandstones occur in the western portion of the basin. Progressive uplift to the northwest along Sverdrup Rim, in Campanian and Maastrichtian time, and in the northeast in Maastrichtian time, supplied sediments to the south and southwest (Fig. 16). By late Maastrichtian time, fluvial-deltaic sedimentation occurred over most of Sverdrup Basin (basal part of Eureka Sound Formation) (Fig. 17). Source areas included the northeastern part of Sverdrup Rim and the Interior Platform and craton to the east and south. Marine equivalents of these strata occur on the continental shelf adjacent to Canada Basin.

Wandel Sea Basin deposition in Late Cretaceous time was controlled by a strike-slip mobile belt in which a series of local, pull-apart basins formed within the frame of a generally NW–SE trending system of anastomosing faults. Rapid infill of very immature sediments characterises all the basins, and they were all variously deformed close to the Cretaceous–Tertiary boundary (Håkansson et al., 1981b). The marine basins (e.g., the Nakkehoved Formation) all accumulated in excess of 450 m of sediments dominated by finegrained greywackes and sandstones with fairly restricted, mostly sparse mollusc faunas. The ages of two of the basins have been determined as Turonian-Coniacian and Santonian, respectively (Håkansson et al., 1981b). The non-marine basins (Herlufsholm Strand Formation) are dominated by quite varied lithologies probably deposited in intertonguing alluvial fan complexes. These basins similarly have accumulated in excess of 450 m of sediments including conglomerates, various sands, and coaly shale, but precise age determinations are still wanting (Håkansson, 1979; Håkansson et al., 1981b). In the westernmost part of the Wandel Sea Basin an additional non-marine sequence is initiated with plant bearing shale and sandstone overlain by at least 5 km of extrusive volcanic rocks with a few intercalations of clastic sediments (Brown and Parsons, 1981). The age of the basal clastics is mid-Cretaceous or younger, whereas sediments high in the sequence contain pollen of Campanian–Maastrichtian age (Batten et al., 1981). This per-alkaline volcanic suite (Kap Washington Group) postdates a mainly Upper Cretaceous N–S trending dyke swarm in the same region, and together

they are considered (Batten et al., 1981; Soper et al., 1981) early expressions of a continental rift-tectonic scenario preceding the onset of spreading in the Eurasian Basin.

In the Baffin Bay region, knowledge of Turonian to Maastrichtian strata is restricted to outcrops from Disko Island to Svartenhuk in central West Greenland and Bylot Island of Canada. In West Greenland the strata, which are at least as thick as 1000 m, display a facies change from fluvial sandstone in the south, through deltaic sandstone, shale and coal (Atane and Pautut formations) to prodelta shale in the north (Henderson et al., 1976). The sea possibly reached its maximum extent in early Campanian time, and the shoreline position was stable until block faulting in the Paleocene caused the formation of an angular unconformity on top of the succession (Hansen, 1981). On Bylot Island the strata of this interval comprise 600 m of Campanian and Maastrichtian grey silty shale (W.S. Hopkins and N.S. Ioannides, pers. comm., 1978). A low angle unconformity truncates the shale. Upper Cretaceous faunas in the Baffin Bay region have close affinities with those of the Western Interior Region of North America, suggesting marine connections with that area across the Arctic and northern Interior Platforms (Birkelund, 1965; Birkelund and Perch-Nielsen, 1976). Connections with the North Atlantic did not become open until perhaps Campanian time (Williams and Stelck, 1975), and it is not until the Lower Paleocene that the faunas and microfloras of the two areas become similar (Rosenkrantz, 1970; Hansen, 1981). The nature and thickness of the Turonian to Maastrichtian strata underlying Baffin Bay are unknown; because marine shales occur on both margins a succession of marine shale is probably present.

REGIONAL TRANSGRESSIVE–REGRESSIVE CONTROLS

We have drawn transgressive–regressive curves (Fig. 18) for the northern Alaska–Yukon region and the Canadian Archipelago, areas which are 800 km apart and which are in somewhat different tectonic settings. These curves cannot be regarded specifically as either sea-level or tectonic movement curves as they simply record transgression and regression as interpreted from the stratigraphic column. They record sediment supply rates which varied areally and tem-

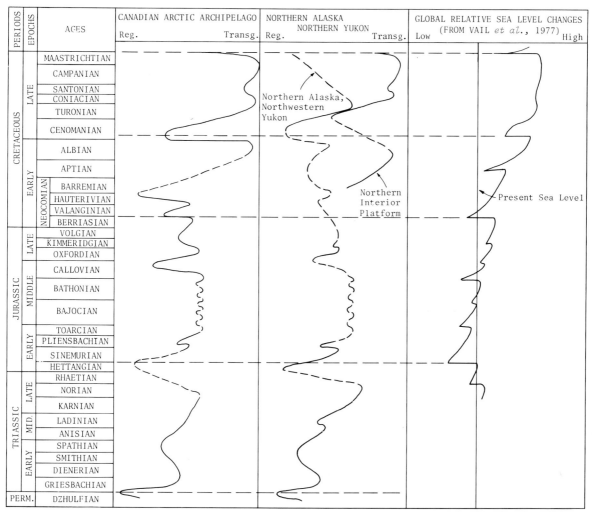

Fig. 18. Transgression—regression curves of Canadian Arctic Archipelago, north Alaska—northwestern Yukon, and Northern Interior of Canada; compared with global sea level fluctuation curves from Vail et al. (1977). Dashed lines indicate extrapolation from relatively meager data.

porally. Transgressions are indicated by onlap relations in basin-marginal areas; regressions by extensive clastic progradation into the basins or by unconformities.

The timing of some major transgressive—regressive events between the two areas during the Triassic and Jurassic periods is similar (Fig. 18). Some major regressive episodes, e.g. latest Triassic and latest Jurassic to Berriasian, coincide with the global sea-level fluctuations suggested by Vail et al. (1977), lending support to the concept of some global sea-level fluctuations.

The effect of orogeny is well expressed in the contrast between the record of the advancing Cordilleran Orogen of northern Alaska and northern Yukon during the Cretaceous, and coincident subsidence of adjacent foredeeps in the northern interior region of Canada (see Figs. 4, 18). The transgressive—regressive record in the foredeeps is similar to that of the Arctic Archipelago where there was no contemporaneous major orogenic event, and also to the global pattern interpreted by Vail et al. (1977). Thus the effect of local orogeny has not entirely obscured the evidence for eustatic sea-level fluctuations as Jeletzky (1978) suggested. Eustatic sea-level changes appear to have left an imprint on the Arctic stratigraphic record despite the complexities of local tectonic movements and sediment supply variation.

PALEOLATITUDES, PALEOCLIMATOLOGY AND FAUNAL AFFINITIES

The interpretation of Mesozoic paleolatitudes of Arctic North America is based on paleomagnetic data that mainly come from the west-central parts of the continent (Irving, 1973, 1977). These data indicate an irregular northward shift of Arctic Canada from approximately 40° to 45°N in Early Triassic time, to a position in the area of 65° to 80°N by Jurassic and Cretaceous time.

Certain climate-related rock types support the concept of general but episodic cooling associated with northward drift since the Triassic, and generally more equable conditions than occur at present. By direct comparison with the location of present-day climatic regions (Miller, 1961) the Triassic paleolatitudes would suggest a cool temperate climate for Arctic North America. However, Lower and Middle Triassic fluvial strata contain abundant red shale and siltstone which suggest a hot, seasonally dry climate (Van Houten, 1973) similar to the savannah areas which presently occur between 15° and 25°N. Coal in Upper Triassic fluvial strata may indicate a more humid climate during the Late Triassic. Relatively warm climates may have extended as far north and south as latitude 45° in the Triassic, as is also suggested by the common occurrence of limestone in Middle and Upper Triassic strata of the North American Arctic.

Triassic marine faunas are essentially cosmopolitan in their affinities, although some paleolatitudinal variation is indicated by the lower diversity of pelagic faunas of the Canadian Arctic compared with those of western Canada (Tozer, 1970a). From Middle Bajocian time onward, many of the Arctic ammonites are significantly different from more southerly ammonites of the same ages, and a succession of distinct boreal faunas has been recognized (see Imlay, 1965; Jeletzky, 1971). Arctic marine bivalve faunas exhibit generally decreasing diversity throughout Jurassic time, and those of the Middle and Late Jurassic are less diverse than equivalent faunas farther south. The boreal Cretaceous ammonites, like those of the Middle and Late Jurassic, are largely different from, and less diverse than, their more southerly equivalents (Jeletzky, 1971). Other striking features of the boreal Cretaceous fauna are a lack of diversity and the absence or rarity of rudistids, colonial corals, orbitoid foraminifera, brachiopods, gastropods, and echinoderms. The decreased diversity of Middle Jurassic through Cretaceous marine Arctic faunas is evidence for Mesozoic paleolatitudinal climatic belts, by analogy with Holocene north–south diversity gradients (Stehli, 1968). The northward decrease in the abundance of carbonate strata in the Jurassic and Cretaceous of North America, and their essential absence in the Arctic, further indicate paleolatitudinal climatic differentiation of the continent during these times. Clark (1977), by comparison with modern distributions, showed that Arctic Upper Cretaceous phytoplankton assemblages indicate a warmer water temperature than exists at present. During much of the Cretaceous, like the Tertiary, the Arctic climate must have been less severe than that of today, because coal beds are common in the fluvial-deltaic strata. At some times, as suggested by palynomorphs in the late Albian and Cenomanian Hassel Formation for example, warm temperate conditions may have prevailed (Hopkins and Balkwill, 1973). However, stellate nodules in Valanginian shale were interpreted by Kemper and Schmitz (1975) to indicate cold marine paleoenvironments.

SUMMARY

Mesozoic strata are present in outcrop and in the subsurface of Arctic North America along a broad belt bordering the Arctic Ocean, from northwestern Alaska to northeastern Greenland, and are present also on the continental margins of Baffin Bay. The strata record the effects of local and regional tectonic activity, eustatic sea-level fluctuations, and changing climate conditions. This interplay created a mosaic of intertonguing sandstone, siltstone and shale, with minor limestone, punctuated by local and regional unconformities.

The Mesozoic rocks include parts of two of the three tectono-stratigraphic assemblages which comprise the Phanerozoic strata of the area: the lower, approximately Mississippian to mid-Early Cretaceous in age; and the upper, mid-Early Cretaceous to latest Cretaceous. Each assemblage is characterized by a distinctive tectonic setting; they are differentiated by tectonic realignment whose occurrence was clearly evident by approximately Barremian time. The realignment was initiated somewhat earlier, in the latest Jurassic, in northern Alaska, where the main significance was a change in source areas from north for the early Mesozoic to south for the late Mesozoic, in association with uplift of Brooks Geanticline. In Arctic Canada, the change of regimes is characterized by onlap of mid- to upper-Lower Cretaceous strata over broad shelf areas beyond previous basin margins.

Triassic to mid-Early Cretaceous strata were deposited in three major basins which had developed in Mississippian time on highly deformed lower Paleozoic and Proterozoic strata: Brooks-Mackenzie Basin of northern Alaska and Yukon, Sverdrup Basin of Canadian Arctic Archipelago, and Wandel Sea Basin of eastern North Greenland. The basins were linked by long-lasting but possibly intermittent marine connections, and their relative dispositions prior to continental drift are not satisfactorily known. They were bounded by land areas to the south (stable craton), and at least in northern Alaska, to the north. These land areas were important sediment source areas during this time interval.

Major tectonic events characterizing the second, mid-Early Cretaceous to latest Cretaceous tectonostratigraphic assemblage are: uplift and expansion of Brooks Geanticline and the Cordilleran Orogen, formation of a series of foredeep troughs in front of these uplifts, submergence of an extensive northern landmass, transgression far onto the stable craton, accelerated subsidence in the Baffin Bay region, and development of a major strike-slip mobile belt in eastern North Greenland.

The Mesozoic Era was initiated in Arctic North America by a regional transgression in earliest Triassic time. Within the basins, clearly defined marginal (sandstone) and basinal (shale and siltstone) facies were deposited. Further transgression from Early Triassic to Norian time resulted in shorelines migrating onto the craton, periodically interrupted by deltaic progradation in some areas. In late Norian and Rhaetian time, sandstones were deposited widely, and regional emergence occurred in the latest Triassic.

During Early and Middle Jurassic time, general regional subsidence resumed, leaving a mosaic of intertonguing sandstone and shale in the basins. Some of the Jurassic sandstone prograded over much of the basinal areas. Intermittent transgressions from the Late Oxfordian to early Valanginian are marked by widespread shales intertongued with coarse clastic sedimentary rocks along the basin margins, and some extensive deltaic lobes. In latest Jurassic time, Brooks Geanticline became a major source of sediment along the southwest margin of Brooks-Mackenzie Basin.

Regional regression in the late Valanginian and Hauterivian resulted in widespread deltaic deposition in the central portions of the basins and erosion of the basin margins.

From Late Hauterivian to Cenomanian time, thick (6000 m) successions of clastic sediments, derived from Brooks Geanticline and the Cordilleran Orogen, prograded into the adjacent foredeeps. The clastic wedges consist of a wide variety of strata from deep marine turbidites to fluvial sandstone and conglomerate. To the east, a thin (max. 300 m) sequence of sandstone and shale was deposited on the Interior Platform of the craton in late Aptian and Albian time. This area became emergent in the late Albian and Cenomanian. A thicker (1200 m) succession of sandstone and shale was deposited in Sverdrup Basin; fluvial-deltaic sedimentation took place in the basin during the low stand of sea level in late Albian and Cenomanian time. Barremian to Albian basal nonmarine sandstones were deposited over Precambrian crystalline rocks in Baffin Bay region.

Regional transgression in Turonian time resulted in widespread deposition of dark, organic-rich shale which characterizes much of the Upper Cretaceous stratigraphic record; coeval coarse clastic deposits are confined to areas near major source terranes. Volcanic ash was spread regionally represented by extensive but very thin bentonite beds, and in central North Greenland several kilometers of volcanic extrusions have been preserved. Late Maastrichtian regression led to widespread fluvial-deltaic sedimentation and emergence at the close of the Mesozoic Era.

Deterioration of climatic conditions in Arctic North America possibly related to northward continental drift may be indicated by the development of a succession of boreal marine faunas and absence of limestones in the later Mesozoic succession.

REFERENCES

Ahlbrandt, T.S., Huffman, A.C., Fox, J.E. and Pasternack, Ira, 1979. Depositional framework and reservoir studies of selected Nanushuk Group outcrops, North slope, Alaska. In: T.S. Ahlbrandt (Editor), *Preliminary Geologic, Petrologic, and Paleontologic Results of the Study of Nanushuk Group Rocks, North Slope, Alaska—U.S. Geol. Surv. Circ.*, 794, pp. 14–31.

Balkwill, H.R., 1978. Evolution of Sverdrup Basin, Arctic Canada. *Am. Assoc. Petrol. Geol., Bull.*, 62, pp. 1004–1028.

Balkwill, H.R. and Roy, K.J., 1977. Geology of King Christian

Island, District of Franklin. *Geol. Surv. Can. Mem.*, 386.

Balkwill, H.R. and Roy, K.J., 1979. Sverdrup Basin response to cratonic events. *Am. Assoc. Petrol. Geol., Bull.*, 63, p. 413.

Balkwill, H.R., Wilson, D.G. and Wall, J.H., 1977. Ringnes Formation (Upper Jurassic), Sverdrup Basin, Canadian Arctic Archipelago. *Bull. Can. Petrol. Geol.*, 25, pp. 1115–1144.

Bamber, E.W. and Waterhouse, J.B., 1971. Carboniferous and Permian stratigraphy and paleontology, northern Yukon Territory, Canada. *Bull. Can. Petrol. Geol.*, 19, pp. 29–250.

Batten, D.J., Brown, P.E., Dawes, P.R., Higgins, A.K., Koch, B.E., Parsons, I. and Soper, N.J., 1981. Peralkaline volcanicity on the Eurasian Basin margin. *Nature*. In press.

Birkelund, T., 1965. Ammonites from the Upper Cretaceous of West Greenland. *Medd. Groenl.*, 179 (7): 192 pp.

Birkelund, T. and Perch-Nielsen, K., 1976. Late Paleozoic–Mesozoic evolution of central East Greenland. In: A. Escher and W.S. Watts (Editors), *Geology of Greenland*. Groenl. Geol. Undersolgelse, pp. 305–339.

Brabb, E.E., 1969. Six new Paleozoic and Mesozoic formations in east-central Alaska. *U.S. Geol. Surv. Bull.*, 1274-I, pp. 111–125.

Brideaux, W.W., 1977. Taxonomy of Upper Jurassic–Lower Cretaceous microplankton from the Richardson Mountains, District of Mackenzie, Canada. *Geol. Surv. Can., Bull.*, 281.

Brideaux, W.W. and Fisher, M.J., 1976. Upper Jurassic–Lower Cretaceous Dinoflagellate assemblages from Arctic Canada. *Geol. Surv. Can., Bull.*, 259.

Brosgé, W.P. and Dutro, T., Jr., 1973. Paleozoic rocks of northern and central Alaska. In: M.G. Pitcher (Editor), *Proceedings of the Second International Symposium on Arctic Geology, San Fransisco, Calif. – Am. Assoc. Petrol. Geol., Mem.*, 19, pp. 361–375.

Brosgé, W.P. and Tailleur, I.L., 1970. Tectonic history of northern Alaska. In: M.G. Pitcher (Editor), *Proceedings of the Geological Seminar on the North slope of Alaska – Am. Assoc. Petrol. Geol., Pacific Sect.*, pp. D1–D17.

Brown, P.E. and Parsons, I., 1981. The Kap Washington volcanics. *Groenl. Geol. Undersogelse Rapp.*, 105. In press.

Campbell, R.H., 1967. Areal geology in the vicinity of the Chariot Site, Lisburne Peninsula, northwestern Alaska. *U.S. Geol. Surv., Prof. Pap.*, 395, 71 pp.

Carey, S.W., 1955. The orocline concept in geotectonics, Part I. *Pap. Proc. R. Soc. Tasmania*, 89, pp. 255–288.

Carey, S.W., 1976. *The Expanding Earth*. Elsevier, Amsterdam, 485 pp.

Chapman, R.M. and Sable, E.G., 1960. Geology of the Utukok–Corwin region, northwestern Alaska. *U.S. Geol. Surv., Prof. Pap.*, 303-C, pp. C47–C167.

Christie, R.L., 1979. The Franklinian Geosyncline in the Canadian Arctic and its relationship to Svalbard. *Saertrykk Nor. Polarinst., Skr.*, 167, pp. 263–314.

Churkin, M., Jr., 1969. Paleozoic tectonic history of the Arctic Basin north of Alaska. *Science*, 165, pp. 549–555.

Churkin, M., Jr., 1973. Geologic concepts of Arctic Ocean Basin. In: *Arctic Geology – Am. Assoc. Petrol. Geol.,*

Mem., 19, pp. 485–499.

Clark, D.L., 1975. Geological history of Arctic Ocean Basin. In: C.J. Yorath, E.R. Parker and D.J. Glass (Editors), *Canada's Continental Margins and Offshore Petroleum Exploration – Can. Soc. Petrol. Geol., Mem.*, 4, pp. 501–524.

Clark, D.L., 1977. Climatic factors of the Late Mesozoic and Cenozoic Arctic Ocean. In: M.J. Dunbar (Editor), *Polar Oceans*. Arctic Institute of North America, Montreal, Que., 682 pp.

Davies, E.H. and Norris, G., 1980. Latitudinal variations in encystment modes and species diversity in Jurassic dinoflagellates. In: D.W. Strangway (Editor), *The Continental Crust and its Mineral Deposits – Geol. Assoc. Can., Spec. Pap.*, 20, pp. 361–373.

Dawes, P.R., 1976. Precambrian to Tertiary of northern Greenland. In: A. Escher and W.S. Watt (Editors), *Geology of Greenland*. Geol. Surv. Greenland, pp. 284–303.

Detterman, R.L., 1956. New and redefined nomenclature of the Nanushuk Group. In: G. Gryc et al., *Mesozoic Sequence in Colville River Region, Northern Alaska – Am. Assoc. Petrol. Geol., Bull.*, 40, pp. 233–244.

Detterman, R.L., 1973. Mesozoic sequence in Arctic Alaska. In: *Proceedings of the Second International Arctic Symposium – Am. Assoc. Petrol. Geol., Mem.*, 19, pp. 376–387.

Detterman, R.L., Reiser, H.N., Brosgé, W.P. and Dutro, J.T., Jr., 1975. Post-Carboniferous stratigraphy, northeastern Alaska. *U.S. Geol. Surv., Prof. Pap.*, 886, 46 pp.

Dixon, J., 1979. The Lower Cretaceous Atkinson Point Formation (new name) on the Tuktoyaktuk Peninsula, N.W.T.: A coastal fan-delta to marine sequence. *Bull. Can. Petrol. Geol.*, 27, pp. 163–182.

Dixon, J., 1981. Jurassic and Lower Cretaceous subsurface stratigraphy of the Mackenzie Delta–Tuktoyaktuk Peninsula, N.W.T. *Geol. Surv. Can., Bull.*, 349. In press.

Dutro, J.T. Jr., 1981. Geology of Alaska bordering the Arctic Ocean. In: A.E.M. Nairn, M. Churkin and F.G. Stehli (Editors), *The Ocean Basins and Margins, 5. The Arctic Ocean*. Plenum, New York, N.Y., pp. 21–36.

Eardley, A.J., 1961. History of geologic thought on the origin of the Arctic basin. In: G.O. Raasch (Editor), *Geology of the Arctic*. Univ. of Toronto Press, Toronto, Ont., pp. 409–436.

Embry, A.F. and Klovan, J.E., 1976. The Middle–Upper Devonian clastic wedge of the Franklinian Geosyncline. *Bull. Can. Soc. Petrol. Geol*, 24, pp. 485–639.

Forsyth, D.A., Mair, J.A. and Fraser, I., 1979. Crustal structure of the central Sverdrup Basin. *Can. J. Earth Sci.*, 16, pp. 1581–1598.

Frebold, H., 1957. The Jurassic Fernie Group in the Canadian Rocky Mountains and foothills. *Geol. Surv. Can., Mem.*, 287.

Frebold, H., 1964. Illustrations of Canadian fossils. Jurassic of western and Arctic Canada. *Geol. Surv. Can., Pap.*, 63-4.

Frebold, H., 1970. Marine Jurassic faunas. In: R.J.W. Douglas (Editor), *Geological and Economic Minerals of Canada – Geol. Surv. Can., Econ. Geol. Rep.*, 1, pp. 641–648.

Frebold, H., 1975. The Jurassic faunas of the Canadian Arctic Lower Jurassic ammonites, biostratigraphy and correlations *Geol. Surv. Can., Bull.*, 243.

Frebold, H. and Poulton, T.P., 1977. Hettangian (Lower Jurassic) rocks and faunas, northern Yukon Territory. *Can. J. Earth Sci.*, 14, pp. 89–101.

Fujita, K., 1978. Pre-Cenozoic tectonic evolution of northern Siberia. *J. Geol.*, 86, pp. 159–172.

Grantz, A., Holmes, M.L. and Kososki, B.A., 1975. Geologic framework of the Alaskan Continental terrace in the Chukchi and Beaufort Seas. In: C.J. Yorath, E.R. Parker and D.J. Glass (Editors) *Canada's Continental Margins and Offshore Petroleum Exploration – Can. Soc. Petrol. Geol., Mem.*, 4, pp. 669–700.

Grantz, A., Eittreim, S. and Dinter, D.A., 1979. Geology and tectonic development of the continental margin north of Alaska. *Tectonophysics*, 59, pp. 263–291.

Håkansson, E. 1979. Carboniferous to Tertiary development of the Wandel Sea Basin, eastern North Greenland. *Groenl. Geol. Undersogelse, Rapp.* 88: 73–83.

Håkansson, E. and Heinberg, C., 1977. Reconnaissance work in the Triassic of the Wandel Sea Basin, Peary Land, eastern North Greenland: *Groenlands Geol. Undersogelse, Rapp.* 85, pp. 11–15.

Håkansson, E. and Pedersen, S.S., 1981. Late Paleozoic to Paleogene development of the continental margin in North Greenland (Abstr.). *Third Int. Symp. Arctic Geology, Abstr.*, p. 57.

Håkansson, E., Birkelund, T., Piasecki, S. and Zakharov, V., 1981a. Jurassic–Cretaceous boundary strata of the extreme arctic (Peary Land, North Greenland). *Geol. Soc. Denmark, Bull.* In press.

Håkansson, E., Heinberg, C. and Stemmerik, L. 1981b. The Wandel Sea Basin from Holm Land to Lockwood Ø, eastern North Greenland. *Groenl. Geol. Undersogelse, Rapp.* 105. In press.

Hamilton, W., 1970. The Uralides and the motion of the Russian and Siberian Platforms. *Geol. Soc. Am. Bull.*, 81: 2553–2576.

Hansen, J.M., 1981. Stratigraphy and structure of the Paleocene in central West Greenland and Denmark. *Groenl. Geol. Undersogelse, Bull.* In press.

Harland. W.B., 1973. Mesozoic geology of Svalbard. In: M.G. Pitcher (Editor), *Arctic Geology – Am. Assoc. Petrol. Geol., Mem.*, 19, pp. 135–148.

Henao-Londoño, D. 1977. Correlation of producing formations in the Sverdrup Basin. *Bull. Can. Petrol. Geol.*, 25 (5), pp. 969–980.

Henderson, G., Rosenkrantz, A. and Schiener, E.J., 1976. Cretaceous–Tertiary sedimentary rocks of West Greenland. In: A. Escher and W.S. Watt (Editors), *Geology of Greenland*. Geol. Surv. Greenland, pp. 338–362.

Herron, E.M., Dewey, J.F. and Pitman, W.C., 1974. Plate tectonics model for the evolution of the Arctic. *Geology*, 2, pp. 227–380.

Hopkins, W.S. and Balkwill, H.R., 1973. Description, palynology, and paleoecology of the Hassel Formation (Cretaceous) on eastern Ellef Ringnes Island, District of Franklin. *Geol. Surv. Can., Pap.*, 72-37, 31 pp.

Imlay, R.W., 1961. Characteristic Lower Cretaceous megafossils from northern Alaska. *U.S. Geol. Surv., Prof. Pap.*, 335, 74 pp.

Imlay, R.W., 1965. Jurassic marine faunal differentiation in North America. *J. Paleontol.*, 39, pp. 1023–1038.

Imlay, R.W., 1976. Middle Jurassic (Bajocian and Bathonian) Ammonites from northern Alaska. *U.S. Geol. Surv., Prof. Pap.*, 854, 19 pp.

Imlay, R.W. and Detterman, R.L., 1973. Jurassic paleogeography of Alaska. *U.S. Geol. Surv., Prof. Pap.*, 801, 34 pp.

Irving, E., 1973. Latitude variation of the Canadian Arctic Islands during the Phanerozoic. In: J.D. Aitken and D.J. Glass (Editors), *Proceedings of the Symposium on the Geology of the Canadian Arctic*. Geol. Assoc. Can. and Can. Soc. Petrol. Geol., Saskatoon, Sask., pp. 1–3.

Irving, E., 1977. Drift of the major continental blacks since the Devonian. *Nature*, 270, pp. 304–309.

Jeletzky, J.A., 1958. Uppermost Jurassic and Cretaceous rocks of Aklavik Range, Northeastern Richardson Mountains, Northwest Territories (107D/4 part of). *Geol. Surv. Can., Pap.*, 58-2.

Jeletzky, J.A., 1961. Upper Jurassic and Lower Cretaceous rocks, west flank of Richardson Mountains between the headwaters of Blow River and Bell River, Yukon Territory. *Geol. Surv. Can., Pap.*, 61-9.

Jeletzky, J.A., 1964. Illustrations of Canadian fossils. Lower Cretaceous marine index fossils of the sedimentary basins of western and Arctic Canda. *Geol. Surv. Can., Pap.*, 64-11.

Jeletzky, J.A., 1966. Upper Volgian (Late Jurassic) Ammonites and Buchias of Arctic Canada. *Geol. Surv. Can., Bull.*, 128, 51 pp., 8 pl.

Jeletzky, J.A., 1967. Jurassic and (?) Triassic rocks of the eastern slope of Richardson Mountains, northwestern District of Mackenzie, 106M and 107B (part of). *Geol. Surv. Can., Pap.*, 66-50.

Jeletzky, J.A., 1970. Cretaceous macrofaunas. In: R.J.W. Douglas (Editor), *Geology and Economic Minerals of Canada – Geol. Surv. Can. Econ. Geol. Rep.*, 1, pp. 649–662.

Jeletzky, J.A., 1971. Marine Cretaceous Biotic provinces of western and Arctic Canada. *Proc. N. Am. Paleontol. Conv., Part L*, pp. 1638–1659.

Jeletzky, J.A., 1975. Jurassic and Lower Cretaceous paleogeography and depositional tectonics of Porcupine Plateau, adjacent areas of northern Yukon and those of Mackenzie District, N.W.T. *Geol. Surv. Can., Pap.*, 74-16.

Jeletzky, J.A., 1977. Porcupine River Formation; a new Upper Jurassic sandstone unit, northwest Yukon Territory. *Geol. Surv. Can., Pap.*, 76-27.

Jeletzky, J.A., 1978. Causes of Cretaceous oscillations of sea level in western and Arctic Canada and some general geotectonic implications. *Geol. Surv. Can., Pap.*, 77-18.

Jeletzky, J.A., 1980. New or formerly poorly known, biochronologically and paleogeographically important gastroplitinid and cleoniceratid (Ammonitida) taxa from Middle Albian rocks of mid-western and Arctic Canada. *Geol. Surv. Can., Pap.*, 79-22.

Keen, C.E. and Hyndman, R.D., 1979. Geophysical review of

the continental margins of eastern and western Canada. *Can. J. Earth Sci.*, 16, pp. 712–747.

Kemper, E., 1977. Biostratigraphy of the Valanginian in Sverdrup Basin, District of Franklin. *Geol. Surv. Can., Pap.*, 76-32.

Kemper, E. and Schmitz, H.H., 1975. Stellate nodules from the Upper Deer Bay Formation (Valanginian) of Arctic Canada. *Geol. Surv. Can., Pap.*, 75-1C, pp. 109–119.

Kent, P.E., 1977. The Mesozoic development of aseismic continental margins. *J. Geol. Soc. Lond.*, 134, pt. 1, p. 1–18.

Kerr, J.W., 1980. A plate tectonic contest in Arctic Canada. In: D.W. Strangway (Editor), *The Continental Crust and its Mineral Deposits – Geol. Assoc. Can., Spec. Pap.*, 20, pp. 457–486.

Kummel, B., 1943. Middle Triassic ammonites from Peary Land. *Medd. Groenl.*, 127 (1): 21 pp.

Lerand, M., 1973. Beaufort Sea. In: R.G. McCrossan (Editor), *The Future Petroleum Provinces of Canada: Their Geology and Potential – Can. Soc. Petrol. Geol., Mem.*, 1, pp. 315–386.

McMillan, N.J., 1973. Labrador Sea and Baffin Bay. In: R.G. McCrossan (Editor), *The Future Petroleum Provinces of Canada: Their Geology and Potential – Can. Soc. Petrol. Geol., Mem.*, 1, pp. 473–518.

Meneley, R.A., Henao, D. and Merritt, R., 1975. The northwest margin of the Sverdrup Basin. In: C.J. Yorath, E.R. Parker and D.J. Glass (Editors), *Canada's Continental Margins and Offshore Petroleum Exploration – Can. Soc. Petrol. Geol., Mem.*, 4, pp. 531–544.

Meyerhoff, A.A., 1973. Origin of Arctic and North Atlantic Oceans. In: M.G. Pitcher (Editor), *Arctic Geology – Am. Assoc. Petrol. Geol., Mem.*, 19, pp. 562–582.

Miall, A.D., 1975. Post-Paleozoic Geology of Banks, Prince Patrick and Eglinton Islands, Arctic Canada. In: C.J. Yorath, E.R. Parker and D.J. Glass (Editors), *Canada's Continental Margins and Offshore Petroleum Exploration – Can. Soc. Petrol. Geol., Mem.*, 4, pp. 557–587.

Miall, A.D., 1979. Mesozoic and Tertiary geology of Banks Island, Arctic Canada: the history of an unstable craton margin. *Geol. Surv. Can., Mem.*, 387.

Miller, A.A., 1961. *Climatology*. Methuen, London, 320 pp.

Mountjoy, E.W., 1967a. Triassic stratigraphy of northern Yukon Territory. *Geol. Surv. Can., Pap.*, 66-19.

Mountjoy, E.W., 1967b. Upper Cretaceous and Tertiary stratigraphy, northern Yukon Territory and northwestern District of Mackenzie. *Geol. Surv. Can., Pap.*, 66-16.

Myhr, D.W. and Young, F.G., 1975. Lower Cretaceous (Neocomian) sandstone sequence of Mackenzie Delta and Richardson Mountains area. *Geol. Surv. Can., Pap.*, 75-1C, pp. 247–266.

Nassichuk, W.W., Thorsteinsson, R. and Tozer, E.T., 1972. Permian–Triassic boundary in the Canadian Arctic Archipelago. *Bull. Can. Petrol. Geol.*, 20, p. 651–658.

Newman, G.W., Mull, C.G. and Watkins, N.D., 1977. Northern Alaska paleomagnetism, plate rotation, and tectonics. *Alaska Geological Society Symposium on Alaskan Geology, Anchorage, 1977*. Abstract in Program, pp. 16–19.

Norris, D.K., 1974. Structural geometry and geological history of the northern Canadian Cordillera. In: A.E. Wren and R.B. Cruz (Editors) *Proceedings of the 1973 National*

Convention of the Canadian Society for Exploration Geophysics, pp. 18–45.

Norris, D.K. (Editor), 1981. The geology, mineral and hydrocarbon potential of northern Yukon Territory and northwestern District of Mackenzie. *Geol. Surv. Can., Mem.* In prep.

Norris, D.K. and Hopkins, W.S., Jr., 1977. The geology of the Bonnet Plume Basin, Yukon Territory. *Geol. Surv. Can., Pap.*, 76-8.

Norris, D.K. and Yorath, C.J. 1981. The North American Plate from the Arctic Archipelago to the Romanzof Mountains. In: A.E.M. Nairn, F.G. Stehli and M. Churkin (Editors), *The Ocean Basins and Margins, 5, The Arctic Ocean*. Plenum, New York, N.Y., pp. 37–104.

Patton, W.W., Jr., 1966. Regional geology of the Kateel River Quadrangle, Alaska. *U.S. Geol. Surv. Misc. Geol. Inv.*, Map I-437.

Patton, W.W., Jr., 1967. Regional geology of the Candle quadrangle, Alaska. *U.S. Geol. Surv. Misc. Geol. Inv.*, Map I-492.

Pedersen, K.R., 1968. Angiospermous leaves from the Lower Cretaceous Kome Formation of northern West Greenland. *Groenl. Geol. Undersogelse, Rapp.*, 15: 17–18.

Plauchut, B.D., 1971. Geology of the Sverdrup Basin. *Bull. Can. Petrol. Geol.*, 19, pp. 659–679.

Poulton, T.P. 1978a. Internal correlations and thickness trends, Jurassic Bug Creek Formation, northeastern Yukon and adjacent Northwest Territories. *Geol. Surv. Can., Pap.*, 78-1B, pp. 27–30.

Poulton, T.P., 1978b. Pre-Late Oxfordian Jurassic biostratigraphy of northern Yukon and adjacent Northwest Territories. *Geol. Assoc. Can., Spec. Pap.* 18, pp. 445–471.

Poulton, T.P., Leskiw, K. and Audretsch, A.P., 1981. Stratigraphy and microfossils of the Jurassic Bug Creek Group of northern Richardson Mountains, northern Yukon and adjacent Northwest Territories. *Geol. Surv. Can., Bull.*; 325. In press.

Reiser, H.H., Lanphere, M.A. and Brosgé, W.P., 1965. Jurassic Age of Mafic Igneous Complex, Christian Quadrangle, Alaska. In: *Geological Survey Research, 1965 – U.S. Geol. Surv., Prof. Pap.*, 525-C, pp. C68–C71.

Richards, H.G., 1974. Tectonic evolution of Alaska. *Am. Assoc. Petrol. Geol., Bull.*, 58, pp. 79–105.

Rickwood, F.K., 1970. The Prudhoe Field. In: *Proceedings of the Geological Seminar on the North Slope of Alaska*. Pacific Sect., Am. Assoc. Petrol. Geol., 1970.

Rosenkrantz, A. and Pulvertaft, T.C.R., 1969. Cretaceous–Tertiary stratigraphy and tectonics in northern West Greenland. In: M. Kay (Editor), *North Atlantic: Geology and Continental Drift – Am. Assoc. Petrol. Geol., Mem.*, 12, pp. 883–898.

Rosenkrantz, A., 1970. Marine Upper Cretaceous and lowermost Tertiary deposits in West Greenland. *Geol. Soc. Denmark, Bull.*, 19: 406–453.

Sainsbury, C.L. and Bloomstein, E., 1977. Compressive interaction of the Siberian and North American plates, Bering Strait region, Alaska. *Alaska Geological Society Symposium on Alaskan Geology, Anchorage, 1977*. Abstract in Program, pp. 19–20.

Smith, A.G. and Briden, J.C., 1977. *Mesozoic and Cenozoic*

Paleocontinental Maps. Cambridge University Press, Cambridge, 63 pp.

Soper, N.J., Dawes, P.R. and Higgins, A.K., 1981. Cretaceous–Tertiary magmatic and tectonic events in North Greenland and the history of adjacent ocean basins. In: *Nares Strait: a Central Conflict in Plate Tectonics Studies of the Arctic. Medd. Groenl.* In press.

Stehli, F.G., 1968. Taxonomic diversity gradients in pole location: The recent model. In: E.T. Drake (Editor), *Evolution and Environment. Yale University Press, New Haven, Conn.*, pp. 163–227.

Sweeney, J.F., 1977. Subsidence of the Sverdrup Basin, Canadian Arctic Islands. *Bull. Geol. Soc. Am.*, 88, pp. 41–48.

Tailleur, I.L., 1973. Probable Rift Origin of Canada Basin, Arctic Ocean. In: M.G. Pitcher (Editor), *Arctic Geology – Am. Assoc. Petrol. Geol., Mem.*, 19, pp. 526–535.

Tailleur, I.L. and Brosgé, W.P., 1970. Tectonic History of Northern Alaska. In: *Proceedings of the Geological Seminar on the North Slope of Alaska.* Am. Assoc. Petrol. Geol., Pacific Sect., pp. E1–E20.

Tassonyi, E.J., 1969. Subsurface geology, lower Mackenzie River and Anderson River area, District of Mackenzie. *Geol. Surv. Can., Pap.*, 68-25, 207 pp.

Tozer, E.T., 1961. Triassic stratigraphy and faunas, Queen Elizabeth Islands, Arctic Archipelago. *Geol. Surv. Can., Mem.*, 316.

Tozer, E.T., 1967. A standard for Triassic time. *Geol. Surv. Can., Bull.*, 156.

Tozer, E.T., 1970a. Marine Triassic faunas. In: R.J.W. Douglas (Editor), *Geology and Economic Minerals of Canada – Geol. Surv. Can., Econ. Geol. Rep.*, 1, pp. 633–640.

Tozer, E.T., 1970b. Mesozoic and Cenozoic. In: R. Thorsteinsson and E.T. Tozer, Geology of the Arctic Archipelago (Chapter X). In: R.J.W. Douglas (Editor), *Geology and Economic Minerals of Canada – Geol. Surv. Can., Econ. Geol. Rep.*, 1, pp. 574–588.

Tozer, E.T. and Thorsteinsson, R., 1964. Western Queen Elizabeth Islands, Arctic Archipelago. *Geol. Surv. Can., Mem.*, 332, 242 pp.

Trettin, H.P., 1973. Early Paleozoic Evolution of Northern Parts of Canadian Arctic Archipelago. In: M.G. Pitcher (Editor), *Proceedings of the Second International Symposium on Arctic Geology, San Francisco, Calif. – Am. Assoc. Petrol. Geol., Mem.*, 19, pp. 57–75.

Trettin, H.P. and Balkwill, H.R., 1979. Contributions to the tectonic history of the Innuitian Province, Arctic Canada. *Can. J. Earth. Sci.*, 16, pp. 748–769.

Trettin, H.P., Frisch, T.O., Sobczak, L.W., Weber, J.R., Niblett, E.R., Law, L.K., de Laurier, I. and Whitham, K., 1972. The Innuitian Province. In: R.A. Price and R.J.W. Douglas (Editors), *Variations in Tectonic Styles in Canada – Geol. Assoc. Can., Spec. Pap.*, 11, pp. 83–180.

Vail, P.R., Mitchum, R.M., Jr. and Thompson, S. III, 1977. Seismic stratigraphy and global changes of sea level, Part 4. Global cycles of relative changes of sea level. In: C.E. Payton (Editor), *Seismic Stratigraphy: Applications to Hydrocarbon Exploration – Am. Assoc. Petrol. Geol., Mem.*, 26, pp. 83–97.

Van Houten, F.B., 1973. Origin of red beds: a review 1961-1972. In: F.A. Donath (Editor), *Annual Review of Earth and Planetary Sciences.* George Banta and Co., Vol. 1, pp. 39–61.

Vogt, P.R. and Avery, O.E., 1974. Tectonic history of the Arctic Basins: Partial solutions and unsolved mysteries. In: Y. Herman (Editor), *Marine Geology and Oceanography of the Arctic Seas.* Springer Verlag, New York, N.Y., pp. 83–117.

Vogt, P.R. and Ostenso, N.A., 1970. Magnetic and gravity profiles across the Alpha Cordillera and their relation to Arctic sea-floor spreading. *J. Geophys. Res.*, 75, pp. 4925–4937.

Williams, G.D. and Stelck, C.R., 1975. Speculations on the Cretaceous paleogeography of North America. *Geol. Assoc. Can., Spec. Pap.*, 13, pp. 1–20.

Yorath, C.J. and Cook, D.G., 1981. The Cretaceous and Tertiary stratigraphy and paleogeography, northern Interior plains, District of Mackenzie. *Geol. Surv. Can., Mem.*, 398. In press.

Yorath, C.J. and Norris, D.K., 1975. The tectonic development of the Southern Beaufort Sea and its relationships to the origin of the Arctic Ocean Basin. In: C.J. Yorath, E.R. Parker and D.J. Glass (Editors), *Canada's Continental Margins and Offshore Petroleum Exploration – Can. Soc. Petrol. Geol., Mem.*, 4, pp. 580–611.

Yorath, C.J., Balkwill, H.R. and Klassen, R.W., 1975. Franklin Bay and Malloch Hill map areas, District of Mackenzie. *Geol. Surv. Can., Pap.*, 74-36.

Young, F.G., 1975a. Stratigraphic and sedimentologic studies in northeastern Eagle Plain, Yukon Territory. *Geol. Surv. Can., Pap.*, 75-1 B, pp. 309–323.

Young, F.G., 1975b. Upper Cretaceous stratigraphy, Yukon Coastal Plain and northwestern Mackenzie Delta. *Geol. Surv. Can., Bull.* 249.

Young, F.G., 1977. The mid-Cretaceous flysch and phosphatic ironstone sequence, northern Richardson Mountains, Yukon Territory. In: *Report of Activities, Part C; Geol. Surv. Can., Pap.*, 77-1C, pp. 67–74.

Young, F.G., Myhr, D.W. and Yorath, C.J., 1976. Geology of the Beaufort-Mackenzie Basin. *Geol. Surv. Can., Pap.*, 76-11, 65 pp.

Ziegler, P.A., 1969. The development of sedimentary basins in western and Arctic Canada. *Alta. Soc. Petrol. Geol., Calgary*, 88 pp.

Chapter 2

THE NORTHWESTERN ATLANTIC OCEAN MARGIN*

JAMES P. OWENS

INTRODUCTION

Investigations of the Mesozoic geology of the western Atlantic continental margin have increased greatly in recent years. The main impetus for these studies has been two-fold: (a) economic, the potential for oil on the submerged shelf; and (b) scientific, the geology of the continental margin in relationship to plate tectonics.

It has now been well established by extensive drilling that the ocean basins are Mesozoic in age. Therefore, in many respects, a summary of the Mesozoic rocks of the western North Atlantic margin is documentation of the tectonic evolution of that ocean basin. The purpose of this paper is to evaluate the Mesozoic geology along the continental margin between Newfoundland on the north and Florida on the south. Because of the enormous area encompassed, and because of page limitations, the major emphasis of this report will be on the area between the continental shelf edge and the inner edge of the Triassic and Jurassic basins. This report is an updated synthesis of the physical stratigraphy of this region but only touches peripherally upon the many types of geophysical or biostratigraphic investigations being conducted throughout the western Atlantic Ocean.

The Mesozoic history of the western North Atlantic margin was initiated by a series of relatively small rift valleys extending from Georgia on the south to the Bay of Fundy on the north. The development of these rift valleys was sporadic through time. The exposed basins of the Carolinas were formed early in the Late Triassic (Carnian) whereas some of the northern basins began later in the Triassic (Norian) and continued to remain active into the Middle Jurassic (Bajocian). Basins apparently of Triassic to Jurassic(?) age lie beneath the Coastal Plain well to the east

of the exposed basins. The width of the zone of all these basins is approximately 400 km along much of the western Atlantic continental margin. The basins were formed on faulted continental crust and typically filled with very feldspathic clastic sediments which were introduced from the surrounding crystalline terrane by streams and/or alluvial fans. Lakes, as evidenced by lacustrine deposits, were common in most of these intermontane basins. Basaltic volcanic rocks were extruded into a few of the basins, and these flows are interstratified with the sediments introduced into the basins from their perimeters. In the northern western Atlantic margin some of the basins had some connection with marine waters as evidenced by a thick evaporite (halite) deposit (Argo Formation). The age of this unit is uncertain but possibly is pre-Early Jurassic (Hettangian). In the Middle and Late Jurassic, the area off Newfoundland and southern Florida was invaded by normal marine waters and thus the western Atlantic Ocean was established at least by this time. Later in the Jurassic much of the continental margin was arched and the crust locally attenuated, faulted, and fractured extensively. Off eastern Canada, the salts were mobilized during the faulting forming extensive diapirs which penetrated the overlying younger sediments. The fractures were extensively intruded by a diabase dike system centered somewhere east of the Carolinas (a hot spot or dome). Both the extrusive rocks and later dike systems are continental-type flows and dikes. Chemically, both these rock groups are tholeiitic (quartz-normative and olivine-normative) or "primitive" in nature. Such igneous rocks are interpreted to have originated deep in the lower crust or upper mantle.

* Publication authorized by the Director, U.S. Geological Survey.

33

Many of these rocks have been dated radiometrically, and the time spanned by these ages have been called the Palisades Event by some workers, for example, by Cornet in 1977. The Palisades Diabase has been dated at 193 m.y. (by Dallmeyer, 1975) or near the Triassic-Jurassic boundary. There is, however, considerable scatter in the radiometric ages so whether or not the Mesozoic dikes and flows were emplaced synchronously is conjectural. Field relationships, for example, suggest that the dikes (diabases) may postdate the flows even though both types seem to have the same age range.

The Early Cretaceous was a time of extensive coastal instability north of South Carolina. Large deltas, localized in many of the embayments, built well out into the oceanic basin, some at least as far as the edge of the present Continental Shelf. Large volumes of these clastic sediments were derived mainly from crystalline, metamorphic and second-cycle Appalachian sedimentary rocks. At the same time, a thick carbonate–anhydrite sequence was accreting on the southern Florida–Bahama platform.

The Upper Cretaceous rocks are spread as a thin blanket over all the older rock sequences. Volumetrically these deposits are much thinner than the underlying sediments but are much more widespread over the continental margin. Characteristically these deposits in the area north of South Carolina are dominantly shelf marine-deltaic deposits. These deposits are cyclic in nature recording several transgressions and regressions during this time interval. Lithologically, the Upper Cretaceous deposits of this region have abundant glauconite sand, and heavy-mineral assemblages indicate a metamorphic rock (Piedmont) provenance for much of the clastic sediments in these units. South of South Carolina these rocks are dominantly sands in the updip areas grading to carbonate sediments downdip. Glauconite, though locally common, is much less abundant in the southern Atlantic Coastal Plain than in the northern.

PHYSIOGRAPHY

Most of the eastern margin of North America, particularly south of New England, is bordered by a low-lying terrane. This terrane is divided into two physiographic provinces, the Piedmont and the Atlantic Coastal Plain (Fig. 1). The Coastal Plain physiographic province may be subdivided into two parts: the emerged and the submerged. Nearly all the

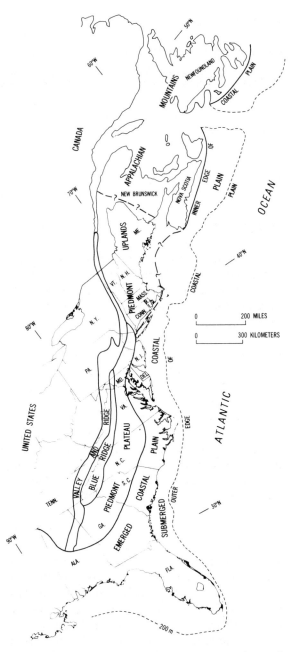

Fig. 1. Map of physiographic provinces bordering the northwestern continental margin of the North Atlantic Ocean.

emerged part of the Coastal Plain is south of New England, and in this region nearly all of the geologic studies have been made. In general, the Coastal Plain is composed largely of unconsolidated sediments which have been shaped into a flat to gently rolling topography.

The Piedmont physiographic province is adjacent

to the Coastal Plain typically at slightly higher elevations. The bulk of this province is underlain by
crystalline rocks. The near-surface rocks in this province have been widely saprolitized, and the surface is
typified by irregular to gently rolling topography.
Most of the exposed Triassic and Jurassic basins are
within this province.

In New England, crystalline rocks similar to those
in the Piedmont to the south are widespread; however,
the terrane is hilly to mountainous. Maher (1971)
referred to this region as the Piedmont uplands
(terminology followed in this report). Northward into
Canada, the character of some of the rocks changes
from crystalline to consolidated unmetamorphosed
sedimentary rocks, and this part of the terrane is
referred to as the Appalachian Mountains (Canada
Geological Survey, 1969). Between New England and
Newfoundland, this mountainous terrane is bordered
on the east by the submerged Coastal Plain.

Physiographic provinces west of the Piedmont —
Piedmont Uplands — Appalachian Mountains terrane
are shown in Fig. 1, but are not discussed in this
study.

STRUCTURAL ELEMENTS

In general, the sub-Coastal Plain continental
margin consists of a series of arches (highs) and
embayments (lows) which divide this region into
broad basins and troughs. Fig. 2 shows the tectonic
elements of the Coastal Plain physiographic province
which will be discussed throughout this report. A
major omission from this diagram is the exposed
rift-valley system ("Triassic") west of the Coastal
Plain physiographic province (see Fig. 3). It should
be stressed here that most of these features have been
defined by the stratigraphic distribution of the Coastal
Plain beds, not by observed folds or faults, particularly basement faults. Off eastern Canada, however, the
basins appear to be controlled by basement faults
(see Jansa and Wade, 1975). It is possible, therefore,
that most of the southern basins have such an origin,
but structural data to support such a contention is
lacking or poorly documented. An interesting feature
of some of the basins or depocenters is their mobility.
Perhaps the most intensively studied basins are the
Salisbury and Raritan embayments (c.f. Petters,
1976; Wolfe, 1976; Doyle and Robbins, 1977; Owens
et al., 1977; among others). Here the constructional

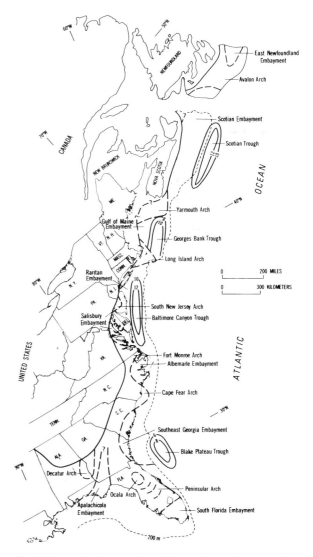

Fig. 2. Tectonic elements found along the northwestern Atlantic continental margin. As shown, the margin consists of
a series of arches and embayments which are flanked by
elongate troughs perpendicular to axes of embayments.
Trough depths are in kilometers. The shelf edge is approximately at the 200-m isobath. Not shown on this diagram are
the Triassic and Jurassic basins (modified from Owens, 1970,
and Maher, 1971).

deltas which filled these embayments migrate from
south to north through time; Barremian in the south
to Cenomanian in the north. The age of basement
deformation then varies through time. The oldest
rocks found in the emerged part of the Salisbury
Embayment are Early Cretaceous in age (Barremian),
whereas the oldest rocks in the southeastern Georgia
embayment are Late Cretaceous (Cenomanian) (R.A.

Christopher, written communication, 1977). Basin mobility is, in fact, the dominant characteristic along this so-called passive margin or trailing edge of the continent along the western North Atlantic margin.

In general, the embayments are flanked seaward by very deep troughs which trend perpendicular to the embayment axial trends. From north to south these troughs, as shown in Fig. 2, are the Scotian Trough, Georges Bank Trough, Baltimore Canyon Trough, and the Blake Plateau Trough.

Brown et al. (1972) proposed a structural model for the Atlantic Coastal Plain primarily derived from the stratigraphic patterns observed between New Jersey and North Carolina. Basically, within this model, the basement is faulted into a series of grabens and half grabens which tilted in different directions through time. Generally, however, the blocks were bounded by faults (termed hinge zones) one set oriented north—south and the other N—25°E. Associated with these faults were complementary northwest- and north-striking right-lateral shear fractures, respectively. Sheridan (1974a, b, 1976) expanded on this concept, and on the basis of a variety of geophysical methods, proposed a model which provided for all the known depositional basins along the North American continental margin. His model, however, required rotation of the blocks to restore the margin to its pre-rifting state. Known basement faults of the type proposed by the above cited authors and their relationship to known depositional basins have yet to be documented. Recently, however, small reverse faults have been observed along the inner edge of the Coastal Plain in southeastern Virginia (Mixon and Newell, 1977) and in eastern Georgia (Powell and O'Connor, 1979). These faults lie along the presumed Fall Line between the Coastal Plain and Piedmont province, but the reverse nature of the observed faults is unusual for a hinge-zone fault. In any case, the known "hinge" zone faults have a general northeast—southwest orientation but the complementary faults proposed by Brown et al. and Sheridan have yet to be found.

STRATIGRAPHY

The Mesozoic rocks of the western Atlantic seaboard are subdivided in this report into four time-stratigraphic units: (1) Late Triassic to Early Jurassic; (2) late Early to Late Jurassic; (3) Early

Cretaceous; and (4) Late Cretaceous. Because data for the younger sediments are much more abundant, particularly those of Late Cretaceous age, these deposits will be discussed in relationship to the various stages of the Late Cretaceous. Paleogeographic maps have been prepared for all the chronostratigraphic units, but because of the small scale of the maps, only generalized lithologic trends are shown (sand—shale—limestone). In some areas, marine sands interfinger with nonmarine sands; but no attempt is made to separate these sands into their genetic types.

Upper Triassic and Lower Jurassic rocks

Distribution

Rocks of this age are exposed along the western margin of the study area or buried beneath the Coastal Plain from the panhandle of Florida to east of Newfoundland (Fig. 3). Typically these rocks are deposited in narrow basins elongated parallel to the Appalachian Mountain trend. Although the individual basins are narrow, collectively they form a belt as much as 400 km wide. The best exposed and most investigated of these are the western basins. Basically, all these studies indicate marked similarities in structural style and sedimentological characteristics within the basins. In most reports, the fill in these basins has traditionally been referred to as the "Triassic red beds or red-bed sequences". Unfortunately, in many areas thick sections of red-colored sediments are also found in younger Coastal Plain units (for example, the Potomac Group of Early to early Late Cretaceous age in the middle Atlantic States), and new fossil data suggest that part of these "Triassic" sections are of Jurassic age (Cornet, 1977).

Thicknesses

Estimates of the total thickness of the sedimentary and extrusive rocks in the basins of this age are difficult to assess. Some authors have added up measured stratigraphic thickness and come to very thick estimates of basin fill (Lee, 1977). Others have argued that this classical stratigraphic technique cannot be used in such a tectonic setting as the basin is filled from the perimeter of the rift valley; the formations in the basin are not sheets but differ greatly in thickness from area to area (Van Houten, 1969; Bain and Harvey, 1977). Additionally, the basin fill has been faulted post-depositionally and complicates the addi-

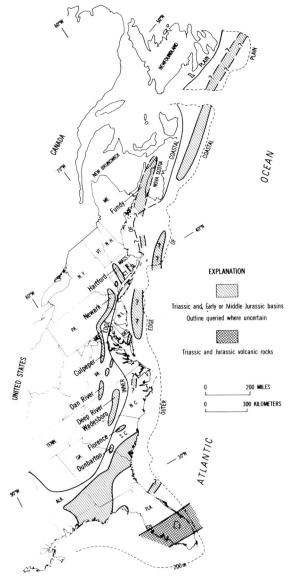

Fig. 3. Approximate outline of the major known or inferred Triassic and Jurassic basins along the northwestern continental margin. The presence of basins east of the continental margins is poorly understood. Note several of the basins lie to the west of the inner edge of the Coastal Plain (modified from Marine and Siple, 1974; Bain and Harvey, 1977; Van Houten, 1977; Hubert et al., 1978; Gohn et al., 1978).

tive stratigraphic-sequence measurement technique (Bain and Harvey, 1977).

The basin fill is reported to be about 1800 m in the Wadesboro–Deep River Basin (Reinemund, 1955), 5000 m in the Dan River Basin (Thayer, 1970), 5000–6000 m in the Newark Basin (Van Houten, 1969), 3600 m in the Hartford Basin (Hubert et al., 1978) and 1100 m in the Fundy Basin (Klein, 1962).

Offshore, because of extensive block faulting and lack of diagnostic age criteria, the thickness of beds of this age is poorly understood. Jansa and Wade (1975) conclude, however, that units of this age in this region may be as much as 1500 m thick. Schlee (1978) estimates that as much as 8 km of strata of this age may be present in the Georges Bank Trough, on the basis of geophysical characteristics.

In the Baltimore Canyon Trough, Poag (1978) in his schematic cross-section, depicts only a thin section of Triassic(?) strata. How much may be equal in volume to the Georges Bank thickness is unknown.

Less is known about the thickness of the red-bed sequences beneath the emerged Coastal Plain. In Georgia one such basin (Dunbarton) has more than 250 m of red beds (Marine and Siple, 1974). Near Charleston, S.C. the interbedded volcanic red-bed sequence is a minimum of 400 m thick (Gohn et al., 1978). In the subsurface of Georgia the red-bed volcanic interbeds may be more than 1000 m.

All the thicknesses quoted are maximum estimates.

Age

The rocks in this chronostratigraphic group have been dated in a variety of ways (vertebrate faunas, K/Ar dating of the intrusive and extrusive rocks; and most recently, pollen and spores). Cornet (1977) summarized much of these data. In particular, he studied the pollen and spores in much detail and part of the data in Table I are based on his findings. In general, he proposed that the basins were younger from south (in his particular study, the Wadesboro–Deep River Basin) to north (Hartford Basin) (Table I). He also noted that the radiometric dates on the flows and intrusive rocks grouped near 190 m.y. which he termed the Palisades Event. He considered this magmatic event to be coeval between the basins and would have erupted in time along the Triassic–Jurassic boundary. Actually, considerable radiometric age is spread within the intrusive and extrusive rocks associated with this rifting in eastern North America. Clearly the flows and diabase dikes and sills are temporally separated, the flows predating the sills and dikes. Unfortunately, the imprecision of the K/Ar method due to a variety of reasons, does not resolve this age date differential. In fact, on the basis of the radiometric results, the red-bed sequence could be as old as Early Triassic or as young as Late Jurassic. The best resolution of age therefore must rely upon the sporomorphs which appear to be the best tool for age zonation.

TABLE I

Stratigraphic nomenclature of the Triassic and Jurassic formations in the western Atlantic region*

Era	System	Epoch	Stage	Southern Florida (1)	Dunbarton, Ga.	Charleston, S.C.	Wadesboro–Deep River, N.C. (2)	Dan River, N.C.–Va. (2)	Culpeper, Va. (2)	Newark–Gettysburg, N.J.–Pa. (2)	Hartford, Conn. (4)	Fundy, N.B., Canada (2)	Scotian, Canada (3)	East New-foundland, Canada (3)
MESOZOIC	JURASSIC	LATE	Tithonian	Fort Pierce Formation (in part)		Absent		Absent	Absent	Absent	Absent	Absent	Mic-Mac and Abenaki Formations	Mic-Mac and Verrill Canyon (in part) Formations
			Kimmeridgian	Unnamed limestone unit										
			Oxfordian	Absent									Mohawk Formation	Mohawk Formation
		MIDDLE	Callovian								? ? ?			
			Bathonian											
			Bajocian											
		EARLY	Toarcian								Portland Formation		Iroquois Formation	Iroquois Formation
			Pliensbachian								?			
			Sinemurian								East Berlin Formation		? Argo Formation	Argo Formation
			Hettangian			Basalt		Stoneville / Cow Branch / Pine Hill	Bull Run Formation / Balls Bluff Siltstone / Manassas Sandstone	Brunswick Shale	? / Shuttle Meadow Formation(?) / New Haven Arkose	Scots Bay Formation		
	TRIASSIC	LATE	Rhaetian		? ?	Unnamed Red beds(?)	Sanford, Cummock and Pekin Formations		? ?	Lockatong and Stockton Formations	Absent	Annapolis Formation(?) / Blomidon Formation	Eurydice Formation ?	Eurydice Formation ?
			Norian			? ?			Absent			Wolfville Formation ?		
			Carnian		Unnamed Red beds									

*Data modified from the following sources: (1) Meyerhoff and Hatten, 1974; (2) Cornet, 1977; (3) Jansa and Wade, 1975; (4) Hubert et al., 1978.

Structural characteristics

Most of the Upper Triassic and Lower Jurassic beds are found within elongate, sharply bounded basins. The long axes of these basins commonly are oriented parallel to the pre-existing trends found in the adjacent older Appalachian terranes. Specifically, the basins have a general northeast trend except for the Hartford Basin which trends nearly north. All the basins were extensively faulted after the deposition of the basin sediments. Typically, block faulting is most common, and the strata within each basin are tilted, suggesting a half-graben form as the dominant structural style. However, not all of the basins have their strata tilted in the same direction; some have a prominent westerly dip component (Newark and Dan River basins, for example) whereas others have a strong easterly dip component (Hartford and Deep River, for example). Most of the basins have one major border-fault system in which the faults are high-angle normal in nature. In most areas the faulted borders are marked by the presence of widespread fanglomerate beds. The notable exception to this small, discrete basin distribution is in the South Carolina to Florida area (Fig. 2). Here the "red-bed"-volcanic sequences have a much greater areal distribution and certainly do not have the "elongate" basin shape common in the north. The reasons for this widespread areal distribution are still uncertain, although some workers postulate that Georgia—Alabama—South Carolina is an aulacogen (failed arm) during a spreading episode. The occurrence of widespread Triassic and Lower Jurassic volcanic rocks in the south Florida Embayment (Fig. 3) however, does not make such an origin probable. Cramer (1975) shows numerous faults through this region associated with a large northeast-trending sub-Coastal Plain graben beneath the Georgia Coastal Plain. These faults which apparently control the distribution of Triassic sedimentary rocks, do not conform to the model shown by Gohn et al. (1978). The reasons for this unusual broad "basin" in Georgia are still poorly known.

Some authors have proposed that Late Triassic and Early Jurassic basins originally also had a much broader distribution in the north, "the broad terrane" hypothesis (Russell, 1880; Longwell, 1922; Sanders, 1960). In this hypothesis the Triassic and Jurassic basins were once part of a very large rift valley which was arched post-depositionally and then was extensively bevelled. The basins as we see them today are only edge remnants of this large arch.

This simplistic structural interpretation, however, has serious deficiencies. If the major troughs beneath the submerged Coastal Plain shown in Fig. 2 are underlain by Triassic rocks as proposed by others (see Minard et al., 1974, or Schultz and Grover, 1974), then a single broadly arched rift valley does not fit the distribution of a large single-rift system. Additionally, as will be discussed shortly, the basins were formed at different times, thus could not be a single megarift system.

In summary, the Late Triassic and Early Jurassic basins are interpreted by most writers to represent the stage of continental fragmentation and early rifting during which North America began to decouple from Europe and Africa. As such, the rift valleys would have resulted from extensional forces. Tectonic activity (subsidence) and the accompanied sedimentation continued however in many of the basins for a long time period after the initial fracturing (according to Cornet 1977, fig. 33) as much as 46 m.y. Later, as subsidence continued, basaltic volcanic rocks for the most part were erupted on the basin floor. In the Newark Basin, three such major eruptive events took place (Watchung basalt flows, Faust, 1975). This eruptive period took place somewhere near the Triassic—Jurassic boundary (ca. 186—192 m.y.). Sedimentation continued after this eruptive period, perhaps well into the Middle Jurassic (Bajocian stage; Cornet, 1977). At some time following this sedimentation period, the basins were warped extensively, and the strata tilted and extensively fractured. Numerous diabase dikes filled the fracture systems which extended well beyond the basin margins and into the surrounding crystalline rocks (see Popenoe and Zietz, 1977).

An igneous rock-age problem, however, is related to this sequence of tectonic events. The few dates from the volcanic rocks and diabases suggest they are approximately coeval. If the structural history above is valid, then a conflict exists with the K/Ar age determinations because the dikes must be considerably younger than the extrusive rocks based on the areal distribution of these rocks. The dikes show essentially the same "primitive" tholeiitic character as the extrusive rocks which are interbedded with the red-bed sequence (Gottfried et al., 1977), although they appear to postdate the extrusive rocks. Additionally, as shown by Popenoe and Zietz (1977), these dikes extend far beyond the basin boundaries in South

Carolina. May (1971) also indicates the widespread distribution of these dikes along the circum-Atlantic regions and postulated that they originally radiated from a zone (hot spot or dome) located somewhere beneath the Blake Plateau east of the South Carolina coast.

Sedimentological characteristics

The sediment fill varies from basin to basin, but certain lithologies are common to all. Basically, these are the red-bed (or red-colored) sequences found in many areas of eastern North America.

A general textural distribution pattern is common to all the basin sediments; that is, coarse clasts typically concentrated near the basin margins grade to finer clasts near the basin center or in topographic lows within the faulted basins. Such a model is shown for the Dan River Basin of North Carolina—Virginia (Fig. 4). This model, however, does not apply to all basins. Some of the basins appear to be composites of more than one basin, and in those the distribution of the facies varies in a time sense between basins. The variation in the spatial relationships of the fine-grained to coarse-grained facies in vertical sections has been demonstrated (Fig. 5) in the composite Deep River—Wadesboro Basin (Reinemund, 1955; Bain and Harvey, 1977).

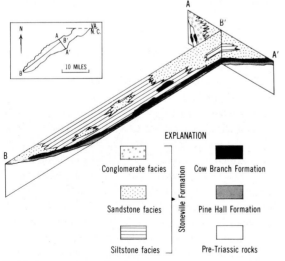

Fig. 4. Cross-sections showing textural variations in Triassic and Jurassic sedimentary rocks from the Dan River Basin (from Thayer, 1970). Note that the fine-grained facies of Thayer's Cow Branch Formation and siltstone facies of Thayer's Stoneville Formation are thickest in the central part of the basin.

Most of the rocks in the basins are consolidated, although locally some are semiconsolidated. The sandstones are red, brown or white and are extensively cross-stratified. The red-colored siltstones or claystones are typically massive. Both these consolidated rock types are considered to be fluviatile in origin (channel sands and overbank siltstone deposits). Root casts are locally very abundant in the red-colored siltstones, but other trace fossils are rare. Thinly bedded, limey, fine-grained beds intercalated with the fluviatile beds are considered to be lacustrine in origin. Root casts are common in some of these lacustrine sequences as well as other trace fossils (cray fish, *Scoyenia?* burrows). A fresh water fauna dominated by benthonic types (mostly fresh water pelecypods and ostracods) are locally common in many of the basins (Bain and Harvey, 1977). The thin-bedded lacustrine sequences occur in a variety of colors (gray-green, reddish-green or black). The red color possibly is dependent upon the amount of time these beds were subaerially exposed during or after the time of deposition.

Compositionally the basin fill from the Fundy Group on the north to the Deep River—Wadesboro Basin on the south varies considerably. The basin fill typically reflects the character of the rocks surrounding the basins. All the known basins were formed on continental crust, and thus the sands typically are first-cycle, immature, and very feldspathic. Arkoses and high-rank graywackes (Krynine, 1948) are common sand types found in many of the basins that reflect a crystalline rock provenance. In other basins, unmetamorphosed Paleozoic limestone clasts in the border fanglomerates of the Newark and Culpeper basins attest to the Appalachian sedimentary source for some of the detritus in such basins. Perhaps the most elegant relationship between basin fill and nearby source rocks is within the Fundy Basin (Klein, 1962). Here, in a single formation (Wolfville), a wide variety of sand composition was observed. The sandstone types were shown to be related directly to the underlying crystalline rocks (producing arkoses) or to Paleozoic sandstones (producing orthoquartzites).

The finer-grained lacustrine beds contain the bulk of the chemical precipitates in the Late Triassic and Early Jurassic basins. These chemically produced sediments are mainly limestones, sometimes dolomite, or as in the Newark Basin, they contain abundant double salts and zeolites (Van Houten, 1969). Layered

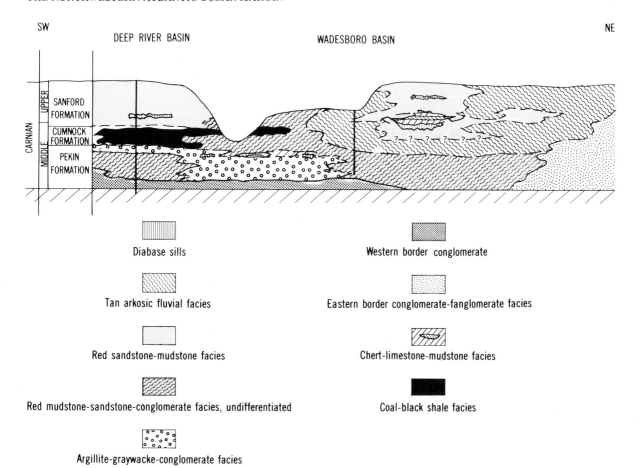

Fig. 5. Cross-section through the Deep River–Wadesboro Basin (Fig. 3) showing the facies variations within and between the two basins. Note the more complex facies patterns in these basins than found in the Dan River Basin (Fig. 4) (modified from Bain and Harvey, 1977).

chert, also thought to be a chemical precipitate, is found in some of the basins (Wadesboro–Deep River, Bain and Harvey, 1977).

Igneous rocks

Igneous intrusive and extrusive rocks, basaltic in composition, also are present in many of the basins. The basalts and diabases are largely tholeitic continental types (Gottfried et al., 1977). Such basalts and diabases, because of their light rare-earth enrichment and high K/Rb ratios, are interpreted to be "primitive" in nature, and closer to the theoretical mantle composition than most other continental basalts.

In some basins, most particularly the northern ones (Culpeper, Newark, Hartford, and Fundy), basaltic flows are interbedded with the basin sedi-

ments. No basaltic flows of Triassic or Jurassic age are known outside the limits of the basins.

The intrusive dike systems which are present in South Carolina, North Carolina, and southern Virginia are abundant and cut across the basin at a high angle to the axial trend of the basins (Fig. 6). These dike systems in many areas extend well beyond the basin limits and are interpreted to postdate the flows. The systems apparently are related to a large upwarping or doming of the basement during the early rifting phase of the Atlantic Ocean (May, 1971).

Paleoenvironmental and tectonic interpretation

The Upper Triassic and Lower Jurassic sedimentary rocks were largely deposited in mixed fluviatile, alluvial fan, and lacustrine environments. The one notable exception is the Argo Formation, which, because of its thick halite concentrations, would

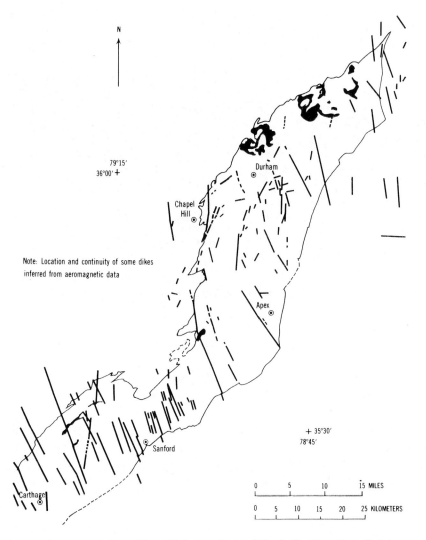

Fig. 6. Dike system in Deep River—Wadesboro Basin of North Carolina. Note that in general the dikes cut across the long axis of the basin and that some are outside the basin margins (modified from Bain and Harvey, 1977).

suggest some marine influence to supply the large quantity of saline water needed to produce such a volume of salt (about 600 m).

In a typical Triassic and Jurassic depositional basin, coarse clastic sedimentary rocks were introduced into the basin either by alluvial fans or by braided streams. The association of fanglomerates with step-faulting along the borders of the basins indicate that at least part of the basin fill was of alluvial fan origin. A similar relationship has been noted commonly in the intermontane basins of the western United States. A large part of the basin fill, however, was introduced by streams which entered the basin (Van Houten, 1969). The typical association of

channel-fill cross-bedded sand and massive overbank deposits point to a fluvial origin for a large part of the fill. The formation of caliche in many of the alluvial overbank beds suggests that a hot, arid climate prevailed at least in many of the basins. Thinly laminated deposits of lacustrine origin are present in all of the exposed basins. Chemical precipitates (mainly carbonate rocks but locally zeolites and some chert) are common in these lacustrine strata. Olson (1978) has shown that lacustrine facies are widespread in most of the basins, and has classified the lakes as to their original depths; deep-to-shallow facies (Newark and Dan River basins) or only shallow facies (Deep River—Wadesboro).

Many of the extrusive volcanic rocks of this age confined to the basins have well-developed pillow structures which may be very vesicular. The well-developed pillows suggest eruption into standing bodies of water. Many of the basalts, particularly in the Newark and Hartford basins, are interbedded with the sedimentary rocks of Early Jurassic or latest Triassic age (Cornet, 1977).

Structurally, the basins appear originally to have been rift valleys. This rifting is presumed to represent the earliest break-up of the supercontinent (Pangaea), but to predate the first marine incursions of the modern Atlantic Ocean. Rift-related fractures were apparently parallel to older Appalachian fracture systems. Post-depositionally, the basins were arched and extensively faulted. Numerous diabase dikes cut the faulted rocks but also extended beyond the limits of the basins (Popenoe and Zietz, 1977). Diabase sills or plugs apparently intruded the basins at the same time. Interestingly, the basalts and the later diabase sills have essentially the same chemical characteristics (Gottfried et al., 1977).

Upper Lower to Upper Jurassic rocks

Distribution

Rocks of late Early to Late Jurassic age have a limited areal distribution. These rocks are best known in the submerged Coastal Plain of eastern Canada and in the southern Florida Embayment (Fig. 7). As illustrated in Fig. 7, rocks of this age are probably present in the Georges Bank and Baltimore Canyon troughs and probably seaward of the Continental Shelf margin along much of the Atlantic seaboard east of the Continental Slope (Poag, 1978). In general, rocks of this age are east of the Upper Triassic and lowermost Jurassic rocks. This is especially true east of the Continental Shelf and south of North Carolina.

As shown in Table I, these rocks are best known in the submerged Coastal Plain off eastern Canada. Here there was an oscillation between nonmarine and marginal marine facies during this time period. The marginal marine sequence primarily consists of abundant salt (Argo?), Lower Jurassic or uppermost Triassic, or anhydrite—carbonate sequence (Iroquois), to mixed marine nonmarine (Mohawk), and to more normal marine (parts) of the upper part of the Mic-Mac, Abenaki and Verrill Canyon in Late Jurassic time. Rocks of probable Late Jurassic age are also present in the southern Florida Embayment and

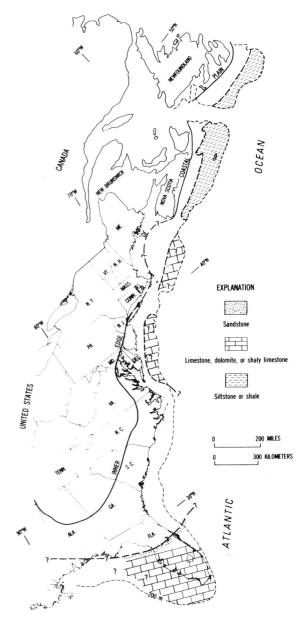

Fig. 7. Distribution of the upper Lower to Upper Jurassic rocks along eastern North America. The presence and lithology of sedimentary rocks of this age in the Georges Bank Trough and Baltimore Canyon Trough is problematical.

probably beneath the submerged shelf west of Florida. Here the section includes parts of the Fort Pierce Formation and possibly includes equivalents of the Louann Salt of the Gulf Coast region off the tip of Florida (Meyerhoff and Hatten, 1974).

Thickness

The upper Lower to Upper Jurassic sedimentary rocks vary considerably in thickness between basins. In the Scotian area, sedimentary rocks of this age are approximately 1600 m thick (Jansa and Wade, 1975). They are established to be as much as 4000 m in the Georges Bank Trough (Schlee, 1978) with perhaps an equal thickness in the Baltimore Canyon Trough and in the Blake Plateau Trough (Scott and Cole, 1975). In the south Florida Embayment, beds of this age are thin or absent, depending upon how much of the Fort Pierce Formation is of Late Jurassic age. In general, the beds of this age are thinner than those of Late Triassic to Early Jurassic age along the continental margin.

Age

The dating of the upper Lower to Upper Jurassic sedimentary rocks is by means of fossils in most areas. In the Scotian Shelf, the zonation of dinoflagellate and spore occurrences was used (Williams, 1975). There, the oldest unit dated by palynomorphs is the Mohawk Formation (Bathonian). In Florida, even though parts of the Fort Pierce Formation contain a rather diverse planktonic forminaferal assemblage (Applin and Applin, 1967; Meyerhoff and Hatten, 1974), not enough diagnostic forms are present to positively establish either a Late Jurassic or an Early Cretaceous age (Table I). As interpreted by Meyerhoff and Hatten (1974), the lower part of the Fort Pierce and an underlying unnamed unit might be considered

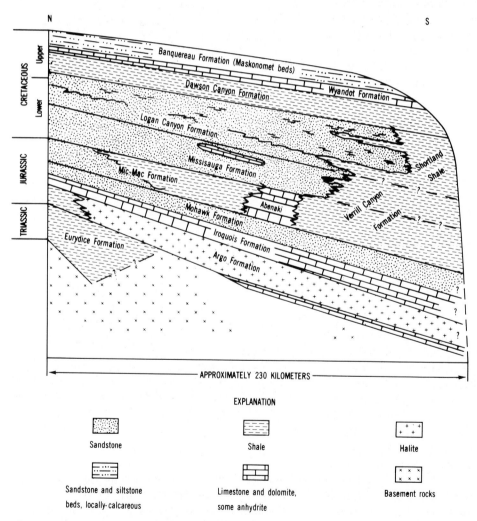

Fig. 8. Idealized cross-section across the Scotian Shelf. Data largely from Jansa and Wade (1975). Note relative abundance of clastic sediments in section, particularly in the Cretaceous units.

to be latest Jurassic in age (Table I). On the basis of the present published information, however, the bulk of known Middle to Upper Jurassic sedimentary rocks are in the north off eastern Canada.

Structural characteristics

Numerous seismic-reflection profiles across the Scotian Shelf and Grand Banks show that the Jurassic sections are block-faulted and that the strata within the fault blocks are extensively intruded by salt diapirs (Jansa and Wade, 1975). The structures in southern Florida are less well known (Applin and Applin, 1965; Meyerhoff and Hatten, 1974). No diapirs are presently known in peninsular Florida but diapirs are reported to be present under the shelf west of Florida (Meyerhoff and Hatten, 1974). Whether the south Florida beds are faulted to the extent shown off eastern Canada is not known. Southward in Cuba, however, strata of this age are faulted extensively (Meyerhoff and Hatten, 1974) and resemble the faulted basement off eastern Canada.

Sedimentary characteristics

The lithologic characteristics of the sedimentary rocks of this age sequence are considerably different from those in the chronostratigraphic unit previously described. Specifically, the upper Lower to Upper Jurassic sequences have more marine characteristics than the older rock sequences in which fluvial–lacustrine depositional environments dominated.

The character of upper Lower to Upper Jurassic beds is best known for the Scotian Shelf area off eastern Canada and the south Florida Embayment where both basins have been extensively drilled. Fig. 8 is an interpretive north–south cross-section through the Scotian Shelf. Because of extensive block faulting, the units in this chronostratigraphic group are only in scattered sections and are not stacked continuously as shown. This illustration therefore is a diagrammatic, composite, restored section.

The oldest unit (Argo Formation) placed in this stratigraphic sequence is a very thick massive halite which interfingers updip with a massive red shale—

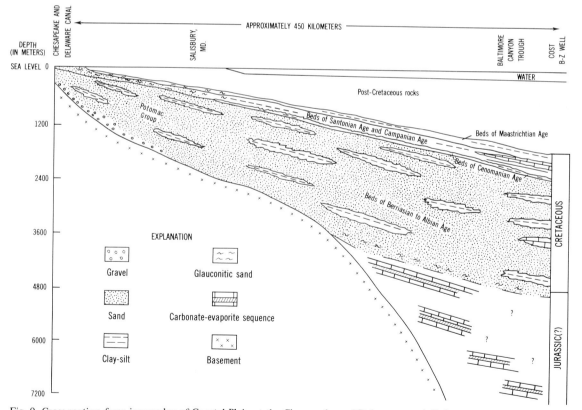

Fig. 9. Cross-section from inner edge of Coastal Plain at the Chesapeake and Delaware canal, Delaware, through the central part of the Salisbury Embayment and into the Baltimore Canyon Trough (data sources: Doyle and Robbins, 1977; Scholle, 1977; Mattick et al., 1974).

anhydrite sequence. This dominantly salt bed is overlain by a thick dolomitic unit. The bedding in this dolomitic unit (Iroquois Formation) varies throughout the Scotian Basin but is characterized by horizontal lamination, locally by bioturbation and graded bedding; small-scale cross-bedding and rare fossils have also been recorded. This unit has been interpreted to be largely of basin-edge (tidal-flat) origin although presumably equivalent beds in the Grand Banks area are considered to be a deeper water marine facies (Jansa and Wade, 1975). The dolomite is overlapped by a clayey medium- to coarse-grained conglomeratic sandstone locally interstratified with thin fossiliferous limestone (Mohawk Formation). The sandstone in this unit is extensively cross-bedded. Jansa and Wade (1975) consider the sandstone to be largely nonmarine (alluvial deltaic plain), but nearshore, because of the interstratified marine limestone.

The youngest Jurassic beds in the Scotian Shelf region are the Mic-Mac, Abenaki, and Verill Canyon units. These units laterally form a complete marine-facies sequence common in many other sedimentary basins. Closest to shore, the Mic-Mac Formation is dominantly a sand-containing carbonaceous, glauconite sand locally interbedded with thin fossiliferous limestone beds. Basinward, the limestones increase in thickness and become the dominant lithology. These beds are called the Abenaki Formation. Near the shelf edge, the Abenaki interfingers with the Verrill Canyon Formation, a dominantly shaly unit, calcareous in proximity to the Abenaki and more of a typical clastic shale where the Abenaki interfingers with the Mic-Mac beds. The Verrill Canyon is moderately fossiliferous throughout.

Fig. 9 is a cross-section through the emerged and submerged Coastal Plain in the middle Atlantic States from near the inner edge of the Coastal Plain to the Baltimore Canyon Trough. The lithologies in the deeper parts of the trough are based on seismic characteristics and thought to be largely a carbonate—evaporite sequence of possible Jurassic age (Mattick et al., 1974). Whether any thick salt beds are present in this sequence is to date unknown. In many respects

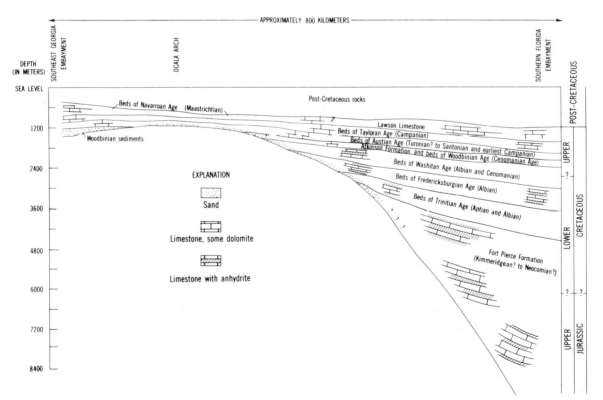

Fig. 10. Cross-section from southeast Georgia Embayment on the north to south Florida Embayment on the south. Note the abundance of carbonate rock in this part of the Atlantic Coastal Plain as compared with the Scotian Shelf area (data from Applin and Applin, 1965, 1967; Meyerhoff and Hatten, 1974).

the lithologies of Jurassic(?) beds in this part of the coast resemble those in the Scotian Trough.

Fig. 10 is a cross-section from southern Georgia to southern Florida. In general, the Jurassic section in peninsular Florida is included in the Fort Pierce Formation. This unit consists of a thin basal red-bed clastic (arkosic) sequence which underlies the formation near the basin edge and perhaps basinward. This thin sand is overlain by a carbonate—evaporite sequence in which limestone is more abundant than dolomite. Thin to thick beds of anhydrite are commonly interstratified with the carbonate rocks (Applin and Applin, 1967). This unit thickens southward into the Florida Keys area. Meyerhoff and Hatten (1974) suggest that Louann Salt equivalents (Jurassic) may be very thick south of Florida. Clearly, the possible Jurassic sequence in this area is a carbonate—evaporite sequence with a thin nearshore sand facies and is similar to the facies in the troughs to the north.

Paleoenvironmental interpretations

The late Early and Late Jurassic time presents the first definite indications that marine waters occupied large parts of the western Atlantic Ocean Basin. Apparently, during this time, particularly during the late Middle Jurassic, the sea occupied isolated block-faulted basins which had restricted water circulation and in which large evaporite deposits were formed (Argo Formation, for example). Broad tidal-flat conditions producing thick supratidal deposits gradually encroached over the northern basins producing thick carbonate sequences. By latest Jurassic time, the Scotian Basin was overlapped by normal open-shelf conditions (Mic-Mac, Abenaki, and Verrill Canyon association) in which clastic dominated.

The geologic record along the southeastern Atlantic States, is less complete. Only the uppermost Jurassic beds are known to be present here and these sediments are shallow-water limestones interbedded with lesser amounts of dolomite and anhydrite. Some have interpreted the full section in the Scotian area and incomplete record to the south as an indication that the Atlantic Ocean opened from north to south. Interestingly, this is exactly the reverse of the early rifting pattern in the Late Triassic and Early Jurassic basins as documented by Cornet (1977), i.e., basins becoming younger from south to north.

Part of this problem may be because of the lack of well-documented data for the age of the thick volume of sediments east of the Continental Shelf off the southeastern Atlantic States in the Blake Plateau Trough. Whatever the case, the Jurassic(?) sediments off the southeastern Atlantic States were deposited on a relatively featureless basement and probably are mostly carbonate bank in origin. Evaporite facies (anhydrite) appear most common in the eastern Gulf of Mexico (Louann Salt) and perhaps in the Bahamas (unnamed unit) (Meyerhoff and Hatten, 1974).

It is now apparent that marine conditions first prevailed in this part of the western Atlantic Ocean Basin (rift zone?) during the Jurassic. The ultimate answers to the exact timing and location of this or these events still remain to be worked out. Tectonic activity seems to have been more intense in the Scotian Shelf area where clastic sediments are much more common than in the southern Florida Embayment.

Lower Cretaceous beds

Distribution

For a long time, the oldest post-Triassic deposits thought to be in eastern North America were Early Cretaceous in age. In part, this concept was the result of early botanical studies on the basal Coastal Plain beds of the middle Atlantic States (Fontaine, 1889). Only in the Salisbury Embayment (Fig. 2), were the oldest Coastal Plain beds exposed along the western Atlantic seaboard and therefore accessible for study.

Subsequent subsurface studies of the emerged Coastal Plain, and dredging and drilling of the submerged Coastal Plain, have shown that rocks of Early Cretaceous age are present in the study area and have the distribution shown in Fig. 11. This diagram shows that the Lower Cretaceous rocks are more widespread than the Upper Triassic and Lower Jurassic or Middle to Upper Jurassic sedimentary rocks, particularly from the Albermarle Embayment to the Scotian Shelf. Although it has been suggested that the Lower Cretaceous rocks are thick and widespread in the Apalachicola Embayment in southwestern Georgia (Cramer, 1975), preliminary palynological studies of the lower Coastal Plain section there indicates the presence of lower Upper Cretaceous (Cenomanian) rocks (R. Christopher, written communication 1977). Thus, the Lower Cretaceous beds in the study area are much more abundant from North Carolina to the Scotian Shelf than to the south. The notable exception in this southern area is the south Florida Embayment where a nearly complete Lower Creta-

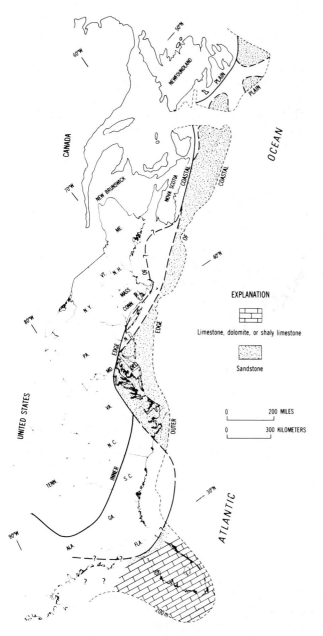

Fig. 11. Distribution of Lower Cretaceous sediments along eastern North America. Note that the bulk of the sediments from North Carolina to Newfoundland is clastic. In southern Florida carbonate sediments dominate.

ceous carbonate-rock section is present (Applin and Applin, 1965).

Thickness

In the Scotian Shelf area the Lower Cretaceous sediments aggregate about 5000 m (Jansa and Wade,

1975) with about 2400 m in the Baltimore Canyon Trough (Scholle, 1977). South of central North Carolina, Lower Cretaceous sediments are absent except in Florida where they are at least 3000 m thick in the southern Florida Embayment (Applin and Applin, 1965).

Age

The age of the Lower Cretaceous rocks north of Florida was determined by dinoflagellates and spores in the Scotian Shelf area (Williams, 1975). All the stages of the Early Cretaceous are represented there (Table II). In the middle Atlantic States, dating from pollen and spores, supported by some microfauna, again showed the entire Lower Cretaceous section to be present (Table II). In both areas, the age of these sediments is better established than the older previously described chronostratigraphic units (Triassic and Jurassic).

In peninsular Florida, the Early Cretaceous age (Neocomian–Albian) (Table II) was determined using foraminifers which are locally abundant in many of the units (Meyerhoff and Hatten, 1974). Apparently, a complete Lower Cretaceous section is present in southern Florida and eastward in the nearby Bahamas.

Structural characteristics

The structures associated with beds of this age are best known in the Scotian Shelf area. Here, block-faulting which produced the major basins or depocenters of the Canadian Continental Shelf, the Scotian and East Newfoundland basins, continued to subside in the Early Cretaceous. The geologic relationships within the Georges Bank Trough during this period are poorly known. On the basis of regional trends, however, Lower Cretaceous sediments would be present here as in the Scotian Trough. Jansa and Wade (1975) proposed that the Early Cretaceous was the time of maximum tectonic activity in the Nova Scotia–Newfoundland area. Many of the major basins or depocenters shown on the tectonic-elements map (Fig. 2) were formed during this period.

Southward, the initial depression of the Salisbury Embayment took place, and its offshore deep depocenter, the Baltimore Canyon Trough, deepened significantly during the Early Cretaceous. What structures controlled subsidence of these basins (Salisbury Embayment and Baltimore Canyon Trough) are still unknown. Jansa and Wade (1975) proposed hinge faults as prominent structures in the formation

TABLE II

Correlation chart of Cretaceous formations from Scotian Shelf on north to southern Florida on south*

Period	Stage	Pollen zone	Florida (4)	Georgia (3)	South Carolina (subsurface) (3)	Central and southern North Carolina (3)	New Jersey—Maryland (2)	Scotian Shelf (1)
LATE CRETACEOUS	Maastrichtian	CA-6/MA-1B	Lawson Limestone	Providence Sand	Peedee Formation	Peedee Formation	Tinton Sand Red Bank Sand Navesink Formation Mount Laurel Sand Wenonah Formation Marshalltown Formation Englishtown Formation Woodbury Clay Merchantville Formation	Banquerean Formation
	Campanian	CA-6/MA-1A CA-5B CA-5 CA-4 CA-3 CA-2	Pine Key Formation, undivided	Ripley Formation Blufftown Formation	Black Creek Formation	Black Creek Formation		Wyandot formation
	Santonian Coniacian Turonian	V		Eutaw Formation	Unnamed beds	Middendorf and Cape Fear Formations	Magothy Formation	Dawson Canyon Formation
	Cenomanian	IV III	Atkinson Formation ?	Tuscaloosa Formation ?	Unnamed beds ?	Unnamed beds? ? ?	Raritan Formation Elk Neck beds	Logan Canyon Formation Eider Member
EARLY CRETACEOUS	Albian	IIC IIB IIA	Naples Bay Group Big Cypress Group Ocean Reef Group					
	Aptian	I	Glades Group	Absent	Absent	Absent	Potomac Group	Mississauga Formation
	Barremian		Fort Pierce Formation (in part)					
	Hauterivian						Absent	
	Valanginian							
	Berriasian							

Sources: (1) Jansa and Wade, 1975; (2) from Owens et al., 1977; (3) modified from Hazel et al., 1977; (4) modified from Applin and Applin, 1967, and from Meyerhoff and Hatten, 1974.

of these basins, but the sausage shape of the Scotian Basin and Georges Bank Trough suggests a reactivated Triassic rift-basin origin for these troughs (Schultz and Grover, 1974). A similar interpretation is proposed for the Baltimore Canyon Trough (Minard et al., 1974).

Lower Cretaceous sediments within the study area are absent between North Carolina and southern Florida; therefore, this probably was a positive area during this time. In the southern Florida Embayment, the Lower Cretaceous appears to onlap the Ocala or Peninsular Arch. Possibly the basin was controlled here by a hinge fault along the basin edge, but the definitive structural evidence for this relationship is lacking. In summary, tectonic activity in the Early Cretaceous was much greater in the northwestern part of the Atlantic margin than along the southwestern continental margin or essentially a similar tectonic-intensity relationship to that which occurred in the late Early to Late Jurassic time for the same area.

Sedimentological characteristics

In the Scotian Shelf area, the Lower Cretaceous rocks are included in the Missisauga Formation (Fig. 8). This formation is largely a sandy unit which interfingers with the upper part of the Verrill Canyon Formation. This clastic unit is interpreted to be largely part of a deltaic sequence with the Missisauga primarily an alluvial delta-plain deposit. The Missisauga represents, according to Jansa and Wade (1975), a massive progradation of coarse clastic sediments into the Scotian Basin during the Early Cretaceous. The sedimentary characteristics of Missisauga include abundant cross-bedding, scours, abundant carbonaceous fragments, shale clasts, and graded sequences. A general lack of microfauna indicates a lessening of marine influence in this part of the western Atlantic during this time.

In the Salisbury Embayment, Lower Cretaceous beds are widely exposed and compose about one half the thickness of the emerged Coastal Plain (Fig. 9). The fluvio-deltaic nature of the beds have been discussed by a large number of investigators (for example, Glaser, 1969). In this area the Lower Cretaceous sediments are included in the Potomac Group, which consists of a northern orthoquartzitic sand facies and southern arkosic facies. Biostratigraphic studies have shown that the Potomac Group consists of three major northeast-trending basins through time (Neocomian to Cenomanian) (Fig. 2).

Lithologically, the Potomac Group consists of interbedded, unconsolidated light-gray sand and red, white, black, and green clay. Carbonaceous matter is very abundant locally, some in the form of large coalified or silicified logs. The sands are extensively cross-bedded and contain very few trace fossils. The unconsolidated nature of the Lower Cretaceous sediments is maintained downdip in the Baltimore Canyon Trough. Here the Lower Cretaceous sediments maintain their dominantly nonmarine lithic characteristics; i.e. interstratified sands and clays containing local concentrations of coalified matter (Scholle, 1977). However, thin beds of dolomite and limestone are also interbedded with the sands suggesting some marine influence during this time.

In the southern Florida Embayment (Fig. 11), the Lower Cretaceous is clearly a carbonate-sediment province and not a clastic-sediment province as to the north. In south Florida, the Lower Cretaceous rocks are an interstratified carbonate—anhydrite facies. The stratigraphic nomenclature assigned to these subsurface sediments is shown in Fig. 10.

Paleoenvironmental interpretation

The Lower Cretaceous strata along the Atlantic margins present a marked contrast in sedimentary regimes. A major regression followed the Late Jurassic transgression in the northern region. North of North Carolina from the Salisbury Embayment to the Scotian Shelf, clastic wedges formed by prograding deltas extended from the inner edge of the Coastal Plain to nearly the shelf edge. Apparently, considerable uplift in the foreland provided a flood of clastic sediments into these basins. In peninsular Florida, in contrast, basin subsidence was on a stable platform with the accumulation of thick carbonate—anhydrite beds throughout the Early Cretaceous.

Thus, the northern region is characterized by an unstable foreland clastic-dominated province whereas the southern Atlantic was characterized by a stable-platform province.

Upper Cretaceous beds

The last chronostratigraphic grouping to be discussed in this chapter is the Upper Cretaceous Series. Because of its widespread distribution in the emerged Coastal Plain, more is known about this chronostratigraphic group than the three groupings discussed above. Therefore, this chronostratigraphic grouping will be separated into beds of Cenomanian, Turonian(?) to earliest Campanian, Campanian, and Maastrichtian age.

Table II shows the stratigraphic nomenclature applied to these beds in the study area. In general the Upper Cretaceous rocks are much thinner throughout the Atlantic Coastal Plain than any of the older Mesozoic chronostratigraphic sequences (Figs. 8–10).

Because of the large number of time-stratigraphic units of this age, the thickness for all the units are contained in Table III.

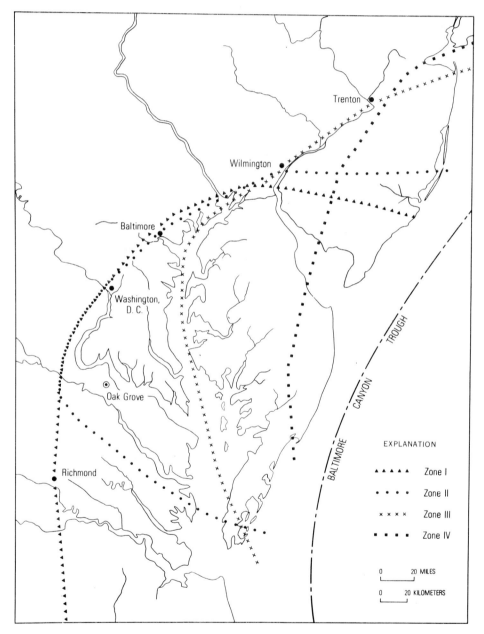

Fig. 12. Map showing distribution of Cretaceous sediments indicating migration of depocenters through time in the middle Atlantic States. Zone I, Early Cretaceous (Barremian to Neocomian); Zone II, Early Cretaceous (Aptian and Albian); Zone III, Early and Late Cretaceous (Albian to early Cenomanian); Zone IV, Late Cretaceous (middle to late(?) Cenomanian) (from Reinhardt et al., 1979).

TABLE III

Maximum thickness reported (m) of time-stratigraphic units

	Scotian Shelf[*1]	Baltimore Through[*2]	Southeast Georgia Embayment[*3]	South Florida Embayment[*4]
Maastrichtian	400	50	300	300
Campanian	200	200	200	200
Turonian(?) to earliest Campanian	1000	600	100	125
Cenomanian	1000	200	300	100

[*1] Jansa and Wade, 1975; [*2] Scholle, 1977; [*3] Gohn et al., 1978; [*4] Meyerhoff and Hatten, 1974.

Beds of Cenomanian age

Distribution. The distribution of Cenomanian beds along the Atlantic Coast is shown in Fig. 13. As can be seen, beds of this age overlap much of the Coastal Plain physiographic province and, for the most part, occupy the major embayments shown in Fig. 13. The notable exception is the absence of Cenomanian beds, particularly those of middle to late Cenomanian age, in one of the largest embayments on the eastern seaboard, the Salisbury Embayment.

In the Scotian Shelf area on the submerged Coastal Plain, the Logan Formation is interpreted to be an upper Lower Cretaceous to lower Upper Cretaceous transgressive sequence (Jansa and Wade, 1975). The upper part of this unit has been locally subdivided into a separate member, the Eider. This unit is dated as Cenomanian and overlaps all the older beds (Albian) of the Logan and most of the adjacent structural highs and foreland.

In the Raritan Embayment, beds of Cenomanian age are widespread. Here the beds of Cenomanian age crop out for a short distance along the inner edge of the Coastal Plain. To the south the Cenomanian beds are found only in the subsurface where they are overlapped by younger beds. The Cenomanian beds reappear at the surface in central and western Georgia (Fig. 13).

Age. The sediments in this sequence have been dated by a variety of fossil types: planktonic foraminifers (Petters, 1976), molluscs (Owens and Sohl, 1973; Owens et al., 1977), and pollen and spores

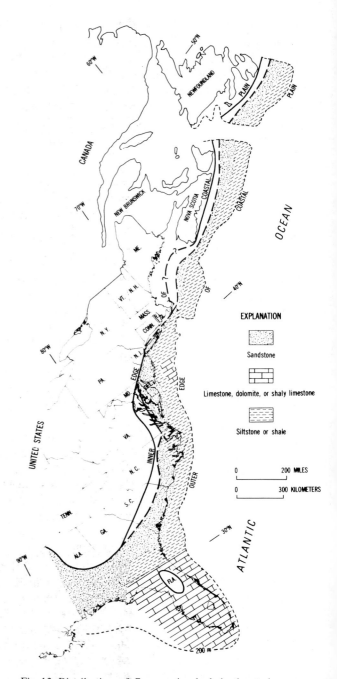

EXPLANATION

Sandstone

Limestone, dolomite, or shaly limestone

Siltstone or shale

0	200 MILES

0	300 KILOMETERS

Fig. 13. Distribution of Cenomanian beds in the study area. Note that the carbonate facies has moved north of Ocala Arch since Early Cretaceous time but is absent on the arch itself. North of Florida, the clastic facies are also more silty and clayey than those in the Lower Cretaceous sediments. Additionally, the Cenomanian beds are more widespread over the continent than any of the older chronostratigraphic groups.

(Owens et al., 1977; Christopher et al., 1979). Because of the wide variety of environments, particularly marine to nonmarine, palynomorphs have been used effectively to establish the age of these beds. The pollen zone for this unit is zone IV (see R.A. Christopher in Owens et al., 1977, for a discussion of the age of this zone). In most studies, this unit is considered to be late Cenomanian in age. The notable exception are the beds in the Raritan Embayment informally called the Elk Neck beds (Table II) which are considered to be early Cenomanian (zone IIC, and III, Doyle and Robbins, 1977).

Paleoenvironmental interpretation. The Cenomanian-age beds represent the first major marine transgression to reach parts of the present inner edge of the emerged Coastal Plain. The Raritan Formation of New Jersey is such a unit. In the southern Coastal Plain in the Georgia—Florida area, the facies change from a thick nonmarine sand—clay unit in outcrop (Tuscaloosa Formation) which grades downdip to a black marine shale, and even farther downdip to a thin limestone, the Atkinson Formation of Applin and Applin (1965). These units are mainly present in the Apalachicola Embayment (Fig. 2) and are thin or absent to the south over the Ocala Arch (Applin and Applin, 1967). South of the arch, the Cenomanian beds have not been separated from a thick chalk sequence which represents the entire Upper Cretaceous in the southern Florida Embayment (Meyerhoff and Hatten, 1974).

Structural characteristics. Block-faulting, which apparently controlled sedimentation patterns in the older units, is not apparent within these beds. Even though beds of this age are in well-defined embayments, surface faults defining these basins have not been discovered to date. Jansa and Wade (1975) do show some hinge-line faulting of the Eider Member of the Logan Canyon Formation in the Scotian basin in addition to disruptions of bedding due to diapiric intrusions.

Sedimentological characteristics. The lithologies of the sediments within this chronostratigraphic sequence vary considerably between areas but show a recognizable large-scale distributional pattern. These units are sandy or silty in the updip areas and calcareous in the downdip areas. The Tuscaloosa and Atkinson Formations in the Georgia—Florida area illustrate this relationship. The Tuscaloosa is found along the inner edge of the Coastal Plain where it consists of interbedded arkosic sands and massive thin to thick mottled red to green clays. The subsurface Atkinson is principally a limestone in Florida.

Northward into the Carolinas, the Cenomanian beds are interstratified nonmarine—marine (clastic-rich part of the lower member of the Atkinson Formation of Applin and Applin, 1967). These beds are marginal-marine facies of this formation. Similar marginal-marine facies are present in the Albermarle and Raritan embayments in the middle Atlantic States.

The same situation is present in the middle Atlantic States between the Raritan Formation (clastic sediments) and limey sediments found beneath the emerged Coastal Plain (Bass River Formation, Petters, 1976) to dominantly carbonate-rich sediments in the Baltimore Canyon (Scholle, 1977). The sand beds in this part of the study area are orthoquartzitic rather than arkosic as to the south.

In the Scotian Shelf area, the sedimentological pattern noted to the south does not apply. Here, beds of this age are included in the sandy and shaley Logan Canyon Formation (Albian and Cenomanian in age). Locally, the Cenomanian beds are lithologically distinct (the Eider Member) from the rest of the formation. In the Scotian Shelf area the Eider Member is a glauconite sandstone.

In summary, therefore, the paleoenvironments in the Cenomanian beds appear to represent depositional conditions similar to those along parts of the Atlantic Ocean margin today (clastic sediments nearshore, clastic sediments over the shelf in the north and in the south, and carbonate sediments offshore).

Beds of Turonian(?) to earliest Campanian age

Distribution. Fig. 14 shows the distribution of Turonian(?) to lowermost Campanian beds in the western Atlantic region. In general, sediments of this age overlap much of the emerged and nearly all the submerged Coastal Plain. The formations composing this group of sediments are shown in Table II. These units have a somewhat different areal distribution than the Cenomanian beds. Specifically, they are more widespread across the Cape Fear Arch (North Carolina—South Carolina boundary) and within the Salisbury Embayment.

Structural characteristics. The formations in this group, although widespread, lie in marked depocenters; the Scotian Shelf Basin and Raritan Embayment are two of the more prominent examples. These depocenters, although known in some detail, have

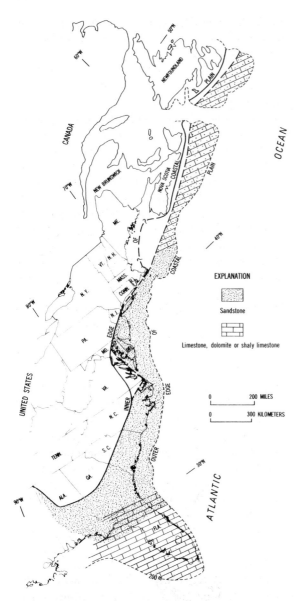

Fig. 14. Distribution of beds of Turonian(?) to earliest Campanian age. The beds of this age are dominantly carbonate sediments in the northern and southern regions and clastic sediments in the central and south-central areas.

not been shown to be fault-bounded to date. In general, the only distortion of these strata seem to be related to piercement structures (diapirs) beneath the submerged Coastal Plain in the Scotian Shelf area (Jansa and Wade, 1975).

Sedimentological characteristics. As in all the Upper Cretaceous beds, the unit contains mainly clastic sediments except in the southern Florida Em-

bayment where carbonate sediments dominate. The best exposures of these beds are along the inner margin of the Coastal Plain (Magothy, Middendorf, and Cape Fear, basal part of the Black Creek and Eutaw, Table II). These basin-edge deposits are largely cross-bedded sands containing beds of abundant dark-gray clay. Carbonaceous matter, mostly finely comminuted, is abundant throughout all of these units.

Most of the sands are quartz-rich. Feldspar and mica are major constituents in the sands, although mica is locally very abundant in the clay interbeds. Pyrite is a common constituent in most of these units.

In the Scotian Shelf (Jansa and Wade, 1975), the Dawson Canyon Formation is primarily a greenish-gray shale and siltstone locally containing thin but persistent beds of limestone (Petril Member). The shales are locally calcareous and contain abundant glauconitic sands.

In the southern Florida Embayment, the beds of this age are included in the Pine Key Formation of Meyerhoff and Hatten (1974). There the beds are dominantly a massive limestone sequence.

Age. The age of these beds is not precisely known mainly because of conflicts between the various faunal or floristic groups used to determine a precise age. The assigning of a Turonian and Coniacian age for this unit is contentious. Using palymorphs, the best zonation suggests a Turonian to Santonian to earliest Campanian age (Table II). In this system, the deposits of this age are classified as zone V (Christopher et al., 1979). The Turonian and Coniacian appear to be absent, especially along the inner edge of the Atlantic Coastal Plain if ammonite zonation is used (Hazel et al., 1977).

Paleoenvironmental interpretations. The deposits in this unit represent a variety of depositional environments. The most common depositional environment between New Jersey and Georgia is a series of small deltaic lobes extending into the embayments along the Coastal Plain margin. Downdip in peninsular Florida, the deposits are largely carbonate bank in origin. In the Scotian Shelf area, the deposits are largely inner-shelf deposits (silts and sands). Thus, the Turonian(?) to lowermost Campanian sands appear to represent what is commonly interpreted as a normal delta grading laterally to shelf deposits, or where sedimentation is least intense, carbonate-bank accumulates (such as in central and southern Florida).

Beds of Campanian age

Distribution. Fig. 15 illustrates the distribution of beds of Campanian age along the western Atlantic margin. These units represent an extensive transgres-

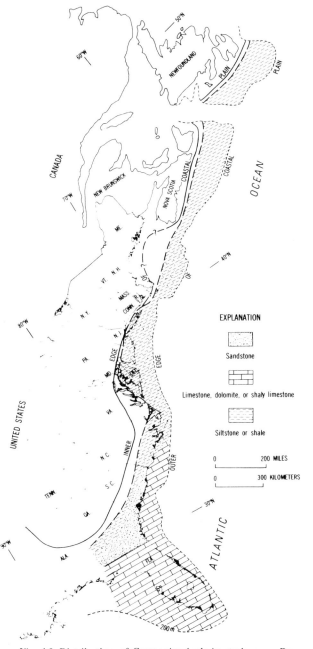

Fig. 15. Distribution of Campanian beds in study area. During this time, the extent of the carbonate facies has increased in the south and decreased north of North Carolina relative to its distribution during Santonian time. In addition, the beds of this age have overlapped most of what now constitutes the Atlantic Coastal Plain.

sion of the continental margin. The chronostratigraphic sequence, however, is absent in most of the Virginia Coastal Plain where it is somewhat to the east or seaward of the Turonian(?) to Santonian and Campanian beds. Nevertheless, beds of this age are extensive over much of the Atlantic Coastal Plain. The formational nomenclature associated with these age beds is shown in Table II. As can be seen, numerous Upper Cretaceous formations are defined in this interval; this is, in part, a reflection of the length of time encompassed by these stages (as much as 17 m.y., Owens and Sohl, 1973).

Age. The age of the Campanian beds are known as well as any of the Upper Cretaceous beds. The upper boundary, however, depends upon whether a planktonic foraminifer zonation or an ammonite zonation is used (see discussion in Owens and Sohl, 1973). Recent pollen and spore zonation (Wolfe, 1976; Christopher et al., 1979) has been found to have wide-spread application throughout the exposed beds and in the subsurface. The pollen zonation is directly comparable and complementary to the ammonite zonation. In any case, the Campanian beds represent the longest period of time available for the deposition of Upper Cretaceous formations in the Atlantic Coastal Plain. Radiometric ages (K/Ar method) on glauconites from the Campanian beds in New Jersey (Table II) ranged from about 81 to 64 m.y. (Owens and Sohl, 1973).

Structural characteristics. The Campanian sediments appear to have accumulated in local depocenters along the inner edge of its outcrop belt. Of particular note are the Campanian sections in New Jersey, northern North Carolina, and the southeast Georgia embayment. The thickness of Campanian sediments in the southern Florida Embayment is poorly understood, but they appear to thicken southward from the Ocala Arch into this basin.

Sedimentological characteristics. As in all the Upper Cretaceous formations, a wide variety of marginal-marine to normal-marine environments is reflected in the beds of this age. Where these beds are exposed between New Jersey and Georgia, they show a characteristically cyclic nature. The cycles typically are asymmetrical commonly having glauconite sand as the basal transgressive unit and a quartz sand as the upper regressive part of the cycle. In the Campanian beds of New Jersey, at least two cycles of sedimentation are present (Owens and Sohl, 1969). As more detailed studies are made on the beds in the southeast Atlantic Coastal Plain, cyclicity may be more widely recognized. In the deep surface at Charleston, S.C., the cycles in rocks of this age are discernible in the calcareous subsurface facies (Gohn et al., 1977).

The sediments of this chronostratigraphic sequence are, in the outcrop areas, some of the most glauconitic beds in the Atlantic Coastal Plain (Merchantville and Marshalltown formations of the middle Atlantic States,. Table II). These beds typically are associated in the outcrop sections with carbonaceous black silts or clays. Downdip in the middle shelf, they are commonly associated with calcareous shales (Scotian Shelf) and in the distal areas with carbonate banks (southern Florida Embayment) (Applin and Applin, 1967).

The deposits of this age tend to thicken downdip although the thickening is not excessive, certainly, if compared with the downdip thickening of the Lower Cretaceous beds.

Paleoenvironmental interpretation. The Campanian beds of the Atlantic Coastal Plain are mainly inner to middle shelf deposits. Locally, nearshore sublittoral to delta-front deposits are also present and these aggregate a considerable sand thickness near the inner margin of the Coastal Plain.

The Campanian beds, because of their thickness and abundance of glauconite, suggest a waning of clastic supply during the Late Cretaceous to the many basins into which they were deposited. Sufficient clastic sediments, however, were supplied to the depocenters to inhibit significant carbonate deposition except in the distal parts of the shelf. Again, the notable exception is in the south Florida embayment where carbonate shelf deposits are well developed throughout this time.

Maastrichtian beds

Distribution. The Maastrichtian beds have a slightly more limited overall distribution than the Campanian strata in the Coastal Plain (Fig. 16). Beds of this age are thin or absent in the Grand Banks area east of Newfoundland. In other areas, the Maastrichtian beds have overlapped most of the Atlantic Coastal Plain.

Age. In all areas of the Atlantic Coastal Plain, most paleontologists consider the youngest beds of the Cretaceous system to be of middle Maastrichtian age. Using the planktonic foraminifer zonation (Pessagno, 1969), the youngest zone present is the *Globotruncana gansseri* zone. An upper Maastrichtian

planktonic zone, the *Abathomphalus mayaroensis* zone, is poorly known and probably absent in the Atlantic Coastal Plain (C. Smith, written communication, 1977). In the ammonite zonation the youngest types present are the *Baculites columna* and *B. carina-*

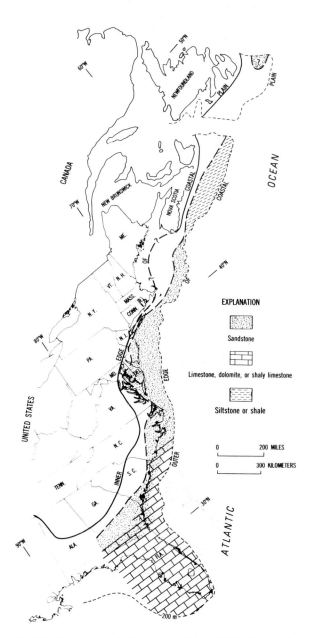

Fig. 16. The uppermost Cretaceous (Maastrichtian) beds are areally less widespread in the Atlantic Coastal Plain than the underlying Campanian beds. Carbonate sediments in the Maastrichtian section are more widespread in the southern parts of eastern North America than in the northern or central region.

trus, both middle Maastrichtian forms (Owens and Sohl, 1973). The upper Maastrichtian beds which are widespread in Mexico are not known in the Atlantic Coastal Plain, even in the deep troughs (for example, in the Baltimore Canyon Trough, Scholle, 1977). In the pollen zonation, Maastrichtian beds contain pollen indicative of zone CA 6/MA-1B (Wolfe, 1976), which is considered early to middle Maastrichtian in age.

Structural characteristics. The Maastrichtian beds show little postdepositional distortion. In general, these beds appear to thicken from the landward margin seaward. These beds, however, are deposited in well-defined depocenters, although again, no pronounced faults have been recognized. In general, the Maastrichtian beds are much thinner than the underlying Campanian beds, and the upper beds seem to be removed by a widespread erosion interval during the latest Cretaceous and earliest Tertiary (Danian).

Sedimentological characteristics. The Maastrichtian beds are lithologically similar to the underlying Campanian beds in most basins. In the Scotian Shelf area (Table II) these beds are included in the Banquereau Formation (Williams, 1975) (Maskonomet beds of others). In the middle Atlantic States, the Maastrichtian beds have the same glauconitic and cyclic characteristics as the underlying Campanian beds (Owens and Sohl, 1973). In the southeast, the Providence Sand is the thick updip marginal-marine clastic facies which interfingers downdip with unnamed thick shaley calcareous beds which grade into the Lawson Limestone in peninsular Florida.

Paleoenvironmental analysis. All the uppermost Cretaceous strata in the Atlantic Coastal Plain were deposited in a marine shelf environment. The nearshore sandy facies is well developed along the inner edge of the Coastal Plain (Providence Sand of Georgia and Red Bank Sand of New Jersey are good examples of thick sand units). Downdip, these sandy units interfinger with shaly sediments, and, in the distal parts of the basins in the southern Coastal Plain, with calcareous limestone facies. In the distal parts of the northern Coastal Plain, however, only sand and dark clay are present in this time interval (Scholle, 1977). These sand and clay beds represent only the earliest Maastrichtian or perhaps late Campanian intervals (C.W. Poag, in Scholle, 1977).

MESOZOIC GEOLOGIC HISTORY

The Mesozoic geologic history of the western North Atlantic Ocean margin began with early rifting. The earliest dated sediments associated with this fracturing are Late Triassic (Carnian, about 210 m.y. old). A series of irregular horst and graben structures were formed on thinned continental crust. In most areas, highly feldspathic sediments derived from crystalline basement were introduced into the rift valleys from all sides, but typically, more from one side to the fault-bounded valleys than from the other. These first-cycle sediments were introduced by alluvial fans and/or extensive braided streams. Most of the coarse sand and gravel clasts were deposited near the basin edges with the finer-grained material being carried to the basin centers where they were deposited in perennial or ephemeral playa-lake environments. In some of the basins, (Deep River–Wadesboro), these lakes were bordered by large marshes or swamps with the vegetation being subsequently compacted to produce coal beds. In other basins, fresh water carbonate deposits accreted in a lacustrine environment either as thinly bedded limestones or, more commonly, as dolomites interbedded with fine clastic sediments. In some of the basins (Newark), the waters were sufficiently hypersaline to produce thick zeolite-rich facies.

The rift valleys were not all formed at the same time nor did they persist for the same length of time. In general, the southernmost rift valleys formed first and received sediment for only a short duration (Table I). Sediments accumulated in northern basins from Pennsylvania to Connecticut for a much longer time. In the Connecticut Basin the sedimentary record represents a time period of about 46 m.y. (Cornet, 1977). The position of sedimentation in the basins offshore is poorly known, so the timing and width of the rift zone is still poorly known. It is apparent, however, that the site of these basins is not associated with a Mesozoic mid-Atlantic fracture system that produced oceanic crust. For reasons as yet unknown, the spreading center lay well east of these basins (Vogt, 1973).

During Early to Middle Jurassic time, the crystalline basement east of Nova Scotia and Newfoundland was block-faulted and the basins were invaded by marine waters. Apparently, the climate was sufficiently arid to produce an evaporite sequence (thick halite beds of the Argo Formation). In late

Middle Jurassic time, a major marine transgression affected at least the northwestern Atlantic margin and also the southern Florida area. The distribution of these beds on the Continental Shelf between eastern Canada and Florida, is conjectural, but beds of this age probably are present in the major troughs or depocenters which parallel the present coast. In the Scotian Shelf area, the beds of this age are largely clastic sediments (Mic-Mac and Verrill Canyon) whereas in the Florida Embayment carbonate sedimentation dominated. This relationship suggests that the forelands were more tectonically active in the Canadian area than in the Florida area.

The Lower Cretaceous beds in the area north of Cape Hatteras are largely nonmarine, suggesting a major regression during which deltas prograded into the Scotian and Baltimore Canyon troughs. The massive outpouring of clastic sediments into these northern troughs suggests major uplift in adjacent uplands. In the Blake Plateau Bahamas southern Florida area carbonate sediment again was accreting during Early Cretaceous time. Evaporite basin accumulation is indicated because of the interbedding of calcite and anhydrite in these Lower Cretaceous sediments. Because of the paucity of terrigenous clastic sediments in these beds, the southern Appalachian region was probably tectonically less active than the area north of Cape Hatteras during this time period.

In Late Cretaceous time, the continental margins were largely inundated several times during a series of well-defined transgressions and regressions. The sedimentation rate, however, was considerably reduced during this time period as compared with Early Cretaceous time. The first major transgression to reach the present inner edge of the Coastal Plain was in the middle Cenomanian (Raritan Formation of New Jersey). In the southeastern Coastal Plain, equivalent beds are nonmarine (Tuscaloosa Formation). The marine equivalents of these units are present in the deep subsurface of Georgia, South Carolina, and North Carolina (Atkinson Formation or its unnamed lateral equivalents).

In Turonian(?) to earliest Campanian time, a series of marine deltas prograded into the oceanic basin along its edge from New Jersey to Georgia. The effects of this clastic influx are not evident in the carbonate rocks of this age in peninsular Florida where carbonate sediment continued to accrete.

The Campanian and Maastrichtian beds in the middle Atlantic States are largely cyclic in nature producing well-defined transgressive (glauconite-rich) and regressive (largely quartz sands) beds. In the southeastern Atlantic States, the spatial arrangement of facies is for clastic sediments to be deposited in the updip areas and carbonate sediments in the downdip areas. As in the older beds, if the clastic deposits are a measure of basin and/or foreland instability, then the northern continental margin was more unstable than the southern, a situation which was established in the Middle to Late Jurassic forming the entire western Atlantic margin. In any case, by Late Cretaceous time, open-marine conditions prevailed along the entire western Atlantic margin.

REFERENCES

Applin, P.L. and Applin, E.R., 1965. The Comanche Series and associated rocks in the subsurface in central and south Florida. *U.S. Geol. Surv. Prof. Pap.*, 447: 84 pp.

Applin, P.L. and Applin, E.R., 1967. The Gulf Series in the subsurface in northern Florida and southern Georgia. *U.S. Geol. Surv. Prof. Pap.*, 524-G: 34 pp.

Bain, G.L. and Harvey, B.W., 1977. Field guide to the geology of the Durham Triassic basin, North Carolina. In: *Carolina Geol. Soc., Field Trip Guidebook*, Oct. 7–9, 1977.

Barnett, R.S., 1975. Basement structure of Florida and its tectonic implications. *Trans. Gulf Coast Assoc. Geol. Soc.*, XXV: 122–142.

Brown, P.M., Miller, J.A. and Swain, F.M., 1972. Structural and stratigraphic framework and spatial distribution of permeability of the Atlantic Coastal Plain, North Carolina to New Jersey. *U.S. Geol. Surv. Prof. Pap.*, 796: 79 pp.

Canada Geological Survey, 1969. *Geological Map of Canada*. Canada Geol. Surv. Map 1251A, scale 1:5.000.000.

Christopher, R.A., Owens, J.P. and Sohl, N.F., 1979. Late Cretaceous palynomorphs from the Cape Fear Formation of North Carolina. *Southeastern Geol.*, 20(3).

Cornet, B., 1977. *The Palynostratigraphy and Age of the Newark Supergroup*. Penn. State Univ., unpubl. thesis, 501 pp.

Cramer, H.R., 1975. Isopach and lithofacies analyses of the Cretaceous and Cenozoic rocks of the central Coastal Plain of Georgia. *Ga. Geol. Surv. Bull.*, 87: 21–86.

Dallmeyer, R.D., 1975. The Palisades sill: A Jurassic intrusion(?) Evidence from $^{40}Ar/^{39}Ar$ incremental release ages. *Geology*, 3(5): 243–245.

Dillon, W.P., Paull, C.K., Buffler, R.T. and Faill, J.P., 1979. Structure and development of southeast Georgia embayment and northern Blake Plateau: Preliminary analysis. *Am. Assoc. Petrol. Geol. Bull.* In press.

Dooley, R.E., 1977. *K-Ar Relationships in Dolerite Dikes of Georgia*. Ga. Inst. Technol., unpubl. thesis, 185 pp.

Doyle, J.A. and Robbins, E.I., 1977. Angiosperm pollen zonation of the continental Cretaceous of the Atlantic

Coastal Plain and its application to deep wells in the Salisbury embayment. *Palynology*, 1: 43–78.

Faust, G.T., 1975. A review and interpretation of the geologic setting of the Watchung Basalt flows, New Jersey. *U.S. Geol. Surv. Prof. Pap.*, 864-A: 42 pp.

Folger, D.W., Hathaway, J.C., Christopher, R.A., Valentine, P.C. and Poag, C.W., 1978. Stratigraphic test well, Nantucket Island, Massachusetts. *U.S. Geol. Surv. Circ.*, 773: 27 pp.

Fontaine, W.M., 1889. The Potomac or Younger Mesozoic flora. *U.S. Geol. Surv. Monogr.*, 15.

Glaser, J.D., 1969. Petrology and origin of Potomac and Magothy (Cretaceous) sediments, middle Atlantic Coastal Plain. *Md. Geol. Surv., Rept. Invest.*, 11: 101 pp.

Gohn, G.S., Higgins, B.B., Smith, C.C. and Owens, J.P., 1977. Lithostratigraphy of the deep corehole (Clubhouse Crossroads Corehole 1) near Charleston, S.C. In: D.W. Rankin (Editor). *Studies Related to the Charleston, South Carolina Earthquake of 1886 – A Preliminary Report. U.S. Geol. Surv. Prof. Pap.*, 1028E: 43–70.

Gohn, G.S., Bybell, L.M., Smith, C.C. and Owens, J.P., 1978. Preliminary stratigraphic cross-sections of the Southeastern states: Cretaceous sediments along the South Carolina coastal margin. *U.S. Geol. Surv. Miscell. Field Invest.*, MF-1015A.

Gottfried, David, Annel, C.S. and Schwarz, L.J., 1977. Geochemistry of subsurface basalt from the deep corehole (Clubhouse Crossroads Corehole 1) near Charleston, South Carolina – magma type and tectonic implication. In: D.W. Rankin (Editor), *Studies Related to the Charleston, South Carolina Earthquake of 1886 – A Preliminary Report. U.S. Geol. Surv. Prof. Pap.*, 1028G: 91–114.

Hazel, J.E., Bybell, L.M., Christopher, R.A., Fredricksen, N.O., May, F.E., McLean, D.M., Poore, R.C., Smith, C.C., Sohl, N.F., Valentine, P.C. and Witmer, R.J., 1977. Biostratigraphy of the Deep Corehole (Clubhouse Crossroads Corehole 1) near Charleston, South Carolina. In: D.W. Rankin (Editor), *Studies Related to the Charleston, South Carolina Earthquake of 1886. U.S. Geol. Surv. Prof. Pap.*, 1028F: 71–89.

Hubert, J.F., Reed, A.A., Dowdall, W.L. and Gilchrist, J.M., 1978. *Guide to the Redbeds of Central Connecticut – 1978 Fieldtrip, Eastern Section, Society of Economic Paleontologists and Mineralogists*. Univ. Mass., Dept. Geol. Geogr. Contrib., 32: 129 pp.

Jansa, L.F. and Wade, J.A., 1975. Geology of the continental margin off Nova Scotia and Newfoundland. In: *Offshore Geology of Eastern Canada. Geol. Surv. Canada, 2, Regional Geol., Pap.*, 74-30, part 11, pp. 51–105.

Klein, G. de Vries, 1962. Triassic sedimentation, Maritime Provinces, Canada. *Geol. Soc. Am. Bull.*, 73 (9): 1127–1146.

Krynine, P.D., 1948. The megascopic study and field classification of sedimentary rocks. *J. Geol.*, 56 (2): 130–165.

Lee, K.Y., 1977. Triassic stratigraphy in the northern part of the Culpeper basin, Virginia and Maryland. *U.S. Geol. Surv. Bull.*, 1422-C: C1–C17.

Longwell, C.R., 1922. Notes on structure of Triassic rocks in southern Connecticut. *Am. J. Sci., 5th Ser.*, 4: 223–236.

Maher, J.S., 1971. Geologic framework and petroleum potential of the Atlantic Coastal Plain and continental shelf. *U.S. Geol. Surv. Prof. Pap.*, 659: 92 pp.

Manspeizer, W., Puffer, J.H. and Cousminer, H.L., 1978. Separation of Morocco and eastern North America: A Triassic–Liassic stratigraphic record. *Geol. Soc. Am. Bull.*, 89: 901–920.

Marine, I.W. and Siple, G.E., 1974. Buried Triassic basin in Central Savannah River area, South Carolina and Georgia. *Geol. Soc. Am. Bull.*, 85: 311–320.

Mattick, R.E., Foote, R.Q., Weaver, N.L. and Grim, N.S., 1974. Structural framework of United States Atlantic continental shelf north of Cape Hatteras. In: *East Coast Symposium – Am. Assoc. Petrol. Geol.*, 58 (6): 1179–1190.

May, P.R., 1971. Pattern of Triassic–Jurassic diabase dikes around the North Atlantic in the context of predrift position of the continents. *Geol. Soc. Am. Bull.*, 82: 1285–1292.

Meyerhoff, A.A. and Hatten, C.W., 1974. Bahama salient of North America, stratigraphy and petroleum potential. In: *East Coast Symposium – Amer. Assoc. Petrol. Geol.*, 58 (6): 1169–1178.

Minard, J.P. and Owens, J.P., 1966. Domes in the Atlantic Coastal Plain east of Trenton, New Jersey. *U.S. Geol. Surv. Prof. Pap.*, 550-B: B16–19.

Minard, J.P., Perry, W.J., Weed, E.G.A., Rhodehamel, E.C., Robbins, E.I. and Mixon, R.B., 1974. Preliminary report on geology along Atlantic coastal margin of northeastern United States. In: *East Coast Symposium – Am. Assoc. Petrol. Geol.*, 58 (6): 1169–1178.

Mixon, R.B. and Newell, W.L., 1977. Stafford fault system: Structures documenting Cretaceous and Tertiary deformation along the Fall Line in northeastern Virginia. *Geology*, (5): 437–440.

Olson, P., 1978. Comparative paleolimnology of Newark Supergroup Lakes. In: J.F. Hubert, A.A. Reed, W.L. Dowall and J.M. Gilchrist. *Guide to the Redbeds of Central Connecticut–1978 Fieldtrip, Eastern Section, Society of Economic Paleontologists and Mineralogists*. Univ. Mass. Dept. Geol. Geogr. Contrib., 32, 129 pp.

Owens, J.P., 1970. Post-Triassic movements in the central and southern Appalachians as recorded by sediments of the Atlantic Coastal Plain. In: *Studies of Appalachian Geology, Central and Southern*. Interscience, New York, N.Y., pp. 417–427.

Owens, J.P. and Sohl, N.F., 1969. Shelf and deltaic paleoenvironments in the Cretaceous–Tertiary formations of the New Jersey Coastal Plain. In: S. Subitzky (Editor), *Geology of Selected Areas in New Jersey and Eastern Pennsylvania and Guidebook to Excursions*. Rutgers University Press, New Brunswick, N.J., pp. 235–278.

Owens, J.P. and Sohl, N.F., 1973. Glauconites from the New Jersey–Maryland Coastal Plains; their K/Ar ages and use in stratigraphic studies. *Geol. Soc. Am. Bull.*, 84: 2811–2838.

Owens, J.P., Sohl, N.F. and Minard, J.P., 1977. A field guide to the Cretaceous and Lower Tertiary beds of the Raritan and Salisbury Embayments, New Jersey, Delaware and

Maryland. *Ann. Am. Assoc. Petrol. Geol.–Soc. Econ. Paleontol. Mineral. Conv., Washington, D.C.*, June 9–11, 111 pp.

Pessagno, E.A., Jr., 1969. Upper Cretaceous stratigraphy of the western Gulf Coast of Mexico, Texas, Arkansas. *Geol. Soc. Am. Mem.*, 111: 139 pp.

Petters, S.W., 1976. Upper Cretaceous subsurface stratigraphy of Atlantic Coastal Plain of New Jersey. *Am. Assoc. Petrol. Geol. Bull.*, 60 (1): 87–107.

Poag, C.W., 1978. Stratigraphy of the Atlantic continental shelf and slope of the United States. *Ann. Rev. Earth Planet Sci.*, 6: 251–280.

Popenoe, P. and Zietz, I., 1977. The nature of the geophysical basement beneath the Coastal Plain of South Carolina and northeastern Georgia. In: D.W. Rankin (Editor), *Studies Related to the Charleston, South Carolina Earthquake of 1886–A Preliminary Report. U.S. Geol. Surv. Prof. Pap.*, 1028-I: 119–138.

Powell, D.C. and O'Connor, B.J., 1979. Belair Fault zone: Evidence of Tertiary fault displacement in eastern Georgia. *Geology*. In press.

Reinemund, J.A., 1955. Geology of the Deep River coal field, North Carolina. *U.S. Geol. Surv. Prof. Pap.*, 246: 159 pp.

Reinhardt, J., Christopher, R.A. and Owens, J.P., 1978. Lower Cretaceous stratigraphy of the Oak Grove core, northeastern Virginia. *Va. Div. miner. Resour. Publ.*, 20, Ch.3.

Russell, I.C., 1880. On the former extent of the Triassic Formation of the Atlantic strata. *Am. J. Sci.*, 14: 703–712.

Sanders, J.E., 1960. Structural history of Triassic rocks of the Connecticut Appalachians. *Geol. Soc. Am. Bull.*, 81 (10): 2993–3013.

Schlee, J.S., 1978. Acoustic stratigraphy of Georges Bank [Abstr.]. *Geol. Soc. Am. Abstr. Progr.*, 1 (2): 84.

Scholle, P.A., 1977. Data summary and petroleum potential. *U.S. Geol. Surv. Circ.*, 750: 8–14.

Schultz, L.K. and Grover, R.L., 1974. Geology of Georges Bank Basin. *Am. Assoc. Petrol. Geol. Bull.*, 58: 1159–1168.

Scott, K.R. and Cole, J.M., 1975. U.S. Atlantic margin looks favorable. *Oil Gas J.*, 73 (1): 95–99.

Sheridan, R.E., 1974a. Atlantic continental margin of North America. In: C.A. Burke and C.L. Drake (Editors) *Geology of Continental Margins*. Springer-Verlag, Berlin, pp. 391–407.

Sheridan, R.E., 1974b. Conceptual model for the block fault origin of the North American Atlantic continental margin geosyncline. *Geology*, 2: 465–468.

Sheridan, R.E., 1976. Sedimentary basins of the Atlantic margin of North America. *Tectonophysics*, 36: 113–132.

Thayer, P.A., 1970. Stratigraphy and geology of Dan River Triassic basin, North Carolina. *Southeastern Geol.*, 12 (1): 1–32.

Van Houten, F.B., 1969. Late Triassic Newark Group in north central New Jersey and Pennsylvania and New York. In: *Geology of Selected Areas in New Jersey and Guidebook of Excursions*. Rutgers University Press, New Brunswick, N.J., pp. 314–347.

Van Houten, F.B., 1977. Triassic–Liassic deposits of Morocco and eastern North America: Comparison. *Am. Assoc. Petrol. Geol. Bull.*, 61 (1): 79–99.

Vogt, P.R., 1973. Early events in the opening of the North Atlantic. In: *Implications of Continental Drift to the Earth Science, 2*. Academic Press, London, New York, pp. 693–712.

Williams, G.L., 1975. Dinoflagellate and spore stratigraphy of the Mesozoic–Cenozoic offshore eastern Canada. In: *Offshore of Eastern Canada–Geol. Surv. Can. Pap.*, 74–30, vol. 2, pp. 107–145.

Wolfe, J.A., 1976. Stratigraphic distribution of some pollen types from the Campanian and lower Maastrichtian rocks (Upper Cretaceous of the middle Atlantic states). *U.S. Geol. Surv. Prof. Pap.*, 977: 18 pp.

Chapter 3

MEXICO

KEITH YOUNG

INTRODUCTION

The Mesozoic history of Mexico is consistent with a model involving the spreading Atlantic Ocean and the subducting, eastern margin of the Pacific Ocean Basin. The spreading floor of the Atlantic Ocean had a greater importance in the Triassic and produced structures that influenced distribution of sediments and organisms through the Jurassic, the Cretaceous, and the early part of the Cenozoic. The subducting, eastern margin of the Pacific Ocean Basin was more important in the later Mesozoic and produced structures and metamorphism that tend to mask previous geologic history. Mesozoic rocks of the central and north-central parts of Mexico overlie a more stable area of Precambrian and Paleozoic rocks.

Mesozoic volcanism was common in the Pacific states of Mexico, and in the Jurassic and Cretaceous included the intrusion of granitic and dioritic masses of considerable extent. Volcaniclastic rocks are particularly abundant in the Middle Cretaceous of those southern states bordering the Pacific Ocean (the states from Sinaloa to Michoacan, inclusive) and through much of the Sierra Madre del Sur, where there are also metamorphic rocks of Middle Cretaceous age. Platform and deeper, but nevertheless within the phototropic zone, pelagic limestones are common in the Late Jurassic and most of the Cretaceous, particularly in the eastern and northeastern parts of Mexico. Cretaceous reefoid limestones also occur on raised blocks between basins of volcaniclastic deposition in the Pacific coast states from Sinaloa through Jalisco, Nyarit, Colima, Michoacan, and into Guerrero. In all, a great variety of geology, representing many kinds of events, characterizes the Mesozoic of Mexico.

STRATIGRAPHY AND PALEOGEOGRAPHY

Triassic

As in many parts of the World the Triassic record in Mexico is quite restricted. There are four prominent geological activities represented in these Triassic rocks: (1) granitic intrusion in the northeast; (2) volcanism in the west-central; (3) marine basins of deposition in the northwest and west-central; and (4) deposition of red beds in the east-central and south. Marine, Triassic deposits of Mexico are restricted to the Late Triassic (Norian and Carnian) (Jaworski, 1929).

Aléncaster (1961) has shown the Upper Triassic of Sonora to have been deposited in two basins (Figs. 1, 2). The Paleobahía del Antimónio is centered around the old mining district of El Antimónio; about 150 km to the southeast is the Cuenca San Marcial, centered just south of Hermosillo, Sonora. (Fig. 3 is a map of Mexico showing the states, for the convenience of those unfamiliar with Mexican political subdivisions.) These two basins were separated by a structural sill over which some Late Triassic sediments were deposited, connecting the two basins with a continuity of sedimentary rocks (Aléncaster, 1961). The sediments are not all marine, but marine deposits are intercalated with continental sediments, mostly rather fine-grained, that contain beds of coal (Wilson and Rocha, 1946; Aléncaster, 1961). The combined basins, including the sill and all of the marine sediments, have been termed the Sonoran Embayment (Fig. 1); it opened to the Pacific Ocean.

In the vicinity of Zacatecas, Zacatecas (Figs. 1, 4), Late Triassic rocks have been described by Burckhardt (1905, 1906, 1930a), Burckhardt and Scalia (1906), Frech (1906), and Vásquez (1949). Cantu

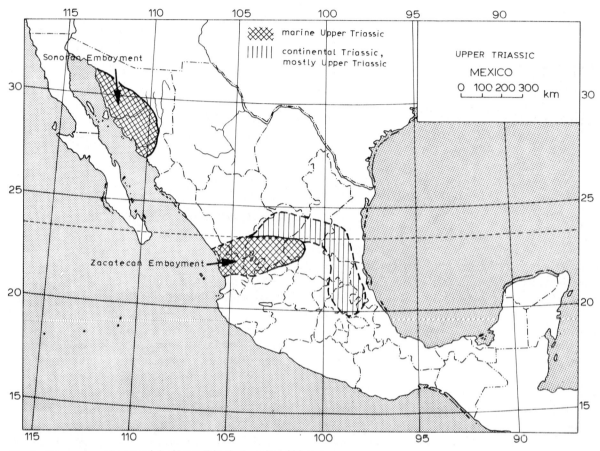

Fig. 1. Paleogeographic map of the Upper Triassic deposits of Mexico.

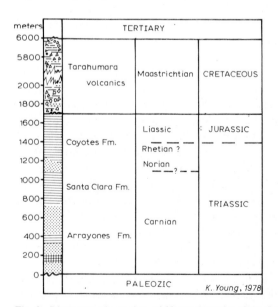

Fig. 2. Diagrammatic section of Mesozoic rocks of the Sonoran Basin, Sonora, Mexico.

Chapa (1969) has described a Late Triassic ammonite, *Juvavites* sp., from Charcas, state of San Luis Potosí, which extends the Zacatecan Basin farther southeast than it had previously been construed. This most recent discovery also emphasizes that much of the post-Triassic to pre-Late Jurassic history of Mexico is little understood, especially in the Sierra Madre Occidental, where most of the older rocks are covered by Cenozoic volcanic rocks. Although De Cserna (1969a) shows an elongated (north–south), Triassic, depositional area paralleling a Cordilleran Geosyncline, such an interpretation requires more evidence before it can be included on paleogeographic maps; even then, the concept may have to be restricted to pre-Upper Triassic rocks. In the meantime a basinal reentrant, the Zacatecan Embayment, from the Pacific Ocean, as proposed by most authors (e.g., Burckhardt, 1930a), seems to explain most simply this isolated sequence of marine Triassic rocks. This section has not been thoroughly studied, to my knowledge, since Burck-

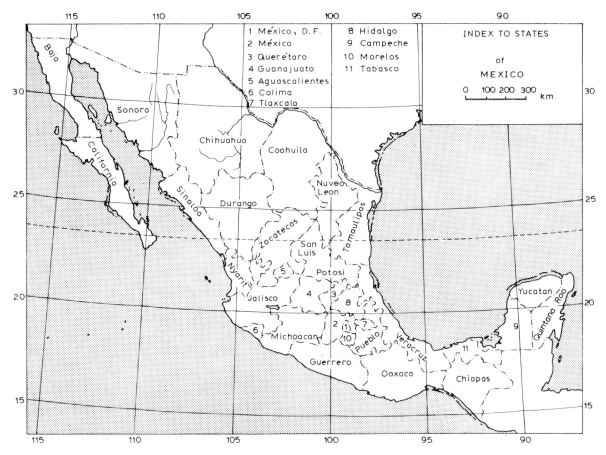

Fig. 3. Index maps to the states of Mexico.

hardt and Scalia (1906), who reported that these rocks lie between older metamorphic rocks and a younger, green, volcanic sequence (Fig. 4) which cannot be related to the green rocks farther north in Chihuahua, rocks interlayered with Permian volcanic rocks (Bridges, 1965; De Cserna et al., 1970). The relation of the Zacatecas section to metamorphic rocks of Mesozoic age (Córdoba et al., 1976) along the eastern Sinaloan boundary is still unclear.

These are the primary marine deposits of the Triassic of Mexico. Rumors of Late Triassic rocks in the Campeche Basin and underlying the Liassic in the Huayacocotlan Basin persist but are difficult to verify. Such rocks, however, should be expected, although they are not indicated by Sansores M. (1956), Gonzalez (1969), or Flores (1974).

Nonmarine, Triassic rocks are more widespread than the marine. Nonmarine, Triassic rocks (Fig. 1) are exposed in the Sierra Madre Oriental from Hidalgo to the north, continually, to Tamaulipas, where

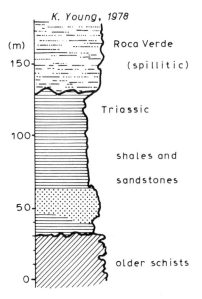

Fig. 4. Diagrammatic section of Upper Triassic rocks at Zacatecas, Zacatecas, Mexico.

they are known as the La Boca Formation of the Huizachal Group (Mixon et al., 1959). They extend through the southern third of Tamaulipas and the southern third of Nuevo Leon. There are red beds of Triassic age which also crop out in a broad belt across northern San Luis Potosí and northern Zacatecas. The overlying Jurassic red beds are spread over a much greater area. In the region of Galeana, Nuevo Leon, volcaniclastic rocks appear within the Triassic red beds. These volcanically derived rocks thicken to the west and in northeastern Zacatecas constitute over half of a section which is over 200 m thick (Belcher, 1978). Red beds to the north and west of those illustrated in Fig. 1 appear to be of Jurassic age (Belcher, 1978). To the south of Hidalgo, especially in Chiapas, Oaxaca, and the subsurface of southern Veracruz and Tabasco, are red beds usually correlated with the Todos Santos Formation of Guatemala and

Honduras (Fig. 19). These extend to the west into Guerrero as the Chapolapa Formation (Klesse, 1970). In Guatemala the Todos Santos Formation is said to be latest Triassic and Jurassic (Mills et al., 1967; Finch, 1972). It extends on deeper into Guatemala and into Honduras as the Atíma Formation (Finch, 1972; Ritchie, 1975). In this report I include these with the continental Liassic (Fig. 5), since Jurassic fossils are definitely known from the Todos Santos Formation in Mexico, and Triassic fossils are unknown, except that they have been described from the Todos Santos Formation to the south of Mexico (Newberry, 1888). However, the probability that lower parts of the Chapolapa and Todos Santos Formations may be Late Triassic in southern Mexico should not be ignored. Triassic plants have been described from northeastern Hidalgo by Silva Pineda (1963), but this far north, in a different basin, these beds are

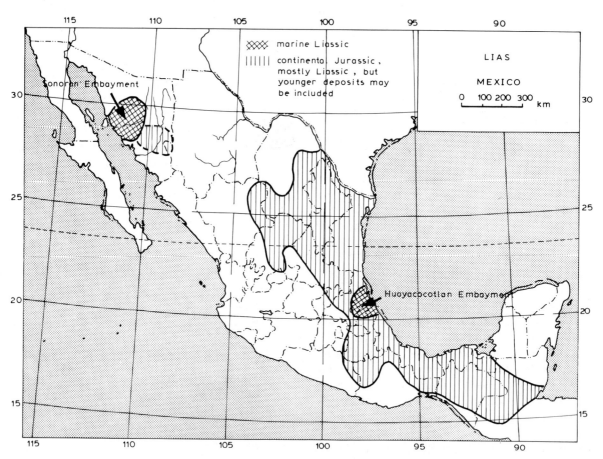

Fig. 5. Index map to the marine Liassic of Mexico. The continental rocks depicted on this map include all of the Jurassic red beds and are not restricted to the Lias.

more often referred to the Huizachal (Sansores and Navarrete, 1969) or just referred to as "los Lechos Rojos" (Ornelas, 1969).

The rocks of the La Boca Formation of the Huizachal Group in the Sierra Madre Oriental and the equivalent red beds of the Huayacocotlan Embayment and surrounding areas often change from a thickness of as much as 1000 m to only a few meters in a distance of only a few kilometers (Toledo-Toledo, 1969; Varela, 1969). That these deposits are thick locally in basins, thinning over intervening highs, seems apparent (Belcher, 1978), but the data are still incomplete.

Jurassic

Jurassic rocks occur throughout northern and eastern Mexico, but in most areas only Late Jurassic rocks are represented. Included in the Late Jurassic rocks are great thicknesses of evaporites, not only in Mexico, but all along the northern part of the Gulf of Mexico. That these evaporitic deposits extend throughout the area of the Gulf of Mexico is indicated by some (e.g., Cook and Bally, 1976), but the determination of actual age is, as yet, unsatisfactory.

Liassic

Known marine Liassic rocks of Mexico are restricted to two areas (Fig. 5). In the Sonoran Embayment the southernmost of two basins containing marine Triassic rocks – the Cuenca San Marcial – also contains marine Liassic (Fig. 2). Deposition was probably continuous from the Carnian into the Lias (Aléncaster, 1961). The Liassic seaway in this area spread farther south, to the Guaymas area, than did the Triassic, and also farther inland (Córdoba et al., 1976). However, it was still a small, marine re-entrant that must have been open to the Pacific Ocean, and it still represents the Sonoran Embayment.

The other small area of marine, Liassic rocks constitutes the Huayacocotlan Embayment of the Veracruz–Hidalgo area (Figs. 5 and 6), a small, marine re-entrant from the ancestral Gulf of Mexico (Burckhardt, 1930a; Erben, 1957b). Erben also questionably shows a small basin during the Liassic, extending from the Pacific Ocean into Guerrero, the Guerreran Embayment (Fig. 7), but there is little evidence, as yet, of its Liassic importance.

Much of the Lower Jurassic of Mexico consists of red beds. They extend from deep into Guatemala and

GUERRERAN EMBAYMENT	SE PUEBLA	HUAYACOCOTLAN EMBAYMENT	
		Pimienta	MALM
Acahuizotla		Tamán	
		Tepexic	
Consuelo	Mapache (?)	Huehuetepec	DOGGER
		Huayacocotla	LIAS
		red beds	TRIASSIC

K. Young, 1981

Fig. 6. Relation of Jurassic strata from the Huayacocotlan Embayment to those of the platform in the State of Pueblo and the Guerreran Embayment. Not all of the formational names utilized by geologists are utilized on this diagram. The Huehuetepec Formation contains both red beds and lagoonal deposits.

Honduras, where they are known as the Todos Santos and Atíma Formations, northward into Chiapas, Tabasco, and Oaxaca. Liassic red beds in eastern Guerrero, Pueblo, and on to the north, may have been continuous with these, but in different areas the red beds are now interpreted to be of different ages (Sanchez Montes de Oca, 1969; Varela, 1969; Cantu Chapa, 1971).

In the states of San Luis Potosí, Tamaulipas, Coahuila, and much of the subsurface adjacent to the Gulf of Mexico, these Liassic red beds are known as the La Joya Formation. The La Joya Formation overlies the Late Triassic La Boca Formation, and together they constitute the Huizachal Group (Mixon et al., 1959). Both are red beds. But the Jurassic red beds are much more widespread than the Triassic red beds, especially to the northwest. Although these red beds are illustrated on the Liassic map (Fig. 5), they are not restricted to the Liassic; some of them are Dogger (Erben 1957a; Silva Pineda, 1969, 1970), and some of them are even Upper Jurassic (Cook and Bally, 1976). If there are similar Cretaceous red beds, as indicated to the south and east of the Campeche Embayment (= Isthmian Embayment of Ewing et al., 1970) by Sanchez Montes de Oca (1969), they have not yet been fully documented.

On the east side of the Aldama Platform in Chihuahua (Fig. 8) there is a fairly thick sequence of red sandstone, red pebbly sandstone, and red silty sandstone (Diaz and De Cserna, 1956); it lies disconformably below marine late lower-Aptian (*Cheloni-*

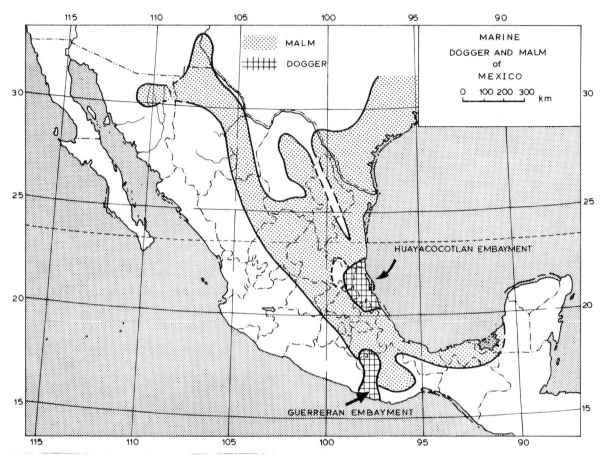

Fig. 7. Paleogeographic map depicting the maximum transgression of Dogger and Malm seaways in Mexico.

ceras sp.) and unconformably over the Permian flysch of that area. Similar beds to the west of Torreon, Durango, the Nazas Formation, underlie marine Upper Jurassic (García Calderón, 1976). Whether the strata east of the Aldama Platform are marginal to the Cretaceous Navarette and Las Vigas Formations (Neocomian and early Aptian) or marginal to the Kimmeridgian cannot be determined at this time, and any correlation with the Jurassic because of their reddish color is unsafe.

In a paper as short as this there is not sufficient space to include and discuss completely all of the stratigraphic nomenclature, but stratigraphic diagrams illustrating the nomenclature in key section are important; most of the references contain such diagrams. As the Triassic red beds, so the Jurassic red beds vary greatly in thickness over short distances, with sediment sources from many greatly divergent directions (Belcher, 1978). For example, the disposi-

tion of red beds would indicate that the extension of a Toarcian seaway over the Miquihuana Platform, as shown by Erben (1957b, fig. 4) to be impossible, since the elements of the Platform were already extant at that time (Fig. 8). But small seaways in small elongate basins aligned from 30 to 20 degrees west are not impossible. Such basins would be parallel to major features of the positive elements of the platform. Near areas of marine deposition, such as the Sonoran and Huayacocotlan Embayments, red beds are often replaced by gray shales, also continental, that may contain plants and thin seams of coal (Wilson and Rocha, 1946; Silva Pineda, 1963, 1969, 1970).

Dogger

The marine rocks representing the Dogger are as restricted as the Liassic rocks (Figs. 4, 7). These rocks are primarily Bathonian, as the Bajocian is known for

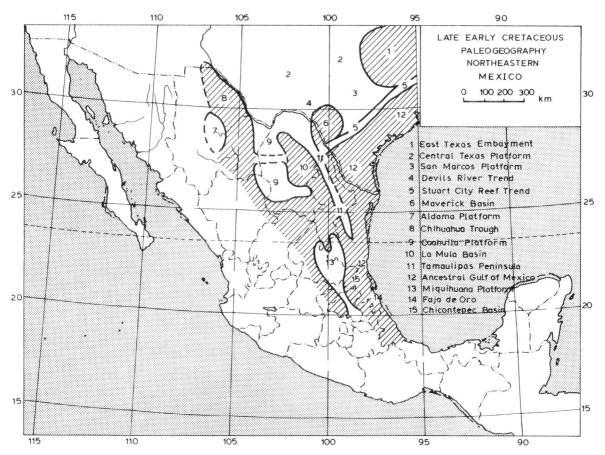

Fig. 8. Index map to the major tectonic elements of northeastern Mexico during the Jurassic and Cretaceous.

certain from only the Guerrreran Embayment in Mexico (Ochoterena, 1963), although interpreted to be present in the Huayacocotlan Embayment. Marine, Bathonian rocks lie above the Toarcian in the Huaya-cocotlan Embayment, occupying about the same geographic areas as the marine Pliensbachian sediments (Erben, 1957b, fig. 5; Figs. 5, 6). Marine Dogger is at times more restricted than the Liassic, and continental red beds intervene between the Liassic and the Callovian (Cantu Chapa, 1971).

The Sonoran Embayment had become inactive by Middle Jurassic, and this paleogeographic feature did not again reappear. The Guerreran Embayment, questionably suggested as appearing in the late Liassic by Erben (1957b), is, in the Middle Jurassic, a distinct, although a small, feature containing marine, Bathonian and Bajocian sediments. Erben believed that some of the Bathonian marine rocks interfinger with Jurassic red beds, and if so, deposition of red beds

was continuous from the Liassic into the Dogger in some areas. Silva Pineda (1970) has described Middle Jurassic plants from north of the northern end of the Guerreran Embayment. However, the red beds usually cannot be separated paleontologically or physically and are therefore included on the map of the Lias (Fig. 5). Cook and Bally (1976) show even a wider distribution of marine, Liassic and Dogger rocks than does Erben, and they indicate that some of the evaporites of northeastern Mexico are probably Middle Jurassic.

Malm

Although the basic elements of Mesozoic paleogeography of Mexico (Fig. 8) appear to have been established before the end of the Triassic, rapid changes in facies, rapid changes in thickness, and lack of data from the outcrop and subsurface, make the details of that paleogeography difficult to document, at least prior to the Aptian. Burckhardt's (1930ab) original interpretations dealt primarily with the western margin of various seaways of different times. Imlay (several papers, but especially 1938 and 1953) developed interpretations that since have only been modified in detail (Carillo Bravo, 1969; Carrasco, 1970; Cook and Bally, 1976; Young, 1978), except along the Pacific margin of Mexico, where much new information in the last decade is still insufficient for definite interpretations.

Perhaps the Liassic and Dogger sea invaded east-central Mexico through a series of elongate basins (grabens?) between intervening, narrow ridges (horsts?) (Toledo Toledo, 1969; Varela, 1969) of a since abandoned, Triassic, spreading rift-area, but the data are not yet clear. During the Malm there is enough evidence to indicate that paleotopography controlled the pathways of expansion of the Late Jurassic seaways. Whether this actually started in the Bajocian, Callovian, or even the Oxfordian, is more a matter of individual aesthetics than a matter of real data obtainable at the present time, but such interpretations depend on the age of most of the outcrops of limestone north of Veracruz. This Jurassic limestone usually has been identified as Zuloaga (Oxfordian), but many of the outcrops now appear to be "Cotton Valley" Limestone (Kimmeridgian), a limestone unit at the base of the La Casita (the Cotton Valley of the Texas Gulf Coast). Nevertheless, the Late Jurassic, primarily Kimmeridgian, seaway underwent a great expansion across Mexico and particularly across northern Mexico, except for the western states adjacent to the then subducting margin of the Pacific Ocean Basin, where there is no known marine invasion of Late Jurassic age. This does not, however, preclude some of the metamorphic rocks, dated as Mesozoic, from being anywhere in the pre-Albian Mesozoic, as De Cserna (1969a, 1970) so frequently indicates.

Furthermore, by late Kimmeridgian the Jurassic seaway had invaded the Chihuahua Trough (Córdoba and Guerrero, 1970) and the La Mula Basin (= Sabinas Basin of many authors) (Salinas, 1969). Cook and Bally (1976) indicate that the La Mula Basin was invaded in the Oxfordian, as is also indicated by Humphrey and Diaz (ms.). Sabinas Basin was first used to describe the intermontane coal basin (Fig. 20) in the vicinity of Nueva Rosita, Coahuila (Robeck et al., 1956), it is not the same feature as the "Sabinas" Basin used by other authors to describe the Late Ju-

rassic—Early Cretaceous basin between the Coahuila Platform on the west and the Tamaulipas Peninsula on the east (Fig. 8); at one time E.J. Guzman termed this basin "Golfo de Mesozoica de Coahuila" (lectures at the University of Texas at Austin, 1956). This name is too long to conveniently use on maps, but Guzman was correct in differentiating that basin from the Sabinas (coal) Basin proper. In order to prevent confusion, I have, for many years, termed this other Mesozoic basin the La Mula Basin.

Whether, at some time during the Late Jurassic, the Miquihuana Platform (or archipelago) was completely covered by sediments or not remains unclear, but the presence of nerineid limestones along some of the southern part of the eastern margin of that platform indicates deposition in very shallow water, and the absence of Jurassic over parts of the platform suggests nondeposition, although its removal by erosion is not ruled out. Imlay (1937, 1943) has interpreted the Miquihuana area as an island, and Böse (letter to J.A. Udden, Univ. of Texas, Archives at Austin) interpreted this particular island as the source area for the Galeana Sandstone (lower Hauterivian).

Certainly the Jurassic seaway did not cover the Aldama Platform or the Coahuila Platform (Fig. 9), and the Tamaulipas Peninsula, although completely covered by a Kimmeridgian sea at times, was probably an archipelago of small islands on a shallow, elongate platform during much of the Tithonian and lower Neocomian, accounting for the restricted accumulation of various deposits of evaporites in basins (La Mula) to its west (Fig. 8). The top of the La Caja Formation (Tithonian?) in the Rincon de Pelillo, Sierra Gomas, west of Potrero, Nuevo Leon, appears to be lagoonal, but on the east side of the Tamaulipas Peninsula; the barrier to the east of this lagoonal facies is unknown.

In the Chihuahua Trough there are also Late Jurassic evaporites, and near Placer de Guadalupe the "green formation" between the Paleozoic and the marine Jurassic, long thought to be Jurassic, was considered to be Permian by Bridges (1965). De Cserna et al. (1970) now report interlayered Permian rhyolites, dated radiometrically, within this formation.

Jurassic rocks between Arteága Canyon, southeast Coahuila, and Valles, southeast San Luis Potosí, crop out in very isolated sections that make interpretation difficult. The limestone identified as "Zuloaga" in the Sierra de Fraile, Nuevo Leon, the Sierra Minas Viejas, Nuevo Leon, and near Galeana, southern Nuevo

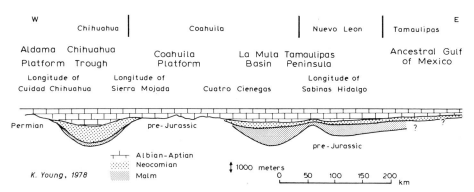

Fig. 9. East—west cross-section illustrating the general relations of Jurassic—Neocomian and Neocomian—post-Neocomian rocks in northern Mexico.

Leon, now appears to be "Cotton Valley Limestone" (Kimmeridgian) and has little to do with the real Zuloaga Limestone (Oxfordian) (Moor, 1978). Therefore, the ages of evaporites underlying this limestone need to be reinterpreted; they are more likely Kimmeridgian (Buckner equivalent in Texas) than Oxfordian.

In both the Campeche Embayment and northeastern Mexico there are thick sequences of evaporites. Those in the La Mula Basin may be Oxfordian at the oldest, and there also appear to be Oxfordian evaporites below the Zuloaga Limestone in areas south of Cuidad Victoria. A younger sequence of evaporites, at least in Nuevo Leon and Coahuila, is Kimmeridgian.

The southern part of the Rio Grande Trench, which in space, if not in time, is a continuation to the north of the Chihuahua Trough, must have been active before the end of the Late Jurassic, because the Grimm, Hunt, Brown, and American Arctic, Ltd., No. 1 Mobil 32 encountered Jurassic rocks, tentatively identified as Oxfordian (but more likely Kimmeridgian in the writer's opinion) at depths of 4210 m (Thompson, 1975). This extends the Chihuahua Trough into New Mexico somewhat north of its most northern limit on the maps of DeFord and Haenggi (1970), De Cserna (1970), Hayes (1970), and Cantu Chapa (1976).

There are metamorphosed rocks of Mesozoic age along the Sinaloa—Durango border (De Cserna, 1970; Córdoba et al., 1976). Cook and Bally (1976) show some of these as Triassic. Slightly metamorphosed flyschoid rocks in northwestern Jalisco, just south of Cabo Corrientes, were long labeled as metamorphosed Mesozoic (Salas et al., 1968), but Córdoba et al., (1976) now show these rocks to be metamorphosed Paleozoic, but the age of the rocks in Jalisco has not

been firmly established. The age of most of the metamorphic rocks in the Sierra Madre del Sur has also not been completely established, although some have yielded fossils as young as Middle Cretaceous (Palmer, 1928a).

That there are Jurassic intrusive and volcaniclastic rocks in western Mexico is doubted by few, but the subduction zone along the western coast of Mexico had proceeded for only a short time by the Late Jurassic, and igneous events of the Jurassic are minor when compared to the igneous events of the Cretaceous. The Sierra Madre Occidental is composed largely of Cenozoic igneous rocks, which not only overwhelm, in volume, those of the Cretaceous and Jurassic, but also conceal most of the pre-Cenozoic rocks of that large area.

Cretaceous

Early students of the Cretaceous of Mexico divided the System into lower, middle, and upper (Böse and Cavins, 1927); the lowermost division consisted of the initial, largely terrigenous rocks, mostly derived from local sources; the middle division consisted of the mostly carbonate part of the System, starting in the later Barremian and ending in the early Campanian, then thought to be Santonian; the upper subdivision included the final, terrigenous part of the System, consisting mostly of pre- and synorogenic deposits from the rising mountains of the Laramide Orogeny to the west; it includes rocks ranging in age from the upper 2/3 of the Campanian to, and including, the Maastrichtian. The base of the third subdivision was younger in the east and older in the west as the carbonate rocks were replaced by pre-orogenic terrigenous facies from the west (De Cserna, 1970).

Neocomian

There was some retreat of marine deposition near the end of the Jurassic, as indicated by the absence of transitional deposits over the Tamaulipas Platform and by the *Leopoldia victoriensis* Imlay zone resting on late Kimmeridgian at Samalayuca, northern Chihuahua (Webb, 1969a, b; Cantu Chapa, 1970) (Fig. 9). Many authors (e.g., DeFord and Haenggi, 1970, p. 181) seem to assume that either Young (in Webb, 1969a) or Cantu Chapa (1970) are in error on the ages of these sandstones. When the area was visited by the members of the A-13 excursion of the International Geological Congress in 1956 (Diaz and De Cserna, 1956), the late Walter Biese was certain of the Jurassic age of the outcrop visited at that time. The specimen of *Leopoldia victoriensis* Imlay, collected by Webb, cannot be doubted. Even the structure in the rocks surrounding the Sierra de Samalayuca is not all that simple (DeFord and Haenggi, 1970; Berg, 1969, 1970). Furthermore, it can never be proved

that Webb collected from exactly the same beds at exactly the same locality from which the specimens described by Cantu Chapa were collected. King (1947) originally mapped the formations as Jurassic–Lower Cretaceous undifferentiated, and it seems that both Kimmeridgian and Neocomian rocks are here represented. After all, the environments of deposition and the source areas would not have been that different.

Valanginian rocks rest on marine Kimmeridgian rocks or on Jurassic red beds across the Miquihuana Platform (Imlay, 1938) and other areas of northern Mexico.

The maximum extent of the Neocomian seaway (Fig. 10) was about the same as that of the Kimmeridgian, perhaps slightly less in the Chihuahua Trough, perhaps a bit more on the Miquihuana Platform.

Wilson and Pialli (1978, fig. 4) show interfingering at the boundary of the La Casita and Taraises Forma-

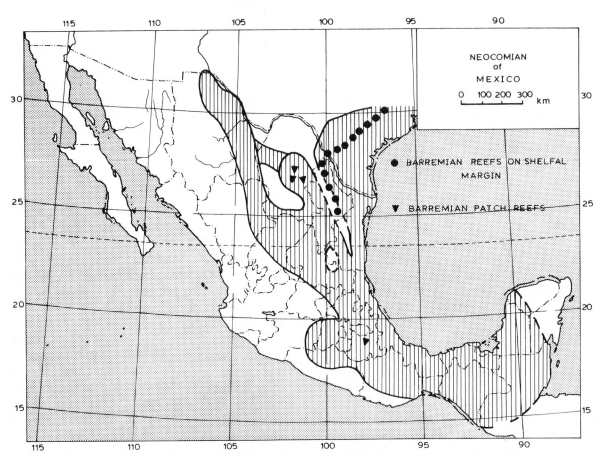

Fig. 10. Paleogeographic map depicting the maximum transgression of the Neocomian seaway in Mexico.

tions, although in the text they use paleontological indices for differentiating these boundaries; thus, there are probably changes in facies along and across the boundary of the Neocomian with the Jurassic in the La Mula Basin northeast of Saltillo, southern Coahuila. The Late Jurassic, subsequent to the Cotton Valley Limestone, is terrigenous (Böse, 1923a). The lower part of the Neocomian is also terrigenous, the terrigenous rocks gradually becoming more calcareous upward and away from the terrigenous sources as the Taraises Formation, or equivalent formations, grades upwardly and eastwardly into the Cupido Limestone. Eastward from the Tamaulipas Peninsula, into the ancestral Gulf of Mexico, the calcareous upper member of the Taraises Formation of the eastern Flank of Sierra Gomas, along the border of Coahuila and Nuevo Leon (Fuentes Fuentes, 1964), becomes more calcareous and is called Cupido Limestone when logged in wells in eastern Nuevo Leon.

It was during the Late Jurassic that the great barrier reef system (Fig. 8), surrounding much of the Gulf of Mexico, first began; it was to dominate the paleogeography of that region for about 30 million years. However, it was during the upper part of the Neocomian that this reef system began to take on the aspect that it would maintain until near the end of the Cenomanian (Young, 1972; Bebout and Loucks, 1974).

Aptian

There is a general break in the continuity of deposition of carbonate rocks during the late Aptian. This resulted in the deposition of an Aptian influx of terrigenous sediments during the *Deshayesites–Cheloniceras* zones, recognized as the La Pena Formation in the north of Mexico and the Otates Formation in the south. These beds become more calcareous to the south of Mexico into Guatemala and Honduras, where they have not been separated from the Coban or Atíma Formations (Mills et al., 1967; Finch, 1972; Ritchie, 1975), but, nevertheless, retain their distinctiveness of thinbedded limestones (7 to 10 beds per meter). Fig. 11 represents the seaway for the latest Aptian (the *Clansayes* horizon), which was the most widespread part of the Aptian seaway. Differences between my maps and those of Hayes (1970b) are largely in the selection of the horizon to be depicted than in disagreement of extent of seaways at any particular time.

Toward the northeast of the Coahuila Platform in Coahuila the La Virgen evaporites represent an Aptian, evaporitic platform-facies behind the upper part of the Cupido Limestone (Böse, 1923a; Smith et al., 1974).

Below the upper Aptian, carbonate formations have not been named according to their facies. Below the *Deshayesites–Cheloniceras*-bearing beds one can usually get by with using Cupido Limestone, or lower Tamaulipas for offshore limestone deposits, in most of the eastern Mexico or Sligo Limestone in the United States or the subsurface of northeastern Mexico. Above the *Cheloniceras* level, on the other hand, a great hodgepodge (variety) of names are used to differentiate the lateral and vertical distribution of the different facies of calcareous, sedimentary rocks. Tracts of grainstones, rudist banks, mudstones, wackestones, inner platform, outer platform, etc., each receives a different formal name.

Albian–Turonian

The Albian–Turonian history of northern and eastern Mexico is primarily centered around the different facies associated with the different platforms and basins (Fig. 8). Up until the Aptian the Coahuila Peninsula was largely exposed, shedding sediments into the Chihuahua Trough on the west to help produce the Navarette (Neocomian) and Las Vigas (Neocomian and lower Aptian) Formations. To the east, along its eastern margin, following the emplacement of late Triassic granites that crop out near Las Delicias and in the Potrero de la Mula (Denison et al., 1970) terrigenous sediments were shed into the La Mula Basin. Arkosic wedges of different ages [San Marcos (Neocomian), Menchaca (Jurassic), Patula (late Neocomian)] extend into the Basin, and some were even washed as far as the Rio Grande Embayment to produce a few red beds prior to the development of the Maverick Basin. Other unique terrigenous deposits, the La Mula Shale (a red deposit of Barremian to late Aptian age) and Barril Viejo Shale (Valanginian and Hauterivian), also have their sources in the Coahuila Platform. By the latest Aptian most of the areas that had been sources of terrigenous sediments, locally, were covered, and the deposition of carbonate rocks became the dominant depositional program of all of northern Mexico. During regressive hemicycles (Cooper, 1977; Kauffman, 1977; Matsumoto, 1977) reefoid rocks prograded away from the platforms. On the Coahuila Platform the late Albian–early Cenomanian transgression was rapid, and reefoid rocks

were covered with calcareous mudstones and wackestones; reefoid rocks did not again appear on that
platform (Fig. 12).

To the south, along the Miquihuana Platform (=
Valles–San Luis Potosí Platform of Carillo-Bravo,
1969, the southern end of which was called the Actopan Platform by Carrasco, 1971), transgression was
less rapid, and there was karstic development on some
parts of the platform following the earliest Cenomanian. Just southeast of Miquihuana large specimens of
Sauvagesia may indicate Turonian. 35 km east of San
Luis Potosí, in the Sierra de Alvarez, *Radiolites* sp. cf.
mullerriedi Bauman and *Hippurites mexicanus* Bar-
cena (De Cserna and Bello-Barrádas, 1963; Bartlett,
1978) from the Cuautla Formation are upper Turonian or younger. Here the Cuautla Formation appears
to rest unconformably on El Abra with a well-developed karstic terra rosa and red conglomerate separating the two formations.

Off, but adjacent to, the Miquihuana Platform cal-
cispheric mudstones and wackestones, which are
probably not of deep water, show little evidence of
the adjacent platform (Waisley, 1978). Most of the
limestones off the platform are black and highly organic. Those of Cenomanian age may represent the
Cenomanian anoxic event of general occurrence
throughout the Atlantic Ocean (Hays and Pitman,
1973; Ryan and Cita, 1977; Schlanger and Jenkyns,
1976; Hallam, 1977), but most of the organic, black
limestones off the platform are older than Cenomanian; some of these probably represent deposition in
local, anoxic basins of poor bottom circulation
(Schlanger and Jenkyns, 1976). But, not all black,
organic limestones are anoxic. The mudstones and
wackestones in the area of Barramberri, in the re-entrant at the north end of the Miquihuana Platform,
which I have drawn as extending farther north than
Carrillo Bravo (1969) or Carrasco (1971), give little

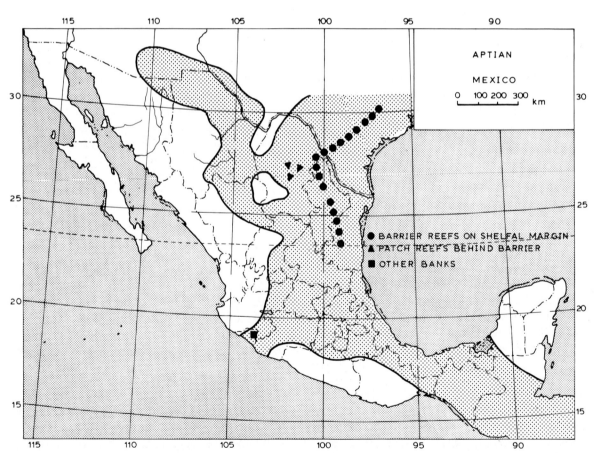

Fig. 11. Paleogeography of the maximum transgression of the Aptian seaway of Mexico. The distribution of known rudist banks
are also indicated.

evidence of the adjacent platform, perhaps indicating that the relief of the platform was not great near that area (Moor, 1978; Waisley, 1978). However, soft sediment flows on the west side of the platform north of San Luis Potosí (Waisley, 1978) and on the east side of the Miquihuana Platform in Peregrina Canyon west of Cuidad Victoria, during deposition of the late Albian and early Cenomanian Cuesta del Cura Formation, indicate an effective paleoslope from the platform. The general Cenomanian transgression is not apparent in the reefoid rocks around the margin of this platform.

Although some of the platform areas north of the Miquihuana Platform received rudist deposits (Figs. 12, 13) into the earliest part of the early Cenomanian (e.g., Stuart City Trend; Sierra de la Gloria near Monclova, Coahuila; northeast corner of the Coahuila Platform; Sierra de la Párra, northeast Chihuahua), all of these, unlike the Miquihuana Platform, foundered

by the end of the early Cenomanian and were covered with mudstones and wackestones (Jones, 1938; Smith, 1970) deposited in the rapidly transgressing early Cenomanian seaway. As in the Albian, the Cretaceous seaway of the early Cenomanian was more widespread in Mexico than at any other time during the Mesozoic (Fig. 12). The rising mountains resulting from the Laramide Orogeny with accompanying volcanism forced retreat of the seaway during the later Cretaceous (post-early Cenomanian) (Fig. 14), even though general transgression elsewhere was more extensive into the Campanian and Maastrichtian (Cooper, 1977; Kauffman, 1977; Matsumoto, 1977).

In northeastern Chihuahua and northern Coahuila the lower part of the early Cenomanian contains the terrigenous Del Rio Claystone, whereas the upper part is represented by the Buda Limestone. In northern and eastern Mexico, during the greatest advance of the Cretaceous seaway, the early Cenomanian is

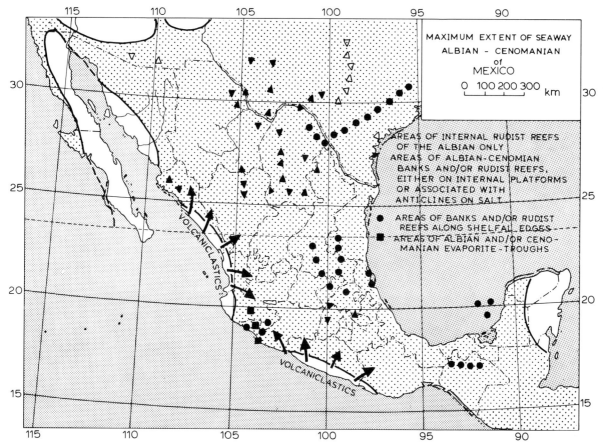

Fig. 12. Paleogeography of the maximum extent of Albian and Cenomanian seaways in Mexico. The maximum extent of the late Albian seaway is almost identical with that of the maximum extent of the early Cenomanian seaway. The seaway was less extensive during the Late Cenomanian. Known rudist banks and deposits are also indicated.

represented by the black, organic limestone of the Cuesta del Cura Formation. To the west in eastern Durango and western Zacatecas the limestone is still called Cuesta del Cura, but here there are 12 to 20 cycles per meter, in contrast to the 3 to 5 cycles per meter in eastern and northeastern Mexico. In this area the base of the Cuesta del Cura is also older than farther east, as this facies replaces the underlying, more massive limestones from east to west (De Cserna, 1976). The Cuesta del Cura Limestone then extends southward along the western side of the Miquihuana Platform, so close to the platform near Matehuala that small fragments of rudistids and other reef debris occur in some of the grainier beds.

Around the edges of the Miquihuana Platform early Cenomanian reefoid rocks replace the Cuesta del Cura Limestone. To the southwest in Colima and Jalisco thinbedded limestones, averaging 16 cycles per meter, may also be early Cenomanian (Palmer, 1928b), but may also include Albian.

Upper Cenomanian rocks are more restricted than lower Cenomanian, and the upper Cenomanian seems to be absent on the Miquihuana Platform and on the Faja de Oro (Coogan et al., 1972). Further north the upper Cenomanian is missing across parts of the Coahuila Peninsula, the zone of *Kanabiceras septem-seriatim* (Cragin) being spotty at Tanque Toribio (Jones, 1938; Young, 1978), with the rest of the upper Cenomanian and the lower part of the lower Turonian being absent. Upper Cenomanian is present in the Chihuahua Trough (Wolleben, 1965, 1967; Powell, 1963, 1965) and at El Paso, across the border from northeastern Chihuahua (Powell, 1967, 1970).

Through part of northeastern Mexico the upper Cenomanian is terrigenous (Agua Nueva Formation) as is the lower-most Turonian. To the east into the Gulf of Mexico and to the south along the eastern front of the Sierra Madre Oriental, Tamaulipas, the Agua Nueva Formation is sometimes replaced by the lower part of the overlying San Felipe Limestone

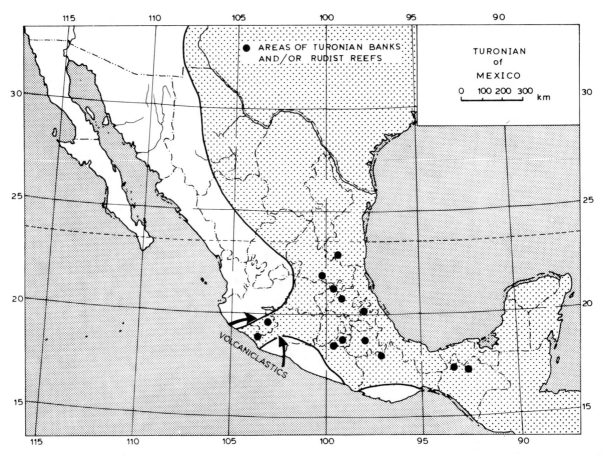

Fig. 13 Paleogeography of the Turonian of Mexico. Areas of known rudist banks and volcaniclastic deposits are also indicated.

(Fuentes Fuentes, 1964; Bishop, 1970, 1972), which also includes the upper Turonian, the Coniacian, the Santonian, and the lower part of the Campanian (e.g., *Parapuzosia boesei* Scott and Moore, species of *Submortoniceras*, and *Scaphites hippocrepis* s. l.)

Although upper Turonian rocks crop out over most of the Coahuila Peninsula in chalky beds, they are generally absent over the Miquihuana Platform, except to the south of the state of San Luis Potosí. An exception is in the Sierra de Alvarez (De Cserna and Bello Barrádas, 1963), just east of San Luis Potosí, where reefoid rocks (Cuautla Formation) contain *Radiolites, Hippurites*, and probably *Bayleoidea* (Bartlett, 1978). Similar rocks extend on south into Morelos and Oaxaca (Fe;rusquía, 1976) and northern Guerrero as the Morelos Formation (Bauman, 1958; Fries, 1960; Klesse, 1970). To the west, in the western Sierra Madre del Sur (Michoacan, Jalisco, and Colima), rocks containing *Bayleoidea, Radiolites*, and caprinids are probably Turonian (Palmer, 1928b).

Senonian

At the present time erosion has proceeded until enough Albian rocks are exposed that the paleogeographic elements of eastern and northern Mexico are reasonably well established (Fig. 8).

The Mesozoic history of eastern and northern Mexico is largely one of carbonate deposition on platforms, or karstic developments on platforms' versus offshore deposition of carbonate rocks. The most widespread marine deposits in Mexico are those of the Albian and Cenomanian (Fig. 12), and the maximum extent of the seaway of these two stages is similar, and somewhat beyond the western extent of the Aptian seaway (Fig. 11). Following the Cenomanian the Late Cretaceous seaway of Mexico steadily receded to the east (Fig. 14) as a result of Laramide orogenic pulses and the resulting terrigenous deposits filling in the seaway and displacing marine water (Fig. 15). The Caracol Formation of western Zacatecas, eastern Durango, and neighboring areas, represents ter-

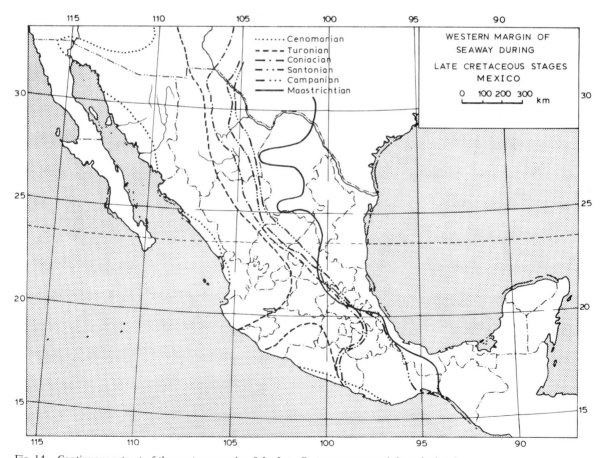

Fig. 14.   Continuous retreat of the western margin of the Late Cretaceous seaway is here depicted.

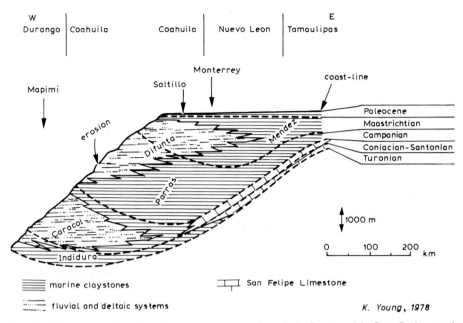

Fig. 15. Eastern migration of terrigenous depocenters through the Parras and La Popa Basins, northern Mexico.

rigenous debris from these rising mountains to the west, and these Turonian, Coniacian, and Santonian deltaic facies can be observed overlying lower Turonian *Mytiloides*-bearing strata in the Sierra de Cadeña and near Mapimí, northeastern Durango.

In northeastern Chihuahua, in the Sobaco Syncline, west of Cuchillo Parado, marine sandstones invade presumably upper Turonian strata overlying beds containing the ammonite, *Spathites*. The history of this marine regression of the Cretaceous seaway in Mexico is recorded in northeastern Chihuahua, near Ojinaga, where deltaic and fluvial systems consecutively replace younger and younger Campanian rocks to the east (Wolleben, 1965, 1966; Powell, 1967; Wilson, 1970; Weidie et al., 1972). The same story is told in the western part of the Párras Basin along the south side of the Laguna de Mayran, southern Coahuila. The upper part of the San Felipe Limestone (Santonian and lowest Campanian) of eastern Coahuila and Nuevo Leon is here represented by the Párras Shale of southwestern Coahuila, and perhaps the deltaic Caracol Formation even farther west (Fig. 15). The Méndez Formation (late Campanian of Tamaulipas) is represented to the west by the deltaic and fluvial systems of the Difunta Formation. The base of the Difunta Formation then rises from west to east, upper Campanian in southwestern Coahuila, until in Nuevo Leon it represents only the Maastrichtian. To the east, in Tamaulipas, the entire Difunta Formation (deltaic and fluvial systems) is replaced by the Méndez Shale.

To the south the more western facies (Fig. 15) have been removed by erosion, and the sequence of events is less clear, although the fluvial and deltaic systems of the Caracol Formation are recorded in northwestern Zacatecas (García-Calderón, 1976; De Cserna, 1976).

Coniacian. Coniacian rocks are of fairly wide distribution through the Caracol, Párras, and San Felipe Formations of the northeast of Mexico (Fig. 15). They extend south through Veracruz and are again widespread in the south of Mexico (Fig. 16). They are likewise present on the west side of the Miquihuana Platform as far south as Matehuala, where they occur in the San Felipe Formation in the mountains east of Sierra del Catorce. Burckhardt (1921) has described Coniacian ammonites from Zumpango del Rio, Guerrero, and Bonet (1969) considers Coniacian to be present in the Guzmantla Formation of the platform facies near Orizaba, Veracruz, and in the Necoxtla and Maltrata Formations of the offshore facies. Martínez (1972a, b) shows Coniacian to be present in a platform facies around the southern margin of the Campeche Basin, Chiapas.

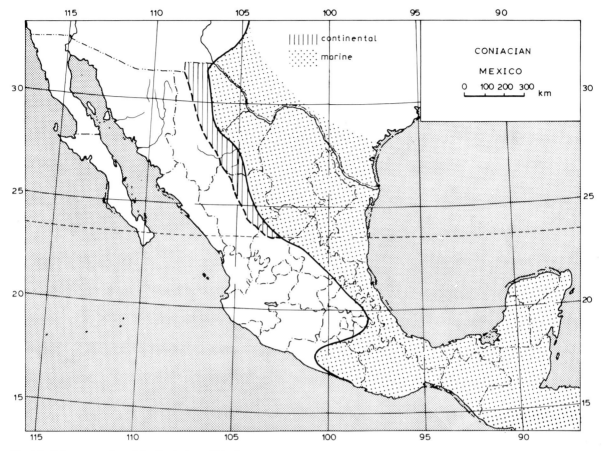

Fig. 16. Paleogeography of the Coniacian of Mexico.

Santonian. Santonian rocks have much the same distribution as Coniacian rocks (Fig. 17). In the north there are fluvial and deltaic facies in the Caracol Formation, and marine pelagic facies in the Párras Shale and San Felipe Limestone. Like the Coniacian, Santonian rocks probably occur in the San Felipe Limestone just east of Sierra del Catorce, west of Matehuala, San Luis Potosí. Santonian rocks also extend to the south on the east side of the Miquihuana Platform, where they appear to spread out over a large area of Chiapas, Oaxaca, and eastern Guerrero. Bonet (1969) considers the Santonian present in platform sections in southwestern Veracruz and adjacent Hidalgo, and Martínez (1972a) shows rocks of this age to be present in platform sections in Chiapas along the southern margin of the Campeche Embayment.

As in Colombia (Bürgl, 1956) there is a problem with the Santonian in Mexico. One of the indices for the Santonian in Colombia (Bürgl, 1956) and southern Mexico (Martínez, 1972b) is *Globotruncana fornicata* Plummer. Plummer (1931) first collected the types of this species from the Bergstrom Formation (Young, 1965), upper Campanian above the *Hoplitoplacenticeras vari* zone, at Moore and Berry's Crossing of Onion Creek, Travis County, Texas, south of Austin. McNulty (1976) records the *Globotruncana fornicata* zone from the upper part of the Austin Chalk (zones of *Submortoniceras tequesquitense* and *Delawarella delawarensis*), lower Campanian. Either *Globotruncana fornicata* Plummer is a long-ranged species that should not be used for dating, or different authors assign separate species to this taxon. A less likely possibility is that beds in Colombia and southern Mexico, considered Santonian, are really Campanian. On the other hand, Sanchez Montes de Oca (1969) indicates that the Coniacian and Santonian are both absent through much of the platform area of central Chiapas.

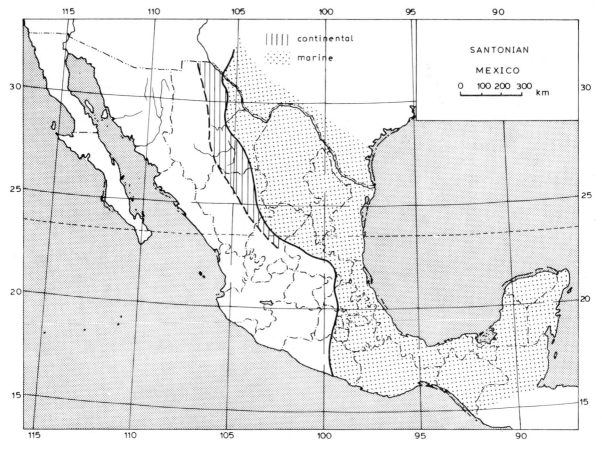

Fig. 17. Paleogeography of the Santonian of Mexico.

Campanian. Campanian rocks have much the same distribution as the Maastrichtian rocks (Fig. 18), although to the north the western margin of the seaway has not been displaced as far east (Figs. 14 and 15). There is probably Campanian in the Méndez Shale between Matehuala and Sierrra del Catorce.

A question arises as to the Maastrichtian or Campanian age of some of the deposits containing rudists. Myers (1968) interpreted the Cárdenas Formation as Maastrichtian. However, the presence of *Arctostrea aguilerae* (Böse) might indicate a Campanian age according to Sohl and Kauffman (1964), although they prefer the Maastrichtian date. Some of the earlier strata of the Cárdenas Formation in the Cárdenas area (Arroyo de la Mula at Rayon) contain a species of *Praebarrettia* closely related to *Praebarrettia sparcilirata* (Whitfield), with which species it has been identified. But the specimens of this species at Rayon contain distinct vertical grooves on the exterior of the

attached valves, are more primitive, and seem to differ from those specimens of *Praebarrettia sparcilirata* from the San Luis Conglomerate in Chiapas.

The Tamazopo Limestone can be interpreted in two ways. The first interpretation is as a series of facies tracts with Tamazopo and Cárdenas being laterally equivalent (Sansores and Girard, 1969), the Tamazopo as the higher part of the bank and the Cárdenas beds as an adjacent lagoon on the west. The second interpretation considers the Tamazopo to be older than the Cárdenas (Böse, 1906). The first interpretation is aesthetically most satisfying, especially since the Tamazopo Limestone contains, generally, the same fossils as the Cárdenas (Böse, 1906), except that boundstones of *Distefanella* sp. and *Biradiolites aguilerae* (Böse) are common in the Tamazopo and absent from the Cárdenas; this occurrence could be reflecting differences in depth of water.

Farther south, in the Campeche Embayment,

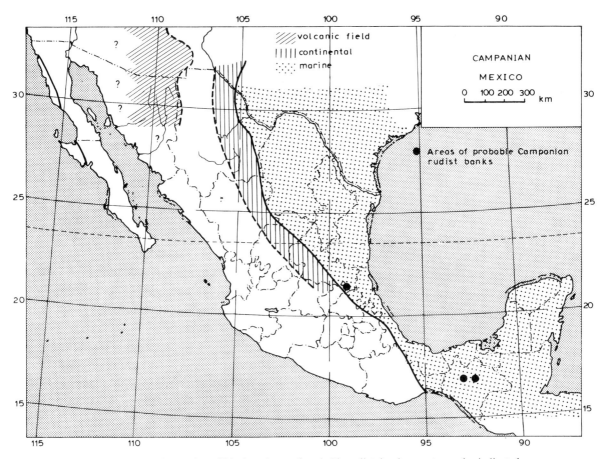

Fig. 18. Paleogeography of the Campanian of Mexico. Areas of probable rudist developments are also indicated.

Chiapas, there are strata with *Chiapasella* and *Titanosarcolites* that are undoubtedly Maastrichtian (Alencáster, 1971). Underlying these strata of the Ocozocuautla Formation (Fig. 19) are other strata of the same formation with *Praebarrettia sparcilirata* (Whitfield). The San Luis Conglomerate contains cobbles composed of *Praebarrettia sparcilirata* (Whitfield) that must have been eroded from older beds. In all, it is probable that on the platforms, at least, Campanian is thin to absent, or else it is represented in rudistaceous rocks that are now assigned to the Maastrichtian. One does gather the impression that basinal facies dated as Campanian on foraminifers are timeequivalent to Maastrichtian rocks on platforms dated by rudists.

Maastrichtian. From Rayon, southern San Luis Potosí, south of Cárdenas, to the north past Cuidad de Mais, El Doctór and Tula, large down-faulted areas contain Maastrichtian (Cárdenas Formation) terrigenous deposits that in some areas are sufficiently carbonate to represent such a low rate of deposition that such rudists as *Tampsia, Durania, Hippurites, Coralliochama, Distefanella, Bournonia,* and *Sauvagesia,* are abundant (Myers, 1968). Where the Rio Verde passes through the eastern limestone ranges of the Sierra Madre Oriental late Campanian or Maastrichtian reefoid rocks contain the genera *Praebarrettia, Biradiolites, Distefanella, Thyrastylon, Durania, Sauvagesia, Tampsia,* and many corals and large bryozoans and algae. These fossils occur in the younger rocks at Tamazopo (Böse, 1906), constituting the type locality of the Tamazopo Limestone, where it rests disconformably on a karstic surface developed on late Albian or early Cenomanian reefoid rocks (El Abra Limestone). Sansores and Girard (1969) have interpreted the Tamazopo Limestone as a Maastrichtian reefoid facies along the eastern margin of the Miquihuana Platform,

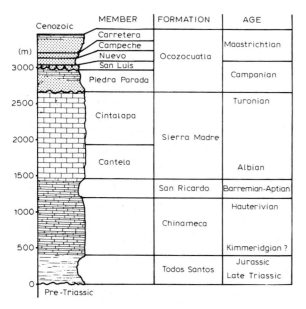

	MEMBER	FORMATION	AGE
Cenozoic			
	Carretera		Maastrichtian
(m)	Campeche	Ocozocuatla	
	Nuevo		
3000	San Luis		Campanian
	Piedra Parada		
2500	Cintalapa		Turonian
2000		Sierra Madre	
	Cantela		Albian
1500		San Ricardo	Barremian-Aptian
1000		Chinameca	Hauterivian
500			Kimmeridgian ?
		Todos Santos	Jurassic
0			Late Triassic
Pre-Triassic			

Fig. 19. Composite section from Central Chiapas along the platform south of the Campeche Basin. The problem of the presence of Coniacian and Santonian rocks in this area is discussed in the text.

with Méndez deposited to the east and Cárdenas Formation on the sagging central part of the platform to the west.

Further south in Pueblo, southern Veracruz, Chiapas, and Tabasco, Maastrichtian rocks contain many of the above rudist genera plus *Chiapasella* and *Titanosarcolites* (Chubb, 1959; Aléncaster, 1971). In the western part of this area the Maastrichtian rocks overlie red conglomerates (the San Luis Conglomerate).

The Maastrichtian rocks are mostly terrigenous, except for the grainstones and rudist banks of the Tamazopo and Ocozocuautla Formations and the larger limestone banks in the Difunta Formation of the La Popa Basin north of Monterrey, Nuevo Leon (Fig. 20) (McBride et al., 1974), which are composed largely of red algae, but contain species of *Coralliochama*, hippuritids, *Durania*, and caprinids.

Cretaceous facies: rudist

Except for areas of Pueblo, where there are Barremian monopleurids, rudistaceous deposits in most areas of Mexico appear to have developed locally in the lowest Aptian. There seems to be no such thing as a continuous rudistid reef, except for some migrating

biostromes, particularly in Texas. Most rudistid developments represent patches of rudistid growth in the proper environment along platform or shelf edges. Stabler and Marquez (1978) interpret early rudistaceous deposits in the La Mula Basin as originating on highs in the middle of the Basin. They consider the highs as mud anticlines, but the underlying mud had not received much load at that time. More likely the great wedges of Cretaceous terrigenous sediments debouching off the Coahuila Platform were overloading the Jurassic evaporites, producing evaporite-anticlines at the toes of these terrigenous wedges. The anticlines resulted in shallow water over their crests, producing small, shallow water platforms (Elliott, 1978), ideal environments for rudistid growth. There is evidence that salt was flowing throughout the Cretaceous in the subsurface of the La Mula and La Popa Basins. Debris flows from such salt-anticlines (highs) occur in deposits that now reveal only the direction of the source, the anticlines probably having collapsed after salt withdrawal; they were buried by younger terrigenous wedges.

In northern and eastern Mexico most of the rudistid reefs are distributed around and over the Coahuila and Miquihuana Platforms and the Faja de Oro during the Albian and early Cenomanian, and around the Miquihuana Platform during the later Cretaceous. In southern Mexico they occur on platforms south and west of the Campeche Embayment.

In the Sierra Madre del Sur, from Jalisco through Michoacan into Guerrero and Morelos, Albian through Turonian reefoid limestones or bank deposits are on narrow, submarine, topographic highs between narrow basins that contain volcaniclastic and/or fine-grained terrigenous sediments. But the rudistaceous reefs in the western Sierra Madre del Sur are only part of the story. The mountains east of Tecopan, on the shore of Colima, which constitute the coast ranges in this area, are composed of gypsum; the gypsum is in fault contact with fine-grained terrigenous rocks and with thinly bedded, cyclic limestone of about 16 cycles per meter. All of these rocks were considered Cenomanian by Palmer (1928b), but that age has not been verified, and similar deposits represent different ages of the Middle Cretaceous in different basins. They probably represent basinal deposits in long grabens separated by horsts upon which the rudistaceous banks developed simultaneously, and collectively probably represent all of the time from Middle Albian through the Turonian, according to the determina-

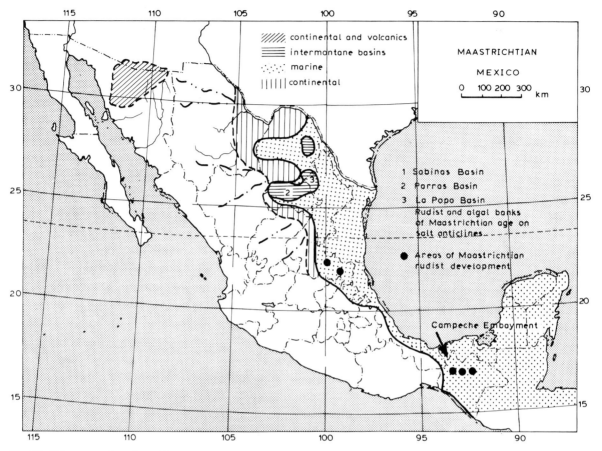

Fig. 20. Paleogeography of the Maastrichtian of Mexico. Areas of rudist development and intermontane basins are also indicated.

tions of the ranges of genera of rudists (Chubb, 1971; Coogan, 1973; 1978). Middle Cretaceous limestones and volcanic rocks continue to the eastward in the Sierra Madre del Sur into Guerrero and Morelos (Bonet, 1971).

Cretaceous facies: rocks of ill-defined age, the Pacific Slope and the Sierra Madre del Sur

Those isolated outcrops of Cretaceous rocks along the Pacific slopes of Mexico, north of Cabo Corrientes, Jalisco, are labeled Lower Cretaceous, Upper Cretaceous, or Cretaceous undifferentiated (Córdoba et al., 1976). Most of these appear to be late Albian, early Cenomanian, or both, and are here so indicated on Fig. 15. The Albian and Cenomanian are shown on the same map, largely because there is little difference in the maximum extent of seaways in Mexico during these two stages, but also because differentiation of

these stages in these western states is not yet possible. Most of the Albian deposits are limestone and marl, but shales begin to appear in east-central Chihuahua, and west of Cuidad Chihuahua a large wedge of shale replaces middle Albian limestone (Guerrero, 1969), even though equivalent beds at Arivechi, northeastern Sonora, are calcareous mudstones. Along the Pacific slope, and even as far east as Durango, in the Sierra de Cadeña, the late Albian and early Cenomanian deposits are mostly thin-bedded limestone, in beds of from 12 to 20 cycles per meter, usually resting on Albian, rudistaceous, reefoid rocks. They are usually mudstones, although some graded beds of siltstone to mudstone may be observed.

Along the Pacific slope and eastward through much of the Sierra Madre del Sur Middle Cretaceous outcrops occur in four ways: (1) as horsts in Cenozoic volcanics; (2) as grabens in metamorphics older than Cretaceous; (3) as ranges of mountains that are

probably horsts exposed through Cenozoic volcaniclastics; and (4) as horsts that do not represent mountains but are enclosed by Cenozoic volcanic rocks.

The ages of some of these rocks are not completely settled. Palmer (1928a,b) has considered those
of Michoacan (Coalcoman), Colima (Paso del Rio and
the mountains east of Tecopan), and Jalisco (Soyatlan de Adentro) as Cenomanian. Redescription of
some of the rudists (Chubb, 1971; Coogan, 1973,
1978) indicates that the rocks at Coalcoman and Soyatlan de Adentro are probably Albian, although the
upper ranges of such rudist genera as *Caprinuloidea,
Texicaprina*, and *Kimbleia* are not well established.
Certainly, the latter two genera seem to range into
the Cenomanian. In addition, the fossiliferous strata
at Soyatlan de Adentro are downslope debris flows
intermingled with volcaniclastic strata.

Turonian deposits in this area, such as those east
of Barra de Navidad, Colima, and at Huescalapa, Jalisco, contain *Bayleoidea clivei* Palmer, *Radiolites* sp. cf.
mullerriedi Bauman, and *Discocyclina* sp., plus caprinids.

VOLCANISM

It is not usually possible to refer to volcanism
within the time scale used for the paleogeographic
maps of this paper. The Triassic and Jurassic volcanism for southern Arizona has been documented
(Hayes, 1970a; Cooper, 1970). Undoubtedly some of
these volcanic fields extend from the Sierrita and
other mountains of southernmost Arizona southward
into northern Sonora (Salas, 1970). Triassic volcanic
rocks of western Mexico extend into western Zacatecas (Belcher, 1978), where they make up as much as a
third of the total Triassic section. Their eastern attenuation extends to the east into southeastern Coahuila.

The exact age of the "Rocas Verdes", green spil-
litic volcanics that overlie the Upper Triassic at Zacatecas, has been debated, in that rocks of similar lithology and stratigraphic position in other areas, except that the underlying Upper Triassic is absent,
have been interpreted as older than Jurassic, probably
because "Rocas Verdes" at Placer de Guadalupae,
northeastern Chihuahua, contain interlayered rhyolites of Permian age (De Cserna et al., 1970). These
differences in interpretation still need to be resolved.

No one will deny that during the Jurassic western

Mexico suffered extensive igneous activity. But, the
extent and effect of such activity, above and beyond
"local" regional metamorphism, has not been defined.

Cretaceous igneous events are much more widespread than Triassic or Jurassic events, but even they
are subdued by the overwhelming cover of Cenozoic
igneous rocks.

There is extensive igneous activity during the Late
Cretaceous and into the Cenozoic through southern
Sonora and Sinaloa and Nyarit (McDowell and Cla-
baugh, 1978). As Nyarit joins Jalisco the Sierra Madre
Occidental joins the Sierra Madre del Sur. But in the
Sierra Madre del Sur (one of the major volcanic trans-
forms of Middle America) the Neogene, including the
Quaternary, volcanic system has dampened, and even
concealed major Cretaceous igneous and metamor-
phic events.

At Paso del Rio, Colima, the single reefoid cycle
with *Immanitas*, a unique rudistid, is intruded by
diorite of unknown origin, as are the conformably
overlying fine-grained volcaniclastics, which are supposedly Cenomanian (Palmer, 1928b). At Coalcoman,
Michoacan, and Soyatlan de Adentro, Jalisco, particularly the latter, rudists are preserved in debris
flows down Middle Cretaceous submarine paleoslopes, interbedded with volcaniclastics, very similar
to the rubble beds with Maastrichtian rudists that
have rolled down an ancient paleoslope north of Ha-
cienda Flor de Alba, Puerto Rico.

At Huescalapa, near Cuidad Guzman, Jalisco, Turonian cycles in carbonate rocks with *Bayleoidea clivi*
Palmer, *Radiolites*, and caprinids, grade upward into
volcaniclastics (Palmer, 1928b). Eastward from Michoacan into Guerrero and Morelos there are volcaniclastic sediments of Middle Cretaceous age, and areas
in which Middle Cretaceous rocks (zone of *Plesiotur-
rilites brazoensis*), basal early Cenomanian (Palmer,
1928a) have been metamorphosed. This is the same
faunal zone as that of the silicified fauna described by
Böse (1923b) from the middle of the Cuesta del Cura
formation of northern Zacatecas; it also occurs at
Sierra del Catorce. The unique silicification of small
ammonites may be associated with silica in the rocks
resulting from the deposition of volcaniclastic rocks.
Siliceous sponge spicules are also abundant in the
fine-grained Middle Cretaceous rocks of the states of
Colima and Jalisco.

Much more is to be learned of metamorphism and
igneous activity of the Mesozoic in western and
southern Mexico.

TECTONISM

Mesozoic tectonism of Mexico, so far as known, is generally apparent in the preceding discussion because it is reflected in all of the sedimentary rocks. The Coahuila Platform is essentially a massif, at least the eastern edge of which is held up in part by Late Triassic granites (Denison, 1970; King et al., 1944).

Eastern Mexico, including the Miquihuana Platform, as best interpreted at present on insufficient evidence, probably is associated with an ancient, later aborted, Triassic spreading system (Belcher, 1978).

The western coast, and adjacent areas, is associated with a subducting margin of the Pacific Ocean, but much of the entire region is covered by Cenozoic volcanics, and the exposures of Mesozoic rocks are few.

Central Mexico and north-central Mexico, as far south as southern San Luis Potosí, at least, is associated with a Paleozoic and Precambrian terrane that has resulted in greater tectonic stability (De Cserna, 1969a, 1970).

Salt tectonics are a field of their own, and in part produce the ridge and valley province (Humphrey, 1956) of the northeast, and they are especially developed through part of the Sierra Madre Oriental (Heim, 1940; De Cserna, 1956). The northeastern part of the Chihuahua Trough demonstrates unusual salt tectonics (Haenggi and Gries, 1970; Gries and Haenggi, 1970).

The Sierra Madre del Sur and the Isthmus of Tehuantepec are associated with and owe their peculiar configurations to their nearness to the northern margin of the Caribbean Plate. From the Sierra Madre del Sur to the south there are dominant, Mesozoic and Cenozoic, transforms that need further study.

Perhaps resulting from the more rapid movement of the American Plate to the west during the Late Cretaceous, the Laramide Orogeny began to affect rocks in the west of Mexico earlier than in the east. Thus the base of preorogenic terrigenous rocks occurs in younger and younger rocks to the east (De Cserna, 1970), rising from the late Turonian in the west into the Maastrichtian to the east.

SUMMARY

To summarize this paper, which in itself is only a summary, requires emphasizing only the briefest highlights. The Mesozoic history of eastern Mexico is dominated by a Triassic spreading system, subsequently aborted, but the tectonics of which influence deposition into the Cenozoic.

The Pacific regions of Mexico are dominated by geologic events associated with a subducting continental margin; in the particular margin in Mexico younger and younger volcanism tends to conceal earlier geologic events.

The Central Plateau area of northern Mexico, including the altiplano to at least as far south as northeastern Jalisco and southern San Luis Potosí, is made more stable by an underlying Paleozoic terrane that was largely unaffected by either subduction to the west or rifting to the east.

The Siera Madre del Sur is unique, with its interlayering of volcaniclastics and marine sedimentary rocks, in some areas dating at least from the Aptian. The thousands of meters of Middle and Late Cretaceous volcaniclastics testify to the tremendous volcanic energies released here in the Mesozoic, a volcanism that is yet continuing.

The accompanying paleogeographic maps, as imperfect as they are, tell a story of intermittent but general transgression from the Middle Jurassic to the middle of the Cretaceous, followed by continual regression to the end of the Cretaceous as a result of Laramide orogenies and progradation of the deltaic and fluvial systems from west to east across Mexico.

In western Mexico volcaniclastic rocks are known from the Triassic to the Pleistocene, and through the Mesozoic represent increasing activity from Triassic to Cretaceous. In the Sierra Madre del Sur Cretaceous volcaniclastic rocks are several thousands of meters thick. Mesozoic intrusives include the late Triassic granites along the eastern margin of the Coahuila Peninsula, and diorite and other intrusives through the Pacific states of Sinaloa, Nyarit, Jalisco, and Colima.

ACKNOWLEDGEMENTS

Much of my work in Mexico during various periods has been supported by the Geology Foundation at the University of Texas at Austin or by the University of Texas at Austin Research Institute. I also wish to thank Bob Belcher, Samuel G. Bartlett, and Calixto Ramírez for relevant information.

REFERENCES

Aléncaster, G., 1961. Estratigrafiá del Triásico superior de la parte central del Estado do Sonora. *Univ. Nac. Auton. Méx., Inst. Geol., Paleontol. Mex.*, 11, pt. 1, 18 pp.

Aléncaster, G., 1971. Rudistas del Cretácico superior de Chiapas; Parte I. *Univ. Nac. Auton. Méx., Paleontol. Mex.*, 34, 91 pp.

Barceló-Duarte, J., 1978. *Estratigrafía y petrografía detallada del area de Tehuacán, San Juan Raya, Edo. de Puebla.* Tesis Professional, Facultad de Ingenieria, Univ. Nacl. Autón. de México, 143 pp.

Bartlett, S.G., 1978. *Carbonate Rocks of the Area around San Francisco, Municipio Zaragoza, State of San Luis Potosi.* Thesis, Univ. of Texas, Austin.

Bauman Jr., C.F., 1958. Dos radiolitidos nuevos de la región de Cuernavaca, Morelos. *Univ. Nac. Auton. Méx., Inst. Geol., Paleontol. Mex.*, 3, 9 pp.

Bebout, D.G. and Loucks, R.G., 1974. Stuart City Trend, Lower Cretaceous, south Texas, a carbonate shelf-margin model for hydrocarbon exploration. *Univ. Texas, Bur. Econ. Geol. Rept. Invest.*, 78, 80 pp.

Bebout, D.G., Coogan, A.H. and Maggio, C., 1969. Golden Lane-Poza Rica Trends, Mexico: an alternate interpretation. *Am. Assoc. Petrol. Geol.*, 53: 706.

Belcher, B., 1978. *Depositional Environments, Paleomagnetism, and Tectonic Significance of Lower Mesozoic Red Beds in Northeastern Mexico.* Thesis, Univ. of Texas, Austin.

Berg, E.L., 1969. Geology of Sierra de Samalayuca, Chihuahua, Mexico. In: D.A. Córdoba, et al. (Editors), *Guidebook to the Border Region.* New Mex. Geol. Soc., pp. 176–181.

Berg, E.L., 1970. *Geology of the Sierra de Samalayuca, Chihuahua, Mexico.* Thesis, Univ. of Texas, Austin, 79 pp., 18 figs., 1 pl., + appendices by Erasmo A. Euarte and Ivan F. Wilson.

Bishop, B.A., 1970. Stratigraphy of Sierra de Picachos. Am. Assoc. Petrol. Geol. Bull., 54: 1245–1270, 25 figs.

Bishop, B.A., 1972. Petrography and origin of Cretaceous limestones, Sierra de Picachos and vicinity, Nuevo Leon, Mexico. *J. Sediment Petrol.*, 42: 270–286, 29 figs.

Bonet, F., 1969. Microfacies de las calizas Cretácicas de la región Córdoba-Orizaba. In: *Seminario sobre exploración petroléra, Mesa Redonda 4. Problemas de exploración de la Cuenca de Papaloapan.* Inst. Mex. Petróleo, 24 pp., 1 fig., 8 pls., 2 tbls.

Bonet, F., 1971. Espeleología de la region de Cacahuamilpa, Gro.: *Univ. Nac. Auton. Méx., Inst. Geol., Bol.*, 90, 98 pp., 16 maps, 2 tbls., 26 pls.

Böse, E., 1906. La fauna de moluscos del Senoniáno de Cárdenas, San Luis Potosí. *Inst. Geol. Méx. Bol.*, 24, 95 pp., 18 pls.

Böse, E., 1923a. Vestiges of an ancient continent in northeast Mexico. *Am. J. Sci., 5th Ser.*, 6: 127–136; 196–214; 310–337.

Böse, E., 1923b. Algunas faunas Cretácicas de Zacatecas, Durango, y Guerrero. *Inst. Geol. Méx. Bol.*, 42: 219 + iv, 19 pls.

Böse, E. and Cavins, O.A., 1927. The Cretaceous and Tertiary of southern Texas and northern Mexico. *Univ. Tex. Bull.*, 2748: 7–142, pl. 19.

Bridges, L.W., II, 1965. Geología del area de Plomosas, Chihuahua. *Univ. Nac. Auton. Méx., Inst. Geol., Bol.*, 74, pt. 1, pp. 1–134, 21 figs., 3 pls., 11 tbls. (tranl. by Diego A. Córdoba).

Burckhardt, C., 1905. La faune marine du Trias supérieur de Zacatecas. *Inst. Geol. Méx. Bol.*, 21, 44 pp., 8 pls.

Burckhardt, C., 1906. Sobre del descubrimiento del Trias marino en Zacatecas. *Inst. Geol. Méx. Bol.*, 2: 43–45.

Burckhardt, C., 1921. El Cretaceo superior de Zumpango Del Rio. *Inst. Geol. Méx. Bol.*, 33, Pt. 2, pp. 81–135, pls. 22–32 (text published in 1919, atlas in 1921).

Burckhardt, C., 1930a. Etude synthétique sur le Mésozoique Mexicain, Première partie (Jurassic et Triassic). *Mém. Soc. Paléontol. Suisse*, 49: 122 pp.; 32 figs.; 11 tbls.

Burckhardt, C., 1930b. Etude synthétique sur le Mésozoique Mexicain, seconde partie (Cretaceous). *Mém. Soc. Paléontol. Suisse*, 50: 125–280; figs. 33–65; tbls. 12–18.

Burckhardt, C. and Scalia, S., 1906. Géologie des environs de Zacatecas. *Geol. Congr. Intern, 10th, Mexico, Guide to Excursions 16*, 26 pp., 10 figs., map.

Bürgl, H., 1956. *Stratigraphic Paleontology in Colombia.* Unpublished ms.

Cantu Chapa, A., 1969. Una nueva localidad del Triásico Superior marino en México. *Rev. Inst. Mex. Petról.*, 1 (2): 71–72, 1 fig.

Cantu Chapa, A., 1970. El Kimeridgiáno Inferior de Samalayuca, Chihuahua. *Rev. Inst. Mex. Petról.*, 2 (3): 40–44, 1 fig., 1 pl.

Cantu Chapa, A., 1971. La serie Huasteca (Jurásico Medio-Superior) del centro este de México. *Rev. Inst. Mex. Petról.*, 3 (2): 17–40, 1 fig., 4 pls.

Cantu Chapa, A., 1976. Nuevas localidades del Kimeridgiáno y Titoniáno en Chihuahua (Norte de México). *Rev. Inst. Mex. Petról.*, 8 (2): 38–49; 3 figs., 2 pls.

Cantu Chapa, C.M., 1969. Los rocas Eocretácicas de Zitacuaro, Michoacan. *Inst. Mex. Petról., Sec. Geol., Monogr.* 2, pp. 1–18, 1 map.

Cantu Chapa, C.M., 1974. Una nueva localidad del Cretácico Inferior en México. *Rev. Inst. Mex. Petról.*, 6 (4): 51–55; 1 fig., 1 pl.

Carrasco V., B., 1970. La Formación El Abra (Formación El Doctór) en la Plataforma Valles-San Luis Potosí. *Rev. Inst. Mex. Petról.*, 2 (3): 97–99; 1 fig.

Carrasco V., B., 1971. Litofacies de la formación El Abra en la Plataforma de Actopan, Hgo. *Rev. Inst. Mex. Petról.*, 3 (1): 5–26; 38 figs.

Carrillo Bravo, J., 1969. Exploración geológica y posibilidades petroléras de la Plataforma Valles–San Luis Potosí (Sierra Madre Oriental–Altiplano Mexicano). 19 pp., il. In: *Seminario sobre Exploración Petroléra. Mesa Redonda 6: Problemas de exploración en areas posiblemente petroliferas de la Republica Mexicana.* Inst. Mex. Petróleo., 19 pp.

Chubb, L.J., 1959. Cretaceous of central Chiapas, Mexico. *Am. Assoc. Petrol Geol. Bull.*, 43: 725–756; 10 figs., 3 tbls.

Chubb, L.J., 1971. Rudists of Jamaica. *Palaeontogr. Am.*, 8 (45): 161–257 + ii; 10 figs.; Pls. 27–57.

Coogan, A.H., 1973. Nuevos rudistas del Albiáno y Cenomaniáno de México y del sur de Texas. *Rev. Inst. Mex. Petról.*, 5 (2): 51–82; 8 figs.; 9 pls.

Coogan, A.H., 1978. Early and Middle Cretaceous Hippuritacea (Rudists) of the Gulf Coast. *Univ. Texas Austin, Bur. Econ. Geol., Rept. Invest.*, 89: 32–70; 7 figs., 18 pls.

Coogan, A.H., Bebout, D.G. and Maggio, C., 1972. Depositional environments and geologic history of Golden Lane and Poza Rica Trend, Mexico, an alternative view. *Am. assoc. Petrol. Geol. Bull.*, 56: 1419–1447; 21 figs.

Cook, T.D. and Bally, A.W., 1976. *Stratigraphic Atlas of North and Central America*. Princeton Univ. Press, Princeton, 272 pp.

Cooper, J.R., 1970. Mesozoic stratigraphy of the Sierrita Mountains, Pima County, Arizona. *U.S. Geol. Surv. Prof. Pap.*, 658-D: 42 pp.; 8 figs.

Cooper, M.R., 1977. Eustacy during the Cretaceous: its implications and importance. *Palaeogeograph, Palaeoclimatol, Palaeoecol.*, 22: 1–60; 18 figs.

Córdoba, D.A. and Guerrero García, J., 1970. Mesozoic stratigraphy of the northern portion of the Chihuahua Trough. In: K. Seewald and D. Sundeen (Editors), *Symposium in Honor of Professor Ronald K. DeFord*. West Tex. Geol. Soc., pp. 83–97, 8 figs., 2 tbls.

Córdoba, D.A. et al., 1976. *Carta Geológica de la Republica Mexicana*. Escala 1/2,000,000, 4th ed. Comité de la Carta Geológica de México, D.F.

De Cserna, E.G. and Bello Barrádas, A., 1963. Geología de la parte central de la Sierra de Alvarez, Municipio de Zaragoza, Estado de San Luis Potosí. *Univ. Nac. Auton. Méx., Inst. Geol. Bol.*, 71 (2): 23–63; 2 figs.; 12 pls.

De Cserna, Z., 1956. Tectonica de la Sierra Madre Oriental de México, entre Torreon y Monterrey. *XX Congr. Geol. Intern.*, iv + 87 pp., 7 pls., 1 tbl., frontispiece.

De Cserna, Z., 1969a. Tectonic framework of southern Mexico and its bearing on the problem of continental drift. *Soc. Geol. Méx. Bol.*, 30 (2): 159–168; 2 figs.; 1 pl. (for 1967).

De Cserna, Z., 1969b. Notas sobre la geología del area del Tecomatlan, Estado de Pueblo. *Univ. Nac. Auton. Méx., Inst. Geol., Paleontol. Mex.*, 27: 79–88; 3 figs.

De Cserna, Z., 1970. Mesozoic sedimentation, magmatic activity and deformation in northern Mexico. In: K. Seewald and D. Sundeen (Editors), *Symposium in Honor of Professor Ronald K. DeFord*. West Tex. Geol. Soc., pp. 99–117, 9 figs.

De Cserna, Z., 1976. Geology of the Fresnillo area, Zacatecas, México. *Geol. Soc. Am. Bull.*, 87: 1191–1199; 8 figs.; 1 tbl.

De Cserna, Z., Rincón-Orta, C., Solorio-Munguía, J. and Schmitter-Villada, E., 1970. Una edad radiometrica Permica temprana de la región de Placer de Guadalupe, noreste de Chihuahua. *Bol. Soc. Geol. Méx.*, 31: 65–73; 2 figs.

DeFord, R.K. and Haenggi, W.T., 1970. Stratigraphic nomenclature of Cretaceous rocks in northeastern Chihuahua. In: K. Seewald and D. Sundeen (Editors), *Symposium in Honor of Professor Ronald K. DeFord*. West Tex. Geol. Soc., pp. 175–196, 7 figs., 2 appendices.

Denison, R.E., 1970. Basement rock framework of parts of Texas, southern New Mexico and northern Mexico. In: K. Seewald and D. Sundeen (Editors), *Symposium in Honor of Professor Ronald K. DeFord*. West Tex. Geol. Soc., pp. 3–14, 2 figs., 3 tbls.

Diaz, T. and De Cserna, Z., 1956. *Guidebook for Excursion A-13*. Congr. Geol. Intern., XX Sess., English translation, 126 pp.

Elliott, T.L., 1978. *Carbonate Slope Deposits of the Lower Cretaceous, Northeastern Mexico*. Thesis, Univ. Texas, Austin.

Erben, H.K., 1957a. New biostratigraphic correlations in the Jurassic of eastern and south-central Mexico: *XX Congr. Geol. Intern., Publ.*, 11: 43–52.

Erben, H.K., 1957b. Paleogeographic reconstructions for the Lower and Middle Jurassic and for the Callovian of Mexico. *XX Congr. Geol. Intern., Publ.*, 11: 35–49.

Ewing, J.I., Edgar, N.T. and Antoine, J.W., 1970. 10. Structure of the Gulf of Mexico and Caribbean Sea. In: A.E. Maxwell (Editor), *The Sea: Ideas and Observations on Progress in the Study of the Seas*. Wiley Interscience, New York, N.Y., pp. 321–358, 18 figs.

Ferrusquia Villafranca, I., 1976. Estudios geológico–paleontológicos en le region Mixteca, Pte. 1: Geología del area Tamazulapan–Teposcolula–Yanhuitlan, Mixteca Alta, Estado de Oaxaca, México. *Univ. Nat. Auton. Méx., Inst. Geol. Bol.*, 97: 160 pp.; 9 figs.; 12 pls.; 20 tbls.

Finch, R.C., 1972. *Geology of the San Pedro Zacapa Quadrangle, Honduras, Central America*. Thesis, Univ. Texas, Austin, 238 + xxiii pp., 12 figs., 10 pls.

Flores, L.R., 1974. Datos sobre le bioestratigrafía del Jurásico Inferior y Medio del subsuelo de la región de Tampico, Tamps. *Rev. Inst. Mex. Petról.*, 6 (3): 6–15; 2 figs.; 2 pls.; 1 tbl.

Frech, F., 1906. Ueber aviculiden von Palaeozoischem habitus aus der Trias von Zacatecas. *C.R. 10th Sess., Congr., Geol. Intern.*, 1: 327–333; 2 pls.

Fries, Jr., C., 1960. Geología del Estado de Morelos y de partes adyacentes de México y Guerrero, region central-meridional de México. *Univ. Nac. Aut. Méx., Inst. Geol. Bol.*, 60: 236 pp.; 4 figs.; 24 pls.; 1 tbl.; 1 map.

Fuentes Fuentes, R.P., 1964. *Stratigraphy of Sierra Santa Clara and Sierra Gomas, Nuevo Leon, Mexico*. Thesis, Univ. Texas, Austin, 216 pp., 10 figs., 5 pls.

García Calderón, J., 1976. Investigación hidrogeológica de la región de "El Cardito", Zacatecas. *Univ. Nac. Auton. Méx., Inst. Geol., Bol.*, 98: 101 pp.; 12 figs.; 2 maps.

Gonzalez García, R., 1969. Areas con posibilidades de produción en sedimentos del Jurásico superior (Caloviáno–Titoniáno). In: *Seminario sobre exploración petroléra. Mesa Redonda 3. Problemas du exploración del distrito de Poza Rica*. Inst. Méx. Petróleo, 11 pp., 8 figs.

Gries, J.C. and Haenggi, W. T., 1970. Structural evolution of the eastern Chihuahua tectonic belt. In: K. Seewald and D, Sundeen (Editors), *Symposium in Honor of Professor Ronald K. DeFord*. West Tex. Geol. Soc., pp. 119–137, 14 figs.

Guerrero, C.J., 1969. Stratigraphy of Sierra Banco de Lucero, state of Chihuahua. In: D.A. Córdoba et al. (Editors), *Guidebook to the Border Region.* New Mexico Geol. Soc., pp. 171–172.

Haenggi, W.T. and Gries, J.C., 1970. Structural evolution of northeastern Chihuahua tectonic belt. pp. 55–69, il., In: D.H. Campbell, D.F. Reaser and B. Jones (Editors), *Geology of the Southern Quitman Mountains Area, Trans-Pecos Texas. Soc. Econ. Paleontol. Mineral., Permian Basin Sect. Publ.*, 70–12, pp. 55–69.

Hallam, A., 1977. Anoxic events in the Cretaceous ocean. *Nature*, 268: 15–16.

Hayes, P.T., 1970a. Mesozoic stratigraphy of the Mule and Huachuca Mountains, Arizona. *U.S. Geol. Surv. Prof. Pap.*, 658-A, 28 pp., 15 figs.

Hayes, P.T., 1970b. Cretaceous paleogeography of southeastern Arizona and adjacent areas. *U.S. Geol. Surv. Prof. Pap.*, 658-B, 42 pp., 6 figs.

Hays, J.D. and Pitman, W.L., 1973. Lithospheric plate motion, sea-level changes and climatic and ecological consequences. *Nature*, 246: 18–22; 4 figs.

Heim, A., 1940. The front ranges of the Sierra Madre Oriental, Mexico, from Cuidad Victoria to Tamazunchale. *Eclogal Geol. Helv.*, 33: 313–362; 10 figs.; sections, map.

Humphrey, W.E., 1956. Tectonic framework of northeast Mexico. *Gulf Coast Assoc. Geol. Soc. Tr.*, 6: 25–35; 1 fig.; 3 correlation charts.

Humphrey, W.E. and Diaz, G., T., [ms.] Jurassic and Lower Cretaceous stratigrapgy and tectonics of northeast Mexico. [unpublished].

Imlay, R.W., 1937. Lower Neocomian fossils from the Miquihuana region, Mexico. *J. Paleontol.*, 11: 552–574; 8 figs.; pls. 70–83.

Imlay, R.W., 1938. Studies of the Mexican Geosyncline. *Geol. Soc. Am. Bull.*, 49: 1651–1694; 6 figs.

Imlay, R.W., 1943. Evidence of Upper Jurassic landmass in eastern Mexico. *Am. Assoc. Petrol. Geol. Bull.*, 27: 524–529; 1 fig.

Imlay, R.W., 1953. Las formaciónes Jurásicas de México. *Soc. Geol. Mex. Bol.*, 16 (1): 1–65; 5 figs.; 1 tbl.

Jaworski, E., 1929. Eine Lias-fauna aus Nordwest-Mexico. *Abh. Schweiz. Paläontol. Ges.*, 84: 1–12; 1 fig.; 1 pl.

Jones, T.S., 1938. Geology of the Sierra de la Peña and paleontology of the Indidura Formation, Coahuila, Mexico. *Geol. Soc. Am. Bull.*, 49: 61–150; 4 figs.; 13 pls.

Kauffman, E.G., 1977. Geological and biological overview; western Interior Cretaceous Basin. *Mountain Geol.*, 14 (3, 4): 75–99; 12 figs.; 1 tbl.

King, P.B., 1947. (maps, 1940) *Cartas geológicas y mineras de la Republica Mexicana, México.* Univ. Nac. Auton. Méx.

King, R.E., Dunbar, C.O., Cloud, P.E. Jr. and Miller, A.K., 1944. Geology and paleontology of the Permian area northwest of Las Delicias, southwestern Coahuila, Mexico. *Geol. Soc. Am. Spec. Pap.*, 52: 172 pp.; 29 figs.; 45 pls.

Klesse, E., 1970. Geology of the El Ocotito-Ixcuinatoyac region and of La Dicha stratiform sulphide deposit, state of Guerrero. *Soc. Geol. Mex. Bol.*, 31: 107–140; 2 figs.; 10 pls.; 4 tbls.

Martínez, R.E., 1972a. Presencia del Turoniáno, Coniaciáno, Santoniáno y ausencia del Campaniáno en el Mesozoica de Chiapas. *Rev. Inst. Mex. Petról.*, 4 (4): 5–15; 1 fig.; 4 pls.

Martínez, R.E., 1972b. Zonificación microfaunistica del Mesozoica de la parte oriental de la Sierra Madre de Chiapas. *Conferencia presentada en la II convención Nacional de la Sociedad Geológica Mexicana, Mazatlan, 5th Mayo*, pp. 1–6.

Matsumoto, T., 1977. On the so-called Cretaceous transgressions. *Palaeontol. Soc. Japan, Spec. Pap.*, 21: 75–84; 3 figs.

McBride, E.F., Weidie, A.E., Wolleben, J.A. and Laudon, R.C., 1974. Stratigraphy and structure of the Parras and La Popa Basins, northeastern Mexico. *Geol. Soc. Am. Bull.*, 85: 1603–1622; 18 figs.; 1 tbl.

McDowell, F.W. and Clabaugh, S.E., 1978. Ignimbrites of the Sierra Madre Occidental and their relation to the tectonic history of western Mexico. 41 ms pp., 9 figs.

McNulty, C.L. Jr., 1976. *Globotruncana fornicata* zone of Upper Austin Group (Cretaceous), northeast Texas. *Am. Assoc. Petrol. Geol. Bull.*, 60: 2058–2062.

Mills, R.A., Hugh, K.E., Feray, D.E. and Swolfs, H.C., 1967. Mesozoic stratigraphy of Honduras. *Am. Assoc. Petrol. Geol. Bull.*, 51: 1711–1786.

Mixon, R.B., Murray, G.E. and Diaz G., T., 1959. Age and correlation of Huizachal Group (Mesozoic), state of Tamaulipas, Mexico. *Am. Assoc. Petrol. Geol. Bull.*, 43: 757–771; 11 figs.

Moor, A., 1978. *Stratigrapgy and Structure of Potosí Anticline, Nuevo Leon, Mexico.* Thesis, Univ. Texas, Austin.

Myers, R.L., 1968. Biostratigraphy of the Cárdenas Formation (Upper Cretaceous), San Luis Potosí, México. *Univ. Nac. Auton. Méx., Paleontol. Mex.*, 24: 89 pp.; 3 figs.; 1 tbl.; 16 pls.

Newberry, J.S., 1888. Rhaetic plants from Honduras. *Am. J. Sci.*, Ser. 3, 36: 342–351.

Ochoterena F., H., 1963. Amonitas del Jurásico Medio y del Calloviáno de México, I. *Parastrenoceras* gen. nov. *Univ. Nac. Auton. Méx., Inst. Geol., Paleontol. Mex.*, 16: 26 pp.; 10 figs.; 5 pls.

Ochoterena F., H., 1966. Amonitas del Jurásico Medio de México, II. *Infrapatoceras* gen. nov. *Univ. Nac. Aut. Méx. Inst. Geol., Paleontol. Mex.*, 23: 18 pp.; 5 figs.; 3 pls.

Ornelas O., F., 1969. Posibilidades petroliferas de los sedimentos subyacentes al Calloviáno en sedimentos de facies lagunales (Huehuetepec) In: *Seminario sobre exploración petroléra, Mesa Redonda 3: Problemas de exploración del distrito de Poza Rica. Inst. Mex. Petróleo*, 7 pp.

Palmer, R.H., 1928a. Geology of southern Oaxaca, Mexico. *J. Geol.*, 36: 718–734.

Palmer, R.H., 1928b. The rudistids of southern Mexico. *San Francisco, Calif. Acad. Sci., Occ. Pap.*, 14: 137 pp.; 8 figs.; 17 pls.

Plummer, H.J., 1931. Some Cretaceous Foraminifera in Texas. *Univ. Texas Bull.*, 3101: 109–203.

Powell, J.D., 1963. Cenomanian–Turonian (Cretaceous) ammonites from Trans-Pecos Texas and northeastern Chihuahua, Mexico. *J. Paleontol.*, 37: 309–322; 3 figs.; pls. 31–34.

Powell, J.D., 1965. Late Cretaceous platform-basin facies,

northern Mexico and adjacent Texas. *Am. Assoc. Petrol. Geol. Bull.*, 49: 511–525.

Powell, J.D., 1967. Mammitine ammonites in Trans-Pecos Texas. *Texas J. Sci.*, 19: 311–322.

Powel, J.D., 1970. Early Upper Cretaceous faunal zones southwest of the Diablo-Coahuila Platform. pp. 96–99, 1 fig.; In: D.H. Campbell, D.F. Reaser and B.R. Jones (Editor), *Geology of the Southern Quitman Mountains Area, Trans-Pecos Texas. Soc. Econ. Paleontol. Mineral., Permian Basin Sect. Publ.*, 70–12: 127 pp.

Rangin, C., 1977. Sobre la presencia de Jurásico Superior con amonitas en Sonora septentrional. *Univ. Nacl. Autón. de México, Inst. Geol. Revista*, 1 (1): 1–4.

Ritchie, A.W., 1975. *Geology of the San Juan Zacatepequez Quadrangle, Guatemala, Central America.* Thesis, Univ. Texas, Austin, xv + 119.; 13 figs.; 3 pls.

Robeck, R.C., Pesquera V., R. and Salvadore U., A., 1956. Geologiá y depositos de carbón de la región de Sabinas, Estado de Coahuila. *XX Congr. Geol. Intern.*, 109 pp.; 2 figs.; 13 pls.; 9 tbls.

Ryan, W.B.F. and Cita, M.B., 1977. Ignorance concerning episodes of ocean-wide stagnation. *Marine Geol.*, 23: 197–215.

Salas, G.A., 1970. Areal geology and petrology of the igneous rocks of the Santa Ana region, northwest Sonora. *Soc. Geol. Mex. Bol.*, 31 (1): 11–64; 3 figs.; 22 pls.; 11 tbls.; 3 appendices [for 1968].

Salas, G.P. et al., 1968. *Carta Geológica de la Republica de México.* Com. Carta Geol. de México, México D.F., 2 sheets.

Salinas E., S., 1969. Golfo de Sabinas Jurássico Superior y su correlación. In: *Seminario sobre exploración petroléra, Mesa Redonda 6. Problemas de exploración en areas posiblemente petroliferas de la Republica Mexicana: Inst. Mex. del Petróleo*, 13 pp.

Sánchez Montes de Oca, R., 1969. Estratigrafiá y paleogeografiá del Mesozoica de Chiapas. In: *Seminario sobre exploración petrolera, Mesa Redonda 5. Problemas de exploración de la zona sur: Inst. Mex. del Petróleo*, 31 pp.; 11 figs.

Sansores Manzanilla, E., 1956. Breves notas sobre la geologia de la Peninsula de Yucatan. *XX Congr. Geol. Intern., Excursion C-7*, pp. 123–129.

Sansores Manzanilla, E. and G. Navarrete, R., 1969. Bosquejo geológico de la zona norte. In: *Seminario sobre exploración petroléra, Mesa Redonda 2. Problemas de exploración de la zona norte. Inst. Mex. del Petróleo*, 37 pp.

Schlanger, S.O. and Jenkyns, H.C., 1976. Cretaceous anoxic events: causes and consequences. *Geol. Mijnb.*, 55: 179–184.

Silva Pineda, A., 1963. Plantas del Triásico superior del Estado de Hidalgo. *Univ. Nac. Auton. Méx., Inst. Geol., Paleontol. Mex.*, 18: 12 pp.; 1 fig.; 7 pls.

Silva Pineda, A., 1969. Plantas fosiles del Jurásico Medio de Tecomatlan, Estado de Pueblo. *Univ. Nac. Auton. Méx., Inst. Geol., Palenontol. Mex.*, 27: 1–76; 19 pls.

Silva Pineda, A., 1970. Plantas fosiles del Jurásico Medio de la region Tezoatlan, Oax. *Libreto Guia excursión México-Oaxaca. Soc. Geol. Mex.*, 129–143; 11 figs.; 1 pl.

Smith, C.I., 1970. Lower Cretaceous stratigraphy, northern Coahuila, Mexico. *Univ. of Texas, Bur. Econ. Geol., Rept. Invest.*, 65: 101 pp.; 20 figs.; 15 pls.; 4 tbls.

Smith, C.I., Charleston A., S., and Brown, J.B., 1974. Lower Cretaceous: shelf, platform reef and basinal deposits, southwest Texas and northern Coahuila. *West Tex. Geol. Soc. and Permian Basin Sect., Soc. Econ. Paleontol. Mineral., Joint field trip, Guidebook*, 13 pp.

Sohl, N.F. and Kauffman, E.G., 1964. Giant Upper Cretaceous oysters from the Gulf Coast and Caribbean. *U.S. Geol. Surv. Prof. Pap.*, 483-H, 22 p.; 3 figs.; 5 pls.; 2 tbls.

Stabler, C.L. and Marquez D., B., 1978. Initiation of Lower Cretaceous reefs in Sabinas Basin, northeast Mexico. *Univ. Texas, Austin, Bur. Econ. Geol. Rept. Invest.*, 89: pp. 299–301.

Thompson, S., III, 1975. Oil and gas exploration wells in Dona Aña County, New Mexico. In: W.R. Seager, R.E. Clemons and J.F. Callender (Editors), *Guidebook of the Las Cruces Country.* New Mex. Geol. Soc., pp. 171–174.

Toledo Toledo, M., 1969. Problemas de exploración de las plataforma continental de la zona norte. In: *Seminario sobre exploración petroléra, Mesa Redonda 2. Problemas de exploración de la zona norte. Inst. Mex. Petróleo*, 35 pp.

Varela Hernandez, A., 1969. Problemas de la exploración petroléra en el homoclinal de San José de las Rusias, Tam. In: *Seminario sobre exploración petroléra, Mesa Redonda 2. Problemas de exploración de la zona norte. Inst. Mex. Petróleo*, 19 pp.

Vásquez, E.M., 1949. Los Criaderos minerales de "El Bote", Estado de Zacatecas. *Inst. Nac. para la Investigación de Recursos Minerales Bol.*, 24: 39 pp.; 2 figs.; 5 pls.; 5 tbls.

Velarde N., P., 1969. Posibilidades petroliferas de las formaciónes Mesozoicas en la Cuenca de Veracruz. In: *Seminario sobre exploración petroléra, Mesa Redonda 4. Problemas de exploración de las Cuenca de Papaloapan. Inst. Mex. Petróleo*, 12 pp.

Waisley, S., 1978. *Depositional History of Cuesta del Cura Formation, Northern Mexico.* Thesis, Univ. Texas, Austin.

Webb, D.S., 1969a. A proposito de la edad de las rocas de las colinas de Samalayuca noreste de Chihuahua. *Soc. Geol. Mex. Bol.*, 30: 155–158.

Webb, D.S., 1969b. Facets of the geology of the Sierra del Presidio area, north-central Chihuahua. In: D.A. Córdoba et al. (Editors), *Guidebook to the Border Region.* New Mex. Geol. Soc., pp. 182–185.

Weidie, A.E., Wolleben, J.A. and McBride, E.F., 1972. Late Cretaceous depositional systems in northeastern Mexico. *Gulf Coast Assoc. Geol. Soc., Trans.*, 22: 323–329.

Wilson, I.F. and Rocha, V.S., 1946. Los yacimientos de carbón de la región de Santa Clara, Municipio de San Javier, Estado de Sonora. *Comité Directivo par la Invest. de los Recursos Minerales de México, Bol.*, 9: 108 pp.; 8 figs.; 9 tbls.; 8 pls.

Wilson, J.L. and Pialli, G., 1978. A Lower Cretaceous shelf margin in northern Mexico. *Univ. Texas, Austin, Bur. Econ. Geol. Rept. Invest.*, 89: 286–294; 5 figs.

Wilson, J.A., 1970. Vertebrate biostratigraphy of Trans-Pecos Texas. In: K. Seewald and D. Sundeen (Editors), *Sym-*

posium in Honor of Professor Ronald K. DeFord. West Texas Geol. Soc., pp. 159–166, 8 figs.

Wolleben, J.A., 1965. Nomenclatúra litoestratigrafiá de las unidades del Cretácico superior en el Oeste de Texas y el noreste de Chihuahua. *Geol. Soc. Mex. Bol.*, 27: 65–74.

Wolleben, J.A., 1966. *Biostratigraphy of the Ojinaga and San Carlos Formations of West Texas and Northeastern Chihuahua.* Thesis, Univ. Texas, Austin, viii + 62 pp.; 8 figs.; 3 pls.; 2 tbls.

Wolleben, J.A., 1967. Senonian (Cretaceous) Mollusca from Trans-Pecos Texas and northeastern Chihuahua, Mexico. *J. Paleontol.*, 41: 1150–1165; 8 figs.; pls., 147–152; 2 tbls.

Young, K., 1965. A revision of Taylor nomenclature, Upper Cretaceous, central Texas. *Univ. Tex., Bur. Econ. Geol. Geo. Circ.*, 65–3: 11 p.; 3 figs.; 2 tbls.

Young, K., 1972. Cretaceous paleogeography: implications of endemic ammonite faunas. *Univ. Texas, Austin, Bur. Econ. Geol. Geo-Cric.*, 72–2: 12 pp.; 4 figs.; 3 tbls.

Young, K., 1978. Middle Cretaceous rocks of Mexico and Texas. *Univ. Texas, Austin, Bur. Econ. Geol. Rept. Invest.*, 89: pp. 325–332; 11 figs.

Chapter 4

THE CARIBBEAN REGION

J. BUTTERLIN

INTRODUCTION

The Caribbean region consists basically of the Caribbean Sea lying to the north of the South American continent and bounded, on the west, by the isthmus of Central America continued by Nuclear Central America (the extension south of the North American craton). Its northern and eastern limits are formed by a complex of islands, the Greater and Lesser Antilles. Thus the Gulf of Mexico and the Bahama Banks lie outside the region discussed here. The geology of the islands generally grouped as the West Indian Islands, the Caribbean Sea and Central America will be briefly reviewed here, in an attempt to generate paleogeographic maps which will serve to show the distribution of facies at few periods during the Mesozoic, the last two readily defined by important orogenic phases. The periods are: Triassic, Jurassic, Lower Cretaceous and Upper Cretaceous. The maps are based upon present geographic location of the lands and seas for at this stage there is no better alternative.

The basin of the Caribbean Sea s.l. can be divided into four sub-basins: in the south, an eastern Venezuelan and a western Colombian Basin, partially separated by the Beata Ridge (Fig. 1), in the north, the Yucatan Basin, bounded by the Yucatan Peninsula in the west, by the Cayman Ridge to the south and Cuba to the northeast. Between the Yucatan Basin and the Colombian Basin lies the Cayman Trough separated from the first by the Cayman Ridge and from the second by the Nicaraguan Rise and its northeast continuation towards Jamaica.

It will be noted finally that the Venezuelan Basin is subdivided by a north—south positive structure, the Aves Swell with the Venezuelan Basin s.str. to the west and the Grenada Trough to the east.

The West Indian Islands form various groups within and around the Caribbean Sea. In the northwest there are the Greater Antilles (Cuba, Haiti, Puerto Rico, Jamaica and Virgin Islands) containing most of the larger islands and the geologically similar Virgin Islands. In the northeast and the east lie the Lesser Antilles (or better the oriental Lesser Antilles) which form two parallel arcs convex to the east, the more westerly *volcanic* Lesser Antilles, and, the *calcareous* Lesser Antilles to the east.

In the south are the Dutch Windward Islands, with the Venezuelan Windward Islands lying to the east. The former are similar geologically to the floor of the Caribbean Sea and the latter together with Trinidad and Tobago are similar to, or a continuation of, the northeast margin of South America (Caribbean Cordillera of Venezuela). Finally, east of the oriental Lesser Antilles lies Barbados isolated on the Barbados Ridge. The origin of this island has been much discussed. The numerous islands of the Bahamas lie east of Florida. They form the foreland of the occidental Greater Antilles to the north.

Central America is a geographical unit but not a structural one. It is composed, in fact, of two units separated by the Nicaragua Trench, a northwest—southeast graben, in which is found the Managua Lake in the northwest and the Nicaragua Lake in the south (Dengo, 1968). The northern unit, or Nuclear Central America, represents the prolongation and the continental end of North America. The southern unit, or Isthmian Central America, appears as the prolongation of and the end of, the northwest part of the South American Andes. This unit was probably only recently (Plio—Quaternary) connected to Nuclear Central America, on the evidence of mammalian migrations (Lloyd, 1963).

The oldest dated rocks found in the Antilles are the Jurassic rocks in Cuba. In the other islands of the

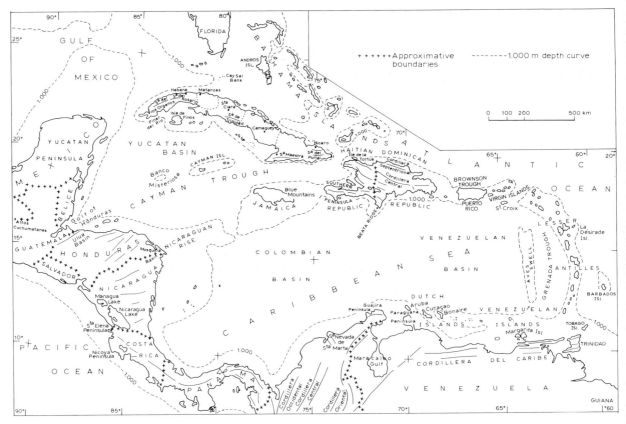

Fig. 1. Geographic features of the Caribbean region.

Greater Antilles and in the southern part of the Antillean Arc (Windward Islands: Dutch and Venezuelan islands), with the exception of Trinidad, no pre-Cretaceous formations have been identified with certainty. In the Bahamas and Trinidad, Upper Jurassic rocks are known and, in Désirade Island, an Upper Jurassic age has been assigned to some magmatic rocks (Fink, 1968). In Central America, Jurassic and probably Upper Triassic marine deposits exist in Honduras. Continental molasse deposits are of about the same age or are more recent.

There is thus very little information available on the history of the Antillean region in pre-Cretaceous times. Indeed there are authors for whom the region did not even exist at the beginning of the Mesozoic, the region being variously born in Middle Triassic (Dietz and Holden, 1970), Upper Triassic (Funnel and Smith, 1968) or during the Jurassic (Le Pichon and Fox, 1971) from the divergent rifting westward of North America and South America. It is thus critical geologically to know whether Paleozoic structural units from the neighbouring continents continue into the Antillean region, i.e. whether there was a pre-Mesozoic basement in the Caribbean.

With the beginning of the Cretaceous, the pattern is very much clearer and, in fact, all the major hypotheses explaining the origin of the Caribbean Sea (permanent ocean, foundered continent or rift zone – Dengo, 1969) are in agreement that, since the end of the Jurassic, the constitution and extent of the Caribbean region was more or less similar to that presently observed.

The Cretaceous, absent only in the Lesser Antilles (save in Désirade Island)[1] is characterized by very important generally mafic volcanism, with serpentinized peridotites and deep marine basalts in the Lower Cretaceous and andesitic to dacitic volcanism emplaced in shallow marine or subcontinental environments and associated with granodioritic in-

_____
[1] Recent studies (P. Bouysse and P. Martin, 1979, Bull. BRGM, No. 3/4) seem to indicate that the northern part of the Lesser Antilles exists since Upper Cretaceous and perhaps uppermost Jurassic.

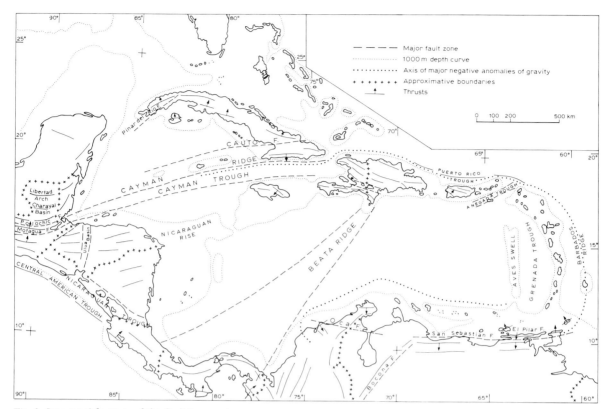

Fig. 2. Structural features of the Caribbean region.

trusions in the Upper Cretaceous. Sedimentary rocks, both calcareous and clastic are abundant particularly in the outer margins (external zones).

These rocks were subjected to orogenic deformation, the main phase being in Middle Cretaceous (Subhercynian phase) with important metamorphism, and in the uppermost Cretaceous and Lower Tertiary (Laramian phase) which was associated with important tangential movements.

STRUCTURAL FRAMEWORK

We shall consider the structural framework in the Caribbean region successively during the pre-Cretaceous and the Cretaceous period.

Pre-Cretaceous

As indicated in the section on stratigraphy, it is difficult to determine the geologic situation of the Caribbean region before the Upper Jurassic or Lower Cretaceous. The majority of the geologists consider that this region was created in the Middle or Upper Triassic or perhaps in the Lower Jurassic by the divergent westerly drift between North and South America. Structurally it is probable that at the *beginning of the Mesozoic* the Hercynian structures of northern Central America continued into Cuba which probably then occupied a more southerly position, for the Cayman Trough and Ridge and the Yucatan Basin did not exist at that time.

In the Jurassic, the Caribbean region expanded and, according to Le Pichon and Fox (1971), it attained about its present extension by the end of the period. Structurally, this period was characterized by a north—south expansion of the Caribbean Sea depression with formation of an oceanic crust, whether by the volcanic activity of a west—east mid-Caribbean ridge or by the penetration of a Pacific plate in the Caribbean (see section p. 109 ff.). It was a period also characterized by the formation of a shallow marginal basin with clastic deposits in the south of Cuba and evaporitic deposits in the north.

During the Upper Jurassic there were orogenic movements (pre-Upper Oxfordian, pre-Middle Kimmeridgian) in Cuba according to some authors. However, as indicated further in the stratigraphy-section (p. 98), the arguments in favour of these orogenic phases are not decisive. There were, however, important tectonic events in the uppermost Jurassic, perhaps of epirogenic character, inasmuch as the Lower Cretaceous paleogeography was completely different from that of the Upper Jurassic.

Cretaceous

In the Lower Cretaceous (or uppermost Jurassic) there began a typical geosynclinal cycle in northern Central America, in the Greater Antilles and along the northern margin of South America (Dutch and Venezuelan Windward Islands, Trinidad and Tobago). The extension of the different parts of this geosynclinal basin will be indicated in the section on paleogeography (p. 104). An important orogenic phase occurred in the Middle Cretaceous (Albian in general – apparently Turonian in Cuba). It is called the Austrian or Subhercynian phase and was characterized by regional metamorphism of blue schist and greenschist facies in the inner zones (eugeosynclinal), excluding the Virgin Islands where the only unmetamorphosed Lower Cretaceous deposits are found.

It is not very easy to determine the intensity of the folding because the intensity of folding in the Laramian phase s.l. (Maastrichtian to Middle Eocene) obliterated the macrofoldings of the Subhercynian phase. No microtectonic study which would permit the reconstruction of the successive phases of folding has so far been carried out. The Laramian phase began in the Maastrichtian and, with periods of rest, continued until the Middle (or Upper) Eocene. Regional metamorphism was weak or absent, but granodioritic intrusions were common. This phase was also characterized by very important tangential movements with thrusting and formation of thrust nappes, particularly in Guatemala, Cuba, Hispaniola, Jamaica, possibly in Puerto Rico, the Venezuelan Windward Islands, Tobago and Trinidad. The ultrabasic rocks (serpentinized peridotites) are often associated in the nappes structures, particularly in northern Central America, in Cuba, Hispaniola, Puerto Rico and the Venezuelan Islands.

The dominant direction of thrusting is outwards from the Antillean Arc and to the north in Central America but there is also thrusting inwards in the Antillean Arc, particularly in the southeastern Cuba (Sierra Maestra), in the central part of the Haitian Republic, in the Venezuelan Islands, Tobago and Trinidad.

Important Campanian to Upper Eocene flysch facies deposits, frequently with wildflysch, confirm the intensity of the Laramian orogenic phases.

The fold axes are generally oriented northwest—southeast to west—east in the north and southwest—northeast to west—east in the southern parts of the Caribbean.

STRATIGRAPHY (Figs. 3, 4, 5, 6, 7)

Triassic

Arguments for the existence of Triassic rocks can only be presented for Cuba and for Nuclear Central America.

In Cuba, the argument for the presence of Triassic rocks is based primarily on the dating of polyphase metamorphism and upon the age of the flora in the lower part of the San Cayetano Formation (cf. below, Jurassic – Vachrameev, 1966).

According to Somin and Millan (1973a) three types of metamorphic complex can be recognized in the island. From the Isla de Pinos to eastern Oriente province type I is a greenschist facies and consists of metamorphosed or only slightly metamorphosed clastic and calcareous rocks similar to the San Cayetano Formation which is dated as Upper Triassic to lower-Upper Jurassic. The supposition that the type I rocks are of the same age gains some support from recrystallized fossils found on the Isla de Pinos which may be of Middle Triassic–Upper Jurassic age (Somin and Millan, 1973b). The type II, ophiolitic complex, shows characteristically high pressure/low temperature metamorphism. It consists of massive basic and tuffaceous metavolcanites and amphibolites associated with glaucophane schists and serpentinized peridotites. This complex was subjected to a later metamorphism of greenschist and even of amphibolite facies (Boiteau et al., 1972a; Boiteau and Campos, 1973; Boiteau and Michard, 1976). It is covered by Upper Cretaceous sedimentary-volcanic formations and, according to Somin and Millan (1973a), is also of Cretaceous age. However, in view of the double metamorphism, it may be older than the Subhercynian (Middle Cretaceous) orogenic phase. Nevertheless, it is probably linked to this orogenic cycle (or even to an older cycle).

The type III complex includes amphibolites re-

Fig. 3. Geologic map of Cuba (after Butterlin, 1977). *1* = Quaternary; *2* = Neogene; *3* = Paleogene; *4* = Upper Cretaceous; *5* = Lower Cretaceous; *6* = Upper Jurassic; *7* = Middle and Lower Jurassic; *8* = base complex (Paleozoic(?) −Mesozoic); *9* = acid intrusive rocks (diorites, quartz, granites); *10* = gabbros, peridotites and especially serpentines.

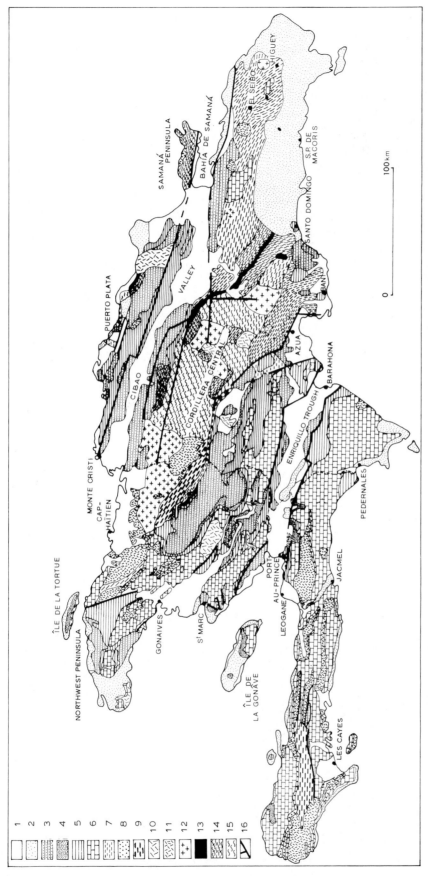

Fig. 4. Geologic map of Hispaniola (after Butterlin, 1977). *1* = Holocene, alluvial deposits; *2* = Pleistocene and Miocene, conglomerates, sandstone, sands, gypsum; *4* = Oligocene and Miocene, marl, sandstone, conglomerates; *5* = Oligocene, limestone, conglomerates; *3* = Pliocene and Pliocene, coral limestones; *6* = Eocene, limestone; *7* = Eocene, detritic rocks; *8* = Paleocene–Lower Eocene, base conglomerate, sandstone; *9* = Cretaceous, tuffs, schists (flysch), limestone; *10* = Cenozoic, volcanic rocks; *11* = Cretaceous, volcanic rocks; *12* = Cretaceous-Eocene, granodiorite intrusive rocks; *13* = Cretaceous(?), serpentinised peridotites; *14* = Paleozoic(?)–Mesozoic(?), metamorphic rocks; *15* = pre-Tertiary, undifferentiated rocks; *16* = folds.

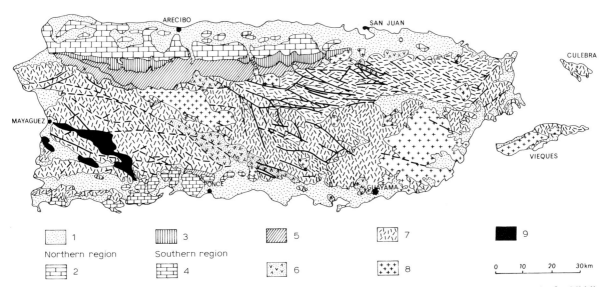

Fig. 5. Geologic map of Puerto Rico (after Butterlin, 1977). *1* = Pliocene–Quaternary, alluvial deposits, dune sands; *2* = Middle and Lower Miocene, marly limestones (Camuy Formation), Aymamon limestone, Aguada limestone; *3* = Middle Oligocene–Lower Miocene, detritic limestone rocks (Cibao Formation, Lares limestone, San Sebastian Formation); *4* = Upper Oligocene–Middle(?) Lower(?) Miocene, Ponce limestone, Angola limestone; *5* = Middle Oligocene, detritic rocks (Juana Diaz Formation); *6* = Paleocene–Eocene, volcanic and sedimentary rocks; *7* = Cretaceous and pre-Cretaceous(?), volcanic and sedimentary rocks, metamorphic rocks; *8* = Late Cretaceous–Eocene, granodiorites and quartz diorites; *9* = Cretaceous, serpentines.

sulting from the metamorphism of basic igneous rocks. According to Boiteau and Campos (1973), in the Sierra del Purial, these rocks suffered a phase of metamorphism prior to that of the ophiolitic complex. The age may be thus Upper Jurassic or older. In the Escambray Mountains, a radiometric measurement on an intrusive granodioritic batholith into the amphibolites gives a Lower Jurassic age (180±10 m.y.-Khudoley, in Hatten and Khudoley, 1967). In this region, the amphibolites are included, in the Trinidad Formation, with gneisses, micaschists, chlorite schists and serpentines.

On the Isla de Pinos, there are outcrops of micaschists with sillimanite, garnet, disthene and staurolite associated with quartzites, serpentines and gabbros and intrusive granodiorites. This complex is considered of Paleozoic age by Meyerhoff (in Khudoley and Meyerhoff 1971) and Skvor (1969), as Precambrian by Tikhomirov and Izquierdo (1968), and Mesozoic by Khudoley.

The author considers the type I metamorphics to be the same age as the San Cayetano Formation. The type II is regarded as linked to the Subhercynian cycle along with the metamorphosed basic igneous rocks, including amphibolites and serpentines classified as type III, for they are lithologically similar to, and associated with rocks of type II.

The highly metamorphosed rocks of type III (gneiss, granites, micaschists with sillimanite and staurolite) may be of Paleozoic age, like similar rocks in Nuclear Central America which are included in the Chuacus Group (McBirney, 1963) and Las Ovejas Group (Schwartz, 1971). In the latter region, the Tambor Formation (McBirney and Bass, 1969), an ophiolite complex associated with glaucophane schists, and the greenschist facies rocks of the Chuacus Group correspond to the types II and I and are probably of Mesozoic age.

On this basis it seems unlikely that metamorphosed Paleozoic rocks extend to other islands of the Greater Antilles. The metamorphic complexes encountered in Hispaniola, Puerto Rico and Jamaica are of type II (in the author's opinion) or I, and are probably of Mesozoic age, linked to the sub-Hercynian cycle.

In Central America, Triassic marine sedimentary rocks are thin or absent. They may be represented by the lower part (Upper Triassic?) of the El Plan Formation of Honduras (Mills et al., 1967) in the Mosquitia Basin, though most of the formation appears to be Jurassic in age. The El Plan Formation marine littoral and deltaic deposits interdigitate with 1.200 m of continental beds, a molasse facies of conglomerates, graywackes, shales, red sandstones, occasio-

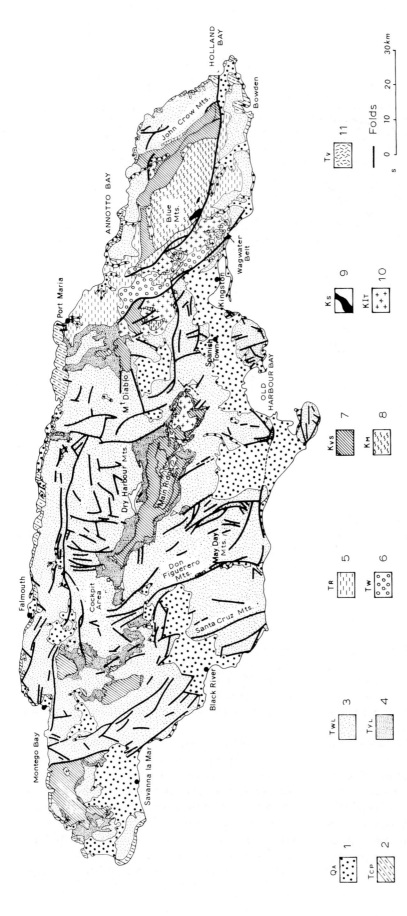

Fig. 6. Geologic map of Jamaica (after Butterlin, 1977). *1* = Quaternary, alluvial deposits; *2* = Middle(?) Miocene–Pleistocene(?), marine detritic rocks (Coastal group); *3* = Middle Eocene–Middle Miocene, white limestone; *4* = Middle Eocene, yellow limestone; *5* = Lower and Middle Eocene, flysch and gypsum (Richmond layers); *6* = Lower Eocene, conglomerates (Wagwater conglomerate); *7* = Cretaceous, volcanic and sedimentary rocks; *8* = Cretaceous, metamorphic rocks; *9* = Cretaceous, serpentines; *10* = Late Cretaceous–Paleocene, Middle Eocene, granodiorites and quartz microdiorites; *11* = Cretaceous and Tertiary, volcanic rocks.

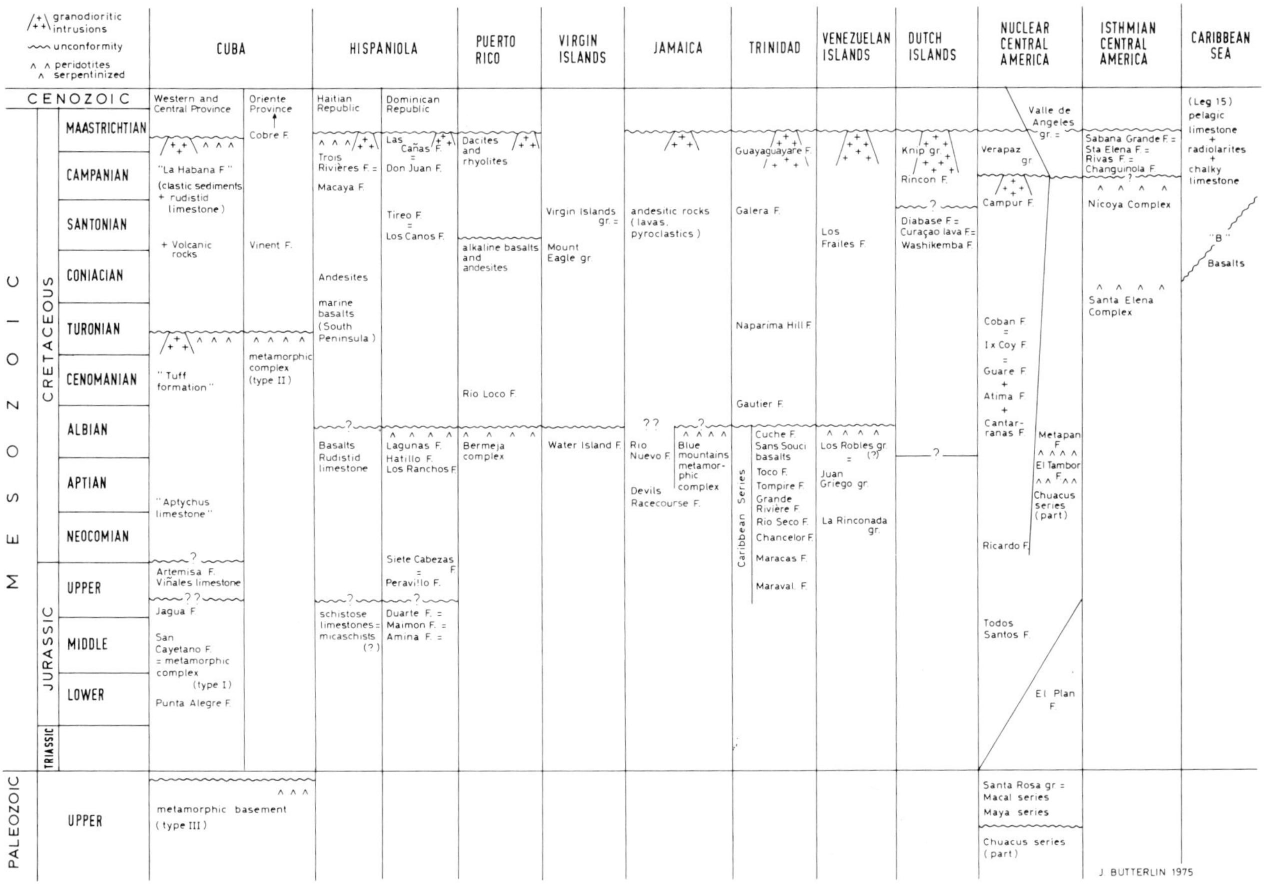

Legend:
- /+\ granodioritic intrusions
- ~~~ unconformity
- ^ ^ peridotites
- ^ serpentinized

			CUBA		HISPANIOLA		PUERTO RICO	VIRGIN ISLANDS	JAMAICA	TRINIDAD	VENEZUELAN ISLANDS	DUTCH ISLANDS	NUCLEAR CENTRAL AMERICA	ISTHMIAN CENTRAL AMERICA	CARIBBEAN SEA
			Western and Central Province	Oriente Province	Haitian Republic	Dominican Republic									(Leg 15) pelagic limestone + radiolarites + chalky limestone
CENOZOIC				Cobre F.										Valle de Angeles gr. / Verapaz gr.	Sabana Grande F. = Sta Elena F. = Rivas F. = Changuinola F.	
MESOZOIC	CRETACEOUS	MAASTRICHTIAN	"La Habana F." (clastic sediments + rudistid limestone)		Las Cañas F. / Trois Rivières F. / Macaya F.	Dacites and rhyolites / Don Juan F.				Guayaguayare F.		Knip gr. / Rincon F.	Campur F.	Nicoya Complex	
		CAMPANIAN						Virgin Islands gr.	andesitic rocks (lavas, pyroclastics)	Galera F.	Los Frailes F.	Diabase F. = Curaçao lava F. = Washikemba F.			
		SANTONIAN	+ Volcanic rocks	Vinent F.	Tireo F. = Los Canos F.										
		CONIACIAN			Andesites	alkaline basalts and andesites / Mount Eagle gr.								Santa Elena Complex	"B" Basalts
		TURONIAN	"Tuff formation"	metamorphic complex (type II)	marine basalts (South Peninsula)				Naparima Hill F.			Coban F. = Ix Coy F. = Guare F. + Atima F. + Cantarranas F.		
		CENOMANIAN					Rio Loco F.			Gautier F.					
		ALBIAN			Basalts Rudistid limestone	Lagunas F. / Hatillo F. / Los Ranchos F.	Bermeja complex	Water Island F.	Rio Nuevo F. / Devils Racecourse F.	Blue mountains metamorphic complex	Cuche F. / Sans Souci basalts / Toco F. / Tompire F. / Grande Rivière F. / Rio Seco F. / Chancelor F. / Maracas F. / Maraval F.	Los Robles gr. (?) / Juan Griego gr. / La Rinconada gr.		Metapan F. / El Tambor F. / Chuacus series (part)	
		APTIAN	"Aptychus limestone"											Ricardo F.	
		NEOCOMIAN				Siete Cabezas = Peravillo F.									
	JURASSIC	UPPER	Artemisa F. / Viñales limestone / Jagua F.										Todos Santos F.	
		MIDDLE	San Cayetano F. = metamorphic complex (type I)		schistose limestones = micaschists (?)	Duarte F. = Maimon F. = Amina F. =									
		LOWER	Punta Alegre F.											El Plan F.	
	TRIASSIC														
PALEOZOIC		UPPER	metamorphic basement (type III)										Santa Rosa gr = Macal series / Maya series / Chuacus series (part)		

Fig. 7. Major Mesozoic formations of the Caribbean region.

J. BUTTERLIN 1975

nally cross-bedded, referred to the Todos Santos Formation. The latter is essentially Jurassic in age but the lower part may be as old as Triassic.

Jurassic

Rocks of Jurassic age, while more widely distributed than Triassic, have only been identified with certainty in a few places. Sedimentary and metamorphic rocks are found in Cuba and Trinidad and igneous rocks dated as Jurassic are known from Désirade Island (calcareous Lesser Antilles) (Table I).

The oldest Jurassic deposits in Cuba are those of the San Cayetano Formation which, as already indicated, may extend down into the Upper Triassic. It has outcrops in the Pinar del Rio Province and consists of dominantly continental, deltaic sediments which may locally be flysch-like and has some intercalations of shallow marine limestones. While probably several thousand meters thick, because of tectonic complications, both intense folding and faulting, its true thickness is in some doubt.

The flora associated with these sediments apparently ranges from Upper Triassic to Upper Jurassic (Vachrameev, 1966) and, in the upper marine beds, some forms indicate a possible Bajocian–Callovian age. It is limited to pre-Upper Oxfordian by the overlying Jagua Formation in some regions and to pre-Tithonian by the overlying Artemisa Formation in others. As mentioned earlier the type I metamorphic complex is also assigned the same age as the San Cayetano Formation.

It is regarded also as the time equivalent of the evaporite sequence of the Punta Alegre Formation (Meyerhoff and Hatten, 1968) found in the northern part of the island. The flora of the latter only indicates a post-Permian–pre-Cretaceous age and it might possibly be correlated with the Louann Salt Formation of the Gulf Coast (Rhetian – Middle Jurassic, Jux, 1961).

The Upper Jurassic is represented by the 300 m thick Jagua Formation (Palmer, 1945) of grey to black sublithographic limestones (= Azucar Formation, Hatten, 1957), followed by sandstone and shale with calcareous concretions and ending in a grey sublithographic limestone. It contains an Upper Oxfordian fauna (Judoley[*1] and Furrazola-Bermudez, 1968) in the middle part.

The Viñales Limestone (De Golyer, 1918), 1000 m thick, begins with a limestone conglomerate or breccia. It consists of thick beds of grey limestone and dolomites, which form a striking karst topography. The ammonite fauna indicates a middle Kimmeridgian to middle Tithonian age (Judoley and Furrazola-Bermudez, 1968). Lower Kimmeridgian is therefore unknown, and relationships between the Viñales Limestones and the Jagua Formation are not clear. According to Palmer (1945) and Furrazola-Bermudez et al. (1964), the Viñales Limestone lies unconformably on the Jagua Formation, and begins with a basal conglomerate. As the San Cayetano Formation is far more folded than the Viñales Limestone, this would seem to indicate an orogenic phase between the deposit of these two formations. However, according to Hatten (1957), Knipper et al. (1967) and the author's observations, the Viñales Formation is conformable with the Jagua Formation, the "basal conglomerate" of the Viñales Limestone being, in fact, a tectonic breccia (see also Rigassi-Studer, 1963). The San Cayetano Formation is much more folded than the Viñales Limestone but the difference may be attributed to a difference in the competence of the beds.

The existence, in Cuba, of a Nevadian orogenic phase is still debated. It would be surprising nevertheless if, after an orogenic phase, deposits would have begun with such pure limestones as these of the Viñales Formation. There are some serpentines, tectonically associated with the Jurassic formations, in the eastern part of Pinar del Rio Province, but this must be the result of Middle Cretaceous, Upper Cretaceous and Lower Tertiary tectonic activity.

The Artemisa Formation conformably overlies the Viñales Limestone in the Sierra de los Organos, but transgresses directly into the San Cayetano Formation in the more eastern Sierra del Rosario. It

---

[*1] Judoley, Spanish transliteration of English Khudoley-spelling retained for bibliographical reasons.

consists of a regularly thin-bedded limestone with chert, radiolarite and shale, indicating deposition in deeper water than the Viñales Limestone. These deposits contain middle and upper Tithonian foraminifers and ammonites.

The varying stratigraphic relationships between the cited formations, in different regions, lead Meyerhoff (in Khudoley and Meyerhoff, 1971) to the conclusion that these formations are simply facies with a varying stratigraphic distribution, and are not the results of a break in deposition owing to tectonic movements, as Khudoley seems to think (same publication).

It has been already indicated, that Jurassic is unknown in the other Greater Antilles. Nevertheless, it must be observed that the base of some formations in these islands, which contain Lower Cretaceous fossils, could belong to the Jurassic.

In the Venezuelan Lesser Antilles where metamorphic rocks similar to these of the Caracas group are encountered, Jurassic may be present, particularly in Margarita (see below). We find the same situation in Tobago.

In the northern mountains of Trinidad, the Upper Jurassic is represented by the lower part of the "Caribbean Series", a thick metamorphic complex (about 6.000 m) of greenschist facies, which may probably be correlated with the Caracas group of the Venezuelan Coastal Cordillera.

At that time, these rocks were being deposited in a Jurassic and Cretaceous marine eugeosyncline trough, the western prolongation of which must eventually be sought in the circum-Mediterranean regions. It may also extend to the Venezuelan Islands and Tobago (see below).

The Caribbean Series began with the *Maraval Formation*, a 400 m (min. thickness) marble-like limestone, encountered in the western part of the mountain range. The base is not exposed. It is overlain by 1.500 m of the Maracas Formation, a flysch facies of alternating sandstone, phyllite and quartzite, indicating the occurrence of an orogenic phase during this period (Upper Jurassic?). The Maracas Formation is overlain in turn by the Chancelor Formation, a sequence of limestone, phyllite and quartzite, cropping out in the western region. The Rio Seco Formation, composed of phyllite and limestone and considering the time equivalent in the eastern region, contains a Tithonian ammonite (Barr and Saunders, 1968). The upper part of the Caribbean Series, which belongs to the Lower Cretaceous, will be considered later.

Finally, Fink Jr. (1968, in Anonymous, 1974) has indicated that the spilites and quartz keratophyres of Désirade Island, east of Guadeloupe, are associated with trondhjemites of a radiometric age between 142 ± 9 m.y. (Upper Jurassic) and 43.4 ± 4 m.y. (Eocene). According to Fink Jr. the first corresponds to the real age of the complex. But the trondhjemites intrude the volcanic complex and are therefore younger.

Several hypotheses have been proposed concerning the origin of this complex (see L.K. Fink in Anonymous, 1974). It has been suggested that it may be a prolongation of the Greater Antilles Mesozoic magmatic arc (Fink, 1968; Meyerhoff and Meyerhoff, 1972), or a fragment of Atlantic crust, later subducted below the Caribbean or, altenatively, a part of autochthonous oceanic crust upon which the eastern Lesser Antilles Arc was built during the Tertiary and subsequently uplifted (Fink et al., 1972). The second hypothesis seems favored by the existence of 90 m.y. The Atlantic crust east of Désirade Island (Pitman and Talwani, 1972), with the age difference between it and the Désirade basement (about 50 m.y.) due to subduction of the Atlantic plate which would thus begin in the Eocene times.*1

On the other hand, in the Bahamas, which can be considered as the foreland of the western Greater Antilles (Cuba, Hispaniola), there exists a thick carbonate series (10.500 m thick, cf. Banks, 1967), formed of sub-horizontal limestones and dolomites, the base of which may be of Upper Jurassic age, by comparison with the deposits of the Atlantic Ocean, east of the island San Salvador. These carbonate deposits overlie clastic formations of Upper Triassic – Lower Triassic age (Dietz et al., 1970) but this hypothesis is not accepted by Glockhoff (1973).

The sinking crust, on which this sedimentary series was deposited, according to the different authors, is either oceanic (Dietz et al., 1970, 1971; Glockhoff, 1973) or continental (Sheridan, 1971; see also Uchupi et al., 1971). According to Khudoley and Meyerhoff (1971) the basement below the sedimentary series of the Bahamas is continental, as in Florida and in Cuba. These series are Paleozoic at least, or even perhaps Precambrian.

In Nuclear Central America, bearing in mind

Isthmian Central America did not exist, the Jurassic sedimentation pattern repeats that presumed established in Late Triassic times (Fig. 2). The marine El Plan Formation, interdigitating with the continental Todos Santos Formation is best seen in Honduras. The age of the Todos Santos Formation varies according to the region, but is essentially Jurassic. Mills et al. (1967) suggest an orogenic phase occurred in Middle Jurassic; they also note that, where the Todos Santos Formation overlies the El Plan Formation, andesitic flows occur sometimes between the two formations.

Cretaceous

Well dated Cretaceous formations are known all over the West Indies except in the Lesser Antilles where the only rocks of this age are a volcanic series which began in the Jurassic found on Désirade Island (see above). In Nuclear Central America, marine shallow water deposits predominate, but it is probable that mafic volcanic and clastic deposits were formed and subsequently metamorphosed. In Isthmian Central America, Upper Cretaceous marine sediments lie unconformably over an ophiolitic complex of probable Lower Cretaceous age.

Lower Cretaceous

In Cuba there are marked regional variations in the Lower Cretaceous deposits. In southern areas thick (2500–4500 m?) deep marine mafic volcanic rocks ("Tuff Formation"), associated with serpentinized peridotites and gabbros (oceanic crust) predominate. They lie on chlorite schists. Some interbedded pelagic limestones are dated as Neocomian (Aptian according to Khudoley) to Lower Turonian. These rocks are slightly metamorphosed.

Metamorphic rocks of type II (see above) encountered in the Sierra del Purial (northeast Oriente Province) and in the Escambray Mountains are probably of Lower Cretaceous age. North of this region, Lower Cretaceous clastic deposits are followed by 1200 m pelagic limestones of Neocomian (or Tithonian) to Cenomanian age. Along the north coastal region the marine deposits are of a shallower water origin. There are some 3300 m of miliolid and rudistid limestones (Urgonian facies).

In the small islands near the northern coast of Cuba the first deposits are evaporites and dolomites (Neocomian?–Aptian?), followed by pelagic lime-

---

*1 See footnote p. 90.

stones. In the Bahamas, the Cretaceous (and Upper Jurassic?) shallow marine carbonate deposits are 1866–2166 m thick of which 726–976 m is of Upper Cretaceous age.

In Hispaniola, the Lower Cretaceous is predominantly volcanic and is represented by thick spilites (Duarte Formation in the Dominican Republic, Bowin, 1975) and keratophyres (Maimon Formation, Los Ranchos Formation, Bowin, 1975) associated with graywackes, tuffs and some serpentines. They have been metamorphosed to a greenschist facies and crop out in the Dominican "Cordillera Central" and Haitian "Massif du Nord". The similar and thick Los Caños Formation of the Dominican "Cordillera Septentrional" is probably equivalent and is associated with glaucophane and actinolite schists, overlying serpentines (Nagle, 1966, 1971). Thus serpentinized peridotites crop out in two parallel belts in the Dominican Republic, one on the northern flank of the Cordillera Central, parallel to its axis and curving to the southeast at the eastern end, the other extending in the Cordillera Septentrional and the Samaná Peninsula (Nagle, 1971). The Los Caños Formation is unconformably overlain by the Imbert Formation of Paleocene–Lower Eocene age.

These rocks are pre-Middle Aptian–Middle Albian age, because they are overlain by the limestones and radiolarites of the Hatillo Formation, with fauna of this age. According to Bowin (1975) they may even be pre-Cretaceous. Hornblendite stocks intruding the Dominican formations gave a $127 \pm 5\%$ m.y. (Neocomian) age – and amphibole schists $91 \pm 10\%$ m.y. (Middle Cretaceous) (Khudoley and Meyerhoff, 1971).

In Île de la Tortue (Haitian Republic) and in the Samaná Peninsula, metamorphic sheared schistose limestones, unconformably overlain by Paleocene–Lower Eocene sedimentary rocks (Butterlin, 1960) may be of Lower Cretaceous age.[*1]

The most highly metamorphosed rocks found in Hispaniola are pebbles of micaschists and garnetiferous quartz schists which occur in the North and Leogane plains of the Haitian Republic (Woodring et al., 1924). They do not occur in situ and could even be ballast from vessels of the colonial epoch. On the southern peninsula of Haiti, a caprinid fauna of Aptian–Albian age has been found in the argillaceous sediments interbedded in basaltic pillow-lavas (Reeside, 1947).

In Puerto Rico, the Bermeja Complex which crops out in the southwest corner of the island (Mattson, 1960, 1968) is of Lower Cretaceous age. It consists predominantly of serpentines, spilites with pillow lavas, gneiss and gneissic amphibolites. The latter rocks have a radiometric age of 110 ± 3 m.y. (Middle Cretaceous; Mattson, 1964; Khudoley and Meyerhoff, 1971). Some highly folded recrystallized radiolarites, cut by keratophyre dikes, rest unconformably upon the Bermeja Complex. Mattson (1973) considers that they belong to a nappe structure, but it is probable that they form part of the same ophiolitic complex. The complex in unconformably overlain by andesitic volcanic rocks, considered equivalent to the Rio Loco Formation of Albian–Cenomanian age.

In the Virgin Islands, argillaceous sediments are overlain by the 4500 m thick Water Island Formation, consisting of keratophyres (80%) and spilites (20%) (with pillow-lavas and some clastic material), extruded at depths greater than 4500 m. They exhibit slight burial metamorphism (Hekinian, 1971). The Water Island Formation is cut by keratophyre dikes which gave radiometric ages of 110 ± 10 and 106 ± 10 m.y. (Aptian–Albian, Donnelly, 1966) unconformably overlain by the Virgin Islands group (see below) the lower part of which has an Albian or Cenomanian age.

In Jamaica, the Blue Mountains in the eastern part of the island are formed principally of volcanic and sedimentary greenschist grade metamorphic rocks (chlorite and sericite schists, amphibolites, marbles) which are associated with a few serpentinized peridotites. Some Cretaceous rudistids (Lower Cretaceous?) have been found in the Blue Mountains (L.J. Chubb, in Zans et al., 1963) and the metamorphic series is unconformably overlain by rudistid limestones, shales and conglomerates of Campanian age.

In the eastern part (Benbow inlier) of the Cornwall-Middlesex Zone (central and western regions of the island), 1000 m of basaltic pillow-lavas and andesitic rocks and radiolarites (Devils Racecourse Formation) are interbedded with calcareous beds containing Valanginian (?) (L.J. Chubb, in Zans et al., 1963), Barremian and Aptian faunas (Burke et al., 1968; N.F. Sohl, in Khudoley and Meyerhoff, 1971; Roobol, 1972). Sandstones and shales with andesitic and basaltic pebbles (Rio Nuevo Formation) are interbedded with limestones of Albian age.

[*1] Upper Senonian fossils were recently encountered (J.M. Villa et al., 1982, in press).

In the Dutch islands, the lower part of the Cretaceous volcanic sedimentary series, mainly Upper Cretaceous (see below) in age contains Albian ammonites (Beets and McGillavry, 1976).

In the Venezuelan islands, probably Jurassic–Lower Cretaceous series exist, particularly in Margarita Island. There, the oldest deposits correspond to the La Rinconada group, consisting of serpentines associated with recrystallized and very sheared granitic rocks, metabasic rocks and micaschists (Maresch, 1972), which have suffered a high pressure/low temperature metamorphism. These rocks are conformably overlain by the Juan Griego group of metavolcanic gneiss, micaschists, marble, associated with serpentinized peridotite, gabbro amphibolite, amphibolite schists of greenschist and amphibolite facies, and cut by pegmatite dikes (Gonzalez de Juana and Vignali, 1972). According to Maresch (1972), this group may be correlated with the Caracas group of the Caribbean Cordillera of Venezuela and the Caribbean series of Trinidad of Upper Jurassic and Lower Cretaceous age. Similar rocks crop out in the Venezuelan islands Orchila, Los Hermanos and probably Blanquilla.

The Los Robles group is in fault contact with rocks of the Juan Griego group. It consists of phyllite, chlorite and graphite schists, limestones in the greenschist facies. It is considered, either partially equivalent to the Juan Griego group (Hess and Maxwell, 1949) or younger than the latter lying upon it unconformably (Taylor, 1960) or conformably (Gonzalez de Juana and Vignali, 1972).

In Trinidad the Lower Cretaceous is represented by the upper part of the Caribbean series (see above). The Rio Seco Formation with a Tithonian fauna is overlain by the Grande Riviere Formation of phyllite and limestone, followed by the phyllitic shales of the Tompire Formation with Barremian ammonites and finally the Toco Formation, formed by shales, sandstones and limestones of Barremian–Lower Aptian age, associated with gypsum possibly moved upwards from Neocomian–Barremian beds, as in the Paria Peninsula of Venezuela (Saunders, 1972).

On the eastern margin of the Central Mountain Range the Sans Souci basalt, a sequence of basaltic flows and breccia, rests upon the Toco Formation and pre-Senonian (pre-Galera Formation, see below) beds. No other volcanic rocks are found in Trinidad. In the eastern part of the Central Mountains the Lower Cretaceous is represented by the Cuche Forma-

tion, shales, quartzites, limestones. These are overlain by the marls of the upper Aptian–lower Albian age Maridale Formation. The two formations can be correlated with the Sucre group of the eastern part of the Caribbean Cordillera (Metz, 1968). They are followed unconformably by the Gautier Formation, shales and black limestones of upper Albian–lower Cenomanian age.

In the Tobago Island the situation is not very clear (Maxwell, 1948). The oldest deposits are the North Coast schist group, a sequence of 2340 m of metavolcanics and metasediments with metatuff, amphibolites, greenstones, sericite schists, quartzite and chert of the greenschist facies, associated with peridotite not much serpentinized. They are very similar to the Robles group of Margarita. This complex is overlain, unconformably, by the Tobago volcanic group of dacitic and andesitic rocks, with pillow-lavas but also with some pyroclastic (tuffs, breccia, conglomerate) rocks indicating both marine and continental influences. The top of this group excluded, it is intruded by a mainly dioritic batholith with a radiometric age of 113 ± 6 m.y. (Middle Cretaceous, Meyerhoff and Meyerhoff, 1972). It is in turn unconformably overlain by Mio–Pliocene deposits.

Finally it should be noted that the radiometric age of the magmatic complex of Désirade Island in the Calcareous Lesser Antilles varies between Upper Jurassic and Eocene so that it may be in part of Cretaceous age.

In Nuclear Central America, the Cretaceous is represented in the central and north part of Guatemala by the Ixcoy group or formation (Anderson et al., 1973, Clemons et al., 1974) of thick limestones and dolomites with chert, and very thick (2400 m, Vinson, 1962) deposits of gypsum although probably locally exaggerated by diapirism. It has a Neocomian (Aptian?)–Campanian (Maastrichtian?) age. The lower part is thus Lower Cretaceous. The Yoyoa group of Honduras and southeastern Guatemala (Mills et al., 1967; Burkart et al., 1973) is probably time-equivalent. Limestones of the middle part of the Metapan Formation (the lower continental part must be equivalent to the Todos Santos Formation – see above) crops out in the southern part of Nuclear Central America. 2000–3300 m of limestones, dolomites and evaporites were also found in wells in the Mexican Yucatan Peninsula (Salas, 1955; Viniegra, 1971).

The ophiolitic complex of the El Tambor Formation and the greenschist metasedimentary rocks of the Chuacus group may be of Lower Cretaceous age and thrust northward over the calcareous Cretaceous rocks of Central Guatemala.

Upper Cretaceous

In southern and central Cuba, above the Tuff Formation which extends into lower Turonian, the remainder of the Upper Cretaceous consists of clastic rocks (conglomerate, graywacke and shale), in part of flysch and wildflysch facies, with some rudistid limestones and of Campanian—Maastrichtian age (La Habana Formation). The lower Senonian is not known and probably not represented. In the northern region as in the Lower Cretaceous the calcareous deposits of shallow- and deep-water origin predominate; these unconformably overlie pre-Turonian deposits. They are found in wells on the north coast where they appear to be very thick (Echevarria and Veliev, 1967). Between the two regions some Upper Cretaceous volcanoclastic, basaltic to rhyolitic deposits with ignimbrites are found. They are intruded by granodiorite batholiths with radiometric age of 61 to 78 m.y. (Khudoley and Meyerhoff, 1971).

In the east (Oriente Province) the Upper Cretaceous is probably represented by the Vinent Formation in the Sierra Maestra. The formation consists of volcanic rocks which have been metamorphosed by numerous Eocene (46—58 m.y.) granodioritic intrusions. This formation is probably overlain by the volcano-sedimentary Cobre Formation of Upper Cretaceous (?)—Paleocene to Lower Eocene and Middle Eocene (?) age, but the contact between the two formations is not very clear.

In Hispaniola, the Upper Cretaceous is represented by volcanic (andesitic and dacitic rocks) and sedimentary rocks. In the Haitian Republic, the andesitic rocks of the Massif du Nord and of the Massif de Terre Neuve (Northwestern Peninsula) are cut by quartz dioritic intrusions which form a large batholith occupying the eastern part of the Massif du Nord. The batholith probably has an upper-Senonian age (Butterlin, 1960; Kesler, 1971).

Sedimentary rocks are represented both by rudistid and foraminiferal limestones (Trois Rivières Formation) of Campanian—Maastrichtian age (possibly also of Cenomanian—Turonian age according to Ayala Castañares, 1959) and by clastic deposits, in part of flysch facies, which crop out extensively on the southern flank of the Massif du Nord and con-

tinue into the Dominican Republic (Cheilletz and Lewis, 1976). In the Southern Peninsula of the Haitian Republic there are Upper Cretaceous pelagic limestones of lower-Santonian (Reeside Jr, 1947) and upper-Senonian (mainly upper Campanian: Ayala Castañares, 1959) age (Macaya Formation) with basaltic and doleritic pillow-lavas, interbedded with the limestones. As indicated earlier the base of this series is of Aptian—Albian age. It is very similar to the rocks found in the floor of the Caribbean Sea, in the Leg 15 well of J.O.I.D.E.S. (Saunders et al., 1974). In the Venezuelan Basin and on the northeastern slope of the Nicaraguan Rise, marine basalts are overlain by upper-Senonian limestones, the contact corresponding to the reflector horizon "B". On the Beata Ridge, pillow-lavas basalts are found, the youngest of them having a radiometric age (Fox et al., 1968, 1970) of 65 m.y. The Cretaceous of the Dutch Leeward Islands shows also much similarily (see below).

Thus all these present continental areas were originally parts of the Caribbean Sea, uplifted by faulting during the Tertiary (Miocene?).

In Puerto Rico, the Upper Cretaceous was marked by an important volcanic episode with deposits 6000—8000 m thick. The evolution of this volcanism was marked by a progression towards more acidic magmas (tholeiitic and andesitic in the Lower Senonian, dacitic and rhyolitic in the Upper Senonian) and by deposition in progressively shallower water with subcontinental pyroclastic deposits becoming more abundant (Lidiak, 1968). However, basaltic rocks continue to prevail (McIntyre et al., 1970) in the northwest. Horizons of fossiliferous limestones are intercalated in these beds (Pessagno, 1960—1966; — Kauffman, 1965). Intrusive granodiorites have a radiometric age of 65 ± 3 to 70 ± 20 m.y. (uppermost Cretaceous).

In the Virgin Islands there is a similar Upper Cretaceous volcano-sedimentary series, several thousands of meters thick which, in St Thomas and St John Islands, unconformably overlies the pre-Albian Water Island Formation (see above). The age of the intercalated sediments varies from upper Albian to Middle Eocene.

The Upper Cretaceous in Ste Croix Island begins with the Mount Eagle group, some 4500 m turbidites, followed by volcano-sedimentary deposits. The upper part has a Campanian to Lower Maastrichtian age. In the Virgin Islands the Cretaceous is intruded by a granodiorite batholith, partly submarine, of Middle Eocene or younger age (Helsley, 1971) for it cuts the Tortola Formation, the upper part of which has a Middle Eocene age.

In Jamaica, Upper Cretaceous rocks are abundant in "inliers" in the central and western regions (Cornwall—Middlesex zone) where they are represented by a volcano-sedimentary series. The volcanic rocks are mainly andesitic, predominantly pyroclastic and are interbedded with rudistid limestones. Intrusions of granodiorite, particularly the Above Rocks granodiorite with a radiometric age of 64 to 75 m.y. (uppermost Cretaceous) northwest of Kingston, cut these beds.

In the eastern part of the island, along the eastern margin of the Blue Mountains, a thick (3600–7000 m) Upper Cretaceous volcano-sedimentary sequence was deposited in a subsiding basin. The volcanic rocks are mainly basaltic but also include andesitic flows and pyroclastics (tuff and conglomerate) rocks and interbedded reef and foraminiferal limestones, sandstones, conglomerates and shales of Turonian to Maastrichtian age (Annual Report of the Geological Survey of Jamaica, 1966).

In the Bahamas, the Upper Cretaceous is represented by about 1000 m limestones and dolomites. These latter must correspond to the Cuban orogenic periods (Goodell and Garman, 1969).

In the Dutch Leeward Islands, the Upper Cretaceous is characterized chiefly by several thousands meters of volcano-sedimentary rocks (diabase and diabase tuff formation in Aruba, Curaçao lava in Curaçao, Washikemba Formation in Bonaire). The volcanic rocks are generally of mafic type: diabase and diabasic pyroclastics (tuff, breccia, conglomerate) with amphibolite schists. These rocks are principally marine but subaerial eruptions occur with ignimbrites in Bonaire (Beets and Lodder, 1967). Fossiliferous interbedded graywackes, limestones, radiolarites are of Albian to Turonian (or lower Senonian?) age (Beets and McGillavry, 1976). The beds are slightly metamorphosed and are intruded by granodioritic rocks dated at 72 ± 7 and 73 ± 3 m.y. (Priem et al., 1966; Priem, 1967). They are unconformably overlain by clastic, calcareous and siliceous rocks of Senonian to Danian age, in part of flysch type (Beets, in Matsumoto and Beets, 1966; — Krijnen, 1967; Krijnen, in Beets and Lodder, 1967).

As already indicated (see above) the mafic volcano-sedimentary series of the Dutch islands is very similar to the Cretaceous of the Caribbean basin and probably represents the marginal parts of this structural unit uplifted by faulting during the Tertiary (Miocene?).

In the Venezuelan islands, there are Upper Cretaceous volcanic andesitic rocks of the Los Frailes Formation, found in the Margarita, Los Frailes and Los Testigos archipelagoes. In Margarita they are unconformably overlain by the sedimentary Punta Carnero group, of Lower-Middle Eocene age (Bermudez and Gomez, 1966; Butterlin, 1970; Caudri, 1974). Granodioritic intrusions, aligned with those of the Dutch Leeward Islands and the southern part of the Aves Ridge (Fox et al., 1971), crop out in the Venezuelan islands. They have a radiometric age of 57 to 89 m.y. (71 ± 5 m.y. in the Matasiete stock of eastern Margarita, Olmeta, 1968) and must represent two intrusive periods, one of Upper Cretaceous age (78 to 89 m.y.) and one of Eocene age (57 to 60 m.y.), corresponding to the orogenic periods in this region.

In Trinidad, in the Northern Mountain Range, the Caribbean series are overlain with marked unconformity, by the Galera Formation of slightly metamorphic phyllite, sandstone and shales of Senonian age. In the Central Mountain Range, the Lower Cretaceous series (see above) are unconformably overlain by the Gautier Formation, shales and black limestones of upper-Albian—lower-Cenomanian age, upon which rests Turonian—Lower-Senonian siltstones of the Naparima Hill Formation. This is in turn overlain by the Maastrichtian shales and black limestones of the Guayaguayare Formation. Similar deposits were found in wells in the southern region of the island. They were folded during an uppermost Cretaceous—Early Tertiary orogenic phase at which time thick flysch deposits were formed.

In Nuclear Central America, thick calcareous deposits of Lower—Upper Cretaceous age are known. An orogenic phase occurred during the Upper Cretaceous (Cenomanian (?), Horne et al., 1974 — upper-Turonian (?), Mills et al., 1967, Senonian). In Guatemala, this period is marked by flysch deposits (red sandstones, shales and calcarenites of the Sepur Formation; Wilson, 1974), followed by another orogenic period at the end of the Cretaceous. These two orogenic periods were accompanied by granodioritic intrusions.

In Honduras, the Upper Cretaceous orogenic period was followed by mainly continental red molasse deposits, but with some littoral beds. These are the

3000 m Valle de Angeles group, of Senonian–Miocene age. In the southern part of Nuclear Central America similar red molasse deposits of Upper Cretaceous age crop out. They are referred to the upper part of the Metapan Formation. Marine Senonian deposits are found in the coastal regions of Honduras (Mosquitia Basin, Mills and Hugh, 1968 – Ulua Basin, Mills et al., 1967 = Esquias Formation).

In Isthmian Central America; the oldest sedimentary formations are of Campanian age. They overlie an ophiolitic series which corresponds to the Nicoya Complex and Santa Elena Complex. The former crops out along the Pacific coastal peninsulas of Costa Rica, particularly in the Nicoya Peninsula and the latter in a region close, to the Santa Elena Peninsula. The Nicoya Complex is an ophiolitic metamorphosed complex, of the greenschist facies (Weyl, 1969). It consists of tholeiitic, sometimes spilitic, metabasalts (pillow-lavas, tuffs, and agglomerates), interbedded with metasedimentary rocks (limestones, silicified lutites, lydites, phtanites and graywackes) with gabbroic intrusions. The Santa Elena Complex consists of serpentinized peridotites, metabasalts, amphibolites and hornblende schists. According to Dengo (1969), it is older than the Nicoya Complex, but the age cannot be very different for the complexes are lithologically very similar and geographically very close. It is overlain, with angular unconformity, by the Santa Elena Formation consisting of a conglomerate with serpentine boulders, followed by Campanian–Maastrichtian limestones, lutites and sandstones. These rocks are correlated with the Sabana Grande Formation (see below). A complex similar to the Nicoya Complex occurs in the Azuero Peninsula of Panama (Ferencic et al., 1971).[*1]

All the complexes are intensely folded. Like the Franciscan Formation of California they may correspond to Pacific oceanic crust, incorporated into the continent as a consequence of its westward drift and interaction with the margin of the Pacific oceanic plate (Pichler et al., 1974). According to Dengo (1968) and Barr and Escalante (1969), these complexes are unconformably overlain, and according to Henningsen (1966a, b) conformably, by marine volcano-sedimentary formations of Campanian–Maastrichtian age (see below). In fact, the contact is probably of a tectonic type. Dengo (1962) claims, but without any conclusive argument, that the ophiolitic complexes are probably of Upper Jurassic–Lower Cretaceous age; however, Barr and Escalante (1969) have obtained a radiometric age of 72 ± 5 m.y. (Campanian) for the upper basalts of the series.

Dengo (1968), Goossens (1973) and Pichler et al. (1974) consider that these complexes may be correlated with the unmetamorphosed diabasic group of the Colombian Cordillera Occidental (Julivert et al., 1969) of Lower (or Middle (?)) Cretaceous to Paleocene age. In fact, although the accretion of the Pacific oceanic crust to the continent is known from California down to Ecuador passing through the Pacific border of the Baja California Peninsula, Isthmian Central America and Colombia, it is not necessarily contemporaneous, according to the origin we have supposed.

The marine volcano-sedimentary deposits in the Nicoya Peninsula begin with the Sabana Grande Formation of silicified limestones and phtanites. According to Dengo (1968), it is unconformably overlain by the Upper Campanian–Maastrichtian Rivas Formation which crops out in southwestern Nicaragua and in Costa Rica with graywackes, arkoses, phyllites, sandstones, tuffs, volcanic lavas. The Rivas Formation is then equivalent for Dengo to the Changuinola Formation of comparable lithology, cropping out in the east of Costa Rica and west of Panama (Fischer and Pessagno, 1965). However, according to Henningsen and Weyl (1967) these two formations must be correlated also with the Sabana Grande Formation. In any case, they are unconformably overlain by Eocene beds, indicating the existence of a Late Cretaceous-Early Tertiary orogenic phase.

PALEOGEOGRAPHY (Figs. 8, 9, 10, 11)

It is difficult to present the paleogeography of the Caribbean region in the Mesozoic because it is probable that the relative position of the continent, islands and seas was not the same as today and because of the lack of information on the geology of the different regions and, in particular, the age of the metamorphic formations.

For these reasons the paleogeographic maps presented are tentative, and based upon the present position of the geographical units, can only give infor-

---

[*1] Recent studies by J. Azema et al. (1979) of the Nicoya Complex indicate the presence of Upper Jurassic and Upper Albian fossils.

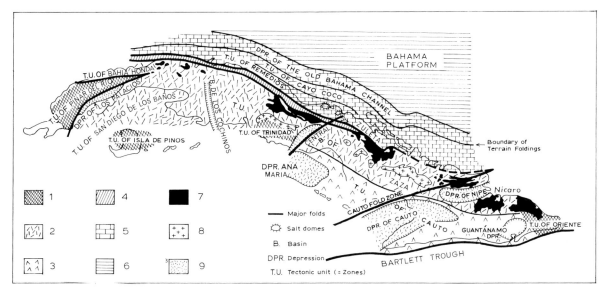

Fig. 8. Structural units of the island of Cuba (after Butterlin, 1977). *1* = Intrageanticlinal, metamorphic and Cretaceous sedimentary rocks; *2* = areas of Cretaceous magmatism (septentrional eugeosyncline); *3* = areas of Cretaceous and Tertiary magmatism, meridional eugeosyncline; *4* = thin layers of mixed Cretaceous deposits (miogeanticline = marginal boundary of the geosyncline), Las Villas unit; *5* = deposits of folded Cretaceous limestones (miogeosyncline); *6* = deposits of unfolded Cretaceous limestones (foreland); *7* = basic and ultrabasic rocks (age?); *8* = granitic rocks (Cretaceous–Eocene); *9* = Tertiary depressions.

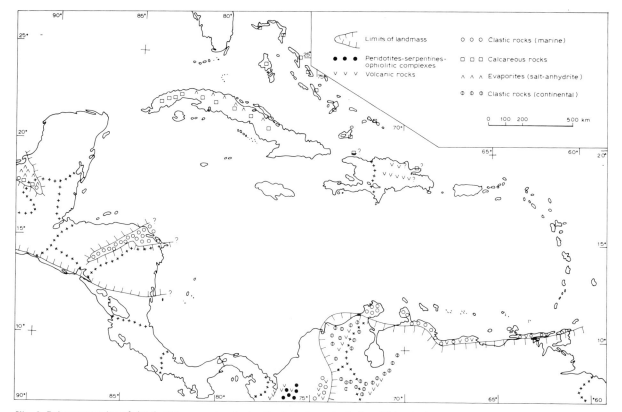

Fig. 9. Paleogeography of the Caribbean region during the Late Jurassic (before Tithonian).

mations on the present status of knowledge on the geology of this region.

Triassic

As indicated before (see above), it is probable that the Caribbean region did not exist before Middle Triassic, Upper Triassic or even Jurassic. It may have been the divergent drift westward of North and South America which created the region.

Yet we have seen that in all probability the Paleozoic structural units of Nuclear Central America continued eastwards into the southern regions of Cuba at the end of the Paleozoic, forming mountains as in Nuclear Central America, totally emergent. The erosion of these mountains was the source of the continental molasse deposits in the western parts of Cuba (San Cayetano Formation) and in Nuclear Central America (Todos Santos Formation). In this last region the sea invaded narrow depressions in Honduras (E1 Plan Formation).

In the Bahamas also clastic deposits occurred in

Upper Triassic and Lower Jurassic, with the clastic material probably derived from the southwestern United States.

Jurassic

The Jurassic was probably marked by the development of the Caribbean region and Le Pichon and Fox (1971) consider that at the end of the Jurassic this region had, more or less, the extension of to-day. Only relative east—west movements of the different parts related to strike-slip faulting, occurred during the Cretaceous.

At the beginning of the Jurassic, the paleogeography was not dissimilar to that of the Upper Triassic. In Cuba, the molasse deposits continue and in the northern regions there was a development of lagunar basins with deposit of evaporites.

In Nuclear Central America the continental molasse deposits resulted from the erosion of the Hercynian mountains. The erosion was probably renewed by epeirogenic movements during the Middle Jurassic and

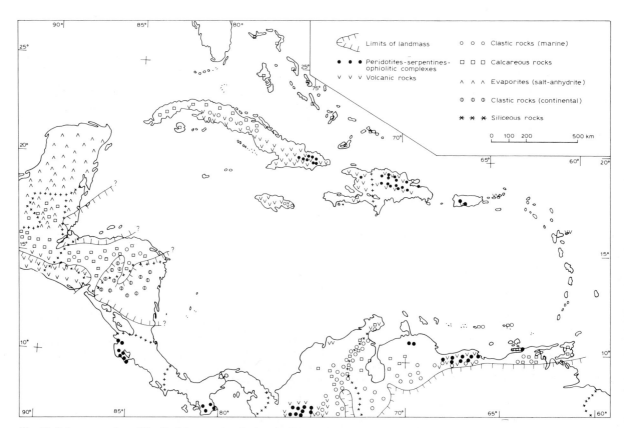

Fig. 10. Paleogeography of the Caribbean region during the Early Cretaceous (Aptian).

continued during Upper Jurassic and probably into the base of the Lower Cretaceous.

During the Upper Jurassic, calcareous sediments formed upon a marine platform in the western regions of Cuba. On the Bahama Platform too the period began with monotonous shallow water marine carbonate sedimentation.

In the southern regions an eugeosynclinal deep basin formed which extended along the Caribbean margin of Venezuela and Trinidad and into the Venezuelan islands and Tobago. Along the margin of this basin there existed a volcanic arc of islands (Maresch, 1974). Thick volcanic and sedimentary deposits occurred during the Upper Jurassic and Lower Cretaceous.

Cretaceous

Lower Cretaceous

In the Lower Cretaceous, for the first time, all the structural units of the Caribbean region appear clearly.

In the Greater Antilles a eugeosynclinal basin extended over the southern parts of Cuba (Zaza zone of Khudoley and Meyerhoff, 1971; Santa Clara zone of Meyerhoff and Hatten, 1968), the northern area of the Haitian Republic (Northwestern Peninsula, Massif du Nord), northern and central areas of the Dominican Republic (Cordillera Septentrional, Cordillera Central) and the southwestern part of Puerto Rico and the Virgin islands. The Oriente Province of Cuba, where more metamorphic Upper Mesozoic rocks are found, must have occupied a more internal position. The SW—NE Cauto Fault Zone which separates the two regions is of strike-slip type so that, originally, the Oriente Province (Cauto zones of Khudoley and Meyerhoff, 1971) must have had a position south of the Zaza zone.

The situation of Jamaica is more doubtful. It is probable that the Cayman Trough appeared during the Tertiary (see Bowin, 1968; Uchuppi, 1973) and the Yucatan Basin may have originated during the Upper Cretaceous. The similarity between the Tertiary rocks of Jamaica and Cayman islands which lie on the

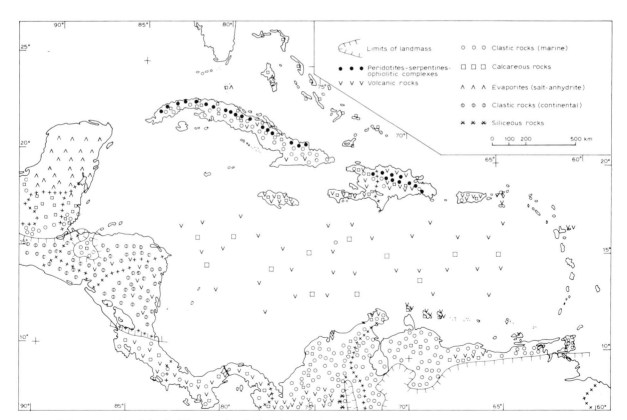

Fig. 11. Paleogeography of the Caribbean region during the Late Cretaceous (Senonian).

Cayman Ridge confirms this hypothesis. If this is the case, the Blue Mountains must be part of the Cauto zone (see above) and the Cornwall–Middlesex zone, part of the Zaza zone, a deduction which implies a subsequent counter-clockwise rotation of the island.

Thick outpourings of submarine mafic lavas, associated with pelagic and neritic calcareous sediments occurred in the eugeosynclinal basin during the Lower Cretaceous. These deposits probably rest on an oceanic, peridotitic, crust. This basin was limited to the north in Cuba by a miogeanticlinal ridge (southern part of the Las Villas structural unit; Khudoley and Meyerhoff, 1971). In this zone the deposits are thin clastics followed by calcareous (100–300 m) rocks. To the north a miogeosynclinal basin developed (northern part of Las Villas structural unit and Remedios structural unit of Khudoley and Meyerhoff, 1971 = Camajuani Meyerhoff and Hatten's zone, 1968). Here thick (1000–3000 m) pelagic and neritic calcareous sediments were deposited. On the north margin of the basin (Cayo Coco structural unit of Khudoley and Meyerhoff, 1971) 1000–2000 m of Neocomian? to Aptain? littoral to lagunal deposits, with dolomites and evaporites were laid down followed by pelagic carbonates.

On the foreland (Bahama Platform) about 1000 m of calcareous, neritic to littoral sediments with limestones and dolomites formed continuing the Upper Jurassic pattern of sedimentation.

Along the northern margin of South America, including the Venezuelan islands, Trinidad and Tobago, the situation paralleled that of Upper Jurassic (see above). Volcano-sedimentary deposits continued in the eugeosynclinal basin formed during the Upper Jurassic. It is probable that in the northern part of this basin important submarine outpourings of mafic magmas, deposited on oceanic crust occured (Maresch, 1974).

In Nuclear Central America, the Lower Cretaceous is characterized by the development of a submarine platform extending over the Yucatan Peninsula, the central part of Guatemala, Honduras and the northern part of Nicaragua. On this subsiding platform, interrupted by deeper basins and with shallow lagoons in the northern part, the sediments laid down, were thick carbonates. It is probable that south of this platform there existed a deep marine basin with volcano-sedimentary deposits, similar to the eugeosynclinal basin of the northern margin of South America with mafic volcanics representing submarine out-

pourings onto an oceanic crust. The same marine eugeosynclinal basin or a parallel basin extended into Isthmian Central America.

Over the whole Caribbean region, the Middle Cretaceous was an important period of orogenic activity in the internal parts of the orogen exactly as in Mexico. It was marked by regional metamorphism of greenschist facies and also metamorphism of high pressure/low temperature (blue schist facies), the latter in this region, and strong deformation.

Upper Cretaceous

In the Greater Antilles the Upper Cretaceous is generally characterized by a continuation of volcanic activity. The volcanic deposits, however, were laid down in shallower water or were subaerial, with abundant pyroclastics. Compositionally they are also more acid. Interbedded sedimentary deposits are common particularly in the upper part (upper Senonian).

In central Cuba, volcanic deposits were subordinate. The deposits are mainly clastic in the south and central regions and calcareous in the northern part of the island. There was an important hiatus in sedimentation, particularly in the lower Senonian, indicating the importance of the Middle Cretaceous orogenic activity (Subhercynian phase). In the South Peninsula of the Haitian Republic there persisted from Lower Cretaceous a deep basin in which mafic submarine lavas poured onto an oceanic crust which also received pelagic calcareous sedimentation. This marine basin extended to the south-central part of the Republic for the Cul-de-Sac depression did not exist at this time.

On the Bahama Platform shallow marine calcareous sedimentation with dolomites probably corresponded to the orogenic movements in Cuba.

In the Dutch Windward Islands sedimentation was similar to that in the South Peninsula of the Haitian Republic, and suggests a deep marine basin existed in this region. Epirogenic movements in the Senonian are indicated by shallower marine sedimentation and flysch deposits.

In the Venezuelan islands, the Upper Cretaceous was also essentially characterized by a volcanic activity of andesitic type and it was probably the same on Tobago. Over all Trinidad, the Upper Cretaceous was marked by dominantly clastic sedimentation linked to the Middle Cretaceous tectonic activity although some black limestones occur. Sedimentation was particularly developed in the central and southern parts of the island.

In Nuclear Central America, the lower Upper Cretaceous was a period of calcareous sedimentation on marine platforms in the northern and central regions. Tectonic activity beginning in the middle Upper Cretaceous resulted in clastic sedimentation, both marine (flysch type deposits) and continental, with red beds in the upper part of the Upper Cretaceous. Over the northernmost regions, however, calcareous sediments and evaporites continued to be deposited indicating a quietly subsiding platform whose movement continued into the Lower Tertiary.

In Isthmian Central America, volcano-sedimentary sedimentation began in the Upper Senonian. A volcanic island arc probably existed, the volcanic products of which accumulated on the marine platforms surrounding the islands. They are interbedded with shallow sedimentary, calcareous and clastic deposits.

The Laramian orogeny s.l. (Upper Cretaceous to Upper Eocene) was very important in the Caribbean region. It was marked by nappe structures which moved to the north-northeast in Cuba and perhaps in Puerto Rico (Mattson, 1973) and the Venezuelan Islands, Trinidad and Nuclear Central America. Intense folding and faulting occurred in the other islands of the Greater Antilles.

INTERPRETATION OF EVOLUTION OF THE CARIBBEAN REGION IN THE MESOZOIC (Fig. 2, 12)

To explain the evolution of the Caribbean region, the existence of a Caribbean plate (Le Pichon, 1968; Molnar and Sykes, 1969) has been proposed. According to this hypothesis the present region of the Caribbean Sea resulted from the diverging westerly motion of North America and South America during the Jurassic (Le Pichon and Fox, 1971). According to Le Pichon and Fox (1971), the Caribbean area was completely formed by the end of the Jurassic, and from this period forth the divergence ended and was replaced by an ordinary parallel westward drift. The rate of motion of North America was faster than that of South America, and resulted in a west—east extension. This in turn produced west—east strike-slip faulting along the northern and southern margins of the Caribbean region, i.e., the border faults of the Cayman Trough and the Oca—San Sebastian—El Pilar faults, respectively.

As yet, no west—east spreading center has been found in the Caribbean Sea, to explain how in such a newly formed area, the Caribbean crust, with its suboceanic characters (i.e., different from those of the typical oceanic crust; Officer et al., 1957) was formed. Christofferson (1973) considers that such a spreading ridge formed in the Upper Cretaceous in the Caribbean area but has since disappeared by subduction below the South American plate. This does

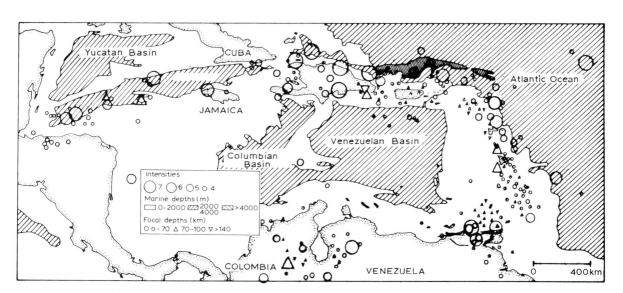

Fig. 12. Seismicity of the Caribbean region (after Molnar and Sykes, 1969, simplified).

not explain the tectonic evolution during the Sub-hercynian cycle which is older than the supposed mid-Caribbean Ridge. Recently, Maresch (1974) has proposed a similar hypothesis for the southern margin of the Caribbean without, however, any clear idea of the source of the tectonic compression and neither author explains the change from a north–south Mesozoic compression to a Cenozoic east–west trend.

Other authors have thus preferred to suppose that the Caribbean crust is a fragment of the Pacific crust. Edgar et al. (1971) and Malfait and Dinkelman (1972) suppose that the Caribbean plate was the result of early crustal spreading in the eastern Pacific, from the East Pacific Ridge, which was wedged between North America and South America, probably during the Upper Jurassic. It would represent the only relict of the pre-Tertiary crust of the eastern Pacific area, the rest of the crust having disappeared by subduction under the westward drifting South American plate (Herron, 1972). This explains, according to these authors, the particular constitution of this crust, intermediate between a typical oceanic crust and a continental crust, although more similar to the former than the latter (Officer et al., 1957). A similar situation is found in the crust of the North-western Pacific, which is probably a contemporaneous and equivalent crustal zone (Den et al., 1969). Wilson (1966), for his part, has supposed that in the Carib-bean area the thrusting of the Pacific oceanic crust over the Atlantic crust occurred but this seems mechanically difficult (Edgar et al., 1971). However, none of these theories explain clearly why the Carib-bean crust is different from normal oceanic crust. Even the definition of the Caribbean crust is a subject of discussion, e.g. the Cretaceous basalts of the Venezuelan Basin are probably submarine flows alter-nating with sedimentary rocks, as in the South Peninsula of the Haitian Republic and in the Dutch Leeward Islands (Butterlin, 1960, 1973; Saunders et al., 1974; Beets and McGillavry, 1978) and not an oceanic crust.

In these hypotheses which suppose a Pacific origin for the Caribbean plate, the Caribbean region during Lower and Middle Mesozoic was a zone of tension and not a region of compression, an opinion expressed by many authors (Bullard et al., 1965; Le Pichon, 1968; Ball and Harrisson, 1969; Ball et al., 1971; Tanner, 1971; Le Pichon and Fox, 1971, etc. ...). Extension was followed by, or associated with, west–east strike-slip faulting, as a result of the eastward

motion of the Caribbean plate, relative to the adja-cent regions. This relative movement is regarded as responsible for the tectonic activity observed on the Caribbean margins (Silver, 1972). The Caribbean plate would be anticipated to have undergone subduc-tion along its northern and southern borders, under both the North and South American continents. On the other hand, along its eastern border, the Atlantic plate was subducted under it. This latter phase of sub-duction probably did not begin before the Eocene period (see the section on stratigraphy). On the oth-er hand the tectonic activity along the northern and southern margins of the Caribbean region occurred in the Upper Mesozoic.[1]

Thus such an hypothesis does not explain all the structural characters observed in the Caribbean region, particularly in Cuba, in the Cordillera del Caribe of Venezuela, the islands of Trinidad, Tobago and the Venezuelan Islands, that is, mountain ranges of geo-synclinal type with thrusting obviously connected to a strong compression, in a SSW–NNE direction in Cuba and N–S along the southern margin of the Caribbean Sea. The W–E faulting and the oblique faulting SW–NE in Cuba, NW–SE along the southern margin appear as subsidiary phenomena, subsequent to the main orogeny. It should also be noted that according to Bellizzia (1972a, b), the entire southern margin of the Caribbean Sea (Cordillera del Caribe, s.l.) constitutes a huge nappe, thrust from north to south. The relative motion of the Caribbean plate from west to east could produce the compression at the northern and southern margins of the Caribbean plate, but were they sufficiently strong to produce the compression observed.

Malfait and Dinkelman (1972) tried to overcome some of these difficulties. Following Edgar et al. (1971), they admit a Caribbean plate derived from the eastern Pacific area which formed an important protuberance in the Caribbean region, during Upper Cretaceous. According to them and to Freeland and Dietz (1971), a convergent motion occurred, between North America and South America during the Creta-ceous. It produced a subduction of the Caribbean plate under northern South America (Bell, 1972). This subduction ceased at the end of the Cretaceous as a consequence of the slackening of the convergent motion (Dietz and Holden, 1970). The Caribbean

[1] See footnote p. 90.

Cordillera were then uplifted by isostatic adjustment. Malfait and Dinkelman consider that the Caribbean plate motion was, at first, directed northeast with subduction under the Atlantic plate, west of the Beata Ridge, the situation being reversed east of the ridge. Walper (1973) expressed the same opinion. So, according to Malfait and Dinkelman, the Beata Ridge represents a transform fault. It was only at the beginning of the Oligocene that the Caribbean plate separated from the east Pacific plate, by the easterly extension of the Central American Trough. Since that time, the direction of its motion changed to west—east.

This ingeneous hypothesis raises some problems.

(1) Not all the authors agree that underthrusting occurred in the western Greater Antilles, south of the islands. Mattson (1973) considers that in the Greater Antilles the underthrusting occurred from north to south, i.e., that the Atlantic plate being more dense in this region was subducted under the Caribbean plate. This is neither the opinion of Walper (1973), nor of Khudoley and Meyerhoff (1971) who consider, for example, the Puerto Rican Trough as an ordinary graben, as is also the opinion of Bunce (1966). The serpentines which bound the "eugeosyncline" in northern Cuba seem to result from a south to north nappe, and provide no argument in support of Mattson's hypothesis, and the existence of a Benioff zone south of Cuba is also unproven, although more probable. It would explain the origin of the serpentines and the mafic volcanism in the Lower Cretaceous and the andesitic volcanism in the Upper Cretaceous by a relative movement of the Benioff zone to the south.

(2) There is nothing in the Beata Ridge structure to lead one to suppose that it corresponds to a transform fault (see Fox et al., 1968, 1970). Some authors, such as Arden (1969), consider that it was formerly connected with the Nicaraguan Rise and that it reached its present position by rift valley expansion which corresponds today to the Colombian Basin, which has separated the two positive zones.

(3) The relative motions supposed by Malfait and Dinkelman (1972) neither provide a better explanation than the earlier hypothesis nor explain the tectonic evolution of the northern margin of South America during the Mesozoic, or that of Cuba prior to the Upper Cretaceous.

To conclude, we can agree with Le Pichon (1968), that the interpretation of the Caribbean region remains one of the least convincing parts of the plate tectonics theory.

Excluding the plate theory, are there other theories which give a better account of the evolution of this region in the Mesozoic?

In fact, there exist two old and opposed theories recently reconsidered. The first that the Caribbean Sea has been oceanic since an early origin with about its present extension. This theory was supported by Schuchert (1935), and more recently proposed by McGillavry (1970) and also by Meyerhoff, in Khudoley and Meyerhoff (1971) and Meyerhoff and Meyerhoff (1972), in opposition to the global tectonics.

McGillavry considered the Caribbean Sea area as an old, stable, oceanic plate cut accross by a spur, prolonging the Hercynian orogen of Nuclear Central America and now incorporated in the nappes of Pinar del Rio Province. During the Lower Cretaceous, according to Kozary (1968) two tensional zones with marine basic volcanism appeared: one in the Greater Antilles, the second in the southern Lesser Antilles (including the Villa de Cura zone then thrust southwards). According to McGillavry they are "abortive" geosynclines without any marked characters and called "of oceanic realm".

During the Upper Cretaceous, this period of extension was followed by a break in the contact between the Caribbean oceanic plate and the thick volcanic zones, followed by a compressive period with outwards thrusting in Cuba (northwest corner of the plate) and in Venezuela (southeast corner of the plate) involving a clockwise rotational movement of the Caribbean plate. At first, subduction of the Caribbean oceanic plate under the margins occurred, and then an obduction on the evaporite plate of Cuba and on the South American continent.

In the eastern Lesser Antilles this would correspond to Cenozoic thrusting of the spreading Atlantic plate under the Caribbean plate. In this way, McGillavry combines the idea of an ancient stable Caribbean plate, which explains the Upper Mesozoic to Lower Tertiary orogenies, with the notion of a continental drift which explains the formation of the eastern Lesser Antilles. His hypothesis, however, does not explain the particular constitution of the Caribbean crust, nor does it take in account the thrusting in the interior of the arc (absent in his fig. 3) and it only touches lightly on the problem of the origin of the tectonic compressions which have produced the structures of the Antillean Arc.

According to Meyerhoff the Caribbean area was always oceanic, joining the Pacific Ocean with the Atlantic Ocean. The zone was bounded northwards by a double island arc convex towards the ocean, i.e., a Cuban segment and a segment formed by the other Greater Antillean Islands. The Cayman Ridge, the Bartlett Trough and the Nicaraguan Rise are regarded as the prolongation of the Paleozoic structures of Nuclear Central America. If we exclude the contradiction between the direct connection of Central American Paleozoic structures and the Sierra Maestra, indicated by Meyerhoff, and the fact that he admits that the basement complex of Cuba is of Paleozoic age in numerous regions, then his hypothesis only differs from that of global tectonics by the supposition that the central and eastern parts of the Caribbean Sea have always existed. This does not fundamentally change the evolution of the region, considering that he considers it since Upper Jurassic, i.e. in a period when according to Le Pichon and Fox (1971), the Caribbean space had reached practically its present extension.

Meyerhoff's conception is no more satisfactory than the global tectonics as an explanation of the particular constitution of the Caribbean crust, even including with it the evolution of "continentalization" by acid intrusions, as for example, the Aves Swell (Meyerhoff and Meyerhoff, 1972). This idea explains Late Mesozoic and Early Cenozoic orogenic cycles, by the compression of the permanent Caribbean plate, against the North and South American continents and the formation of the Aves Swell and then the eastern Lesser Antilles, by a west—east compression. But this concept does not explain the origin of the tectonic forces which produced these compressions. It only supposes that the existence of a rotational movement of unknown origin is in the adjacent continents. Meyerhoff and Meyerhoff (1972) really only highlight the difficulties of the global tectonics to explain evolution of the Caribbean region. Yet, Malfait and Dinkelman (1972) think that geophysical data, and particularly the paleomagnetic (Steinhauser et al., 1972), oppose to a static conception of the Caribbean region such as that of Meyerhoff and Meyerhoff.

The second hypothesis, also an old one, and also static is the supposition that the Caribbean is a submerged continental block subsequently "oceanized" or with the basification of its floor. This hypothesis was supported by Bucher (1947), Eardley (1954, 1962), Butterlin (1956), Chubb (1960, and in Zans et al., 1963) and recently this concept was reintroduced by Judoley and Furrazola-Bermudez (1967, 1971) and by Skvor (1969).

The hypothesis can give an explanation of the particular constitution of the Caribbean crust. According to Skvor, it corresponds to a submerged Paleozoic orogen subjected to an oceanization or basification from the mantle. This caused an isostatic disequilibrium with the adjacent zones of Greater Antilles where the sialic basement did not undergo the same evolution and resulted in tectonic motions (known under the name of "regeneration of continent"), observed during Mesozoic and Cenozoic. The hypothesis cannot explain either the basic volcanism in the "eugeosyncline" of the Greater Antilles during the Lower Cretaceous nor the abundance of serpentinized peridotites in Cuba, except by supposing the existence of parallel deep faults striking NW—SE, the presence of which remains to be proved since they are not observed in the field.

It is clear that, in fact, the three major groups of hypotheses to explain the origin and the evolution of the structures of the Caribbean region (Dengo, 1969) agree in that since the end of the Jurassic the constitution and the extension of the Caribbean Sea crust was quite similar to those presently observed in geophysical studies.

The concepts differ principally in the following aspects: (1) the origin, the constitution and the extension of the Caribbean floor before the Late Jurassic; and (2) the origin of the tectonic compression which affected the region particularly during the Late Mesozoic and Early Cenozoic.

These are the points for which the present observations give the least certain data and for that reason, it is very difficult to choose between these three concepts. The answers to these two questions will serve then to discriminate between the various hypotheses.

It seems sure nevertheless that something occurred between the Upper Mesozoic and the Tertiary, probably in the Eocene which changed completely the structural tectonic framework of the Caribbean region.

Before this event, it is probable that after the extension of the region during the Jurassic which permitted the formation of the Caribbean region, there occurred a north—south compression in the Cretaceous (Subhercynian orogenic phase) followed by an extension (deep marine basalts of the Caribbean Sea) followed again by another compression phase (Laramian orogenic phase) which preceded the east–west drift of the Caribbean plate beginning in the Eocene.

These successive movements remain to be explained; a task for future geologists.

REFERENCES

Adamovich, A. and Chejovich, V., 1964. Principales caracteristica de la geologia y de los minerales utiles de la region Noroeste de la Provincia de Oriente. *Nuestra Ind., Rev. Tecnol.*, 2 (1): 14–20.

Alencaster, G., 1977. Moluscos braquiopodos del Jurásico superior de Chiapas. *Inst. Geol. Rev. UNAM*, 1 (2): 151–166.

Anderson, P.S., Gebhard, J.A., Gomez, A., Pedreira, A. and Roseman, H.L., 1973. Contribucion al conocimiento geologico de las localidades de Apipe, Neiva, Palermo y Yaguara, departamento del Huila, Colombia. In: *Resumenes Congr. Latinoamericano Geol., 2°, Caracas, Nov. 1973*, pp. 136–137.

Anonymous, 1967. Activities of the Geological Society of Jamaica. *J. Geol. Soc. Jamaica, Geonotes*, 9: 42–57.

Anonymous, 1974. Livret guide d'excursions dans les Antilles françaises. *B.R.G.M.*, 205 pp.

Anonymous, 1976. Transactions de la VIIème Conférence géologique des Caraïbes, Guadeloupe, 1974. *Impr. Réunies, Chambéry, France*, 617 pp.

Anonymous, 1978. *Mapa geologica de la República de El Salvador, 1/100.000*. Bundesanstalt fur Geowissenschaften und Rohstoffe, Hannover.

Anonymous, 1980. *Geologia de Venezuela y de sus cuencas petroliferas*. Ediciones Foninoes, Caracas, 1031 pp.

Arden Jr., D.D., 1969. Geologic history of the Nicaraguan rise. *Trans. Gulf. Coast. Assoc. Geol. Soc.*, 19: 295–309.

Aubouin, J., 1975. De la Méditerranée aux Caraïbes: éléments d'une comparaison. *C.R. Acad. Sci. Paris*, Sér. D, 281: 215–218.

Ayala-Castañares, A., 1959. Estudio de algunos microfosiles planctonicos de las calizas del Cretacico superior de la Republica de Haïti. *Paleontol. Mex.*, 4: 41 pp.

Azema, J., 1979. Découverte d'Albien supérieur à Ammonites dans le matériel volcano-sédimentaire du "complexe de Nicoya" (province de Guanacaste, Costa Rica). *C.R. Somm. Soc. Géol. Fr.*, 3: 129–131.

Azema, J. and Tournon, J., 1980. La péninsule de Santa Elena, Costa Rica: un massif ultrabasique charrié en marge pacifique de l'Amérique Centrale. *C.R. Acad. Sci. Paris*, 290, Sér. D, pp. 9–12.

Ball, M.M., Harrison, C.G.A., Supko, P.R., Bock, W. and Maloney, N.J., 1971. Marine geophysical measurements on the Southern Boundary of the Caribbean Sea. In: T.W. Donnelly (Editor), *Caribbean Geophysical, Tectonic and Petrologic Studies. Geol. Soc. Am. Mem.*, 130, pp. 1–33.

Ball, M.M. and Harrison, C.G.A., 1969. Origin of the Gulf and Caribbean and implications regarding ocean ridge extension, migration and shear. *Trans. Gulf. Coast Assoc. Geol. Soc.*, 19: 287–294.

Banks, J.E., 1967. Geologic history of the Florida–Bahama platform. *Trans. Gulf. Coast. Assoc. Geol. Soc.*, 17: 261–264.

Barr, K.W. and Escalante, G., 1969. Contribucion al esclarecimiento de la Edad del Complejo de Nicoya, Costa Rica. *Inst. Centroamericano. Invest. Tecnol. Ind., (I.C.A.I.T.I.)*, 11: 43–47.

Barr, K.W. and Saunders, J.B., 1968. An outline of the geology of Trinidad. *Trans. Caribb. Geol. Conf., 4th, Port-of-Spain, 1965*, pp. 1–10.

Beckmann, J.P. et al., 1956. *Lexique Stratigraphique International, V. Amérique latine, Fasc. 2b. Antilles (sauf Cuba et Antilles vénézuéliennes)*. C.N.R.S., Paris, 494 pp.

Beets, D.J., 1968. Stratigraphy and sedimentary petrography of Cretaceous and lower Tertiary strata of Curacao. *Fifth Caribb. Geol. Conf., Abstr. Pap.*

Beets, D.J. and Lodder, W., 1967. Indications for the presence of ignimbrites in the Cretaceous Washikemba Formation of the isle of Bonaire, Netherlands Antilles. *K. Ned. Akad. Wetensch., Proc.*, Ser. B, 70 (1): 63–67.

Beets, D.J. and McGillavry, H., 1976. New data on the stratigraphic succession of Bonaire (Netherlands Antilles) (abstract). *Trans. VII Conf. Geol. Caraïbes, Guadeloupe, 1974*, p. 495.

Bell, J.S., 1972. Geotectonic evolution of the southern Caribbean area. In: R. Shagam et al., *Studies in Earth and Space Sciences. Geol. Soc. Am. Mem.*, 132: 369–386.

Bellizzia, G.A., 1972a. Is the entire Caribbean Mountain Belt of northern Venezuela allochthonous? In: R. Shagam et al. *Studies in Earth and Space Sciences. Geol. Soc. Am. Mem.*, 132: 363–368.

Bellizzia, G.A., 1972b. Sistema montanoso del Caribe, Borde Sur de la Placa Caribe. Es una Cordillera aloctona? *Memorias. Trans. VI Conf. Geol. Caribe, Margarita 1971*, pp. 247–258.

Bellon, H. and Tournon, J., 1978. Contribution de la géochronométrie K-Ar à l'étude du magmatisme de Costa Rica, Amérique Centrale. *Bull. Soc. Géol. Fr.*, 7ème Ser., XX (6): 955–9.

Bermudez, P.J. and Gomez, H.A., 1966. Estudio paleontologico de une seccion del Eoceno grupo Punto Carnero de la Isla Margarita, Venezuela. *Mem. Soc. Sci. Nat. La Salle*, 26 (75): 205–259.

Bermudez, P.J. and Hoffstetter, R., 1959. *Lexique stratigraphique international, Vol. V. Amérique latine, fasc. 2c. Cuba et îles adjacentes*. C.N.R.S., Paris, 140 pp.

Berryhill, H.L. Jr., 1966. Lithology and tectonic implications of Upper Cretaceous clastic volcanic rocks, Northeast Puerto Rico. *Trans. 3rd Caribb. Geol. Conf., Kingston, 1962*, pp. 34–38.

Blesch, R.R., 1966. Mapa geologico preliminar, Republica Dominicana, 1/250.000. Reconocimiento y evaluation de los recursos naturales de la Republica Dominicana. *Union Panamericana*, 2.

Bertrand, J., Delaloye, M., Fontignie, D. and Vuagnat, M., 1978. Ages (K-Ar) sur diverses ophiolites et roches associées de la cordillère Centrale du Guatemala. *Schweiz. Mineral. Petrogr. Mitt.*, 58 (3): 405–13.

Boiteau, A. and Campos, M., 1973. *Datos preliminares sobre la geologia de la parte sur de la Sierra del Purial (Oriente, Cuba)*. Preprint, 21 pp.

Boiteau, A. and Michard, A., 1976. Données nouvelles sur le socle métamorphique de Cuba. Problèmes d'application de la tectonique des plaques. *VII° Caribb. Geol. Conf., Guadeloupe*, 1974, pp. 221–226.

Boiteau, A., Michard, A. and Saliot, P., 1972a. Métamorphisme de haute pression dans le complexe ophiolitique du Purial (Oriente, Cuba). *C.R. Acad. Sci. Paris, Sér. D*, 274: 2137–2140.

Boiteau, A., Butterlin, J., Michard, A. and Saliot, P., 1972b. Le complexe ophiolitique du Purial (Oriente, Cuba) et son métamorphisme de haute pression problémes de datation et de corrélation. *C.R. Acad. Sci. Paris, Sér. D*, 275: 895–898.

Bowin, C.O., 1966. Geology of central Dominican Republic (A case history of part of an Island Arc). *Geol. Soc. Am. Mem.*, 98: 83 pp.

Bowin, C.O., 1968. Geophysical study of the Cayman Trough. *J. Geophys. Res.*, 73 (16): 5159–5173.

Bowin, C.O., 1975. The geology of Hispaniola. In: A.E.M. Nairn and G. Stehli (Editors), *The Ocean Basins and Margins, 3. The Gulf of Mexico and the Caribbean*. Plenum Press, New York, N.Y.

Bronnimann, O., 1955. Microfossils *incertae sedis* from the Upper Jurassic and Lower Cretaceous of Cuba. *Micropaleontology*, 1 (1): 28–49.

Bronniman, P. and Rigassi, D., 1963. Contribution to the geology and paleontology of the area of the city of La Habana and its surroundings. *Eclogae Geol. Helv.*, 56 (1): 193–480.

Bucher, W.H., 1947. Problems of earth deformation illustrated by the Caribbean Sea Basin. *Trans. N. Y. Acad. Sci.*, Ser 9, 2.

Bucher, W.H., 1950. Geologic-tectonic map of the United States of Venezuela (1/1.000.000). *Geol. Soc. Am. Mem.*

Bullard, E., Everett, J.E. and Smith, A.G., 1965. The fit of the continents around the Atlantic. In: *A Symposium on Continental Drift. Prog. Soc. Philos. Trans. G.B.*, 1088: 41–51.

Bunce, E.T., 1966. The Puerto Rico Trench. In: *Continental Margins and Islands Arcs. Geol. Surv. Can., Pap.*, 66 (15): 165–176.

Burkart, B., Clemons, R.E. and Crane, D.C., 1973. Mesozoic and Cenozoic stratigraphy of southeastern Guatemala. *Bull. Am. Assoc. Petrol. Geol.*, 57 (1) 63–73.

Burke, K., Coates, A.G. and Robinson, E., 1968. Geology of the Benbow Inlier and surrounding areas, Jamaica. *Trans. 4th. Carib. Geol. Conf., Port of Spain, Trinidad*, 1965, pp. 299–307.

Butterlin, J., 1956. *La constitution géologique et la structure des Antilles*. C.N.R.S. Paris, 453 pp.

Butterlin, J., 1960. Géologie générale et régionale de la République d'Haïti. *Inst. Htes Etudes Amér. Latines, Paris, Trav. Mém.*, 6: 194 pp.

Butterlin, J., 1970. Macroforaminiferos y Edad de la Formacion Punta Mosquito (Grupo Punta Carnero) de la Isla de Margarita (Venezuela). *Bol. Inf. Asoc. Venez. Geol. Min. Pet.*, 13 (10): 273–317.

Butterlin, J., 1973. Regard sur l'origine et l'évolution des unités structurales de la région des Caraibes. *Bull. Soc. Géol. Fr.*, 14 (5): 46–54.

Butterlin, J., 1977. *Géologie structurale de la région des Caraïbes et de l'Amérique de Sud*. Masson, Paris, 276 pp.

Carfantan, J.C., 1977. La cobijadura de Motozintla. Un paléoarco volcanico en Chiapas. *Inst. Geol. U.N.A.M., Rev.*, 1 (2): 133–137.

Case, J.E., 1975. *Geologic Framework of the Caribbean Region*. Nat. Sci. Found., (preprint) 24 pp.

Case, J.E. and Holcombe, T.L., 1980. *Geologic-Tectonic Map of the Caribbean Region*. U.S. Geol. Surv., Washington D.C., 1/2.500.000; 2 sheets.

Caudri, B.C.M., 1974. The larger Foraminifera of Punta Mosquito, Margarita Island, Venezuela. In: *Contributions to the Geology and Paleobiology of the Caribbean and Adjacent Areas. Verhandl. Naturf. Ges. Basel*, 84 (1): 293–318.

Chancellor, G.R.C., Reyment, R.A. and Talt, E.A., 1978. Notes on Lower Turonian ammonites from Loma El Macho, Coahuila, Mexico. *Bull. Geol. Inst. Univ. Uppsala*, 7: 85–101.

Cheilletz, A. and Lewis, J., 1976. Contribution à l'étude de la bordure méridionale du massif du N-NE d'Haïti. *Trans. VII Conf. Geol. Caraïbes. Guadeloupe, 1974*, pp. 243–247.

Cheilletz, A., Kachrillo, J.-J., Sonet, J. and Zimmermann, J.-L., 1978. Pétrographie et géochronologie de deux complexes intrusifs à porphyres cuprifères d'Haiti. Contribution à la connaissance de la province cuprifère laramienne de l'arc insulaire des Grandes Antilles. *Bull. Soc. Géol. France*, 7ème sér. 20 (6): 907–914.

Christofferson, E., 1973. Linear magnetic anomalies in the Colombia Basin, Central Caribbean Sea. *Geol. Soc. Am. Bull.*, 84 (10): 3217–3230.

Chubb, L.J., 1960. The Antillean Cretaceous geosyncline. *Trans. 2nd Caribb. Geol. Conf., Mayaguez, 1959*, pp. 17–26.

Clemons, R.E., Anderson, Th.H., Bohnenberger, O.H. and Burkhart, B., 1974. Stratigraphic nomenclature of recognized Paleozoic and Mesozoic rocks of Western Guatemala. *Am. Assoc. Petrol. Geol., Bull.*, 58 (2): 313–320.

Coates, A.G., 1968. The geology of the Cretaceous Central Inlier around Arthurs Sea Clarendon, Jamaica. *Trans. 4th Caribb. Geol. Conf., Port of Spain, Trinidad*, 1965, pp. 309–315.

Cooke, D.L. and Bailey, B.V., 1966. Marble at Mt Hibernia, Jamaica. *Trans. 3rd Geol. Conf., Kingston, 1962*, pp. 123–128.

Cooper, G.A., 1979. Tertiary and Cretaceous Brachiopods from Cuba and the Caribbean. *Smithson. Contrib. Paleobiol.*, 37: 50 pp.

De Golyer, E.L., 1918. The geology of Cuban petroleum deposits. *Bull. Am. Assoc. Petrol. Geol.*, 2: 133–167.

Den, N.W. et al., 1969. Seismic refraction measurements in the northwest Pacific basin. *J. Geophys. Res.*, 74 (6): 1421–1434.

Dengo, G., 1962. Tectonic-igneous sequence in Costa Rica. In: *Petrologic Studies, a Volume to Honor A.F. Buddington. Geol. Soc. Am.*, pp. 133–161.

Dengo, G., 1968. *Estructura geologica, historica tectonica y morfologia de America Central*. Inst. Centramericano de Investig. y Tecnologia Industrial (I.C.A.I.T.I.), 50 pp.

Dengo, G., 1969. Problems of tectonic relations between Central America and the Caribbean. *Trans. Gulf. Coast Assoc. Geol. Soc.*, 19: 311–320.

Dietz, R.S. and Holden, J.C., 1970. The breakup of Pangaea. *Sci. Am.*, 223 (4): 30–41.

Dietz, R.S., Holden, J.C. and Sproll, W.P., 1970. Evolution and subsidence of the Bahama Platform. *Geol. Soc. Am. Bull.*, 81, (7): 1915–1928.

Dietz, R.S., Holden, J.C. and Sproll, W.P., 1971. Geotectonic evolution and subsidence of the Bahama Platform: Reply. *Geol. Soc. Am. Bull.*, 82 (3): 811–814.

Dinkelman, M.G. and Brown, J.F., 1977. K-Ar geochronology and its significance to the geologic setting of La Desirade, Lesser Antilles. *Carib. Geol. Conf, 8th, Curaçao Abstracts*, pp. 38–39.

Draper, G. et al., 1976. Low-grade metamorphic belt in Jamaica and its tectonic implications. *Geol. Soc. Am. Bull.*, 87 (9): 1283–1290.

Ducloz, C. and Vuagnat, M. 1962. A propos de l'âge des serpentinites de Cuba. *Arch. Sci., Genève*, 15 (2): 309–332.

Eardley, A.J., 1954. Tectonic relations of North and South America. *Bull. Am. Assoc. Petrol. Geol.*, 38 (5): 707–773.

Eardley, A.J., 1962. *Structural Geology of North America*. 2nd ed. Harper and Row, New York, N.Y., 743 pp.

Echevarria, G. and Veliev, M., 1967. La perforacion de los pozos produndos "Francés 5" y "Fragoso": La Habana. *Minist. Ind. Rev. Tecnol.*, 5 (1): 49–54.

Edgar, N.T., Ewing, J.L. and Hennion, J., 1971. Seismic refraction and reflection in the Caribbean Sea. *Bull. Am. Assoc. Petrol. Geol.*, 55 (6): 833–870.

Ferencic, A., Del Giudice, D. and Recchi, G., 1971. Tecto-magmatic and metallogenic relationships of the Region Central Panama–Costa Rica. *Trans. 5th Carib. Geol. Conf., Geol. Bull.*, No. 5. Queens College Press, pp. 189–197.

Fink, L.K. Jr., 1968. *Geology of the Guadeloupe Region, Lesser Antilles Island Arc*. Dissert., Univ. Miami, 121 pp.

Fink, L.K. Jr., Harper, C.T., Stipp, J.J. and Nagle, F., 1972. The tectonic significance of la Desirade-possible reflect sea floor crust. (abstr.) *Mem. Trans. VI Conf. Geol. Caribe, Margarita*, 1971, p. 302.

Fischer, S.P. and Pessagno, E.A. Jr., 1965. Upper Cretaceous strata of northwestern Panama. *Bull. Am. Assoc. Petrol. Geol.*, 49 (4): 433–444.

Fox, P.J., Ruddiman, W., Heezen, B.C. and Ryan, W.B.F., 1968. Mesozoic igneous oceanic crust from the Caribbean (abstract). *Prog. Ann. Mtg., Geol. Soc. Am.*, 1968, p. 101.

Fox, P.J., Ruddiman, W.F., Ryan, W.B.F. and Heezen, B.C., 1970. The geology of the Caribbean crust, 1. Beata Ridge. *Tectonophysics*, 10 (5/6): 495–513.

Fox, P.J., Heezen, B.C., Ruddiman, W.F. and Ryan, W.B., 1971. Igneous rocks from the Beata Ridge. *Trans. 5th Carib. Geol. Conf., Bull. No. 5*, Queens College Press, p. 65.

Franco Rubio, M., 1978. Estratigrafía del Albiano–Cenoma-niano en la región de Naica, Chihuahua. *Rev. Inst. Geol. U.N.A.M.*, 2 (2): 132–149.

Freeland, G.L. and Dietz, R.S., 1971. Plate tectonic evolution of the Caribbean – Gulf of Mexico region. *Nature*, 232 (5305): 20–23.

Funnell, B.M. and Smith, A.G., 1968. Opening of the Atlantic Ocean. *Nature*, 219 (5161): 1328–1333.

Furrazola-Bermudez, G. et al., 1964. *Geologia de Cuba*. Inst. Cubano Recur. Miner., Dpto Cient. Geol., 239 pp.

Galli, C. and Schmidt-Effing, R., 1977. Estratigrafía de la cubierta sedimentaria supra-ofiolítica cretácica de Costa Rica. *Rev. Cienc. Tecn. Univ. Costa Rica*, 1: 87–96.

Glockhoff, C., 1973. Geotectonic evolution and subsidence of Bahama Platform: Discussion. *Bull. Geol. Soc. Am.*, 84 (10): 3473–3476.

Gonzalez de Juana, C. and Vignali, C.M., 1972. Rocas meta-morficas e igneas en la peninsula de Macanao, Margarita, Venezuela. *Mem. Trans. VI Conf. Geol. Caribe, Margarita*, 1971, pp. 63–68.

Goodell, H.G. and Garmann, R.K., 1969. Carbonate geochemistry of superior deep test well, Andros Island, Bahamas. *Bull. Am. Assoc. Petrol. Geol.*, 53 (3): 513–536.

Goosens, P.J., 1973. *Late Mesozoic–Early Tertiary Volcanic Island are along the northwestern South American Continental Margin*. II Congr. Latinoamercano Geol., 1973, 15 pp. (mimeogr.).

Hatten, C.W., 1957. *Geologic report on Sierra de los Organos*. La Habana, Ministeria de Industrias Archives, Unpub. Rept., 140 pp.

Hatten, C.W. and Khudoley, K.M., 1967. Principal features of Cuban geology: discussion and reply. *Bull. Am. Assoc. Petrol. Geol.*, 51 (5): 780–791.

Hatten, C.W. et al., 1958. *Geology of Central Cuba, eastern Las Villas and western Camagüey Provinces, Cuba*. La Habana, Ministerio Industrias Archives, Unpub. Rept., 250 pp.

Hekinian, R., 1971. Petrological and geochemical study of spilites and associated rocks from St John, U.S. Virgin Islands. *Geol. Soc. Am. Bull.*, 82 (3): 659–682.

Helsley, C.E., 1971. Geology of the British Virgin Islands. *Fifth Caribb. Geol. Conf., Abstr. Pap*.

Henningsen, D., 1966a. Estratigrafia y Paleogeografia de los Sedimentos del Cretacico Superior y del Terciario en el sector Sureste de Costa Rica. *Publ. Inst. Centroamericano Investig. Tecnol. Ind. Geol. (I.C.A.I.T.I.)*, 1: 53–57.

Henningsen, D., 1966b. Notes on stratigraphy and paleontology of Upper Cretaceous and Tertiary sediments in Southern Costa Rica. *Bull. Am. Assoc. Petrol. Geol.*, 50 (3): 562–566.

Henningsen, D., 1966c. Die pazifische Küstenkordillera Costa Ricas und ihre Stellung innerhalb des süd-zentral-amerikanischen Gebirges. *Geotekt. Forsch.*, 23: 3–66.

Henningsen, D. and Weyl, R., 1967. Ozeanische Kruste im Nicoya-Komplex von Costa Rica (Mittelamerika). *Geol. Rundsch.*, 57: 33–47.

Herron, E.M., 1972. Sea-floor spreading and Cenozoic history of the East-Central Pacific. *Geol. Soc. Am. Bull.*, 83 (6): 1671–1692.

Hess, H.H. and Maxwell, J.C., 1949. Geological reconnaissance of the Island of Margarita, I. *Bull. Geol. Soc. Am.*, 60: 1857–1868.

Hoffstetter, R., Dengo, G., Dixon, C.G., Meyer-Abich, H., Weyl, R., Woodring, W.P. and Zoppis Bracci, L., 1960. *Lexique Stratigraphique International, Vol. V. Amérique Latine, Fasc. 2a. Amérique Centrale*. C.N.R.S. Paris, 366 pp.

Horne, G.S., Pushkar, P. and Shafiqullah, M., 1974. Laramide plutons on the Londward continuation of the Bonacca Ridge, Northern Honduras. *Trans. 7th Carib. Geol. Conf., Guadeloupe*, 1974, 9 pp.

Horne, G.S., Clark, G.S. and Pushkar, P., 1976. Pre-cretaceous rocks of northwestern Honduras: basement terrane in Sierra de Omoa. *Bull. Am. Assoc. Petrol. Geol.*, 60 (4): 566–583.

Housa, V., 1969. Neocomian Rhyncholites from Cuba. *J. Paleontol.*, 43 (1): 119–124.

Iturralde-Vinent, M.A., 1975. Problems in application of modern tectonic hypotheses to Cuba and Caribbean region. *Bull. Am. Assoc. Petrol. Geol.*, 59 (5): 838–855.

Judoley, C.M. and Furrazola-Bermudez, G., 1967. La posicion de Cuba en la estructura geologica de la region del Caribe. *Rev. Technol.*, 5 (6): 10–19.

Judoley, C.M. and Furrazola-Bermudez, G., 1968. *Estratigrafia y Fauna del Jurasico de Cuba*. Acad. Cienc. Cuba, Dep. Geol., 126 pp.

Judoley, C.M. and Furrazola-Bermudez, G., 1971. *Geologia del Area del Caribe y de la Costa del Golfo de Mexico*. Inst. Cubano del Libro, 286 pp.

Julivert, M. et al., 1969. *Lexique Stratigraphique Internacional, Vol. V. Amérique Latine, Fasc. 4a. Colombie*. C.N.R.S., Paris, 650 pp.

Jux, U., 1961. The palynologic age of diapiric and bedded salt in the Gulf coastal province. *La. Geol. Surv. Bull.*, 38: 46 pp.

Kauffman, E.G., 1965. The Upper Cretaceous *Inoceramus* of Puerto Rico. *4th Carib. Geol. Conf.*, 20 pp.

Kesler, S.E., 1968. Igneous rocks of the Terre Neuve mountains and Massif du Nord, Haïti (abstr.) In: *Abstracts, 5th Carib. Geol. Conf. Virgin Islands, July 1–5, 1968*. Puerto Rico Univ., Mayagüez, P.R., pp. 41–43.

Kesler, S.E., 1971. Petrology of the Terre-Neuve Igneous Province, Northern Haiti. *Geol. Soc. Am. Mem.*, 130: 119–137.

Khudoley, K.M., 1967. Principal features of Cuban geology. *Bull. Am. Assoc. Petrol. Geol.*, 51 (5): 668–677.

Khudoley, K.M. and Meyerhoff, A.A., 1971. Paleogeography and geological history of Greater Antilles. *Geol. Soc. Am. Mem.*, 129: 199 pp.

Knipper, A.L., Puscharovsky, Y.M. and Puig, M., 1967. Estructura tectonica de las montanas de la Sierra de los Organos en la zona del pueblo de Viñales y situacion en ellas de los cuerpos de serpentinitas. *La Habana, Acad. Cienc. Cuba, Rev. Geol. Ano*, 1 (1): 138–146.

Kozary, M.T., 1968. Ultramafic rocks in thrust zones of Northwestern Oriente province Cuba. *Bull. Am. Assoc. Petrol. Geol.*, 52 (12): 2298–2317.

Krijnen, J.P., 1967. Pseudorbitoid foraminifera from Curaçao, I, II. *Proc. K. Ned. Akad. Wet.*, Ser. B, 70 (2): 144–164.

Krommelbein, Von K., 1963. Beitrage zur geologischen Kenntnis der Sierra de los Organos (Cuba). *Z. Dtsch. Geol. Ges.*, 114 (1): 92–120.

Kumpera, O., 1968. Contribucion a la geologia de la Sierra de Nipe. *Tec. Cien., Univ. Oriente*, 1, 23 pp.

Ladd, J., 1976. Relative motion between North and South America and Caribbean tectonics. *Trans. VII Conf. Geol. Caraibes, Guadeloupe*, 1974, pp. 63–68.

Ladd, J.W. and Watkins, J.S., 1979. Tectonic development of trench arc complexes on the northern and southern margins of the Venezuelan Basin. *Mem. Am. Assoc. Petrol. Geol.*, 29: 363–371.

Lau, W. and Rajpaulsingh, W., 1976. A structural review of Trinidad, West Indies in the light of current tectonics and wrench fault theory. *Trans. VII Conf. Geol. Caraibes, Guadeloupe*, 1974, pp. 473–483.

Le Pichon, X., 1968. Sea-floor spreading and continental drift. *J. Geophys. Res.*, 73 (12): 3661–3697.

Le Pichon, X. and Fox, P.J., 1971. Marginal offsets, fracture zones and the early opening of the North Atlantic. *J. Geophys. Res.*, 76 (26): 6294–6308.

Lidiak, E.G., 1968. Variation in calc-alkalic and tholeiitic volcanism in the Puerto Rico orogen (abst.) In. *Abstracts of Papers. 5th Carib. Geol. Conf., Virgin Islands, July 1–5, 1968, Mayaguez, Puerto-Rico*, pp. 45–46.

Lidz, B. and Nagle, F. (Editors), 1980. *Hispaniola: Tectonic Focal Point of the Northern Caribbean—Three Geologic Studies in the Dominican Republic*. Miami Geol. Soc., 97 pp., cartes.

Lloyd, J.J., 1963. Tectonic history of the South Central American orogene. In: Childs and Beebe (Editors), *Backbone of the Americas. Mem. Am. Assoc. Petrol. Geol.* 2: 88–100.

MacDonald, W.D., Doolan, B.L., Cordani, U.G., 1971. Cretaceous-Early Tertiary Metamorphic K-Ar age values from the South Caribbean. *Geol. Soc. Am. Bull.*, 82 (5): 1381–1388.

Malfait, B.T. and Dinkelman, M.G., 1972. Circum-Caribbean tectonic and igneous activity and the evolution of the Caribbean Plate. *Geol. Soc. Am. Bull.*, 83 (2): 251–272.

Maresch, W.V., 1971. Mesozoic high-P low-T metamorphism on Isla Margarita, Venezuela and its significance in the development of the Venezuelan coast ranges (abstr.). *Mem. Trans. VI Conf. Geol. Caribe, Margarita*, 1971, p. 366.

Maresch, W.V., 1972. Eclogitic—amphibolitic rocks on Margarita Island, Venezuela: a preliminary account. In: R. Shagam et al. (Editors), *Studies in Earth and Space Sciences, Geol. Soc. Am. Mem.*, 132: 429–437.

Maresch, W.V., 1974. Plate tectonics origin of the Caribbean Mountain System of northern South America: Discussion and proposal. *Geol. Soc. Am. Bull.*, 85: 669–682.

Maresch, W.V., 1975. The geology of northeastern Margarita Island, Venezuela: a contribution to the study of Caribbean plate margins. *Geol. Rundsch.*, 64 (3): 846–883.

Matsumoto, T. and Beets, D.J., 1966. A Cretaceous Ammonite from the Island of Curaçao, Netherlands, Antilles (with appendix: Stratigraphic position and age of a Cretaceous Ammonite from Curaçao, Netherlands, Antilles). *Mem. Fac. Sci., Kyushu Univ., Ser. D., Geol.*, 17 (3): 277–286.

Mattson, P.H., 1960. Geology of the Mayagüez area, Puerto Rico. *Geol. Soc. Am. Bull.*, 71 (3): 319–362.

Mattson, P.H., 1964. Petrography and structure of serpenti-

Mattson, P.H., 1960. Geology of the Mayagüez area, Puerto Rico. *Geol. Soc. Am. Bull.*, 71 (3): 319–362.

Mattson, P.H., 1964. Petrography and structure of serpentinite from Mayagüez, Puerto Rico. In: C.A. Burk, *A Study of Serpentinite. Nat. Acad. Sci.-Nat. Res. Counc. Publ.*, 1188: 7–24.

Mattson, P.H., 1966. Geological characteristics of Puerto Rico. In: *Continental Margins and Islands Arcs. Geol. Surv. Can., Pap.*, 66-15: 124–138.

Mattson, P.H., 1966. Unconformity between Cretaceous and Eocene rocks in Central Puerto Rico. *Trans. 3rd Carib. Geol. Conf., 1962, Geol. Surv. Dept., Kingston*, pp. 49–51.

Mattson, P.H., 1968. Basements Rocks, Vulcanism and Deformation in the Greater Antilles (abstr.). *Prof. Ann. Mtg., 1968, Geol. Soc. Am.*, p. 191.

Mattson, P.H., 1968. Serpentinite, gneiss and bedded chert in Puerto Rico (abstr.). *Abstr. Pap., 5th Carib. Geol. Conf., St. Thomas, Virgin Island*.

Mattson, P.H., 1973. Middle Cretaceous nappe structures in Puerto Rican ophiolites and their relations to the tectonic history of the Greater Antilles. *Geol. Soc. Am. Bull.*, 84 (1): 21–38.

Mattson, P.H., 1976. Mesozoic and Tertiary lithospheric plate interactions in the Northern Caribbean. *Trans. VII Conf. Geol. Caraibes, Guadeloupe*, 1974, pp. 79–85.

Mattson, P.H. and Pessagno, E.A., 1979. Jurassic and Early Cretaceous radiolarians in Puerto-Rican ophiolite–Tectonic implication. *Geology*, 7 (9): 440–444.

Maurrasse, F. et al., 1979. Upraised Caribbean sea floor below acoustic reflecto B at the Southern Peninsula of Haiti. *Geol. Mijnb.*, 58 (1): 71–83.

Maxwell, J.C., 1948. Geology of Tobago, British West Indies. *Geol. Soc. Am. Bull.*, 59: 801–854.

McBirney, A.R., 1963. Geology of a part of the Central Guatemalan Cordillera. *Univ. Calif. Publ., Geol. Sci. Berkeley*, 38 (4): 177–242.

McBirney, A.R. and Bass, M.N., 1969. Structural relations of the pre-Mesozoic rocks of Northern Central America. In: *Tectonic Relations of Northern Central America and the Western Caribbean, the Bonacca Expedition. Geol. Soc. Am. Mem.*, 11: 269–280.

McIntyre, D.H., Aaron, J.M. and Tobisch, O.T., 1970. Cretaceous and Lower Tertiary stratigraphy in northwestern Puerto Rico. *U.S. Geol. Surv. Bull.*, Part D, 1294: 1–16.

Mercier de Lepinay, B., Labesse, B., Sigal, J. and Vila, J.M., 1979. Sédimentation chaotique et tectonique tangentielle maestrichtiennes dans la presqu'ile du Sud d'Haiti (Ile d'Hispaniola, Grandes Antilles). *C.R. Acad. Sci., Paris*, 289, Sér. D, pp. 887–890.

Metz, H.L., 1968. Stratigraphy and geologic history of extreme northeastern serrania del Interior, state of Sucre, Venezuela. *Trans. 4th Carib. Geol. Conf. Port-of-Spain*, 1965, pp. 275–292.

Meyerhoff, A.A., 1964. Las formaciones geologicas de Cuba. *Int. Geol. Rev.*, 6 (1): 149–156.

Meyerhoff, A.A. and Hatten, C.W., 1968. Diapiric structures in central Cuba. In: J. Braustein and G.D. Brien. *Diapirism and Diapirs. Mem. Am. Assoc. Petrol. Geol.*, 8: 315–357.

Meyerhoff, A.A. and Meyerhoff, H.A., 1972. Continental drift, IV. The Caribbean "plate". *J. Geol.*, 80 (1): 34–60.

Meyerhoff, H.A., 1933. Geology of Puerto Rico. *Univ. P.R. Monogr.*, Ser. B., 1: 306 pp.

Millan, G., 1974. El complejo cristalino mesozoico de Isla de Pinos. Su metamorfismo. *Acad. Cienc. Cuba, Ser. Geol.*, 23: 1–16.

Millan, G. and Somin, M., 1975. El metamorfismo del complejo vulcanogeno cretacico en los alrededores del Escambray. *Acad. Cienc. Cuba, Ser. Geol.*, 18: 1–8.

Mills, R.A. and Hugh, K.E., 1968. Reconnaissance geologic map of Mosquitia region, Honduras and Nicaragua Caribbean coast. *5th Carib. Geol. Conf., St. Thomas, Virgin Islands*, 2 pp.

Mills, R.A. and Hugh, K.E., 1974. Reconnaissance geologic map of Mosquitia region, Honduras and Nacaragua caribbean coast. *Bull. Am. Assoc. Petrol. Geol.*, 58 (2): 189–207.

Mills, R.A., Hugh, K.E., Feray, D.E. and Swolfs, H.C., 1967. Mesozoic stratigraphy of Honduras. *Bull. Am. Assoc. Petrol. Geol.*, 51 (9): 1711–1786.

Molnar, P. and Sykes, L.R., 1969. Tectonics of the Caribbean and Middle America Regions from focal mechanisms and seismicity. *Geol. Soc. Am., Bull.*, 80 (9): 1639–1684.

Moticska, P., 1972. Geologia del archipielago de los Frailes. *Mem. Trans. VI Conf. Geol. Caribe, Margarita*, 1971, pp. 69–73.

Nagle, F., 1966. *Geology of the Puerto Plata area, Dominican Republic*. Nat. Sci. Found. Grant G 14217, Dept. Geol., 171 pp.

Nagle, F., 1971. Caribbean geology. *Bull. Marine Sci.*, 21 (2): 375–439.

Nagle, F., 1979. Geology of the Puerto Plata area, Dominican Republic. In: B. Lidz and F. Nagle (Editors), *Hispaniola, tectonic focal point of the northern Caribbean*. Miami Geol. Soc., pp. 1–28.

Officer, C.B., Ewing, J.I., Edwards, R.S. and Johnson, H.R., 1957. Geophysical investigations in the eastern Caribbean: Venezuelan Basin, Antilles island arc and Puerto Rico trench. *Geol. Soc. Am. Bull.*, 68 (3): 359–378.

Olmeta, A.M., 1968. Determinacion de edades radiometricas en rocas de Venezuela y su procedimiento por el metodo K/Ar. *Bol. Geol. Venez.*, 10 (19): 339–344.

Palmer, R.H., 1945. Outline of the geology of Cuba. *J. Geol.*, 53 (1): 1–34.

Palmer, H.C., 1963. *Geology of the Moncion–Jarabacoa Area, Dominican Republic*. Dept. Geol., Princeton University, 256 pp.

Pardo, G., 1969. Geologia de la parte central de la isla de Cuba. *Congr. Geol. Venez., 4th, Caracas, 1969*.

Pessagno, E.A. Jr., 1960. Stratigraphy and micropaleontology of the Cretaceous and Lower Tertiary of Puerto Rico. *Micropaleontology*, 6 (1): 87–110.

Pessagno, E.A. Jr., 1962. The Upper Cretaceous stratigraphy and micropaleontology of south central Puerto Rico. *Micropaleontology*, 8 (3): 349–368.

Pessagno, E.A. Jr., 1965. Eaglefordian (Cenomanian–Turonian) stratigraphy in Mexico, Texas and the West Indies. *4th Carib. Geol. Conf. Port-of-Spain*, Preprint, 19 pp.

Pessagno, E.A. Jr., 1966. Upper Cretaceous Radiolaria from Puerto Rico. *Trans. 3rd. Caribb. Geol. Conf. 1962, Geol. Surv. Dept., Kingston*, pp. 160–162.

Pichler, H., Stibane, F.R. and Weyl, R., 1974. Basischer Magmatism und Krustenbau im südlichen Mittelamerika, Kolumbien und Ecuador. *Neues Jahrb. Geol. Palaontol. Monatsh.*, 2: 102–126.

Pinet, P.R., 1975. Structural evolution of the Honduras continental margin and the sea floor south of the western Cayman Trough. *Geol. Soc. Am. Bull.*, 86 (6): 830–838.

Pitman, N.C. and Talwani, M., 1972. Sea-floor spreading in the North Atlantic. *Geol. Soc. Am. Bull.*, 83: 619–646.

Potter, H., 1976. Type sections of the Maraval, Maracas and Chancellor formations in the Caribbean group of the Northern range of Trinidad. *Trans. VII Conf. Geol. Caraibes, Guadeloupe*, 1974, pp. 505–527.

Priem, H.N.A., 1967. *Beknopt Verslag over de Werkzaamheden in het Jaar van September 1966 tot September 1967*. Z.W.O. Laboratorium voor Isotopengeologie, Amsterdam.

Priem, H.N.A., Boelrijk, N.A.I.M., Verschure, R.H., Hebeda, E.H. and Lagaay, R.A., 1966. Isotopic age of the Quartz-Diorite Batholith on the Island of Aruba, Netherlands Antilles. *Geol. Mijnb.*, 45 (6): 188–190.

Pyle, T.E., Antoine, J.W., Fahlquist, D.A. and Bryant, W.R., 1969. Magnetic anomalies in Straits of Florida. *Bull. Am. Assoc. Petrol. Geol.*, 53 (12): 2501–1505.

Rangin, C., 1978. Speculative model of Mesozoic geodynamics, Central Baja California to Northeastern Sonora. In: Howell, McDougall (Editors), Mesozoic Symposium. pp. 85–106, 10 fig.

Reeside, J.B. Jr., 1947. Upper Cretaceous Ammonites from Haiti. *U.S. Geol. Surv. Prof. Pap.*, 214-A: 1–11.

Richards, H.G., 1963. Stratigraphy of earliest Mesozoic sediments in southeastern Mexico and western Guatemala. *Bull. Am. Assoc. Petrol. Geol.*, 47 (10: 1861–1870.

Rigassi-Studer, D., 1963. Sur la géologie de la Sierra de Los Organos, Cuba. *Arch. Sci., Genève*, 16 (2): 339–350.

Roberts, R.J. and Irving, E.M., 1957. Mineral deposits of Central America (with a section on manganese deposits of Panama, by F.S. Simons). *U.S. Geol. Surv. Bull.*, 1034: 205 pp.

Roobol, M.J., 1972. The volcanic geology of Jamaica. *Mem. Trans. VI Conf. Geol. Caribe, Margarita*, 1971, pp. 100–107.

Salas, G.P., 1955. Posibles futuras provincias petroligeras en Mexico. *Bol. Asoc. Mex. Geol. Petrol.*, VII (3-4): 1–34.

Sapper, K., 1937. Mittelamerika (Unter Mitarbeit von W. Staub). *Handbuch Region. Geol.*, Bern, 8, 160 pp.

Saunders, J.B., 1972. Recent paleontological results from the Northern Range of Trinidad. *Mem. Trans. VI Conf. Geol. Caribe, Margarita*, 1971, pp. 455–460.

Saunders, J.B., Edgar, N.T., Donnelly, T.W. and Hay, W.W., 1974. Cruise Synthesis. Leg 15. *Initial Reports of the Deep Sea Drilling Project*. U.S. Govt. Print. Office, Washington, XV, pp. 1077–1111.

Schmidt-Effing, R., 1979. Alter und Genese des Nicoya-Komplexes, einer ozeanischen Paläokruste (Oberjura bis Eozän) im südlichen Zentralamerika. *Geol. Rundsch.*, 68 (2): 457–494.

Schuchert, C., 1935. *Historical Geology of the Antillean Caribbean Region*. Wiley, New York, N.Y., 811 pp.

Schwartz, D.P., 1971. Petrology and structural geology along the Motagua fault zone, Guatemala (abstr.) *VI Conf. Geol. Caribe, Margarita, Abstr.*, 1971, pp. 5–6.

Sheridan, R.E., 1971. Geotectonic evolution and subsidence of Bahama Platform: Discussion. *Geol. Soc. Am. Bull.*, 82 (3): 807–810.

Silver, E.A., 1972. Geophysical study of the Venezuelan Borderland (abstr.). *Abstr. Progr. Geol. Soc. Am.*, 4 (3): 237.

Skvor, V., 1969. The Caribbean area: A case of destruction and regeneration of continent. *Geol. Soc. Am. Bull.*, 80 (6): 961–968.

Sohl, N.F., 1967. On the Trechmann-Chubb Controversy regarding the age of the "Carbonaceous Shale" of Jamaica. *J. Geol. Soc. Jamaica, Geonotes*, 9: 1–10.

Somin, M.L. and Millan, G., 1973a. *Generalidades sobre la geologia de los complejos metamorficos de Cuba*. Preprint, 2 p.

Somin, M.L. and Millan, G., 1973b. *Sobre del descubrimiento de microfauna mesozoica dentro de las secuencias metamorficas de Isla de Pinos*. Preprint, 2 p.

Speed, R.C., Gerhard, L.C. and McKee, E.H., 1979. Ages of deposition, deformation and intrusion of Cretaceous rocks, eastern St Croix, Virgin Islands. *Bull. Geol. Soc. Am.*, 90 (7): 629–632.

Steinhauser, P., Vincenz, S.A. and Dasgupta, S.N., 1972. Paleomagnetism of some Lower Cretaceous Lavas on Jamaica. *Trans. Am. Geophys. Union (E.O.S.)*, 53 (4): 356–357.

Stephan, J.-F., Beck, C.M., Bellizzia, A. and Blanchet, R., 1980. La chaîne caraibe du Pacifique à l'Atlantique. *26ème Congr. Géol. Intern., Paris, 1980. Colloque C 5: Géologie des chaînes issues de la Téthys*, pp. 38–59.

Tanner, W.F., 1971. Growth rates of Venezuelan beach ridges. *Sed. Geol.*, 6: 215–220.

Tardy, M., 1978. Essai sur la reconstitution de l'évolution paléogéographique et structurale de la partie septentrionale du Mexique au cours du Mésozoique et du Cénozoique. *Bull. Soc. Géol. Fr.*, 7ème sér., XIX (6): 1297–1308.

Tardy, M., 1980. *Contribution à l'étude géologique de la Sierra Madre Orientale du Mexique*. Thèse Doctorat d'Etat, Univ. Pierre et Marie Curie, Paris, 459 pp. (miméographiée).

Taylor, G.C., 1960. Geologia de la isla de Margarita, Venezuela. In: *Mem. Tercer Congr. Geol. Venez. Bol. Geol. Caracas Publ. Espec.*, 3, t.II, pp. 838–893.

Tikhomirov, I.N. and Izquierdo, J.M., 1968. The magmatic formations of Cuba (abstr.). *Int. Geol. Congr., 23rd, Rep., Czechoslovakia*, p. 67.

Tobisch, O.T., 1968. Gneissic amphibolite at Las Palmas, Puerto Rico, and its significance in the early history of the Greater Antilles Island Arc. *Geol. Soc. Am. Bull.*, 79 (5): 557–574.

Uchupi, E., 1973. Eastern Yucatan Continental Margin and western Caribbean Tectonics. *Bull Am. Assoc. Petrol. Geol.*, 57 (6): 1075–1085.

Uchupi, E., Milliman, J.D., Luyendyk, B.P., Bowin, C.O. and Emery, K.O., 1971. Structure and origin of Southeastern

Bahamas. *Bull. Am. Assoc. Petrol. Geol.*, 55 (5): 687–704.

Vachrameev, V.A., 1966. Primer descubrimiento de flora del Jurasico en Cuba. *Rev. Tecnol., La Habana, Cuba*, 4 (2): 22–25.

Versey, H.R., 1966. Annual report of the Geological Survey Department for the year ended 31st March 1966. *Geol. Surv. Jamaica*, 1966, 17 pp.

Villa, J.M., Bourgois, J., Llinas, R. and Tavares, I., 1980. Tectoniques superposées à Hispaniola (Grandes Antilles) (résumé). *Réunion annuelle Sci. Terre, Marseille*, 1980, 1 p.

Viniegra, O.F., 1971. Age and evolution of salt basins of Southeastern Mexico. *Bull. Am. Assoc. Petrol. Geol.*, 46 (4): 425–456.

Vinson, G.L., 1962. Upper Cretaceous and Tertiary stratigraphy of Guatemala. *Bull. Am. Assoc. Petrol. Geol.*, 46 (4): 425–456.

Vinson, G.L. and Brineman, J.H., 1963. Nuclear Central America, Hub of antillean transverse belt. In: Childs and Beebe et al. (Editors), *Backbone of the Americas. Mem. Am. Assoc. Petrol. Geol.*, 2: 101–112.

Walper, J.L., 1973. Middle Cretaceous nappe structures in Puerto Rican ophiolites and their relation to the Tectonic History of the Greater Antilles: Discussion. *Geol. Soc. Am. Bull.*, 84 (11): 3755–3756.

Wassall, H., 1957. The relationship of oil and serpentine in Cuba. *Congr. Geol. Int. XX Sess. Mexico*, 1956, sect. 3, pp. 65–77.

Weyl, R., 1969. Magmatische Förderphasen und Gesteinschemismus in Costa Rica (Mittelamerika) *Neues Jahrb. Geol. Palaontol. Monatsh.*, 7: 423–446.

Weyl, R., 1980. *Geology of Central America* (2nd, completely revised edition) Borntraeger, Berlin, 371 pp.

Williams, M.D., 1975. Emplacement of Sierra de Santa Cruz, Eastern Guatemala. *Bull. Am. Assoc. Petrol. Geol.*, 59 (7): 1211–1216.

Wilson, H.H., 1974. Cretaceous sedimentation and orogeny in Nuclear Central America. *Bull. Am. Assoc. Petrol. Geol.*, 58 (7): 1348–1396.

Wilson, J.T., 1966. Are the structures of the Caribbean and Scotia arc region analogous to ice rafting? *Earth Planet. Sci. Lett.*, 1: 335–338.

Woodring, W.P., Brown, J.S. and Burbank, W.S., 1924. *Geology of the Republic of Haiti*. Dept. Public Works, Port-au-Prince, 631 pp.

Zans, V.A., Chubb, L.J., Versey, H.R., Williams, J.B., Robinson, E. and Cooke, D.L., 1963. *Synopsis of the Geology of Jamaica.* (An explanation of the 1958 provisional geological map of Jamaica.) Geol. Surv. Dept. Jamaica., 72 pp.

Chapter 5

THE NORTHERN ANDES

W.V. MARESCH

INTRODUCTION

The Northern Andes area in this chapter is taken to represent that part of the Andean System of South America that extends from the Huancabamba deflection near the Peru–Ecuador border along the western, northwestern, and northern edges of the Guayana Shield to include the Caribbean Mountain System of Venezuela. The mountain ranges of Trinidad, a geological and structural extension of the Caribbean Mountain System, have been described by Butterlin elsewhere in this volume (Chapter 4).

At first sight, this area presents itself as a heterogeneous complex of anastomosing structural elements that point to a complicated orogenic history of the region. It is important to note, therefore, that our interest in this volume lies in the geology and geological history of the Mesozoic Era only; for, although the present geography and structural classification is used for orientation (Fig. 1), it must be kept well in mind that the breath-taking grandeur of the Andes and the intricate structural relationships of northern Venezuela are due mainly to tectonic processes that did not begin to leave their mark until the closing stages of the Mesozoic Era, and in many cases much, much later.

It is for this reason that the maps presented in this chapter may not be regarded at face value as paleogeographic representations. They attempt to show for the most part only the depositional environment of Mesozoic rocks as they are distributed today. Our knowledge of the paleotectonic history of this region has evolved to such an extent that a majority of investigators involved in the area suspect that major paleotectonic movements have rearranged the gross structural elements by dismemberment and/or collage. Since the exact nature of these movements is still unclear, true paleogeographic representations are still equivocal.

It is not possible to present all points of view in a short summary such as this. An attempt is made to reflect the most recent majority opinion – this by no means indicates that the final solution to the various problems has been found. Generally only the most recent literature, which is itself based on fundamental older work, is cited, but this should suffice as a starting point for the more deeply interested reader. Pertinent regional summaries of much more detailed scope than is possible here have recently been presented by Bell (1974), Campbell (1974b, c), and Shagam (1975). Relevant stratigraphic lexica are those by Hoffstetter (1956), Julivert (1968), and Venezuela (1970). This chapter was submitted in its present form in December 1978.

REGIONAL SETTING AND GEOLOGICAL PROVINCES

Although the Northern Andes region nominally begins at the Huancabamba deflection (Gerth, 1955; Gansser, 1973), this description concerns itself only with latitudes north of Guayaquil.

In Ecuador a distinct morphological and structural tripartite division is possible into the Litoral, the Sierra, and the Oriente region (Fig. 1). The Litoral is characterized by a broad, low platform known as the Daule Platform or Ecuadorian Coastal Depression. It separates low hills and structural arches directly along the coast from the dominating gigantic wall of the Sierra, which towers over 4,000 m above the lowlands to the east and west. The Sierra, topped by inactive and sparsely active volcanoes (e.g., Chimborazo, elev. 6,267 m), is divided into two parallel chains, the Cordillera Occidental to the west, and the

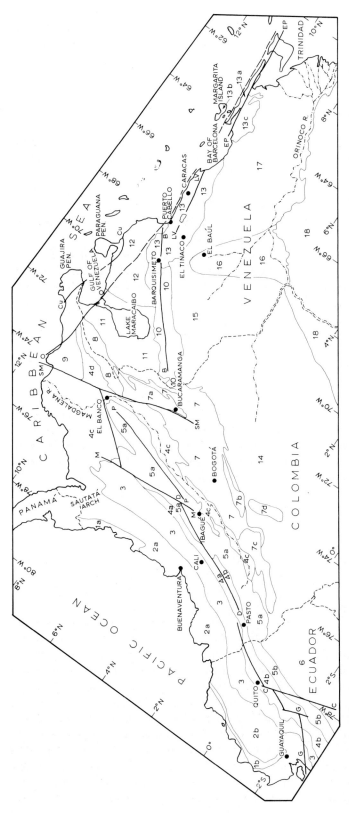

Fig. 1. Index map with structural classification of Northern Andes employed in this chapter (after Venezuela, 1970; Sauer, 1971; Bell, 1972; Campbell, 1974a).

1 = Pacific Coast Range; *a* = Serranía de Baudó, and *b* = Coastal Hills; *2a* = Bolívar Geosyncline, *2b* = Daule Platform or Coastal Depression of Ecuador; *3* = Cordillera Occidental; *4* = Intramontane Depressions, *a* = Cauca Basin, *b* = Quito-Cuenca Basin, *c* = Magdalena Valley, and *d* = César Basin; *5* = North Andean Cardinal Cordillera, *a* = Cordillera Central of Colombia, and *b* = Cordillera Real of Ecuador; *6* = Oriente; *7* = Cordillera Oriental, *a* = Santander Massif, *b* = Quetame Massif, *c* = Garzón Massif, and *d* = Macarena Block; *8* = Sierra de Perijá; *9* = Sierra Nevada de Santa Marta; *10* = Mérida (or Venezuelan) Andes; *11* = Maracaibo Basin; *12* = Falcón Basin; *13* = Caribbean Mountain System, *a* = Araya–Paria, *b* = Margarita Island, and *c* = Eastern Interior Range (= *17a*); *14* = Llanos Basin; *15* = Barinas-Apure Basin; *16* = El Baúl Arch; *17* = East Venezuela Basin; *18* = Guayana Shield.

Transcurrent faults (D = dextral; S = sinistral): *B* = Boconó (D); *O* = Oca (D); *EP* = El Pilar (D); *LV* = La Victoria (D); *Cu* = Cuisa (D); *SM* = Santa Marta (S); *P* = Palestina (D); *M* = Medellín (S); *D* = Dolores (or Romeral) (D); *C* = Cotopaxi-Baños (S); *G* = Guayaquil (D).

Cordillera Real (or Cordillera Central) to the east. The two are separated by a broad, north–south graben structure of alternating basins (hoyas) and cross-ridges (nudos) in trellis-like pattern. Another structural break separates the Cordillera Real from the structurally simple lowlands of the Oriente region to the east. Several apparently simple uplifts along the foot of the Cordillera Real in this area might be viewed morphologically as precursors of the Cordillera Oriental in Colombia. In Ecuador, the Andean System is about 100 km wide.

At the Ecuador–Colombia border, the north–south Cordillera Occidental and Cordillera Real join morphologically at the "Cumbre" or "Nudo de Pasto", and then diverge as the Cordillera Occidental and Cordillera Central of Colombia, respectively, with a new trend of N30°–45°E (Fig. 1). Along the Caribbean coast, the Cordillera Occidental splits into two diverging, subdued branches, while the Cordillera Central dies out morphologically and gives way to a complex of fault-bounded basins and uplifts. The Pacific Coast Ranges (or Serranía de Baudó), which extend into Panamá from northwestern Colombia, are, according to Campbell (1974a), a continuation of the coastal hills and uplifts west of the Daule Platform in Ecuador. The latter is represented in Colombia by the Bolívar Geosyncline (Nygren, 1950) along the Atrato Valley. As in Ecuador, the Cordilleras Occidental and Central of Colombia are separated by a major tectonic depression which forms the Cauca Valley, and is the site, for a great part of its length, of the Dolores-Romeral and Cauca Faults (Fig. 1).

At the latitude of the "Nudo de Pasto", a third major range, the Cordillera Oriental with elevations up to 5,500 m, diverges to the northeast thus that the maximum width of the orogenic belt in Colombia reaches 450 km. The Magdalena Valley, a huge topographic and structural basin, separates the Cordillera Oriental from the Cordillera Central.

In northeastern Colombia and western Venezuela (Fig. 1), the trend of the main orographic elements arcs to the northeast to connect with the east–west tectonic belts of north-central and northeast Venezuela and Trinidad. This complicated hinge zone, composed mainly of isolated massifs and ranges cropping out within younger Tertiary basins and fold-belts, has led to a variety of interpretations regarding the correlation of structural elements between the northeast–southwest Colombian Andes, and the east–west Caribbean Mountain System of Venezuela (e.g., Gerth,

1955; Alvarez, 1971; Butterlin, 1972; Gansser, 1973; Campbell, 1974a). Despite the many tantalizing similarities stressed by these authors in the orogenic system of the Pacific and Caribbean margins, there are, however, also clear differences that may be based on distinctly different orogenic modes (e.g., Mencher, 1963; Maresch, 1974, 1976).

The northernmost massifs of this "hinge" zone are the Sierra Nevada de Santa Marta, and the low hills of the Guajira and Paraguaná Peninsulas. The Sierra Nevada de Santa Marta is a spectacular, fault-bounded uplift of 5,800 m maximum elevation, separated from the Cordillera Central to the southwest and the Guajira Peninsula to the northeast by major Tertiary to Recent basins.

Near Bucaramanga, the Cordillera Oriental splits into two branches that envelop the Maracaibo Basin of western Venezuela. The Western branch is known as the Sierra de Perijá north of 9°N latitude, and reaches elevations of 2,200–2,600 m. It is, like the Sierra Nevada de Santa Marta, abruptly terminated by a major east–west fault (the Oca Fault). The Sierra de Perijá forms the divide between the Maracaibo Basin and the Magdalena Valley, and represents the international boundary between Colombia and Venezuela. In contrast to this range, the northeasterly striking Mérida (or Venezuelan) Andes represent a chain distinct from the Cordillera Oriental with elevations up to 5,000 m. The junction with the latter is an intricate network of northeast and northwest, criss-crossing structures (Gansser, 1973). Between the Sierra de Perijá and Mérida Andes lie the low, featureless, oil-rich Maracaibo Basin, and the subdued hills and valleys of the Falcón Basin.

The Mérida Andes give way, beyond Barquisimeto (Fig. 1), to the structurally complex Caribbean Mountain System of Venezuela, which follows, interrupted only by the Bay of Barcelona, the Caribbean coast as far as Trinidad. The Caribbean Mountain System may be divided into a northern Coast Range (Cordillera de la Costa) with elevations up to 2,765 m, and a southern, more subdued, but geomorphologically more complex Interior Range (Serranía del Interior). The large Island of Margarita may be reckoned as part of the Cordillera de la Costa. For the Caribbean Mountain System, specific reference is made to the terminology as defined by Venezuela (1970, pp. 11–13), for many authors have informally used different names for this mountain system, often employing the term Cordillera de la Costa, or translations

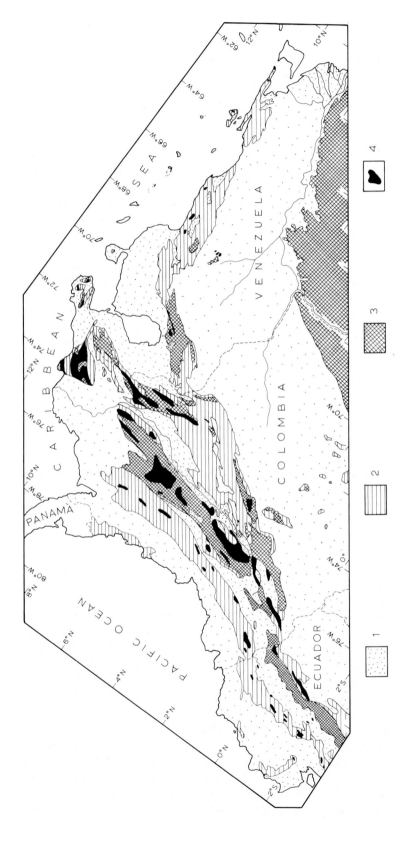

Fig. 2. Simplified geological sketch map of the Northern Andes to indicate the areal extent of Mesozoic rocks. Mainly after Blondel et al. (1964), with modifications from Case et al. (1971), Sauer (1971), and Tschanz et al. (1974). *1* = Post-Mesozoic; *2* = Mesozoic, may include Permian and Paleogene rocks in some regions; *3* = pre-Mesozoic basement; *4* = Mesozoic to Lower Tertiary acid intrusives, probably includes some intrusives of Upper Paleozoic age.

thereof, to designate this entire orogenic belt.

Broad, structurally simple lowlands separate all these mountain ranges from the Guayana Shield to the south, southeast, and east, respectively.

STRATIGRAPHY

Pre-Mesozoic basement

Pre-Mesozoic rocks have been reported from the Cordillera Occidental, Cordillera Real, and the Oriente region of Ecuador (Fig. 2). Unfossiliferous schists and gneisses of variable metamorphic grade make up most of the Cordillera Real. Unmetamorphosed, in part fossiliferous, sediments are known from the Oriente region. All of these rock series are generally considered to be predominantly of pre-Mesozoic age (Sauer, 1971; Faucher and Savoyat, 1973; Herbert, 1977), although contact relations indicate a possible Cretaceous age for some of the volcanics exposed in the inter-Andean depression. A possible Lower Mesozoic age for the Macuma Fm. of the Oriente region has been mentioned (Herbert, 1977, p. 153). An entirely different point of view is taken by Feininger (1975), who considers the major part of the metamorphic rocks of the Cordillera Real to be a lateral and contemporaneous equivalent of well-dated Cretaceous strata in the Oriente region to the east. Because the implications for the age of metamorphism in the Cordillera Real and therefore the paleotectonic history of the Ecuadorian Andes are enormous; new data on this problem should be expected in the near future (T. Feininger, personal communication, 1979).

Dark pelites similar to the Paleozoic sediments of the Oriente region have been described from the Cordillera Occidental, and although not fossiliferous, have been interpreted to be outcrops of the Paleozoic basement underlying the Cordillera Occidental and Daule Platform (Sauer, 1971). Recent geophysical investigations in western Colombia (Case et al., 1971; Meissner et al., 1976) make it appear more likely, however, that the nature of the crust west of the Cordillera Real is actually oceanic (and Mesozoic in age), rather than continental in character.

Pre-Mesozoic rocks are found in a number of locations in Colombia, but are restricted to the Cordillera Central and regions to the east. The Cordillera Central is a framework of several belts of pre-Mesozoic crystalline schist, gneiss, marble, amphibolite, migmatite, and granulite (Botero, 1963; Radelli, 1967; Feininger, 1970; Göbel, 1978), as well as younger batholithic intrusions. The lower-grade rocks tend to predominate. Ages are thought to vary from Devonian (Stibane, 1967, 1968) to Precambrian (Case et al., 1971).

In various massifs of the Cordillera Oriental, as well as the "hinge zone" of northeastern Colombia and western Venezuela, pre-Mesozoic rocks ranging in age from Precambrian to Permian have been reported. Precambrian schists and gneisses are known from the Garzón Massif (Stibane, 1968), the Macarena Block (Gansser, 1973), and the Santander Massif (Goldsmith et al., 1971) of the Cordillera Oriental. They are further exposed in the Sierra Nevada de Santa Marta (Gansser, 1955; MacDonald and Hurley, 1969; Tschanz et al., 1974), the Sierra de Perijá (Martín B., 1968), and the Mérida Andes (Venezuela, 1970, pp. 305—306). The subsurface continuity between these exposures and the Guayana Shield appears likely (Gansser, 1973).

Tschanz et al. (1974) stress the fact that granulite of Guayana Shield type is found in the Andean System only in the Sierra Nevada de Santa Marta, the Cordillera Central, and the Garzón Massif of the Cordillera Oriental. These granulites are dated or inferred to be older than 1,000 m.y. All other exposures of Precambrian rocks between these occurrences and the Guayana Shield are less than 1,000 m.y. old. Tschanz et al. (1974) surmise that these granulite exposures are blocks tectonically isolated from the main Guayana Shield by younger tectonic events.

Paleozoic strata of non-metamorphic to low-grade type, ranging in age from Cambrian to Permian (in contrast to the lack of Upper Paleozoic sediments in the Cordillera Central), are known from the various massifs of the Cordillera Oriental (e.g., Stibane, 1967, 1968). Other Paleozoic strata have been reported from the Sierra de Perijá (Miller, 1962; Venezuela, 1970), and Mérida Andes (e.g., Shagam, 1972a, b). Some of the more strongly metamorphosed rocks and associated acid plutons exposed in the Sierra Nevada de Santa Marta (Tschanz et al., 1974) and the Guajira (Alvarez, 1967, 1971) and Paraguaná Peninsulas (Feo-Codecido, 1971) may also be of Paleozoic age.

East of Barquisimeto, the problem of defining a pre-Mesozoic basement is more acute, although it is generally inferred to extend as far north as the Caribbean coast in the Caribbean Mountain System (e.g., Menéndez, 1967; Bell, 1971, 1974; Maresch, 1974).

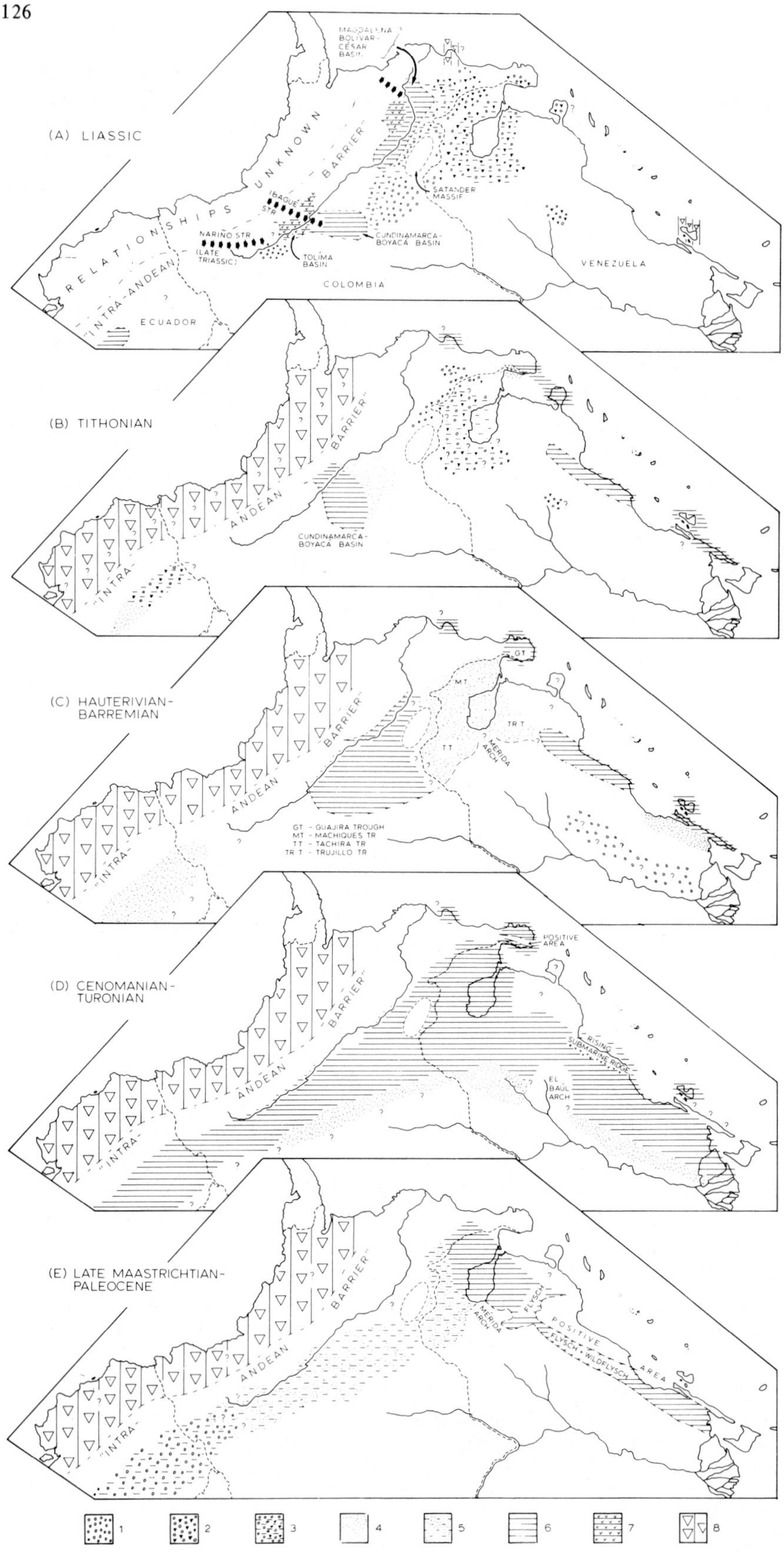

(A) LIASSIC
MAGDALENA BOLIVAR-CESAR BASIN
"INTRA-ANDEAN" BARRIER
RELATIONSHIPS UNKNOWN
IBAGUE STR
NARIÑO STR (LATE TRIASSIC)
TOLIMA BASIN
CUNDINAMARCA-BOYACA BASIN
SATANDER MASSIF
COLOMBIA
VENEZUELA
ECUADOR
(B) TITHONIAN
CUNDINAMARCA-BOYACA BASIN
(C) HAUTERIVIAN-BARREMIAN
GT - GUAJIRA TROUGH
MT - MACHIQUES TR
TT - TACHIRA TR
TR T - TRUJILLO TR
MERIDA ARCH
(D) CENOMANIAN-TURONIAN
POSITIVE AREA
RISING SUBMARINE RIDGE
EL BAUL ARCH
(E) LATE MAASTRICHTIAN-PALEOCENE
MERIDA ARCH
FLYSCH
POSITIVE AREA
FLYSCH, WILDFLYSCH
1
2
3
4
5
6
7
8

Pre-Mesozoic ages seem certain for metamorphic rocks and intrusives in the El Baúl and likely in the El Tinaco areas (Venezuela, 1970, p. 237, pp. 210–211; Menéndez, 1965; Martín B., 1968), although González Silva (1972) has at one time interpreted the contact between the El Tinaco Complex and the overlying Jurassic metasediments to be transitional. Stephan et al. (1977) present sedimentological evidence to support the concept of an extensive "Paleocordillera de El Tinaco" shedding Mesozoic and older debris and extending from Barquisimeto to the Gulf of Barcelona during the Albian. Some workers (e.g., A. Bellizzia, oral communication, 1975) have speculated that the El Tinaco Complex may indeed be pre-Mesozoic in age, but may represent only a thin allochthonous slice derived from an area further north. A possible analog would be a very small allochthon of Precambrian(?) granulite found approximately 100 km west of Puerto Cabello. The Sebastopol Basal Complex (or Sebastopol gneiss) of the Cordillera de la Costa is commonly interpreted to represent pre-Mesozoic basement (Bell, 1974), but some confusion on this question also still exists (e.g., González Silva, 1971, pp. 366–367; Venezuela, 1970, pp. 563–564, p. 641; Martín B., 1968, p. 347). Pre-Mesozoic rocks are probably not exposed in the Caribbean Mountain System east of the Caracas region, although certain similarities between the Dragon Gneiss of Paria and the Sebastopol Basal Complex exist (H.H. Hess, personal communication, in Kugler, 1972).

Triassic–Jurassic

In describing the Mesozoic geology of the Andes of Colombia, a fundamental difference between the areas to the east and west of the Cordillera Central immediately becomes apparent (Bürgl, 1961, 1967). As Bürgl (1961) states: "From Triassic onward, the western and eastern Andes were manifested as distinct sedimentary, magmatic, and tectonic provinces" (translation by Case et al., 1971). This feature can, without difficulty, be recognized in Ecuador as well (e.g. Sauer, 1971). The Cordillera Central thus is

postulated to have formed an "Umbral Andino" — that is, a positive region or barrier. Stibane (1967) in fact considers this barrier system to have been in effect as early as the Late Paleozoic.

It is not entirely clear to what extent this apparent barrier was submarine or subaerial. It was probably completely inundated in Cretaceous time, and subject, as will be discussed below, to local breaching from the west during the earlier Mesozoic. Jacobs et al. (1963) contend that if it was a land barrier, it contributed surprisingly little to the sediments to the east or west. On the other hand, Sauer (1971) seems to leave no doubt that the continuation of the Cordillera Central in Ecuador, the Cordillera Real, formed a predominantly continental barrier throughout the Mesozoic.

Nevertheless, the distinct possibility exists that the effect ascribed to this so-called "barrier" may actually be a tectonic artifact. Recent geophysical investigations in northwestern Colombia (Case et al., 1971; Case, 1974; Meissner et al., 1976) have led to the alternate hypothesis that the present juxtaposition of geologically different elements along the Cordillera Central is due to the megatectonic suturing of an oceanic sequence to the west with a continental sequence to the east (Faucher and Savoyat, 1973; Pichler et al., 1974; Feininger, 1977). The effects of such a model on the paleogeographical interpretation of the area, as deduced by investigations to date, are still being explored. In any case, it seems advisable to separate an east-Andean region from a west-Andean region at least as far north as the Sierra de Perijá.

East-Andean region

The Triassic–Jurassic periods east of the Cordillera Real — Cordillera Central barrier (termed the "North Andean Cardinal Cordillera" by Sauer, 1971) were characterized by predominantly continental sedimentation, interrupted by local marine incursions, apparently through gaps in the "barrier" to the west (Fig. 3A).

Stibane (1968) and Geyer (1973) regard the east-Andean region as an intramontane basin characterized

Fig. 3. Observed areal distribution of facies types of Mesozoic strata in the Northern Andes. Due to the possibility of major post-Mesozoic tectonic rearrangement, these maps should not be considered as unqualified paleogeographic representations. Villa de Cura Klippe of Caribbean Mountain System has been disregarded. *1* = predominantly terrestrial sedimentation, red-bed development typical; *2* = as in *1*, with abundant volcanism; *3* = as *1* and 2, with strong intermittent marine influence; *4* = littoral/terrestrial nearshore conditions; *5* = brackish and deltaic conditions, coal common; *6* = marine sedimentation, predominantly continental shelf and slope environment; *7* = as in *6*, with abundant volcanism; *8* = abyssal floor environment.

mainly by block faulting (= "continental geosyncline" of Bürgl, 1967), in which the source of sediments is to be found in the Guayana Shield to the east, as well as in local uplifted areas such as the Quetame, Garzón, and Santander Massifs, and probably in the intra-Andean barrier itself (see, however, Jacobs et al., 1963).

Cediel (1969) speaks of "typical molasse sediments" in the Santander area. One may appreciate the difficulties involved in correlating and dating these poorly fossiliferous continental sediments with their rapid and complicated lateral facies variations (Trümpy, 1943; Bürgl, 1964; Cediel, 1969; Geyer, 1973; Rabe, 1977).

The studies of Tschopp (1948, 1953, 1956) do not indicate the existence of Triassic strata in Ecuador, although Sauer (1971) and Geyer (1974) consider their presence likely. The oldest Mesozoic sediments in Ecuador thus are the limestones and interbedded shales and calcareous sandstones of the marine, flysch-like Santiago Formation, which crops out only in the Cordilleras de Cutucú and Condor in the southern Oriente region (Tschopp, 1948, 1953; Geyer, 1974). The Santiago Formation yields Lower Sinemurian ammonites (Geyer, 1974), but could very likely contain older and younger beds.

The Santiago Formation is unconformably overlain by the far more extensive Chapiza Formation, which is known in outcrop and from drilling along the entire eastern edge of the Cordillera Real in a strip at least 100 km wide (Tschopp, 1953; Sauer, 1971). The Chapiza Formation represents an abrupt facies change from the Santiago Formation, in that the environment of sedimentation now varies among terrestrial, lacustrine, and shallow marine conditions. Gray, brown, and red sandstones, siltstones, and clays comprise the main part of the formation. Evaporite deposits are common in the lower part, while the upper part (known as the Misahuallí member) shows the effects of intense volcanism. Here massive intrusions and pyroclastics may occur to the exclusion of normal clastic sediments (Tschopp, 1956; Sauer, 1971). The age of the poorly fossiliferous Chapiza can be only indirectly estimated: the formation must be younger than the Liassic Santiago Formation and older than the overlying Lower Cretaceous Hollín sandstone. Thus the Misahuallí member is probably of Upper Jurassic to Lower Cretaceous age.

The pre-Cretaceous Mesozoic strata of the east-Andean region of Colombia are divisible into two

main series: the Payandé Group (as used by Geyer, 1973) and the Girón Group (as used by Rabe, 1977). The Payandé Group is exposed intermittently along the eastern flank of the Cordillera Central. Near Ibagué (see Fig. 1), the Payandé Group may be divided into the Luisa Formation (defined by Geyer, 1973; = pre-Payandé Formation or Pre-Payandé Red Beds of Nelson, 1957, and Trümpy, 1943, respectively), the Payandé Formation, and the El Salitre Formation (defined by Geyer, 1973; = post-Payandé Formation or Post-Payandé Red Beds of Nelson, 1957, and Trümpy, 1943). The Luisa Formation is a "red bed" series with minor marine influence and overlies, with slight disconformity where the base is exposed, a series of rhyodacitic extrusives of possible Permo-Triassic age (Nelson, 1957; Bürgl, 1964). Note, however, that Barrero (1968) prefers to consider the rhyodacites as younger intrusives into the Luisa Formation. Overlying the Luisa Formation are the cherty limestones of the Upper Norian, marine Payandé Formation (Trümpy, 1943; Geyer, 1973), and the red beds of the El Salitre Formation composed predominantly of acid to intermediate porphyritic flows, tuffs, and volcanic breccias, with intercalated, thin limestone beds (Trümpy, 1943; Geyer, 1973). Because of the generally unfossiliferous nature of the El Salitre Formation, it is not known how far these rocks extend into the Jurassic. The upper limit may be bracketed between overlying marine Lower Cretaceous and the Sinemurian age of the ammonite-bearing, marine Morrocoyal "subformation" of the El Salitre Formation, which is locally exposed at the northern limit of the Cordillera Central near El Banco (Fig. 1).

In exposures along the Rio Saldaño, about 50—75 km south of the area studied by Geyer, Cediel et al. (1976) have introduced the term Saldaña Fm. for Nelson's "post-Payandé Fm.". Fossils recently found in the Saldaña Formation (Cediel et al., 1978) point to a Norian age for at least part of this unit.

In general, the Girón Group strata of the Cordillera Oriental have been much less influenced by marine incursions than the Payandé Group to the west (Geyer, 1973), and consequently continental sediments predominate. The historical development of the complex stratigraphy involving the use of the term Girón is beyond the scope of this summary, but Geyer (1973) and Rabe (1977) provide a good up-to-date discussion.

The Girón Group comprises, in the main, a variety

of conglomerates, medium- to coarse-grained sandstones, siltstones, and shales. The siltstones and shales are commonly red in colour, while the sandstones usually exhibit more subdued grayish tints. Limestones and marls, as well as rocks of volcanic origin, may be locally important. In the type area near Bucaramanga, the Girón Group exceeds 5,500 m in thickness (Cediel, 1969; Geyer, 1973; Rabe, 1977).

Rocks of Girón type are known in the Cordillera Oriental from the northwestern margin of the Garzón massif through to the Santander Massif, the Sierra de Perijá, the Sierra Nevada de Santa Marta, and the Guajira and Paraguaná Peninsulas. Such rocks are also found in western Venezuela, where they are referred to as the La Quinta Formation (Venezuela, 1970).

A consideration of marine intercalations in the Payandé Group and Girón Group has led Geyer (1973) to postulate a distinct pattern for the marine incursions from the west through the intra-Andean "barrier" during the Triassic and Jurassic (Fig. 3A). He envisions three main marine basins during the Late Triassic–Early Jurassic: (1) one in the upper Magdalena Valley (Tolíma Basin), fed through a break in the barrier near Pasto ("Nariño Strait"); (2) the Cundinamarca–Boyacá Basin around Bogotá, fed through the "Strait of Ibagué"; and (3) a basin along the lower to middle reaches of the Magdalena Valley (Magdalena-Bolívar-César Basin), fed through a break between the present Cordillera Central and the Sierra Nevada de Santa Marta. Connections between these centers of marine deposition are probable. The Lower Liassic, marine Santiago Formation of southern Ecuador probably outlines a similar marine basin (Sauer, 1971; Geyer, 1974).

Of these marine basins, the Cundinamarca–Boyacá Basin is the most significant as far as the later Mesozoic history of the east-Andean region is concerned. It is from this basin that, beginning in the Latest Jurassic (Fig. 3B), the Cretaceous transgression of the Cordillera Central spread to the north and south. Stibane (1967, 1968) refers this periodic, localized, vertical tectonic activity in the Ibagué–Bogotá region (he reports a similar situation in the Paleozoic) to the intersection of north–south and east–west fault systems in this area.

West-Andean region

As noted in an earlier section, the rocks of the west-Andean region are now accepted by most authors as having originated in an oceanic realm. A lively debate has evolved, however, on the origin and nature of the volcanic rocks which form a large part of the sequence. Because recent opinion seems to favour a maximum age of Lower Cretaceous, discussion of these units is deferred to the appropriate section below.

Hinge zone: Sierra de Perijá, Sierra Nevada de Santa Marta, Guajira and Paraguaná Peninsulas, Maracaibo Basin, Falcón Basin, Mérida Andes, Barinas–Apure Basin

Rocks of Girón type, that is predominantly terrestrial and limno-fluviatile deposits with common red beds, are known from the entire hinge zone. However, in contrast to the studied sections of the Cordillera Oriental of Colombia (Cediel, 1969; Geyer, 1973; Rabe, 1977), these strata usually include *abundant* lavas and pyroclastics. Again, the development of a useful stratigraphy has been difficult and complex due to the lack of determinable fossils and rapid lateral facies changes.

Traditionally, the term La Quinta (Kündig, 1938) has been used to describe the Triassic–Jurassic red beds of western Venezuela and the Sierra de Perijá (Miller, 1962; Venezuela, 1970; Geyer, 1973). In the Sierra de Perijá and western part of the Maracaibo Basin, the La Quinta Formation is combined with underlying terrestrial strata, which also contain abundant volcanic material but generally lack the development of typical red beds, into a La Gé Group (Hea and Whitman, 1960; Venezuela, 1970). The thickness of these strata varies widely. They may be entirely absent locally, or reach several thousand meters. Gansser (1955) has estimated 4,000 m of red beds in the northwestern Sierra de Perijá. The La Gé Group as a whole may exceed 5,000 m (Venezuela, 1970).

Bowen (1972) has recently suggested that the La Gé Group may be much younger than previously thought, varying from Permian to Upper Jurassic, and that the La Quinta Formation is restricted entirely to the Jurassic. Thus two interpretations are possible: either the Girón Group of the Cordillera Oriental of Colombia is younger than assumed by Geyer (1973), as Rabe (1977) has indicated in the Santander area, or it is the La Gé Group rather than the La Quinta Formation that represents the Girón Group time equivalent in the Sierra de Perijá. Nevertheless, it appears likely that some of these problems may be resolved in the near future, for the stratigraphy of the Upper Paleozoic to Jurassic continental deposits of western Venezuela is at present being revised (O. Macsotay, personal communication, 1975).

Similar problems of stratigraphy beset the red beds and associated, predominantly terrestrial sediments and volcanics of the Sierra Nevada de Santa Marta (Hedberg, 1942; Gansser, 1955), which, according to Gansser, may have at one time covered vastly greater areas of this uplifted massif.

According to Tschanz et al. (1974), these deposits, which are restricted to the southeastern edge of the massif, may be divided into the Permian to Lower Triassic spilitic volcanics and graywackes of the Corual Formation, the Triassic red beds and red spilitic volcanic rocks of the Guatapuri Formation, and unnamed Jurassic volcanic rocks and ignimbrites extending from the Upper Jurassic into the Cretaceous. These ages are based on geological inference. However, radiometric dates also reported by Tschanz et al. (1974) for the Guatapuri Formation are more in accord with the interpretation of Geyer (1973), who places the Corual–Guatapuri sequence in the Upper Triassic to Middle Jurassic. Both Geyer (1973) and Gansser (1973) stress the importance of a tiny fault-bounded occurrence of Liassic shales and quartzitic sandstones (the "El Indio" beds of southwestern Sierra Nevada de Santa Marta), which Geyer and Gansser correlate with deposits of marine incursions of approximately similar age in the east-Andean region of Ecuador and Colombia to the south.

To the southeast, rocks of La Quinta type have been variously described as isolated remnants in the Maracaibo Basin (e.g., Toas Island; Pimentel, 1976), in the Táchira Basin between the Cordillera Oriental and the Mérida Andes (Ramírez and Campos, 1972), and along the northwestern and southeastern flanks of the Mérida Andes themselves (Useche and Fierro, 1972). Scattered occurrences from the north-central part of the latter are also mentioned by Ramírez et al. (1972).

In general, the sporadic occurrence and greatly varying thickness of these strata are related to the predominance of vertical tectonics during their deposition (e.g., Stainforth, 1969; Shagam, 1972a). Although the predominantly continental, limno-fluviatile environment of deposition of the Venezuelan Triassic–Jurassic deposits thus corresponds closely to that noted for eastern Colombia, it should be emphasized that intermittent marine influence appears to have been much more common in the Venezuelan case (e.g. Ramírez and Campos, 1972; Useche and Fierro, 1972; Geyer, 1973).

The geographically isolated occurrences of Triassic–Jurassic rocks of the Guajira and Paraguaná Peninsulas exhibit basic similarities as well as important differences with respect to other occurrences in the hinge zone. Geyer (1973) has recently summarized the Guajira occurrences based on own observations and the work of Renz (1956, 1960), Rollins (1960, 1965), MacDonald (1964), Lockwood (1965), and Radelli (1962, 1967). His interpretation is followed here (Fig. 4).

The strata in question are the approximately coeval (according to Geyer) Cojoro and Cocinas Groups of the narrow "Guajira Trough" in the southern half of the Guajira Peninsula. The Cocinas Group, of at least 3,300 m thickness, is composed predominantly of marine shales and sandstones with minor red beds in the lower part (Cheterló Formation) and limestones (in part as reefs) in the upper section (Jipi Formation). With the exception of the Tithonian of Colombia, a thick, continuous section of non-metamorphic marine sediments of Triassic–Jurassic age is found nowhere else in Colombia and Venezuela (Alvarez, 1971).

While the northern margin of this marine trough is unknown due to extensive faulting, the southern margin is probably represented by the deposits of the Cojoro Group, of which at least 832 m have been

---

Fig. 4. Generalized correlation chart of representative areas in the Northern Andes. Time scale after Van Eysinga (1975).

Source: 1 = Sauer (1971); 2 = Julivert (1968); 3 = Geyer (1973); 4 = Goossens and Rose (1973); 5 = Case et al. (1971); 6 = Venezuela (1970); 7 = Rollins (1965); 8 = Petzall (1972); 9 = Miller (1962); 10 = Bowen (1972); 11 = Bellizzia and Rodriguez (1968); 12 = Rabe (1977); 13 = Maresch (1974); 14 = Maresch (1975); 15 = Seiders (1965); 16 = Santamaría and Schubert (1974); 17 = Cediel et al. (1976); 18 = E. Colmenares (oral communication, 1978).

Symbols: a = contact poorly defined; b = hiatus; c = large uncertainties in age possible; d = true stratigraphic extent unknown; e = predominantly terrestrial sediments, red beds common; f = as in e, with abundant volcanics; g = as in e and f, with strong intermittant marine influence; h = littoral/terrestrial near-shore sediments, transgressive/regressive sandstone and conglomerate typical; i = predominantly brackisch and deltaic deposits, coal common; j = marine continental shelf and rise sediments; k = as in j, with abundant volcanics; l = predominantly abyssal, siliceous sediments with associated basic volcanics; m = predominantly basic volcanics with associated siliceous sediments.

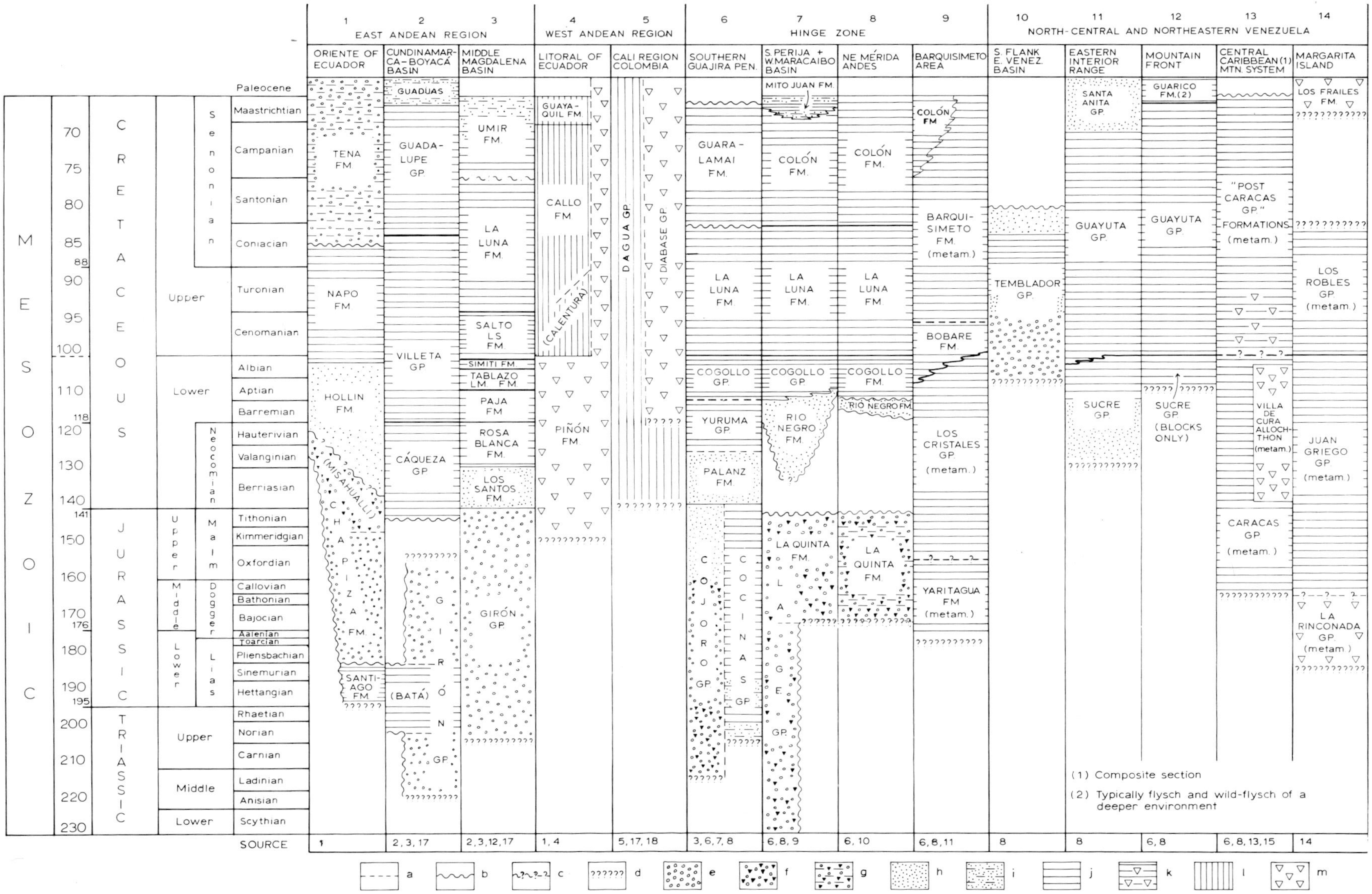

Stratigraphic correlation chart — Northern South America

Legend markers (bottom): a, b, c, d, e, f, g, h, i, j, k, l, m

(1) Composite section
(2) Typically flysch and wild-flysch of a deeper environment

Time scale (right to left): Paleocene, Maastrichtian, Campanian, Santonian, Coniacian, Turonian, Cenomanian, Albian, Aptian, Barremian, Hauterivian, Valanginian, Berriasian, Tithonian, Kimmeridgian, Oxfordian, Callovian, Bathonian, Bajocian, Aalenian, Toarcian, Pliensbachian, Sinemurian, Hettangian, Rhaetian, Norian, Carnian, Ladinian, Anisian, Scythian

System / Series: CRETACEOUS (Senonian, Upper, Lower — Neocomian), JURASSIC (Malm / Upper, Dogger / Middle, Lias / Lower), TRIASSIC (Upper, Middle, Lower)

Age scale (Ma): 70, 75, 80, 88, 90, 95, 100, 110, 118, 120, 130, 140, 141, 150, 160, 170, 176, 180, 190, 195, 200, 210, 220, 230

MESOZOIC

Col	Region	Formations (young → old)	Source
1	ORIENTE OF ECUADOR (EAST ANDEAN REGION)	TENA FM, NAPO FM, HOLLIN FM (MISAHUALLI), SANTIAGO FM, CHAPIZA FM	1
2	CUNDINAMARCA-BOYACÁ BASIN	GUADUAS, GUADALUPE GP, VILLETA GP, CÁQUEZA GP, (BATÁ), GIRÓN GP	2, 3, 17
3	MIDDLE MAGDALENA BASIN	UMIR FM, LA LUNA FM, SALTO LS FM, SIMITI FM, TABLAZO LM FM, PAJA FM, ROSA BLANCA FM, LOS SANTOS FM, GIRÓN GP	2, 3, 12, 17
4	LITORAL OF ECUADOR (WEST ANDEAN REGION)	GUAYAQUIL FM, CALLO FM (CALENTURA), PIÑÓN FM	1, 4
5	CALI REGION COLOMBIA	DIABASE GP, DAGUA GP	5, 17, 18
6	SOUTHERN GUAJIRA PEN.	GUARA-LAMAI FM, LA LUNA FM, COGOLLO GP, YURUMA GP, PALANZ FM, COJORO GP, COCINAS GP	3, 6, 7, 8
7	S. PERIJA + W. MARACAIBO BASIN (HINGE ZONE)	MITO JUAN FM, COLÓN FM, LA LUNA FM, COGOLLO GP, RIO NEGRO FM, LA QUINTA FM, LA GE GP	6, 8, 9
8	NE MÉRIDA ANDES	COLÓN FM, LA LUNA FM, COGOLLO FM, RIO NEGRO FM, LA QUINTA FM	6, 10
9	BARQUISIMETO AREA	COLÓN FM, BARQUISIMETO FM (metam.), BOBARE FM, LOS CRISTALES GP (metam.), YARITAGUA FM (metam.)	6, 8, 11
10	S. FLANK E. VENEZ. BASIN	TEMBLADOR GP	8
11	EASTERN INTERIOR ANDES	SANTA ANITA GP, GUAYUTA GP, SUCRE GP	8
12	MOUNTAIN FRONT (EASTERN INTERIOR RANGE)	GUARICO FM (2), GUAYUTA GP, SUCRE GP (BLOCKS ONLY)	12
13	CENTRAL CARIBBEAN MTN SYSTEM	"POST CARACAS" FORMATIONS (metam.), VILLA DE CURA ALLOCHTHON (metam.), CARACAS GP (metam.)	6, 8, 13, 15
14	MARGARITA ISLAND	LOS FRAILES FM, LOS ROBLES GP (metam.), JUAN GRIEGO GP (metam.), LA RINCONADA GP (metam.)	6, 8

Column group headings:
- Cols 1–3: EAST ANDEAN REGION
- Cols 4–5: WEST ANDEAN REGION
- Cols 6–8: HINGE ZONE
- Cols 10–14: NORTH-CENTRAL AND NORTHEASTERN VENEZUELA

measured. The lower section (Ranchogrande Formation) is a typical continental red bed complex with abundant rhyodacitic volcanics, thus corresponding generally to the La Quinta strata elsewhere in the hinge zone. Upwards in section, the red beds grade into well-sorted, coarsely bedded marine sandstone, which is interpreted to represent deposits of the marine beach environment contemporaneous with the formation of reefs of the upper Cocinas Group just to the north.

The identification of Late Jurassic ammonites in the upper Cocinas Group, as well as the likely association of some of the rhyodacites of the Ranchogrande Formation with dated Triassic granodiorite intrusions, makes it appear probable that these sediments represent most of the Triassic—Jurassic interval.

A very similar Triassic—Jurassic sequence is probable from Paraguaná Peninsula. Lightly metamorphosed shales, with limy, sandy, and conglomeratic intercalations (the Pueblo Nuevo Formation of MacDonald, 1968, or "semi-metamorphic series" of Feo-Codecido, 1971) yield Tithonian ammonites and therefore correspond both in age and lithology to the upper Cocinas Group of Guajira (MacDonald, 1968; Geyer, 1973). The Pueblo Nuevo Formation is also considered part of the Ruma Metamorphic zone (see p. 141. Red beds of La Quinta type are known from bore-holes (Engleman, 1935; MacDonald, 1964; Feo-Codecido, 1971). Fragments of dacitic extrusives (La Quinta?) in the lower part of the Pueblo Nuevo Formation (Feo-Codecido, 1971) tend to support the concept that the La Quinta is older, in contrast to the relationship shown on official Venezuelan correlation charts (Venezuela, 1970; Petzall, 1972).

Whereas the divisions Triassic—Jurassic (mainly continental conditions) and Cretaceous (mainly marine) allow a logical framework for the description of rock units in Colombia, Ecuador, and western Venezuela, this is no longer true of the areas east of Barquisimeto. With one small exception from the El Baúl area (Fig. 1), no unmetamorphosed Triassic—Jurassic sediments or extrusive volcanics are known with certainty from the entire region. On the other hand, many of the metamorphic rocks of the Caribbean Mountain System are postulated to extend from Cretaceous back into the Jurassic or even Triassic (e.g., Venezuela, 1970; Petzall, 1972). Nevertheless, these pre-Cretaceous ages are based in large part on indirect evidence at best, and hardly allow an age subdivision of the metamorphic column to be made

with confidence. The Mesozoic geology of the rest of Venezuela will thus be treated on the basis of metamorphic and non-metamorphic successions on p. 137.

Cretaceous

The geological differences between the west-Andean and east-Andean regimes become apparent in the Cretaceous, although both areas are characterized by marine sedimentation. To the east, epicontinental seas predominated, with local subsidence to bathyal depths, and led to "tranquil" sedimentation and very little or no volcanism. This east-Andean geosyncline, considered a "typical miogeosyncline" by Campbell and Bürgl (1965), covered an immense area from Venezuela to Bolivia. By analogy, the west-Andean region has been referred to as a eugeosyncline (e.g., Bürgl, 1967; Case et al., 1971), but this interpretation has been disputed (Pichler et al., 1974). Regardless of terminology, however, it is clear that the west-Andean area was characterized by abundant basic to intermediate volcanism and predominantly deep-water conditions (Fig. 3, C, D, E).

East-Andean region

The Cretaceous of the east-Andean area is one of the most complete and paleontologically interesting sections in the world (Bürgl, 1963). The rich ammonite fauna present has allowed the detailed recognition of stages and substages.

According to Campbell and Bürgl (1965), the axis of the east-Andean geosyncline lay along the present western flank of the Cordillera Oriental. Nevertheless, it is the north—south variation in thickness that is most striking (Bürgl, 1963, fig. 7). Bürgl (1963) and Campbell and Bürgl (1965) mention a stratigraphic thickness of at least 11,000 m to possibly more than 16,000 m for the Cretaceous section in the Cundinamarca—Boyacá Basin of the Bogotá area (Fig. 3). In southern Colombia and the northern Oriente of Ecuador, however, this section thins to only several hundred meters. North of Bucaramanga (Fig. 1), less than about 3,000 m of marine Cretaceous are found. Local thinning also occurs around the Paleozoic massifs (Fig. 1) throughout the Cordillera Oriental.

The section is composed predominantly of shale, along with limestone and sandstone — all deposited under "tranquil" conditions (Campbell and Bürgl,

1965). Euxinic shale predominates in the Bogotá area. Sandstone becomes important to the east near the sediment source region represented by the Guayana Shield, and limestone is typical to the west and north as well as in northern Ecuador.

Bürgl (1962, 1963) recognizes a distinct cyclical pattern in the sedimentation history of the geosyncline due to cyclical subsidence and uplift, producing alternating deepening and shoaling conditions. One such pattern of cycles is represented by the alternation of ammonite-bearing euxinic shales with shallow-water conglomerates, sandstones, marls, etc., which in the Cundinamarca–Boyacá Basin, according to Bürgl, can be correlated with the Cretaceous stages and substages. A second type of cycle, of more regional importance, is the megacycle ("Grosszyklus") of Bürgl (1962, 1963). Although an oversimplification, because these megacycles are not everywhere contemporaneous, they may be used as a guide in following the transgressive and regressive history of the Cretaceous sea in the east-Andean region.

Cundinamarca cycle. As had occurred before in the Late Triassic to Liassic, an ingress of marine conditions reached the Cundinamarca–Boyacá Basin, presumably from the west through the "Straits of Ibagué" (Fig. 3A), in Tithonian time. From Tithonian to Valanginian time, marine sedimentation was generally restricted to this basin (Fig. 3B).

In northeastern Ecuador, the volcanic Misahuallí "subformation" of the continental Chapiza Formation may have extended into the Lower Cretaceous. It is overlain by the sandstones of the Hollín Formation (Sauer, 1971), which transgressed from the southwest during this time (Tschopp, 1956). Both in Colombia and Ecuador, fauna of Andean character ("Pacific type") predominated (Olsson, 1956; Tschopp, 1956; Bürgl, 1963).

Santander cycle. Beginning in the Hauterivian (Fig. 3C), the geosyncline extended north from Bogotá to the Caribbean coast along the "intra-Andean barrier" (Bürgl, 1963; Geyer, 1973). Tethyan (or "European type") fauna became predominant. Contrary to the paleogeographic representations of these two authors, at least a minor connection with the Oriente of Ecuador must have already existed at this time, for the sandstones of the Hollín Formation of this region indicate a change in direction of transgression (now from the northwest). Furthermore, the replacement of Pacific fauna by European types took place in Ecuador at about this time (Tschopp, 1956; Sauer, 1971).

Tolíma cycle. The time-span Albian to early Senonian (Fig. 3D) probably marks the greatest extent of the east-Andean geosyncline (from the Caribbean Sea to Bolivia, according to Weeks, 1947), and extensive filling of the trough between the North Andean Cardinal Cordillera (Sauer, 1971) and the Guayana Shield continued. In Ecuador, the littoral to sub-littoral Hollín Formation was replaced by the deep neritic Napo limestone (Sauer, 1971). The Tethyan character of the fauna gradually waned, and, in the Senonian, the Andean character again took over.

Magdalena cycle. This is essentially a regressive cycle extending from the early Senonian into the Tertiary, with its greatest marine extent in the early Maastrichtian (Fig. 3E). It is characterized by marked facies development, indicating a division into distinct basins of sedimentation (Campbell and Bürgl, 1965). The sea apparently retreated to the north, for while littoral to terrestrial sedimentation (Tena red beds) commenced in Ecuador in the Santonian, marine sedimentation continued in the Cordillera Oriental until the Maastrichtian (Campbell and Bürgl, 1965; Sauer, 1971).

West-Andean region

The Mesozoic stratigraphy of the west-Andean region is only imperfectly known. In Colombia, these rocks have been studied in most detail in the Buenaventura–Cali area around the 4°N latitude (Nelson, 1957; Bürgl, 1961; Colmenares, 1978). Traditionally, the sequence has been divided into a lower Dagua Group of Triassic(?) to Cretaceous age (Nelson, 1957; Bürgl, 1961; Geyer, 1973) and an overlying Upper Cretaceous Diabase Group (Hubach, 1957; Nelson, 1957; = "Grupo Porfirítico-Diabásico of Bürgl, 1961). However, the age of the Dagua Group is based mainly on inference, since no usable fossils have yet been identified. Recently (e.g. Cediel et al., 1976; Colmenares, 1978, and oral communication, 1978), the Dagua and Diabase Groups have come to be considered time-equivalent and probably no older than Cretaceous. This interpretation would be more in accord with similar sequences in Ecuador (see below).

The Dagua Group consists of about ?10,000 m (Campbell, 1974b) of phyllitic slate, mica schist, basaltic volcanic rocks, and some limestone as well as siliceous slates. The lower part of the Chita Group from southern Colombia (Kehrer, 1939) may be correlated with the Dagua Group.

The Diabase Group is composed of diabase, spilite,

intercalated tuff and agglomerate, and minor ultra-mafic rocks (Nelson, 1957). Campbell (1974b) reports a possible thickness of ?10,000 m also for the Diabase Group. Thin beds of black chert and siliceous shale have yielded fossils indicating a minimum range of Barremian to Senonian (Case et al., 1971). Neverthe-less, the oldest known dated rocks overlying the Dia-base Group are the black cherts and dirty greenish sandstones of the Paleocene Nogales Formation (Nel-son, 1957). Similarly, Paleocene fossils have been found in the marine Chita Formation of southern Colombia (Nelson, 1957), the upper part of which is commonly correlated with the Diabase Formation (Geyer, 1973; Pichler et al., 1974).

Detailed stratigraphic correlation with other parts of the west-Andean region is difficult due to sparse regional data and poor age control. Nevertheless, rocks of this type are known to underlie the Coastal Cordillera (Serranía de Baudó) and Sautatá Arch (the connection between the Cordillera Occidental and the Serranía de Darién of Panamá) of northwestern Co-lombia, much of the Cordillera Occidental of Colom-bia and Ecuador, and most of the Litoral area of Ecuador, though exposed here mainly in the low coastal hills (Hubach and Radelli, 1962; Case et al., 1971; Sauer, 1971). In northwestern Colombia, gray-wackes, shales, and andesites have also been reported associated with this volcanic complex. Ages in this region are estimated to range from Jurassic to Eocene or younger (Case et al., 1971).

The question whether the low-grade Cajamarca Group of the Cordillera Central is Paleozoic or Meso-zoic (and thus equivalent to the Dagua Group—Diabase Group association) in age has occupied geologists for some time (see Butterlin, 1972, and Geyer, 1973, for a historical account of this discussion). Both Radelli (1967) and Butterlin (1972) have strongly supported the concept of a Mesozoic eugeosyncline in the posi-tion of the Cordillera Central (rather than the "intra-Andean swell or barrier" discussed earlier) in which the Cajamarca Group accumulated and was metamor-phosed in Late Mesozoic time. This view appears to be clearly refuted by the work of Stibane (1970), who makes a strong case for the existence of Caja-marca rock fragments in the Triassic Luisa Forma-tion. Thus a Paleozoic age appears at present to be most plausible for the Cajamarca Group.

Nevertheless, it should be borne in mind, as Sha-gam (1975) has pointed out, that the name Caja-marca has become something of a "waste-basket" designation and may possibly encompass a variety of strata of heterogeneous ages. A similar problem emerges in Ecuador, where Herbert (1977) views the metamorphic rocks of the Cordillera Real (the con-tinuation of the Cordillera Central of Colombia) as Paleozoic in age, while Feininger (1975) points to a Cretaceous age with metamorphism at the close of the Mesozoic Era.

The Diabase Group finds its Ecuadorian counter-parts, via the Chita Group of southwestern Colombia, in the Piñón Formation (Tschopp, 1958) of the Li-toral area and the Diabase-Porphyry Formation (cor-responds to "basalt-andesite formation" in common American usage) of Sauer (1965, 1971) in the Cor-dillera Occidental. The Piñón Formation and the Diabase—Porphyry Formation have also been com-bined into a Piñón Group s.l. (Servicio Nacional de Geología y Minería, 1969).

Outcrops of the Piñón Formation in the Litoral region are relatively restricted in comparison with those of the Diabase-Porphyry Formation in the Cor-dillera Occidental, but drilling has indicated that the Piñón Formation may underlie much of the pre-dominantly Tertiary cover of the Ecuadorian coastal depression (Sauer, 1971, pp. 47—48).

Two distinct units have been recognized by Goos-sens (1973) in the rocks of the Piñón Formation s.str., that is those exposed in the Ecuadorian Litoral region — "Lower unit: a series of phaneritic mafic and ultramafic rocks including harzburgite, serpen-tine, gabbro and diabase; Upper unit: aphanitic and porphyritic basalts occurring as dikes and flows with-in the overlying marine sediments". The Diabase—Porphyry Formation consists of pillow-lavas, massive diabase, siliceous sedimentary intercalations, and locally cumulate gabbro and peridotite (Juteau et al., 1977).

The ages of these units can be only indirectly determined. The lower unit of the Piñón Formation s.str. is overlain in the low hills of the Litoral region by the siliceous, tuffaceous marine lutites of the Cal-lo Formation (whose "subformation" Calentura re-presents the Cenomanian to Senonian), concordantly overlain in turn by the Maastrichtian Guayaquil For-mation (Thalmann, 1946; Sauer, 1971; see also Faucher and Savoyat, 1973). Lithologically corre-sponding units of the Sierra region (termed "Com-plexe Piñón-Cayo[*1] de la Sierra" by Faucher and

---

[*1]   The spellings "Callo" and "Cayo" are both commonly found in literature on Ecuador.

Savoyat, 1973) have yielded Upper Cretaceous ages, but do not reach the Maastrichtian, which here is represented by flysch.

K/Ar determinations by Goossens and Rose (1973) on volcanic flows from the Litoral region yielded ages between 110 ± 10 and 54 ± 5 m.y.

The similarity of these volcanic sequences with the Diabase Group of Colombia and correlative strata in Central America led to the definition of a "Basic Igneous Complex" in the entire region in question by Goossens and Rose (1973). The obvious similarity of the Basic Igneous Complex to ophiolites throughout the world, thus suggesting that this unit might be a slice of obducted Pacific oceanic crust, has led to a variety of attempts to determine the petrochemistry of the Complex and compare it to Pacific crust (e.g., Goossens, 1973; Goossens and Rose, 1973, 1975; Pichler et al., 1974; Goossens et al., 1977; Herbert, 1977; Juteau et al., 1977; P.K. Hörmann, oral communication, 1978). The results have been equivocal. Part of the problem appears to be one of definition, for the quartz tholeiites of one author may be the basaltic andesites of another.

Goossens, as author and co-author in a series of publications cited above, has stressed the overall tholeiitic nature of the Basic Igneous Complex and concludes that the volcanics are "more like ocean-ridge tholeiites than other varieties", with a generally bimodal population of predominant basalt and lesser basaltic andesite. The chemistry could be compared with either an ocean-floor tholeiite or an early island-arc tholeiite setting. Pichler et al. (1974) similarly favour an ocean-floor tholeiite interpretation.

Other authors stress the calc-alkalic affinities of this volcanic suite. Herbert (1977) concludes that the "Basic" Igneous Complex exposed in the Cordillera Occidental of Ecuador is in fact almost exclusively calc-alkalic in character. Hörmann (oral communication, 1978) extends this interpretation to the Basic Igneous Complex of the Litoral region of Ecuador and in a preliminary form also to Colombia.

Clearly, the problem is far from being resolved. This state of affairs has, however, not deterred most workers in the region from accepting that the west-Andean region originated in an oceanic realm that has subsequently been welded onto continental South America. These recent views stand in contrast to the classical interpretation of "early Andean magmatism" in an in-situ west-Andean eugeosyncline.

Hinge zone: Sierra de Perijá, Sierra Nevada de Santa Marta, Guajira and Paraguaná Peninsulas, Maracaibo Basin, Falcón Basin, Mérida Andes, Barinas-Apure Basin[1]

The Cretaceous deposits of the hinge zone may be regarded lithologically and paleogeographically as a continuation of the so-called "miogeosyncline" of the east-Andean region. Nevertheless, the thickness of the Cretaceous only locally reaches a few thousand meters, and the environment of deposition is generally referred to as of "platform type" (e.g., Zambrano et al., 1971), rather than "geosynclinal". Due to great thicknesses of Tertiary sediments, almost nothing is known of the Cretaceous record in the Falcón Basin. In the following, the continuation or transition from the Guajira Cretaceous section to the Cretaceous to the south in the Gulf of Venezuela is largely hypothetical, for as yet little is known of the stratigraphy underlying the Gulf of Venezuela region. Metamorphosed strata of probable Cretaceous age from the Barquisimeto region, the northern tip of the Sierra Nevada de Santa Marta, and the Guajira Peninsula (Fig. 5) are discussed in conjunction with the metamorphic complex of the Caribbean Mountain System in a subsequent section (see p. 139).

Essentially, the Cretaceous represents a marine transgression over probably the entire hinge zone, covering a peneplaned Paleozoic and Triassic—Jurassic terrane. Local basins developed in the Guajira, Machiques, Táchira, and Trujillo areas (Fig. 3) that allowed greater thicknesses of sediment to accumulate without, however, affecting the bathymetry of the depositional environment to any great extent (Zambrano et al., 1971).

The oldest Cretaceous rocks are the sandstones, with intercalated conglomerates, variegated shales and rare limestones of the Río Negro Formation (Hedberg, 1931). The strata are predominantly continental with marine intercalations. They are generally pre-Aptian in age, probably mainly Barremian (Venezuela, 1970, p. 527), but could conceivably extend

---

[1]   The following is based largely on a synthesis of this important oil-producing area by Zambrano et al. (1971), who have incorporated and interpreted both published and extensive unpublished data. A discussion of the complex development of nomenclature and existing differences of opinion is beyond the scope of this summary. For a discussion of surface—subsurface stratigraphy of the Barinas—Apure Basin see Feo-Codecido (1972).

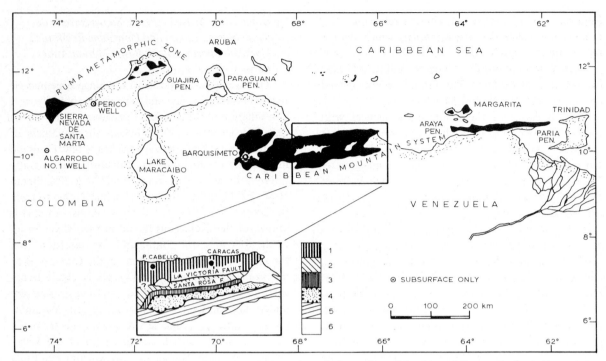

Fig. 5. Distribution (shaded areas) of metamorphosed Mesozoic rocks in northern Colombia and Venezuela. *Inset*: Tectonic belts of central part of Caribbean Mountain System (includes both metamorphosed, belts *1–4*, and unmetamorphosed, belts 5 and 6, strata): *1* = Coast Range Tectonic Belt; *2* = Caucagua – El Tinaco Tectonic Belt; *3* = Paracotos Tectonic Belt; *4* = Villa de Cura Tectonic Belt; *5* = Foothills Tectonic Belt, Thrust Fault Zone, and Overturned Belt; *6* = Gentle Dips Belt. After Bell, 1971.

back into the Tithonian (e.g., Geyer, 1973). As such, the Río Negro Formation represents initiation of the Cretaceous transgression and is essentially analogous to the transgressive sandstones and conglomerates of the Los Santos Formation and Arcabuco sandstone of Colombia (Geyer, 1973), and the Hollín Formation of Ecuador (Sauer, 1971). The same interval on Guajira is represented by the wholly marine sandstones and limestones of the Palanz Formation (arenaceous) and Yuruma Group (predominantly limestone). While the Río Negro Formation may vary in thickness from a maximum of 1,000 m, such as in the Táchira Trough, to only 17 m on the main Maracaibo Platform, a thickness of 1,200 m is typical for the Palanz–Yuruma interval (Zambrano et al., 1971). As might be expected, the source of sediments was the Guayana Shield to the southeast. The 0-isopach contour for the Río Negro Formation at the end of the Barremian generally follows the present-day position of the Mérida Andes, looping to the northwest in the central part because of the Mérida Arch (Fig. 3C).

The Aptian to lower Senonian interval is dominated by marine limestone, with shale and cherty slate. The

lower part of the Cogollo Group (Sutton, 1946), the Apón Formation, conformably overlies the Río Negro Formation and is generally similar in geographic extent. These nodular limestones and black shales are restricted to the Aptian, and rarely exceed several hundred meters in thickness. The middle part of the Cogollo Group shows a distinct facies development: it is divisible along a line approximately parallel to 10°N latitude into the Lisure Formation (glauconitic sandstone, sandy limestone, 55–180 m thick) to the north, and the Aguardiente Formation (sandstone, conglomerate, shale, 30–560 m thick) to the south. The Guayana shoreline in the Albian retreated to the southeast to allow deposition of the basal clastics of the Barinas–Apure Basin. In some interpretations, these basal clastics are referred to the Río Negro Formation (e.g. Venezuela, 1970, correlation chart). The upper part of the Cogollo Formation is represented by a thin, 30–40 m interval of shallow-water marine limestone, which covered the entire region.

Again in the Cenomanian, Turonian, and lower Coniacian, a typical northwest–southeast facies distribution, related to the sediment source from the Guayana Shield to the southeast, can be recognized. Epineritic, predominantly arenaceous sediments marked the southeastern part of the Barinas–Apure Basin. The northwestern part formed a neritic, transitional province characterized mainly by 300–800 m of shale and decreasing amounts of arenaceous sediments. Northwest of the present-day southeast flank of the Mérida Andes, sedimentation was of "La Luna" type. The La Luna Formation (Garner, 1926) is composed of concretionary limestone and lesser shale. It is locally cherty and has a high content of organic matter. Coarse clastics are missing, and euxinic conditions apparently prevailed. Thickness varies from 100 to 250 m. Locally, on the southwestern part of the depositional platform, the Táchira area, shale predominated during the Cenomanian–Turonian and has given rise to the term Capacho Formation. The Cogollo Group and the La Luna Formation are the reservoir and source rocks of much of Venezuela's petroleum.

During the remainder of the Late Cretaceous, most of the area was the site of deposition of the homogeneous gray to black foraminiferal shales of the Colón Formation (Liddle, 1928), indicating uniform subsidence and deep, open-sea conditions. During the Maastrichtian, intercalations of limestone and sandstone are found in the west-central part of the Maracaibo Platform, giving rise to the definition of the shallow-water Mito Juan Formation (Garner, 1926). Together, these strata reach a thickness of 700–800 m. In the Barinas–Apure Basin, the proximity of the source rocks of the Guayana Shield again led to the deposition of 100–400 m of arenaceous sediments. On Guajira, a slight hiatus in sedimentation after the deposition of the La Luna Formation was followed by the shallow-water limestones, in part arenaceous, and calcareous shales of the Guaralamai Formation (530 m thick).

The end of the Mesozoic was marked by a profound change in the paleogeography of the hinge zone (Fig. 3E). The Barinas–Apure Basin and the Mérida Arch were subjected to subaerial erosion. The latter continued to the northwest to form a broad platform characterized by shallow-water conditions. To the southwest, on the Colombia flank, deltaic and freshwater conditions prevailed. On the northeastern flank, however, thick units of flysch and wild-flysch accumulated, often containing huge olistostromes of Cretaceous metamorphic and non-metamorphic for-

mations (Bellizzia and Rodríguez, 1968; Zambrano et al., 1971).

The Jurassic–Cretaceous of the Caribbean Mountain System, El Baúl Uplift, and East Venezuela Basin

As noted on p. 132 the logical division of the Mesozoic into pre-Cretaceous and Cretaceous intervals in western Venezuela, Colombia, and Ecuador is no longer useful for descriptive purposes in the area east of Barquisimeto.

Neglecting for the moment the metamorphic complex of the Caribbean Mountain System (Fig. 5), we are left with a vast region between the latter and the Guayana Shield, which is exposed along the Orinoco River. This region is called the East Venezuela Basin, and is separated from the Barinas–Apure Basin by the El Baúl Uplift (Fig. 1). It represents the easternmost portion of the extensive Andean sedimentary belt that borders the Guayana Shield.

Although Mesozoic strata are exposed only in the foothills of the Caribbean Mountain System and in the Eastern Interior Ranges, extensive drilling has shown them to underlie most of the thick Tertiary section of the East Venezuela Basin.

The only occurrence of pre-Cretaceous unmetamorphosed Mesozoic rocks has been described from a small outcrop on the El Baúl Uplift. Termed the Guacamayas Group (Martín B., 1961) they consist of stratified volcanics (lava flows, tuffs, volcanic sandstones and conglomerates) with a minimum thickness of 325 m, and resemble the El Totumo volcanics of the La Quinta Formation in the Sierra de Perijá (Venezuela, 1970, p. 274; Bowen, 1972). On the basis of this resemblance, and stratigraphic relationships within the El Baúl area, the Guacamayas Group is placed in the Triassic–Jurassic.

The Lower Cretaceous sedimentary sequence points to deposition near the borders of a shallow sea encroaching on the low-relief Guayana Craton from the north and east (Hedberg, 1956). The oldest Lower Cretaceous deposits (Neocomian to Aptian, Venezuela, 1970, p. 84) exposed are the continental and shallow-water marine sandstones, shales, and limestones of the Barranquín Formation (Liddle, 1928), the lowermost part of the Sucre Group (Hedberg, 1950).[1] The base of the formation is un-

---

[1]   For a detailed discussion and departure from traditional views on intra-Sucre stratigraphy, see Guillaume et al., 1972.

known, however, and correlation with metamorphosed equivalents on Araya-Paria (González de Juana et al., 1965) suggests that an even older age for the lower parts is likely. The Barranquín Formation is known mainly from the Eastern Interior Ranges, where more than 1690 m are exposed. Although differences in facies exist, it may be correlated with the Río Negro Formation of western Venezuela (Fig. 4).

The upper part of the Sucre Group in the Eastern Interior Ranges is composed of marine reefal limestone, glauconitic sandstone, and deeper-water shale of the Aptian to Cenomanian El Cantil, Chimaná, and Borracha Formations (Petzall, 1972). The El Cantil Formation may be correlated with the Cogollo Formation of western Venezuela (Fig. 4). Allochthonous blocks of El Cantil limestone in younger formations suggest the existence of a similar sequence along the mountain front west of the Gulf of Barcelona as well.

In the remaining regions of the East Venezuela Basin to the south, Cretaceous rocks are known only from the subsurface. The Lower Cretaceous here is represented by the Canoa Formation (Dusenbury, 1960a) of the Temblador Group (Hedberg, 1942). The Canoa Formation is composed of less than 100 m of non-marine, mottled sandstone, mudstone, and claystone, discordantly overlying pre-Mesozoic basement. As such, it is considered the non-marine southern lateral equivalent of the marine upper Sucre Group to the north.

In late Albian to Cenomanian time, a distinct paleogeographical change resulted in the restriction of this Lower Cretaceous shelf sea to a "deep and subsiding inland seaway" (Hedberg, 1956), with poor bottom circulation and the deposition of organic-rich sediments (Hedberg, 1956; Menéndez, 1967). The resulting strata are represented by the carbonaceous shales, limestones, and cherts of the Cenomanian to Coniacian (Venezuela, 1970, correlation chart) Querecual Formation (Hedberg, 1937a), which are exposed along the entire northern edge of the East Venezuela Basin. The thickness is a uniform 750 m. The Querecual Formation finds a facial and temporal equivalent in the La Luna Formation of western Venezuela (Fig. 4).

The Querecual Formation represents the lower part of the Guayuta Group (Liddle, 1928) which extends into the Maastrichtian. The upper part is known as the San Antonio Formation (Hedberg,

1937a) east of the Bay of Barcelona, and differs lithologically from the Querecual only through the presence of interbedded sandstone and chert of variable quantity. West of the Bay of Barcelona, the San Antonio Formation is renamed by some authors as the Mucaria Formation (Renz and Short, 1960), because siliceous shale predominates.

The Late Cretaceous sea must have extended almost as far south as the Orinoco (Fig. 3D), for Upper Cretaceous glauconitic sandstone and mudstone, as well as carbonaceous shale, and thin beds of dolomite or dolomitic limestone are known from the subsurface in wells from the El Baúl Arch to the Orinoco Delta (Fig. 1). These sediments comprise the Tigre Formation (Dusenbury, 1960a), of which about 600 m conformably overlie the non-marine Canoa Formation, and together with the latter forms the Temblador Group. The sandstone of the upper part of the Tigre Formation indicates a regressive phase (Venezuela, 1970, p. 588) that effectively terminated the Cretaceous transgression in the southern part of the East Venezuela Basin.

Marine sedimentation continued, however, along the southern margin of the Caribbean Mountain System west of the Bay of Barcelona, and in the area of the Eastern Interior Ranges, although markedly different sedimentary conditions characterized these two regions (Fig. 3E).

The Maastrichtian to Lower Tertiary interval in the Eastern Interior Ranges is represented by the Santa Anita Group (Hedberg, 1937b), of which only the lowest members, the San Juan Formation (Hedberg, 1937a) and the Vidoño Formation (Hedberg and Pyre, 1944) are relevant here. The San Juan Formation represents a wedge of shallow-water, well-sorted, coarsely bedded to massive sandstone which grades laterally northward as well as upward in section into the foraminiferal, black, deep-water shales of the Vidoño Formation. These deposits mark the beginning of an interval of transgressive–regressive environments along the southern margin of a deep-water basin (Venezuela, 1970; Saizarbitoria, 1972). West of the Bay of Barcelona, the San Juan Formation is missing, and here the Guayuta Group extends into the Maastrichtian to be replaced near the Mesozoic–Tertiary boundary by the Guárico Formation (Mencher, 1950). The latter is a thick (more than 2,000 m) flysch section which grades to the southeast and east into the Santa Anita Group. The turbidites of the Guárico Formation develop "wildflysch" character with the inclusion of numerous exotic blocks near the present contact with the metamorphic section (e.g., Beck, 1978). Rocks of this formation border the metamorphic section in a belt about 80 km wide from the Bay of Barcelona to the Mérida Andes. Similar Maastrichtian to Lower Tertiary flysch deposits are found under a variety of formational names in the Barquisimeto Trough, west and southwest of Barquisimeto.

Clearly, west of the Bay of Barcelona, a narrow, deep basin of rapid deposition was located along the present southern and western margin of the Caribbean Mountain System in the Maastrichtian to Early Tertiary, and was connected with the flysch basins of the hinge zone of western Venezuela (Fig. 3E).

The metamorphic complex of the Caribbean Mountain System

No attempt can be made here to discuss the controversial stratigraphy of the metamorphic rocks of the Caribbean Mountain System in detail. Various recent opinions have been expressed and summarized in Menéndez (1967), Venezuela (1970), Bell (1971, 1974), Petzall (1972), Harvey (1972), and Maresch (1974).

It should be emphasized in addition that the boundary between the metamorphic and non-metamorphic sections in Venezuela is not everywhere easily recognizable. Along the southern and western margin of the Caribbean Mountain System, the effects of metamorphism in general seem to die out fairly gradually. Nevertheless, the metamorphic–nonmetamorphic boundary appears to approximate the dividing line between so many other basic differences that the subdivision seems justified:

(1) The sedimentary, non-metamorphic section of the East Venezuela Basin is notably free of volcanic influence. The metamorphic section contains abundant volcanic material.

(2) The sedimentary, non-metamorphic section is, with the exception of the narrow Foothills Tectonic Belt and Thrust Fault Zone (Fig. 5), autochthonous. The metamorphic section is in large part, if not entirely, allochthonous.

(3) The metamorphic section contains marine Jurassic and Cretaceous strata. The non-metamorphic strata are restricted to the Cretaceous and Tertiary, though minor Jurassic terrestrial strata may locally be present.

As indicated above, the metamorphic grade in-

creases from south to north, from the west near Barquisimeto, and from the east on Araya-Paria. The highest-grade rocks may be referred to the epidote–amphibolite facies, and occur directly along the Caribbean coast west of the Bay of Barcelona, the western tip of Araya Peninsula, and on Margarita Island (Bellizzia, 1969). In situ eclogite occurrences are known from the highest-grade zone (e.g., Morgan, 1970; Maresch, 1974). For the purpose of this chapter, the metamorphic section west of the Bay of Barcelona may be divided into the basal Caracas Group (Aguerrevere and Zuloaga, 1937), a sequence of younger formations informally grouped together as the post-Caracas Group (Seiders, 1965; Menéndez, 1967), and the Villa de Cura Group (Aguerrevere and Zuloaga, 1937) with associated metavolcanics (Fig. 4).

The ages of these metamorphic strata can, for the most part, be ascertained only indirectly, for fossils are rare and often of no biostratigraphic value. On the basis of these sparse occurrences (Wolcott, 1943; Dusenbury and Wolcott, 1949; Dusenbury, 1960b; Bermúdez and Rodriguéz, 1962; Urbani, 1969; Furrer, 1972; Bellizzia, 1972a; González de Juana and Vignali, 1972; Vignali, 1972; Macsotay, 1972a, b; Asuaje, 1972; González de Juana et al., 1968, 1972), contact relationships of dated intrusives and extrusives (González de Juana et al., 1974), and correlation with dated strata in Trinidad (e.g., Venezuela, 1970, p. 120), the Caracas Group and its equivalents are estimated to be of Jurassic (or less likely, Triassic) to Lower Cretaceous age. The post-Caracas Group and its equivalents are then Upper Cretaceous in age, while the Villa de Cura Group must be older than mid-Cretaceous. The probable correlation of the Villa de Cura Group with similar dated volcanic rocks of the Dutch Leeward Islands and Venezuelan Caribbean Islands (Maresch, 1974; Beets and Mac Gillavry, 1977) points to a Lower Cretaceous age (Santamaría and Schubert, 1974; Wiedmann, 1978).

The Caracas Group represents a stable-shelf, southwardly transgressive sequence, 3,000–4,000 m thick, which was derived from a granitic source and unconformably overlies granitic basement (Menéndez, 1967; Bell, 1971; González Silva, 1971; Maresch, 1974). The lower part of the Caracas Group consists of quartzose metaclastics and metaconglomerates as well as horizons of marble lenses of probable biohermal origin (Dengo, 1953; Smith, 1953). Rocks of volcanic origin are subordinate. The upper part of the section is graphitic and calcareous, with numerous limestone lenses (e.g., Smith, 1953), and suggests the advent of a euxinic environment over large parts of the depositional shelf. Lithologically similar rocks, though in part more siliceous, are also exposed on Margarita Island (Taylor, 1960; González de Juana and Vignali, 1972; Maresch, 1971, 1972, 1975) and on the Araya-Paria Peninsula (Schubert, 1971; González de Juana et al., 1972). On Margarita Island, these Caracas Group equivalents overlie a thick section of basic metavolcanics, rather than granitic basement (Maresch, 1971, 1972, 1974, 1975). To the west, in the Barquisimeto area and environs to the southeast, equivalents may be found in the Yaritagua Formation and Los Cristales Group (Bellizzia and Rodriguez, 1968), as well as the Araure, Agua Blanca, and Cojedes Formations (Renz and Short, 1960).

In the mid-Cretaceous, abundant volcanic products appear in the stratigraphic column. The post-Caracas Group, which is in general exposed further south than the Caracas Group (see Fig. 5), and is metamorphosed to a lower grade, therefore comprises abundant greenstone metatuffs and basaltic pillowed metalavas in its lowest parts. This volcanic activity declined in the Upper Cretaceous, and was replaced by the sedimentation of a shale–graywacke sequence with only minor metavolcanic content. Metamorphic rocks of this age are essentially absent in eastern Venezuela, with the possible exception of the Los Robles Group on Margarita Island (Hess and Maxwell, 1949; Maresch, 1974), and occur only as huge olistostromes of the Barquisimeto Formation within Tertiary flysch formations in the Barquisimeto area (e.g., Bellizzia and Rodriguez, 1968).

The term Villa de Cura Group refers to a block of water-laid tuffs and lavas, predominantly spilitic but also andesitic, and minor graphitic phyllites and quartz-albite schists (Shagam, 1960; Piburn, 1968). The block is approximately 250 km long, up to 28 km wide, and 3 to 6 km thick. According to Piburn (1968), the Villa de Cura Group has been metamorphosed in the "blueschist facies", but "greenschists" also occur. Overlying and closely associated with the sequence is the Tiara Formation (Smith, 1953), which is lithologically similar but of lower metamorphic grade (prehnite–pumpellyite facies, according to Piburn, 1968). The Tiramuto Formation (Menéndez, 1965), of similar lithology and low metamorphic grade, is a probable correlative from the El Tinaco region.

The importance of the Villa de Cura Group and associated metavolcanics goes beyond the mere anomalous presence of abundant rocks of this type and grade of metamorphism in the Caribbean Mountain System. Whereas geologists had predominantly attempted to correlate the metamorphic Mesozoic sequence of Venezuela with the unmetamorphosed sequence, H.H. Hess (oral communication, in Oxburgh, 1965) was the first to introduce the concept of allochthony on a large scale to the regional geology of the area. He suggested that the entire Villa de Cura Group was allochthonous, and had originated to the north of the present Caribbean coast. The concept of allochthony was developed and elaborated by, among others, Seiders (1965), Menéndez (1967), Bell (1968, 1971, 1972), Bellizzia (1972a, b), Maresch (1974), Skerlec (1976), Skerlec and Hargraves (1977), Stephan (1977a, b), and Beck (1978).

In summary, the following features stand out.

(1) The Villa de Cura Group, the Tiara Formation, and the Tiramuto Formation are klippen or truncated nappes which were originally deposited and metamorphosed prior to the Late Cretaceous north of the present Caribbean coast. Movement of these allochthons towards the south occurred intermittently from the Late Cretaceous or later until the Oligocene.

(2) Regional metamorphism in the Caribbean Mountain System took place in the Late Cretaceous and suggests activation of the orogenic mechanism at this time (Maresch, 1974).

(3) Gravity sliding or thrusting towards the south of various units of the Caribbean Mountain System commenced possibly in the Late Cretaceous and continued to post-Eocene time:

(a) The post-Caracas Group of the Caucagua–El Tinaco Tectonic Belt (Fig. 5) is allochthonous (Bellizzia, 1972b), and is itself covered by allochthonous blocks of older Caracas Group metasediments (Bellizzia, 1972a).

(b) The Paracotos Tectonic Belt, the Foothills Tectonic Belt ("Piemontine nappe" of Beck, 1978), and the imbricated Thrust Fault Zone (Fig. 5) are all allochthonous. In the latter two, non-metamorphic rocks are thus also involved (the Querecual, Mucaria, Guárico sequence; see p. 137.

(c) Stephan (1977a, b) postulates the existence of a major "Lara nappe" in the Barquisimeto area.

(d) The entire Caribbean Mountain System is to be viewed as a huge allochthon (Bellizzia, 1972a, b).

When all these suggestions are taken together, a picture of an orogenic belt emerges that has been internally dismembered, and on the whole has suffered a net transport towards the south.

One of the indications supporting the concept of major disruptions involving the Caribbean Mountain System is the presence of a belt of metamorphosed Mesozoic rocks (Fig. 5) in the northern Sierra Nevada de Santa Marta, the Guajira and Paraguaná Peninsulas, and Aruba (see Butterlin, this volume Chapter 4) of the Dutch Leeward Islands. This belt, termed the Ruma Metamorphic Zone by MacDonald et al. (1971), is approximately parallel to the Caribbean Mountain System metamorphics, but is offset in the Barquisimeto region by about 200–250 km (Bellizzia, 1972b).

This Ruma Metamorphic Zone represents the only other belt of metamorphosed Mesozoic rocks in Venezuela and northern Colombia and therefore deserves special mention. The occurrence in the extreme northwestern tip of the Sierra Nevada de Santa Marta (the Santa Marta and Sevilla Metamorphic Belts of Tschanz et al., 1974) has been described in detail by Doolan (1970, 1972), MacDonald et al. (1971), and Tschanz et al. (1974). Several belts of metabasic, metapelitic, and quartzo-feldspathic rocks are exposed, ranging in grade from lower greenschist southward to the amphibolite facies. On the basis of well-cuttings, these belts appear to continue (with offsets on the order of 100 km or so) to the southwest and northeast beneath Tertiary sediments.

The interpretation of the polymetamorphic history of these schists and gneisses by MacDonald et al. (1971) and Tschanz et al. (1974), on the basis of radiometric age determinations, is complex. A Permian–Triassic regional metamorphic episode is inferred, as well as several more in the time span Jurassic to Paleocene. Not all of these episodes can be recognized in the various metamorphic belts of the region, however, and interpretations are clearly hampered by the resetting of radiometric clocks by younger Tertiary intrusions. Moreover, some of these belts, especially in the far northwest, are definitely allochthonous. Nevertheless, the Mesozoic metamorphic history indicated here appears to be more complex and to have commenced earlier than in the Caribbean Mountain System. The true age of the metamorphic strata is open to debate.

Detailed geologic investigation by MacDonald (1964), Rollins (1965), Lockwood (1965), and Alvarez (1967) has indicated the existence of phyllite, quartzite, schist, and lesser amounts of chert and metavolcanic rocks, as well as minor marbles, in the northern part of the Guajira Peninsula (Fig. 5). Alvarez (1967) has proposed the term Bahía Honda Group for these metamorphics of greenschist grade. Fossils indicating a stratigraphic range from Upper Aptian to Maastrichtian (Lockwood, 1965; Alvarez, 1967) strongly suggest that most of these rocks are Cretaceous in age. Folding and metamorphism occurred in latest Cretaceous to Early Tertiary time (Alvarez, 1971). Interestingly, the metaclastics of the Guajira are postulated to have been derived from a northern source (Alvarez, 1971).

GEOLOGICAL HISTORY

One important aspect of the Mesozoic geological history of the Northern Andes, the environment of deposition of the rocks involved, is summarized in Fig. 3. Apparently continental red bed sedimentation characterized the Triassic–Jurassic periods, although this time interval is not recorded in the west-Andean region. Marine sediments are typical of the Cretaceous.

A unified picture of orogenic activity in the Northern Andes is, however, difficult to deduce for a number of reasons. For example, the quality of geological data available is highly variable. On the one hand, excellent paleontological control in the Cretaceous section of the Cordillera Oriental of Colombia allows the recognition of relatively minor disturbances during the course of sedimentation. On the other, even the gross age of voluminous rock series such as the Cordillera Occidental of Colombia and the Caribbean Mountain System of Venezuela is only poorly known. Tectonic disturbances affecting such rocks are therefore difficult to bracket in time. In addition, as Shagam (1975) has also pointed out, it is often doubtful whether the true stratigraphy of metamorphosed sequences has been satisfactorily corrected for the effects of metamorphic zonation.

Descriptions of orogenic activity are understandably biased by the particular author's concept of orogenic theory. While, for instance, older work attempts to view the geological history using Stillean concepts, many workers have in recent years come to view the Andean system of Colombia and Ecuador as a prime example of an orogenic mode in the sense of plate tectonics.

Although the tenets of plate tectonics appear to signal the right approach to the problem, much more basic data will be required before a satisfactory,

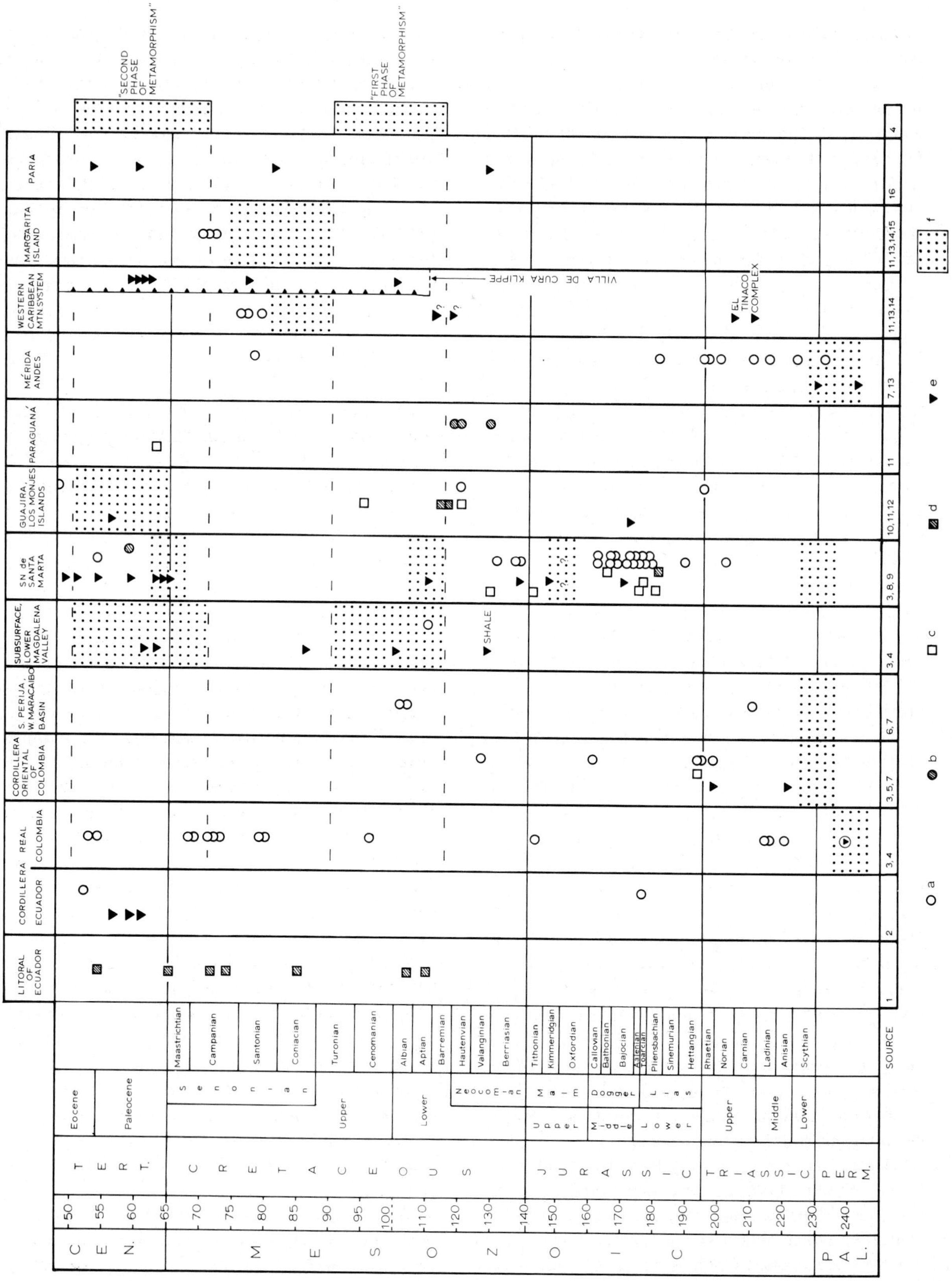
"SECOND PHASE OF METAMORPHISM"
"FIRST PHASE OF METAMORPHISM"
PARIA
MARGARITA ISLAND
WESTERN CARIBBEAN MTN SYSTEM
VILLA DE CURA KLIPPE
EL TINACO COMPLEX
MÉRIDA ANDES
PARAGUANÁ
GUAJIRA, LOS MONJES ISLANDS
SN de SANTA MARTA
SUBSURFACE, LOWER MAGDALENA VALLEY
SHALE
S. PERIJA, W MARACAIBO BASIN
CORDILLERA ORIENTAL OF COLOMBIA
CORDILLERA REAL COLOMBIA
CORDILLERA ECUADOR
LITORAL OF ECUADOR
SOURCE
4
16
11,13,14,15
11,13,14
7,13
11
10,11,12
3,8,9
3,4
6,7
3,5,7
3,4
2
1
Maastrichtian
Campanian
Santonian
Coniacian
Turonian
Cenomanian
Albian
Aptian
Barremian
Hautenvian
Valanginian
Berriasian
Tithonian
Kimmeridgian
Oxfordian
Callovian
Bathonian
Bajocian
Aalenian
Toarcian
Pliensbachian
Sinemurian
Hettangian
Rhaetian
Norian
Carnian
Ladinian
Anisian
Scythian
Eocene
Paleocene
Senonian
Upper
Lower
Neocomian
Malm
Dogger
Lias
Upper
Middle
Lower
Upper
Middle
Lower
TERT.
CRETACEOUS
JURASSIC
TRIASSIC
PERM.
50 55 60 65 70 75 80 85 90 95 100 110 120 130 140 150 160 170 180 190 200 210 220 230 240
CEN.
MESOZOIC
PAL.
a
b
c
d
e
f

unified model can evolve. There is little doubt, however, that major tectonic rotations and relative translations of large crustal blocks have occurred in the Northern Andes during the Late Mesozoic and Tertiary. For instance, MacDonald and Opdyke (1972) conclude, on the basis of paleomagnetic evidence, that Guajira Peninsula was located approximately 10° south of its present latitude in the Cretaceous, perhaps in a position offshore of present-day Peru. Scott (1978) has presented paleomagnetic evidence for the Triassic Luisa Formation of Colombia (see p. 128 and concludes that the Cordillera Central block was originally located along the Paleozoic South American margin of southern Peru to northernmost Chile. This translation, which is coupled with a 90° clockwise rotation, must have occurred in post-Triassic time. Skerlec (1976) and Skerlec and Hargraves (1977) have similarly shown that the Dutch Leeward Islands (see Butterlin, Chapter 4 this volume), as well as a number of rock units of the Caribbean Mountain System of Venezuela, show evidence of a 90° clockwise rotation in Late Mesozoic to Tertiary time.

A variety of paleotectonic models have been considered which attempt to take into account such large-scale rotations and translations. Several major themes are common to almost all of these models and have already been touched upon in other sections of this summary. The west-Andean region may be confidently considered to be of oceanic character. Whether the volcanics of the area are interpreted to be predominantly Pacific mid-ocean ridge basalts or basic to intermediate island-arc derivatives will depend to some extent on a better knowledge of the geochemistry of the East Pacific ocean floor. The west-Andean region must have subsequently been sutured onto the continental edge of South America. There may indeed have been several such accreting slivers. A net northeastward translation of such slivers as well as slices of the sialic margin appears to have taken place and is probably still in progress. In the Venezuelan section of the Andean belt, there is ample evidence for large-scale southward thrusting or gravity sliding. The distribution and type of metamorphic rocks exposed here suggest that a collision between the continental margin and the island arc now north of the present coast has occurred in Late Cretaceous time (see below).

Clearly, a unified paleotectonic model for the Northern Andes will be possible only when the configuration of such mobile elements can be reconstructed with confidence for any given interval of geological time. As mentioned earlier, it is for this reason that Fig. 3 can not be thought of as a simple paleogeographic summary.

MAGMATISM AND METAMORPHISM

Two further important aspects of geological history in the Northern Andes concern acid intrusive igneous activity, which is an integral part of the classical orogenic concept (see for instance Shagam, 1975), and regional metamorphism, from which important clues regarding the style of tectonic activity may be deduced. Even if we have to admit considerable ignorance concerning the Mesozoic paleogeography of the Northern Andes, it is instructive to summarize, area by area, the pertinent data available.

In Fig. 6 known radiometric dates and ages of metamorphism, where the latter can be confidently assigned to a relatively short time interval, have been compiled. Most of the ages were determined by K/Ar methods, so that caution must be exercised in interpreting them. No attempt has been made to influence the distribution by subjectively singling out "unrealistic" determinations. In general, the given dates will be "minimum" ages.

A number of authors have cited evidence for a Late Permian to Early Triassic age of regional metamorphism and strong deformation in the Cordillera

Fig. 6. Summary of published radiometric age determinations for the Northern Andes. Most have been obtained by the K/Ar method (see pertinent "Source" for further information). Ages with a reported relative error exceeding ± 10% have been neglected. Time scale after Van Eysinga (1975).

Source: 1 = Goossens and Rose (1973); 2 = Herbert (1977); 3 = Irving (1975); 4 = Thery et al. (1977); 5 = MacDonald and Opdyke (1974); 6 = Mencher (1963); 7 = Shagam (1975); 8 = MacDonald et al. (1971); 9 = Tschanz et al. (1974); 10 = Lockwood (1965); 11 = Santamaría and Schubert (1974); 12 = MacDonald and Opdyke (1972); 13 = Martín B. (1968); 14 = Maresch (1974); 15 = González de Juana and Vignali (1972); 16 = González de Juana et al. (1974).

Symbols: Age determined on: *a* = acid intrusive; *b* = basic intrusive; *c* = acid extrusive; *d* = basic extrusive; *e* = metamorphosed rock; *f* = probable age of metamorphism.

Central of Colombia, the Sierra de Perijá and Santander Massif of the Cordillera Oriental, the Sierra Nevada de Santa Marta, and the Mérida Andes. A pre-Triassic, Late Paleozoic event also appears likely on Guajira Peninsula (e.g. Lockwood, 1965) and for the El Tinaco Complex of the Caribbean Mountain System (Martín B., 1968). On the other hand, others prefer a Caledonian age for metamorphism in the Cordillera Central/Real of Colombia (Stibane, 1970) and Ecuador (Herbert, 1977) and the Santander Massif of the Cordillera Oriental of Colombia (Irving, 1975). Plausible arguments have been advanced for both views. A thoughtful discussion of the problem is given by Shagam (1975). These Paleozoic metamorphic events may generally be considered to be of the low- to intermediate-pressure type (in the sense of Miyashiro, 1972).

By contrast, regions subjected to Mesozoic or earliest Tertiary regional metamorphism are restricted to the Pacific–Caribbean margins and show evidence of high P/T metamorphic gradients. In addition, the ultramafic rocks of the Northern Andes occur almost exclusively in these belts. Specifically, these are the west-Andean region, the Ruma Metamorphic Zone and the Caribbean Mountain System of Venezuela. In the west-Andean region, the grade of metamorphism is generally very low, although it must be admitted that the metamorphic petrology here is still very poorly known. The age of the metamorphism, inasmuch as recrystallization is probably related to ocean-floor processes (e.g., Colmenares, 1978) and simple burial (e.g., Nelson, 1957) is of only minor interest. Of considerable importance, however, are isolated exposures of blueschists, which have been reported from the Cauca Valley (Orrego et al., 1980).

The Ruma Metamorphic Zone has already been described in some detail (see p. 140). Metamorphic events are placed in the mid-Cretaceous and again the Maastrichtian–Paleocene in northwestern Sierra Nevada de Santa Marta and in the Maastrichtian–Paleocene on Guajira. Only the low-pressure type of metamorphism has been regionally recognized (Lockwood, 1965; Doolan, 1970), but, locally, boulders of eclogite are known from Guajira (Green et al., 1968).

Such eclogites are much more common as in-situ constituents along the Caribbean coast from Puerto Cabello to Margarita Island of the Caribbean Mountain System. They appear to be a normal consequence of prograde metamorphism from south to north (Morgan, 1970; Maresch, 1974, 1977) and yield in-

valuable information on the metamorphic history of this mountain range. The required high P/T ratio necessary during metamorphism may at present best be explained by calling upon a collision between the Venezuelan continental margin and the Aruba–Blanquilla island arc now to the north of the coast (Maresch, 1974; Beets and Mac Gillavry, 1977). The Villa de Cura klippe can then logically be considered as a former fragment of this island arc. Constraints on the timing of the metamorphic event suggest an early Senonian age for the collision.

Thery et al. (1977) have compared radiometric ages from northern South America (see also Fig. 6) and have concluded that two metamorphic episodes, a mid-Cretaceous phase and a Maastrichtian to Eocene phase, may be recognized from northern Colombia to Paria Peninsula. If Thery et al.'s generalizations are indeed correct, and probably more radiometric data will be needed to prove their point, then the early Senonian age of the island-arc collision is clearly discordant. This may be a further indication that this collision mechanism is applicable only to one segment of the Northern Andes mountain belt (Maresch, 1976).

One of the most impressive features of the Andes as a whole are the many large acid to intermediate plutons exposed (Gansser, 1973). The unevenness of data distribution and quality make the recognition of distinct age provinces for these plutons a potentially hazardous exercise in the Northern Andes (see for example Irving, 1975). Nevertheless, the distribution of plutons in time and space does not appear to be uniform (Fig. 6).

Apparent Upper Paleozoic to Lower Triassic plutons have been dated mainly in the Cordillera Central of Colombia, the Sierra de Perijá, the Mérida Andes, and the El Baúl Uplift (see Fig. 3). Triassic–Jurassic plutons are known in the Cordillera Oriental of Colombia, the eastern side of the Cordillera Central of Colombia and the Cordillera Real of Ecuador, the Sierra Nevada de Santa Marta, Guajira Peninsula, the Mérida Andes and probably Paria Peninsula. Cretaceous acid intrusives are concentrated along the axis of the Cordillera Central of Colombia and the Coast Ranges of the Caribbean Mountain System, but have been recorded to a minor extent from almost all other areas as well. Lower Tertiary plutons are generally restricted to the west-Andean region, but are reported as far east as Guajira Peninsula from the Ruma Metamorphic Zone. Explanations for such an apparent distribution in pluton ages remain speculative at present (see Tschanz et al., 1974; Irving, 1975; Shagam, 1975, for recent contributions to the problem).

From our present knowledge of magmatism, metamorphism, and deformation, we are forced to conclude that, while tantalizing new ideas have evolved on the geological history of the Northern Andes, an overall paleotectonic and paleogeographic synthesis of the region in Mesozoic to Early Tertiary time is not yet possible.

REFERENCES

Aguerrevere, S.E. and Zuloaga, G., 1937. Observaciones geológicos en la parte central de la Cordillera de la Costa, Venezuela. Bol. Geol. Min., 1: 3–22.

Alvarez, W., 1967. Geology of the Simarua and Carpintero Areas, Guajira Peninsula, Colombia. Ph. D. Thesis, Princeton Univ., Princeton, N. J., 168 pp.

Alvarez, W., 1971. Fragmented Andean belt of northern Colombia. In: T.W. Donnelly (Editor), Caribbean Geophysical, Tectonic, and Petrologic Studies, Geol. Soc. Am. Mem., 130: 77–96.

Asuaje, L.A., 1972. Geología de la región de Guatire – Cabo Codera (Abstr.). Mem. IV Venez. Geol. Congr., Bol. Geol., Publ. Espec., 5 (3): 1289–1290.

Barrero L., D., 1968. Geology of the Area West of Payandé, Tolíma, Colombia. M.A. Thesis, Indiana Univ., Bloomington, Ind., 60 pp.

Beck, C.M., 1978. Polyphasic Tertiary tectonics of the interior range in the central part of the Western Caribbean Chain, Guárico State, northern Venezuela. Geol. Mijnbouw, 57: 99–104.

Beets, D.J. and Mac Gillavry, H.J., 1977. Outline of the Cretaceous and Early Tertiary history of Curaçao, Bonaire and Aruba. GUA Pap. Geol., Ser. 1, 10: 1–6.

Bell, J.S., 1968. Geología de la región de Camatagua, Estado Aragua, Venezuela. Bol. Geol., 9: 292–440.

Bell, J.S., 1971. Tectonic evolution of the central part of the Venezuelan Coast Ranges. In: T.W. Donnelly (Editor), Caribbean Geophysical, Tectonic, and Petrologic Studies, Geol. Soc. Am. Mem., 130: 107–118.

Bell, J.S., 1972. Geotectonic evolution of the southern Caribbean area. In: R. Shagam et al. (Editors), Studies in Earth and Space Sciences, Geol. Soc. Am. Mem., 132: 369–386.

Bell, J.S., 1974. Venezuelan Coast Ranges. In: A.M. Spencer (Editor), Mesozoic–Cenozoic Orogenic Belts. Geol. Soc. (London) Spec. Publ., 4: 683–703.

Bellizzia G., A., 1969. Mapa Geológico de la República de Venezuela. Ministerio de Minas e Hidrocarburos, Dirección de Geología, Caracas, Venezuela.

Bellizzia G., A., 1972a. Sistema Montañoso del Caribe, borde sur de la placa Caribe ¿Es una cordillera alóctona? Mem. VI Carib. Geol. Conf., Margarita, 1971, pp. 247–258.

Bellizzia G., A., 1972b. Is the entire Caribbean Mountain Belt of northern Venezuela allochthonous? In: R. Shagam et al. (Editors), Studies in Earth and Space Sciences, Geol. Soc. Am. Mem., 132: 363–368.

Bellizzia G., A. and Rodríguez G., D., 1968. Consideraciónes sobre la estratigrafía de los Estados Lara, Yaracuy, Cojedes, y Carabobo. Bol. Geol., 9: 515–563.

Bermúdez, P.J. and Rodríguez G., D., 1962. Notas sobre la presencia de tintínidos o calpionelas en Venezuela. Asoc. Venez. Geol. Min. Petról. Bol. Inform., 5: 51–57.

Blondel, F. et al. (Comision de la Carta Geológico del Mundo) 1964. Mapa Geológico de América del Sur.

Botero A., G., 1963. Contribución al conocimiento de la geología de la zona central de Antioquia. An. Fac. Min., 57: 1–110.

Bowen, J.M., 1972. Estratigrafía del Precretáceo en la parte norte de la Sierra de Perijá. Mem. IV Venez. Geol. Congr., Bol. Geol., Publ. Espec., 5 (2): 729–761.

Bürgl, H., 1961. Historia geológica de Colombia. Rev. Acad. Colomb. Cienc. Exact. Fís. Nat., 11: 137–191.

Bürgl, H., 1962. Sedimentación cíclica en el geosinclinal cretáceo de la Cordillera Oriental de Colombia. Bol. Geol., 7: 85–118.

Bürgl, H., 1963. Die rhythmischen Bewegungen der Kreidegeosynklinale der Ostkordillere Kolumbiens. Geol. Rundsch., 53: 706–731.

Bürgl, H., 1964. El "Jura-Triásico" de Colombia. Bol. Geol., 12: 5–31.

Bürgl, H., 1967. The orogenesis of the Andean system of Colombia. Tectonophysics, 4: 429–443.

Butterlin, J., 1972. La posición estructural de los Andes de Colombia. Mem. IV Venez. Geol. Congr., Bol. Geol., Publ. Espec., 5 (2): 1185–1200.

Campbell, C.J., 1974a. Structural classification of northwestern South America. Verh. Naturforsch. Ges. Basel, 84: 68–79.

Campbell, C.J., 1974b. Colombian Andes. In: A.M. Spencer (Editor), Mesozoic–Cenozoic Orogenic Belts. Geol. Soc. (London) Spec. Publ., 4: 705–724.

Campbell, C.J., 1974c. Ecuadorian Andes. In: A.M. Spencer (Editor), Mesozoic–Cenozoic Orogenic Belts. Geol. Soc. (London) Spec. Publ., 4: 725–735.

Campbell, C.J. and Bürgl, H., 1965. Section through the Eastern Cordillera of Colombia, South America. Geol. Soc. Am. Bull., 76: 567–590.

Case, J.E., 1974. Oceanic crust forms basement of eastern Panamá. Geol. Soc. Am. Bull., 85: 645–652.

Case, J.E., Durán S., L.G., López R., A., and Moore, W.R., 1971. Tectonic investigations in western Colombia and eastern Panamá. Geol. Soc. Am. Bull., 82: 2685–2711.

Cediel, F., 1969. Die Girón-Gruppe. Eine früh-mesozoische Molasse der Ostkordillere Kolumbiens. Neues Jahrb. Geol. Palaeontol. Abh., 133: 111–162.

Cediel, F., Ujueta, G. and Cáceres, C., 1976. Mapa Geológico de Colombia (+ accompanying memoir, 22 pp.), 1 : 1,000,000. Ediciones Geotec, Bogotá.

Cediel, F., Colmenares, F. and Mojica, J., 1978. Sobre la edad de la Formación Saldaña en el Mesozoico Colombiano. 6. Geowiss. Lateinamer.-Koll. Stuttgart, Tagungsh., Nov. 1978, p. 11.

Colmenares, F., 1978. Geología de la sección Buga–Buenaventura, Cordillera Occidental, Colombia. *6. Geowiss. Lateinamer.-Koll. Stuttgart, Tagungsh.*, Nov. 1978, p. 11.

Dengo, G., 1953. Geology of the Caracas region. *Geol. Soc. Am. Bull.*, 64: 7–40.

Doolan, B.L., 1970. *The Structure and Metamorphism of the Santa Marta Area, Colombia, South America*. Ph. D. Thesis, New York State Univ., Binghamton, N.Y., 200 pp.

Doolan, B.L., 1972. The structure and metamorphism of the Santa Marta area, Colombia, South America. *Mem. VI Carib. Geol. Conf., Margarita*, 1971, pp. 239–240.

Dusenbury, A.N. Jr., 1960a. The stratigraphy of the Cretaceous Temblador Group of the eastern Venezuela basin. *Asoc. Venez. Geol. Min. Petról. Bol. Inform.*, 3: 246–257.

Dusenbury, A.N. Jr., 1960b. Revision of the microfauna described from the Cretaceous metamorphics in Quebrada Yaguapa. *Asoc. Venez. Geol. Min. Petról. Bol. Inform.*, 3: 316–317.

Dusenbury, A.N. Jr. and Wolcott, P.P., 1949. Rocas metamórficas cretácicas en la Cordillera de la Costa, Venezuela. *Asoc. Venez. Geol. Min. Petról. Bol. Inform.*, 1: 17–26.

Engleman, R., 1935. Geology of Venezuelan Andes. *Am. Assoc. Petrol. Geol. Bull.*, 19: 769–792.

Faucher, B. and Savoyat, E., 1973. Esquisse géologique des Andes de l'Equateur. *Rev. Géogr. Phys. Géol. Dyn.*, 15: 115–142.

Feininger, T., 1970. The Palestina fault, Colombia, *Geol. Soc. Am. Bull.*, 81: 1201–1216.

Feininger, T., 1975. Origin of petroleum in the Oriente of Ecuador. *Am. Assoc. Petrol. Geol. Bull.*, 59: 1166–1175.

Feo-Codecido, G., 1971. Guía de la excursión a la Península de Paraguaná, Estado Falcón. *Mem. IV Venez. Geol. Congr., Bol. Geol., Publ. Espec.*, 5 (1): 304–315.

Feo-Codecido, G., 1972. Contribución a la estratigrafía de la Cuenca Barinas–Apure. *Mem. IV Venez. Geol. Congr., Bol. Geol., Publ. Espec.*, 5 (2): 773–792.

Furrer, M.A., 1972. Fossil tintinnids in Venezuela. *Mem. VI Carib. Geol. Conf., Margarita*, 1971, pp. 451–454.

Gansser, A., 1955. Ein Beitrag zur Geologie und Petrographie der Sierra Nevada de Santa Marta (Kolumbien, Südamerika). *Schweiz. Miner. Petrogr. Mitt.*, 35: 209–279.

Gansser, A., 1973. Facts and theories on the Andes. *J. Geol. Soc. Lond.*, 129: 93–131.

Garner, A.H., 1926. Suggested nomenclature and correlation of geological formations in Venezuela. *Am. Inst. Min. Metall. Trans.*, 1926: 677–684.

Gerth, H., 1955. *Der geologische Bau der südamerikanischen Kordillere*. Borntraeger, Berlin, 264 pp.

Geyer, O.F., 1973. Das präkretazische Mesozoikum von Kolumbien. *Geol. Jahrb.*, Ser. B, 5: 1–155.

Geyer, O.F., 1974. Der Unterjura (Santiago-Formation) von Ekuador. *Neues Jahrb. Geol. Paläontol. Monatsh.*, 1974: 525–541.

Göbel, V.W., 1978. Über Bau und geologische Entwicklung der südlichen Zentralkordillere Kolumbiens. *Münster. Forsch. Geol. Paläontol.*, 44/45: 127–141.

Goldsmith, R., Marvin, R.F. and Mehnert, H.H., 1971. Radiometric ages in the Santander Massif, Eastern Cordillera,

Colombian Andes. *U.S. Geol. Surv. Prof. Pap.*, 750-D: 44–49.

González de Juana, C., Muñoz, N.G. and Vignali, M., 1965. Reconocimiento geológico de la parte oriental de Paria. *Asoc. Venez. Geol. Min. Petról. Bol. Inform.*, 8: 255–279.

González de Juana, C., Muñoz, N.G. and Vignali C., M., 1968. On the geology of Eastern Paria (Venezuela). *Trans. IV Carib. Geol. Conf., Port-of-Spain, Trinidad*, pp. 25–29.

González de Juana, C., Muñoz, N.G. and Vignali C., M., 1972. Reconocimiento geológico de la Península de Paria, Venezuela. *Mem. IV Venez. Geol. Congr., Bol. Geología, Publ. Espec.*, 5 (3): 1549–1588.

González de Juana, C. and Vignali C., M., 1972. Rocas metamórficas e ígneas en la Península de Macanao, Margarita, Venezuela. *Mem. VI Carib. Geol. Conf., Margarita*, 1971, pp. 63–68.

González de Juana, C., Santamaría, F. and Navarro F., E., 1974. A few considerations on the age, origin and relations of the Dragon Gneiss, Paria Peninsula, Venezuela. *Verh. Naturforsch. Ges. Basel*, 84: 153–163.

González S., L.A., 1971. Guía de la excursión a la parte central de la Cordillera de la Costa: Caracas–Valencia–El Palito–Morrón. *Mem. IV Venez. Geol. Congr., Bol. Geol., Publ. Espec.*, 5 (1): 363–379.

Gonzáles S., L.A., 1972. Geología de la Cordillera de la Costa, zona centro-occidental. *Mem. IV Venez. Geol. Congr., Bol. Geol., Publ. Espec.*, 5 (3): 1589–1618.

Goossens, P.J., 1973. Arco de islas volcanicas durante el Mesozoico superior al Tertiary inferior a lo largo del margen continental noroccidental de la America del Sur. *Congr. Latinoamericano de Geología, 2nd, Caracas, Resumenes*, pp. 67–68 (available from author in form of English synopsis of presentation).

Goossens, P.J. and Rose, W.I. Jr., 1973. Chemical composition and age determination of tholeiitic rocks in the Basic Igneous Complex, Ecuador. *Geol. Soc. Am. Bull.*, 84: 1043–1052.

Goossens, P.J. and Rose, W.I. Jr., 1975. Geochemistry of basalts of the Basic Igneous Complex of northwestern South America and Panama. *Trans. Am. Geophys. Union EOS*, 56: 474.

Goossens, P.J., Rose, W.I. Jr. and Flores, D., 1977. Geochemistry of tholeiites of the Basic Igneous Complex of northwestern South America. *Geol. Soc. Am. Bull.*, 88: 1711–1720.

Green, D.H., Lockwood, J.P. and Kiss, E., 1968. Eclogite and almandine–jadeite–quartz rock from the Guajíra Peninsula, Colombia, South America. *Am. Miner.*, 53: 1320–1335.

Guillaume, M.A., Bolli, H.M. and Beckmann, J.P., 1972. Estratigrafía del Cretáceo Inferior en la Serranía del Interior, Oriente de Venezuela. *Mem. IV Venez. Geol. Congr., Bol. Geol., Publ. Espec.*, 5 (3): 1619–1659.

Harvey, S.R.M., 1972. Origin of the southern Caribbean Mountains. In: R. Shagam et al. (Editors), *Studies in Earth and Space Sciences, Geol. Soc. Am. Mem.*, 132: 387–400.

Hea, J.P. and Whitman, A.B., 1960. Estratigrafía y petrografía de los sedimentos pre-Cretácicos de la parte norte central de la Sierra de Perijá, Estado Zulia, Venezuela. *Mem. III Venez. Geol. Congr., Bol. Geol., Pub. Espec.*, 3: 351–376.

Hedberg, H.D., 1931. Cretaceous limestones as petroleum source rocks in northwestern Venezuela. *Am. Assoc. Petrol. Geol. Bull.*, 15: 229–244.

Hedberg, H.D., 1937a. Stratigraphy of the rio Querecual section of northeastern Venezuela. *Geol. Soc. Am. Bull.*, 48: 1971–2024.

Hedberg, H.D., 1937b. Stratigraphy of the rio Querecual section of northeastern Anzoátegui, Venezuela. *Bol. Geol. Min.*, 1: 239–250.

Hedberg, H.D., 1942. Mesozoic stratigraphy of northern South America. *Proc. Am. Sci. Congr., 8th, Wash., D.C., 1940*, 4: 195–227.

Hedberg, H.D., 1950. Geology of the eastern Venezuela basin (Anzoátegui – Monagas – Sucre – eastern Guárico portion). *Geol. Soc. Am. Bull.*, 61: 1173–1216.

Hedberg, H.D., 1956. Northeastern Venezuela. In: W.F. Jenks (Editor), *Handbook of South American Geology, Geol. Soc. Am. Mem.*, 65: 337–340.

Hedberg, H.D. and Pyre, A., 1944. Stratigraphy of northeastern Anzoátegui, Venezuela. *Am. Assoc. Petrol. Geol. Bull.*, 20: 1–28.

Herbert, H.J., 1977. *Die Grünschiefer der Ost-Kordillere Ecuadors und ihr metamorpher Rahmen*. Doctoral Dissertation, Eberhard-Karls-Universität, Tübingen, 191 pp.

Hess, H.H and Maxwell, J.C., 1949. Geological reconnaissance of the Island of Margarita, I. *Geol. Soc. Am. Bull.*, 60: 1857–1868.

Hoffstetter, R., 1956. Ecuador – Equateur. *Lexique Stratigraphique Internationale*, V. 5a. Centre Nat. Rech. Sci., Paris, 191 pp.

Hubach, E., 1957. Contribución a las unidades estratigráficos de Colombia. *Inst. Geol. Nac. Inf.*, 1212: 1–166.

Hubach, E. and Radelli, L., 1962. *Mapa Geológico de Colombia*. Minist. Minas Petról., Serv. Geol. Nac., Bogotá.

Irving, E.M., 1975. Structural evolution of the northernmost Andes, Colombia. *Geol. Surv. Prof. Pap.*, 846: 1–47.

Jacobs, C., Bürgl, H. and Conley, D.L., 1963. Backbone of Colombia. In: O.E. Childs and B.W. Beebe (Editors), *Backbone of the Americas, Am. Assoc. Petrol. Geol. Mem.*, 2: 62–72.

Julivert, M., 1968. Colombie (première partie). *Lexique Stratigraphique Internationale*, V, 4a. Centre Nat. Rech. Sci., Paris, 650 pp.

Juteau, T., Mégard, F., Raharison, L. and Whitechurch, H., 1977. Les assemblages ophiolitiques de l'occident équatorien: nature pétrographique et position structurale. *Bull. Soc. Géol. Fr.*, 7, XIX, 5: 1127–1132.

Kehrer, W., 1939. Zur Geologie der südlichen Zentral- und Ostkordillere der Republik Kolumbien. *Neues Jahrb. Geol. Paläontol., B*, 80: 1–30.

Kugler, H.G., 1972. The Dragon Gneiss of Paria Peninsula (Eastern Venezuela). *Mem. VI Carib. Geol. Conf., Margarita, 1971*, pp. 113–116.

Kündig, E., 1938. Las rocas precretáceas de los Andes Centrales de Venezuela con algunas observaciones sobre su tectónica. *Bol. Geol. Min.*, 2: 21–43.

Liddle, R.A., 1928. *The Geology of Venezuela and Trinidad*. MacGowan, Fort Worth, Texas, 552 pp.

Lockwood, J.P., 1965. *Geology of the Serranía de Jarara, Guajira Peninsula, Colombia*. Ph. D. Thesis, Princeton University, Princeton, N. J., 237 pp.

MacDonald, W.D., 1964. *Geology of the Serranía de Macuira Area, Guajira Peninsula, Colombia*. Ph. D. Thesis, Princeton University, Princeton, N. J. 167 pp.

MacDonald, W.D., 1968. Estratigrafía, estructura y metamorfísmo, rocas del Jurassico Superior, Peninsula de Paraguaná, Venezuela. *Bol. Geol.*, 9: 441–458.

MacDonald, W.D. and Hurley, P.M., 1969. Precambrian gneisses from northern Colombia, South America. *Geol. Soc. Am. Bull.*, 80: 1867–1872.

MacDonald, W.D. and Opdyke, N.D., 1972. Tectonic rotations suggested by paleomagnetic results from Northern Colombia, South America. *J. Geophys. Res.*, 77: 5720–5730.

MacDonald, W.D. and Opdyke, N.D., 1974. Triassic paleomagnetism of northern South America. *Am. Assoc. Petrol. Bull.*, 58: 208–215.

MacDonald, W.D., Doolan, B.L. and Cordani, U.G., 1971. Cretaceous–Early Tertiary metamorphic K-Ar age values from the South Caribbean. *Geol. Soc. Am. Bull.*, 82: 1381–1388.

Macsotay, O., 1972a. Observaciones acerca de la edad y paleoecología de algunas formaciones de la región de Barquisimeto, Estado Lara, Venezuela. *Mem. IV Venez. Geol. Congr., Bol. Geol., Publ. Espec.*, 5 (3): 1673–1701.

Macsotay, O., 1972b. Significado cronológico y paleoecológico de los amonites desenrolladas de la Formación Chuspita del Grupo Caracas. *Mem. IV Venez. Geol. Congr., Bol. Geol., Publ. Espec.*, 5 (3): 1703–1714.

Maresch, W.V., 1971. *The Metamorphism and Structure of Northeastern Margarita Island*. Ph.D. Thesis, Princeton University, Princeton, N. J., 278 pp.

Maresch, W.V., 1972. Eclogitic-amphibolitic rocks on Isla Margarita, Venezuela: A preliminary account. In: R. Shagam et al. (Editors), *Studies in Earth and Space Sciences, Geol. Soc. Am. Mem.*, 132: 429–437.

Maresch, W.V., 1974. Plate tectonic origin of the Caribbean Mountain System of northern South America: Discussion and proposal. *Geol. Soc. Am. Bull.*, 85: 669–682.

Maresch, W.V., 1975. The geology of northeastern Margarita Island, Venezuela: A contribution to the study of Caribbean plate margins. *Geol. Rundsch.*, 64: 846–883.

Maresch, W.V., 1976. Implications of a Mesozoic to Early Tertiary, collision-type, plate-tectonic model in northern Venezuela for the southern Caribbean region. *Trans. VII Carib. Geol. Conf., Saint-François, Guadeloupe, 1974*, pp. 485–491.

Maresch, W.V., 1977. Similarity of metamorphic gradients in time and space during metamorphism of the La Rinconada Group, Margarita Island, Venezuela. *GUA Pap. Geol.*, Ser. 1, 9: 110–111.

Martín B., C., 1961. Geología del macizo de El Baúl. *Mem. III Venez. Geol. Congr., Bol. Geol., Publ. Espec.*, 3 (4): 1453–1530.

Martín B., C., 1968. Edades isotópicas de rocas Venezolanas. *Bol. Geol.*, 10: 356–379.

Meissner, R.O., Flueh, E.R., Stibane, F. and Berg, E., 1976. Dynamics of the active plate boundary in southwest Colombia according to recent geophysical measurements. *Tectonophysics*, 35: 115–136.

Mencher, E., 1950. Sucesos cretácicos–eocénicos en el norte de Venezuela. *Asoc. Venez. Geol. Min. Petról Bol. Inform.*, 2: 91–99.

Mencher, E., 1963. Tectonic history of Venezuela. In: O.E. Childs and B.W. Beebe (Editors), *Backbone of the Americas, Am. Assoc. Petrol. Geol. Mem.*, 2: 73–87.

Menéndez, A., 1965. Geología del área de El Tinaco, centro-norte del Estado Cojedes, Venezuela. *Bol. Geol.*, 6: 417–453.

Menéndez, A., 1967. Tectonics of the central part of western Caribbean Mountains, Venezuela. *Stud. Trop. Oceanogr.*, 5: 103–130.

Miller, J.B., 1962. Tectonic trends in Sierra de Perijá and adjacent parts of Venezuela and Colombia. *Am. Assoc. Petrol. Geol. Bull.*, 46: 1565–1595.

Miyashiro, A., 1972. Metamorphism and related magmatism in plate tectonics. *Am. J. Sci.*, 272: 629–656.

Morgan, B.A., 1970. Petrology and mineralogy of eclogite and garnet amphibolite from Puerto Cabello, Venezuela. *J. Petrol.*, 11: 101–145.

Nelson, H.W., 1957. Contribution to the geology of the Central and Western Cordillera of Colombia in the sector between Ibagué and Cali. *Leidse Geol. Meded.*, 22: 1–75.

Nygren, W.E., 1950. Bolívar geosyncline of northwestern South America. *Am. Assoc. Petrol. Geol. Bull.*, 34: 1998–2006.

Olsson, A.A., 1956. Colombia. In: W.F. Jenks (Editor), *Handbook of South American Geology, Geol. Soc. Am. Mem.*, 65: 294–326.

Orrego, A., Cepeda, H. and Rodríguez, G.I., 1980. Esquistos glaucofánicos en el área de Jambaló, Cauca (Colombia). *Geol. Norandina*, 1: 5–10.

Oxburgh, E.R., 1965. Geología de la región oriental del Estado Carabobo, Venezuela. *Bol. Geol.*, 6: 113–208.

Petzall, C., 1972. Cuadro sinóptico de unidades estratigráficas norte de Venezuela. *Mem. VI Carib. Geol. Conf., Margarita*, 1971, p. 491.

Piburn, M.D., 1968. Metamorfísmo y estructura del Grupo Villa de Cura, norte de Venezuela. *Bol. Geol.*, 9: 183–290.

Pichler, H., Stibane, F.R. and Weyl, R., 1974. Basischer Magmatismus und Krustenbau im südlichen Mittelamerika, Kolumbien und Ekuador. *Neues Jahrb. Geol. Palaeontol. Monatsh.*, 1974: 102–126.

Pimentel M., N., 1976. Geology of Toas Island. *Mem. 2nd Latinamerican Congr., Bol. Geol. Publ. Espec.*, 7 (2): 523–525.

Rabe, E.H., 1977. Zur Stratigraphie des ostandinen Raumes von Kolumbien. *Giessener Geol. Schr.*, 11: 1–223.

Radelli, L., 1962. Las dos granitizaciones de la Península de La Guajira (Norte de Colombia). *Geol. Colombiana*, 3: 5–19.

Radelli, L., 1967. Géologie des Andes colombiennes. *Trav. Labor. Géol. Fac. Sci. Grenoble Mém.*, 6: 1–457.

Ramírez, C., García, R. and Campos C., V., 1972. Geología de la región de Timotes, Estados Mérida, Barinas y Trujillo. *Mem. IV Venez. Geol. Congr., Bol. Geol., Publ. Espec.*, 5 (2): 898–934.

Ramírez C., C. and Campos C., V., 1972. Geología de la región de La Grita – San Cristobal, Estado Táchira. *Mem. IV Venez. Geol. Congr., Bol. Geol., Publ. Espec.*, 5 (2): 861–897.

Renz, O., 1956. Cretaceous in western Venezuela and the Guajira (Colombia). *Int. Geol. Congr., 20th, Mexico*, p. 342.

Renz, O., 1960. Geología de la parte sureste de la peninsula de La Guajira (República de Colombia). *Mem. III Venez. Geol. Congr., Bol. Geol., Publ. Espec.*, 3 (1): 317–439.

Renz, O. and Short, K.C., 1960. Estratigrafía de la región comprendida entre El Pao y Acarigua, Estados Cojedes y Portuguesa. *Mem. III Venez. Geol. Congr., Bol. Geol., Publ. Espec.*, 3 (1) 277–315.

Rollins, J.F., 1960. *Stratigraphy and Structure of the Goajira Peninsula, Northwestern Venezuela and Northeastern Colombia*. Ph.D. Thesis, University of Nebraska, Lincoln, Nebr.

Rollins, J.F., 1965. Stratigraphy and structure of the Goajira Peninsula, northwestern Venezuela and northeastern Colombia. *Univ. Nebraska Stud.*, 30: 1–102.

Saizarbitoria, I., 1972. Guía de la Excursión PC-4, segunda parte: geología del área de Pertigalete y autopista Puerto La Cruz–El Tigre. *Mem. VI Carib. Geol. Conf., Margarita*, 1971, pp. 58–62.

Santamaría, F. and Schubert, C., 1974. Geochemistry and geochronology of the Southern Caribbean–Northern Venezuela plate boundary. *Geol. Soc. Am. Bull.*, 85: 1085–1098.

Sauer, W., 1965. *Geología del Ecuador*. Edit. Minist. Educación, Quito, 383 pp.

Sauer, W., 1971. *Geologie von Ekuador*. Borntraeger, Berlin, 316 pp.

Schubert, C., 1971. Metamorphic rocks of the Araya Peninsula, eastern Venezuela. *Geol. Rundsch.*, 60: 1571–1600.

Scott, G.R., 1978. Translation of accretionary slivers: Triassic results from the Central Cordillera of Colombia. *Trans. Am. Geophys. Union EOS*, 59: 1058–1059.

Seiders, V.M., 1965. Geología de Miranda central, Venezuela. *Bol. Geol.*, 6: 289–416.

Servicio Nacional de Geología y Minería, 1969. *Mapa Geológico de la República del Ecuador*. 1:1,000,000, Quito.

Shagam, R., 1960. Geology of central Aragua, Venezuela. *Geol. Soc. Am. Bull.*, 71: 249–302.

Shagam, R., 1972a. Evolución tectónica de Los Andes. *Mem. IV Venez. Geol. Congr., Bol. Geol., Publ. Espec.*, 5 (2): 1201–1261.

Shagam, R., 1972b. Andean research project, Venezuela: Principal data and tectonic implications. In: R. Shagam et al. (Editors), *Studies in Earth and Space Sciences, Geol. Soc. Am. Mem.*, 132: 449–463.

Shagam, R., 1975. The northern termination of the Andes. In: A.E.M. Nairn and F.G. Stehli (Editors), *The Ocean Basins and Margins*, Plenum Press, New York, 3: 325–420.

Skerlec, G., 1976. The western termination of the Caribbean Mountains: A progress report. *Trans. VII Carib. Geol. Conf., Saint-François, Guadeloupe*, 1974, p. 492.

Skerlec, G.M. and Hargraves, R.B., 1977. Tectonic significance of paleomagnetic data from the islands of the
southern Caribbean boundary and northern Venezuela.
GUA Pap. Geol., Ser 1, 9: 190–191.

Smith, R.J., 1953. Geology of the Los Teques – Cua region.
Geol. Soc. Am. Bull., 64: 41–64.

Stainforth, R.M., 1969. The concept of seafloor-spreading
applied to Venezuela. *Asoc. Venez. Geol. Min. Petról.
Bol. Inform.*, 12: 257–274.

Stephan, J.F., 1977a. El contacto Cadena Caribe–Andes
Merideños entre Carora y El Tocuyo (Estado Lara, Venezuela). *GUA Pap. Geol.*, Ser. 1, 9: 197–198.

Stephan, J.F., 1977b. Una interpretación de los Complejos
con Bloques asociados a los flysch Paleoceno-Eoceno de la
Cadena Caribe Venezolana: El emplazamiento submarino
de la Napa de Lara. *GUA Pap. Geol.*, Ser. 1, 9: 199–200.

Stephan, J.F., Beck, C. and Macsotay, O., 1977. Reflexiones
sobre unas facies conglomeráticas marinas en el Albiense
del norte de Venezuela. *GUA Pap. Geol.*, Ser. 1, 9: 201–
202.

Stibane, F.R., 1967. Paläogeographie und Tektogenese der
kolumbianischen Anden. *Geol. Rundsch.*, 56: 629–642.

Stibane, F.R., 1968. Zur Geologie von Kolumbien, Südamerika. Das Quetame und Garzón Massiv. *Geotekt.
Forsch.*, 30: 1–85.

Stibane, F.R., 1970. Beitrag zum Alter der Metamorphose
der Zentral-Kordillere Kolumbiens. *Mitt. Inst. Colombo-
Alemán Invest. Cient.*, 4: 77–82.

Sutton, F.A., 1946. Geology of Maracaibo Basin, Venezuela.
Am. Assoc. Petrol. Geol. Bull., 30: 1621–1741.

Taylor, G.C., 1960. Geología de la Isla de Margarita, Venezuela. *Mem. III Venez. Geol. Congr., Bol. Geol., Publ.
Espec.*, 2: 838–893.

Thalman, H.E., 1946. Micropaleontology of Upper Cretaceous and Paleocene in western Ecuador. *Am. Assoc.
Petrol. Geol. Bull.*, 30: 337–347.

Thery, J.-M., Esquevin, J. and Menéndez, R., 1977. Significa-
tion géotectonique de datations radiométriques dans des
sondages de Basse Magdalena (Colombie). *Bull. Cent.
Rech. Explor.-Prod. Elf-Aquitaine*, 1: 475–494.

Trümpy, D., 1943. Pre-Cretaceous of Colombia. *Geol. Soc.
Am. Bull.*, 54: 1281–1304.

Tschanz, C.M., Marvin, R.F., Cruz B., J., Mehnert, H.H. and
Cebula, G.T., 1974. Geologic evolution of the Sierra
Nevada de Santa Marta, northeastern Colombia, *Geol.
Soc. Am. Bull.*, 85: 273–284.

Tschopp, H.J., 1948. Geologische Skizze von Ekuador. *Ver.
Schweiz. Petrol.-Geol. -Ing. Bull.*, 15: 14–45.

Tschopp, H.J., 1953. Oil exploration in the Oriente of Ecuador, 1938–1950. *Am. Assoc. Petrol. Geol. Bull.*, 37:
2303–2347.

Tschopp, H.J., 1956. Ecuador: Upper Amazon Basin geological province. In: W.F. Jenks (Editor), *Handbook of
South American Geology, Geol. Soc. Am. Mem.*, 65:
253–267.

Urbani, F., 1969. Primera localidad fosilífera del Miembro
Zenda de la formación Las Brisas: Cueva El Indio, La
Guairita, Estado Miranda. *Asoc. Venez. Geol. Min. Petról.
Bol. Inform.*, 12: 446–453.

Useche, A. and Fierro, I., 1972. Geología de la región de
Pregonero, Estados Táchira y Mérida. *Mem. IV Venez.
Geol. Congr., Bol. Geol., Publ. Espec.*, 5 (2): 963–998.

Van Eysinga, F.W.B., 1975. *Geological Time Table*. Elsevier,
Amsterdam, 3rd ed.

Venezuela, Dirección de Geología, 1970. Léxico estratigráfico de Venezuela. *Bol. Geol., Publ. Espec.*, 4: 1–756.

Vignali C., M., 1972. Análisis estructural y eventos tectónicos
de la Península de Macanao, Margarita, Venezuela. *Mem.
VI Carib. Geol. Conf., Margarita*, 1971, pp. 241–246.

Weeks, L.G., 1947. Paleogeography of South America. *Am.
Assoc. Petrol. Geol. Bull.*, 31: 1194–1241.

Wiedmann, J., 1978. Ammonites from the Curaçao Lava Formation, Curaçao, Caribbean. *Geol. Mijnbouw*, 57: 361–
364.

Wolcott, P.P., 1943, Fossils from metamorphic rocks of the
Coast Ranges of Venezuela. *Am. Assoc. Petrol. Geol.
Bull.*, 27: 1632.

Zambrano, E., Vásquez, E., Dural, B., Latreille, M. and Cof-
finieres, B., 1971. Síntesis paleogeográfica y petrolera
del occidente de Venezuela. *Mem. IV Venez. Geol. Congr.,
Bol. Geol., Publ. Espec.*, 5 (1): 483–552.

Chapter 6

BRAZIL

S. PETRI AND J.C. MENDES

INTRODUCTION

Brazil (Fig. 1) which occupies most of South America east of 65° W, can be characterized geologically as a great Precambrian shield to the south bordered by the Amazon depression to the north, the Atlantic to the east and the Andean chain to the west. The most striking feature of the Mesozoic geology of this vast area is the change in the sedimentation pattern which occurred during the latter half of the Mesozoic. The early Mesozoic sedimentation was a continuation of that established during the Paleozoic, of deposition of relatively thin but widely distributed beds in a number of slowly subsiding intracratonic basins. At the end of the Jurassic and continuing into the Cretaceous, tectonic activity led to the development of many grabens or semigrabens, roughly parallel to the present Atlantic coast line, which received a thick wedge of sediments. This pattern of sedimentation dominated the remainder of the Mesozoic although the intracratonic basins continued to receive relatively thin deposits. No geosynclinal sedimentation occurs within Brazil, for the Andean geosyncline lies within neighbouring territories to the west.

North of the Precambrian shield lies the great Amazonas Basin covering an area of some 1,000,000 km². It is conventionally divided into three subbasins by two arches, the Purus and Monte Alegre arches (Fig. 2) and by a third flexure, the Iquitos Arch, from the pericratonic Acre Basin which communicates westwards with the Andean geosyncline through the Pastaza Basin of Peru (Morales, 1959; Miura, 1972). In the east the Parnaiba Basin penetrates deeply into the shield area, for with its extension the São Francisco Basin it is separated from the huge Paraná Basin to the south by the Canastra Arch (Fig.

2). The Paraná Basin is the largest of the intracratonic basins with an area of 1,600,000 km² of which one quarter lies in Argentina and one eighth in Paraguay and Uruguay. It is limited to the west by the Asuncion and the Pampean arches and to the south by the Martim Garcia Arch. There is no limiting arch to the east but the outcrops along that margin show a sigmoidal curvature due to the penetration of the basin by northwest–southeast trending positive areas (e.g., Ponta Grossa Arch) (Northfleet et al., 1969; Fulfaro, 1971).

Along the eastern margin a large number of relatively small, intensely faulted grabens are found, often containing a very thick younger Mesozoic–Cenozoic sequence (Fig. 2) as well as two platform basins, the Ceará–Potiguar and the Recife–João Pessoa, which are not rifted (Asmus and Ponte, 1973). A Guyanan rift valley system, Takutu, running east–west, reaches Brazil at its western tip, in the extreme north of the country (Fig. 3, col. 1).

THE MESOZOIC SEDIMENTARY SEQUENCE

The Brazilian Mesozoic sequences are generally incomplete with numerous unconformities (Figs. 3, 4). At the base, the Permian beds of the intracratonic basins are always regressive culminating with a general emergence and widespread erosion. Thus, no Early Triassic sediments are known, and in the Amazon Basin the only Triassic rocks are a few radiometrically dated dykes. In the coastal basins no Triassic is known.

The only Jurassic sediments are of Late Jurassic age. They are present in the Acre and the Northeast and always separated by unconformities from the Triassic beds when present. There was thus an ubiqui-

Fig. 1. Geographical map of Brazil for the localities cited in the text.

tous Early Jurassic hiatus. In most of the areas of sedimentation there is only one pre-Late Jurassic unconformity. Jurassic beds are poorly fossiliferous and always non-marine, fluvio-lacustrine or arid. As in the Triassic no marine beds are known.

In Guyana, some beds of the Takutu rift valley have yielded pollen flora which could be of Jurassic (Pre-Purbeckian) age (Van der Hammen and Burger, 1966). The Late Jurassic beds are conformably covered by Early Cretaceous, Valanginian to Barremian

beds. This sequence is also of non-marine origin and covered unconformably by late Aptian beds, the first marine sediments of Mesozoic age known in Brazil.

In Guyana, the Valanginian–Barremian beds of the Takutu bear a rather rich pollen flora which would indicate near-shore shallow water environments (Van der Hammen and Burger, 1966).

The Purbeckian–Barremian sequence is well developed in the rift valleys of the northeast and is also present in the southern coastal basins. The Paraná and

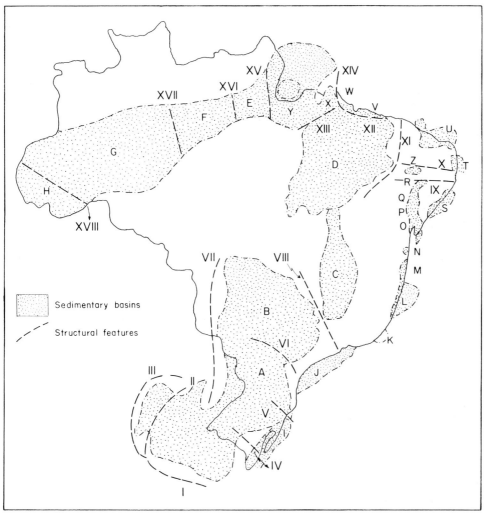

Fig. 2. Map of Brazil with the indication of the sedimentary basins and some of the structural features. A = lower section of the Paraná; B = upper section of the Paraná; C = São Francisco; D = Parnaíba; E = lower section of the Amazonas; F = middle section of the Amazonas; G = upper section of the Amazonas; H = Acre; I = Pelotas; J = Santos; K = Campos; L = Espírito Santo; M = Jequitinhonha; N = Almada; O = Recôncavo; P = South Tucano; Q = North Tucano; R = Jatobá; S = Sergipe–Alagoas; T = Recife–João Pessoa; U = Potiguar (and Ceará); V = Barreirinhas; W = São Luiz; X = Bragança–Vizeu; Y = Amazonas mouth; I = Martim Garcia Arch; II = East Pampean Arch; III = West Pampean Arch; IV = Uruguay–Río Grande shield; V = Torres depressed area; VI = Ponta Grossa Arch; VII = Assuncion Arch; VIII = Canastra Arch; IX = Pernambuco lineament; X = Patos lineament; XI = Serra Grande Arch; XII = Ferrer–Urbano Santos Arch; XIII = Tocantins Arch; XIV = Guamá Arch; XV = Gurupá Arch; XVI = Monte Alegre Arch; XVII = Purus Arch; XVIII = Iquitos Arch.

Parnaíba intracratonic basins exhibit a contemporaneous volcanic sequence with a few sandstone lenses. This sequence is not present in the northern basins indicating one great unconformity covering Triassic, Jurassic and Early Cretaceous times. The Aptian unconformity may not be present in the southern basins but this is not known with certainty for the basins are still not well known.

The Aptian hiatus involved considerable structural readjustments in the coastal areas since the heavily faulted Purbeckian–Barremian sequences exhibit strong dips generally toward the continent while younger beds dip gently toward the ocean. A short time but a widespread hiatus separates the early Aptian from latest Aptian sediments. The Aptian sequence is present in the northeastern and southern

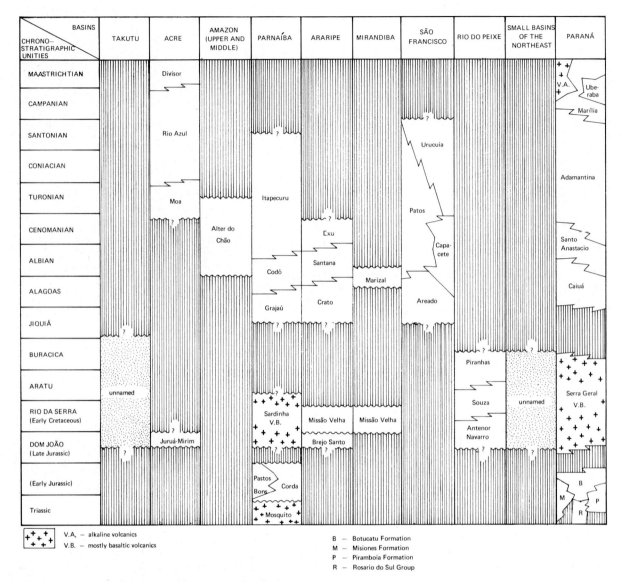

Fig. 3. Correlations of the Mesozoic formations of the interior basins.

basins but not in the northern basins. The late Aptian to Santonian transgressive—regressive sequence marks an important event for the whole eastern half of Brazil. During the middle Albian the marine transgression reached its maximum extent for the whole Mesozoic. It was then present in the interior of the northeast (Chapada do Araripe, Fig. 3, col. 5) and Parnaíba Basin (Codó Sea). Contemporaneous non-marine sediments are known in the São Francisco and Paraná basins.

A different situation is found in the Barreirinhas Basin (Fig. 4, col. 4). This basin was formed only in

late Aptian as a result of an intense taphrogenesis. The erosion of rising block faults permitted the accumulation of 4,000 — 5,000 m of sediments on the sinking block in a short time (late-Aptian—Albian) (Pamplona, 1969; Lima, 1972). The high rate of sedimentation prevented the advance of the sea. The sea was present, however, over the most subsident blocks, on the continental shelf. The Codó Sea which probably entered the Parnaíba Basin (Fig. 3, col. 4) through the Barreirinhas (Fig. 4, col. 4) rift valley formed at the same time was restricted in extent.

The oldest Mesozoic sediments from the Potiguar

The northeast is characterized by: (a) syntectonic Purbeckian–Barremian very thick sedimentation; (b) Aptian evaporites even in inland basins; (c) presence of a Cenomanian unconformity (Fig. 7).

The east-central and middle-north includes the São Francisco, Parnaíba, Potiguar, Barreirinhas, São Luiz and Bragança-Viseu basins. It is characterized by: (a) syntectonic late-Aptian–Santonian very thick sedimentation; (b) no Aptian evaporite sequence in the coastal basins, but anhydrite and gypsum present in inland basins (Figs. 8, 9).

The north includes the Amazon Mouth and Taku-tu rift valleys, the Amazon Intracratonic Basin and

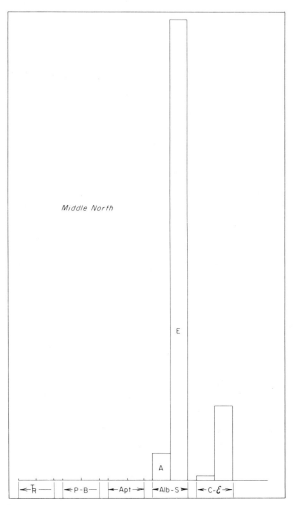

Fig. 8. Mesozoic sequences of the mid-northern region. (For legend see Fig. 5.)

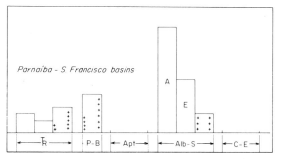

Fig. 9. Mesozoic sequences of the Parnaíba and São Francisco basins. (For legend see Fig. 5.)

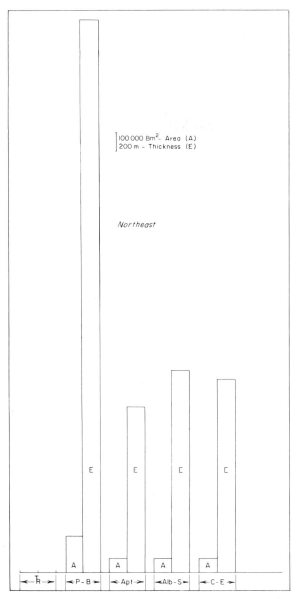

Fig. 7. Mesozoic sequences of the northeastern region. (For legend see Fig. 5.)

the Acre Basin. The Purbeckian—Barremian sequence in Acre and Takutu would eventually be connected to the Andean Geosyncline and the Albian—Santonian of the Amazon Intracratonic Basin would eventually belong to the Barreirinhas and Parnaíba Complex. Only the latest Cretaceous sequence (Turonian—Maastrichtian going up to Eocene) is known in the Amazon Mouth Basin.

No Late Jurassic and pre-Aptian Early Cretaceous marine sediments are known in Brazil. Even the Aptian seas were restricted and therefore with few fossils. The non-marine fossils are not easily correlated with the established world index fossils for those times. The Brazilian Purbeckian—Barremian and Aptian sequences have been intensively studied because most of the country's petroleum comes from these beds. It was then proposed to introduce national time-stratigraphic stages for this interval of time. These stages and their probably international equivalents are (Viana et al., 1971): Alagoas stage

(Late Aptian); Jiquiá (late Barremian to early Aptian); Buracica (early Barremian); Aratu (late Valanginian—Hauterivian); Rio da Serra stage (early Valanginian); Dom João stage (Purbeckian).

The marine deposits which were laid down since Albian times are easier to correlate with the international time-stratigraphic stages.

The Paraná (Fig. 3, col. 10), Santos (Fig. 4, col. 14) and Parnaíba (Fig. 3, col. 4) basins were also the site of enormous outpourings of lava during the late Jurassic—Early Cretaceous. These were fissure eruptions and at Presidente Epitacío in São Paulo piled up a thickness of 1,500 m and at Torres in Río Grande do Sul they poured on to the continental shelf. A later, Upper Cretaceous phase of alkaline volcanics also developed in the Paraná Basin; the earliest lavas occur in rocks of Jurassic age in the Parnaíba Basin. (Fig. 3, col. 4).

The deposits which were laid down in the rift valleys can be grouped according to the tectonic environment under which they were formed (Fig. 13). During the initial stage of active fault tectonism deposition was characterized by a complex interfingering of facies. The phase ended with uplift and erosion to be succeeded by a phase during which the presence of evaporites along with clastics provides the first evidence of restricted marine ingression. It forms a transgressive—regressive cycle and as a consequence the development of a wide-spread unconformity separates the overlying normal open sea marine sequence. This sequence, however, while present in some basins, is not present in all, e.g. in the Recôncavo, Tucano and Jatoba basins (Fig. 4, col. 8, 9) only the beds of the first stage occur.

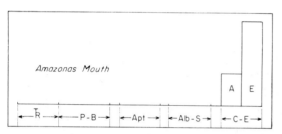

Fig. 10. Mesozoic sequences of the Amazon Mouth rift valleys. (For legend see Fig. 5.)

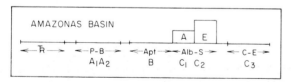

Fig. 11. Mesozoic sequences of the Amazon Intracratonic Basin. (For legend see Fig. 5.)

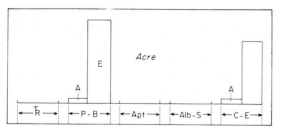

Fig. 12. Mesozoic sequences of the Acre Basin. (For legend see Fig. 5.)

MESOZOIC VOLCANISM

The first Mesozoic volcanism took place in the Amazon where the first dikes and sills were radiometrically dated around 230 m.y. or earliest Triassic. This was the first evidence of igneous activity since the end of Ordovician times. Volcanic activity was sporadically felt in the Amazon up to Jurassic times. This volcanism occurred during a long period of non-deposition and erosion so wherever lavas eventually poured out at the surface they were later eroded and for this reason only dikes and sills are now preserved. Triassic basaltic flows are also known in the Parnaíba Basin.

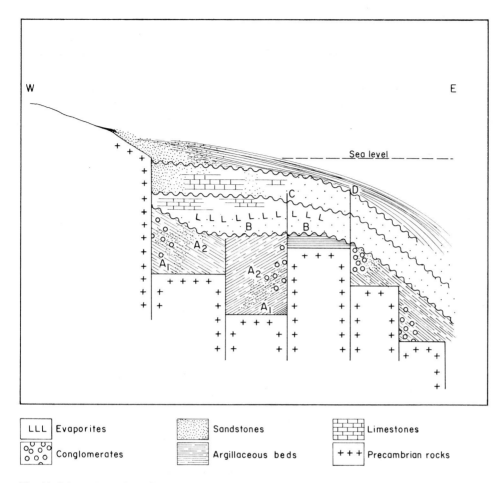

Fig. 13. Schematic section of coastal rift valleys. A_1 = syntectonic sediments under mild conditions; A_2 = syntectonic sediments under strong tectonic influence; B = sediments formed in restricted seas under milder tectonic conditions; C = open marine sediments
(A, B, C = Cretaceous deposits); D = Cenozoic deposits.
D = non-marine and marine deposits;
C = marine and transitional deposits;
B = restricted marine and non-marine deposits;
A = non-marine deposits.

The greatest volcanic manifestation in Brazil started about 150 m.y. B.P., its climax being at about 120 m.y. It affected practically all eastern Brazil including the present continental shelf. Most of the lavas are of basaltic nature, extruded along major tectonic tensional fractures or fissures (Leinz, 1949).

The most affected region is the Paraná Basin both from the point of view of the area of the basalt flows (around 1,000,000 km²), and of the thickness (which reach 1,500 m at Presidente Epitácio of São Paulo, where 32 flows are piled up). These figures make the

volcanism one of the largest outpourings of plateau basalts in the world. Contemporaneous diabase dikes cut the Precambrian outside the Paraná Basin. Dikes are so crowded in certain regions, as for instance in the Ponta Grossa Arch, that more than one per kilometer may be found. They are vertical in attitude and their regional strike is remarkably constant. Some dikes, which probably fed the flows, are almost 1,000 m thick. Contemporaneous basalt flows are also known in the Parnaíba Basin. They are less important than in the Paraná Basin, occupying smaller areas and

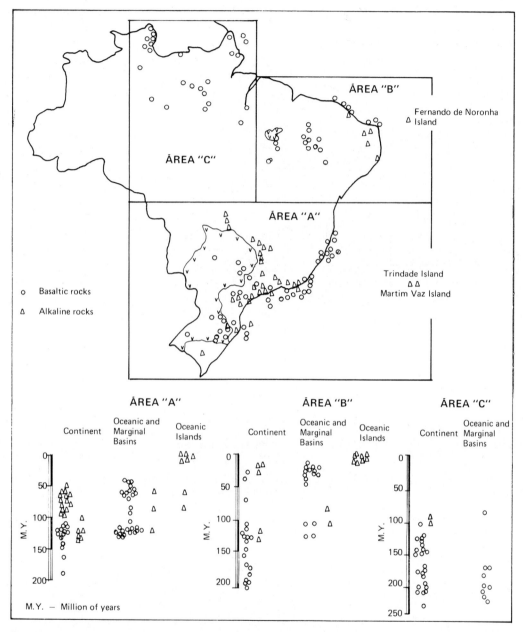

Fig. 14. Mesozoic and Cenozoic volcanism in Brazil (after Asmus and Porto, 1980).

with a total thickness not exceeding 200 m. Diabase dikes are also known at many northeastern localities outside the basin.

Acid and alkaline lavas are partly contemporaneous to the basaltic lavas and alkaline ones are partly younger. Some are found along the Ponta Grossa Arch but most of the alkaline lavas occur along the Canastra Arch. They belong to two age groups, one

130–120 m.y. old and the other 50–80 m.y. A trend in the geographical distribution of volcanism can be distinguished in the Brazilian craton, for the volcanics generally become progressively younger from west and south to east and north. As mentioned earlier the oldest are in the Amazon Basin. The volcanics at the mouth of the Amazon are Triassic and Purbeckian–Barremian in age. The post-Barremian alkaline rocks in the Paraná Basin lie east and north of the basaltic flows. The basaltic rocks in northeastern Brazil are Cretaceous and Cenozoic in age. Most of the volcanics from eastern coastal basins along the present continental shelf are Late Cenozoic in age. Presently there is no active volcanism in Brazil.

TECTONICS

The tectonic history of the Brazilian platform during the Phanerozoic can be divided into three broad stages. The first stage reflects the change in character from the Precambrian geosynclinal phase to the Phanerozoic platform phase. It developed during Cambro-Ordovician times with the infilling of depressions by post-tectonic molasses. The second stage, one of stability, covers the history of the now consolidated Brazilian platform from the Silurian to the Jurassic. Sedimentation was restricted to the intracratonic basins enumerated earlier. During the latest Jurassic and Early Cretaceous major tension faults and rift valleys (mostly coastal) were born and impressive lava flows extruded from the huge tension fissures mostly located in the Paraná Basin but also present in Parnaíba Basin. Great numbers of contemporaneous diabase dikes scattered through the Precambrian shield of eastern Brazil attest the extent of this volcanic activity. This phase is referred to as the phase of platform reactivation or 'Wealdian reactivation' by the Brazilian geologists (Almeida, 1969). It began rather slowly in the Late Jurassic, reached its peak of activity in the Early Cretaceous and then settled gradually down with the re-establishment of platform conditions.

The beginning of the Mesozoic thus lies within the stable stage. However, the passage from Paleozoic to Mesozoic was not so calm as it might be supposed. The intracratonic basins which were sites of intense sedimentation during late Paleozoic were uplifted and subject to erosion at the beginning of the Mesozoic. The three intracratonic basins (Paraná, Parnaíba and Amazon) have a definite transgressive–regressive Late Carboniferous–Permian sequence over 1,000 m thick. The sea is clearly present in the Late Carboniferous and sometimes in the Early Permian, but the Late Permian is always regressive and non-marine. In contrast no Lower Triassic deposits are known in Brazil and the Triassic sedimentation when present is poorly developed and always non-marine.

The sharp change of the nature of sedimentation and the differences in the behaviour of some of the basins when late Paleozoic and early Mesozoic sequences are compared lead to the conclusion that the passage from the first era to the second was accompanied by some tectonic activity.

Amazon Basin

During the Triassic there was some igneous activity in the Amazon Basin. As the region was undergoing continuous erosion during Mesozoic up to Albian times even if lava flows were extruded they were long since destroyed by erosion. Only dikes and sills are known. The diabase intrusions were responsible for the Monte Alegre structural high.

The Iquitos structural high, probably formed during Early Carboniferous but which later disappeared as a consequence of the great Late Carboniferous subsidence, was rejuvenated during the Mesozoic. As a rising land area it was submitted to erosion and became a source of the sediments laid down in the Acre Basin and the middle and lower sections of the Amazon Basin. Later a set of blockfaults uplifted the Acre Cretaceous beds towards the Iquitos Arch.

The Purus Arch was active only in Paleozoic times. The Gurupa Arch was probably formed during the Paleozoic, but it was affected by the Cretaceous fault systems responsible for the Amazon Mouth rift-valley.

The Amazon Basin is cut by an orthogonal system of fractures striking northeast and northwest. This system continues north and south for about a hundred kilometers into the Precambrian, but then changes abruptly to a singe north–south striking system. This abrupt change of strikes is believed to mark the original border of the sedimentary basin.

Parnaíba Basin

The simplified isopach maps of the late Paleozoic and Triassic in the Parnaíba Basin show a change in the structural behaviour of the basin during the interval of non-deposition between the two eras. The depocenter which was symmetrically located during the Late-Carboniferous–Permian cycle became asymmetric during the Triassic located towards the western margin (Fig. 15; Mesner and Woolridge, 1964).

The most striking tectonic feature of the basin is the Ferrer–Urbano Santos Arch which separates the

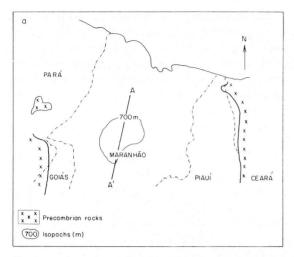 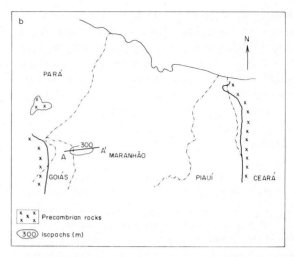

Fig. 15. Change in location of the late Paleozoic (Carboniferous—Permian) (a) and Triassic (b) depocenters in the Parnaíba Basin.

Parnaíba Basin and the São Luiz—Barreirinhas coastal rift valleys and which only developed during the Aptian—Santonian times. The beds in the Parnaíba Basin have a regional northward dip both in the Cretaceous and older sediments. In the coastal margins of Barreirinhas and São Luiz the regional dip is the same. The Cretaceous section thickens gradually northward as far as Ferrer—Urbano Santos and then rapidly increases in the coastal basins (Fig. 16). This arch was affected by later north-west striking transcurrent fault movements of the later Cretaceous subsequently rejuvenated during the Cenozoic.There is also a northeast strike, though less important. Both fault systems affected the Parnaíba and coastal basins.

The Rosario horst lies between the Barreirinhas and São Luiz basins. The Barreirinhas Basin, however, formed during the latest Aptian while the Rosario horst and São Luiz Basin only developed in Late Cretaceous times.

Paraná Basin

The northeastern, northern, western and southern borders of this basin are structural highs with their axis running parallel to the borders. The eastern border, however, crosses the structural features (Fig. 2). These dispositions suggest an age as old as the basin itself for the former borders while the eastern margin appears to be younger; the original border presumably lay further to the east. The main structural features of the eastern border are: (a) Ponta Grossa Arch; (b) Torres downfaulted region; (c) Uruguay—

Río Grande do Sul shield. The region of the present Ponta Grossa Arch has generally been considered as a less negative area than the nearby regions. The least negative areas within it changed several times during the Paleozoic. During the Triassic the arch was positive, that is an emergent land mass extended down to Santa Catarina State separating two negative areas, one in the State of São Paulo and the other in the State of Río Grande do Sul. After a general emergence of the Paraná Basin during Early and Middle

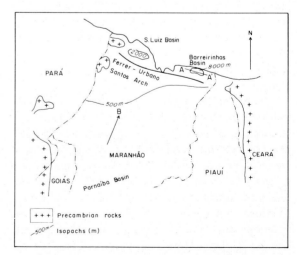

Fig. 16. Isopach map of the Cretaceous thicknesses in the Parnaíba, São Luis and Barreirinhas basins. There is a gradual thickening northwards in the Parnaíba Basin and a very much more rapid increase north of the Ferrer—Urbano Santos Arch in the Barreirinhas Basin. The Ferrer—Urbano Santos Arch developed during Aptian—Santonian times.

Jurassic, the arch appeared again in Late Jurassic times, but its function during the outpouring of the Early Cretaceous basaltic flows is not clear. In Cretaceous times it formed the southern limit to the Bauru deposition but to the south its limit is not clear as the lower section of the Paraná Basin became a land area. In fact, from Maastrichtian times the Paraná Basin in general became a rising land area.

The Torres downfaulted region is a relatively new, post-basaltic tectonic feature. The faulted blocks were downthrown eastward towards the ocean in a series of steps so part of the sedimentary succession of the basin now forms part of the continental shelf. The Uruguay–Río Grande do Sul shield remained a positive area throughout the entire history of the Paraná Basin. The eastern Precambrian border of the

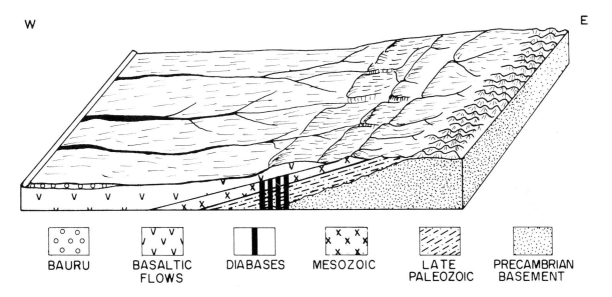

Fig. 17. Hypothetic situation of the upper section of the Paraná Basin at the time of Cretaceous sedimentation (after Fulfaro, 1971).

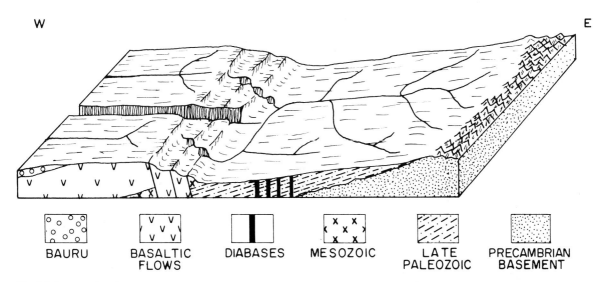

Fig. 18. Present situation of the geological formations in the upper section of the Paraná Basin (after Fulfaro, 1971).

Paraná Basin is formed by upthrown blocks forming the Serra do Mar range which terminates abruptly toward the ocean but slopes smoothly inland. This is a Cenozoic feature but during the Paleozoic and Mesozoic the region may have had a gentle topography rising smoothly toward the east. The characteristics of the sediments suggest that the most efficient barrier lay to the West where the Assuncion Arch formed the source of sediment.

The basaltic flows and the sediments of the Bauru Group now cover the plateaus with eastern escarpments known as 'Serra Geral'. This feature probably dates from the end of Cretaceous or perhaps from the Cenozoic.There were no plateaus nor Serra Geral during the Mesozoic;the region rather was a depressed, subsiding zone in which the Bauru Group was deposited. Faults now cut the basalts and sediments of the Bauru Group. The initial displacements were small but by reactivation the fault throws increased. The slow growth of the fault displacements allowed the rivers to retain their courses crossing the Serra Geral through deep canyons (Fig. 18).

The Northeastern basins

This region was submitted to a complex faulting during the Mesozoic and a number of small isolated Cretaceous basins scattered on the Precambrian basement is all that remains after this faulting and later erosion. The whole area may have once belonged to a fluvio-lacustrine complex draining southward into the Jatobá–Tucano–Recôncavo fluvio-lacustrine system. The faults that disrupted this unity are post-Barremian in age.

The Chapada do Araripe area exhibits a complicated tectonism. A late Aptian–Albian sea penetrated the region probably through the Parnaíba Basin. Block-faulting disrupted this connection and now Precambrian and Paleozoic rocks crop out in this region, and the Chapada do Araripe is a mesa elevated above the surrounding Precambrian rocks. This situation shows an inversion of the relief due to geologically young faults for otherwise the sedimentary record would have been destroyed. The E–W Patos and Pernambuco lineaments are Precambrian in age but were rejuvenated during the Late Cretaceous and possibly also during the Cenozoic. They controlled to a great extent the distribution of the Chapada do Araripe and Jatobá outliers.

The Patos lineament marks the northern limit of

the Chapada do Araripe. Every Cretaceous basin south of it has northerly or easterly dips while every Cretaceous basin to the north has southerly dips. There has therefore been a tendency for negative movements along this lineament during Late Cretaceous and (or) Cenozoic times.

The Pernambuco lineament marks the northern limit of the Jatobá Basin, the area to the north was uplifted and eroded. The lineament reaches the coast south of the city of Recife; to the north, the coastal Cretaceous sediments were preserved while those to the south were uplifted and generally stripped off. The last movements along this lineament are pre-Pliocene since Pliocene beds (Barreiras) which cover it are undisturbed.

Rift-valleys and marginal basins

The rift valleys in the northeast are crossed by predominant northeasterly trending faults and less important northwesterly faults. The northeastern trend is as old as the Precambrian but rejuvenated during the Cretaceous and Tertiary. It is the main structural trend in southern and northeastern Brazil, parallel to the coast line. The Recôncavo rift valley for instance is limited by two northeast striking fault zones. The eastern zone, the Salvador, has a total throw of about 4,000 m while the western, the Maragogipe, has only about 300 m of total throw. The basin is cut by a great number of normal tension faults by striking northeast and some antithetic northwest faults. The Recôncavo rift valley has a shorter history than other rift valleys for its development ceased after the first Purbeckian–Barremian episode.

Other basins in the northeast show further developments. The main period of fault activity when the rift valleys were born was during the Valanginian–Barremian. The tectonic activity continued with progressively lower intensity up to the Early Cenozoic times. The fault blocks generally behaved in much the same manner during their history. The pre-Aptian non-marine succession generally dips rather strongly toward the continent, that is in the opposite direction to the Aptian and later beds which dip oceanwards. The dip toward the ocean was caused by a general Cenozoic tilting of the basins which was not strong enough to reverse the dips of older beds.

The coastal basins may have a present emergent area in the coastal plains while the greater part lies on

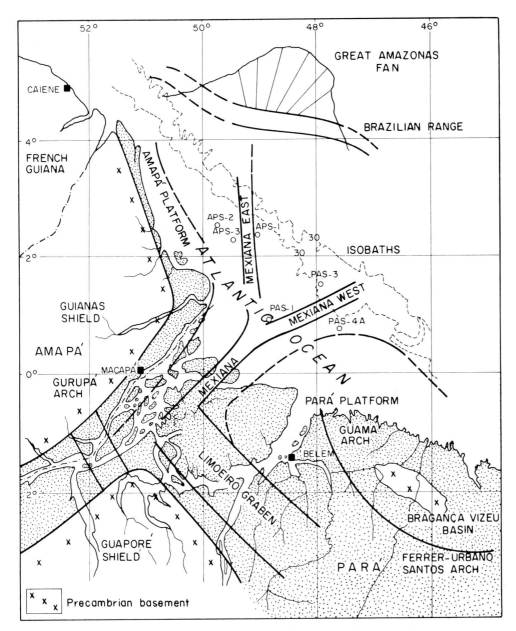

Fig. 19. Structural features of the Amazon Mouth Basin. The Brazilian Range and the great Amazon Fan are Cenozoic features. The grabens, arches and platforms are Late Cretaceous features. (Modified from Rezende and Ferradais, 1971.)

the continental shelf. Some basins have only this submerged part. The line between the continental shelf and the continental slope is considered the limit of the coastal basins. The tectonic structure of the submerged area continues the structure of the emerged part.

The Cenozoic fault systems of the southern coastal basins from Espirito Santo to Pelotas are more im-

portant than the Cenozoic faults of the northeast. They are responsible for important geomorphological features such as, for example, the Serra do Marrange, Paraíba rift valley, São Paulo and Curitiba basins. While the Precambrian of the northeast forms lowland razed by erosion, the Precambrian of the south down to the State of Santa Catarina is a region of young relief. The continental shelf of the Espirito

Santo is enlarged and known as Abrolhos Bank. Volcanic activity which began during the Cretaceous but developed mainly during the Cenozoic was responsible for this enlargement. The Abrolhos Bank may be a tectonic hinge separating the northern and southern coastal basins which have different tectonic characteristics (Asmus et al., 1971).

The Pelotas Basin is located in the Río Grande do Sul continental shelf. Most of the sediments present there are Cenozoic in age. Cretaceous and older sediments may be present in at least some of them and may possibly be connected to the Paraná Intracratonic Basin through the downfaulted blocks of Torres.

The Recife—João Pessoa and Potiguar basins north of the Sergipe—Alagoas, are not so strongly faulted and cannot be classified as rift valleys. From the northeast toward the north, the principal trends of the faults change from northeast to east—west and later to north-northwest, always parallel to the outline of the coast except the trends connected to the Amazon Mouth Basin. The east—west Ferrer—Urbano Santos structural high marks the inland limit of the Barreirinhas and São Luiz basins. The pre-Cenomanian beds dip generally southward against the continent. Cenomanian and later beds dip northward to the ocean. The Sobral lineament (Serra Grande in Fig. 2), between the Potiguar and Barreirinhas basins, is important as a tectonic limit of provinces. West of it, from Ceará to the Barreirinhas continental shelf extensions of the corresponding basins, compression forces exceptionally originated anticlinal and synclinal structures.

The Guamá Arch is a north-northwestern continuation of the Ferrer—Urbano Santos through the State of Pará coastal strip. It separates the Bragança-Viseu Basin from the Pará Platform of the Amazon Mouth Basin (Fig. 19). The region between this and the Amapá Platform is cut by a series of aligned grabens beginning at Grajau, Maranhão State and extending in a northwest direction as far as Marajó Island. The trend changes to northeastward with a bifurcation on the continental shelf into an eastern and western branch (Rezende and Ferradais, 1971; Schaller et al., 1971).

There are numerous small basins on the Pará and Amapá platforms with 1,000 to 2,000 m of Cretaceous sediments covered by Tertiary sediments without any structural relation to these basins. Some unstability prevailed therefore during the Cretaceous, follow-

ed by relative stability during the Tertiary. Some disturbances were felt during Triassic and Cretaceous with some volcanic activity but the syntectonic Jacarezinho and Limoeiro formations in the grabens indicate that a climax of the fault tectonics was reached only in the Late Cretaceous.

The migration of tectonism of the coastal region

The Sergipe—Alagoas is one of the most completely studied of the Brazilian coastal basins and has the most complete Cretaceous sedimentary record. It may be taken therefore as a basis of reference for the other basins. This basin exhibits the following sequences:

Sequence C_3 — late Santonian—Eocene,
Sequence C_2 — late Cenomanian—Early Santonian
Sequence C_1 — Albian—Early Cenomanian,
Sequence   B — Aptian,
Subsequence A_2 — Valanginian—Barremian,
Subsequence A_1 — Purbeckian.

The Recôncavo, Tucano, Jatobá and Almada basins exhibit only the subsequences A_1 and A_2 with the same ages.

The Barreirinhas Basin exhibits a basal sequence tectonically equivalent to sequence A and an upper succession tectonically equivalent to C of the northeastern basins. The Aptian evaporitic sequence B is not present in the Barreirinhas.

The basal sequence A of the Barreirinhas Basin was formed in the same way as the sequence A of the northeastern basins, that is thick conglomerates close to rising fault scarps grading laterally to fluvio-lacustrine deposits reaching considerable thicknesses on the down-faulted blocks (4,000 to 5,000 m).

The sequence A in the Barreirinhas Basin is Albian in age, therefore younger than sequence A of the northeastern basins. The sequences C_1, C_2 and C_3 of the northeastern basins are correlated with only one sequence of the Barreirinhas Basin. The sequence A of the Amazon Mouth Basin was again formed in the same way as the sequence A of the northeastern basins, that is thick conglomerates close to rising fault slopes (Jacarezinho Formation) grading laterally to braided channels fluvial deposits of the Limoeiro Formation. The sequence A of the Amazon is Late Cretaceous to Paleocene in age, younger than in both Barreirinhas and the northeastern basins.

It is evident therefore that a tectonic migration

occurred from the Recôncavo and Sergipe—Alagoas to the Amazon basins. The principal tectonic phase of the northeastern basins is Valanginian—Barremian in age; this phase in the middle-north basins is Albian in age and in the Amazon Mouth Basin it is Late Cretaceous to Tertiary .

The first marine sediments are late Aptian in the northeast, Cenomanian in the Barreirinhas and Tertiary in the Amazon Mouth Basin.

Late Aptian marine deposits with evaporites are present in the coastal basins south of Recôncavo. There is also a pre-Aptian non-marine syntectonic succession in those basins though not reaching the thicknesses known in the northeast. It seems that the Valanginian—Barremian tectonism was not so energetic in the southern coastal basins as it was in the northeast. The most energetic tectonic activity in the south was felt more inland in the intracratonic basin of Paraná, where enormous tension fractures opened and huge lava flows occurred. On the northeastern coast, on the other hand, post-Barremian tectonic activity was not extensive and diminished with time. Late Cretaceous and Tertiary movements here are mostly in the form of growth faults. In the south in contrast the Late Cretaceous and Tertiary Abrolhos volcanics of the Espirito Santo Basin are evidence of important post-Barremian tectonic events and the faults of the southern basins at this time cannot be interpreted as growth faults. The Serra do Mar range and faults in the Santos Basin striking parallel to the range are examples of such faults.

This is further evidence that the main tectonic activity migrated both to the north and to the south from the Recôncavo—Sergipe—Alagoas region reaching the extreme north in the Late Cretaceous to Tertiary times and south to Santa Catarina State during the Tertiary. In the extreme south of the country, in Río Grande do Sul State, the Precambrian is again levelled by erosion. The shallow Pelotas Basin dips uniformly toward the ocean and there are no dip inversions or any other indications of stronger tectonism. No pre-Tertiary sediments other than those deposited in the downfaulted Paraná Intracratonic Basin are known.

The strong Early Cretaceous fault tectonics of Brazil was interpreted as related to the postulated rift leading to separation of South America from Africa. The later presence of the sea on the northern coast was assumed to be the result of a rotational process of the South American and African blocks during their separation.

The situation in the south is more difficult to explain since the Early Cretaceous taphrogenesis was felt mostly inland, in the Paraná Basin, instead of on the coast. The main tectonism migrated with time eastward reaching the coast only in the Tertiary. Asmus and Ferrari (1978) suggested that a hot spot had developed in the south where the more active magmatism would have pushed up the crust. The eastern Brazilian coastal basins, having similar geologic evolution, would be located between this southern hot spot and a weaker one, developed in the extreme northeast, the latter causing a weaker subsidence of the Recife—João Pessoa Basin.

PALEOENVIRONMENTS AND PALEOGEOGRAPHIC EVOLUTION DURING THE MESOZOIC

Triassic (Fig. 20)

The regressive nature of the Upper Permian beds in the three intracratonic basins (Paraná, Parnaíba and Amazon) heralded the general emergence of Brazil during Early Triassic time. No sediments of this age are recognized anywhere in Brazil. Even in Middle and Late Triassic times evidence of deposition is relatively scarce and restricted to the intracratonic basins. There are no sediments in the Amazon Basin, only diabase sills and dikes, and only fluvio-lacustrine sediments are known from the Parnaíba and Paraná basins.

Middle to Late Triassic sediments of the Paraná Basin are found in separate areas in the states of São Paulo and Río Grande do Sul. They are channel and overbank sandstones forming recurring cycles. They show textural and mineralogical uniformities and poor heavy mineral associations indicating that they passed through several sedimentary cycles. Bimodality in cross-beddings points to meandering channels, and ventifacts along foreset beds, found in some places in the State of São Paulo suggest aquatic reworking of older eolian deposits.

The isopachs of the São Paulo sediments suggest a northwest drainage toward the confluence of the Río Grande and Paranaíba Rivers, bordered northeastward by the Canastra Arch and southeastward by the Ponta Grossa Arch (Soares, 1974). There is no information about the nature of the drainage beyond that point, although despite the postulated dry climate the drainage is believed to be exorheic. The most likely

Fig. 20. Schematic presentation of the Triassic paleogeography of Brazil.

continuation of the drainage pattern is southward roughly parallel to the present Paraná River, a region now covered by younger beds. Some sandstones found by drillings may thus belong to the Middle Triassic instead of Late Triassic (Botucatu) as supposed.

During the time of deposition of the Triassic sediments in Río Grande do Sul, a fluctuation toward a wetter climate led to the development of relatively extensive, long-lives lakes associated with flood plains (Gammermann, 1973). In these environments lived a rich fauna of reptilians cotylosaurs, dicynodonts and rynchocephalids, conchostracans, insects and plants (*Dicroites* flora). The Triassic fluvial system of Río Grande do Sul seems to be directed toward the south into Uruguay.

The oldest Triassic rocks in the Parnaíba Basin are

volcanics. Thin sandstone beds may be present between the flows (Aguiar, 1971). A fluvio-lacustrine cycle of sedimentation rests unconformably on the lava flows. The climate was drier at the end of the cycle than in the beginning. Black lacustrine shales contain Late Triassic fossil fishes.

The Brazilian Triassic, therefore, as far as can be surmised by the evidences at hand had a generally dry climate though not arid, with important wetter time intervals.

The Botucatu Formation of the Paraná Basin is made up of well to regularly sorted fine sandstones with scarce matrix and large cross-beds with foresets dipping at around $30°$ tangential at the base. Their 0.5 mm grains exhibit good roundness and sphericity with pitted and frosted surfaces. They are aeolian sediments built up by accretion in front of dunes. A desert therefore was present in the area. The prevailing paleowinds in the north (Minas Gerais and São Paulo) came from the north and north-northeast; in the south (Río Grande do Sul) they came from the southwest; in the State of Paraná there was an important deflexion with the winds coming from the east (Bigarella and Salamuni, 1961).

The homegeneity of the fine sandstones is only locally broken by subaqueous deposits, conglomerates and conglomeratic sandstones laid down under conditions of high energy, and siltstones and mudstones laid down under conditions of low energy. Some foresets of cross-beds exhibit ventifacts along the stratification. The coarse clastics are more common in the lower part of the formation while the fine clastics are more common in the upper part.

The desert sandstones extend over an area of about 1,500,000 km². Erosion must have stripped off the sandstones over a large area along the border of the present basin and it seems therefore likely that the Botucatu may have been a desert larger than the present Sahara. The absence of salt might be explained either by a small salt content of underground water or by dissolution during diagenesis.

Only animal tracks are known as fossils in the Botucatu Formation so only its stratigraphical position below Late Jurassic–Early Cretaceous basaltic lavas and above the Middle to Late Triassic Piramboia and Rosário do Sul formations, defines its age. According to Gamermann (1973), Rosário do Sul grades upward to Botucatu. There could even be a partial synchronism evolving the uppermost Rosário do Sul beds generated in a wet climate, and Botucatu beds generated in a dry climate. Gordon (1947), Leinz (1949) and Eick et al. (1973) have written about unconformable relationships between Botucatu and the overlying Serra Geral lavas. These lavas sometimes ride upon older Botucatu dunes without destroying them, suggesting that the dunes were already consolidated before basalt outpouring. Diabase sills intruded Botucatu beds signifying a thicker cover later destroyed by erosion. A latest Triassic or earliest Jurassic age may therefore be assigned to the Botucatu Formation.

Jurassic

The Juruá Mirim red beds from Acre (Fig. 3, col. 2) contain evaporites suggesting a desert passing westward to coastal environments which in turn pass westward outside of Brazil to marine environments of the Andean Geosyncline.

In Guyana, some beds of the Takutu rift valley (Fig. 3, col. 1) have yielded a pollen flora which could be of Jurassic (pre-Purbeckian) age.

The first sediments of the Aliança Formation, Late Jurassic of the Recôncavo Basin (Fig. 4, col. 9) are red beds with anhydrite, and in the extreme southwest even halite is present.

Sedimentary evidence of post-Botucatu Late Jurassic deposits is very scarce in the Paraná Basin, being restricted to sandstone beds very like the Botucatu, between the first basaltic lava flows. These flows span the time between Jurassic and Cretaceous and continue up to the Barremian.

The post-evaporitic Late Jurassic sediments in the Recôncavo and other rift valleys are made up of clastics with sedimentary structures indicating flood plains of meandering rivers with lakes and oxbow stagnant waters. Primary red beds were formed in this environment. Red argillaceous beds are frequent in the lower part suggesting derivation from red lateritic argillaceous soils. The overbank and the oxbow lakes facies retain evidence of oxidizing conditions whereas the facies of the larger lakes exhibit reducing conditions imprinting greenish hues at the sediments. These lacustrine sediments may contain limestone beds and concentrations of ostracod shells.

Red sandstone beds are more and more frequent toward the top of the sequence suggesting deep erosion in source area resulting in the denudation of the regolith and erosion of fresh rocks. The sandstones at the top of the sequence contain rock fragments and the cross-bedding is of the torrential type. Silicified conifer trunks lie on bedding planes in such a way as to suggest the action of transport. Red hues are seen only in the matrix of the sandstones. The evidence suggests a change from meandering to braided channels.

Late Jurassic sediments of the rift valley basins therefore provide evidence of a change in the climatic conditions from dry to humid with alternation of dry and wet seasons. The Recôncavo and Tucano basins were part of a Late Jurassic tectonic valley which drained toward the south through rivers with headwaters in the Jatobá, Mirandiba, São José do Belmonte and Chapada do Araripe areas (Fig. 21). The sediments of these areas now isolated were probably part of a single basin disrupted by erosion. The thick-

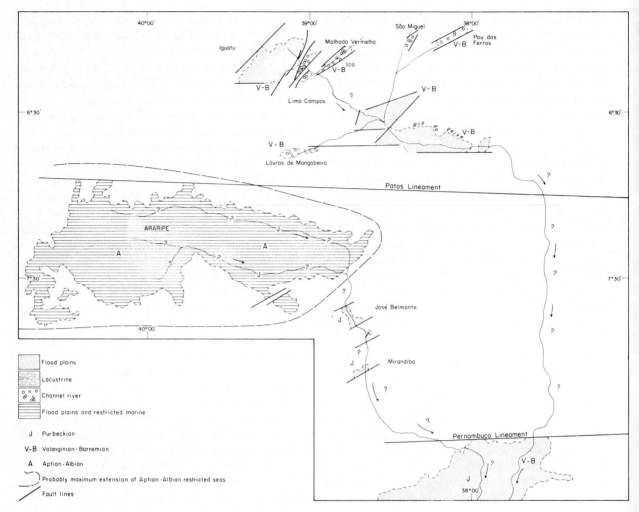

Fig. 21. The Cretaceous basins of the Interior of the northeast.

ness of the formations increases gradually from north to south; the coarseness of sandstones, on the other hand, increases from south to north. The greatest subsidence is found in the southern Recôncavo Basin in the proposed direction of transport.

The south-north band of the Recôncavo—Tucano—Jatobá basins is broken at its northern extremity by a fault zone called Pernambuco lineament which strikes east-west. This lineament of Precambrian age was rejuvenated in Wealdian times. The Jatobá Jurassic and Cretaceous formations were downthrown south of the lineament and Precambrian rocks were upthrown to the north so erosion erased the northern Mesozoic sediments except in a few places where sediments were preserved in downfaulted blocks. This fluvial system extended southward to the Almada Basin (Fig. 4, col. 10).

The Sergipe—Alagoas Late Jurassic deposits though lithologically and paleontologically similar to those of the Recôncavo—Tucano—Jatobá basins probably belonged to one or more independent river systems. The formations now cover discrete areas separated by erosion according to some geologists (Ojeda and Bisol, 1971); however, it may be considered that they were formed in separate river systems.

Cretaceous (Figs. 22—25)

The Paraná Basin was the scene of tremendous volcanic activity which reached its climax in Early Cretaceous times. Basaltic lavas poured out of tension fissures. No volcanoes were built up. A few sandstone horizons occur between the first flows and rather thin aeolian sandstone may cover the last flow.

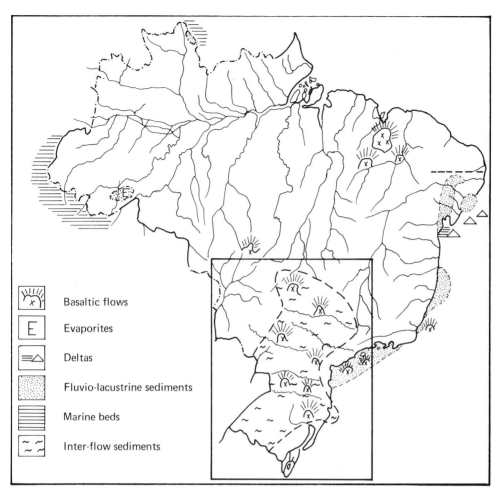

Fig. 22. Schematic presentation of the Late Jurassic–Early Cretaceous paleogeography of Brazil.

This period of Purbeckian–Barremian volcanism was one of the greatest in the world occuping the Paraná Basin, an area of 1,200,000 km². The sequence of flows reaches 1,500 m around Presidente Epitacio, State of São Paulo, without any intercalated sediments. A great number of diabase dikes chronologically correlated to the flows cut through Precambrian rocks beyond the border of the Paraná Basin. The Parnaíba Basin and the coastal basins of Santos Campos and Espirito Santo were also affected by this volcanism though on a smaller scale.

The Paraná Basin and Precambrian bordering areas also underwent a less conspicuous contemporaneous alkanline volcanism with the lavas thrown out from volcanic edifices. Some differentation of basaltic lavas in the Paraná Basin resulted in acid lavas.

As tension fractures were developing in the south, strong fault tectonism was developing in the north-

east. The Late Jurassic Recôncavo–Tucano–Jatobá valley persisted during Early Cretaceous but with a change in the disposition of the 'headwaters'. These headwaters no longer passed through Mirandiba, São José do Belmonte and Araripe region but through Rio de Peixe, Iguatu, Malhada Vermelha, Lima Campos, Icó, São Miguel and Pau dos Ferros (Fig. 21). These small basins were probably part of a once continuous basin extending from Almada to Ceará and Río Grande do Norte.

The lithology, fossils and sedimentary structures of Rio do Peixe point to a flood plain environment evolving to lacustrine and back to a flood plain. These beds have yielded conchostracans, ostracods, fishes and dinosaur footprints. The rhythmic deposition of mudstone and marl beds shows sudden lateral variations to sandstones at Iguatu. Conglomeratic sandstones are the predominant lithology at Icó,

Lavras de Mangabeira, Lima Campos and Pau dos Ferros, suggesting a river channel environment. The marl deposits at Lima Campos with reptilian bones and fish scales (*Lepidotus*) may have been formed in a lake.

The thickness of the sediments increases gradually southward from these small basins to Jatobá and from Jatobá to Recôncavo. This fluvial valley was subjected to faulting with a tendency for greater sub-

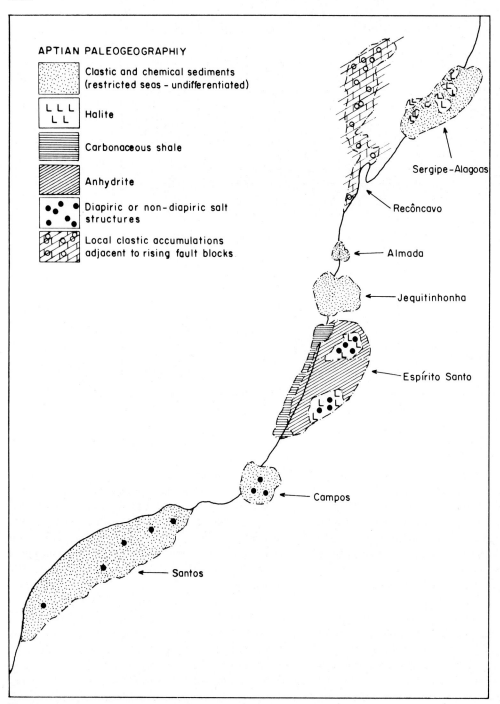

Fig. 23. Schematic presentation of the Aptian paleogeography of Brazil.

sidence in the southern downfaulted blocks. During the time of greatest subsidence the Itaparica and Candeias Formations of Tucano and Recôncavo were laid down. They were formed in large lakes some of which with depths enough for the inception of reducing conditions causing the formation of bituminous shales. Siltstones, mudstones, dolomites and thin beds of limestones with conchostracans, ostracods, fishes, gastropods and plant remains are also present. The lakes passed laterally to flood plains where medium to very fine to silty sandstones with clay galls were formed. The interfingering of the facies indicates fluctuations with time of the geographic distributions of the two environments.

During the time of deposition of the Candeias Formation the intensification of fault movement resulted in the deposition of poorly sorted immature fanglomeratic beds with a high percentage of unstable minerals in front of rising block faults. The coarseness of these deposits decreases rapidly away from the fault escarpments passing gradually to the flood plain and lake deposits. The fanglomerates and talus deposits are over 1,000 m thick in the Recôncavo Basin disappearing at distances of 10–15 km from the fault scarps. They are Valanginian to Barremian in age. There are similar deposits in the Tucano, Jatobá, Almada and Sergipe–Alagoas basins. The lake-flood plain complex lasted from the Purbeckian to the Barremian.

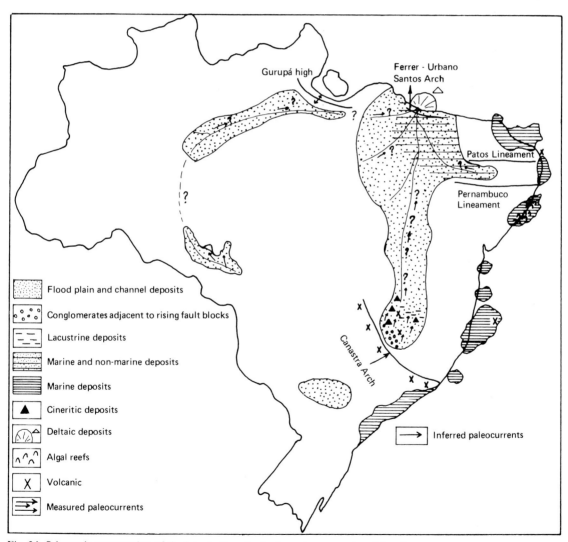

Fig. 24. Schematic presentation of the late-Aptian–Santonian paleogeography of Brazil.

The river discharges into these lakes produced constructive deltas. Three deltaic sub-environments are distinguished: (a) deltaic plain consisting of braided channels and interdistributary basins; (b) front delta recognized by subaqueous slumping; (c) lower channels recognized by structures made by mud flows and perhaps turbidites. The distribution of these three sub-environments suggest a southern progradation of the delta front of the Recôncavo Basin with fluctuations due to tectonic causes.

The final progradation over the lacustrine sedi-

ments of deltaic deposits produced a flood plain environment typified by the deposits of the São Sebastião Formation, made up of coarse to fine feldspathic sandstones and argillaceous variegated siltstones. The coarseness of the clastics tends to increase upward. This formation contains conchostracans, ostracods, gastropods, bivalves and fishes.

The Purbeckian–Barremian cycle of the Recôncavo–Jatobá Basin is unconformably covered by the fluvial Marizal Formation. Its lithology varies from conglomerates to shales and even contains rare lime-

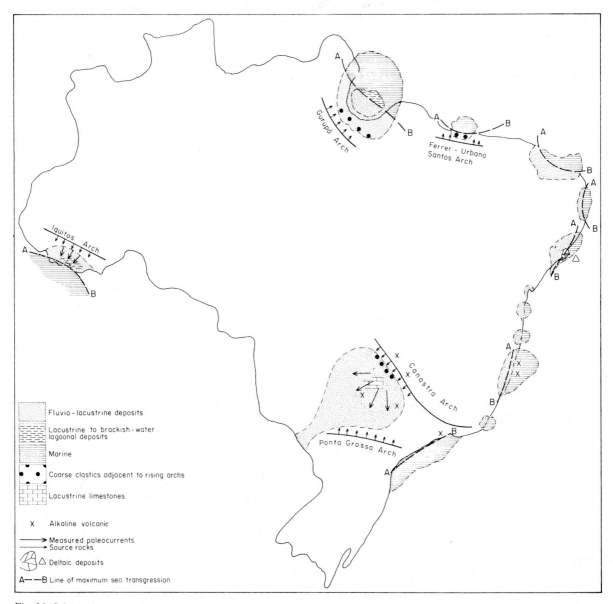

Fig. 25. Schematic presentation of the Campanian–Eocene paleogeography of Brazil.

stone beds. The predominant lithology is a poorly sorted argillaceous sandstone commonly cross-bedded with cut-and-fill structures. The conglomerates contain boulders of varied origins. The Marizal Formations contrary to all other Recôncavo–Jatobá formations, thickens northward to Tucano but it thins again northward from Tucano, showing, however, considerable local variations. The lithological characteristics, distribution and irregular thickness suggest a close association with the uplifted fault blocks with local sources formed by fault escarpments. The transport was torrential or by way of mud flows near the slopes and by laminar flow on the plains away from the steep slopes.

Although no evaporites are known, the climate was dry and an internal drainage might have occurred at least during part of the time of deposition of this formation. The nature of the Marizal Formation and the geographical proximity to contemporaneous evaporitic deposits of Sergipe–Alagoas and Jequitinhonha favour this interpretation. The Marizal Formation is distributed over a much smaller area than the beds of the previous cycle of deposition showing a tendency towards the consolidation of the Recôncavo–Jatobá rift valley to the Brazilian craton. Such a consolidation occurred during the Aptian after the deposition of the Marizal Formation. This unit corresponds therefore to a type of molasse deposit in a rift valley basin.

In the southern and northeastern coastal basins there are formations correlated to the Marizal which contain evaporites among other lithologies. These sediments were formed initially in lagoonal environments passing later to restricted marine conditions where dolomites, anhydrite, halite and potash salts accumulated together with clastics.

The presence of contemporaneous evaporites in Jequitinhonha, Espirito Santo, Campos, Chapada do Araripe and in the Codó Formation of Parnaíba Basin attest the extension of dry climates along the coast from southern to northeastern Brazil and in the north and northeastern interior. The presence of suprabasaltic aeolian sandstones in the states of Paraná and São Paulo also suggests that these regions had contemporaneous dry climates. The Areado Formation of the São Francisco Basin is slightly younger than the evaporite formations. It contains ventifacts reworked by water in the lower beds (Grossi Sad et al., 1971), suggesting dry climates during times preceding the time of deposition of this formation, i.e., possibly

during the time of evaporitic deposition (Aptian). According to these considerations extensive dry climates prevailed over most of eastern Brazil during late Aptian. The Santos Basin evaporites are slightly older than the evaporites from other coastal basins; this fits into the accepted theory of the drifting apart of South America and Africa, starting from the Santos Basin (Asmus, 1975).

The Parnaíba Basin subsided during late Aptian and Albian times and a transgressive–regressive cycle beginning in late Aptian and extending up to the Santonian is recorded in the lithology. The transgressive phase consists of the Grajaú and Codó formations. The Grajaú beds accumulated on flood plain-braided channel environment. The sea then encroached over the land and deposited the beds of the Codó Formation. This sea, however, occupied a restricted basin indicated by the occurrence of anhydrite and by the fossils for besides fishes and ostracodes there are oyster biostromes. The Grajaú flood plains and the restricted marine environments of Codó are recorded beyond the border of the present Parnaíba Basin in Chapada do Araripe (Fig. 3, col. 5).

The Crato, first formation of this cycle in Araripe (Fig. 3, col. 5) was formed in a lake occasionally deep enough for the inception of reducing conditions and the deposition of bituminous shales. Fishes, conchostracans, ostracods and bivalves lived in this lake and plant remains were also fossilized here. This lake was probably connected to the Grajaú flood plain. When marine conditions reached Chapada do Araripe, the Santana Formation was formed. At times when the salt-content of the water of this restricted sea was too high, gypsum precipitated; when the salt-content was nearly normal open marine life could develop with bivalves and echinoids. If the salt content dropped too low, non-marine fish and ostracods were represented and even a crocodile, *Araripesuchus*, is known from these beds. The Santana Formation is famous for its calcareous fish concretions in marls and mudstones.

In Albian–Santonian times, a large river, the predecessor of the present São Francisco, ran from west-central Minas Gerais State, near 20° latitude, to the coast of Maranhão State at less than 2°. While the Codó Formation was accumulating in the Parnaíba Basin, the fluvial Areado Formation was accumulating in Minas Gerais.

The conglomerates of the Areado Formation are genetically connected to the rising Canastra Arch.

They disappear with increasing distance from the arch giving way, gradually, to a rythmic sequence of thin red beds, sandstones, siltstones and mudstones which accumulated on flood-plains.

The colours of these sediments were derived from the source area. The flood plain deposits give way to lacustrine sediments containing oil shales, rich in fish, conchostracans and conifer leaves. The lakes were connected downstream to further flood pains where siltstones and thick cross-bedded sandstones were formed. Gradually the lakes were filled up by sediments.

The regressive phase of the Aptian—Santonian sequence in the Parnaíba Basin started in the middle of the Codó Formation and went on through the Itapecuru Formation, a sequence of flood plains of red beds. Diastems are common in this unit and for this reason the formation despite its relatively low thickness (300 m) may span a long interval of time. The beds dip and thicken northward. Dinosaurs, crocodiles, fishes, invertebrate and plant remains are known in this formation.

The regressive phase is also present at Chapada do Araripe where beds of similar lithology, also formed in flood plain environments, cover the sediments of the Santana Formation.

The first sediments of the Albian—Santonian sequence of the Barreirinhas Basin contain fanglomerates accumulated at the foot of slopes of rising block faults; they grade laterally to flood plain and lacustrine deposits. No marine sediments are known among these basal deposits. These beds are characterized by abrupt lateral facies variations. Fluvial deposits exhibit high energy distributary channels and overbank facies and grade into deltaic facies with low energy distributary channels. The rivers were of a braided channel type frequently covered by fine clastics of the flood plain facies and by swamp deposits where lignites developed. The deltaic sub-environments recognized are delta plain, swamp delta front and prodelta. The prodelta sediments are 2,000—3,000 m thick.

Argillaceous diapir structures are present. Fine clastics are interfingered with fine to coarse sandstones with gradational structures thought to be turbidites, accumulated during times of tectonic instability. Marine sediments only appear in the upper part of the Albian deposits.

The chronohorizons cut the lines of separation of these facies, from the river facies to delta plain, delta front, and prodelta. The piling up of these facies and

the absence of destructive phases lead to the idea of a prograding constructive delta.

This delta was possibly a composite one the product of several rivers converging into an area of high subsidence. The most important of these rivers was probably the predecessor of the São Francisco River which drained a large area of the interior. The non-marine sediment of the Barreirinhas Basin is about 5,000 m thick. The accumulation of such an enormous thickness in such a short time, during the Albian, could not be accomplished, unless a major river was actively transporting material (the Recôncavo Purbeckian—Barremian sequence has a thickness of 6,500 m but accumulated over a greater interval of time). The eastern deviation of the São Francisco River to the East Atlantic Ocean was probably caused by the Late Cretaceous reactivation of the Pernambuco lineament.

During the regressive phase in the Parnaíba, São Luiz and Barreirinhas basins an intensification of uplifts of the Canastra Arch resulted in the appearance of a number of alkaline and ultrabasic volcanoes along and adjacent to the arch. The lavas give way distally to tuffaceous green cross-bedded sandstones and conglomerates; the green colour is caused by weathering to nontronite of the volcanic material. The tuffaceous sandstones grade to argillaceous sandstones rich in montmorillonite derived from volcanic sources. The sandstones are well exposed on the Tocantins—São Francisco divide. Foreset measurements of the cross-bedding of deposits outcropping near the border of the São Francisco Basin, adjacent to the Canastra Arch, indicated paleocurrents flowing in a northernly direction.

The middle-Albian to Turonian Alter-do-Chão Formation of the Amazon Intracratonic Basin is of fluvial origin. It contains palynomorphs similar to those found in chronologically equivalent beds of the Barreirinhas Basin (Daemon, 1974). As no similar beds are found in the Amazon Mouth Basin it is supposed that this fluvial system was drained to the Barreirinhas deltaic system. The Chapada-dos-Parecis beds at the Paraguay—Amazon divide may be contemporaneous. Their thickness increases northward. They were laid down under fluvial (channel and overflood) environments and their cross-beddings suggest a northward drainage which bespeak a connection to the Alter-do-Chão river system.

Marine sediments contemporaneous with the non-marine late-Aptian—Santonian sequences of the São

Francisco and Parnaíba basins are found in the coastal basins. A widespread Cenomanian unconformity occurs in the southern and northeastern coastal basins. This unconformity, however, is not present in the middle north coastal basins (Potiguar to Barreirinhas) where there is continuous sedimentation from late Aptian to Eocene. The beds formed in warm tropical seas as evidenced by widespread limestone beds and the presence of algal bioherms associated with oolitic and pisolitic limestones. Albian was the period of maximum transgression and in a geologically short time the Albian seas covered large areas of the northeast and middle north. The fact that the Albian transgressions have progressively covered a lesser part of the continent suggests that they represent minor episodes in a dominantly regressive movement caused by the continuous rise of the Brazilian craton.

The Potiguar Basin although not being a rift valley has a non-marine late Aptian and Albian sequence extending up to the Cenomanian as in the Barreirinhas Basin. Only during the Turonian did the sea encroach upon the basin.

The latest Cretaceous sediments of the interior, excluding the Acre and the Upper Paraná sub-basins, are Santonian in age. In the Acre Basin there is a continous sequence from Cenomanian to Maastrichtian without any unconformity. These sediments form a transgressive–regressive cycle but the sea never reached the Acre Basin, marine sediments being only found in Peru. In the Acre Basin the transgressive cycle is represented by flood plain deposits grading westward to lacustrine beds which probably in turn graded into lagoonal sediments still further to the west. The beginning of a slow regression probably occurred between Coniacian and Santonian as a consequence of tectonic movements of the Andean orogenesis. The regressive fluvio-lacustrine deposits gradually covered the lake and lagoon beds.

The Canastra alkaline volcanism persisted through the Campanian–Maastrichtian and made its imprints in the sediments of the upper section of the Paraná Basin. This group was intensively studied after 1976. According to Soares and Landin (1976) its depocenter was originally located at the northwestern side of the Ponta Grossa Arch (see Fig. 2), in the northwestern Paraná State and southwestern São Paulo State. The relief would be very uneven and the drainage chaotic due to disturbance of the old drainage by the basaltic lava flows. Later there were migrations of

the depocenter first to the western-central part of São Paulo State and then to the northwest of this state and then southwest of the State of Minas Gerais. Precambrian limestones could be the source rocks for the limestone present in Bauru either as discrete rocks or with alkaline volcanics as the source of the cement in the sandstones. Foreset measurements of cross-bedding in rocks cropping out close to the Canastra Arch indicated predominantly southwestward and southeastward flowing paleocurrents (Suguio, 1973; Mezzalira, 1974). Dinosaurs (both sauropods and theropods), crocodiles, turtles, fish, bivalves, conchostracans, ostracods and algae (charophytae) are present in these beds.

The rising Ponta Grossa Arch in the south and the Canastra Arch in the north limited the Bauru sedimentation to the upper section of the Paraná Basin. The movements were stronger along the Canastra Arch and here the volcanic activity and the coarsest sediments are to be found. The drainage of the river system responsible for the deposition of the Bauru Group continued southwestward beyond the Ponta Grossa Arch much as the present Paraná River. It seems therefore that the predecessor of the Paraná River was born in Upper Cretaceous times as a result of the reorganization of the drainage disrupted by the internal drainage of the Botucatu and by the lava flows. The span of time of deposition of the Bauru Group would fill the interval late Barremian (or early Aptian) to Maastrichtian (from the Caiuá to the Marília Formation; see Fig. 3). The Bauru Group marks the end of the Mesozoic sedimentation in the Paraná Basin. Subsequent alkaline volcanic activity persisted through the latest Cretaceous and into Early Tertiary.

The oldest Cretaceous sediments in the basin at the mouth of the Amazon are Campanian or late Santonian in a sequence which without unconformity extends up to the Paleocene. Two partially interfingered formations make up this cycle: the Jacarezinho Formation which is characterized by immature, poorly sorted coarse clastics with boulders, cobbles and pebbles of different lithologies and reaching the maximum thickness of 1,000 m (slope deposits and fanglomerates adjacent to fault scarps), grading laterally to the fine to coarse sandstones of the 1,500 m thick Limoeiro Formation, with frequent gradational structures starting with conglomerates and thinning upward to dark argillites. The latter seems to have originated in flood plains with braided channels. The Jacarezinho and Limoeiro formations were therefore deposited in syntectonic conditions.

Campanian–Eocene volcanics are known in the Paraná Basin and in Precambrian areas adjacent to this basin. Contemporaneous volcanism is also displayed in the Espírito Santo Basin where it kept pace with Campanian–Eocene sedimentation.

Marine Campanian–Eocene sediments are known in the northeastern and southern coastal basins from the Recife–João Pessoa Basin to the Santos Basin. They are known also on the continental shelf of the Potiguar and Barreirinhas basins. These Late Cretaceous–Eocene seas were warm and tropical. The Campanian–Eocene beds of the Sergipe–Alagoas Basin were formed by a high constructive delta marked by the successive prograding of shallow-water coarse clastics over the deeper-water fine grained clastics. The presence of this delta is also evidenced by the geometry of the lithosomes (Schaller, 1969). Similar deltas occurred in Jequitinhonha, Espírito Santo and Campos basins.

A synthesis of the climatic conditions of the Brazilian Mesozoic and its evolution from the available evidence suggest that the climate was probably always hot, tropical. There are many evaporites, limestones and less frequently organic reefs. The Triassic evidence, though fragmentary, point to dry but not desert climates with important fluctuations towards humid conditions at least as far as the Paraná and the Parnaíba basins are concerned.

Deserts were widespread during the Late Jurassic from the south to the northeast and to Acre. Earliest Cretaceous to Barremian witnessed more humid climates at least in the northeast, but dry climates returned during the Aptian and evaporites were widespread in the coastal basin from southern to northeastern Brazil. Extremely dry climates probably never recurred. No post-Albian evaporites are known even though Late Cretaceous and Tertiary regressive sequences occurred which would set proper conditions for the development of these chemical sediments.

REFERENCES

Aguiar, G.A., 1971. Revisão geológica da bacia paleozóica do Maranhão. *An. XXV Congr. Bras. Geol.* 3: 113–122.

Almeida, F.F.M., 1969. Origem e evolução da plataforma brasileira. *Dept. Nacl. Prod. Miner. Div. Geol. Min. Bull.*, 241 - 36 pp.

Asmus, H.E., 1975. Controle estrutural da deposição mesozoica nas bacias da margem continental brasileira. *Rev. Bras. Geogr.* 5(3): 160-175.

Asmus, H.E., and Ferrari, A.L., 1978. *Hipótese sobre a causa do tectonismo cenozóico na região Nordeste do Brasil.* Proj. Remac n° 4, Petrobrás – Cenpes, pp. 75–88.

Asmus, H.E., and Ponte, F.C., 1973. The Brazilian marginal basins. In: A.E.M. Nairn and F.G. Stehli, (Editors), *The Ocean Basins and Margins, 1, The South Atlantic.* Plenum Press, New York, London, pp. 87–133.

Asmus, H.E. and Porto, R., 1980. *Diferença nos estágios iniciais da evolução da margem continental brasileira: possíveis causas e implicações.* An. XXXI Congr. Bras. Geol., 1: 225–239.

Asmus, H.E. Gomes, J.B. and Pereira, A.C.B., 1971. Integração geológica Regional da Bacia do Espirito Santo. *An. XXV Congr. Bras. Geol.* 3: 235–252.

Bigarella, J.J. and Salamuni, R., 1961, Early Mesozoic wind patterns as suggested by dune bedding in the Botucatu Sandstone of Brazil and Uruguay. *Bull. Geol. Soc. Am.*, 72: 1089-1106.

Daemon, R.F., 1974, *Contribuição à datação da Formação Alter-do-Chão, Bacia do Amazonas.* Unpublished report.

Eick, N.C., Gamermann, N. and Carraro, C.C., 1973. A discordância pre-Formação Serra Geral. *Inst. Geol. Univ. Fed. Rio Grande do Sul, Pesqui.*, 2: 73–77.

Fulfaro, V.J., 1971. A evolução tectônica e paleogeografica da bacia sedimentar do Paraná pelo 'trend surface analysis'. *Esc. Eng. São Carlos, Geol.*, 14: 112 pp.

Gammermann, N., 1973. Formação Rosário do Sul. *Pesqui. Inst. Geol. Porto Alegre*, 2: 5–35.

Gordon, M. Jr. 1947. Classificação das formações gondwânicas do Paraná, Santa Catarina e Rio Grande do Sul. *Dept. Nacl. Prod. Miner., Div. Geol. Min., Notas Prel. Estud.*, 38: 20 pp.

Grossi Sad. J.H., Cardoso, R.N. and Costa, M.T., 1971. Formações cretáceas em Minas Gerais, uma revisão. *Rev. Bras. Geocinêc.*, 1(1): 2–12.

Hasui, Y., 1968. A Formação Uberaba, *An. XXII Congr. Bras. Geol.*, pp. 167–179.

Leinz, V., 1949. Contribuição à Geologia dos derrames basálticos. *Bol. Fac. Fil. Ciên. Letras, Univ. S. Paulo, 103 (Geol. 5)* 61 pp.

Lima, E.C., 1972. Bioestratigrafia da bacia de Barreirinhas. *An. XXVI Congr. Bras. Geol.*, 3: 81-92.

Mendes, J.C. and Petri, S., 1971. Geologia do Brasil. *Inst. Nacl. Livor, Geol.*, 9: 207 pp.

Mesner, J.C. and Woolridge, L.C., 1964. Maranhão basin and Cretaceous Coastal basins. *Bull. Am. Assoc. Petrol. Geol.*, 48 (9): 1475–1512.

Mezzalira, S., 1974. Contribuição ao conhecimento da Estratigrafia e Paleontologia do Arenito Bauru. *Inst. Geográf. Geol., S. Paulo*, 51: 163 pp.

Miura, K., 1972. Possibilidades petrolíferas da bacia do Acre. *An. XXVI Congr. Bras. Geol.*, 3: 15–20.

Morales, L.G., 1959. General geology and oil possibilities of the Amazon Basin, Brazil. *Proc. 5th World Petrol. Congr., Sec. 7, Pap. 51*, pp. 925–942.

Northfleet, A.A., Medeiros, R.A. and Muhlmann, H., 1969. Reavaliação dos dados geológicos da bacia do Paraná. *Bol. Tec. Petrobrás*, 12(3): 291–346.

Ojeda, H.A. and Bisol, D.L., 1971. Integração geológica regional da extensão submarina da bacia sedimentar de Sergipe–Alagoas. *An. XXV Cong. Bras. Geol.*, 3: 215–226.

Pamplona, H.R.P., 1969. Litoestratigrafia da bacia cretácea de Barreirinhas. *Bol. Téc. Petrobrás* 12(3): 261–290.

Rezende, W.M. and Ferradais, J.O., 1971, Integração geológica regional da bacia sedimentar da foz do Amazonas. *An. XXV Congr. Bras. Geol.*, 3: 203–214.

Schaller, H., 1969. Revisão estratigráfica da bacia Sergipe–Alagoas. *Bol. Téc. Petrobrás*, 12(1): 21–86.

Schaller, H., Vasconcellos, D.N. and Castro, J.C., 1971. Estratigrafia preliminar da bacia sedimentar da foz do rio Amazonas. *An. XXV Congr. Bras. Geol.*, 3: 189–202.

Soares, P.C., 1974. *O Mesozóico Gondwânico no Estado de São Paulo*. Ph.D.Thesis, School of Geology of Rio Claro, São Paulo. Unpublished.

Soares, P.C. and Landim, P.M.B., 1976. Comparison between the tectonic evolution of the intracratonic and marginal basins in South Brazil. *An Acad. Bras. Ciên.* 48: 313–324. (Supl. – Proc. Int. Symp. Continental Margins of Atlantic Type).

Soares, P.C., Landim, P.M.B., Fulfaro, V. and Sobreiro Neto A.F., 1980. Ensaio de caracterizaçao estratigráfica do cretáceo no Estado de São Paulo: Grupo Bauru. *Rev. Bras. Geocienc.* 10(3): 177–185.

Suguio, K., 1973. *Formação Bauru, Calcários e sedimentos detríticos associados*. Ph.D. Thesis, University of São Paulo. Unpublished.

Van der Hammen, T. and Burger, D., 1966. Pollen flora and age of the Takutu Formation (Guyana). *Leidse Geol. Meded.*, 38: 173–180.

Viana, C.F., Gama E.G. Jr, Simões, I.A., Moura, J.R. and Fonseca, J.R., 1971. Revisão estratigráfica da bacia Recôncavo–Tucano. *Bol. Téc. Petrobrás*, 14(3/4): 157–192.

Chapter 7

THE TRIASSIC OF ARGENTINA AND CHILE

PEDRO N. STIPANÍCIC

INTRODUCTION

Mesozoic formations are well developed in Argentina and Chile and over the past century an extensive bibliography has been generated. Groeber (1953, 1959) and Leanza (1969) provide important data sources for Argentina while for Chile Munoz (1950), Hofstetter et al. (1957) and Corvalán (1965) may be consulted.

If more emphasis is placed upon the Mesozoic formations of Argentina it is because, in general, they are better known. To facilitate reference, names have been assigned to unnamed units (*nov. form., nov. group*), and others corrected to conform to nomenclatural rules (*nov. nom.*).

The Mesozoic of Argentina and Chile developed on a cratonic foreland and in the Andean orogen. The Andean geosyncline began its development in Triassic times and has a full Jurassic–Cretaceous history. A segment with no strictly geosynclinal conditions (= "liminar" Andes) extended north–south to about 40°S, south of which true geosynclinal conditions pertain and the orogen curves to trend essentially west–east (Vicente, 1970).

Two long troughs separated by a volcanic ridge characterize the liminar Andes. The western ("euliminar") trough extends in Chile as far as the Argentinian border; the eastern ("mioliminar") trough lies in Argentina only penetrating into Chile north of 31°S. The euliminar trough contains a volcano-sedimentary complex in excess of 15,000 m which has suffered slight regional metamorphism (Levi, 1969) and is intruded by granodiorite. There are Middle Triassic, Lower and Middle Jurassic and Lower Cretaceous marine horizons. Marine horizons of the same age are also found in the generally continental, non-volcanic sediments of the mioliminar trough.

The geosynclinal Andes may be divided into three zones, the Western or Pacific (the Patagonian Archipelago), the Patagonian (or main) Cordillera and the Eastern Precordillera, along the Patagonian foreland. The first and second correspond to a typical geosynclinal trough (Katz, 1964) with internal ophiolitic magmatic activity, sub-Hercynian metamorphism and a post-orogenic intrusive stage. The third belt, after an initial volcanic phase developed as a miogeosynclinal basin over the Patagonian Shelf (Vicente, 1970). The Neuquenian–Tarapaquean sector corresponds to most of the liminar Andes, while the geosynclinal Andes equate almost exactly the "Magellanian" sector.

Extensive intracratonic basins received a fill of continental sediments over most of Argentina during Mesozoic times. A shallow epicontinental sea, however, penetrated northwestern Argentina from the Bolivian embayment (liminar Andes) during Triassic times, and later during Cretaceous times parts of Northern Argentina and Chile were invaded in the same way.

The more complete stratigraphic record in the eastern mioliminar trough serves to provide the general scheme of geological events and acts as a standard of comparison for the euliminar, geosynclinal and even continental zones. Sedimentation in the liminar troughs was orogenically controlled and diastrophic phases which correspond to the established global events can be recognized, with some having local orogenic foci. The sedimentary cycles so delimited (Table I) are marked by characteristic paleogeographies reflected in facies and regional slope changes. This scheme, based upon Groeber's ideas (1945, 1953) augmented and modified by Stipanícic (1957) and by Stipanícic and Rodrigo (1970) is closely similar to those of Jurassic and Cretaceous times in

TABLE I

Mesozoic sedimentary cycles and diastrophic phases in the Andes of Argentina and Chile

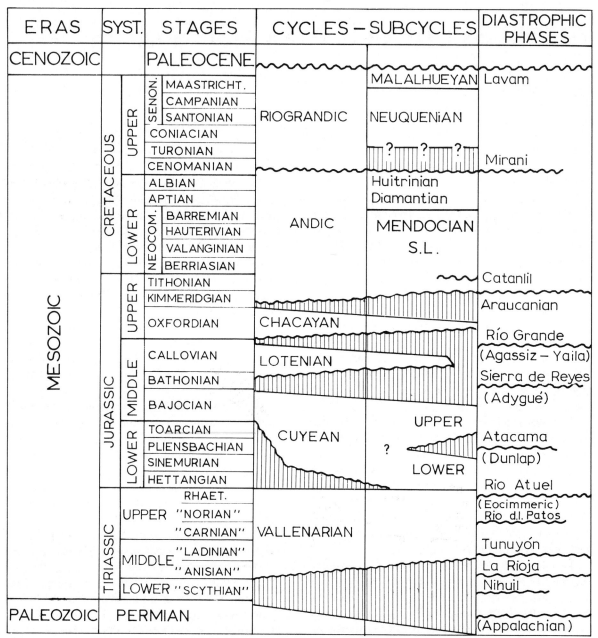

North America or the Russian platform, etc. ...

    Triassic sediments of the first Mesozoic cycle rest upon an erosional surface (the "Platform" of Corvalán, 1965). Active subsidence and intense vulcanism in the Pacific zone suggest an embryonic geosyncline. Three depositional subcycles, separated by unconformities or stratigraphic gaps are recognized during the

Jurassic. They are the Cuyean, Lotenian and Chacayan. The third subcycle ended with Upper Oxfordian rocks and is unconformably[1] overlain by lower Kimmeridgian to Tithonian beds which are

_____

[1] Araucanian diastrophism, which was earlier than and independent of Nevadan activity.

closely associated with the Neocomian sequence as a part of the Mendocian (Mendociano) subcycle of the Andic cycle[1]. This Andic cycle lasted until Coniacian in the external realm (Argentina) but only up to the Cenomanian in the internal realm in Chile (Vicente et al., 1973), the time of closing the geosyncline. To the south in the Magellanian sector (= austral geosyncline) Meso-Neocretaceous, if correctly identified, would define a final Mesozoic cycle, the Neocretaceous ("Supracretácico"). This is separated at its base from the Andean cycle by "Intersenonian" diastrophism ("Peruvian" of Steinmann, "Sub-Hercynian" of Stille) and above from Danian deposits by a strong unconformity in Chile resulting from Laramide activity (= first phase of the First Andean movement of Groeber, 1959). In Argentina an inconspicuous hiatus masks this level.

STRATIGRAPHY

Both igneous and sedimentary rocks occur. The volcanic rocks are dominantly andesites and rhyolites[2]. Some granites are also found. In Argentina with the exception of a small transgression in the northwest, all of the sedimentary rocks are of continental origin. In Chile Triassic marine beds are found at several stages in the deepest basins of the subsiding western area.

The Triassic sedimentary rocks rest, with angular unconformity, upon Neopermian igneous or Eopermian sedimentary rocks (Muñoz, 1950; Groeber and Stipanícic, 1953; Corvalán, 1965). There is no evidence that beds as young as the *Cistecephalus* (J.F. Bonaparte, pers. comm.) zone of Brazil exist here.

The top of the Triassic is readily defined in Argentina and in many parts of Chile by an unconformity, generally angular, resulting from late Norian (Rio Atuel–Valparaiso diastrophism) orogenic activity. The resulting Triassic–Jurassic hiatus may comprise the whole Liassic or parts of it (Stipanícic and Rodrigo, 1970). In the Chilean marine embayment, sedimentation was continuous and the hiatus is not recognized. In the marine beds, the Alpine zonal sequence can be used; however, for the continental sequence the terms Eo, Meso and Neo (Lower, Middle and Upper) are used since comparisons have to be based upon continental faunas and floras of different composition. Recently (Bonaparte, 1973) a biochronological standard based upon the evolution of the reptilian fauna has been proposed.

Chile

Juan de Morales Basin (20°10′S 69°20′W)

The homonymous formation (*nov. form.*) consists of Upper Triassic continental red sandstones and siltstones with *Dicroidium odontopteroides* (Morr.) Goth similar to that found at Ornuni, Peru (Corvalán, 1965) (see Table II).

Pacific Basin (25°30′–40°S)

During Triassic times the western margin of Chile was a zone of subsidence. It may be referred to as the Pacific Basin (= embryonic Mesozoic Andean geosyncline) and is divided by the Los Vilos Peninsula at 33°–34°S into a northern Vallenar–Los Vilos, and a southern Talca–Bío Bío Sub-basin. Marine sediments in the deeper part of the basin interfinger with continental and transitional deposits in the eastern, shallower, part of the basin. There was important igneous activity (Corvalán, 1965; Muñoz, 1950) and in the western part of the basin the flows were sub-aqueous (J.C. Vicente, pers. comm.).

(1) *Vallenar-Los Vilos Sub-basin.* The marine embayment of the Vallenar–Los Vilos Sub-basin penetrates eastwards nearly as far as the present Argentinian border at 29°S. The most complete section occurs between Los Vilos and Los Molles (32°S 71°30′W) (see Table II). It begins with the Middle Anisian–Ladinian El Quereo Formation (*s. str.*) which consists of 300 m of breccias, deltaic sandstones and conglomerates followed by marine greywackes and turbidite shales with *Trematoceras* sp., *Daonella dubia* (Gabb.) and 300 m of sandstones and conglomerates with intercalations of tuff and rhyolitic flows ("keratophyres"). Igneous activity became more intense and the overlying Pichidangui Formation consists of 2,000 m of acid breccias, tuffs and rhyolitic flows. Towards the top there are sandstone horizons with *Dicroidium odontopteroides*, *Yabeiella mareyesiaca* (Kurtz) Oishi. This is followed by relatively thin (50 m) marine shales with upper-Norian fossils (eg. *Sandlingites* ex. gr. *lissoni*) and in excess of 300 m of continental sandstones and conglomerates with *Chiropteris* aff. *copiapensis* Stein.

---

[1] Andean (Andino) for the geosyncline, Andic (Andico) for the sedimentary cycle.

[2] Triassic andesites are commonly referred to as porphyrites in Argentina. In Chile rhyolites are commonly called keratophyres, a term which can also include trachytes.

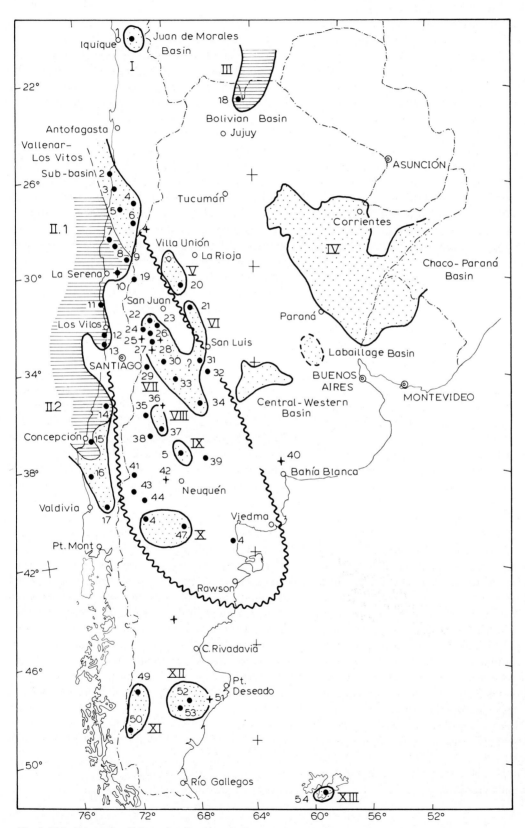

Fig. 1. Main Triassic localities and sedimentary basins.

TABLE II

Correlations of the main Triassic sequences in Chile

SYSTEM PERIOD	JUAN DE MORALES	QUEBRADA CIFUNCHO	LA TERNERA COPIAPÓ	VICUÑA	HUASCO ELQUI	LOS VILOS LOS MOLLES LA LIGUA	CUREPTO TALCA	BÍO-BÍO GOMERO	QUILLÉN NIELOL	TRALCÁN	DIASTROPHIC PHASES & CLIMATES
OVERLIE	SINEMUR.	LIASSIC	JURASSIC	MIDDLE JURASSIC	TOARCIAN	HETTANG.	HETTANG. Mb. A. Tr.	CRETACEOUS	TERTIARY	CRETACEOUS	VALPARAÍSO PHASE / Warm, oxidant in dry seasons / Uniform and more humidity
NORIAN	JUAN DE MORALES Fm.	CIFUNCHO Fm.	LA TERNERA Fm.	LAS BREAS Fm.	HUASCO Fm.	LOS MOLLES Fm. / EL PUQUÉN Fm. / PICHIDANGUI / EL QUEREO Fm. / PASTOS BLANCOS Fm. / TRÁNSITO Fm	EL GUINDO FM.	GOMERO Fm. / UNIHUE Fm. / QUILACOYA Fm. (BÍO-BÍO GROUP)	NIELOL Fm.	TRALCÁN Fm.	ELQUI PHASE
CARNIAN / LADINIAN / ANISIAN / SCYT.											
UNDERLIE	LOW. PERM. UP. CARB.	UPPER PALEOZOIC	PALEOZOIC	UPPER PALEOZOIC	PALEOZOIC	PERMIAN CARBONIF.	PALEOZOIC PRECAMBR.?	PALEOZOIC PRECAMBR.	PALEOZOIC PRECAMBR.	PALEOZOIC PRECAMBR.	

TRIASSIC — LOWER / MIDDLE / UPPER

Legend:
- ◊ Red beds
- ● Coal beds
- ⊘ Marine invertebrates
- ➤ Dicroidium flora
- Y Dicroidium not identified
- Marine
- Continental (or mainly contin.)
- Granites + + +
- Vulcanites v v v
- Absent

and Solms, *Dicroidium odontopteroides, D. zuberi* (Szaj.) Archan. (= *D. feistmanteli* (John) Goth.), *Podozamites elongatus* (Morr.) Feist, *Y. mareyesiaca, Pseudoctenis ctenophylloides* Bonetti, which together form the El Puquén Formation. The overlying marine late-Norian–Rhaetian Lower Los Molles Formation is a sequence of several hundred metres of shale with some sandstone intercalations. It contains *Arcestes* cf. *andersoni* Hyatt & Smith, *Cladiscites* sp., *Pinnacoceras metternichi* (Hauer), *Oxytoma* cf. *inaequivalvis* (Sow.), etc. ... Typical Rhaetian is characterized by *Choristoceras marschi* (Hauer). The succeeding 50 m of Hettangian sandstones and shales which have a conglomeratic base are assigned to the Upper Los Molles Formation (Fuenzalida, 1938). In this region the Keratophyre Series of Chilean authors (Pichidangui Formation) is Carnian in age, but elsewhere may comprise the late Ladinian.

Revision of the marine faunas identified below the volcanic horizon in nearby locations is necessary. In Quereo, Quevedo and Cerro Talimai, Meso-Triassic forms such as *Palaeoneillo* aff. *triangularis* Burck., *Halobia* aff. *zitteli* Linds., *H. rugosa* Smith, *Dinarites* aff. *desertorum* Smith., *Beyrichites* aff. *hannibalis* Toula, *Ceratites* cf. *montisvobis* Smith., *Hungarites* sp., *Arcestes* aff. *ciceronis* Mojs., *Ptychites* sp., etc. ... occur (Muñoz, 1942).

The La Caleta Formation in the same region, a 200 m thickness of chaotic breccias with shaley–limy–sandy intercalations formerly assigned to Triassic or Paleozoic is now known to be (Middle ?) Triassic in the upper part at least. This is based upon the discovery of *D. zuberi* (G. Cecioni, pers. comm.). Middle Triassic sequences in the surrounding areas (La Ligua, Cerro Colorado, 32°30′S 71°15′W) largely repeat the Quereo succession. Post-volcanic beds containing possible Carnian and Norian horizons are here disconformably overlain by the Sinemurian (Muñoz, 1938). Similar facies are also found at Chañaral (29°S 71°20′W), Las Palmas (31°20′S 71°40′W).

Closer to the Argentine border in the Elqui Cordillera (29°30′S 71°10′W), the volcanic beds, rhyolites, tuffs and breccias, together with shales and continental sandstones, 500 to 1,000 m thick, are assigned to the Pastos Blancos Formation. This is disconformably overlain by Toarcian. In neighbouring areas, such as Alto del Carmen, Rio Transito and Huasco the volcanic horizon is referred to as the La Totota Formation and overlies the San Felix Forma-

tion of lower to upper Anisian age, and is succeeded by the Carnian Huasco Formation (*nov. form.*) which consists of 250 m of shale and 800 m of sandstone conglomerate and shale. The Anisian beds contain *Trigonia tabacoensis* Barthel, *Daonella* sp. ex. gr. *sturi* (Ben.), *Cuccoceras* sp., *Beyrichites* sp.. The lower, shaley part of the Huasco Formation is essentially marine with *Halobia*, ammonites and plants; the upper dominantly arenaceous succession contains vertebrates (Barthel, 1958; Zeil and Ichikawa, 1958).

In the non-marine sectors of the basin typically the Triassic consists of conglomerates with tuffs and lavas (1,200 m Cinfucho Formation south-southwest of Taltal, 25°40′S 70°35′W) or conglomerates and red shales (Copiapó area, 27°20′S 70°40′W). At La Ternera (27°10′S 69°45′W) there is a classical sequence, 2,300 m thick, of shales, sandstones and conglomerates, with coal seams and a flora comprising *Dictyophyllum carlsoni* Nath., *Chiropteris copiapensis, Copiapea plicatella* Solms, *Dicroidium lancifolium, Linguifolium steinmanni* Solms, *Yabeiella mareyesiaca, Baiera munsteri* Presl., etc. (Solms-Laubach and Steinmann, 1899; Steinmann, 1921; Segerström, 1962; Corvalán, 1965).

Similar lithologies are found in Rio Turbio–Ramadas (28°S 69°30′W) and at Vicuña (30°S 70°45′W) where the Late Triassic continental sequence, the 230 m Las Breas Formation consists of sandstones, conglomerates, carbonaceous shales and rhyolitic tuffs with *Dicroidium lancifolium, Equisetites* sp. and *Dipteridaceae*, unconformably overlain by Middle Jurassic beds.

(2) *Talca–Bío Bío Sub-basin.* Between 34°20′S and 37°30′S, the Triassic seas penetrated very few kilometres east of the line of the present coast, continental conditions reigning over the rest of the basin. At Curepto-Talca (35°20′S 71°50′W) the sequence of 700 m of shales with sandstone and conglomerate intercalations are here referred to the El Guindo Formation (*nov. form.*). It includes Anisian (*Protacrochordioceras*) and upper Norian (*Cochloceras suessi* Mojs.) horizons. In addition there are beds with *Oxytoma inaequivalvis, Chlamys mojsisovicsi* Kob. and Ich., *Monotis typica* Kipar., *M. subcircularis* Gabb., and continental and transition horizons which have yielded *Clathropteris platyphylla* Brong. and *Baiera* cf. *paucipartita* Nath. Above these follows 200 m of a transition sandstone member below Hettangian shales. There is no paleontological evidence of Ladinian or Carnian.

Further to the south (37°S 72°50'W) the Bío Bío Group (*nov. group*) consists of three thick formations deposited during Carnian and Norian times. The lowest, Quilacoya Formation is composed of 1,500 m of conglomerate, coarse arkoses with shales and coal intercalations, where *Clathropteris platyphylla*, *Cladophlebis rosserti* Presl., *C. australis* Morr., *Chiropteris copiapensis*, *Sagenopteris nilssoniana* Brong., *Linguifolium lilleanum* Arb., *Dicroidium odontopteroides*, *Yabeiella mareyesiaca*, etc., are found. The following 2,500–3,000 m of the Unihue Formation of shales and alternating sandstones and shales contains *Halobia neumayri* Bitt., *Phaenodesmia* sp., *Clathropteris* sp. and *Dicroidium lancifolium*. The final formation of the group, the 3,000 m thick Talcamávida–Gomero Formation, is made up of sandy shales and sandstones with black shale intercalations with *Chiropteris copiapensis, Dicroidium lancifolium*, etc. (Steinmann, 1921; Tavera, 1960). The fauna found by Gomero and studied by Jaworski (1922) may be derived from the Unihue Formation. This included *Myophoria* gr. *vestitae* Alb., *Palaeoneillo elliptica* Gold., *Arcestes* aff. *cheilostoma* Mojs.

In other regions, at Quillen–Nielol (38°30'S 72°45'W) the Triassic is represented by 1,000 m of graywacke with conglomerate, shale and thin coal intercalations (Nielol Formation) with plants similar to those found at La Ternera (Fritzsche, 1921), and at its southernmost extent by the Upper Triassic red conglomerates of the Tralcán Formation (*nov. form.*) (see Table II). Sandstones and shales within the conglomerate yield *Cladophlebis tenua* Oishi (Davis and Karzulovic, 1961; Aguirre and Levi, 1964).

Argentina

As indicated earlier the only marine and transitional facies in Argentina are found in the northwestern region. The Triassic consists essentially of igneous and sedimentary rocks deposited in intracratonic basins. Some beds were formerly assigned to the Rhaetic, later to the Upper Triassic. At the present time the presence of Middle and Upper Triassic is well established.

The sedimentary basins are widely spread, especially across western Argentina. A correlation chart of the Triassic formations from different areas is attempted in Table III. It shows that there is no continuous Triassic section and that in the Bolivian and Chaco–Paraná basins only the top of the Upper Triassic is present, and over Argentina as a whole horizons below the Ladinian are only sporadically present. Mention of the principal basins follows.

The Bolivian Basin

Norian marine strata (with *Monotis* aff. *subcircularis*) are better developed in Bolivia and penetrate into northwestern Argentina along the Upper Bermejo River; in Bolivia the Vitiacuá Formation is represented by 300 m of sandstones which become calcareous upwards passing finally to about 100 m of siliceous limestone and shaley beds (see Table III). *Pytyoporites* (Leanza, 1969; Mingramm and Russo, 1972; Reyes and Salfity, 1973) occurs at several horizons.

The Chaco–Paraná Basin

Cross-bedded red sandstones found in the subsurface of northeastern Argentina and southern Brazil and in outcrops in Uruguay, have been referred to the Upper Triassic. While they are younger than the Middle Triassic Santa Maria Formation of southern Brazil, there is no other evidence of age. These continental and aeolian sandstones are assigned to the Buena Vista Formation (Falconer, 1931).

The Laboulage Basin

Reddish brown sandstones and brown tuffs found in the subsurface in northern Buenos Aires, southern Córdoba and Santa Fé provinces have been referred to the Upper Triassic by analogy with the Buena Vista Formation.

The basins of the central-western region

There are many outcrops of Triassic rocks in the western part of La Rioja, San Juan and Mendoza provinces between 29° and 37°S, and they are known in subsurface as far east as San Luis (66°30'W). They can be discussed in terms of three basins, the Ischigualasto–Villa Union, Marayes–Desaguadero and Cuyo, the last of which is sufficiently large and complex to require more detailed treatment. Igneous activity was widespread and consisted of thick andesitic flows with tuffs and breccias. These are sometimes intruded by granite and both may have been cut by rhyolitic bodies. The andesites ("porphyrites") are assigned to the Choiyoi Group (formerly "Serie Porfiritica Supratriasica") and separated from a heterogeneous complex formerly including Carboniferous, Permian and Triassic rocks (Groeber and Stipanícic,

TABLE III

Correlations of the main Triassic sequences in Argentina

Correlation chart of Triassic sequences across Argentine basins (Bolivian, Chaco–Paraná, Ischigualasto–Villa Unión, Marayes–Desaguadero, San Luis, Cuyo, Malargüe, Puesto Viejo, La Pampa–Buenos Aires–Neuquén, North-Patagonian, Deseado and Southern basins), plotted against System/Period (Triassic: Lower–Scythian; Middle–Anisian, Ladinian; Upper–Carnian, Norian) and Diastrophic Phases & Climates (Río Atuel Phase, Río Los Patos Ph., Tunuyán Phase, La Rioja Phase, Nihuil Phase).

Legend: Marine; Continental; Vulcanites (v v v); Granites (+ + +); Coal beds; Red beds; Bituminous shales; Absent; Marine invertebrates; Continental vertebrates; Micro floras; "Dicroidium" floras; Fossil plants ("Dicroidium" not identified).

1953; Polanski, 1966). The group is still poorly defined for it contains Middle Triassic andesites at Agua Negra and Cordón de Olivares (30°20'S 70'W) which are essentially continuous with the Chilean El Elqui–Alto del Carmen volcanics, and older rocks at Uspallata, Cordillera del Tigre and Cordón del Plata (32°20'S 69°50'W) where radiometric dating indicates Middle, Lower Triassic and Permo-Triassic horizons (Rocha et al., 1971).

The Cordón del Plata and Cerro Chihuíu (or Chihuido) granites (35°20'S 69°40'W) with ages of 204 m.y. and 194 m.y. are Upper Triassic in age (Stipanícic and Linares, 1969; Caminos, 1972). The Cerro Cacheuta stock is referred to the same age (Rolleri and Criado, 1968). Triassic pegmatites in San Juan province date at 202 and 220 m.y. (Valle Fertil, 30°40'S 67°40'W) while the Cerro Come Caballos granite in the La Rioja province (28°10'S 69°10'W) at 224 m.y. dates close to the Permo-Triassic boundary.

Choiyoi extrusives occur as far south as Cordón del Portillo (34°S 69°W), San Rafael-Nihuil (34°10' to 35°S 70°W) and Cerro Puchenque, Barda Blanca (36°S 70°W), Sierra Azul, Sierra de Reyes (37°S 69°40'W) etc. (Rolleri and Criado, 1968, 1970). They also occur in the La Pampa and Neuquén provinces, providing a link with the Rio Negro effusives.

The Triassic sediments can usually be divided into a four-part sequence, basal agglomerates and conglomerates (commonly red) followed by tuffaceous sandstones (sometimes bentonitic) upon which rests a homogeneous section of fine clastics (up to bituminous shales). The succession is frequently terminated by a return to the deposition of red conglomerates and sandstones. Divisions of these sediments were formerly based upon either fossil flora or vertebrate remains. The results did not always agree although they now are essentially the same.

Abundant paleobotanic work on the macroflora has been published by Menendez (1951), microfloral studies have begun and results of the studies of the vertebrate fauna appear in Rusconi (1951), Romer (1960, 1967–72) and Bonaparte (1966a, 1971a, 1973).

(1) *Ischigualasto–Villa Union Basin* (= "Ischichuca–Ischigualasto Basin", 29°15'–30°30'S and 67°30'–68°30'W). The Triassic with an average thickness of 1,500 m (and maximum of 4,000 m) rests with angular unconformity on rocks ranging in age from Precambrian to Eopermian (Frenguelli, 1948; Groeber and Stipanícic, 1953; Andreis, 1969). Two groups are recognized, the lower Paganzo III Group and the upper Agua de la Peña Group (Table III).

The Paganzo III Group is made up of the 400 m Talampaya Formation, a sequence of fine red sandstones with *Chirotherium*-type footprints, and occasional pebble horizons in which are intercalated amygdaloid basalt flows, followed by the Tarjados Formation. The latter, 250 m thick, follows an erosional unconformity and consists, above a basal conglomerate, of red cross-bedded sandstones with brown shale and lighter coloured sandstone intercalations. Kannemeyerid-type dicynodonts have been found in beds of the Tarjados Formation (Groeber and Stipanícic, 1953; Romer and Jensen, 1966; Andreis, 1969; Bonaparte, 1969).

The Agua de la Peña Group (Ischigualasto–Ischichuca) consists of five formations separated by an unconformity from the Tarjados Formation or older formations. The basal Chañares Formation, contemporaneous over the whole basin consists of primary tuffs in which are found rich vertebrate remains. The two following formations interfinger eastwards, and a similar relationship is found between the upper two formations (see Table III). The Ischichuca Formation (0–750 m) composed of conglomerates, sandstones, tuffaceous carbonaceous and bentonitic shales is conformably overlain by and interfingers with the 200–1,000 m Los Rastros Formation, a rhythmic sequence of 50 m units of coarse sandstones and conglomerates, fine grained sandstones and carbonaceous shales.

The facies of the Chañares Formation contains dicynodonts (*Chanaria, Dinodontosaurus*), cynodonts (*Massetognathus, Probelesodon, Chiniquodon*), and ornitosuchians (*Gracilisuchus*). Within the Ischichuca Formation the flora contains *Neocalamites carrerei* (Zeiller) Halle, *Dicroidium zuberi, D. stelznerianum* (Gein.) Town., while the flora of the Los Rastos Formation includes *Phyllotheca australis* Brong., *Neocalamites carrerei, Cladophlebis mesozoica* Kurtz, *Dicroidium odontopteroides, D. lancifolium, D. stelznerianum, D. coriaceum* (John.) Town., *D. zuberi, Xylopteris densifolia* (Du Toit) Freng., *X. argentina* (Kurtz) Freng., *X. elongata* (Carr.) Freng. var. *rigida* Stip. & Bonetti, *Yabeiella mareyesiaca, Y. spathulata* Oishi, *Y. brackebuschiana* (Kurtz) Oishi, etc. A palynomorph assemblage M 2[1] is found in the Los Ras-

---

[1] Three main palynological assemblages have been identified in Argentinian Triassic sediments: M1 and M2, with an Upper Triassic pattern, M3 of the Middle–Upper Triassic boundary (Yrigoyen and Stover, 1970).

tros Formation (Yrigoyen and Stover, 1970) and some of the elements (*Cingularisporites, Clavatriletes, Discosporites, Verrucosisporites, Cycadopites, Punctamonocolpites* and *Monosulcites*) were described by Herbst (1970, 1972). *Estheria*, freshwater fish and thecodont tracks (*Rigalites* of Frenguelli, 1948; Bonaparte, 1966a) are also found.

The Ischigualasto Formation (500–900 m) comprises a basal conglomerate followed by sandstones, with dominant carbonaceous shales and tuffs. Plant fossils are common, such as *Neocalamites carrerei, Dicroidium lancifolium, D. odontopteroides, D. zuberi, D. coriaceum, Xylopteris densifolia, X. elongata, X. argentina, Podozamites elongatus*, along with fish (*Semionotus*) stereospondyla tetrapods (*Promastodonsaurus, Pelorocephalus*), rhynchocephalids (*Scaphonix*), thecodonts (*Proterochampsa, Argentinosuchus*), saurischians (*Ischisaurus*), ornithischians (*Pisanosaurus*), dicynodonts (*Ischigualastia*), cynodonts (*Chiniquodon, Exaeretodon*). The M1 palynological assemblage occurs at the base of the formation with *Leiotriletes, Lophotriletes, Concavisporites, Phyllothecotriletes*, etc. (Frenguelli, 1948; Reig, 1963; Cox, 1965; Bonaparte, 1966a, 1969, 1971a, 1973; Casamiquela, 1967; Yrigoyen and Stover, 1970).

The Los Colorados Formation, a red bed sequence, passes from sandstones with conglomerate lenses below to siltstones above. Depending upon the location within the basin it may be conformable or disconformable upon the Ischigualasto Formation. It contains fossil wood, thecodonts (*Riojasuchus*), saurischians (*Riojasaurus*), crocodilians (*Hemiprotosuchus*), etc. (Frenguelli, 1948; Groeber and Stipanícic, 1953; Bonaparte 1966a, 1969, 1971a, b, 1973; Yrigoyen and Stover, 1970).

These formations of the Agua de la Peña Group cover horizons from the lowermost part of the upper Middle Triassic to the uppermost Triassic (excluding Rhaetian) (Romer, 1960, 1969; Reig, 1963; Cox, 1965, 1968; Bonaparte, 1966c, 1969, 1973; Stipanícic and Bonetti, 1969). The age range of the individual formations is not known with any certainty. Macro and microflora from the lower formations and lithological correlations suggest an age not older than upper Middle Triassic, but the evolutionary character of the vertebrates suggests a somewhat greater antiquity, lower Middle Triassic. While the Paganzo III Group is referred to upper Lower Triassic–lower Middle Triassic the stratigraphic relations of the Ta-

lampaya Formation with some Paganzo II beds referred to the Permian are not well defined.

(2) *The Marayes–Desaguadero Basin*. This basin is separated from the Ischigualasto–Villa Union Basin described above, by the crystalline Paganzo and Valle Fertil Hills. In Marayes in the southeastern San Juan province (31°25'S 67°20'W) Triassic outcrops of the Marayes Group have been subdivided into four formations. The basal, Esquina Colorada Formation, 450 m thick, consists of red conglomerates overlain by shales and tuffs. This is followed, conformably or unconformably, depending upon location in the basin, by 90 m of sandstones with gravels, shales and coal bearing beds of the Quebrada de la Mina Formation and 280 m of sandstones and conglomerates of the Carrizal Formation. The Triassic sequence ends with the 900 m Quebrada del Barro Formation, a sequence of reddish brown sandstones and immature conglomerates indicative of renewed uplift in the source area. The middle part of the Marayes Group contains *Dicroidium odontopteroides, D. lancifolium, D. coriaceum, D. stelznerianum, Pteruchus rhaeticum* (Gein.) Freng. (= *Pteruchus africanus* Thomas) and the M1 palynological association (Geinitz, 1876, 1923; Groeber and Stipanícic, 1953; Yrigoyen and Stover, 1970). The group is assigned to the upper Middle Triassic and Upper Triassic.

Southwards sediments of the San Luis Group (Groeber and Stipanícic, 1953; Stipanícic, 1957) also known as the Gigante Group, once referred to the Triassic on the basis of the vertebrate ichnites and fish fauna (Flores and Criado, 1972), are now known to be mostly Lower Cretaceous in age on the basis of their palynomorphs (M. Yrigoyen, pers. comm.). True Triassic strata closely correlated with beds in the Cuyean Basin containing typical palynomorphs and *Dicroidium odontopteroides* have been found in the subsurface (33°40'S 66°40'W) and in Alto Pencoso (33°S 67°W). Their correlation with the Marayes Triassic seems highly probable and both may be sub-basins, regarded as a part of the marin Cuyean Basin.

At Cerro Varela (34°S 66°30'W) 500 m of andesites and tuffs are referred to the Choiyoi Group (Flores and Criado, 1972).

(3) *The Cuyean Basin* (= Cuenca Cuyana). This is a large continental Triassic basin in western Argentina extending over 500 km from 31°30'S 69°15'W to 36°30'S 66°30'W (Rolleri and Criado, 1968; Yrigoyen and Stover, 1970). Some parts of the basin are much better known than others because of their eco-

nomic or stratigraphic interest. There are syntheses by Frenguelli (1948), Groeber and Stipanícic (1953) with revisions by Rolleri and Criado (1968) and Stipanícic (1972). In the following account the data of Rolleri and Criado (1968) will be used, but alternative correlations and nomenclature will be presented (see also Yrigoyen and Stover, 1970). Within the basin there are marked lithological changes from one part of the basin to another, so that the understanding of the stratigraphy requires description of five sections.

Hilario–Sorocayense-Barreal section. Rocks of the Sorocayense Group (ex. Barreal Group, including the Hilario Group) lying unconformably upon Paleozoic rocks as young as Eopermian, crop out between Tamberiás in the north to Barreal in the south. To the west, in the Ansilta Cordillera thick andesite and rhyolitic flows of the Choiyoi Group are found (Table III).

The Sorocayense Group is divided into three formations of which the lowest is the Barreal Formation. This formation, 150–270 m thick, normally consists of light coloured, fine to medium grained tuffs and tuffaceous sandstones. Only in the southwest do red conglomerates and tuffs appear in the lowest part of the formation. There is an erosional unconformity between the Barreal and the overlying Cortaderita Formation, a 200 m sequence of variegated tuffs, and tuffaceous sandstones with thick gray and green-gray shales and yellow bentonitic beds. A further erosional unconformity separates the Cepeda Formation from the Cortaderita Formation. The Cepeda Formation has a basal, red, andesitic conglomerate followed by 250 m of red sandstones (Stappenbeck, 1910; Du Toit, 1927; Frenguelli, 1944, 1948; Groeber and Stipanícic, 1953; Stipanícic, 1957, 1972; Stipanícic and Bonetti, 1965, 1969).

The facies of the Cortaderita Formation occur at the same horizon over a wide area and are known at Potrerillos 200 km to the south. The Barreal section lies at the margin of the Cuyean Basin and as a result the black bituminous shales characteristic of most of the basin (Potrerillos, Cacheuta, etc., in Mendoza province) are not developed. They do appear 15 km to the east at Rincón Blanco.

The upper part of the Barreal Formation and the Lower Cortaderita Formations contain a rich flora which includes: *Phylloteca australis, Hausmannia dentata* Oishi, *Thaumatopteris dunkeri* Oishi & Yam., *T. pusilla* (Nath.) Oishi & Yam., *Chiropteris zeilleri*

Sew., *C. copiapensis, C. cuneata* (Carr.) Sew., *Dicroidium odontopteroides, D. lancifolium, D. zuberi, D. stelznerianum, D. coriaceum, Pteruchus dubius* Thomas, *Umkomasia macleani* Thomas, *Xylopteris argentina, X. elongata* var. *rigida, Thinnfeldia rhomboidalis* Ett., *Pseudoctenis ctenophylloides, P.* cf. *falconeriana* (Morr.) Bonetti, *P. fissa* du Toit, *Anomozamites nilssoni* (Phyll.) L. & H., *A. gracilis* Nath., *A. insconstant* Braun, *Taeniopteris carruthersi* Ten.-Woods, *T. mc'clellandi* O. & M., *Yabeiella mareyesiaca, Y. brackebuschiana, Y. spathulata, Lepidanthium microrhombeum* (Braun) Schimp., *Copiapaea plicatella, Saportaea dichotoma* (Freng.) Stip. & Bonetti, *S. flabellata* (Freng.) Stip. & Bonetti, *S. intermedia* Stip. & Bonetti, *Sphenobaiera stormbergensis* (Sew.) Freng., *S. argentinae* (Kurtz) Freng., *Podozamites elongatus*, etc. (Stappenbeck, 1910; Du Toit, 1927; Frenguelli, 1942–46, 1944, 1948; Stipanícic and Menéndez, 1949; Groeber and Stipanícic, 1953; Stipanícic and Bonetti, 1965, 1969; Bonetti, 1968; Stipanícic, 1972). The flora most closely resembles the upper Middle Triassic and Upper Triassic of the Northern Hemisphere and these are the ages assigned to the Sorocayense Group.

Rincón Blanco section. Some 15 km east of Hilario this section displays lithologies more typical of the central part of the basin. The succession referred to the Rincón Blanco Group (Borrello and Cuerda, 1965, emend., Stipanícic, 1972) comprises five formations, of which the lowest, a 0–250 m thickness of conglomerates and massive sandstones, is named the Ciénaga Redonda Formation. It is followed by the Cerro Amarillo Formation (150–500 m), a succession of sandstones and stratified conglomerates. The Portezuelo Formation rests upon the Cerro Amarillo Formation with para-unconformity. It is 100–300 m thick with sandstones, conglomerates and shale intercalations from which the M2 palynological association and dicynodont remains (J.F. Bonaparte, pers. comm.) have been obtained. Above follows 150 m of black laminar bituminous shales of the Carrizalito Formation again yielding the M1-M2 palynological association. It is separated by an erosional unconformity from the Casa de Piedra Formation, a thickness of more than 200 m of conglomerates and cross-bedded sandstones from which *Neocalamites* sp., *Dicroidium* sp. and the M1 palynological association (Borrello and Cuerda, 1965; Yrigoyen and Stover, 1970; Stipanícic, 1972) have been extracted.

The Rincón Blanco Group correlates closely with

the Sorocayense, Uspallata and Peñasco groups (Yrigoyen and Stover, 1970; Stipanĭcic, 1972). This permits its assignment to the upper Middle Triassic and Upper Triassic and throws into doubt correlation of the whole group with the lower part of the Uspallata Group proposed by Borrello and Cuerda (1965) and Rolleri and Criado (1968). The correlation proposed is borne out by the microfloral evidence.

The Potrerillos–Cacheuta–Uspallata section. It is composed of nearly continuous outcrop from Cerro Cacheuta, Potrerillos, Agua de la Zerra, Paramillos de Uspallata, Villavicencio, San Isidro to Salagasta (from 32°20'S to 33°S and between 68°45'W and 69°15'W). The whole is referred to the Uspallata Group, a term replacing Cacheuta Group, Cacheuta Series (sensu Frenguelli, 1944, 1948; Groeber and Stipanĭcic, 1953, etc.). Rocks of the Uspallata Group show relatively constant facies types, changes in facies becoming more common towards the central part of the basin (Table III). They can be traced in the subsurface to the south and southeast and may be detected with some facies change and loss of some horizons in the Tupungato, Lunlunta, Barrancas, Carrizal and Punta de Bardas oil fields (33°30'S 68°40'W).

The lowest unit of the group, the Río Mendoza Formation, rests unconformably upon Choiyoi andesites, rhyolites and breccias or even Paleozoic rocks. At the type locality it consists of 450 m of red conglomerate with angular fragments which pass upwards to well bedded sandstones and shales. In other localities only conglomerates are present and in places the thickness may dwindle to zero. Some vertebrate remains have been found. The overlying Las Cabras Formation is not well developed in this section but in the Potrerillos area it consists of 0–600 m of sandstones and fine red conglomerates with rounded pebbles, tuffs and light coloured sandstone and calcareous sandstone intercalations. Shales and carbonaceous horizons occur towards the top. In subsurface the tuff and shale components become more important (Rolleri and Criado, 1968). To the north in Paramillos de Uspallata and Río de Las Peñas tuffaceous sandstones and light coloured tuffs are dominant with some intercalated andesite flows which have yielded a date of 204 m.y. (Groeber and Stipanĭcic, 1953; Valencio and Mitchell, 1972).

An erosional or weak angular unconformity separates the Potrerillos Formation from the Las Cabras (Bracaccini, 1945; Borrello, 1962; Rolleri and

Criado, 1968; Yrigoyen and Stover, 1970). The succession of 700 m is made up of tuffs, light coloured shales, bentonite beds similar to these of the Cortaderita Formation in the Barreal region, sandstone and conglomerate and some carbonaceous levels. Plants and vertebrate remains occur. The Potrerillos Formation is conformably overlain by homogeneous fine sandstones followed by well bedded dark shales, black and bituminous towards the top. These beds of the Cacheuta Formation total 200 m and contain an abundant flora and vertebrate fauna. The Triassic sequence ends with the 700 m Río Blanco Formation, a succession of gray and dark shales and siltstones passing up into red siltstones and conglomerates (Frenguelli, 1948; Groeber and Stipanĭcic, 1953; Rolleri and Criado, 1968). There is a suggestion that the fossiliferous lower part of the Río Blanco Formation should be more appropriately included in the Cacheuta Formation restricting the former to the red beds. In some localities there is a para-unconformity at the base of the red beds.

Two formations formerly included in the Triassic sequence (Rolleri and Criado, 1968), the Barrancas and Punta de Bardas Formations are excluded. The first is separated by an angular unconformity from the Triassic and is of Lower Cretaceous age, while the Punta de Bordas with basalt flows dated at 129 m.y. is related to the magmatism of the Serra Geral (Stipanĭcic and Linares, 1969; Regairaz, 1970).

The Uspallata Group has an important plant and vertebrate flora and fauna of which many paleobotanical studies are made (e.g. Frenguelli, 1942–1946, etc.). A study by Jain and Delevoryas (1967) requires considerable revision[1]. The vertebrate fauna was described by Rusconi, whose data was discussed by Romer (1960) and Bonaparte (1973).

The Las Cabras Formation contains the M3 paly-

---

[1] Identifications of *Cladophlebis* by Jain and Delevoryas have been corrected by Herbst (1971). Other species by the present author, e.g., first identification J+D, second Herbst or Stipanĭcic: *Cladophlebis kurtzi* = *C.* sp (non *kurtzi*); *C. johnstoni* = *C. mendozaensis*; *C. australis* = *C.* cf. *mesozoica*; *C. wielandi* = *C.* aff. *kurzi*; *Dicroidium odontopteroides* (tab. 89, fig. 6; tab. 90, fig. 2) = *D. lancifolium*; *D. feistmanteli* (tab. 90, fig. 9c) = *D.* aff. *acutum*; *D. coriaceum* (tab. 91, fig. 1 left) = *D. stelznerianum*; *Xylopteris rigida* = *Dicroidium coriaceum*; *Sphenobaiera tenuifolia* (tab. 96, fig. 6–8) = *Xylopteris argentina*; *Sph. tenuifolia* (tab. 96, fig. 9) = *Xylopteris* cf. *elongata var. rigida*; *Baiera cuyana* = *Xylopteris argentina*; *Noeggerathiopsis* sp. = *Ginkgoidium* aff. *bifidum* Freng.

nological assemblage with *Sulcatisporites, Striatites, Cadargasporites*. The provenance of the common plants *Neocalamites carreri, Asterotheca fuschi* (Zeiller) Kurtz, *Dicroidium remotum, D. stelznerianum, D. zuberi, Xylopteris spinifolia* (Ten.-Woods) Freng. due to tectonic disturbance is uncertain, and some elements may belong to Potrerillos beds. The flora of the latter includes *Neocalamites carreri, Phyllotheca australis, Asterotheca fuchsi, Cladophlebis integra* (Oishi & Tak.), *D. mendozaensis* (Gein.) Freng., *Dicroidium stelznerianum, D. coriaceum, Pteruchus simmondsi* (Shirley) Thomas (= *Stachiopitys anthoides* Freng.), *Xylopteris argentina, X. elongata var. rigida, Lepidopteris stormbergensis* (Sew.) Tonw. (= *Callipteridium argentinum* Freng.), *Yabeiella mareyesiaca, Y. brackebuschiana, Baiera cuyana, B. bidens* (Ten.-Woods) Freng., *Sphenobaiera argentina, Podozamites elongatus, Cycadocarpidium minus* (Wiel.) Freng., and *C. majus* (Wiel.) Freng., etc. The Cacheuta Formation includes *Neocalamites carrerei, Cladophlebis kurtzi* Freng., *Chiropteris copiapensis, Dicroidium odontopteroides, D. lancifolium, D. cacheutense* (Kurtz) Bonetti, *D. coriaceum, Pteruchus simmondsi, Umkomasia macleani, Xylopteris argentina, Yabeiealla du Toiti, Y. brackebuschiana, Lepidopteris stormbergensis, Sphenobaiera argentinae, Podozamites elongatus, Cycadocarpidium andium*, Freng., with M1 and M2 palynomorphs (Frenguelli, 1942–46; Yrigoyen and Stover, 1970).

The vertebrates include the kannemeyerid *Vinceria* and the traversodontidae *Andesynodon, Rusconiodon* in the Río Mendoza Formation. *Colbertosaurus* is known in the Potrerillos Formation, *Pelorocephalus* and *Cuyusuchus* from the Cacheuta Formation (Bonaparte, 1966a, 1969, 1971a, 1973). Freshwater fish are abundant with *Gyrolepidoides* in the Las Cabras and with *Gyrolepidoides, Semionotus, Challaia, Neochallaia, Mendocinia* and *Pholidophorus* in the Potrerillos and Cacheuta Formation (Rusconi, 1951).

Lithologic, mega- and microfloral correlation, radiometric dating, paleogeographic and paleoclimatic considerations are all consistent with an assignment of the Uspallata Group to the upper Middle and Upper Triassic (Stipanícic and Bonetti, 1969), thus making it essentially contemporaneous with other Argentine Triassic sequences (Table III). According to Romer (1960, 1966) and Bonaparte (1966a, 1969) the evolutionary level of the vertebrates is more appropriate to Scythian–Anisian levels, however, more

recently Bonaparte (1973) accepted a lower Middle Triassic age for the Río Mendoza and Las Cabras Formation and an upper Middle Triassic age for the combined Potrerillos and Cacheuta formations (with a possible continuation into the Upper Triassic).

The Santa Clara–Río de Las Peñas–Las Higueras section. This section occupies a more central position in the Cuyean Basin than the preceding succession. The Peñasco Group (*nov. nom.* for ex. "Santa Clara" Group) consists of six formations. The basal Cielo Formation, 600 m of red conglomerates and tuffaceous coarse-grained sandstones, contains the M3 palynological assemblage. Above lies successively 250 m of red, tuffaceous, fine grained sandstones with rare bituminous shale horizons (Mollar Formation) and 500 m of variegated coarse-grained sandstones with shaley sandstone and shale intercalations, of the Montaña Formation. Then follows the Santa Clara Abajo Formation, some 500 m of alternating greenish shales, siltstones, mudstones and fine sandstones with a few conglomeratic lenses, in which M2 palynomorphs and fish remains are found. The overlying Santa Clara Arriba Formation with a similar thickness also contains M2 palynomorphs in an alternation of black bituminous shales, shales, siltstones and light tuffaceous sandstones. The Triassic succession comes to an end with red conglomerates and sandstones of the Los Alojamientos Formation (Yrigoyen and Stover, 1970).

The former identification of Permian *Glossopteris, Walkomia*, etc. (Polanski, 1970) with fish referred to Middle Triassic forms of *Pseudobeaconia* and *Cheithrolepis* (Bordas, 1944) has been resolved by a revision of the flora which indicates an absence of Paleozoic forms (Rolleri and Criado, 1968) and the presence of typical Middle to Upper Triassic *Lepidopteris* (Baldoni, 1972), the presence of *Cladophlebis* (Rolleri and Criado, 1968) and M2 and M3 palynomorphs. In the Río de Las Peñas Formation ichnites have a Middle Triassic aspect (Romer, 1966) and "*Chirotherium barthi*" tracks formerly regarded as Lower Triassic are now assigned to Middle to Upper Triassic (Romer, 1960; Casamiquela, 1964; Bonaparte, 1966a) with the correlation of the Las Higueras Formation (containing *Neocalamites* cf. *Carrerei*) with the Las Cabras (Rolleri and Criado, 1968; Romer, 1960). In the circumstances the basalt flows in the lower Peñasco Group have a radiometric age of 200 m.y., which appears to be too young.

The Peñasco Group thus correlates closely with

the Uspallata Group (Yrigoyen and Stover, 1970) and must be upper Middle and Upper Triassic in age. It is yet another typical continental assemblage. There are no glacial or marine horizons as reported by Polanski (1970).

The San Rafael section. This section on the San Rafael structural block (34°–36°S 67°30′–69°W) rests unconformably upon Middle Permian volcanics dated at 263 m.y. (González, 1966; Valencio and Mitchell, 1972). The lowest beds belong to the Puesto Viejo Formation, a 300-m succession of siltstones, shales and tuffs containing plants and dicynodonts capped by a basalt flow upon which rests the upper part of the formation consisting of conglomeratic sandstones with some tetrapods and a few basalt flows (González, 1966; Criado, 1972).

According to Casamiquela (in González, 1966) and Bonaparte (1966a, b, 1969, 1973) the tetrapods in the Upper Puesto Viejo Formation (*Kannemeyeria, Pascualgnathus, Cynognathus*) belong to the upper Lower Triassic–lower Middle Triassic. A younger age cannot be excluded for these vertebrates are closely related to forms known in the Rio Mendoza Formation (basal Uspallata Group) the age of which is still under discussion.

A succession similar to the Potrerillos and Cacheuta Formations has been traversed in boreholes in the region of General Alvear (35°S 67°30′W).

The Malargüe Basin

South of Malargüe (35°40′S 69°40′W) andesites, tuffs and rhyolites of the Choiyoi Group are separated by an erosional unconformity from the rocks of the overlying Llantenes Group. The Choiyoi Group volcanics are intruded by granite for which a radiometric age of 194 m.y. has been determined. The lower unit of the Llantenes Group, the Chihuíu Formation, consists of a basal 50 m angular conglomerate followed by 250 m of fine to medium conglomerate alternating with torrential sandstones and green-gray shales. Within the conformably overlying Tronquimalal Formation are two black-shale horizons (80 and 90 m thick) separated by 45 m of fine- to coarse-grained sandstones and capped by 25 m of sandstone and coarse conglomerate (Menendez, 1951; Groeber and Stipanícic, 1953; Stipanícic, 1957).

The flora includes *Cladophlebis oblonga* Halle, *C. denticulata* Brong., *C. antarctica* (Nath.) Halle, *Copiapea plicatella, Chiropteris copiapensis, Linguifolium diemenense, Taeniopteris stenophylla* Kryst., *Sphe-*

nopteris membranosa Feist., *Yabeiella mareyesiaca, Y. brackebuschiana, Dicroidium odontopteroides, D. lancifolium, D. zuberi, Xylopteris argentina, X. elongata* var. *rigida, X. e.* var. *irregularis* Stip. & Bonetti, *Ginkgoidium nathorsti* Yok., *Ctenis takamiana* Oishi & Hutz, *Nilssonia princeps* (O. & M.) Sew. etc. (Menéndez, 1951; Stipanícic, 1957). Floral evidence thus suggests an upper Middle Triassic–Upper Triassic age for the Llantenes Group; the Permian age suggested by Rolleri and Criado (1970) is excluded by the presence of *Dicroidium zuberi* (M.I.R. Bonetti, pers. comm.)

Southern La Pampa and Buenos Aires province and North Patagonia

The Lihuel Calel rhyolites and tuffs in the La Pampa province (38°S 65°40′W), the granite porphyry in subsurface in the Neuquén province (Challaco, 39°S 69°W) and the Perez Lacube granite in Buenos-Aires province (30°15′S 62°45′W) all date at about the Permian–Triassic boundary, with ages of 226 m.y. (Halpern et al., 1971) and 227 m.y. (Cingolani and Varela, 1973). The last lies on the northern margin of the Colorado Basin in which neither Permian nor Triassic rocks have been found in boreholes although their presence on the western margin of the basin and on the Argentinian Shelf is suggested by seismic velocities of 4.5 to 5.5 km/S (Zambrano, 1972).

In the Neuquén province in the Cordillera del Viento (= Choiyoi Mahuida) (37°S 70°30′W) younger Paleozoic beds including Lower Permian are intruded by granites and unconformably overlain by 1,500 m of Choiyoi Formation andesites, rhyolites and basalts, which are in turn unconformably overlain by the Lower Lias (Groeber and Stipanícic, 1953; Zöllner and Amos, 1955). The Choiyoi Group, unquestionably Triassic in the type area, can be traced southwards to Pino Hachado, Aluminé, Collón Curá, Chacaicó and in subsurface in oil wells in the Neuquén, north Río Negro and south Mendoza Provinces (Lambert, 1946; Lambert and Galli, 1950; Galli, 1969).

In the Catriel oil field (37°50′S 68°W) black shales within the Choiyoi Group yield gymnosperm pollen (*Allisporites*) and monocolpates of Triassic age. The Choiyoi is overlain by 200 m of brown shales and tuffaceous siltstones assigned to the Planicie Morada Formation (*nov. nom.*) which must also be of Triassic age for it is unconformably overlain by Jurassic sediments.

Over the broad North Patagonian Massif, 40°–43°S 65°–70°W, extensive Triassic igneous and sedimentary rocks unconformably overlie a Paleozoic substratum, which contains granites as young as 232 m.y. These include upper Lower Triassic and lower Middle Triassic andesites (226 and 220 m.y., resp.) and uppermost Triassic Sierra Grande andesites (180 m.y., Halpern et al. 1971; Stipanícic et al., 1971), found at 41°30'S 65°20'W.

At Paso Flores (40°30'S 70°30'W) Triassic sediments rest unconformably upon Choiyoi andesites. The sequence (= Paso Flores Formation) consists of 200 m of conglomerates, sandstones and tuffs. It contains *Dictyophyllum tenuifolium*, *D. spectabile* Nath., *Clathropteris australis*, *Dicroidium lancifolium*, *Xylopteris argentina*, *Yabeiella wielandi*, *Sphenobaiera stormbergensis*, *Ginkgoites crassipes* Feist., etc. which indicate an Upper Triassic age (Frenguelli, 1937, 1948; Groeber and Stipanícic, 1953; Stipanícic and Bonetti, 1969; Stipanícic et al., 1971).

A further Upper Triassic succession which may include some upper Middle Triassic crops out west of Los Menucos (40°50'S 68°10'W) and consists of 400 m tuff, sandstone and conglomerate with some silicified tuff towards the top. Vertebrate tracks, *Dicroidium lancifolium* and *D. zuberi* are found (Casamiquela, 1964).

The conglomerates and sandstones of Cerro Negro, Chubut province, in the southern part of the Patagonian Massif (43°50'S 69°30'W), referred to the Triassic by Lesta and Ferello (1972) are Jurassic in age. A Triassic microdiorite dated at 195–218 m.y. (Ferello and Lesta, 1973) occurs in an exploration well (LPI) at Lagunas Polacios (44°40'S 69°10'W).

Austral Cordillera and Southern Patagonia

In the San Martin (49°S 72°30'W) and Pueyrredón (47°30'S 72°W) Lakes is the Austral Cordilleran Basin, in which a sequence of conglomerates and sandstones with red tuffaceous and shaly intercalations, resting unconformably upon Carboniferous and unconformably overlain by Kimmeridgian, was assigned to the Triassic although no paleontological control was available (Leanza, 1972).

There are Middle to Upper Triassic granites intruding Lower Permian sediments at La Juanita (47°40'S 67°20'W) and La Leona (48°05'S 67°20'W) in the Deseado Massif according to age determination (200–208 m.y., Stipanícic et al., 1971). West of the massif a Triassic basin developed and the El Tranquilo

Group, named from El Tranquilo (48°S 69°W), which includes two formations, was deposited. The lower may be of Middle Triassic age, the vertebrate of the upper has a Keuper, Upper Triassic aspect (*Plateosaurus*, Bonaparte 1973; Casimiquela, 1964). The lower consists of 650 m of light shales and sandstones, the upper of 250 m of red shales and sandstones with boulder conglomerate lenses. The flora of the El Tranquilo Group includes *Cladophlebis mesozoica*, *C. indica*, *Linguifolium lilleanum*, *L. diemenense*, *Dicroidium odontopteroides*, *D. lancifolium*, *D. cacheutense*, *Xylopteris elongata*, *X. argentina*, *Yabeiella mareyesiaca*, *Y. brackebuschiana*, *Lepidopteris stormbergensis*, etc. (Stipanícic, 1957; Stipanícic and Bonetti, 1969; Baldoni, 1972) and large *Dicroidium* ("zuberi" type) and new forms of *Linguifolium*.

At La Enriqueta, 90 km WSW of El Tranquilo, contemporaneous deposits with *Dicroidium* occur, with yet another at La Juanita (Baldoni, 1972).

Falkland Islands (Islas Malvinas)

On Solitude Island (52°S 59°30'W) 2,000 m of sandstones, shales and mudstones overlie Permian beds. These are referred to the Saint Charles Strait Formation of the Falkland Basin, and because of the presence of *Neocalamites* cf. *carrerei* are regarded as Triassic in age (Borrello, 1972).

FLORA, FAUNA AND CLIMATE

There is a rich homogeneous flora between 29°S and 40°30'S which has a typical Gondwana aspect. The dominant forms are *Dicroidium*, *Xylopteris* and *Yabeiella*. Cosmopolitan forms such as *Equisetites*, *Neocalamites*, *Cladophlebis*, *Pseudoctenis*, *Linguifolium*, *Baiera* occur.

The Barreal (San Juan) flora shows a close affinity with those of the Northern Hemisphere and surprisingly contains *Saportaea*, a ginkgophyte of Chinese provenance. The *Dipteridaceae* in both Argentina and Chile also show relationship with northern forms known in Japan, Sweden and Germany. The prevalent large *Dicroidium* and *Linguifolium* in the El Tranquilo flora (Santa Cruz province) show possible affinities with Australian forms.

The Chilean marine invertebrate faunas have a Pacific aspect.

Recent studies of the tetrapods define four evolutionary groups, with the youngest three giving ages compatible with those from other lines of evidence. The older, the "Puesto Viejo" reptile age of J.F. Bonaparte suggests an older age than that given by the macro- and micro-floras, absolute age and geological correlations. The relationships of the Argentinian tetrapods was established by comparison with South African and Madagascan faunas.

Paleoclimate studies are in their early stages in both countries. During the Upper Triassic a warm oxidising climate existed; the desert conditions suggested by Padula and Mingramm (1969) scarcely seem reasonable given the floral and vertebrate-faunal evidence (J.F. Bonaparte, pers. comm.). The same argument may be applied to the uppermost Triassic red beds and the lower-middle Middle Triassic of midwest Argentina where alternating wet and dry seasons are proposed (J.F. Bonaparte and G. Bossi, pers. comm.). More humid conditions resulting in the formation of lakes may typify upper Middle and lower Upper Triassic times. The same conditions hold for the Triassic of southern Patagonia.

MAGMATISM AND DIASTROPHISM

Middle Permian diastrophic movements were very important. They were responsible for the angular unconformity which exists below the Triassic strata. These movements were followed by an erosional phase lasting into the Lower Triassic reinforced by positive vertical movement. During this phase, the Nihuil phase, clastic deposition was virtually absent in the region.

Other diastrophic phases occurred during Triassic times, all of a synepeirogenic, not orogenic character.

Magmatic activity was marked by the intrusion of Early Triassic granites (e.g. La Rioja, 224 m.y.; southern Buenos Aires, 227 m.y.; and Río Negro, 232 m.y.) and andesite effusives (e.g., Lihuel Calel, 226 m.y., and the North Patagonian Massif, 226—220 m.y.). The effusive activity of the Choiyoi Group, well developed in western Argentina, began at about this time.

The basalt flows of the lower Paganzo III and Puesto Viejo Formation, and the andesites, tuffs and breccias of the Choiyoi Group represent an activity occurring after the Permian diastrophic phase and the Nihuil phase.

The La Rioja diastrophic phase in mid-lower Middle Triassic resulted in a widespread unconformity

with an erosional contact between pre- and post-Rioja beds. In Chile this time interval is represented by the El Querco turbidites.

A further conspicuous oscillation in upper Middle Triassic times, the Tunuyan phase (Table I), resulted in a regional unconformity in the northern Mendoza and southern San Juan provinces. This is the unconformity between the Las Cabras and Potrerillos formations, and between the Barreal and Cortaderita formations.

Magmatic activity was intense at the time of the La Rioja movements (up to middle Upper Triassic) with a plutonic phase affecting the Deseado Massif (200—208 m.y.), Cordón del Plata (204 m.y.), Cacheuta, Chihuiu (194 m.y.) and a more important effusive phase of andesite and rhyolite flows with tuffs and breccias and occasional basalts (e.g., El Quereo, Pichindangui, Cinfucho, Pastos Blancos, Alto del Carmen, Tránsito, Huaco in Chile, and Agua Negro, Olivares, Barreal, Cabras, Mollar in Argentina). Some of the Chilean andesites and rhyolites were submarine. In Argentina the explosive volcanic facies was common and tuffaceous products play a large role in the Triassic sediments. They covered, and destroyed, rich floras now found fossil as for example in the Cortaderita Formation at Barreal.

Moderate, oscillatory movements (Rio de Los Patos phase) in middle Upper Triassic times resulted in erosional and local unconformities at the base of the youngest Triassic red beds, and in the occurrence of coarse conglomerates. The terminal diastrophic phase of the Triassic (Río Atuel—Valparaiso) had a strong negative component which made possible the very rapid marine ingression in Early Jurassic times (as early as Rhaetian in the deeper Chilean basins) over a peneplaned surface. Marine conditions reached San Juan, Mendoza, Neuquén and Chubut provinces at somewhat different times.

ACKNOWLEDGEMENTS

The author wishes to express his gratitude to Drs. J. Bonaparte, G. Bossi, R. Pascual and M. Yrigoyen, from Argentina, and G. Cecioni and R. Thiele Cartagena, from Chile, for their help in the revision of the manuscript, and to the same colleagues and to M. Bonetti, J.M. Deguisto, S. Hogg and F. Nullo, who have kindly furnished new unpublished data.

REFERENCES

Aguirre, L. and Levi, B., 1964. Geología de la Cordillera de los Andes de las provincias de Cautín, Valdivia, Osorno y Llanquihue. *Chile Inst. Invest. Geol., Bol.*, 17: 37 pp.

Andreis, R.R., 1969. Los basaltos olivínicos del Cerro Guandacol (Sierra de Maz, Provincia de La Rioja) y su posición estratigráfica. *Cuartas Jorn. Geol. Arg., Actas*, I: 15–33.

Baldoni, A.M., 1972. El género *Lepidopteris* (Pteridosperma) en el Triásico de Argentina. *Ameghiniana*, 9 (1): 1–16.

Barthel, K.W., 1958. Eine Marine Faunula aus der mittleren Trias von Chile. *Neues Jahrb. Geol. Palaeontol., Abh.*, 106 (3): 352–382.

Bonaparte, J.F., 1966a. Cronología de algunas formaciones triásicas de Argentina basada en restos de Tetrápodos. *Asoc. Geol. Arg. Rev.*, 21 (1): 20–38.

Bonaparte, J.F., 1966b. *Chiniquodon* Huene (Therapsida-Cynodontia) en el Triásico de Ischigualasto, Argentina. *Acta Geol. Lilloana*, 8: 157–169.

Bonaparte, J.F., 1966c. Una neuva "fauna" Triásica de Argentina (Therapsida: Cynodontia, Dicynodontia). Consideraciones filogenéticas y paleobiogeográficas. *Ameghiniana*, 4 (8): 243–296.

Bonaparte, J.F., 1969. Dos nuevas "faunas" de reptiles triásicos de Argentina. *Gondwana Stratigraphy, IUGS Symposium, Buenos Aires, 1967*, pp. 283–284.

Bonaparte, J.F., 1971a. Annotated list of the South American Triassic Tetrapods. *Second Gondwana Symp., Proc. Pap., S. Africa*, pp. 665–682.

Bonaparte, J.F., 1971b. Los Tetrápodos del sector superior de la Formación Los Colorados, La Rioja, Argentina (Triásico superior), I Parte. *Opera Lilloana*, 22.

Bonaparte, J.F., 1973. Edades/Reptil para el Triásico en Argentina y Brasil. *Quinto Congr. Geol. Arg., Actas*, 3: 93–130.

Bonetti, M.I.R., 1968. Las especies del género *Pseudoctenis* en la flora triásica de Barreal. (San Juan). *Ameghiniana*, 5 (10): 433–446.

Bordas, A., 1944. Peces Triásicos de la Quebrada de Santa Clara (Mendoza y San Juan). *Physis*, 19: 433–460.

Borrello, A.V., 1962. Fanglomerado Río Mendoza (Triásico, Prov. de Mendoza). *Comis. Invest. Cient. Prov. Buenos Aires*, 1 (3).

Borrello, A.V., 1972. Islas Malvinas. In: A.F. Leanza (Editor), *Geología Regional Argentina*. Acad. Nac. Cienc. Cba., Córdoba, pp. 755–770.

Borrello, A.V. and Cuerda, A.J., 1965. Grupo Rincón Blanco (Triásico, San Juan). *Comis. Invest. Cient. Prov. Buenos Aires, Notas*, 2 (10).

Bowes, W., Knowles, P.H., Moraga, A. and Serrano, M., 1961. Reconnaissance for uranium in the Chañaral–Taltal area, Provinces of Antofagasta and Atacama, Chile. *U.S. At. Energy Comm., RME-4565 (Rev.)*, 21 pp.

Bracaccini, I.O., 1945. Acerca de los movimientos intertriásicos en Mendoza Norte. *Inst. Panam. Ing. Minas y Geol., Sec. Arg., Prim. Reun. Com.*

Caminos, R., 1972. Cordillera Frontal. In: A.F. Leanza (Editor), *Geología Regional Argentina*. Acad. Nac. Cienc. Cba., Córdoba, pp. 305–343.

Casamiquela, R., 1964. *Estudios icnológicos. Problemas y métodos de la icnología ... etc.* Publ. Gobierno Prov. Río Negro, Minist. Asuntos Sociales.

Casamiquela, R., 1967. Un nuevo dinosaurio ornitisquio Triásico (*Pisanosaurus mertii*, Ornithopoda) de la Formación Ischigualasto, Argentina. *Ameghiniana*, 4 (2): 47–64.

Cingolani, C.A. and Varela, P., 1973. Exámen geocronológico por el método rubidio-estroncio de las rocas ígneas de las Sierras Australes bonaerenses. *Quinto Congr. Geol. Arg., Actas*, 1: 349–371.

Corvalán, J., 1965. Geología general. In: *Geografía Económica de Chile*, Corp. Fomento y Produc., Santiago, pp. 35–82.

Cox, C.B., 1965. New Triassic Dicynodonts from South America. Their origins and relationships. *Phil. Trans. R. Soc. (Lond.)*, B 248 (753): 457–516.

Cox, C.B., 1968. The Chañares (Argentina) Triassic reptile fauna. IV, The Dicynodont fauna. *Breviora*, 295: 1–27.

Criado Roque, P., 1972. Bloque de San Rafaél. In: A.F. Leanza (Editor), *Geología Regional Argentina*. Acad. Nac. Cienc. Cba., Córdoba, pp. 282–295.

Davis, S. and Karzulovic, J., 1961. Deslizamientos en el valle del río San Pedro, Provincia de Valdivia, Chile. *Inst. Invest. Geol. Chile, Publ.*, 20: 108 pp.

Du Toit, A.L., 1927. A geological comparison of South America with South Africa. *Carnegie Inst. Publ.*, 381: 1–157.

Falconer, J.D., 1931. Terrenos Gondwáanicos del Departamento de Tacuarembó. *Inst. Geol. Perf. Uruguay, Bol.*, 15: 3–17.

Ferello, R. and Lesta, P., 1973. Acerca de la existencia de una dorsal interior en el sector central de la serranía de San Bernardo (Chubut). *Quinto Congr. Geol. Arg., Actas*, 4: 19–26.

Flores, M.A. and Criado Roque, P., 1972. Cuenca de San Luis. In: A.F. Leanza (Editor), *Geología Regional Argentina*. Acad. Nac. Cienc. Cba., Córdoba, pp. 567–579.

Frenguelli, J., 1937. La flórula jurásica de Paso Flores en el Neuquén con referencia a la de Piedra Pintada y otras floras jurásicas Argentinas. *La Plata, Univ. Nac., Mus., Rev., Paleontol.*, 1 (3): 67–108.

Frenguelli, J., 1942–6. Contribución al conocimiento de la flora del Gondwana superior en la Argentina. *La Plata, Univ. Nac., Mus., Notas Paleontol.*, 42–51 (1942, Pt. 1–11), 57–60 (1943, Pt. 11–14), 63–68 and 70–80 (1944, Pt. 15–31), 87 (1946, Pt. 33).

Frenguelli, J., 1944. La serie del llamado Rético en el Oeste Argentino. *La Plata, Univ. Nac., Mus. Notas, Geol.*, 9 (30): 261–270.

Frenguelli, J., 1948. Estratigrafía del llamado Rético en la Argentina, In: *Geografía de la República Argentina*, 8 (2): 159–309. Soc. Arg. Est. Geogr., GAEA.

Fritzsche, C., 1921. La jeolojia de la rejión comprendida entre los ríos Cautín, Cholchol i Quillén, en la Provincia de Cautín, i los yacimientos de carbón antracitosos de Nielol. *Soc. Nac. Min., Bol. Miner.*, 33 (271–2): 595–626.

Fuenzalida, H., 1938. Las Capas de Los Molles. *Museo Nac. Hist. Nat. Chile*, 16: 67–98. Santiago.

Galli, C.A., 1969. Descripción Geológica de la Hoja 35a, Lago Aluminé (Provincia del Neuquén). *Arg., Dir. Nac. Geol. Minería, Bol.*, 108: 5–45.

Geinitz, H.B., 1876. Ueber rhaetische Thier- und Pflanzenreste in den argentinischen Provinzen La Rioja, San Juan und Mendoza. *Palaeontographica, Suppl.*, 3 (2): 14 pp.

Geinitz, H.B., 1923. Plantas y animales réticos de la República Argentina. *Acad. Nac. Cienc. Cba., Actas*, 8 (3–4): 335–347.

Gonzalez, E., 1966. El hallazgo de infra- meso- triásico continental en el sur del área pedemontana mendocina. *Acta Geol. Lilloana*, 8: 101–134.

Groeber, P., 1945. Lista de terrenos a distinguirse en el mapa geológico de América del Sur. *Seg. Reun. Com. Inst. Panam. Ing. Min. Geol., Sec. Arg., Buenos Aires*.

Groeber, P., 1953. Mesozoico. In: *Geografía de la República Argentina*, 2 (1): 9–541. Soc. Arg. Est. Geogr., GAEA.

Groeber, P., 1959. Supracretácico. In: *Geografía de la República Argentina*, 2 (2): 1–165. Soc. Arg. Est. Geogr., GAEA.

Groeber, P. and Stipanicic, P.N., 1953. Triásico. In: *Geografía de la República Argentina*, 2 (1): 13–141. Soc. Arg. Est. Geogr., GAEA.

Halpern, M., Linares, E. and Latorre, C., 1971. Edad rubidio-estroncio de rocas volcánicas e hipabisales (?) del área Norte de la Patagonia, República Argentina. *Asoc. Geol. Argent., Rev.*, 26 (2): 169–174.

Herbst, R., 1970. Estudio palinológico de la Cuenca Ischigualasto-Villa Unión (Triásico), provincias de San Juan–La Rioja, I. Introducción, II. Monoaperturados. *Ameghiniana* 7 (1): 83–97.

Herbst, R., 1971. Paleophytologia Kurtziana, III.7. Revisión de las especies argentinas del género Cladophlebis. *Ameghiniana*, 8 (3–4): 265–281.

Herbst, R., 1972. Estudio palinológico de la Cuenca Ischigualasto–Villa Unión (Triásico), provincias de San Juan–La Rioja, III. Esporas Triletes. *Ameghiniana*, 9 (3): 280–288.

Hoffstetter, R., Fuenzalida, H. and Cecioni, G., 1957. Chile-Chili. In: *Lexique Stratigraphique International*, V (7): 1–444. Paris.

Jain, R.K. and Delevoryas, T., 1967. A Middle Triassic flora from the Cacheuta Formation, Mina de Petróleo, Argentina. *Palaeontology*, 10 (4): 564–89.

Jaworski, E., 1922. Die marine Trias in Südamerika. *Neues Jahrb. Min. Geol. Palaeontol.*, 47: 93–200.

Katz, H.R., 1964. Conceptos nuevos sobre el desarrollo geosinclinal y del sistema cordillerano en el extremo austral del continente. *Soc. Geol. Chile, Resumenes*, 7: 1–8.

Lambert, L.R., 1946. Contribución al conocimiento de la Sierra de Chacay Co (Neuquén). *Soc. Geol. Argent., Rev.*, 1 (4): 231–252.

Lambert, L.R. and Galli, C.A., 1950. Observaciones geológicas en la región situada entre Piedra del Aguila y Paso Flores (Neuquén). *Asoc. Geol. Arg. Rev.*, 5 (4): 227–232.

Leanza, A.F., 1969. Sistema de Salta. Su edad, sus peces voladores, su asincronismo con el Horizonte Calcáreo Dolomítico y con las Calizas de Mirafloras y la hibridez del Sistema Subandino. *Asoc. Geol. Arg., Rev.*, 24 (4): 393–407.

Lesta, P.J. and Ferello, R., 1972. Región Extraandina de Chubut y Norte de Santa Cruz. In: A.F. Leanza (Editor), *Geología Regional Argentina*. Acad. Nac. Cienc. Cba., Córdoba, pp. 601–653.

Levi, B., 1969. Burial metamorphism of a Cretaceous volcanic sequence west from Santiago, Chile. *Contrib. Mineral. Petrol.*, 24: 30–49.

Menendez, C.A., 1951. La flora de la Formación Llantenes, Prov. de Mendoza. *Rev. Inst. Invest. Mus. Arg. Cienc. Nat., Bot.*, 2 (3): 147–261.

Mingramm, A., and Russo, A., 1972. Sierras Subandinas y Chaco Salteño. In: A.F. Leanza (Editor), *Geología Regional Argentina*. Acad. Nac. Cienc. Cba., Córdoba, pp. 184–211.

Muñoz, J., 1938. Geología de la región de Longotoma y Guaquén en la provincia de Aconcagua. *Bol. Minas Petról.*, 8 (81): 222–284.

Muñoz, J., 1942. Rasgos generales de la constitución geológica de la Cordillera de la Costa, especialmente en la provincia de Coquimbo. *An. Primer. Congr. Panam. Ing. Minas Geol.*, 2: 285–318.

Muñoz, J., 1950. Geología de Chile. In: *Geografía Económica de Chile*, T. 1: 55–187. Corp. Fomento Produc., Santiago.

Padula, E. and Mingramm, A., 1969. Sub-surface Mesozoic Red-Beds of the Chaco–Mesopotamian Region, Argentina, and their relatives in Uruguay and Brazil. *Gondwana Stratigraphy, IUGS Symposium, Buenos Aires, 1967*, pp. 1053–1071.

Polanski, J., 1966. Edades de eruptivas suprapaleozoicas asociadas con el diastrofismo varíscico. *Asoc. Geol. Arg. Rev.*, 21 (1): 5–19.

Polanski, J., 1970. *Carbónico y Pérmico de la Argentina*. Edit. EUDEBA, Buenos Aires, 216 pp.

Regairaz, A.C., 1970. Contribución al conocimiento de las discordancias en el área de las Huayquerías (Mendoza, Argentina). *Cuartas J. Geol. Arg., Actas*, 2: 243–254.

Reig, O.A., 1963. La presencia de dinosaurios saurisquios en los "estratos de Ischigualasto" (mesotriásico superior) de San Juan. *Ameghiniana*, 3: 3–20.

Reyes, F.C. and Salfity, J.A., 1973. Consideraciones sobre la estratigrafía del Cretácico (Subgrupo Pirgua) del noroeste argentino. *Quinto Congr. Geol. Arg., Actas*, 3: 355–385.

Rocha Campos, A.A., Amaral, G. and Aparicio, E.P., 1971. Algunas edades K-Ar de la "Serie Porfirítica" de la Precordillera y Cordillera Frontal de Mendoza, República Argentina. *Asoc. Geol. Arg. Rev.*, XXVI (3): 311–316.

Rolleri, E.O. and Criado Roque, P., 1968. La Cuenca Triásica del Norte de Mendoza. *Terceras J. Geol. Arg., Actas*, 1: 1–76.

Rolleri, E.O. and Criado Roque, P., 1970. Geología de la Provincia de Mendoza. *Cuartas J. Geol. Arg., Actas*, 2: 1–46.

Romer, A.S., 1960. Vertebrate-bearing continental Triassic strata in Mendoza region, Argentina. *Geol. Soc. Am. Bull.*, 71: 1279–1294.

Romer, A.S., 1966. The Chañares (Argentina) Triassic reptile fauna, I. Introduction. *Breviora*, 247: 1–14.

Romer, A.S., 1967. The Chañares (Argentina) Triassic reptile fauna, III. Two New Gomphodonts, *Massetognathus pas-*

cuali and *M. teruggi*. *Breviora*, 264: 1–25.

Romer, A.S., 1969. The Chañares (Argentina) Triassic reptile fauna, V. A New Chiniquodontid cynodont, *Probelesodon lewisi* – Cynodont ancestry. *Breviora*, 333: 1–24.

Romer, A.S., 1971. The Chañares (Argentina) Triassic reptile fauna, XI. Two new long-snouted thecodonts, *Chañaresuchus* and *Gualosuchus*. *Breviora*, 379: 1–22.

Romer, A.S., 1972a. The Chañares (Argentina) Triassic reptile fauna, XIII. An early ornithosuchid pseudosuchian, *Gracilisuchus stipanicicorum* gen. et sp. now. *Breviora*, 389: 1–24.

Romer, A.S., 1972b. The Chañares (Argentina) Triassic reptile fauna, XVI. Thecodont classification. *Breviora*, 395: 1–24.

Romer, A.S. and Jensen, J.A., 1966. The Chañares (Argentina) Triassic reptile fauna, II. Sketch of the geology of the Río Chañares–Río Gualo region. *Breviora*, 252: 1–20.

Rusconi, C., 1951. Laberintodontes triásicos y pérmicos de Mendoza. *Rev. Mus. Hist., Nat. Mendoza*, 5 (1–5): 33–158.

Segerström, K., 1962. Regional geology of the Chañarcillo Silver Mining District and adjacent areas, Chile. *Econ. Geol.*, 57 (8): 1247–1261.

Solms-Laubach, H. and Steinmann, G., 1899. Das Auftreten und die Flora de rhätischen Kohlenschichten von La Ternera, Chile. *Neues Jahrb. Miner. Geol.*, 12: 581–609.

Stappenbeck, R., 1910. La Precordillera de San Juan y Mendoza. *Direc. Nac. Geol. Min., Anal.*, 4 (3): 1–187.

Steinmann, G., 1921. Rhätische Floren und Landverbindungen auf der Südhalbkugel. *Geol. Rundsch.*, 11: 350–354.

Stipanícic, P.N., 1957. El Sistema Triásico en la Argentina. *XX Congr. Geol. Intern., Sec. II*, pp. 73–112.

Stipanícic, P.N., 1972. Cuenca Triásica de Barreal (provincia de San Juan). In: A.F. Leanza (Editor), *Geología Regional Argentina*. Acad. Nac. Cienc. Cba., Córdoba, pp. 537–566.

Stipanícic, P.N. and Bonetti, M., 1965. Las especies del género "Saportae" del Triásico de Barreal (San Juan). Museo Argent. Cienc. Nat. *"B. Rivadavia", Rev., Paleontol.* 1 (4): 81–114.

Stipanícic, P.N. and Bonetti, M., 1969. Consideraciones sobre la cronología de los terrenos Triásicos Argentinos. *Gondwana Stratigraphy, IUGS Symposium, Buenos Aires, 1967*, pp. 1081–1119.

Stipanícic, P.N. and Linares, E., 1969. Edades radimétricas determinadas para la República Argentina y su significado geológico. *Acad. Nac. Cienc. Cba., Bol.*, 42 (1): 51–96.

Stipanícic, P.N. and Menendez, C.A., 1949. Contribución al conocimiento de la flora de Barreal (Prov. de San Juan), I. Dipteridacea. *Bol. Inf. Petrol.*, 291: 44–73.

Stipanícic, P.N. and Rodrigo, F., 1970. El diastrofismo Jurásico en Argentina y Chile. *Cuartas J. Geol. Arg., Actas*, 2: 353–368.

Stipanícic, P.N., Toubes, R.O., Spikerman, J.P. and Halpern, M., 1971. Sobre composición y edad de algunas plutonitas del NE de la provincia de Santa Cruz, Patagonia, República Argentina. *Asoc. Geol. Arg. Rev.*, 26 (4): 459–467.

Tavera, J., 1960. 1. El Triásico del Valle Inferior del río Bío Bío, 2. El Plioceno de Bahía Horcon en la Provincia de Valparaiso. *Univ. Chile, Inst. Geol., Publ.*, 18: 320–362.

Valencio, D.A. and Mitchell, J.G., 1972. Edad potasio-argón y paleomagnetismo de rocas ígneas de la Formación Quebrada del Pimiento y Las Cabras, provincia de Mendoza, República Argentina. *Asoc. Geol. Arg., Rev.*, 27 (2): 170–178.

Vicente, J.C., 1970. Tectónica. In: *Conferencia sobre Problemas de la Tierra Sólida* (Proyecto Internacional del Manto Superior), Vol. 1. *Seminario sobre el Programa Geofísico Andino y Problemas Geológicos y Geofísicos relacionados*. Buenos Aires, pp. 162–188.

Vicente, J.C., Charrier, R., Davidson, J., Mpodozis, A. and Rivano, S., 1973. La orogénesis Subhercínica: Fase mayor de la evolución paleogeográfica y estructural de los Andes Argentino-Chilenos Centrales. *Quinto Congr. Geol. Arg., Actas*, 5: 81–98.

Yrigoyen, M. and Stover, L., 1970. La palinología como elemento de correlación del Triásico de la Cuenca Cuyana. Cuartas J. Geol. Arg., Actas, 2: 427–447.

Zambrano, J.J., 1972. Cuenca del Colorado. In: A.F. Leanza (Editor), *Geología Regional Argentina*. Acad. Cienc. Cba., Córdoba, pp. 419–438.

Zeil, W. and Ichikawa, K., 1958. Marine Mittel-Trias in der Hochkordillere der Provinz Atacama (Chile). *Neues Jahrb. Geol. Palaeontol., Abh.*, 106 (3): 339–351.

Zöllner, W. and Amos, A.J., 1955. Acerca del Paleozoico superior y Triásico del Cerro La Premia, Andacollo (Neuquén). *Asoc. Geol. Arg., Rev.*, 10 (2): 127–135.

Chapter 8

THE JURASSIC OF ARGENTINA AND CHILE

A.C. RICCARDI

INTRODUCTION

The geological study of the Jurassic in Argentina and Chile began with Gottsche's (1878) monograph on the fossils found by Stelzner (1873) in the Espinacito Pass (see Fig. 1). This fauna was later revised by Tornquist (1898). During the last decades of the 19th century and the beginning of the 20th there were several important contributions to the stratigraphy and paleontology of the Jurassic of Argentina and Chile. In Chile, Steinmann (1881) studied the area of Caracoles, and Möricke (1894) Copiapó and several other localities. In Argentina, Bodenbender (1892) extended Stelzner's observations to southern Mendoza and Neuquén and Burckhardt (1900, 1902, 1903) made important studies in the Andes of Neuquén and Mendoza.

In the first half of the present century the stratigraphy of the Jurassic of Neuquén and Mendoza was thoroughly studied by Groeber (1918, 1929, 1933, 1946, 1951) and Weaver (1931). At the same time Feruglio (1933, 1942) and Piatnitzky (1932, 1936) made important contributions to the knowledge of the Patagonian Jurassic. A number of paleontological papers were also published, followed by the detailed reviews by Feruglio (1949) and Groeber et al. (1953).

The last 25 years have been characterized by the publication of a number of reviews and new studies, mostly of a specialized nature, that have improved and refined the knowledge of the Jurassic of Argentina and Chile. A review on the advances in Argentina was published by Stipanicic (1969), as well as an analysis of the diastrophic phases (Stipanicic and Rodrigo, 1970a, b). The biostratigraphy of the Chilean Jurassic has been reviewed by Hillebrandt (1970, 1971) and many new geological and paleontological papers, most of which will be mentioned below, have contributed to the subject. General accounts, covering a part or the whole Jurassic of this region, can also be found in Arkell (1956), Harrington (1956, 1962), Muñoz Cristi (1956), Hoffstetter et al. (1957), Groeber (1963), Hölder (1964), Zeil (1964), Ruiz et al. (1965), and Hallam (1975).

Jurassic rocks in Argentina and Chile crop out over extensive areas (see Fig. 1) and comprise a great variety of marine and continental facies. South of 39°S most of the Jurassic is represented by continental volcanic rocks, associated with sediments in some areas. Marine rocks in the region are represented by Liassic strata of central Patagonia and some uppermost Jurassic of southern Patagonia. North of 39°S the Jurassic, mostly marine, is restricted to a north–south belt roughly coincident with the boundary between Argentina and Chile. Between 39° and 31°S most outcrops are confined to Argentina, whereas north of 31°S the Jurassic is only exposed in Chile. To the east of this belt, in the Principal Cordillera of Argentina and Chile, the Jurassic is chiefly marine, while to the west, in the Coast Cordillera of Chile, it is developed as a thick volcanic sequence which includes some marine intercalations. The extensive basaltic flows of the Serra Geral or Arapey Formation as well as the stratigraphically related sandstones of the Botucatu or Tacuarembo Formation, known from northeastern Argentina, Uruguay, Paraguay and southern Brazil, previously regarded as Jurassic (see Caorsi and Goñi, 1958; Putzer, 1962), were later included in the Lower Cretaceous, although minor igneous activity was present in the Upper Jurassic (Amaral et al., 1966; McDougall and Rüegg, 1966; Stipanicic, 1967; Padula and Mingramm, 1969; Padula, 1972).

The Jurassic generally rests unconformably on Upper Triassic vulcanites, e.g. most of Neuquén and Mendoza, or on Upper Paleozoic sediments, e.g. Chubut. In some parts of the Coast Cordillera of Chile, however, the Jurassic is transitional to marine

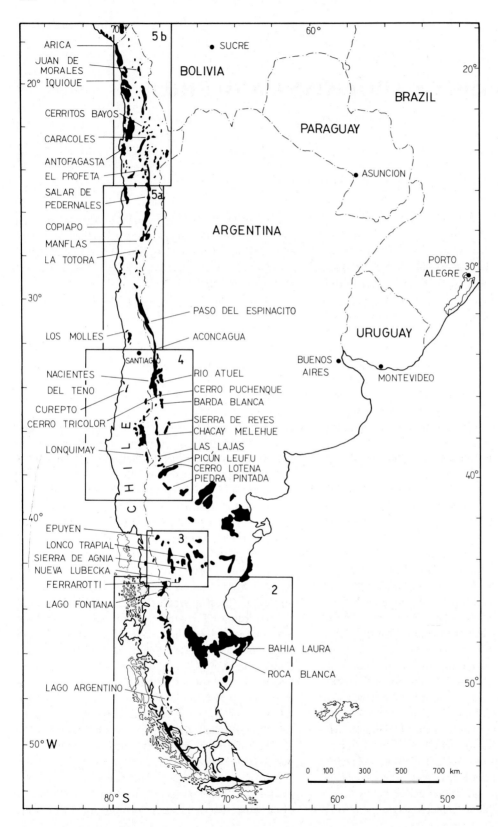

Fig. 1. Distribution of Jurassic outcrops in Argentina and Chile. Rectangles indicate Figs. 2–5.

Triassic, while in other areas, such as Chacay Melehue, Manflas and Cordillera Domeyko, it seems to
pass down into Triassic volcanics.

Within the Jurassic of Argentina and Chile there
has been recognized a major sedimentary cycle extending from the Hettangian to the Oxfordian, followed by Kimmeridgian continental vulcanites and
sediments. The sequence was subdivided by Groeber
(cf. Groeber et al., 1953) into three sedimentary sub-
cycles, named in ascending order: Cuyan, Lotenian
and Chacayan. The Cuyan was said to comprise the
marine sedimentary succession that began in the Lias
and reached its maximum expansion in the Bajocian.
The name Lotenian was applied to the Callovian sedimentary sequence. The Chacayan was said to include
the sediments of the Oxfordian–Kimmeridgian.
Within the Jurassic five diastrophic phases (see the
tectonism-section, p. 234) were recognized, of which
only the first and last are clearly evident. Contemporaneous volcanic activity was also represented,
especially during the Late Triassic–Early Jurassic and
Middle Jurassic.

A second major cycle began in the Tithonian and
extended well into the Early Cretaceous. The Tithonian of Argentina and Chile is therefore more closely
related to the Lower Cretaceous than to the Jurassic
succession, in such a way that, as Harrington (1962)
pointed out, 'there is little doubt that if geology had
been born in South America instead of Europe, the
Jurassic–Cretaceous boundary would have been
drawn between the Kimmeridgian and Tithonian'. On
that basis Groeber (see Groeber et al., 1953) seggre-
gated the Tithonian from the Jurassic and included it
in the beginning of the Cretaceous sedimentary cycle
to which he applied the name 'Andean'. For that reason the present synthesis deals chiefly with the Hettangian–Kimmeridgian sequence, and the Tithonian
is treated in some aspects only. For the paleogeography of the Tithonian the reader is refered to Chapter
9 dealing with the Cretaceous of Argentina and Chile
(N. Malumian).

The present chapter is a synthesis of the existing
knowledge on the Jurassic of Argentina and Chile. Its
aim is to give a general account on the stratigraphy
and paleogeography and to provide access to more
detailed information. A description of the Jurassic
rocks is provided on the basis of generalized sections of
the most important areas of exposures. In order to
achieve that goal the Jurassic outcrops have been
divided into five major areas, from south to north

(Figs. 1–5): (1) Patagonian Cordillera; (2) Deseado
Massif; (3) central-west Patagonia; (4) Principal Cordillera of Argentina and Chile; (5) Coast Cordillera of
Chile. This is followed by a discussion of the geological history and paleogeographic synthesis, and sections
containing information on diastrophic phases, igneous
activity, paleontology and paleoclimatology. The
bibliography is far from exhaustive but an effort has
been made to include the most important relevant
literature.

GENERAL STRATIGRAPHY

Patagonian Cordillera

The Patagonian Cordillera (Fig. 2, Table I) is located on the western border of a structurally depressed area named the Magallanes Basin. The latter
extends over most of southern Patagonia south of
47°S and was formed during the Late Jurassic. The
older recognized units of this area include sedimentary and metamorphic rocks of the Paleozoic and
Jurassic volcanics. The basin was filled with sediments, chiefly marine, during Late Jurassic–Tertiary
times (see Riccardi and Rolleri, 1980).

The Jurassic volcanics, named El Quemado Complex or Quemado Formation or Tobifera Series (see
Feruglio, 1949; Hoffstetter et al., 1957; Riccardi,
1971) cover an extensive area but crop out in a narrow but almost continuous belt along the Patagonian–Fueguian Andes (Fig. 2). To the northeast of
the Magallanes Basin, in extra-Andean Patagonia,
there are also extensive outcrops of rocks of similar
composition and age which have been included in the
Bahia Laura and Lonco Trapial Groups (see descriptions of the Deseado Massif and central-west Patagonia). El Quemado Complex consists of porphyrites,
rhyodacites and andesites and their breccias and tuffs,
which rest on a basal sedimentary breccia or conglomerate. It has a thickness ranging from zero to
2,000+ m within short distances and unconformably
overlies sedimentary metamorphic rocks of Paleozoic
age. These volcanics are for the greater part continental but grade upward into marine tuff and shale. At
Lake Argentino (Fig. 2) the upper levels are interbedded with marine rocks of the Zapata Formation,
which have yielded a Late Jurassic–Early Cretaceous
invertebrate fauna (Feruglio, 1936–37, 1944, 1949;
Leanza, 1968; Riccardi, 1970). In Lake Fontana they

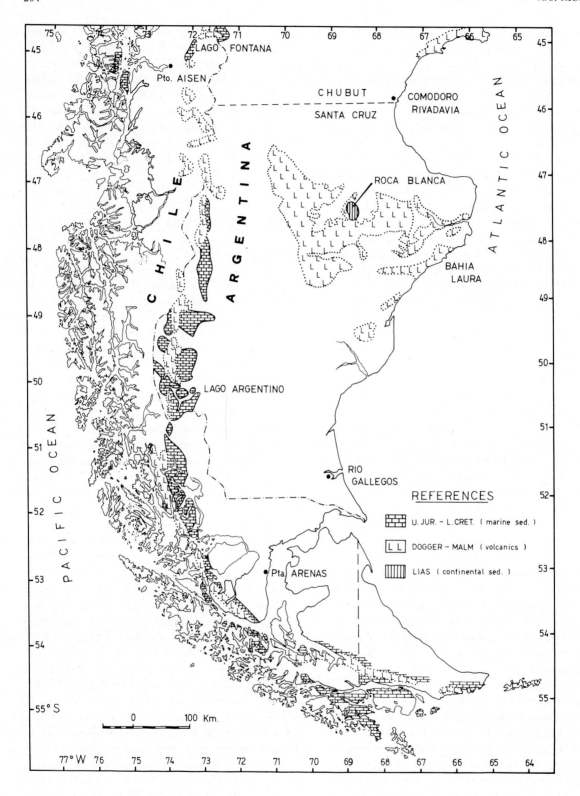

Fig. 2. Major facies distribution of the Jurassic and Lower Cretaceous of southern Patagonia.

TABLE I

Stratigraphy of the Patagonian Cordillera

TIME UNIT	ROCK UNIT	MAX. THICKNESS (m.)	LITHOLOGY	FOSSILS	ENVIRONMENT
LOWER CRETACEOUS	ZAPATA FM	1200	Dark pelites, partly banded, marly		Marine, probably bathyal
UPPER JURASSIC	SPRINGHILL FM	150	Brown sandstones with shales intercalated		Paralic to marine
	COTIDIANO FM	30	Reef limestones, lumachelles		Carbonate shelf and reef
MIDDLE JURASSIC	EL QUEMADO COMPLEX	2100	Porphyrites, rhyodacites, andesites, breccias and tuffs, generally light colored		Continental volcanism

conformably underlie the Oxfordian–Kimmeridgian limestones of the Cotidiano Formation (Ramos, 1976). The presence of Late Jurassic fossils at the top and the superposition on marine Toarcian to the east of Lake Fontana (Malumian and Ploszkiewicz, 1977; Ploszkiewickz and Ramos, 1978) indicate that most of this unit belongs to the Middle–Late Jurassic. Rb-Sr and K-Ar analysis gave values ranging between 158 ± 10 My and 136 My (Halpern, 1973; Nullo et al., 1979; Charrier et al., 1979).

In the northeastern part of the basin, the platform area, the weathering of volcanic rocks resulted in a basal deposit of sandstones, i.e. the Springhill Formation, which is conformably overlain by Cretaceous pelites. This unit paraconformably overlies the Quemado Complex and is generally restricted to the topographic depressions of its irregular surface. It has

an average thickness of 30–40 m with a maximum of 130–150 m. Locally the basal part or even the entire formation may be missing. Part of this unit is buried and it is mainly known from borehole data. In different wells and outcrops the Springhill Formation has yielded fossil remains. The plants, possibly of Tithonian age, have been described by Archangelsky (1977; see Table XVII), whereas the invertebrates, of Tithonian–Berriasian age, have been illustrated by Riccardi (1976, 1977) and Blasco et al. (1979).

The Springhill Formation has been dated on the basis of microfossils as Oxfordian–Kimmeridgian (Sigal et al., 1970; Natland et al., 1974), and on the basis of megafossils as Tithonian–Berriasian (Archangelsky, 1977; Riccardi, 1977). The wide range is attributed to diachronism by some authors (Cecioni and Charrier, 1974; see also Riccardi, 1976).

Deseado Massif

In central-east Patagonia (Fig. 2, Table II) is an area that has remained structurally positive throughout most of its geologic history. A major part of that geologic region, the Deseado Massif, is covered by Middle to Upper Jurassic volcanic and pyroclastic rocks. Older units are concealed or restricted to small and rare outcrops. The oldest known Jurassic rocks in the Deseado Massif, named Roca Blanca Formation (Herbst, 1965), unconformably overlie the Triassic rocks of the El Tranquilo and La Leona Formations. The Roca Blanca Formation consists of grey and green tuffs, siltstones and coarse sandstones of continental origin. It has a thickness of 990–1200 m and at several levels contains fossil plants studied by Herbst (1965; see Table XVI). The Roca Blanca For-

mation also yielded the oldest known Anura, i.e. *Vieraella herbstii* (Reig, 1961; Casamiquela, 1965). The fossils and the stratigraphic relationships of this formation indicate Toarcian–Aalenian age (see Stipanicic and Bonetti, 1970b).

The Roca Blanca Formation is (?) conformably overlain by 200–600 m of basalts and volcanic agglomerates, and in some areas is interbedded with sandstones, tuffs and conglomerates, all of which have been ascribed to the Bajo Pobre Formation (see De Giusto et al., 1980). These two continental units are covered unconformably by the volcanic and pyroclastic rocks of the Bahia Laura Group, which is the extra-Andean equivalent of El Quemado Complex extensively developed in the Patagonian Cordillera.

The Bahia Laura Group has been divided into at least two units, the Chon Aike Formation, chiefly

TABLE II

Stratigraphy of the Deseado Massif

TIME UNIT	ROCK UNIT		MAX. THICKNESS (m.)	LITHOLOGY	FOSSILS	ENVIRONMENT
UPPER JURASSIC						
MIDDLE JURASSIC	BAHIA LAURA GROUP	LA MATILDE FM. / CHON AIKE FM.	+ 1000	Tuffs, sandstones, conglomerates, shales / Acid or mesosilicic flows, or intrusions, coarse to fine pyroclastics		Continental volcanism, lacustrine, fluviatile
		BAJO POBRE FM.	600	Basalts, sandstones, tuffs, conglomerates		Continental volcanism, fluviatile
LOWER JURASSIC		ROCA BLANCA FM.	1200	grey, green, yellow tuffs, siltstones and sandstones		Continental volcanism, fluviatile

volcanic and La Matilde Formation, consisting mainly of pyroclastics. Rocks assigned to these two units seem to alternate and interfinger. Unconformities mentioned in the literature appear therefore to be only of local significance.

La Matilde Formation has yielded plants, anurans and vertebrate tracks (see Stipanicic and Bonetti, 1970b; Stipanicic and Reig, 1955, 1957; Casamiquela, 1964). K-Ar analysis of these rocks gave values of 160.7 My (Cazeneuve, 1965) and 166 ± 5 My (Creer et al., 1972). The Bahia Laura Group is therefore of Middle–Late Jurassic age.

Central-west Patagonia

In central-west Patagonia, north of the Deseado Massif, is a structurally negative area where marine Lias and continental Dogger–Malm overlie an Upper Paleozoic-?Triassic sequence (Fig. 3, Table III). In this region are found the southernmost Lower Jurassic marine outcrops known in South America.

The stratigraphy and paleontology of the Jurassic succession have been described by Piatnitzky (1932, 1936), Feruglio (1933, 1946), Wahnish (1942), Herbst (1961, 1966), Herbst and Anzotegui (1968),

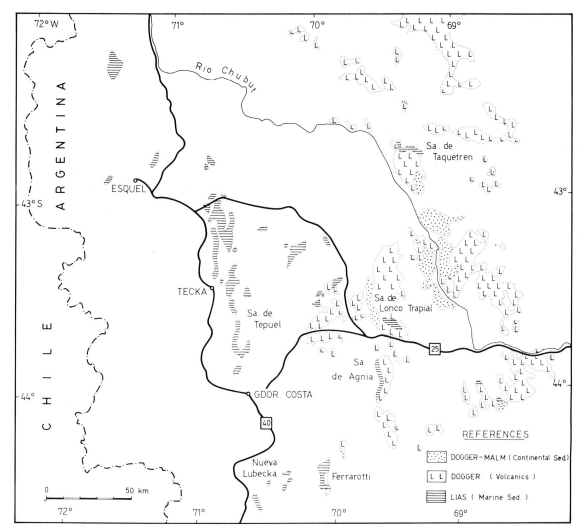

Fig. 3. Major facies distribution of the Jurassic of central-west Patagonia (Lias outcrops also include continental sediments; see Fig. 6).

TABLE III

Stratigraphy of central-west Patagonia

TIME UNIT	ROCK UNIT		MAX. THICKNESS (m.)	LITHOLOGY	FOSSILS	ENVIRONMENT
UPPER JURASSIC						
	CAÑADON ASFALTO FM.		300	Grey, green, yellow tuffs, conglomerates, limestones, bituminous shales, dark siltstones, sandstones		Limnic to continental, arid to semiarid
MIDDLE JURASSIC	LONCO TRAPIAL GROUP	PAMPA DE AGNIA FM.	+1000	Red-violet porphyrites, grey tuffs		Continental volcanism, fluviatile
		CERRO CARNERERO FM.	340	Pink, green, red tuffs, breccias, conglomerates, sandstones		Supra littoral to continental
		OSTA ARENA FM.	190	Ligth colored tuffites, sandstones and conglomerates.		Marine, littoral to continental.
	?					
LOWER JURASSIC	PUNTUDO ALTO FM.		C. 100	Tuffites, ignimbrites, sandstones		Continental
	?					

Tasch and Volkheimer (1970), Robbiano (1971), Musacchio and Riccardi (1971), Nakayama (1973), Nullo (1974), Musacchio (1975), Nullo and Proserpio (1975), Proserpio (1976), Fernandez Garrasino (1977), Blasco et al. (1979, 1980). General accounts can be found in Feruglio (1949), Groeber et al. (1953), Stipanicic and Methol (1972), Lesta and Ferello (1972), and Lesta et al. (1980).

Most of the Jurassic succession is of continental origin and exhibits a wide range of facies change, interfingering and unconformities of doubtful significances. Thus, opinions have long differed about the validity of the units, their nomenclature and relationships. These problems are further enhanced by the lack of a proper taxonomic and biostratigraphic study of the ammonite fauna found in the marine facies.

The Jurassic sequence begins with ignimbrites, tuffs,

tuffites and arkoses of the Puntudo Alto Formation, which has a thickness of about 100 m and has furnished plants of Liassic age (see Stipanicic and Bonetti, 1970a; Table XVI), and could be an equivalent of the Piedra del Aguila Formation of the Neuquén Embayment (see p. 214). The Puntudo Alto Formation is (?) conformably overlain by andesitic conglomerates, the El Cordoba Formation, which could be considered as part of the Lonco Trapial Group (see below).

In the northwestern part of this basin, mostly between 43°–45°S and 70°–72°W (see Fig. 6), follows a marine succession of tuffs, tuffites, sandstones, siltstones and conglomerates, which reaches a maximum thickness of about 300 m in the Sierra de Tepuel. To the north, east and south, the marine strata thin out and the Jurassic becomes totally continental. According to Lesta and Ferello (1972) a narrow sedimentary basin with an almost N–S trend extended from c. 41°S to c. 49°S (See Fig. 6).

This marine Jurassic has received several formational names, e.g. Osta Arena, Lomas Chatas and Mulanguiñeu. The Sinemurian could be represented by the lower levels since, according to some reports, the ammonite genus *Oxynoticeras* is present in the Sierra de Tepuel. The easternmost outcrops contain dactylioceratid and harpoceratid ammonites which indicate a Toarcian age (Musacchio and Riccardi, 1971; Blasco et al., 1979, 1980). An abundant flora has been studied by Herbst (1966; see Table XVI). The westernmost outcrops in Epuyen and farther north, ascribed to the Piltriquitron, Millaqueo, Epuyen–Cholila or Montes de Oca Formation (see Fig. 1; Cazau, 1972; Gonzalez Bonorino, 1974; Gonzalez Diaz and Nullo, 1980; Cucchi and Baldoni, 1980; Lizuain Fuentes, 1980), suggest a possible connection with the Pacific Ocean, or alternatively (?) with the Neuquén Embayment (see Fig. 12), where similar facies are represented within the Los Molles Formation (see p. 214). Marine (?) Bajocian is also known between 43° and 43°45'S, within Chile but near the international boundary (Thiele et al., 1979).

The marine Lias grades upwards, and in part laterally, into continental tuffs, breccias and conglomerates which in the Sierra de Agnia (Figs. 1, 3) have furnished the sauropod *Amygdalodon patagonicus* Cabrera (1947). This unit reaches a thickness of several hundred meters and grades laterally into the volcanic facies of the Lonco Trapial Group. The Lonco Trapial Group has been divided into two units, the Pampa de Agnia Formation, chiefly volcanic, and the Cerro Carnerero or Cajon de Ginebra Formation, of mainly pyroclastic deposits (see Table III). Both units seem to interfinger and alternate. The Lonco Trapial Group is an equivalent to part of the Bahia Laura Group of the Deseado Massif (see p. 206) and of the El Quemado Complex of the Patagonian Cordillera (see p. 203). The stratigraphic relationships of the Lonco Trapial Group suggest a Toarcian–Middle Jurassic age. This conclusion is confirmed by an isotopic age determination that gave 158 ± 6 My for the volcanic Pampa de Agnia Formation (Stipanicic and Rodrigo, 1970a; see also Franchi and Page, 1980).

Unconformably above the Lonco Trapial Group follows a sequence of tuffs, conglomerates, sandstones, siltstones, limestones and bituminous shales that has yielded plants (see Table XVII), fresh water invertebrates (see Tasch and Volkheimer, 1970) and fish (see Table XX). The sequence has a thickness of about 300 m and all evidences point to a lacustrine environment. On the basis of its fossils this unit, named the Cañadón Asfalto Formation, has been assigned to the Callovian–Oxfordian. Its limestones could be equivalent to those of the Cotidiano Formation of the Patagonian Cordillera (see p. 205) and the La Manga Formation of Neuquén and Mendoza (see pp. 215).

Further north of central Chubut a succession of volcanic and pyroclastic rocks, the Taquetren Formation, has furnished a fossil flora similar to that found in the Cañadón Asfalto and La Matilde Formations (see Table XVII) although probably younger in age, i.e. latest Jurassic–earliest Cretaceous. An isotopic age determination gave a value of 136 ± 6 My (Nullo and Proserpio, 1975; see also Franchi and Page, 1980).

Principal Cordillera

The Principal or High Cordillera extends without interruption from southern Neuquén (c. 40°S) to northern Chile (c. 19°S) and further north, along the international boundary of Chile with Argentina and Bolivia (see Harrington, 1956; also Aubouin et al., 1973; Gansser, 1973). It consists fundamentally of Mesozoic and Cenozoic rocks, and the Jurassic System is well represented along its entire extension at the 70°W meridian. From c. 39°S to about 31°S the Jurassic outcrops are well developed in Argentina but from 31°S to about 20°S they are restricted to the western side of the Principal Cordillera within Chile (see Figs. 1, 4, 5).

Neuquén Province

During Jurassic times the sea invaded the area from the west and northwest and formed an embayment with a large eastward extension (see Figs. 12, 13). The Jurassic in Neuquén is therefore well known from a number of outcrops in the west (see Fig. 4) and from borehole data in the east. Together with the exposures in Mendoza Province (see p. 216) it is one of the best known Jurassic areas in South America owing to the studies of Burckhardt (1900, 1902, 1903), Jaworski (1914, 1915), Groeber (1918, 1929, 1946), Weaver (1931), Leanza (1942), Stipanicic (1966, 1969), Marchese (1971), Westermann and Riccardi (1972a, b; 1975), and the geologists of Yacimientos Petrolíferos Fiscales (YPF), the government oil agency (see Digregorio, 1972, 1978; Dellape et al., 1979; Digregorio and Uliana, 1980).

In this region Groeber originally divided the 'Jurassic' into three sedimentary cycles, i.e. Cuyan, Lotenian and Chacayan, ranging in age from the Sinemurian to the Oxfordian–Kimmeridgian (see Groeber et al., 1953; Fig. 7).

The Cuyan is a sedimentary cycle produced by subsidence during which a maximum of 2,500 m of sediments of Liassic–early Callovian age were deposited. It begins with a continental sedimentary infilling of topographic depressions developed on the surface of the Triassic vulcanites. These sediments, sandstones, tuffs and conglomerates are restricted to some isolated areas and to the margins of the embayment.

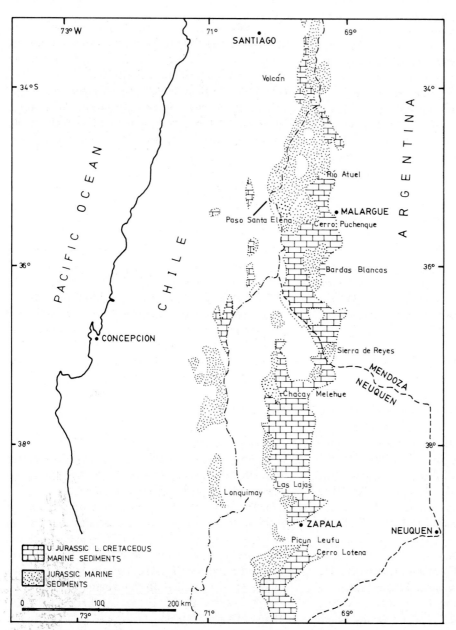

Fig. 4. Distribution of Jurassic and Lower Cretaceous marine sediments in the Principal Cordillera of Argentina and Chile between c. 39°S and 33°S.

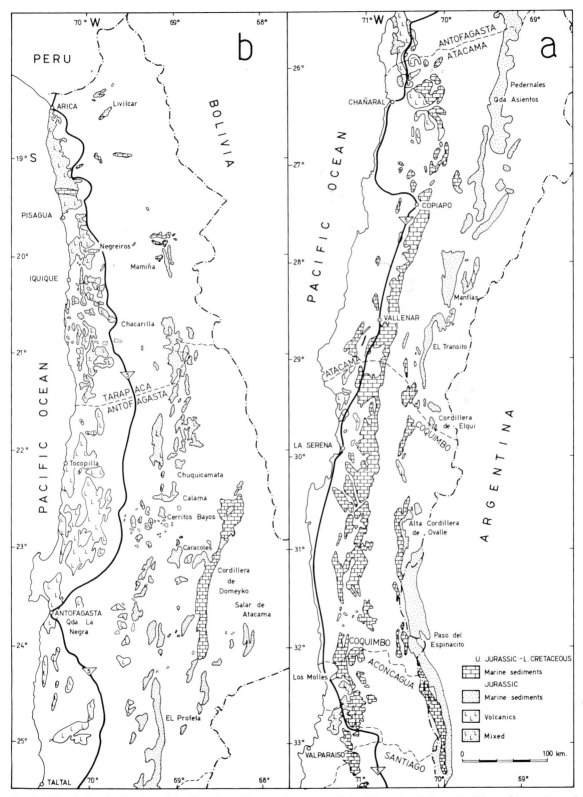

Fig. 5. Major facies distribution of the Jurassic and Lower Cretaceous of the Principal and Coast Cordillera of Argentina and Chile between c. 33°S and 18°30'S.

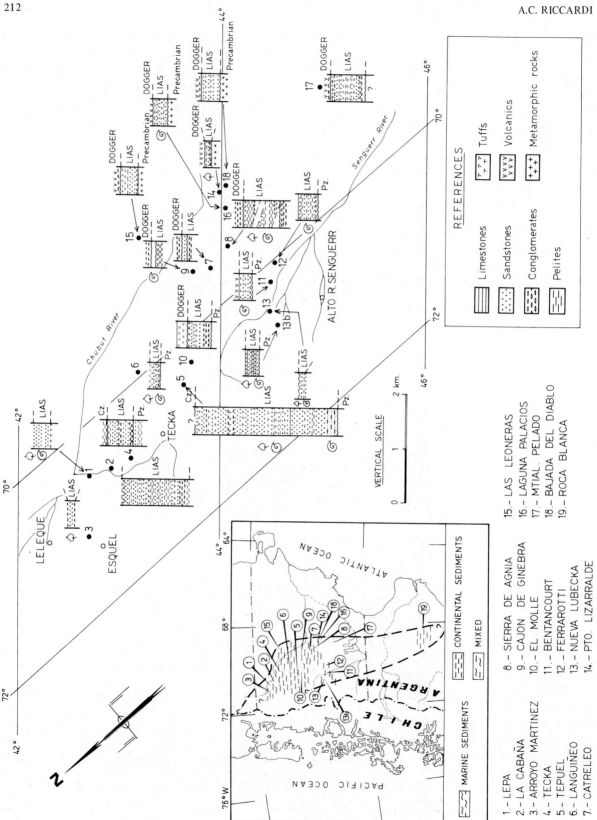

Fig. 6. Comparative sections of the Lias in central-west Patagonia (modified from Lesta et al., 1980).

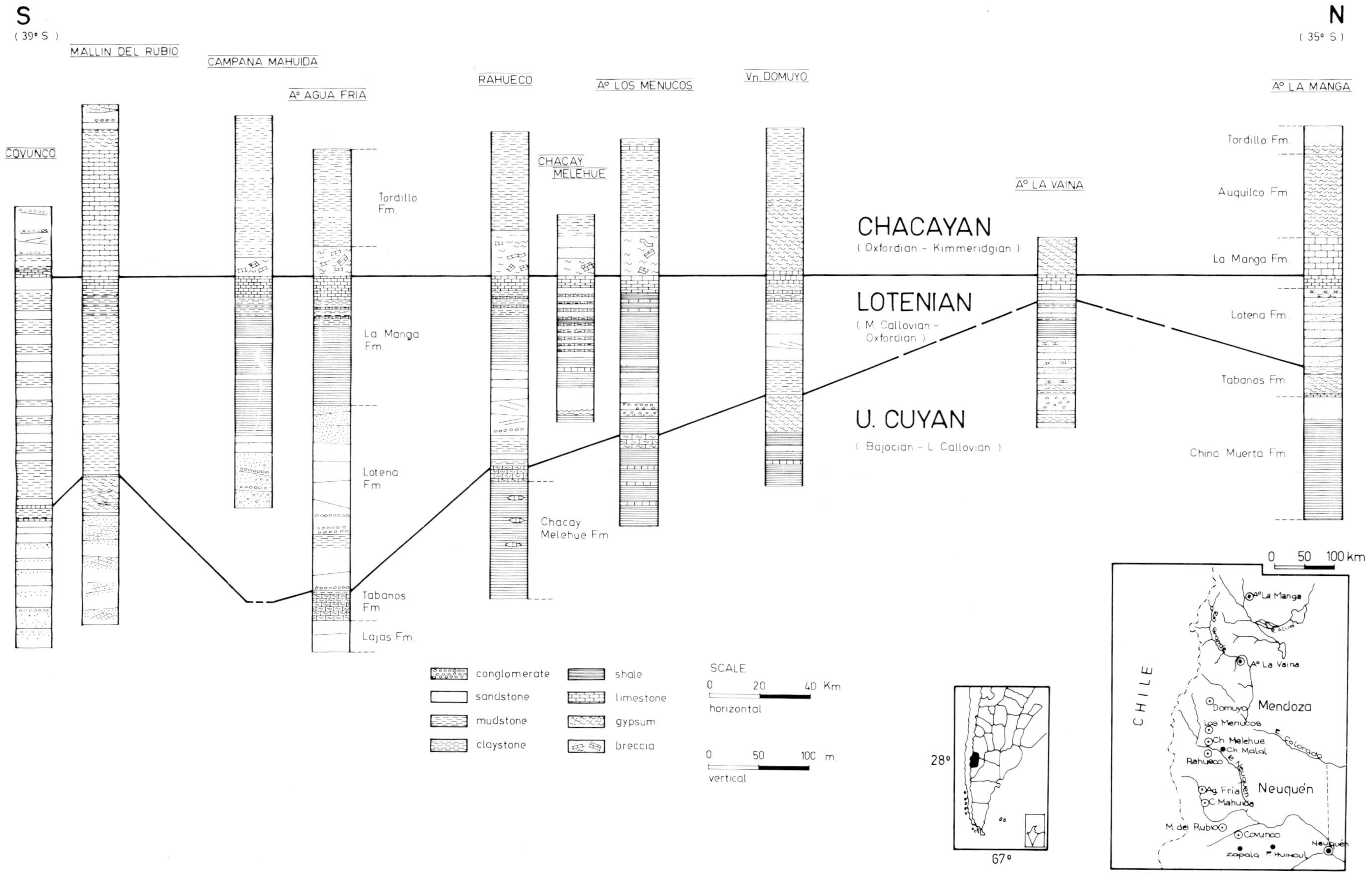

Fig. 7. Comparative sections for the Cuyan, Lotenian and Chacayan of Neuquén and Mendoza (modified from Dellape et al., 1979).

In the subsurface of eastern Neuquén they have been named Planicie Morada Formation (see Table IV). This unit is probably equivalent to the Piedra del Aguila Formation of southwestern Neuquén (see Fig. 15) and the Remoredo Formation of southern Mendoza. Resting on, or interfingering with, these continental sediments or directly upon the Triassic vulcanites, in the western and central part of the embayment is a marine sequence ranging in age from late Sinemurian to early Callovian. It begins with the black shales and limestones of the Los Molles Formation which have furnished fossils indicating ages from late Sinemurian to the early Bajocian. The Los Molles Formation is followed by marine littoral sandstones, of the Lajas and Loma Negra Formations which contain fossils of Bajocian–Callovian age (see Fig. 15, Tables IV, V). This sedimentary sequence ends with

evaporites ascribed to the Tabanos Formation. At Chacay Melehue, in northern Neuquén, the marine shales of the Los Molles Formation continue upwards into Bathonian–Callovian levels which have been named Chacay Melehue Formation (see Figs. 7, 15). In that region the marine shales of the Los Molles and Chacay Melehue Formations are directly overlain by the evaporites of the Tabanos Formation.

On the eastern and southern borders of the embayment the Los Molles, Lajas and Loma Negra Formations grade laterally into continental deposits included in the Lotena (lower part), Challaco and Punta Rosada Formations (see Figs. 8, 15). On the southeastern margin of the embayment, in Piedra Pintada (see Fig. 1), marine Sinemurian–Pliensbachian beds of the Los Molles Formation are interbedded with tuffs and ignimbrites of the Sañico Formation (see Fig. 15).

TABLE IV

Stratigraphy of eastern Neuquén (slightly modified from Digregorio and Uliana, 1980)

TIME-UNIT	ROCK UNIT		MAX. THICK-NESS (m)	LITHOLOGY			FOSSILS	ENVIRONMENT		
L. CRETACEOUS	LOMA MONTOSA FM.		350	Grey-yellowish limestones, shales and calcareous sandstones			⬡ ⧗	Shallow marine, littoral and supralittoral, occasionally sublittoral		
UPPER JURASSIC	VACA MUERTA FM.		200	Dark-greyish shales and limestones			⬡ ⧗ ⬡	Marine, low wave energy, euxinic		
	QUEBRADA DEL SAPO FM.	CATRIEL FM.	50	Grey-greenish and red conglomerates	Fine-grained greenish sandstones and siltstones			Littoral to continental		Littoral to sublittoral
	BARDA NEGRA FM.	SIERRAS BLANCAS FM.	200	Dark grey shales and sandstones, limestones and marls	Green or pink conglomeratic sandstones		⬡ ⧗	Marine littoral to sublittoral		Fluviatile, littoral to shalow marine
MIDDLE JURASSIC	CHALLACO FM. / LOMA NEGRA FM.	PUNTA ROSADA FM.	2300	Red sandstones, mudstones / grey sandstones, mudstones	Green and red sandstones, conglomerates, mudstones and siltstones		⬠ / ⬡	Alluvial	Deltaic, shallow marine	Fluviatile to ? deltaic
LOWER JURASSIC	LOS MOLLES FM.	PLANICIE MORADA FM.		Dark grey shales with sandy and tuffaceous intercalations	Whitish sandstones, reddish conglomerates and mudstones		⧗ ⊗	Marine to continental		Continental
UPPER TRIASSIC	PASO FLORES FM.	BARDA ALTA TUFFS	300	Brownish sandy conglomerates, tuffs and tuffites	Ignimbrites, andesites, tuffs, sandstones, conglomerates		⬠	Continental volcanism, lacustrine, fluviatile		

The Tabanos Formation and its equivalents are overlain by the littoral to continental sandstones and conglomerates of the Lotena Formation (see Figs. 7, 15 and Table V). Within this unit there are, however, marine intercalations with Callovian ammonites. This unit belongs to the lower part of the transgressive—regressive Lotenian–Chacayan sedimentary cycle (see Digregorio and Uliana, 1980). The Lotenian transgression is documented by the shales of the Barda Negra Formation which (?) unconformably overlie the Loma Negra or the Lotena Formation. The Barda Negra Formation contains ammonites of Oxfordian age and towards the embayment margins grades laterally into the clastic sediments of the Sierras Blancas Formation and (?) upper part of the Punta Rosada Formation (see Fig. 8).

In the central part of the embayment, the Lotena and the Barda Negra Formations are overlain by two intergrading sedimentary units (see Fig. 8). The lower unit consists of 0–220 m of limestones and is known as 'blue limestones with *Gryphaea*' or La Manga Formation. It has yielded Oxfordian ammonites probably corresponding to the upper part of the Cordatum to the lower part of the Canaliculatum (= Transversarium) zones (see Table XV) and includes corals, mud mounds, and towards the top, exhibits regressive features. The upper unit consists of up to 300 m of gypsum and/or anhydrite and interbedded limestones and has been named 'Principal Gypsum' or Auquilco Formation. This together with the upper part of the La Manga Formation with which it intergrades laterally, defines the Chacayan regressive cycle. On the eastern margin of the embayment, in subsurface, the Auquilco Formation changes into sandstones and

TABLE V

Stratigraphy of western Neuquén (slightly modified from Digregorio and Uliana, 1980)

TIME UNIT	ROCK UNIT		MAX. THICKNESS (m.)	LITHOLOGY	FOSSILS	ENVIRONMENT
L. CRETACEOUS	MENDOZA GROUP			Black shales and limestones, grey-greenish mudstones and claystones		Marine to continental
		VACA MUERTA FM.	1200			
UPPER JURASSIC	TORDILLO FM.		100	Green sandstones and shales		Littoral to sublittoral
			500	Red sandstones and claystones, tuffs and gypsum		Continental flood plains, channels, swamps
	AUQUILCO FM.		300	White gypsum		Evaporite basin
	LA MANGA FM.		130	Grey micritic limestones, collapse breccias, bioherms		Shallow marine: carbonate platform
MIDDLE JURASSIC	LOTENA FM.		200	Brown and grey conglomerates and conglomeratic sandstones; black grey shales		Littoral to continental
	CUYO GROUP	TABANOS FM.	40	Gypsum, dolomites, oolitic limestones, red claystones		Evaporite basin
		LAJAS FM.	400	Brown and grey sandstones, dark grey shales and limestones		Deltaic: continental, littoral and sublittoral
LOWER JURASSIC		LOS MOLLES FM.	1300	Sandstones, tuffs, dark pelites, limestones		Littoral to restricted basin
UPPER TRIASSIC	CHACAYCO FM.			Tuffs, tuffites, basalts		Continental volcanism, alluvial deposits

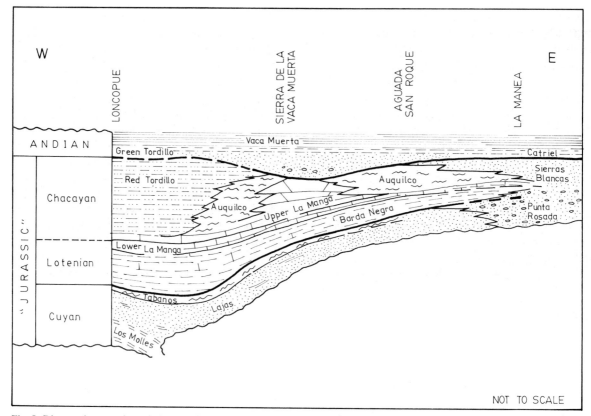

Fig. 8. Diagram interpreting relations in the Jurassic rocks of the Neuquén Embayment (after Digregorio and Uliana, 1980).

conglomerates of the Sierras Blancas Formation, while to the west it grades into the lower part of the Tordillo Formation. On the basis of some gaps in the ammonite record when compared with the European sequence (see Table XIV), a hiatus, attributed to the Rio Grande Diastrophic Phase (see p. 335), was recognized below the La Manga Formation (Stipanicic, 1966; Stipanicic and Rodrigo, 1970b). However, other authors (see Digregorio and Uliana, 1980) noted the transitional character of the sedimentary sequence (see Fig. 7).

The Chacayan sedimentary succession is in turn unconformably overlain (Araucanian Diastrophic Phase) by lower to upper Kimmeridgian or even Tithonian rocks. The Kimmeridgian is represented by the continental sediments of the Tordillo Formation. This unit consists in its lower part of red sandstones and mudstones, and in its upper part of green sandstones and conglomerates. According to some authors (see Digregorio and Uliana, 1980) the lower part grades downwards into the Auquilco Formation whereas the upper part is transitional to the marine

Tithonian of the Vaca Muerta Formation. The unconformity is therefore placed between the red and green sections of the Tordillo Formation (see Fig. 8). The Tordillo Formation increases in thickness and grain size from east to west.

On the northeastern and southwestern margin of the Neuquén Embayment the Upper Jurassic is represented by littoral to continental sandstones (see Table IV). In the former region, the Sierras Blancas and Catriel Formations rest on top of the continental Punta Rosada Formation and both units are in turn covered by Tithonian marine shales of the Vaca Muerta Formation. Along the southwestern border, the Quebrada del Sapo Formation, an equivalent of the Catriel Formation, also underlies the Vaca Muerta Formation.

Rio Atuel–Nacientes del Teno

In Mendoza province, Argentina, the Jurassic sequence is similar to that of the Neuquén Embayment (see Gerth, 1914; Lahee, 1927; Yrigoyen, 1979). In Sierra de Reyes and Cerro Tricolor (see Figs. 1, 4)

PLATE I

A. View of the western flank of Sierra de Reyes, Mendoza province, Argentina. In the nucleus of the anticline are exposed dark volcanic beds of the Remoredo Formation (L. Jurassic). Covering the Remoredo Formation are Middle Jurassic marine sediments, and capping the sequence, in the upper part of the photograph, is the gypsum of the Auquilco Formation (Oxfordian–? Kimmeridgian). In the foreground, the western flank of the anticline is capped by the Vaca Muerta Formation (Tithonian–L. Cretaceous). (Photograph: A.C. Riccardi.)

B. View of the Jurassic section at Bardas Blancas, Mendoza province, Argentina. To the right volcanics of the Choiyoi and Remoredo Formations (U. Triassic–L. Jurassic). In the middle uppermost Lower Jurassic to Middle Jurassic beds, capped to the left by the limestones of the La Manga Formation (Oxfordian). (Photograph: A.C. Riccardi.)

PLATE II

A. View of Jurassic beds exposed to the north of Cerro Puchenque, Mendoza province, Argentina. In the foreground Pliens-bachian marine sediments. In the middle of the photograph an intrusive that penetrated Toarcian–Bajocian marine sediments. To the right follows a Callovian sequence capped by the white gypsum of the Auquilco Formation (Oxfordian–? Kimmeridgian). (Photograph: A.C. Riccardi.)

B. View of Middle Jurassic strata looking north of Paso del Espinacito (4476 m), San Juan province, Argentina. In the upper right edge of the photograph marine Aalenian–Bajocian, covered to the left by marine Callovian. (Photograph: A.C. Riccardi.)

TABLE VI

Stratigraphy of the Atuel River area

TIME UNIT	ROCK UNIT		MAX. THICKNESS (m.)	LITHOLOGY	FOSSILS	ENVIRONMENT
L. CRETACEOUS	MENDOZA GROUP					
		VACA MUERTA FM.	470	Dark grey shales , limestones		Marine
UPPER JURASSIC	TORDILLO FM.		800	Red conglomerates , sandstones , tuffs .		Continental
	AUQUILCO FM.		200	White gypsum		Evaporite basin
	LA MANGA FM.		45	Grey limestones , siltstones		Shallow marine
MIDDLE JURASSIC	LOTENA FM.		200	Grey , green sandstones , with interc siltstones and limestones		Marine to fluviatile
	TABANOS FM.		30	White gypsum		Evaporite basin
	CHINA MUERTA FM.		250	Dark grey sandy shales , siltstones , limestones		Marine
LOWER JURASSIC	RIO ATUEL GROUP	EL CHOLO FM.	600	Green sandstones , siltstones , shales , limestones		Paralic to marine
		EL FRENO FM.	900	Coarse sandstones , conglomerates		Littoral to continental

most of the Lias is represented by the continental tuffs and tuffites of the Remoredo Formation, and the marine sediments begin with the (Pliensbachian) Toarcian–Aalenian. The most complete sequences are exposed in the Cerro Puchenque and Rio Atuel areas, to the west and northwest of Malargüe (see Fig. 4).

In Rio Atuel (see Stipanicic, 1966; Stipanicic and Bonetti, 1970a; Volkheimer, 1970a) the Jurassic begins (see Table VI) with the Rio Atuel Group, which ranges in age from the Hettangian to the Toarcian. It includes two units: the lower El Freno Formation consists of 900 m of coarse sandstones and conglomerates resting unconformably on the Triassic Choiyoi Group and the upper El Cholo Formation formed of sandstones, calcarenites, siltstones and shales. It has furnished plants and invertebrates ranging in age from the Sinemurian to the Toarcian. The El Cholo Formation grades upwards into the Bajocian China Muerta Formation. This unit consists of sandy shales with some interbedded sandstones and has yielded ammonites of Aalenian–Bajocian age.

Above follow 30 m of gypsum, the Tabanos Formation, of Callovian age since the subjacent beds contain macrocephalitid ammonites and the superjacent sandstones of the Lotena Formation have yielded *Reineckeia* ex gr. *anceps* (Reinecke). On top of the Lotena Formation rest the limestones of the La Manga Formation with an Oxfordian ammonite fauna described by Stipanicic (1951). They grade upwards into the 200 m thick gypsum of the Auquilco Formation. The Jurassic sequences end with the continental sediments of the Tordillo Formation which are followed by the marine sediments of the Tithonian–Barremian Mendoza Group.

A similar Middle–Upper Jurassic sequence is exposed in the Chilean part of the Principal Cordillera

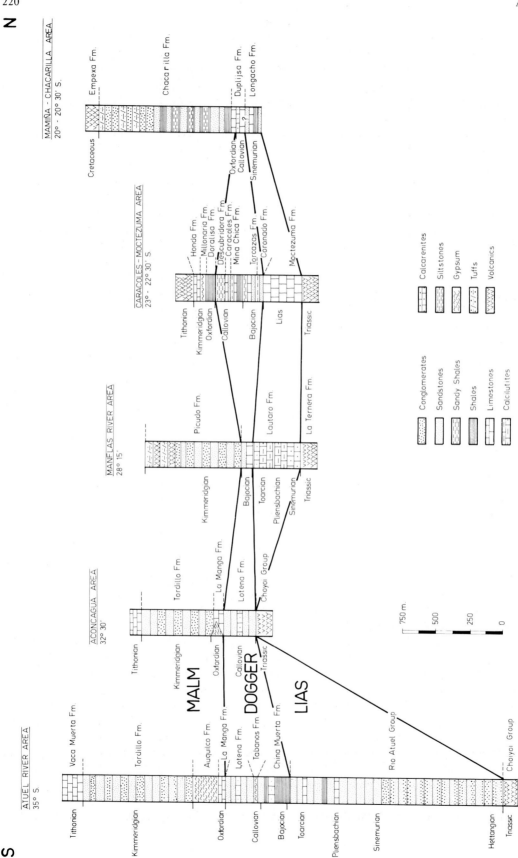

Fig. 9. Comparative sections along the Principal Cordillera of Argentina and Chile between 35°S and 20°S.

(see Fig. 10), the stratigraphic scheme of which has been established by Klohn (1960) and Gonzalez and Vergara (1962), and the paleogeographic analysis across the Argentine–Chilean boundary at 35°S presented by Davidson and Vicente (1973). The southernmost outcrops, located at Lonquimay (see Figs. 1, 4), and their relationships with those of the Neuquén Embayment have been studied by Chotin (1969).

Between 38° and 33°15'S the Jurassic begins with a comprehensive unit, i.e. the Nacientes del Teno or Valle Grande Formation, which can be readily divided into a number of lithologic members ranging in age from the Aalenian to the Oxfordian. The lowest (unnamed) beds rest unconformably on Triassic volcanics and consist of 1,000 m of conglomerates, sandstones, shales, tuffs and tuffites, containing a marine fauna of Aalenian–Bajocian age. Above follows the Rinconada Member, 450 m of sandstones, limestones and shales bearing in the upper part a Callovian–Oxfordian fauna. The Nacientes del Teno Formation ends with 100 m of gypsum, the Santa Elena Member, missing at Lonquimay (Chotin, 1969), conformably overlain by the Rio Damas or Vega Negra Formation. This latter unit is 2,000–3,000 m thick and is chiefly formed by red sandstones, conglomerates, breccias and shales of continental origin. At the top of the Jurassic sequence lie the Leñas Espinoza and Baños del Flaco Formation. They consist of 400–1,500 m of calcareous sandstones and limestones which have yielded Tithonian–Neocomian ammonites (see Corvalan, 1959; Covacevich et al., 1976; Biro, 1980a).

Between 32°45'S and 33°S only the upper part of the Nacientes del Teno Formation and the Rio Damas Formation crop out (see Aguirre, 1960). Similarly, in the Argentine Cordillera at the same latitude and in the Aconcagua area (see Figs. 1, 5), the Jurassic (see Fig. 9) is represented by the La Manga, Auquilco and Tordillo Formations. The oldest exposed levels belong to the Callovian Lotena Formation (see Groeber, 1951, 1963; Yrigoyen, 1976). However, further north at the same longitude, in the Espinacito–Los Patos area (see Figs. 1, 5) marine Sinemurian–Callovian rests paraconformably on Upper Triassic–Lower Jurassic volcanics and is overlain in turn by upper Callovian conglomerates, limestones of the La Manga Formation and gypsum of the Auquilco Formation (Gottsche, 1878; Tornquist, 1898; Schiller, 1912; Rigal, 1930; Lambert, 1943; Groeber, 1951; Stipanicic, 1966; Hillebrandt, 1970; Westermann and Riccardi, 1972b, 1979; Volkheimer et al., 1978a, b;

Herbst, 1980). Marine Callovian is still present at 31°S in the High Cordillera de Ovalle (see Fig. 5) where it overlies marine Toarcian–Bajocian, but the Oxfordian gypsum is missing (Mpodozis et al., 1973; Mpodozis and Rivano, 1976).

Further north, between 30°S and 30°15'S, the marine Jurassic is restricted to Lias and the Dogger–Malm is represented by a continental sequence. According to Dedios (1967), Jurassic rocks rest unconformably on the Upper Triassic Las Breas Formation. The sequence begins with fine grained conglomerates, calcareous sandstones, fossiliferous limestones and shales with a total thickness of about 390 m. This unit, named the Tres Cruces Formation, has yielded Sinemurian–Pliensbachian invertebrate fossils. Above follows unconformably the Algarrobal Formation. It consists of 100–200 m of andesitic rocks with intercalated red colored shales, sandstones and conglomerates of continental origin. This unit has been considered as an equivalent of the La Negra Formation of Atacama and Antofagasta (see p. 227) and is thus assigned to the Middle–Upper Jurassic.

Similarly, between 29°33'S and 29°50'S in the Elqui Cordillera (see Fig. 5), Thiele (1964) identified 1,800 m of limestones, sandstones, conglomerates and volcanics, i.e. the Punilla Formation, containing Pliensbachian–Bajocian invertebrates (see Hillebrandt, 1973a). This unit, which according to Hillebrandt (1973a) is only 220–250 m thick, is limited at top and bottom by volcanics of Cretaceous and Triassic age.

La Totora–Copiapo

At about 29°S (see Fig. 5), Reutter (1974) recognized the presence of a Jurassic marine sequence (see Table VII) formed by conglomerates, sandstones and marls, lying unconformably on Triassic volcanics. This Jurassic unit, named the Lautaro Formation by Segerström (1959, 1968) in the Copiapó area, reaches its maximum thickness of 750 m at Quebrada La Totora, north of El Tránsito. The invertebrate fossils indicate a Sinemurian–Bajocian age (see Hillebrandt, 1970, 1973a). Following unconformably is the continental Picudo Formation of red conglomerates, breccias, tuffs and volcanics. It has a thickness of 1,500 m at Cerro Picudo and its age could be Late Jurassic.

In the Manflas area (see Figs. 5, 9) the presence of Lias and Dogger was first noticed by Bayle and Coquand (1851). Fossil material from this area was

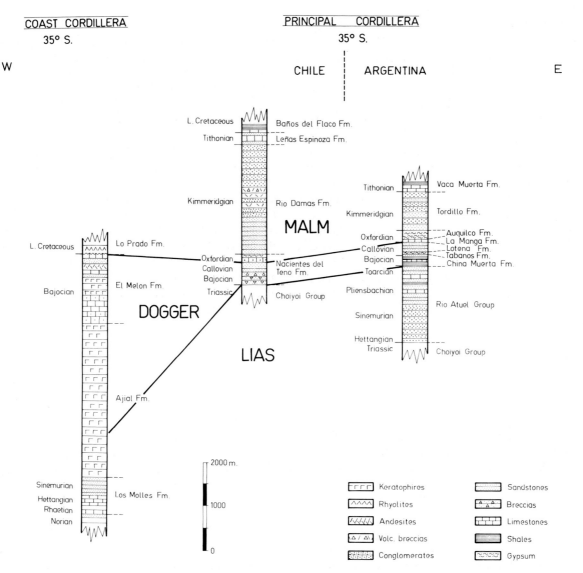

Fig. 10. Comparative Jurassic sections of the Coast and Principal Cordilleras at about 35°S (modified from Aubouin et al., 1973).

described by Burmeister and Giebel (1861), Stein-mann (1881), Möricke (1894), Philippi (1899), Wes-termann and Riccardi (1972b) and Hillebrandt (1977). The stratigraphic nomenclature was establish-ed by Segerström (1959, 1968) and the general geology of the area has recently been reviewed by Jensen and Vicente (1977). An account of the Juras-sic sequence was also given by Hillebrandt (1971, 1973a).

The marine Jurassic ranges in age from the Lias to the Bajocian. The base is transitional to the Upper Triassic volcanic and pyroclastic rocks. The marine sequence is overlain by an Upper Jurassic continental

succession which, although exhibiting extensive facies variation, was named Lautaro Formation. It consists of calcarenites, calcilutites in the lower part and sand-stones and fine grained conglomerates in the upper part. Its thickness decreases from the west (1,200 m) to the east. The ammonite fauna shows that the sea transgressed from west to east from Sinemurian to Toarcian times and that the regression occurred most-ly during the late Toarcian–late Bajocian. Marine Cal-lovian beds are only known north of 28°S and be-come more important between 27° and 26°S, where they are transgressive above Pliensbachian–Bajocian, and farther north (see Hillebrandt, 1973a; Cisternas

TABLE VII

Stratigraphy of La Totora area

TIME UNIT	ROCK UNIT	MAX. THICKNESS (m.)	LITHOLOGY	FOSSILS	ENVIRONMENT
UPPER JURASSIC	PICUDO FM	1500	Red conglomerates, sandstones, andesites, tuffs.	·	Continental volcanism, fluviatile
MIDDLE JURASSIC					
LOWER JURASSIC	LAUTARO FM.	750	Calcarenites, calcilutites, sandstones, shales, conglomerates		Marine, littoral

and Vicente, 1976, Cisternas, 1979).

The marine Jurassic is conformably overlain by 700 m of continental red conglomerates, sandstones, andesites and tuffs of the Upper Jurassic Picudo Formation.

Domeyko Cordillera

The Domeyko Range (see Fig. 5) extends between 22°20'S and 25°30'S and includes an important Jurassic sequence with both marine and continental facies. Recently, and after the pioneering work of Harrington (1961), the general stratigraphy of this range has become better known through the work of Chong (1973, 1976, 1977) and Hillebrandt (1970, 1973a).

To the south a well exposed section is present at Quebrada Asientos, west of Salar de Pedernales (see Figs. 1, 5). According to Harrington (1961; see Fig.

15) the Jurassic succession begins with bituminous grayish black limestones and shales that rest on top of a basal conglomerate. This unit, the Montandon Formation, is c. 400 m thick and contains Sinemurian–Bajocian ammonites (see Garcia, 1967; Mercado, 1978). It is followed by the Callovian Asientos Formation. The Bathonian seems to be missing. The Asientos Formation comprises 570 m of massive blue limestones and brown sandy limestones with intercalations of shales, and has yielded ammonites (Garcia, 1967). It is unconformably covered by the Tithonian–Lower Cretaceous marine sediments of the Pedernales Formation.

In the central and northern parts of the Domeyko Range are several well known localities such as El Profeta, Moctezuma, Limon Verde, Cerritos Bayos and Caracoles (see Figs. 1, 5), some of which were studied by Biese (1957, 1961) and Harrington

TABLE VIII

Stratigraphy of the Caracoles–Moctezuma area

TIME UNIT	ROCK UNIT		MAX. THICKNESS (m.)	LITHOLOGY	FOSSILS	ENVIRONMENT
UPPER JURASSIC				Andesites		Continental volcanism
	HONDA FM.		50	Grey massive limestones		Shallow marine
	MILLONARIA FM.		20	Massive beds of anhydrite		Evaporite basin
	DORALISA FM.		64	Yellow, green blue shales, with intercalated limestones	⬡ ⧖	Marine
MIDDLE JURASSIC	CARACOLES GROUP	DESCUBRIDORA FM.	28	Light colored sandy shales with intercalated limestones and gypsum	⬡ ⧖	Marine
		CARACOLES FM.	20	Grey sandstones, limestones, sandy shales	⬡ ⧖	Marine
		MINA CHICA FM.	75	Brown sandy shales, with intercalated sandstones, limestones	⬡ ⧖	Marine
	TORCAZAS FM.		96	Brown sandy shales, calcareous sandstones	⬡ ⧖	Marine
	CORONADO FM.		40	Calcareous sandstones, fine-grained conglomerates	⬡ ⧖	Marine
LOWER JURASSIC	MOCTEZUMA FM.		290	Blue, brown, grey limestones and marls, andesitic tuffs intercalated	⧖	Marine

(1961). There the Jurassic (Figs. 9, 15; Table VIII) begins with volcanics that interfinger with continental deposits or marine facies bearing an Hettangian ammonite fauna (see Chong, 1977; Naranjo and Covacevich, 1979). Sinemurian to Oxfordian marine facies are widespread, although the Pliensbachian levels are interbedded with conglomerates and evaporites and the Bajocian–Callovian strata with volcanics. This succession, within which a late Bathonian fauna has been reported (see Hillebrandt, 1970, 1973a; Chong 1973, 1977), reaches a thickness of about 700 m in the area of Caracoles–Moctezuma. It is mostly formed by limestones, shales and sandstones and has been divided by Harrington (1961) into several formations (see Table VIII). Some minor changes in the stratigraphic nomenclature were later introduced by Garcia (1967).

The Oxfordian limestones with perisphinctids grade upwards into evaporites of the Millonaria Formation. This in turn is overlain by unfossiliferous limestones and red continental sandstones assigned to the Kimmeridgian–Tithonian. Equivalent continental rocks + 500 m thick which crop out north of the Salar de Atacama (22°45′–23°S) have been named Tonel Formation by Dingman (1963). East of the Salar de Atacama (see Moraga et al., 1974) is exposed a succession c. 3,250 m thick consisting of sediments of continental to littoral origin intercalated with volcanics. This unit, named El Peine Formation, could be an equivalent of the Chacarilla Formation of Tarapaca.

Chacarilla–Mamiña area

The geology of the central region of Tarapaca province (see Figs. 1, 5) in northern Chile, has been studied mainly by Galli (1957, 1968), Galli and Dingman (1962), Dingman and Galli (1965) and Thomas (1967).

The Jurassic (see Figs. 9, 11; Table IX) exposed in the Chacarilla–Mamiña area, between 20°45′S and 20°S, ranges from Lias to Malm. The Liassic rocks, represented only in the southern part of the area, have been called Longacho Formation and consist chiefly of shales, mudstones, fine grained sandstones and limestones of marine origin. The exposed section is 150 m thick, although the base of the formation is not exposed and the upper surface is eroded. It has furnished ammonites identified as *Arietites* spp. which indicate a Sinemurian age. In the region of the Longacho Formation, the succeeding unit is the Chacarilla Formation. The contact between these formations is not exposed. The Chacarilla Formation is more than 1,127 m thick, however, its base is not

exposed and the top has been eroded. According to Garcia (1967) it reaches c. 3,640 m in Huatacondo, south of Chacarilla, consisting of mudstones and sandstones interbedded with few trachyte flows. The lithology shows an upward gradation from near-shore to continental facies which Garcia (1967) named the Majala Formation (marine) and Huatacondo Formation (continental). The fossils include poorly preserved ammonites identified as Oxfordian taxa, bivalves, plants (see Galli and Menendez, 1968) and dinosaur footprints.

East of the Mamiña area, in Juan de Morales, the Jurassic is represented by the Chacarilla Formation and by a succession at least 90 m thick of limestones with intercalated calcareous sandstones. This contains ammonites of Callovian–Oxfordian age and has been named Duplijsa Formation. The contact between the

TABLE IX

Stratigraphy of the Chacarilla–Mamiña area

TIME UNIT	ROCK UNIT	MAX. THICKNESS (m.)	LITHOLOGY	FOSSILS	ENVIRONMENT
UPPER JURASSIC	CHACARILLA FM ?	+ 1127	Breccias , tuffs , shales , siltstones , sandstones		Marine to continental
MIDDLE JURASSIC	DUPLIJSA FM ?	+ 90	Red, grey limestones , calcareous, sandstones		Marine infraneritic
LOWER JURASSIC	LONGACHO FM ?	+ 150	Grey shales , mudstones , sandstones , limestones		Marine sublittoral

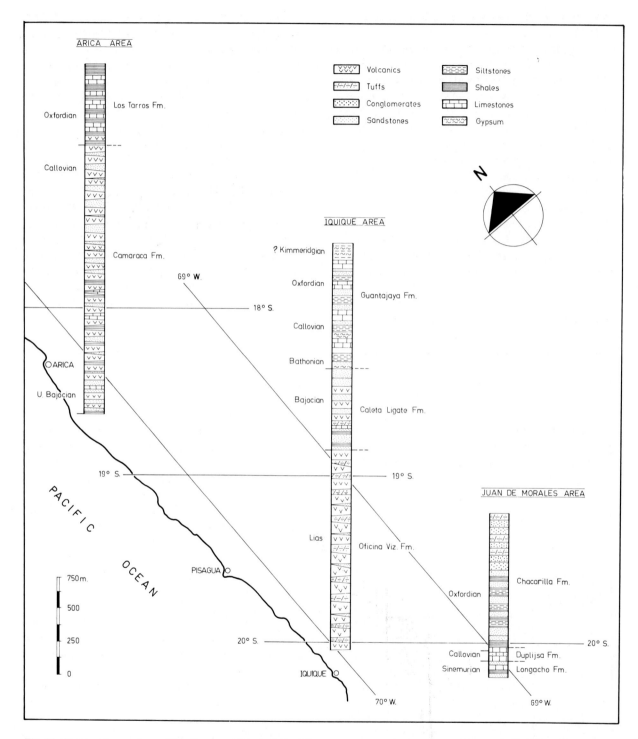

Fig. 11. Comparative sections of the Coast and Principal Cordilleras in northern Chile between 20°S and 18°30'S.

Dupiijsa and Chacarilla Formations is not exposed.

In the Chacarilla–Mamiña region the Upper Jurassic Chacarilla Formation is overlain by Cretaceous volcanic and continental rocks of the Cerro Empexa Formation with angular unconformity.

North of this area (see Fig. 5), in Chile, the Jurassic is exposed in Chismiza, 120 km to the northeast of Iquique and in Livílcar, east of Arica. These are small outcrops which in Chismiza consist of Liassic beds, while most of the others are Oxfordian (see Zeil, 1964; Ruiz et al., 1965; Garcia, 1967).

Coast Cordillera

The Coast Cordillera rises with a north–south trend along the Pacific coast of Chile, from Patagonia to Arica. It consists chiefly of Paleozoic rocks, although on its eastern part, north of 35°S, a very thick volcanic and volcaniclastic sequence with some marine intercalations developed during the Jurassic. North of 21°30'S, the intercalated marine beds have yielded Liassic to Oxfordian fossils, whereas south of 26°S these beds seem to represent only the Lias–Bajocian.

Curepto–Los Vilos

The southernmost outcrops of marine Jurassic in the Coast Cordillera are known from the area of Curepto (c. 35°S; see Fig. 1). In that region, according to Thiele (1965), is a succession of 3,500 m thick shales, sandstones and conglomerates, with interbedded volcanics in its upper part. It has furnished ammonites ranging in age from Hettangian to Toarcian. The strata grade downwards to marine Triassic (see Escobar, 1980).

In the same area but slightly to the north, an Upper Triassic–Lower Jurassic succession has also been reported (Corvalan, 1976). The Lower Jurassic consists of about 930–1,250 m of sandstones and shales and bears fossils of Hettangian–Sinemurian age (see Minato, 1977). It has been included in the Laguna de Tilicura and Rincon de Nuñez Formations.

The classical area is, however, located in the Limache–La Ligua–Los Molles–Los Vilos region (32°–33°S; see Figs. 1, 5). The Jurassic stratigraphy and paleontology are known through the works of Möricke (1894), Fuenzalida (1937), Thomas (1958), Levi (1960), Carter (1963), Cecioni and Westermann (1968), Vicente (1974), Piraces (1976) and Covacevich and Piraces (1976).

The Jurassic section is similar to those known from the Curepto area. Marine beds of Hettangian to Sinemurian age have been assigned to the Los Molles or the Quebrada del Pobre Formation. They consist of 800 m sandstones and shales. The upper part of the formation is interbedded with keratophyres. The Ajial plus Melon Formations which are thousands of meters thick consist mainly of volcanics with sandstones, shales and limestones, follow conformably. This sequence has furnished a Bajocian fauna at several levels. Above are continental volcaniclastic rocks of Early Cretaceous age. The stratigraphic nomenclature and boundaries of this succession have recently been reviewed by Piraces (1976).

Chañaral–Taltal area

Between Chañaral and Taltal (c. 26°S; see Figs. 1, 5) the Jurassic sequence (Cecioni, 1960; Zeil, 1960; Davidson et al., 1976; Davidson and Godoy, 1976; Mercado, 1978; see Table X) begins with grey-yellowish calcareous shales intercalated with grey-greenish fine to coarse sandstones which have been ascribed to the Pan de Azucar Formation. This unit has a thickness of 690 m and has yielded ammonites of Hettangian–Sinemurian (? and Pliensbachian) age (see Cecioni, 1960; Zeil, 1960, 1964). It rests unconformably on the Triassic Cifuncho Formation or on Paleozoic rocks and grades upwards into the Posada Los Tres Hidalgos Formation. The Pan de Azucar Formation seems also to be present north of Taltal at about 24°55'S (see Ferraris, 1978).

The Posada Los Tres Hidalgos Formation is 100 m thick and consists in its lower part of breccias, conglomerates and andesites, and in its upper part of grey-greenish sandstones with intercalated shales and limestones. It has furnished fossil plants. Above follow unconformably andesitic flows with interbedded marine sediments with Bajocian ammonites (see Davidson et al., 1976). The volcaniclastic sequence belongs to La Negra Formation (see Garcia, 1967), a unit formed by thousands of meters of andesites, breccias and tuffs which crop out along the Coast Cordillera for more than 450 km, mostly between Taltal and Iquique (see Fig. 5). To the north and east, the La Negra Formation interfingers with marine Dogger and underlies Tithonian–Neocomian continental sediments (see Ferraris, 1978) but, in some areas, it extends into the Upper Jurassic and Lower Cretaceous (see Davidson et al., 1976). The mineralogy, geochemistry and copper mineralization of this

TABLE X

Stratigraphy of the Chañaral–Taltal area

TIME UNIT	ROCK UNIT	MAX. THICKNESS (m.)	LITHOLOGY	FOSSILS	ENVIRONMENT
UPPER JURASSIC MIDDLE JURASSIC	LA NEGRA FM.	C. 3000	Andesitic flows with intercalated marine sandstones	⑥ ⏳	Volcanism marine to continental
LOWER JURASSIC	POSADA LOS TRES HIDALGOS FM.	100	Grey-greenish sandstones interc. shales and limestones. Breccias, conglomerates, andesites	♤	Marine to Continental
	PAN DE AZUCAR FM.	690	Grey-yellowish calcareous shales intercalated with grey greenish sandstones	⑥	Marine

unit in the Tocopilla area (see Fig. 5) have been studied by Losert (1973) and Palacios (1974).

Iquique–Arica area

The Coast Cordillera between Iquique and Arica in northern Chile has been studied by Biese (1956, 1957), Cecioni and Garcia (1960a, b), Cecioni (1961), Salas et al. (1966), Tobar et al. (1968), Thomas (1970), Silva (1976) and Vila (1976).

The Jurassic strata range from (? Lias)–Bajocian to Oxfordian and were given a number of formational names, mostly by Cecioni and Garcia (1960a, b) and Garcia (1967). Since then the nomenclature has been simplified, so that one set of names can be applied to the stratigraphic succession between Iquique and Pisagua (see Fig. 5) (cf. Thomas, 1970; Silva, 1976), and another set to that around Arica (cf. Salas et al., 1966; Vila, 1976).

Between Iquique and Pisagua (see Table XI) the Jurassic comprises three lithologic units. The lowest unit consists of about 1,000–2,000 m of green marine andesites that have been named the Oficina Viz Formation. The middle unit, i.e. the Caleta Ligate Formation, consists of 500–600 m of glauconitic sandstones, sedimentary breccias, limestones and black shales with a few volcanics intercalated in the lower part. The upper part has furnished fossils of middle Bajocian–Callovian age, but no Bathonian ammonite has been identified. The Caleta Ligate Formation conformably overlies the Oficina Viz Formation. Upwards and laterally it grades into the upper unit, the Guantajaya Formation. The latter consists of about 1,000–1,500 m of dark limestones and sandstones passing upwards into glauconitic sandstones and oolitic limestones with evaporites. The Guantajaya Formation bears middle Bajocian–Late Oxfor-

TABLE XI

Stratigraphy of the Iquique–Pisagua area

TIME UNIT	ROCK UNIT	MAX. THICKNESS (m.)	LITHOLOGY	FOSSILS	ENVIRONMENT
UPPER JURASSIC					
	GUANTAJAYA FM.	1000	Dark and light colored limestones, sandstones, evaporite levels	⚭ ⧖	Marine infraneritic and epineritic
MIDDLE JURASSIC	CALETA LIGATE FM.	600	Green, yellow limestones, sandstones, interc. breccias and andesites	⚭ ⧖	Marine, volcanic
	OFICINA VIZ. FM.	2000	Dark green andesites, volcanic breccias		Marine volcanism
LOWER JURASSIC					

dian fossil invertebrates. The reported presence of Tithonian ammonites (Cecioni, 1961) remains doubtful (see Thomas, 1970; Silva, 1976). Above the Jurassic follow the (?) Cretaceous volcanic rocks of the Punta Barranco Formation or the Lower Miocene–Quaternary Altos de Pica Formation.

Farther north, in the Arica area (see Fig. 11, Table XII), the oldest Jurassic unit is the Camaraca Formation. It comprises 1,900 m of dark andesites with interbedded marine calcareous sandstones, limestones and shales. These levels include an invertebrate fauna with ammonites of late Bajocian–Callovian age. The Camaraca Formation grades upwards into the Los Tarros Formation. The latter consists of about 600 m of dark shales with calcareous concretions intercalated with limestones, quartzites and andesites. It includes Oxfordian–(? early Kimmeridgian) am-

monites and is unconformably overlain by the Lower Cretaceous continental rocks of the Atajaña Formation.

GEOLOGICAL EVOLUTION AND PALEOGEOGRAPHIC SYNTHESIS

General aspects of the paleogeography of the Jurassic of Argentina and Chile have been previously dealt with, among others, by Gerth (1932, 1955), Weaver (1942), Groeber et al. (1953), Harrington (1962), Herrero Ducloux (1963), Cecioni (1964, 1970), Zeil (1964, 1979), Ruiz et al. (1965), Stipanicic (1966), Aubouin and Borrello (1966), Bracaccini (1970), Hillebrandt (1971), Aubouin et al. (1973), Davidson and Vicente (1973), Jensen et al.

TABLE XII

Stratigraphy of the Arica area

TIME UNIT	ROCK UNIT	MAX. THICKNESS (m.)	LITHOLOGY	FOSSILS	ENVIRONMENT
UPPER JURASSIC	LOS TARROS FM.	600	Dark shales , limestones , quartzites andesites	☾	Marine and volcanic
MIDDLE JURASSIC	CAMARACA FM.	1900	Grey to dark andesites , interc. grey calcareous sandstones , limestones , shales	☾	Volcanic and marine
LOWER JURASSIC	?				

(1976), Chotin (1976, 1977), Charrier and Covace-vich (1980), and Zambrano (1980).

Due to the fact that the most important and most studied successions are those of the Principal Cordillera, it would be advisable to subdivide the exposition into two parts according to the two transgressive—regressive sedimentary cycles which characterize that region, i.e. the Cuyan (Hettangian—lower Callovian) and the Lotenian—Chacayan (middle Callovian—Kimmeridgian). In order to simplify the issue, however, the limit between the two parts will be drawn at the Bajocian—Bathonian boundary, i.e. after the maximum extent of the Jurassic sea or before the Tarapaquean and Aconcaguan—Neuquenian Basins began to be delineated (see Figs. 12 and 13).

Lias–Bajocian

Due to the Hercynian diastrophism most of Argentina and Chile remained emergent during Triassic time and underwent extensive erosion and/or continental deposition. At the end of the Triassic and into the beginning of the Jurassic, however, two marine basins began to develop along the western margin of this part of the South American subcontinent: one extending from central-west Argentina to northern Chile and the other in central-west Patagonia (see Fig. 12). Both basins trended NNW—SSE and were arranged *en echélon*, the intervening structurally positive area is the Chubut Dorsal of Aubouin and Borrello (1966), Aubouin et al. (1973) or the Concepcion Land of Cecioni (1964, 1970). At the same time a calcalkaline volcanism began in the west coeval with

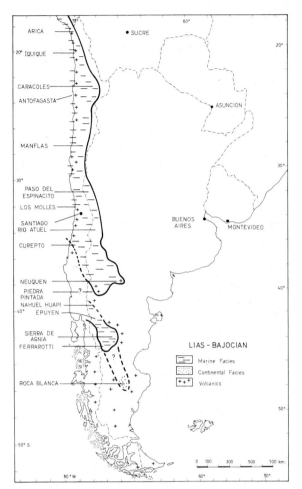

Fig. 12. Paleogeographic map of Argentina and Chile for the Lias—Bajocian.

the last phases of the Hercynian volcanism.

In the northern basin, the Mesozoic sea invaded in the Late Triassic time a restricted area of the central coast of Chile (Fig. 5) between Vallenar (28°30′S) and Curepto (35°S), and began to spread eastwards in Liassic times. During the Hettangian the sea covered the area of the Coast Cordillera from 26°S to 36°S, and the first eastward advance occurred at the latitude of Chañaral—Taltal (see p. 227 et seq.; Figs. 1, 5, 15) where the sea reached the meridian of the Domeyko Cordillera (see Hillebrandt, 1973a; Chong, 1973, 1977).

At the same time thick sequences of conglomerates and sandstones were deposited in the topographic depressions of the continental regions. Those include the El Freno Formation of the Rio Atuel area (see p. 219; Figs. 1, 4, 15) or the Planicie Morada

Formation of eastern Neuquén (see p. 214; Fig. 15).

During the Sinemurian the sea advanced widely over the eastern areas, covering almost the entire Principal Cordillera. However, Liassic rocks seem to be absent north of Taltal (see Fig. 5) in the Coast Cordillera and south of 31°—32°S in the Principal Cordillera of Chile. In the latter area (see Figs. 10, 15), the marine succession begins with Bajocian sediments, although Sinemurian—Pliensbachian are present to the east in Neuquén and Mendoza (see pp. 214, 219). It seems that in this region a ridge remained above sea level throughout the whole Lias (see Davidson and Vicente, 1973). To the east, only a narrow arm of the late Sinemurian—Pliensbachian sea extended in NW—SE direction, lapping against the margin of the Northern Patagonian Massif.

In general, it can be said that at the beginning of the Jurassic the sedimentation was mostly restricted to relatively small and narrow basins originated by fault-controlled subsidence. Thus, the transgressive character of the Liassic sea on the uneven topography of the pre—Jurassic basement (see Bracaccini, 1970) produced an irregular distribution of sediments (see Fig. 15), with marked facies changes even within short distances. Only in Bajocian times existed a major basin, as evidenced by the lateral continuity of the sedimentary units.

North of 31°S the advance of the Toarcian—Bajocian sea was restricted to Chilean territory, where it reached its maximum eastward extent at the same time as in the whole basin. This advance is clearly seen in the eastward overlap, with decreasing thickness, of successively younger units (Hillebrandt, 1973a; Reutter, 1974; Jensen and Vicente, 1977; Cisternas, 1979).

Submarine volcanism lasted in the Coast Cordillera from the Late Triassic times throughout the Early Jurassic. At the end of that epoch, a volcanic ridge or island arc was clearly differentiated in the west, and most marine sedimentation was restricted to a marginal or external basin in the east. The island arc and the marginal basin have been interpreted as the internal (or euliminal) and external (or mioliminal) zones of a Geoliminal Belt, respectively (see Aubouin and Borrello, 1966; Charrier and Vicente, 1972; Aguirre et al., 1974; for other hypotheses see: Ramos, 1978; Digregorio and Uliana, 1980), a belt which became fully developed during the Middle Jurassic.

Most of Patagonia, since the upper Paleozoic—Lower Triassic a continental area, continued to be exposed to erosion throughout Liassic time. Some continental deposits also developed, such as the Roca Blanca Formation in the Deseado Massif (see p. 206) and the Puntudo Alto Formation in central-west Patagonia (see p. 208). In the latter area a shallow (?) late Sinemurian–Pliensbachian sea extended from the northwest and reached its maximum extension during Toarcian times. This central Patagonian embayment was apparently connected with the Pacific Ocean by a northwestern channel. There is, however, no conclusive evidence for such a connection in Chile, although marine sediments of (?) Bajocian age seem to be present near the international boundary between 43°S and 43°45′S. Alternatively, the marine Lias in the Epuyen area (see Fig. 12) and (?) immediately to the south of Nahuel Huapi Lake (González Bonorino and González Bonorino, 1979), may suggest a direct connection with the Neuquén Embayment, with more evidence perhaps concealed under a thick cover of Cenozoic volcanics.

During the Bajocian, continental beds with dinosaur remains were deposited with volcanics in central-west Patagonia. At that time the general uplift continued in the Coast Cordillera of northern Chile, where active volcanism (La Negra Formation and equivalents) changed from submarine to continental. The sediments with Bajocian invertebrates intercalated in these volcanics represent the last known marine record in the Coast Cordillera south of 26°S. The volcanic facies of La Negra Formation spread from the island arc towards the marginal basin in the east-northeast, where it become thinner and interfingered with marine sediments. The absence of late-Middle and Late Jurassic marine sediments in the Coast Cordillera south of 26°S attests to the definitive installation of the volcanic arc as a positive land with subaerial volcanism. Although the island arc separated the Pacific Ocean and the marginal basin, and the principal connection between these two marine domains lay probably only north of 23°–24°S, a number of channels of variable importance must have existed throughout the arc. This also applies to the rest of the Jurassic.

During the Bajocian the transgression reached its maximum extension north of 28°S and south of 31°S, while a regression occurred in the intermediate area from the Toarcian (see Jensen et al., 1976). Bajocian sediments were deposited on Triassic volcanics in the Chilean part of the Principal Cordillera south of 31°S, e.g. Lonquimay and Nacientes del Teno (see Fig. 1;

Chotin, 1970; Davidson and Vicente, 1973). The regressive sequences present in some restricted areas south of 31°S were probably due to the existence of positive relief related to some active volcanos (M. Uliana, personal communication, 1978).

Bathonian–Kimmeridgian (Tithonian)

In central-west Argentina and northern Chile a general regression took place during the Bathonian, although it was not as extensive as previously assumed. Bathonian marine sediments are lacking in most marginal areas and between 26° and 37°S where marine sediments of Callovian (and Oxfordian) age rest directly on Bajocian or even older strata, but to

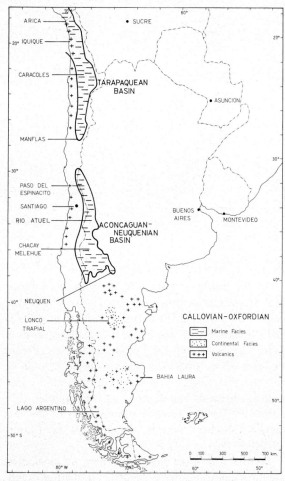

Fig. 13. Paleographic map of Argentina and Chile for the Callovian–Oxfordian. Solid lines around marine facies correspond to generalized boundaries of basins and do not imply absence of connections to the west with the open sea.

the north and south, in areas more centrally located (see Fig. 13) such as Chacay Melehue in northern Neuquén and the northern part of the Domeyko Cordillera a complete or partial Bathonian record exists.

This regression began an irreversible trend that resulted in the delineation of two basins (see Fig. 13), the Tarapaquean to the north and the Aconcaguan–Neuquenian to the south separated by an emergent area, called Antofagasta Land by Cecioni (1964, 1970).

In the Aconcaguan–Neuquenian Basin, this regressive pattern is clearly represented by the northward advance of coarse clastic material, e.g. the Lajas Formation, and the deposition of the gypsum of the Tabanos Formation superposed on finer-grained sediments, e.g. the Los Molles and Chacay Melehue For-

mations. Furthermore, to the south of the Neuquén province the Tabanos Formation is replaced by continental facies (p. 214; Figs. 7, 9).

Although the regressive phase extended into the Callovian, it was immediately followed by renewed transgression. This is clearly evident in the northern part of the Aconcaguan–Neuquenian Basin where the Callovian is transgressive on top of the pre-Jurassic basement (see Fig. 9). To the south evidence of the transgression is given by marine intercalations within the Lotena Formation and the deposition of shales ascribed to the Barda Negra Formation (see Fig. 14). Similarly, in the southern part of the Tarapaquean Basin, the Callovian sea transgressed over areas that during the Bathonian had been above sea level, although it did not extend over the whole region previously covered by the Bajocian sea. Therefore, no record exists of marine Callovian between 28°S and 30°–31°S (see Figs. 9, 13) and at that time the Tarapaquean and Aconcaguan–Neuquenian Basins were definitively separated by the Antofagasta Land. As the sea receded, this positive area became connected with the Coast Cordillera and began to expand to the north and south.

The fact that most of the marginal basin of central-west Argentina and northern Chile became progressively shallower during the Callovian–Oxfordian was due in part to the continuous uplift of the Coast Cordillera which acted as a source of clastic and volcanic material. This process, with deposition from the west and displacement of the central axis of the basin to the east, had its climax during the Late Jurassic when the volcanism reached the Principal Cordillera.

The basins were, however, being filled along their entire margins, including the northern and southern extremities (see Figs. 7, 9). Thus, in the northern sector of the Tarapaquean Basin at the latitude of Iquique (see Fig. 11), the succession became paralic to continental. Only in the Coast Cordillera of Iquique, definite marine conditions persisted for a longer time. To the south in the Domeyko Cordillera, however, small restricted euxinic basins with deposition of gypsiferous and bituminous sediments still persisted. Similar circumstances characterized the Aconcaguan–Neuquenian Basin, which at that time was developing into a closed basin. It is evident, however, from the distribution of Middle Callovian–Oxfordian bio- and lithofacies of this basin (see Fig. 14) that the paleogeographic evolution is far more complex than that outlined here and that a number of

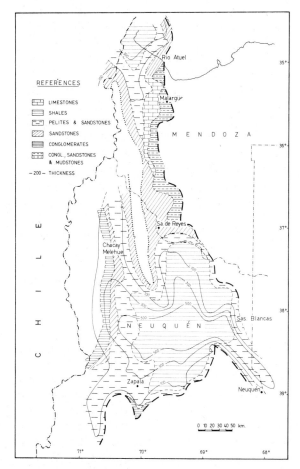

Fig. 14. Isopach and facies distribution of the Lotenian (middle Callovian–Lower Oxfordian) in the Principal Cordillera of Argentina (modified from Dellape et al., 1979).

connections must have existed with the Pacific Ocean.

The general regressive trend sketched above culminated at the end of the Oxfordian or the beginning of the Kimmeridgian with the deposition of thick evaporite sequences in the central part of the basins, and of continental successions on most marginal areas and in Antofagasta Land. This general pattern is clearly present in the southern part of the Aconcaguan–Neuquenian Basin (see Fig. 8) where coarse clastic material coming from the west (Tordillo Formation) and east (Sierras Blancas Formation) interdigitates towards the axial zone of the basin with the evaporites of the Auquilco Formation. During Kimmeridgian time even the evaporites were covered in most areas by terrigenous continental deposits.

Meanwhile, in central and southern Patagonia intensive volcanism occurred throughout most of the Middle and Late Jurassic and produced thick deposits of volcanics and pyroclastics. The Lower Jurassic Central Patagonian embayment was obliterated by these rocks, which spread over most of Patagonia and seem to be composed, in the west of andesites (+ basalts) and to the east by rhyolitic ignimbrites. The Patagonian volcanism was probably related to different causes, e.g. the opening of the South Atlantic, subduction along the western coast of southern South America, and/or the opening of a marginal basin (see Dalziel and Elliot, 1973; Dalziel, 1974; Suarez and Pettigrew, 1976; Barker et al., 1976). Continental sediments, some of them of lacustrine origin, are also represented in central Patagonia. The restriction of the Cañadon Asfalto Formation (see p. 209) to about the same area occupied by the Lower Jurassic marine embayment and the changes in thickness and facies suggest the existence of block faulting subsidence followed by differential erosion of the volcanic basement (M. Uliana, personal communication, 1978).

During the Upper Jurassic large-scale extensional tectonics produced, south of 46°S, a NNW–SSE oriented basin, i.e. the Magallanes Basin. Subsidence, followed by marine transgression, resulted in continued sedimentation throughout the latest Jurassic and Cretaceous (p. 203; Fig. 15). This basin has been interpreted as a typical geosynclinal belt, with marine sedimentary sequences including flysch, ophiolites and high-grade metamorphism. It remained separated from the Geoliminal Belt which lacked those features by the Chubut Dorsal or Concepcion Land (see Aubouin and Borrello, 1966; Vicente, 1970, 1974;

Aubouin et al., 1973; Aguirre et al., 1974). It has also been interpreted as a marginal basin or back-arc marine basin (see Dalziel, 1974; Suarez and Pettigrew, 1976).

During the Tithonian a new transgression, i.e. the Andean sedimentary cycle, began in the Geoliminal Belt of central-west Argentina and northern Chile. This cycle extended well into the Lower Cretaceous, and although in general a repetition of the 'Jurassic Cycle', it was mostly absent in the Coast Cordillera. In the marginal or external basin the Andean is characterized by calcareous and pelitic sediments with scarce volcanics and a rich invertebrate fauna.

TECTONISM

The existence of different diastrophic phases has been pointed out by several authors (see Groeber et al., 1953), and the subject has been dealt with by Stipanicic in a number of papers (see Stipanicic, 1969). A general survey was published by Stipanicic and Rodrigo (1970a, b).

According to these authors, five diastrophic phases occurred between the Late Triassic and the Late Jurassic which they named differently for central-west Argentina, northern Chile and Patagonia. They are, in ascending order: The Rio Atuel (Valparaiso, Austral), the Charahuilla (Atacama, Sureña), the Sierra de Reyes (El Godo, El Molle), the Rio Grande (Maipo, San Jorge), and the Araucana (Chiza, Santa Cruz) phases.

Of all these proposed diastrophic phases, clear evidence exists only for the first and the last. The Rio Atuel diastrophic phase is expressed by an unconformity, sometimes angular, between most Upper Triassic and Lower Jurassic rocks; whereas the Araucana or Araucanian phase produced a general uplift with extensive marine regression at the end of the Oxfordian and deposition of thick sequences of conglomerates during the succeeding Kimmeridgian time. The origin of the Magallanes Basin of southern Patagonia is probably related to this second diastrophic phase.

The other phases were inferred mainly from the apparent hiatuses observed in the ammonite record when compared to the European standard sequence and also from sedimentary discontinuities. Thus, the Charahuilla phase is evidenced by the absence of lower Pliensbachian–Toarcian units in the Chañaral–Taltal area, and of the fauna of these ages in other localities of Argentina and Chile, e.g. Rio Atuel area. The absence of marine index fossils, however, is negative evidence in an area like Chañaral–Taltal where the Upper Lias-Dogger is a thick volcanic sequence with only minor marine intercalations, and cannot be regarded as conclusive. Furthermore, the supposed Pliensbachian hiatus in the Rio Atuel area has been discounted by Hillebrandt (1973a).

The Sierra de Reyes phase is considered the cause of the 'Bathonian hiatus'. This phase should be evident in most marginal areas where the Callovian rests on top of marine Bajocian or even older units, while Bathonian sediments are present in the central parts of the Tarapaquean and Aconcaguan–Neuquenian Basins. However, some uncertainties about the taxonomy of the Bathonian–Callovian Andean macrocephalitid ammonites have some bearing in this problem. Finally, the Rio Grande diastrophic phase supposedly occurred at the Callovian–Oxfordian boundary. Its existence is inferred based on the absence of ammonites indicating the Athleta and Lamberti zones of Europe (see Tables XIV, XV), although the sedimentary sequence seems to be continuous. The existence of an intra-Callovian diastrophic phase separating the Cuyan and Lotenian–Chacayan cycles has also been suggested by Dellape et al. (1979).

IGNEOUS ACTIVITY

The Jurassic volcanism of Argentina and Chile is chiefly developed along the Pacific Coast. It represents the initial magmatism of the Andean orogenic cycle which began simultaneously with, and to the west of, the Jurassic sedimentary basins (see Aguirre et al., 1974; Vicente, 1974). Some Early Jurassic volcanism also occurred within those basins, e.g. Bajo Pobre (see p. 206), Sañico (p. 214).

In the internal zone of the Geoliminal Belt (see p. 231) of central-west Argentina and northern Chile, this volcanism was related to the emplacement of an island arc. It produced thicknesses of thousands of meters of volcanics, ranging in age from the Late Triassic to the Late Jurassic, and consisting of rhyolites, ignimbritic rhyolites, andesitic lavas and pyroclastics. The volcanism at most places was submarine until the Middle Jurassic, but later became subaerial. It spread to the east reaching the external or marginal basin during the Upper Jurassic when andesitic flows were intercalated with continental sediments of Kimmeridgian age.

In Patagonia the Jurassic volcanism is widely present from the Pacific to the Atlantic Ocean. It consists of porphyrites, rhyodacites and andesites and their breccias and tuffs, and was mostly restricted to the Middle and Late Jurassic. The volcanism may have been related to various causes (see p. 234). During the latest Jurassic some minor igneous activity also existed in northeastern Argentina, Uruguay, Paraguay and southern Brazil. It represented the beginning of the volcanic activity that during the Cretaceous would give origin to the extensive basaltic flows of the Serra Geral Formation (see p. 201).

It seems that three maxima of volcanic activity can be recognized, i.e. the Upper Triassic–Lower Jurassic, Middle Jurassic and Upper Jurassic, in coincidence with the Rio Atuel, intra-Callovian and Araucanian diastrophic phases.

In the Upper Jurassic occurred the first Andean plutonic cycle. It was related to the Araucanian diastrophic phase and chiefly restricted to the innermost zone of the Geoliminal Belt (Ruiz et al., 1960; Levi et al., 1963; Charrier and Vicente, 1972; Aguirre et al., 1974; Montecinos, 1979; see also Zeil, 1980, and Zeil et al., 1980).

This plutonism, mainly of adamellitic to tonalitic composition, is also present in the Principal Cordillera. In the Coast Cordillera of northern Chile these rocks, named the Camarones Granodiorite (Garcia, 1967; Chong, 1977), make up a batholith and intrude into the Middle to Upper Jurassic volcanics. Isotopic analysis gave values of 137 ± 4.4 My to 156 ± 4.8 My (see Mercado, 1978). Most other isotopic analyses performed on igneous rocks of the Coast Cordillera confirm a Late Jurassic age, although in some cases Early Jurassic and Early Cretaceous values were also obtained (see Corvalan and Munizaga, 1972; Montecinos, 1979). Evidence of this plutonic cycle has also been found in the Patagonian Cordillera (see Halpern, 1973; Halpern and Carlin, 1971; Toubes and Spikerman, 1974).

PALEONTOLOGICAL FEATURES

Flora

Only in Argentina has the Jurassic flora been studied in detail (see Tables XVI–XVIII). A general survey of the megaflora then known from Argentina, together with a stratigraphic review, was published by

TABLE XIII

Assemblages of Liassic ammonites compared with the European chronozones (assemblage zones and chronozones in roman; modified from Hillebrandt, 1973a, 1981b; Hillebrandt and Schmidt-Effing, 1981)

	EUROPEAN AMMONITE ZONES AND SUBZONES		ARGENTINA AND CHILE	
TOARCIAN	*Dumortieria levesquei*	Pleydellia aalensis	*Pleydellia* cf. *fluitans, Dumortieria* sp.,*Hammatoceras* cf. *porcarellense*	
		Dumortieria moorei	*Pleydellia* cf. *lotharingica, Dumortieria* cf. *pusilla, Hammatoceras* cf. *clavatum, Sphaerocoeloceras brochiiforme*	
		Dumortieria levesquei Phlyseogrammoceras dispansum	*Phlyseogrammoceras tenuicostatum, Hammatoceras insigne, H. speciosum, Sphaerocoeloceras* cf. *brochiiforme* *Phlyseogrammoceras* ex gr. *aalense, Brodieia* cf. *alticarinata*	
	Grammoceras thouarsense		*Phymatoceras copiapense*	
	Hauqia variabilis		*Phymatoceras fabale, P.* ex gr. *lilli*	
	Hildoceras bifrons	Catacoeloceras crassum	*Collina chilensis*	*Peronoceras moerickei, P.* cf. *bolitoense, P.* cf. *planiventer, P.* cf. *crassicostatum, Catacoeloceras* sp.,*Harpoceras* cf. *subexaratum, Phymatoceras* cf. *erbaense, Maconiceras* sp.,*Polyplectus* sp., *Osperlioceras* sp.
				Peronoceras bolitoense, P. cf. *vortex, P.* ex gr. *verticosum, Harpoceras* cf. *subexaratum, Phymatoceras* ex gr. *erbaense, Maconiceras* sp., *Polyplectus* sp., *Frechiella* cf. *helvetica.*
		Peronoceras fibulatum	*Peronoceras pacificum, P.* cf. *verticosum, Maconiceras* sp., *Polyplectus* sp.	
		Dactylioceras commune	*Peronoceras largaense, P.* cf. *subarmatum, P.* cf. *renzi, P.* cf. *desplacei, P.* cf. *choffati, Harpoceras* cf. *chrysanthemum,H.* cf. *falcifer, Mercaticeras* sp.	
	Harpoceras falcifer		*Dactylioceras (Orthodactylites) hoelderi, D. (O.) directum, Nodicoeloceras* cf. *crassoides, Hildaites* cf. *serpentiniformis, H.* cf. *levinsoni, H.* cf. *serpentinus, Harpoceratoides* cf. *alternatus, Eleganticeras* cf. *elegantulum, Tauromeniceras chilense*	
	Dactylioceras tenuicostatum		*Dactylioceras (Orthodactylites) tenuicostatum chilense, D. (O.)* cf. *directum, D. (?O.)* cf. *helianthoides, Nodicoeloceras* sp.*	
			Dactylioceras (Eodactylites) simplex, D. (Orthodactylites) directum, D. (O.) anguinum, Nodicoeloceras cf. *eikenbergi, N.* cf. *pseudosemicelatum, Tauromeniceras* cf. *chakdallaense*	
PLIENS-BACHIAN	Pleuroceras spinatum		*Fanninoceras* cf. *lowrii, Paltarpites* cf. *argutus, Atractites* sp.	
	Amaltheus margaritatus		*Fanninoceras* cf. *lowrii, F. oxyconum. F. disciforme, Arieticeras* cf. *fucinii, A.* cf. *simplex, Reynesocoeloceras (Bettoniceras)* cf. *colubriforme, R. (B.)* cf. *mortilleti, Atractites* sp.	
	Prodactylioceras davoei		*Fanninoceras fannini, Protogrammoceras* ex gr. *normanianum*	
			Fanninoceras behrendseni, Polymorphitinae gen. et sp. indet.	
	Tragophylloceras ibex		*Uptonia* ex gr. *jamesoni, U.* cf. *obsoleta, U.* cf. *angusta, U.* cf. *ignota, Gemmellaroceras* cf. *granuliferum, Tragophylloceras* cf. *wechsleri, Eoamaltheus meridianus*	
	Uptonia jamesoni		*Tropidoceras* cf. *stahli, T.* ex gr. *flandrini*	
SINEMURIAN	Echioceras raricostatum		*Miltoceras* cf. *sellae, Eoderoceras* cf. *pinguecostatum, E.* cf. *unimacula, E.* ex gr. *armatum, Pseudoskirroceras wiedenmayeri, Paltechioceras* cf. *aureolum*	
	Oxynoticeras oxynotum		*Oxynoticeras* cf. *lymense, Cheltonia* cf. *retentum*	
	Asteroceras obtusum		*Epophioceras* cf. *cognitum, Eparietites* cf. *undaries, Oxynoticeras* sp.,*Phylloceras* sp.	
	Caenisites turneri		*Arnioceras* sp.,*Asteroceras* cf. *obtusum*	
	Arnioceras semicostatum		*Arnioceras* cf. *ceratitoides, Angulaticeras* cf. *ventricosum, Agassiceras* sp.	
	Arietites bucklandi		*Megarietites* sp., *Paracoroniceras* cf. *nudaries*	
HET-TAN-GIAN	Schlotheimia angulata		*Schlotheimia* sp.	
	Alsatites liasicus		*Caloceras peruvianum, C. canadense, Alsatites* cf. *platystoma, Ectocentrites* cf. *petersi*	
	Psiloceras planorbis		*Psiloceras planorbis, P. plicatulum*	

TABLE XIV

Assemblages of Middle Jurassic ammonites of Argentina and Chile compared with the European chronozones (assemblage zones and chronozones in roman; information from Hillebrandt, 1970; Westermann and Riccardi, 1979; and unpublished work)

		EUROPE	ARGENTINA AND CHILE
CALLOVIAN	UPPER	Quenstedtoceras lamberti	
	UPPER	Peltoceras athleta	———————————— ? ————————————
	MID	Erymnoceras coronatum	*Reineckeia* spp., *Oxycerites* spp., *Hecticoceras* spp., *Perisphinctes* spp, *Chanasia* sp.
	MID	Kosmoceras jason	
	LOWER	Macrocephalites gracilis	*Eurycephalites vergarensis, E.* spp., *Xenocephalites* spp.
	LOWER		*"Indocephalites" gerthi, Xenocephalites gottschei*
	LOWER	Macrocephalites macrocephalus	*Eurycephalites rotundus, E.* spp., *Xenocephalites* spp., *Neuqueniceras* sp.
	LOWER		*Lilloettia steinmanni, Imlayoceras* spp., *Xenocephalites neuquensis, Bullatimorphites* sp., *Choffatia* sp.
BATHONIAN	UPPER	Clydoniceras discus	*Prohecticoceras* cf. *retrocostatum, Epistrenoceras paracontrarium, Bullatimorphites* sp., *Choffatia* sp., *Parapatoceras* spp.
	UPPER	Prohecticoceras retrocostatum	*? Bullatimorphites* ex gr. *latecentratus, ? Gracilisphinctes* sp.
	MID	Tulites subcontractus	*Eurycephalitinae, ? Procerites* sp.
	L	Zigzagiceras zigzag	
BAJOCIAN	UPPER	Parkinsonia parkinsoni	*? Megasphaeroceras rotundum, ? Eocephalites* sp., *Cadomites* cf. *daubenyi, C.* aff. *deslongchampsi, Leptosphinctes* spp., *Spiroceras orbignyi.*
	UPPER	Garantiana garantiana	
	UPPER	Strenoceras subfurcatum	*Lupherites dehmi, Stephanoceras chilense, Teloceras* aff. *blagdeni, T. crickmayi chacayi.*
	LOWER	Stephanoceras humphriesianum	**Stephanoceras humphriesianum** — *Stephanoceras chilense, S. aricostum caracolense, S.* cf. *allani, Teloceras* aff. *blagdeni, T. crickmayi chacayi* / ? / *Dorsetensia romani, D. liostraca tecta, D.* aff. *deltafalcata, Stephanoceras (S.) pyritosum, S. (Stemmatoceras)* aff. *frechi*
	LOWER	Otoites sauzei	**Emileia giebeli** — *Dorsetensia blancoensis, D. mendozai* / *E. multiformis, E. (Chondromileia) giebeli* s.s., *Stephanoceras (Skirroceras)* cf. *macrum, Chondroceras* spp., *Sonninia alsatica, S. (Papilliceras) espinazitensis* s.s. / *E. giebeli submicrostoma, Pseudotoites sphaeroceroides, P. singularis, P.* spp., *Chondroceras recticostatum, Sonninia (Papilliceras) espinazitensis* s.s.
	LOWER	Witchellia laeviuscula	
	LOWER	Sonninia ovalis	*Pseudotoites singularis, P. sphaeroceroides, Sonninia (Fissilobiceras) zitteli, S. (Papilliceras) espinazitensis altecostata, S. (Euhoploceras) amosi, Bradfordia* aff. *costidensa*
	LOWER	Hyperlioceras discites	*Puchenquia malarguensis Podagrosiceras* spp., *(?) Fontannesia, Sonninia (Euhoploceras) amosi, (?) Eudmetoceras klimakomphalum moerickei*
AALENIAN		Graphoceras concavum	*Zurcheria groeberi Eudmetoceras* cf./aff. *klimakomphalum, E. gerthi, Tmetoceras* spp., *? Podagrosiceras* spp.
		Ludwigia murchisonae	*Planammatoceras* cf. *planinsigne, Bredya* aff. *crassornata, (?) Leioceras opalinum, L. comptum.*
		Leioceras opalinum	

TABLE XV

Assemblages of Upper Jurassic ammonites of Argentina and Chile (information from Hillebrandt, 1970; Stipanicic et al. 1975; Wiedmann 1980a, b)

			EUROPE	ARGENTINA AND CHILE
TITHONIAN	UP-PER		Virgatosphinctes transitorius	*Windhauseniceras internispinosum, V.* ex aff. *transitorius, Wichmanniceras* ex aff. *mirum*
	MID-DLE		Pseudolissoceras zitteli	*Pseudolissoceras zitteli, Aulacosphinctes proximus*
	LOWER		Franconites vimineus	*Virgatosphinctes andesensis, V. choicensis, V.* spp.
			Neochetoceras mucronatum	*Virgatosphinctes pseudolictor, Torquatisphinctes* sp.
			Hybonoticeras hybonotum	
KIMMERIDGIAN	LOWER		Hybonoticeras beckeri	
			Aulacostephanus eudoxus	
			Aspidoceras acanthicum	
			Crussoliceras divisum	
			Ataxioceras hypselocyclum	
			Sutneria platynota	[*Nebrodites pressulus, Idoceras* spp., *Rasenia* sp., *Simaspidoceras* sp.]
OXFORDIAN	UPPER		Idoceras planula	
			Epipeltoceras bimammatum	*Campylites* cf. *mexicanum, Ochetoceras* cf. *hispidum, Decipia* (?) *desertorum, Idoceras* sp., *Euaspidoceras* spp.
			Dichotomoceras bifurcatum	*Discosphinctes* cf. *lucingae, Decipia* (?) *gottschei, Euaspidoceras* cf. *perarmatum*
	MIDDLE		Gregoryceras transversarium	*Peltoceras* sp., *Euaspidoceras* aff. *waageni, Perisphinctes (Kranaosphinctes)* sp., *P. (Prososphinctes)* sp., *P (Arisphinctes)* sp., *Mayaites (Araucanites)* spp.
			Perisphinctes plicatilis	
	LOWER		Cardioceras cordatum	*Gregoryceras* cf. *iteni, Campylites* cf. *henrici*.
			Quenstedtoceras mariae	[*Peltoceras (Parawedekindia) torosum* cf. *fraasi, P. (Peltoceratoides)* cf. *athletoides*]

Stipanicic and Bonetti (1970a, b). Archangelsky (1978) gave an account of the Jurassic megaflora known from Neuquén, Argentina, and information about the Argentina flora was included in the general reviews prepared by Menendez (1969), Archangelsky (1970), Volkheimer and Baldis (1975), and Baldoni (1980). A similarity analysis, using multivariate techniques, was carried out on several taphofloras from Triassic and Lower Jurassic localities of Argentina (Damborenea et al., 1975).

Several changes occurred at the Triassic—Jurassic boundary. The Jurassic megaflora is more cosmopolitan, although the presence of Podocarpaceae and *Athrotaxis* (Toxodiaceae) and the relative scarcity of Ginkgoales indicate differences from Northern Hemisphere floras. *Dicroidium* dies out and the Bennettitales begin to diversify, specially the genus *Otozamites*. Among the ferns, the Dipteridaceae and the genus *Cladophlebis* are quite well represented. The conifers become relatively abundant, mostly in the Middle Jurassic, with genera such as *Elatocladus* and *Brachyphyllum*. As a whole, the Jurassic megaflora

TABLE XVI

Liassic fossil plants described from Argentina (for references see Damborenea et al., 1975)

FOSSILS	FORMATIONS			
	El Cholo	Piedra Pintada	Puntudo Alto + Osta Arena	Roca Blanca
Neocalamites carrerei (Zeiller) Halle	●			
Equisetites frenguellii Orlando		●		
E. patagonica Herbst				●
Marattiopsis munsteri (Goeppert)	●			
Coniopteris leucopetraea Herbst				●
C. meschiana Herbst			●	
Clathropteris obovata Oishi		●		
Dictyophyllum apertum Frenguelli		●		
D. atuelense Herbst	●	●		
D. rothi Frenguelli	●	●		
Thaumatopteris eximia Frenguelli		●		
T. rocablanquensis Herbst				●
Goeppertella macroloba Herbst			●	
G. neuqueniana Herbst		●		
Cladophlebis antarctica Nathorst	●			
C. grahami Frenguelli		●		
C. kurtzi Frenguelli	●			
C. oblonga Halle		●		●
C. pintadensis Herbst		●		
C. ugartei Herbst	●			
Scleropteris vincei Herbst	●	●	●	
Archangelskya protoloxsoma (Kurt.)	●			
Yabeiella mareyesiaca (Geinitz) Oishi	●			
Sagenopteris rhoifolia Presl		●	●	●
Nilssonia princeps (Oldham and Morris) Seward	●			
Otozamites albosaxatilis Herbst			●	●
O. ameghinoi Kurtz		●		
O. barthianus Kurtz		●		
O. bechei Brongniart	●			
O. bunburyanus Zigno		●		
O. chubutensis Herbst			●	
O. hislopi (Oldham) Feistmantel	●		●	
O. sueroi Herbst			●	
Ptylophyllum acutifolium Morris	●			
P. cutchense Morris			●	
Elatocladus conferta (Oldham) Halle	●		●	

shows a gradual change without notable extinctions.

The Jurassic microflora is quite well known, mostly from Neuquén province, Argentina, through a series of papers by Menendez (1968), Volkheimer (1968, 1972, 1974) and Volkheimer and Quatrocchio (1975a, b). A stratigraphic survey of all taxa recorded in Neuquén was published by Volkheimer (1978) and is summarized in Table XVIII. According to Volkheimer (1978) the Lower and Middle Liassic microfloras are clearly different from those of the Upper Triassic and Middle Jurassic. The same distinctive character applies to the Dogger assemblages. All studied Upper Jurassic material, however, comes from Tithonian strata.

TABLE XVII

Middle–Upper Jurassic fossil plants described from Argentina (for references see Stipanicic and Bonetti, 1970b)

FOSSILS	FORMATIONS			
	Los Molles	Cañadón Asfalto + Taquetren	La Matilde	Springhill
Phellinites degiustoi Singer and Archangelsky			•	
Equisetites approximatus Nathorst		•		
Osmundacaulis patagonica (Archangelsky and de la Sota)			•	
Todites williamsoni (Brongniart) Seward	•			
Ruffordia goepperti (Dunker)			•	
Gleichenites juliensis Herbst			•	
G. taquetrensis Herbst and Anzotegui		•		
Clathropteris cfr. *kurtzii* Frenguelli		•		
Thaumatopteris sp.	•	•		
Hausmannia de - ferrariisii Feruglio			•	
Sphenopteris bagualensis Menendez	•			
S. hallei Frenguelli		•		
S. nordenskjoldii Halle		•		
S. sanjulianensis Feruglio			•	
S. patagonica Halle		•	•	
Cladophlebis australis (Morris) Seward			•	
Cladophlebis denticula (Brongniart) Fontaine	•	•		
C. cf. *antarctica* Nathorst		•		
C. grahami Frenguelli		•		
C. patagonica Frenguelli			•	
Scleropteris furcata Halle		•		
Sagenopteris rhoifolia Presl	•	•		
Ticoa magallanica Archangelsky				•
Williamsonia cfr. *gigas* (Lindley and Hutton)		•		
Otozamites sanctae-crucis Feruglio		•	•	•
Ptilophyllum hislopi (Oldham) Seward	•			
P. antarcticum Halle				•
P. patagonicum Berry			•	
Zamites aff. *gigas* Lindley and Hutton				•
Z. pusillus Halle		•		
Cycadolepis sp.				•
Podozamites aff. *gracilis* Arber			•	
Pararaucaria patagonica Wieland			•	
Araucarites sp.				•
Araucarites cutchensis Feistmantel		•		
A. mirabilis (Spegazzini) Florin			•	
A. sanctaecrucis Calder			•	
Masculostrobus altoensis Menendez			•	
Podocarpus (?) *palyssifolia* (Berry) Florin			•	
Elatocladus casamiquelensis Herbst and Anzotegui		•		
E. conferta (Oldham) Halle		•		
E. heterophylla Halle	•			
E. jabalpurensis (Feistmantel) Halle		•		
Athrotaxis ungeri (Halle) Florin		•		
Brachyphyllum ? sp.	•			
B. feistmantelii (Halle)		•		•
Pagiophyllum divaricatum (Bunb.) Seward		•		

TABLE XVIII

Jurassic microflora found in the Neuquén province, Argentina (see Volkheimer, 1978)

	PLIENSBACHIAN	TOARCIAN-AALENIAN	BAJOCIAN	U. BAJOCIAN-L. CALLOVIAN	M.-U. CALLOVIAN	TITHONIAN
Auritulinasporites sp. A	•					
Biretisporites sp.	•				•	•
Concavisporites spp.	•					
C. laticrassus Volkheimer			•			
C. semiangulatus Menéndez				•	•	
Deltoidospora sp. A						•
D. australis (Couper) Pocock		•	•	•	•	
D. minor (Couper) Pocock	•		•	•	•	
Dictyophyllidites sp. A		•				
D. mortoni de Jersey, Playford & Dettmann			•			
Hymenophyllidites mortoni (de Jersey)			•			
Hymenophyllumsporites cf. deltoidus Rouse			•			
H. grandis Volkheimer			•			
Todisporites major Couper	•		•	•	•	
T. minor Couper				•	•	•
Divisisporites sp. A			•			
D. cf. maximus Pflug			•			
Stereisporites sp.						
Calamospora mesozoica Couper			•			
Ceratosporites spp.	•					
Granulatisporites spp.			•			
Rugulatisporites sp. A	•					
R. neuquenensis Volkheimer			•			
Osmundacidites sp. A	•					
O. araucanus Volkheimer			•			
O. diazii Volkheimer				•	•	
Trilites densiverrucosus Menéndez			•			
Apiculatisporites charahuillaensis Volkheimer			•			
Verrucosisporites spp.						
V. cf. opimus Manum			•			
V. varians Volkheimer	•			•	•	
V. cf. walloonensis de Jersey			•			
Leptolepidites cf. macroverrucosus Schulz						
Anapiculatisporites cf. dawsonensis Reiser and Williams						
Hamulatisporis sp. A						
Uvaesporites minimus Volkheimer			•			
Baculatisporites bagualensis Volkheimer			•			
B. tenuis Volkheimer			•			
Lycopodiumsporites austroclavatidites (Cookson)	•		•	•	•	•
L. semimurus (Danzé - Corsin and Laveine)	•					
Foveotriletes microfoveolatus Menéndez			•			
Staplinisporites caminus (Balme)				•	•	
Appendicisporites sp. A						
Ischyosporites cf. crateris Balme	•					
I. marburgensis de Jersey		•				
I. labiatus Volkheimer		•				
I. pachydictyus Menéndez			•			
I. volkheimeri Filatoff				•	•	
Klukisporites variegatus Couper				•	•	
Trilobosporites sp. A						
Nevesisporites sp.	•					
N. vallatus de Jersey and Paten	•					
Gleicheniidites spp.	•					
G. argentinus Volkheimer				•	•	
Interulobites variabilis Volkheimer and Qattrocchio						
Duplexisporites spp.						
Polycingulatisporites reduncus (Bolkhovitina)						
P. sp. A						
Taurocusporites sp.						
T. chlonovae Döring		•	•			
Contignisporites cooksonii (Balme)						
Polypodiaceoiporites neuquenensis Volkheimer			•			
Antulsporites spp.			•	•	•	
Punctatosporites (?) sp. A						
Aratrisporites (?) sp.	•					
Marattisporites scabratus Couper	•		•	•	•	
Cirratriradites minor Volkheimer	•	•				
Peromonolites ? sp.	•			•	•	
P. pehuenche Volkheimer	•			•	•	
Laevigatosporites sp.				•	•	
Callialasporites dampieri (Balme)		•	•	•	•	
C. segmentatus (Balme)		•	•	•	•	•

	PLIENSBACHIAN	TOARCIAN-AALENIAN	BAJOCIAN	U. BAJOCIAN-L. CALLOVIAN	M.-U. CALLOVIAN	TITHONIAN
C. trilobatus (Balme)						•
Tenuisaccites sp.				•	•	•
Alisporites spp.		•		•	•	•
A. cf. robustus Nilsson	•	•		•	•	•
Podocarpidites cf. ellipticus Cookson	•	•		•	•	•
P. verrucosus Volkheimer				•	•	•
Vitreisporites sp.						•
V. pallidus (Reissinger)	•	•	•	•	•	•
Phrixipollenites sp. A						•
Phrixipollenites cf. otagoensis (Couper)				•		
Microcachryidites antarcticus Cookson				•		
Microcachryidites castellanosii Menéndez				•	•	
Trisaccites microsaccatus (Couper)				•	•	•
Dacrycarpites cf. australiensis Cookson and Pike						•
Laricoidites cf. desquematus Goubin				•		
Inaperturopollenites indicus Srivastava				•		
I. microgranulatus Volkheimer					•	
I. turbatus Balme		•			•	•
I. velatus Volkheimer		•				
I. sp.		•			•	
Araucariacites australis Cookson	•			•	•	•
A. fissus Reiser and Williams	•					
A. cf. ghoshii Srivastava				•		
A. pergranulatus Volkheimer				•	•	
Inapertisporites sp.						•
Classopollis cf. classoides Pflug	•			•	•	•
C. intrareticulatus Volkheimer	•	•		•	•	•
C. itunensis Pocock				•		
C. major Groot and Groot						•
C. simplex (Danzé-Corsin and Laveine)	•	•	•	•	•	•
C. sp. A	•					
Perinopollenites sp.						•
P. ? elatoides Couper	•	•				•
Exesipollenites tumulus Balme						•
Cycadopites adjectus de Jersey						•
C. granulatus (de Jersey)	•	•		•	•	•
C. cf. deterius Balme					•	•
C. nitidus (Balme)	•	•		•	•	•
C. punctatus Volkheimer				•	•	
C. cf. tectus (Nilsson)				•		
Cycadaceaelagella nana Volkheimer	•					
Monosulcites angustus Jain	•					
Monosulcites carpentieri Delcourt and Sprumont						•
M. aff. minimus Cookson				•		
M. subgranulosus Couper						•
Equisetosporites sp.						•
E. caichiguensis Volkheimer and Quattrocchio				•	•	
E. menendezii Volkheimer				•	•	
Lagenella sp.						•
Eucommiidites cf. minor Groot and Penny						•
Gonyaulacysta cf. jurassica (Delfl.)						•
Microdinium sp. A						•
Pareodinia cf. ceratophora Deflandre						•
Baltisphaeridium spp.						•
B. cf. debilispirum Wall and Downie	•					•
Campenia austroamericana Volkheimer	•	•				•
Lancettopsis sp.	•	•				•
Pterospermella spp.	•					•
P. cf. goslarensis Mädler						•
Pleurozonaria picunensis Quattrocchio						•
P. suevica (Eisenack)		•				•
Pterosphaeridia volkheimeri Quattrocchio						•
Comasphaeridium sp.						•
Filisphaeridium sp.						•
Hyalosphaera sp.						•
Leiosphaeridia sp. A						•
Leiosphaeridia cf. hyalina (Deflandre)					•	•
L. cf. staplinii Pocock				•		•
Cymatiosphaera sp. A						•

Invertebrates

As in the case of the flora so most studies of Jurassic invertebrates have been carried out on Argentinian material, particularly from the Neuquén and Mendoza provinces. A list, including references, of the described and figured taxa found in Neuquén has been published by Camacho and Riccardi (1978; see Table XIX). These faunas can be considered representative for most of the Jurassic of Argentina and Chile.

The first important paper was by Gottsche (1878), who studied the Dogger material collected by Stelzner (1873) at the Espinacito Pass. Subsequent general studies on the Jurassic invertebrates were published by Philippi (1899), Burckhardt (1900, 1903) and Weaver (1931), the last publication being an outstanding contribution to the knowledge of the fauna of all the Mesozoic of the Argentinian Principal Cordillera. Specialized studies were also carried out, on the ammonite biostratigraphy and paleobiogeography (see Tables XIII–XV), mainly of Chile, by Hillebrandt (1970, 1979), trigonids by Lambert (1944), Perez and Reyes (1977), and Reyes and Perez (1979, 1980), the nautiloids by Cecioni (1963), Early Jurassic brachiopods by Manceñido (1978, 1981), and the Middle–Late Jurassic aptychi by Closs (1962). Other papers are more limited stratigraphically. Most of them deal chiefly with the ammonites, although sometimes other groups are also included.

The Liassic invertebrate fauna is poorly known. The Triassic–Jurassic transition present in Chile is mostly known through the papers of Fuenzalida (1937), Cecioni and Westermann (1968), and Escobar (1980). With the exception of Jaworski (1914, 1915, 1925, 1926a, b), Rigal (1930), Hillebrandt (1973b, 1979), Blasco et al. (1979, 1980) and Hillebrandt and Schmidt-Effing (1981) who studied some Liassic ammonites (Table XIII; see also Minato, 1977), other papers dealt with the rich bivalve fauna (Burckhardt, 1902; Jaworski, 1915; Leanza, 1942; Wahnish, 1942; Covacevich and Escobar, 1979; Damborenea and Manceñido, 1979). The affinities of the Liassic invertebrates (mostly cosmopolitan) are largely with those from the Tethys, although there are some absences and variation in the relative abundance of taxa (see Hillebrandt, 1970, 1979, 1980, 1981a).

Dogger invertebrates, mostly ammonites (Table XIV), are known through several works. Gottsche (1878) and Tornquist (1898) described the fauna from the Espinacito Pass; Steinmann (1881) and

Möricke (1894) studied material from Chile, Stehn (1923) from Chacay Melehue and Caracoles, and Leanza (in Herrero Ducloux and Leanza, 1943) from southern Neuquén and Mendoza. In the last decade, new studies on a regional scale have been carried out by Westermann and Riccardi (see Westermann, 1967; Westermann and Riccardi, 1972a, b, 1975, 1979, 1980), and Hillebrandt (1970, 1977). The ammonite fauna exhibits close similarities to that from the western Tethys (see Westermann and Riccardi, 1976).

Oxfordian invertebrates are poorly known, although some ammonites have been described by Steinmann (1881), Stehn (1923), Leanza (1947b), Stipanicic (1951) and Westermann and Riccardi (in Stipanicic et al., 1975). A decapod crustacea was figured by Chong and Förster (1976). Affinities for the ammonites are clearly Tethyan and the first direct connection with the Indo-Malagasy province is indicated. The Kimmeridgian has previously believed to be represented by some ammonites described by Leanza (1947a) but this has been rejected (see Dellape et al., 1979). Beds of that age seem, however, to be present in Chile (Chong, 1977). A great deal of work has been done on the Tithonian and Lower Cretaceous fauna, probably due to its abundance and good preservation. Papers, mostly on ammonites (Table XV), have been published by Behrendsen (1891–92), Steuer (1897), Haupt (1907), Douville (1910), Krantz (1928), Feruglio (1936–37), Leanza (1945, 1968), Indans (1954), Corvalan (1959), Biro (1980a, b) and Leanza (1980). The most important zonal fossils in Chile were listed by Corvalan and Perez (1958), and the ammonite zonal sequence has been dealt with by Leanza (1947c), Leanza and Hugo (1978) and Wiedmann (1980a, b). The Tithonian ammonite fauna is somewhat differentiated from those of other parts of the world, but its affinities are with the Himalayan and Mediterranean provinces (see Enay, 1972; Wiedmann, 1980a), respectively, in the south and north.

Foraminiferids and ostracods have also been recorded from the Callovian of Neuquén (see Musacchio, 1978, 1980), and the Upper Jurassic of southern Patagonia (see Sigal et al., 1970; Natland et al., 1974).

Vertebrates

The Jurassic vertebrates described and figured from Argentina and Chile are listed on Table XX,

TABLE XIX

Invertebrates (exclusive of ammonites) described from the Neuquén province, Argentina (for references see Camacho and Riccardi, 1978)

LIAS	DOGGER
CORALS *Cyathophora decamera* Gerth *Andenipora liasica* Gerth	**CORALS** *Stylina weaveri* (Gerth) *Montlivaltia delabechii* Edwards *andina* Gerth *Convexastraea* sp.
BRACHIOPODS *Spiriferina rostrata* (Schlotheim) *Lobothyris punctata* (Sowerby)	**BRACHIOPODS** *Lingula beanii* Phillips *Rhynchonella pinguis* Roemer *pectunculoides* (Etalon)
BIVALVES *Palaeoneilo patagonidica* (Leanza) *Cucullaea jaworskii* Leanza *Cucullaea rothi* Leanza *Grammatodon costulatus* (Leanza) *Mytilus scalprum* Bayle and Coquand *Modiolus scalprus* Sowerby *Modiolus gigantoides* Leanza *Gervillia? turgida* Leanza *Gervillaria pallas* (Leanza) *Inoceramus apollo* Leanza *Isognomon isognomonoides* (Stahl) *Isognomon jupiter* (Leanza) *Oxytoma (Oxytoma) inequivalvis* (Sowerby) *Entolium disciformis* (Schuebler) *Entolium* cf. *hehlii* (d'Orbigny) *Propeamussium (Parvamussium) coloradoensis* (Weaver) *Camptonectes lens* (Sowerby) *Pectinula cancellata* Leanza *Chlamys (Chlamys) textoria* (Schlotheim) *Chlamys (Chlamys) textoria* (Schlotheim) *torulosa* Quenstedt *Weyla (Weyla) alata* (v. Buch) *Weyla bodenbenderi* (Behrendsen) *Antiquilima succincta* (Schlotheim) *Limea duplicata* Sowerby *Ctenostreon paucicostatum* Leanza *Gryphaea darwini* Forbes *Lopha longistriata* (Jaworski) *Trigonia catanlilensis* Weaver *Trigonia (Frenguelliella) tapiai* (Lambert) *Trigonia (Frenguelliella) inexpectata* (Jaworski) *Myophorella araucana* (Leanza) *Jaworskiella burckhardti* (Jaworski) *Myophorigonia neuquensis* (Groeber) *Lucina payllalefi* Leanza *Lucina huayquimili* Leanza *Myoconcha neuquena* Leanza *Myoconcha neuquena torulosa* Leanza *Neocrassina aureliae* (Feruglio) *Cardinia andium* (Giebel) *Cardinia* cf. *andium* (Giebel) *Cardinia densestriata* Jaworski *Jurassicardium? asaphum* (Leanza) *Arctica?* sp. *Isocyprina ancatruzi* (Leanza) *Pholadomya corrugata* Koch and Dunker *Pholadomya* cf. *fortunata* Dumortier *Pholadomya* cf. *hemicardia* Roemer *Pholadomya* cf. *plagemanni* Möricke *Homomya neuquena* Leanza *Pachymya (Arcomya?) rotundocaudata* Leanza *Pleuromya striatula* Agassiz	**BIVALVES** *Palaeonucula jaworskii* (Weaver) *Palaeonucula leufuensis* (Weaver) *Isoarca* sp. *Modiolus mollensis* Weaver *Modiolus imbricatus* (Sowerby) *Gervillia leufuensis* Weaver *Isognomon nanus* (Behrendsen) *Inoceramus galoi* Boehm *Entolium disciformis* (Schuebler) *Bositra ornati* (Quenstedt) *Lima laeviuscula* Sowerby *Ctenostreon chilense* Philippi *Ctenostreon neuquensis* Weaver *Ostrea* cf. *roemeri* Quenstedt *Gryphaea neuquensis* Weaver *Gryphaea calceola* Quenstedt *Gryphaea leufuensis* Weaver *Trigonia stelzneri* Gottsche *Trigonia cassiope* d'Orbigny *Trigonia mollesensis* Lambert *Trigonia radixscripta* Lambert *Trigonia densestriata* Behrendsen *Trigonia chacaicoensis* Lambert *Trigonia corderoi* Lambert *Vaugonia covuncoensis* (Lambert) *Myophorella leanzai* (Lambert) *Lucina laevis* Gottsche *Astarte subtetragona* (Münster) *Astarte puelmae* Steinmann *Opis (Trigonopsis) similis* Sowerby *Protocardia striatula* (Sowerby?) Phillips *Tellina leufuensis* Weaver *Pholadomya plagemanni* Möricke *Pholadomya fidicula* (Sowerby) *Homomya gracilis* Agassiz *Pleuromya* cf. *alduini* (Brongniart) *Pleuromya gottschei* Behrendsen
GASTROPODS *Lithotrochus humboldti* v. Buch *Natica catanlilensis* Weaver	**MALM** **BRACHIOPODS** *Lacunosella* cf. *arolica* (Oppel) **BIVALVES** *Cucullaea securis* Leymerie *Cucullaea lotenoensis* Weaver *Ostrea roemeri* Quenstedt *Ostrea lotenoensis* Weaver *Retroceramus? retrorsus* (Keyserling) *Inoceramus backlundi* Sokolov *Inoceramus argentinus* Sokolov *Inoceramus* cf. *galoi* Boehm *Pecten* aff. *rypheus* d'Orbigny *Buchia* aff. *scythica* (Sokolov) *Buchia fischeri* (d'Orbigny) "*Plicatula*" *vacaensis* Weaver *Anditrigonia carrincurensis* (Leanza) *Anditrigonia frenguellii* (Mariñelarena) *Steinmanella splendida* (Leanza) *Trigonia mirandaensis* Lambert *Vaugonia pichimoncolensis* (Lambert) *Eriphyla argentina* Burckhardt *picunleufuensis* Weaver *Astarte* aff. *reginae* Loriol **GASTROPODS** *Capulus argentinus* Haupt

TABLE XX

Jurassic vertebrates described and/or figured from Argentina (A) and Chile (CH) (courtesy of Z.B. de Gasparini)

VERTEBRATES	LIAS A	LIAS CH	DOGGER A	DOGGER CH	MALM A	MALM CH	REFERENCES
OSTEICHTHYES							
Lepidotes cf. *L. maximus* Wagner					•		Weaver, 1931
Pholidophorus argentinus Sáez					•		Sáez, 1939; Bocchino (in revision)
Pholidophorus domeykanus Arratia, Chang & Chong						•	Arratia et. al., 1975 a
Leptolepis australis Sáez					•		Sáez, 1939; Bocchino (in revision)
Leptolepis patagonicus Sáez					•		Sáez, 1940 a; Bocchino 1967 a
Leptolepis argentinus Sáez					•		Sáez, 1940 a; Bocchino (in revision)
Leptolepis sprattiformis Saint - Seine					•		Sáez, 1939; Saint - Seine, 1949
Leptolepis dubius ? (Blainv.)					•		Sáez, 1939; Bocchino (in revision)
Leptolepis opercularis Arratia, Chang & Chong						•	Arratia, et. al., 1975 b
Neolycoptera gracilis Sáez					•		Sáez, 1939; Bocchino (in revision)
Notodectes argentinus Sáez					•		Sáez, 1949; Bocchino (in revision)
Luisiella inexcutata Bocchino					•		Bocchino, 1967 b
Protoclupea chilensis Arratia, Chang & Chong						•	Arratia, et. al., 1975 c
? Bunoderma baini Sáez					•		Sáez, 1940 b; Bocchino, 1967 a
AMPHIBIA							
Vieraella herbstii Reig	•						Reig, 1961; Casamiquela, 1965, 1970
Notobatrachus degiustoi Reig			•				Reig in Stipanicic & Reig, 1955; Reig in Stipanicic & Reig, 1957; Casamiquela, 1961 a, 1970
REPTILIA							
Teleosauridae	•	•					Huene, 1927; Gasparini, 1973 a; Gasparini & Chong (in preparation)
Metriorhynchus casamiquelai Gasparini & Chong				•			Gasparini & Chong, 1977
?Purranisaurus potens Rusconi					•		Rusconi, 1948 a; Gasparini, 1973 b
Geosaurus araucanensis Gasparini & Dellapé					•		Gasparini & Dellapé, 1976; Gasparini, 1978
Notoemys laticentralis Cattoi & Freiberg					•		Cattoi & Freiberg, 1961; Wood & Freiberg, 1977
Thalassemyidae					•		Pascual et al., 1978
Pterodaustro guiñazui Bonaparte						?•	Chong & Gasparini, 1976; Chong, 1976; Casamiquela & Chong, 1980
Wildeichnus navesi Casamiquela			•				Casamiquela, 1964, 1966, 1975 c
Sarmientichnus scaglai Casamiquela			•				Casamiquela, 1964, 1975 c
Delatorrichnus goyenechei Casamiquela			•				Casamiquela, 1964, 1966, 1975 c
Herbstosaurus pigmaeus Casamiquela					?	•	Casamiquela, 1975 c; Pascual et. al., 1978
Amygdalodon patagonicus Cabrera						•	Cabrera, 1947, 1948; Casamiquela, 1960, 1963; Bonaparte, 1980
Theropoda						•	Galli & Dingman, 1962; Dingman & Galli, 1965
Iguanodonichnus frenkii Casamiquela						•	Casamiquela in Casamiquela & Fasola, 1968, Chong & Gasparini, 1976
? Stegosauria						•	Galli & Dingman, 1962; Dingman & Galli, 1965
Archosauria indet.			•				Casamiquela, 1964
Sauria	•						Casamiquela, 1962; Pascual et. al., 1978
Prolacerta patagonica Casamiquela	•						Casamiquela, 1975 a
Plesiosauria		•					Burmeister & Giebel, 1861; Huene, 1927; Chong & Gasparini, 1976
Plesiosauria				•			Biese, 1961; Chong & Gasparini, 1976
Plesiosauria					•		Rusconi, 1943, 1967; Casamiquela, 1970
Ichthyosauria	•	•					Burmeister & Giebel, 1861; Huene, 1927; Casamiquela, 1970; Chong & Gasparini, 1976
Ichthyosauria						•	Chong & Gasparini, 1976; Gasparini & Chong (in preparation)
Ichthyosauria	•		?•			•	Dames, 1893; Huene, 1927; Weaver, 1931; Rusconi, 1948 b, 1949; Pascual et. al., 1978
? Stenopterygius grandis Cabrera			•				Cabrera, 1939; Pascual et al., 1978
Macropterygius ? sp.					?•		Rusconi, 1948 b; Pascual et. al., 1978
? Ophthalmosaurus mendozana (Rusconi)					•		Rusconi, 1940, 1942, 1948, 1967; McGowan, 1972
MAMMALIA							
Ameghinichnus patagonicus Casamiquela			•				Casamiquela, 1960, 1961 a, 1961 b, 1964, 1975 b
Vertebrata indet.			•				Casamiquela, 1964

where all pertinent literature has also been included. Some additional comments seem, nevertheless, relevant. The fish record will be modified in the near future due to new research being carried out on both Argentinian and Chilean material (e.g. Arratia and Chong, 1979). The anuran amphibians, which occur in very well preserved material in Argentina, are the oldest known to date in the world. Terrestrial reptiles which were dominant in Argentina are reliably known only with one sauropod, i.e. *Amygdalodon patagonicus* Cabrera (1947). Evidence for other coelurid dinosaurs and enigmatic saurids is based on scarce material and tracks. The record of marine reptiles is more abundant in Chile but, like the Argentinian, it has not been studied. The ichthyosaurs need revision and the study of crocodilean material is relatively recent. The probable presence of mammal tracks has also been reported from Argentina. A general review on the Jurassic stratigraphy of Argentina in relation to the vertebrates has been published by Pascual et al. (1969), and the Mesozoic tetrapods of the whole South America have been dealt with by Bonaparte (1978, 1979).

CLIMATE

Some general aspects of the Jurassic climate of Argentina have been dealt with in a number of papers by Volkheimer (e.g. 1970a, b). Similar evidence was also considered by Bowen (1961, 1963) who performed several O^{18}/O^{16} paleotemperature analyses on belemnites from the Neuquén and Santa Cruz provinces, Argentina.

In the Lias, the existence of thin coal seems intercalated with clastic rocks, together with the scarcity of limestones, suggests a temperate and humid climate. An increase in temperature during the Dogger is inferred from the prevalence of limestones and the existence of sauropods even in Patagonia, e.g. *Amygdalodon patagonicus* Cabrera. The presence of gypsum and thin coal seams in central-west Argentina and northern Chile also attest to arid and tropical or subtropical conditions. The O^{18}/O^{16} isotope analyses on belemnites from Neuquén indicate that the Late Lias and Middle Bajocian water temperatures were similar, although they were more variable in the Lias. The values obtained range respectively from $16.6°C$ to $29.6°C$ and from $19.7°C$ to $28.6°C$.

The abundant limestone and gypsum and the existence of bioherms in the Oxfordian suggest increasingly warm, dry and arid conditions. The fossil flora indicates that the climate was quite equitable over extensive areas. Only in Patagonia there may have existed a short-lived interval with humid conditions (see Volkheimer, 1970b) as deduced for the Cañadon Asfalto flora. The Upper Jurassic sea was also warm according to the presence of crocodiles and ichthyosaurs. Water temperatures for the Tithonian–Berriasian of southern Patagonia ranged from $23.7°C$ to $25.7°C$ (Bowen, 1961).

All evidence shows that the Jurassic of this region 'was a period which though characterized by climatic differentiation and seasonal change nevertheless was warmer than later times' (Bowen, 1963).

ACKNOWLEDGEMENTS

I am greatly indebted to Prof. G.E.G. Westermann, McMaster University, Canada, and Dr. M. Uliana, Petrobras, Brazil, for valuable criticism, to Dra. Z.B. de Gasparini, La Plata University, Argentina, for providing all included information on Jurassic vertebrates, to Dr. G. Cecioni, Chile University, for the photographs of Chilean stratigraphic sections and to Prof. A.E.M. Nairn, University of South Carolina, and Prof. J. Mills-Westermann, McMaster University, for help with the English text.

REFERENCES

Aguirre, L., 1960. Geología de los Andes de Chile Central, Provincia de Aconcagua. *Chil. Inst. Invest. Geol., Bol.*, 9: 5–70.

Aguirre, L., Charrier, R., Davidson, J., Mpodozis, A., Rivano, S., Thiele, R., Tidy, E., Vergara, M. and Vicente, J.C., 1974. Andean magmatism: its paleogeographic and structural setting in the central part ($30°–35°S$) of the southern Andes. *Pac. Geol.*, 8: 1–38.

Amaral, G., Cordani, V.G., Kawashita, K. and Reynolds, J.H., 1966. Potassium-argon dates of basaltic rocks from southern Brasil. *Geochim. Cosmochim. Acta*, 30: 155–189.

Archangelsky, S., 1970. Fundamentos de paleobotanica. *Univ. Nac. La Plata, Fac. Cienc. Nat. Mus., Ser. Tec. Didact.*, 10: 1–347.

Archangelsky, S., 1977. Vegetales fósiles de la Formación Springhill, Cretácico, en el Subsuelo de la Cuenca Magallánica, Chile. *Ameghiniana*, 13 (2): 141–158.

Archangelsky, S., 1978. Megafloras Fósiles. In: *Relatorio Geología y Recursos Naturales del Neuquén, Septimo Congr. Geol. Argent.*, pp. 187–192.

Arkell, W.J., 1956. *Jurassic Geology of the World*. Oliver and Boyd, Edinburgh, 806 pp.

Arratia, G. and Chong, G., 1979. Los Peces Jurásicos de Chile. *Segundo Congr. Geol. Chileno, Actas*, 3: H125–H144.

Arratia, G., Chang, S. and Chong, G., 1975a. *Pholidophorus domeykanus* n.sp. del Jurásico de Chile. *Rev. Geol. Chile*, 2: 1–9.

Arratia, G., Chang, S. and Chong, G., 1975b. *Leptolepis opercularis* n.sp., from the Jurassic of Chile. *Ameghiniana*, 12 (4): 350–358.

Arratia, G., Chang, S. and Chong, G., 1975c. Sobre un pez fósil del Jurásico de Chile y sus posibles relaciones con clupeidos americanos vivientes. *Rev. Geol. Chile*, 2: 10–21.

Aubouin, J. and Borrello, A.V., 1966. Chaines Andines et Chaines Alpines: Regard sur la géologie de la Cordillère des Andes au parallèle de l'Argentine Moyenne. *Soc. Geol. Fr., Bull.*, 7 (8): 1050–1070.

Aubouin, J., Borrello, A.V., Cecioni, G., Charrier, R., Chotin, P., Frutos, J., Thiele R. and Vicente, J.C., 1973. Esquisse Paléogéographique et structurale des Andes Méridionales. *Rev. Geogr. Phys. Geol. Dyn.*, (2) XV (1–2): 11–72.

Baldoni, A.M., 1980. Análisis de algunas Tafofloras Jurásicas y Eocretácicas de Argentina y Chile. *Segundo Congr. Argent. Paleontol. Bioestratigr., Buenos Aires, 1978, Actas*, V, pp. 41–65.

Barker P. et al., 1976. Evolution of the southwestern Atlantic Ocean Basin: Results of Leg 36, Deep Sea Drill Project. *Initial Rep. DSDP*, 36: 993–1014.

Bayle, E. and Coquand, H., 1851. Mémoire sur les fossiles secondaires recueillis dans le Chili par M. Ignace Domeyko et sur les terrains auxquels ils appartiennent. *Soc. Geol. Fr. Mem.*, (1) 2 (4): 47 pp.

Behrendsen, O., 1891–92. Zur Geologie des Ostabhanges der argentinischen Codillere. *Z. Dtsch. Geol. Ges.*, 43 (1891): 369–420; 44 (1892): 1–42.

Biese, W., 1956. Zur Verbreitung des Jura in chilenischen der andiner Geosynclinale. *Geol. Rundsch.*, 45: 877–919.

Biese, W., 1957. Der Jura von Cerritos Bayos, Calama. *Geol. Jahrb.*, 72: 439–485.

Biese, W., 1961. El Jurásico de Cerritos Bayos. *Univ. Chile, Fac. Cienc. Fis. Mat., Inst. Geol. Publ.*, 61: 7–61.

Biro, L., 1980a. Algunos ammonites nuevos en la Formación Lo Valdés, Titoniano–Neocomiano, Provincia de Santiago (33°50' lat. Sur), Chile. *II Congr. Argent. Paleontol. Bioestr., Buenos Aires*, 1978, I, pp. 223–235.

Biro, L., 1980b. Estudio sobre el límite entre el Titoniano y el Neocomiano en la Formación Lo Valdés, Provincia de Santiago (33°50' lat. Sur) Chile, Principalmente sobre la base de Ammonoideos. *Segundo Congr. Argent. Paleontol. Bioestr., Buenos Aires, 1978, Actas*, V, pp. 137–152.

Blasco, G., Levy, R. and Nullo, F., 1979. Los amonites de la Formación Osta Arena (Liásico) y su posición estratigráfica-Pampa de Agnia (Provincia del Chubut). *Septimo Congr. Geol. Argent., Actas*, II, pp. 407–429.

Blasco, G., Levy, R. and Ploszkiewicz, V., 1980. Las Calizas Toarcianas de Loncopan, Depto. Tehuelches, Provincia

del Chubut. *II Congr. Argent. Paleontol. Bioestr., Buenos Aires, 1978*, I, pp. 191–200.

Blasco, G., Nullo, F. and Proserpio, C., 1979. *Aspidoceras* en Cuenca Austral, Lago Argentino, Provincia de Santa Cruz. *Asoc. Geol. Argent., Rev.*, 34 (4): 282–293.

Bocchino, A., 1967a. Designación de Lectotipos de peces fósiles argentinos y observaciones críticas. *Ameghiniana*, 5 (5): 179–180.

Bocchino, A., 1967b. *Luisiella inexcutata* gen. et sp. nov. (Pisces, Clupeiformes, Dussumieriidae) del Jurásico Superior de la Provincia de Chubut, Argentina. *Ameghiniana*, 5 (2): 91–100.

Bodenbender, G., 1892. Sobre el terreno Jurásico y Cretácico en los Andes Argentinos entre el río Diamante y el río Limay. *Acad. Nac. Cienc. Cba., Bol.*, 13: 5–44.

Bonaparte, J.F., 1978. El Mesozoico de America del Sur y sus Tetrápodos. *Opera Lilloana*, 26: 596 pp.

Bonaparte, J.F., 1979. Faunas y paleobiogeografiá de los tetrápodos Mesozoicos de América del Sur. *Ameghiniana*, 16 (3–4): 217–238.

Bonaparte, J.F., 1980. Los vertebrados tetrápodos del límite Jurásico–Cretácico. *II Congr. Argent. Paleontol. Bioestr., Buenos Aires, 1978, Actas*, V: 77–88.

Bowen, R., 1961. Paleotemperature analysis of Belemnoidea and Jurassic paleoclimatology. *J. Geol.*, 69 (3): 309–320.

Bowen, R., 1963. O^{18}/O^{16} paleotemperature measurements on Mesozoic Belemnoidea from Neuquén and Santa Cruz provinces, Argentina. *J. Paleontol.*, 37 (3): 714–718.

Bracaccini, O.I., 1970. Rasgos tectónicos de las acumulaciones Mesozoicas de las Provincias de Mendoza y Neuquén, República Argentina. *Rev. Asoc. Geol. Argent.*, 25 (2): 275–284.

Burckhardt, C., 1900. Profils géologiques transversaux de la Cordillère argentino-chilienne: *Mus. La Plata, An.*, II: 1–136.

Burckhardt, C., 1902. Le Lias de la Piedra Pintada (Neuquen), III. Sur les fossiles marines du Lias de la Piedra Pintada avec quelques considérations sur l'age et l'importance du gisement. *Mus. La Plata, Rev.*, 10: 243–249.

Burckhardt, C., 1903. Beiträge zur Kenntniss der Jura- und Kreideformation der Cordillere. *Palaeontographica*, 50: 1–144.

Burmeister, H. and Giebel, C., 1861. Die Versteinerungen von Juntas um Thal des Rio Copiapo. *Naturforsch. Ges. Halle, Abh.*, 6: 122–132.

Cabrera, A., 1939. Sobre un nuevo ictiosaurio del Neuquén. *Mus. La Plata, Notas*, 4 (21): 485–491.

Cabrera, A., 1947. Un Nuevo saurópodo del Jurásico de Patagonia. *Mus. La Plata, Notas*, 12 (95): 1–17.

Cabrera, A., 1948. El primer dinosaurio jurásico argentino. *Cienc. Invest.*, 4 (1): 37.

Camacho, H.H. and Riccardi, A.C., 1978. Invertebrados, Megafauna. *Relatorio Geología y Recursos Naturales del Neuquén, Septimo Congr. Geol. Argent.*, pp. 137–144.

Caorsi, J.H., and Goñi, J.C., 1958. Geología Uruguaya. *Inst. Geol. Uruguay, Bol.*, 37: 9–73.

Carter, W.D., 1963. Unconformity marking the Jurassic-Cretaceous boundary in the La Ligua area, Aconcagua province, Chile. *U.S. Geol. Surv. Prof. Pap.*, 450-E: 61–63.

Casamiquela, R., 1960. El hallazgo del primer elenco (icno-lógico) Jurásico de vertebrados terrestres de Latinoaméri-ca (Noticia). *Rev. Asoc. Geol. Argent.*, 15 (1–2): 5–14.

Casamiquela, R., 1961a. Nuevos materiales de '*Notobatra-chus degiustoi*' Reig. La significación del anuro jurásico patagónico. *Univ. Nac. La Plata, Mus. Rev., Paleontol.*, 4 (21): 35–69.

Casamiquela, R., 1961b. Sobre la presencia de un mamífero en el primer elenco (icnológico) de Vertebrados del Jurá-sico de la Patagonia. *Physis*, 22 (63): 225–233.

Casamiquela, R., 1962. Sobre la pisada de un presunto Sauria aberrante en el Liásico del Neuquén (Patagonia). *Ameghi-niana*, 2 (10): 183–186.

Casamiquela, R., 1963. Consideraciones acerca de *Amygda-lodon* Cabrera (Sauropoda, Cetiosauridae) del Jurásico medio de la Patagonia. *Ameghiniana*, 3 (3): 79–92.

Casamiquela, R., 1964. *Estudios icnológicos. Problemas y métodos de la icnología con aplicación al estudio de pisadas mesozoicas (Reptilia, Mammalia) de la Patagonia.* Buenos Aires, 229 pp.

Casamiquela, R., 1965. Nuevo material de '*Vieraella herbstii*' Reig. Reinterpretación de la ranita liásica de la Patagonia y consideraciones sobre filogenia y sistemática de los anu-ros. *Univ. Nac. La Plata, Mus. Rev., Paleontol.*, 4 (27): 265–317.

Casamiquela, R., 1966. Algunas consideraciones teóricas sobre los andares de los dinosaurios saurisquios. Implica-ciones filogenéticas. *Ameghiniana*, 4 (10): 373–385.

Casamiquela, R., 1970. Los Vertebrados Jurásicos de la Ar-gentina y de Chile. *IV Congr. Latinoam. Zool., Caracas, Actas*, II: 873–890.

Casamiquela, R., 1975a. La presencia de un Sauria (Lacer-tilia) en el Liásico de la Patagonia Austral. *I Congr. Argent. Paleont. Bioestr., Tucumán, Actas*, II: 57–70.

Casamiquela, R., 1975b. Sobre la significación de *Ameghini-chnus patagonicus*, un mamífero brincador del Jurásico medio de Santa Cruz (Patagonia). *I Congr. Argent. Paleont. Bioestr., Tucumán, Actas*, II: 71–85.

Casamiquela, R., 1975c. *Herbstosaurus pigmaeus* (Coeluria, Compsognathidae) n. gen. n. sp. del Jurásico medio del Neuquén (Patagonia septentrional). Uno de los más pequeños dinosaurios conocidos. *I Congr. Argent. Paleont. y Bioestr., Tucumán, Actas*, II: 87–103.

Casamiquela, R. and Chong, G., 1980. La presencia de *Ptero-daustro* Bonaparte (Pterodactyloidea) del Neojurásico (?) de la Argentina, en los Andes del Norte de Chile. *II Congr. Argent. Paleontol. Bioestr., Buenos Aires, 1978*, I, pp. 201–209.

Casamiquela, R. and Fasola, A., 1968. Sobre pisadas de dino-saurios del Cretácico inferior de Colchagua (Chile). *Univ. Chile, Fac. Cienc. Fis. Mat., Inst. Geol., Publ.*, 30: 1–24.

Cattoi, N. and Freiberg, M.A., 1961. Nuevo hallazgo de Che-lonia extinguidos en la República Argentina. *Physis*, 22 (63): 202.

Cazau, L., 1972. Cuenca del Ñirihuau–Ñorquinco–Cusha-men. In: A.F. Leanza (Editor), *Geología Regional Argen-tina*. Acad. Nac. Cienc. Cba., Córdoba, pp. 727–740.

Cazeneuve, H., 1965. Datación de una Toba de la Formación Chon Aike (Jurásico de Santa Cruz, Patagonia) por el método de Potasio-Argón. *Ameghiniana* 4 (5): 156–158.

Cecioni, G., 1960. La Zona de Psiloceras planorbis en Chile. *Univ. Chile, Fac. Cienc. Fis. Mat., Inst. Geol., Comunic.*, 1 (1): 1–19.

Cecioni, G., 1961. El Titónico inferior marino en la provincia de Tarapacá y consideraciones sobre el arqueamiento cen-tral de los Andes. *Univ. Chile, Fac. Cienc. Fis. Mat., Inst. Geol., Comunic.*, 1 (3): 19 pp.

Cecioni, G., 1963. Nautiloideos Jurásicos del Norte Grande de Chile. *Univ. Chile, Bol.*, 38: 8 pp.

Cecioni, G., 1964. Ingolfamenti marini giurassici nel Cile set-tentrionale. *Boll. Soc. Nat. Napoli*, 72: 177–206.

Cecioni, G., 1970. *Esquema de Paleogeografía Chilena*. Edit. Universitaria de Chile, Santiago, 143 pp.

Cecioni, G. and Charrier, R., 1974. Relaciones entre la Cuen-ca Patagónica, la Cuenca Andina y el Canal de Mozam-bique. *Ameghiniana*, 11 (1): 1–38.

Cecioni, G. and García, F., 1960a. Observaciones geológicas en la Cordillera de la Costa de Tarapacá. *Chile, Inst. Invest. Geol., Bol.*, 6: 5–28.

Cecioni, G. and García, F., 1960b. Stratigraphy of Coastal Range in Tarapacá Province, Chile. *Am. Assoc. Petrol. Geol., Bull.*, 44 (10): 1609–1620.

Cecioni, G. and Westermann, G.E.G., 1968. The Triassic/ Jurassic marine transition of Coastal Central Chile. *Pac. Geol.*, 1: 41–75.

Charrier, R. and Vicente, J.C., 1972. Liminary and geosyn-cline Andes: Major orogenic phases and synchronical evo-lution of the central and austral sectors of the Southern Andes. *Symp. Upper Mantle Invest., Buenos Aires, 1970*, II: 451–470.

Charrier, R. and Covacevich, V., 1980. Paleogeografía y Bio-estratigrafía del Jurásico superior y Neocomiano en el Sector Austral de los Andes Meridionales Chilenos (42°– 56° Latitud Sur). *Segundo Congr. Argent. Paleontol. Bioestr., Buenos Aires, 1978, Actas*, V: 153–174.

Charrier, R., Linares, E., Niemeyer, H. and Skarmeta, J., 1979. Edades Potasio-Argón de vulcanitas mesozoicas y cenozoicas del sector chileno de la meseta Buenos Aires, Aysen, Chile y su significado geológico. *Septimo Congr. Geol. Argent., Actas*, II: 23–41.

Chong, G., 1973. El Sistema Jurásico en la Cordillera de Domeyko (Chile) entre 24°30' y 25°30' de Lat. Sur. *II Congr. Latinoam. Geol., Caracas*, pp. 765–785.

Chong, G., 1976. Las relaciones de los Sistemas Jurásico y Cretácico en la zona preandina de Chile. *Primer Congr. Geol. Chileno Santiago, Actas*, I: A21–A42.

Chong, G., 1977. Contribution to the knowledge of the Do-meyko Range in the Andes of Northern Chile. *Geol. Rundsch.*, 66 (2): 374–404.

Chong, G. and Förster, R., 1976. *Chilenophoberus atacamen-sis* a new decapod crustacean from the Middle Oxfordian of the Cordillera de Domeyko, northern Chile. *Neues Jahrb. Geol. Palaeontol., Monatsh.*, 3: 145–156.

Chong, G. and Gasparini, Z.B. de, 1976. Los Vertebrados Mesozoicos de Chile y su aporte Geo-Paleontológico. *Sexto Congr. Geol. Argent., Actas*, I: 45–67.

Chotin, P., 1969. Le Jurassique du Lonquimay (Chili). Ses relations avec le Jurassique du Neuquen (Argentine). *Soc. Géol. Fr., Bull.*, (7) 11: 710–716.

Chotin, P., 1976. Essai d'interpretation du Bassin Andin chileno-argentin mésozoique en tant que bassin marginal. *Soc. Géol. Nord, Ann.*, 96 (3): 177–184.

Chotin, P., 1977. Les Andes Méridionales a la Latitude de Concepcion (Chili, 38°S): Portion Intracratonique d'une Chaine développée en bordure de la Marge active Est-Pacifique. *Rev. Géogr. Phys. Géol. Dyn.*, (2) 19 (4): 353–376.

Cisternas, M.E., 1979. Litofacies de transición marino-continental en el Jurásico del Sector La Ola, al sur del Salar de Pedernales. *Segundo Congr. Geol. Chileno, Actas*, 1: A65–A85.

Cisternas, M.E. and Vicente, J.C., 1976. Estudio Geológico del Sector de Las Vegas de San Andres (Prov. de Atacama-Chile). *Primer Congr. Geol. Chileno, Santiago, Actas*, I: A227–A252.

Closs, D., 1962. Los 'Aptychi' (Cephalopoda–Ammonoidea) de Argentina. *Asoc. Geol. Argent., Rev.*, 16 (3–4): 117–141.

Corvalan, J., 1959. El Titoniano de Rio Leñas, Prov. de O'Higgins. *Chile, Inst. Invest. Geol., Bol.*, 3: 5–65.

Corvalan, J., 1976. El Triásico y Jurásico de Vichuquen-Tilicura y de Hualañe, Prov. de Curicó, Implicaciones Paleogeográficas. *Primer Congr. Geol. Chileno, Actas*, I: A137–A154.

Corvalan, J. and Munizaga, F., 1972. Edades radiométricas de rocas intrusivas y metamórficas de la Hoja Valparaiso–San Antonio. *Chile, Inst. Invest. Geol., Bol.*, 28: 5–40.

Corvalan, J. and Perez, E., 1958. Fósiles guías chilenos Tithoniano–Neocomiano. *Chile, Inst. Invest. Geol., Manual*, 1: 1–48.

Covacevich, V. and Escobar, F., 1979. La presencia del género *Otapiria* Marwick, 1935 (Mollusca: Bivalvia) en Chile y su distribución en el ámbito circumpacífico. *Segundo Congr. Geol. Chileno, Actas*, 3: H165–H187.

Covacevich, V. and Piraces, R., 1976. Hallazgo de ammonites del Bajociano superior en la Cordillera de la Costa de Chile Central entre la Cuesta Melón y Limache. *Primer Congr. Geol. Chileno, Actas*, I: C67–C85.

Covacevich, V., Varela, J. and Vergara, M., 1976. Estratigrafía y sedimentación de la Formación Baños del Flaco al sur del Río Tinguiririca, Cordillera de los Andes, Provincia de Curicó, Chile. *Primer Congr. Geol. Chileno, Actas*, I: A191–A211.

Creer, K.M., Mitchell, J.G. and Abou Deeb, J., 1972. Palaeomagnetism and radiometric age of the Jurassic Chon-Aike Formation from Santa Cruz province. *Earth Planet. Sci. Lett.*, 14 (1): 131–138.

Cucchi, R.J. and Baldoni, A., 1980. Hallazgo de plantas mesozoicas en la Formación Epuyén–Cholila, prov. de Chubut. *Asoc. Geol. Argent., Rev.*, 34 (1): 155–156.

Dalziel, I.W.D., 1974. The evolution of the margins of the Scotia Sea. In: C.A. Burke and C.L. Drake (Editors), *The Geology of Continental Margins*. Springer-Verlag, New York, N.Y., pp. 567–579.

Dalziel, I.W.D. and Elliot, D.H., 1973. The Scotia arc and Antarctic margin. In: A.E.M. Nairn and F.G. Stehli (Editors), *The Ocean Basins and Margins, I*. Plenum Press, New York, N.Y., pp. 171–246.

Damborenea, S.E. and Manceñido, M.O., 1979. On the palae-

ogeographical distribution of the pectinid genus *Weyla* (Bivalvia, Lower Jurassic). *Palaeogeogr., Palaeoclimatol. Palaeoecol.*, 27 (1): 85–102.

Damborenea, S.E., Manceñido, M.O. and Riccardi, A.C., 1975. Biofacies y estratigrafía del Liásico de Piedra Pintada, Neuquén, Argentina. *I Congr. Argent. Paleontol. Bioestr., Tucumán, Actas*, II: 173–228.

Dames, W., 1893. Über das Vorkommen von Ichthyopterygiern im Tithon Argentiniens. *Z. Dtsch. Geol. Ges.*, 45: 23–33.

Davidson, J.M. and Godoy, E., 1976. Observaciones sobre un perfil Geológico de los Andes Chilenos en la Latitud 25°40'S. *Sexto Congr. Geol. Argent., Actas*, I: 69–87.

Davidson, J. and Vicente, J.C., 1973. Características Paleogeográficas y Estructurales del Area fronteriza de las Nacientes del Teno (Chile) y Santa Elena (Argentina) (Cordillera Principal, 35° a 35°15' de Latitud Sur). *Quinto Congr. Geol. Argent., Actas*, V: 11–55.

Davidson, J., Godoy, E. and Covacevich, V., 1976. El Bajociano marino de Sierra Minillas (70°30'L.O–26°L.S) y Sierra Fraga (69°50'L.O–27°L.S), Provincia de Atacama, Chile: Edad y Marco Geotectónico de la Formación La Negra en esta latitud. *Primer Congr. Geol. Chileno, Actas*, I: A255–A272.

Dedios, P., 1967. Cuadrángulo Vicuña, Provincia de Coquimbo. *Carta Geol. Chile*, 16: 5–65.

De Giusto, J.M., DiPersia, C.A. and Pezzi, E.E., 1980. Nesocratón del Deseado. *Segundo Simposio de Geología Regional Argentina*. Acad. Nac. Cienc., Cordoba, II: 1389–1430.

Dellape, D.A., Mombrú, C., Pando, G.A., Riccardi, A.C., Uliana, M.A. and Westermann, G.E.G., 1979. Edad y correlación de la Formación Tabanos en Chacay Melehue y otras localidades de Neuquén y Mendoza. Con consideraciones sobre la distribución y significado de las sedimentitas Lotenianas. *Obra Centenario Museo La Plata*, V: 81–105.

Digregorio, J.H., 1972. Neuquén. In: A.F. Leanza (Editor), *Geología Regional Argentina*. Acad. Nac. Cienc., Córdoba, pp. 439–505.

Digregorio, J.H., 1978. Estratigrafía de las Acumulaciones Mesozoicas. *Relatorio Geología y Recursos Naturales del Neuquén. Septimo Congr. Geol. Argent.*, pp. 37–65.

Digregorio, J.H. and Uliana, M.A., 1980. Cuenca Neuquina. *Segundo Simposio de Geología Regional Argentina*. Acad. Nac. Cienc., Córdoba, II: 985–1032.

Dingman, R.J., 1963. Cuadrángulo Tulor, Provincia de Antofagasta. *Carta Geol. Chile*, 11: 5–35.

Dingman, R.J. and Galli, C., 1965. Geology and ground water resources of the Pica Area, Tarapaca Province, Chile. *U.S. Geol. Surv. Bull.*, 1189: 1–113.

Douvillé, R., 1910. Céphalopodes Argentins. *Soc. Géol. Fr., Mém.*, 17 (4): 5–24.

Enay, R., 1972. Paleobiogeographie des Ammonites du Jurassique terminal (Tithonique/Volgien/Portlandien s.l.) et mobilité continentale. *Geobios*, 5 (4): 355–407.

Escobar, F., 1980. Paleontología y bioestratigrafía del Triásico superior y Jurásico inferior en el área de Curepto, Provincia de Talca. *Chile, Inst. Invest. Geol., Bol.*, 35: 1–78.

Fernandez Garrasino, C.A., 1977. Contribución a la estratigrafía de la zona comprendida entre Estancia Ferrarotti, Cerro Colorado y Cerrito Negro, Departamento de Tehuelches, Provincia del Chubut, Argentina. *Asoc. Geol. Argent. Rev.*, 32 (2): 130–144.

Ferraris, F., 1978. Cordillera de la Costa entre 24° y 25° Latitud Sur, Región de Antofagasta. *Carta Geol. Chile*, 26: 3–16.

Feruglio, E., 1933. Fossili liassici della valle del Rio Genua (Patagonia). *Giorn. Geol., Ann. R. Mus. Geol. Bologna*, 9: 1–64.

Feruglio, E., 1936–37. Palaeontographia Patagonica. *Ist Geol. Univ. Padova, Mem.*, 11: 1–384.

Feruglio, E., 1942. La flora del valle del Rio Genoa (Patagonia): Gingkoales et Gymnospermae incertae sedis. *Mus. La Plata, Notas*, 7 (40): 93–110.

Feruglio, E., 1944. Estudios Geológicos y Glaciológicos en la región del lago Argentino (Patagonia). *Acad. Nac. Cienc. Bol.*, 37 (1): 3–255.

Feruglio, E., 1946. La flora liásica del valle del Río Genua (Patagonia). Semina incertae sedis. *Asoc. Geol. Argent., Rev.*, 1 (3): 209–218.

Feruglio, E., 1949–50. Descripción Geológica de la Patagonia. Yac. Petr. Fisc., Buenos Aires, (1949): 1–334; II (1949): 1–349; III (1950): 1–431.

Franchi, M.R., and Page, R.F.N., 1980. Los Basaltos Cretácicos y la evolución magmática del Chubut Occidental. *Asoc. Geol. Argent., Rev.*, 35 (2): 208–229.

Fuenzalida, H., 1937. Las Capas de Los Molles. *Museo Nac. Hist. Nat. Chile*, 16: 67–92.

Galli, C., 1957. Las formaciones geológicas en el borde occidental de la Puna de Atacama, sector de Pica, Tarapacá. *Minerales*, 56: 3–15.

Galli, C., 1968. Cuadrángulo Juan de Morales, Provincia de Tarapacá. *Carta Geol. Chile*, 18: 5–53.

Galli, C., and Dingman, R.J., 1962. Cuadrángulos Pica, Alca, Matilla y Chacarilla, Provincia de Tarapacá. *Carta Geol. Chile*, 3 (2–5): 7–125.

Galli, C., and Menendez, C.A., 1968. Geología de la Quebrada Juan de Morales, Tarapacá, Chile y su flora Jurásica. *Terceras Jorn. Geol. Argent., Actas*, I: 163–171.

Gansser, A., 1973. Facts and theories on the Andes. *J. Geol. Soc. London*, 129 (2): 93–131.

Garcia, F., 1967. Geología del Norte Grande de Chile. *Simp. Geosinclinal Andino*, 1962. Edic. ENAP, Santiago, 138 pp.

Gasparini, Z.B. de, 1973a. *Revision de los Crocodilia (Reptilia) fósiles del territorio argentino. Su evolución, sus relaciones filogenéticas, su clasificación, y sus implicancias estratigráficas.* Univ. Nac. La Plata, Ph.D. Thesis (unpublished).

Gasparini, Z.B. de, 1973b. Revision de '? *Purranisaurus potens*' Rusconi, 1948 (Crocodilia, Thalattosuchia). Los Thalattosuchia como un nuevo Infraorden de los Crocodilia. *Quinto Congr. Geol. Argent. Actas*, 3: 423–431.

Gasparini, Z.B. de, 1978. Consideraciones sobre los Metriorhynchidae (Crocodilia, Mesosuchia): sus origenes, taxonomía y distribución geográfica. *Obra Centenario Museo La Plata*, V: 1–9 (1979).

Gasparini, Z.B. de, 1979. Comentarios críticos sobre los Vertebrados Mesozoicos de Chile. *Segundo Congr. Geol. Chileno, Actas*, 3: H15–H32.

Gasparini, Z.B. de, and Chong, G., 1977. Metriorhynchus casamiquelai n.sp. (Crocodilia, Thalattosuchia), a marine crocodile from the Jurassic (Callovian) of Chile, South America. *Neues Jahrb. Geol. Palaeontol., Abh.*, 153 (3): 341–360.

Gasparini, Z.B. de, and Dellapé, D., 1976. Un nuevo cocodrilo marino (Thalattosuchia, Metriorhynchidae) de la Formación Vaca Muerta (Jurásico, Tithoniano) de la provincia de Neuquen (República Argentina). *Primer Congr. Geol. Chileno, Actas*, I: C1–C21.

Gerth, H., 1914. Stratigraphie und Bau der argentinischen Kordillere zwischen dem Rio Grande und Rio Diamante. *Z. Dtsch. Geol. Ges.*, 65: 568–575.

Gerth, H., 1932. *Geologie Südamerikas.* Borntraeger, Berlin, 389 pp.

Gerth, H., 1955. *Die Geologische Bau der südamerikanischen Kordillere.* Borntraeger, Berlin, 253 pp.

Gonzalez, O. and Vergara, M., 1962. Reconocimiento Geológico de la Cordillera de los Andes entre los paralelos 35° y 38° latitud Sur. *Univ. Chile, Fac. Cienc. Fis. Mat., Inst. Geol., Publ.*, 24: 121 pp.

González Bonorino, F., 1974. La Formación Millaqueo y la 'Serie Porfirítica' de la Cordillera Nordpatagónica: Nota Preliminar. *Asoc. Geol. Argent., Rev.*, 29 (2): 145–153.

González Bonorino, F. and González Bonorino G., 1979. Geología de la Región de San Carlos de Bariloche: Estudio de las Formaciones Terciarias del Grupo Nahuel Huapi. *Asoc. Geol. Argent., Rev.*, 33 (3): 175–210.

Gonzalez Diaz, E.F. and Nullo, F.E., 1980. Cordillera Neuquina. *Segundo Simposio de Geología Regional Argentina.* Acad. Nac. Cienc. Córdoba, II: 1099–1147.

Gottsche, C., 1878. Über jurassische Versteinerungen aus der Argentinischen Cordillere. *Palaeontographica, Suppl.*, 3, 2 (3): 1–50.

Groeber, P., 1918. Estratigrafía del Dogger en la República Argentina. *Argent., Dir. Nac. Geol. Min., Bol.*, 18B: 1–81.

Groeber, P., 1929. Líneas fundamentales de la geología del Neuquén, sur de Mendoza y regiones adyacentes. *Argent., Dir. Nac. Geol. Min., Publ.*, 58: 1–109.

Groeber, P., 1933. Confluencia de los ríos Grande y Barrancas (Mendoza y Neuquén). *Argent., Dir. Nac. Geol. Min., Bol.*, 38: 1–72.

Groeber, P., 1946. Observaciones Geológicas a lo largo del Meridiano 70, I. Hoja Chos Malal. *Asoc. Geol. Argent., Rev.*, 1: 177–208.

Groeber, P., 1951. La Alta Cordillera entre las latitudes 34° y 29°30'. *Mus. Argent. Cienc. Nat. 'Bernardino Rivadavia'. Inst. Nac. Invest. Cienc. Nat., Rev. Geol.*, 1, 5: 235–352.

Groeber, P., 1963. La Cordillera entre las latitudes 22°20' y 40°S. *Acad. Nac. Cienc. Cba., Bol.*, 43 (2–4): 111–175 (posth. publ. by A.F. Leanza).

Groeber, P., Stipanicic, P.N. and Mingramm, A., 1953. Jurásico. In: *Geografía de la República Argentina.* Buenos Aires, II: 143–347.

Hallam, A., 1975. *Jurassic Environments.* Cambridge Univ. Press, Cambridge, 269 pp.

Halpern, M., 1973. Regional Geochronology of Chile south of 50° Latitude. *Geol. Soc. Am. Bull.*, 84: 2407–2422.

Halpern, M. and Carlin, G.M., 1971. Radiometric chronology of crystalline rocks from southern Chile. *Antarct. J. U.S.*, 6 (5): 191–193.

Harrington, H.J., 1956. Argentina. In: W.F. Jenks (Editor), *Handbook of South American Geology. Geol. Soc. Am. Mem.*, 65: 131–165.

Harrington, H.J., 1961. Geology of parts of Antofagasta and Atacama Provinces, Northern Chile. *Am. Assoc. Petrol. Geol. Bull.*, 45 (2): 169–197.

Harrington, H.J., 1962. Paleogeographic development of South America. *Am. Assoc. Petrol. Geol. Bull.*, 46 (10): 1773–1814.

Haupt, O., 1907. Beiträge zur Fauna des oberen Malm un der unteren Kreide in den Argentinischen Kordillere. *Neues Jahrb. Geol. Palaeontol., Abh.*, 23: 187–236.

Herbst, R., 1961. La flora liásica de C. Meschio, Prov. de Chubut, Patagonia. *Ameghiniana*, 2 (9): 337–349.

Herbst, R., 1965. La flora fósil de la Formación Roca Blanca (Prov. de Santa Cruz, Patagonia), con consideraciones geológicas y estratigráficas. *Opera Lilloana*, 12: 3–11.

Herbst, R., 1966. La flora liásica del Grupo Pampa de Agnia, Chubut, Patagonia. *Ameghiniana*, 4 (9): 337–347.

Herbst, R., 1980. Flórula fósil de la Formación Los Patos (Sinemuriano) del Río Los Patos, Provincia de San Juan, República Argentina. *Segundo Congr. Argent. Paleontol. Bioestr., Actas*, I: 175–189.

Herbst, R. and Anzotegui, L.M., 1968. Nuevas plantas de la flora del Jurásico medio (Matildense) de Taquetren, Prov. de Chubut. *Ameghiniana*, 5 (6): 183–190.

Herrero Ducloux, A., 1963. The Andes of Western Argentina. In: The Backbone of the Americas–Tectonic History from Pole to Pole, A Symposium. *Am. Assoc. Petrol. Geol. Mem.*, 2: 16–28.

Herrero Ducloux, A., and Leanza, A.F., 1943. Sobre los Ammonites de la 'Lotena Formation' y su significado geológico. *Mus. La Plata, Notas*, 8 (54): 281–304.

Hillebrandt, A.v., 1970. Zur Biostratigraphie und Ammoniten-Fauna des südamerikanischen Jura (insbes. Chile). *Neues Jahrb. Geol. Palaeontol., Abh.*, 136 (2): 166–211.

Hillebrandt, A.v., 1971. Der Jura in der chilenisch-argentinischen Hoch-kordillere (25°–32° südl. Breite). *Münster Forsch. Geol. Palaeontol.*, 20/21: 63–87.

Hillebrandt, A.v., 1973a. Neue Ergebnisse über den Jura in Chile und Argentinien. *Münster Forsch. Geol. Palaeontol.*, 31/32: 167–199.

Hillebrandt, A.v., 1973b. Die Ammonitengattungen *Bouleiceras* und *Frechiella* im Jura von Chile und Argentinien. *Eclogae Geol. Helv.*, 66: 351–363.

Hillebrandt, A.v., 1977. Ammoniten aus dem Bajocien (Jura) von Chile (Südamerika), Neue Arten des Gattungen Stephanoceras und Domeykoceras n. gen. (Stephanoceratidae). *Mitt. Bayer. Staatssamml. Palaeontol. Hist. Geol.*, 17: 35–69.

Hillebrandt, A.v., 1979. Paleobiogeografía de los ammonites del Lias de la Argentina y áreas vecinas. *Ameghiniana* 16 (3–4): 239–246.

Hillebrandt, A.v., 1980. Paleozoogeografía de Jurásico mari-

no (Lías hasta Oxfordiano) en Suramérica. In: *Nuevos Resultados de la Investigación Geocientífica Alemana en Latinoamérica*. Deutsche Forschungsgemeinschaft, Bonn, pp. 123–134.

Hillebrandt, A.v., 1981a. Kontinentalverschiebung und die paläozoogeographischen Beziehungen des südamerikanischen Lias. *Geol. Rundsch.*, 70 (2): 570–582.

Hillebrandt, A.v., 1981b. Faunas de ammonites del Liásico inferior y medio (Hettangiano hasta Pliensbachiano) de América del Sur (excluyendo Argentina). *Segundo Congr. Latinoamericano de Paleontol. (Porto Alegre, 1981), Simp. Jurásico–Cretácico*, 2: 499–538.

Hillebrandt, A.v. and Schmidt-Effing, R., 1981. Ammoniten aus dem Toarcium (Jura) von Chile (Südamerika). *Zitteliana*, 6: 3–74.

Hoffstetter, R., Fuenzalida, H. and Cecioni, G., 1957. Chile–Chili. In: *Lexique Stratigraphique International* V (7): 1–444.

Hölder, H., 1964. *Jura*. Ferdinand Enke Verlag, Stuttgart, 603 pp.

Huene, F.v., 1927. Beitrag zur Kenntnis mariner mesozoischer Wirbeltiere in Argentinien. *Centralb. Min. Geol. Palaeontol.*. B (1): 22–29.

Indans, J., 1954. Eine Ammonitenfauna aus dem Untertithon der Argentinischen Kordillere in Süd-Mendoza. *Palaeontographica*, A105 (3–6): 96–132.

Jaworski, E., 1914. Beiträge zur Kenntnis der Jura in Südamerika, I. *Neues Jahrb. Geol. Palaeontol., Abh.*, 37: 285–342.

Jaworski, E., 1915. Beiträge zur Kenntnis der Jura in Südamerika, II. *Neues Jahrb. Geol. Palaeontol., Abh.*, 40: 364–456.

Jaworski, E., 1925. Contribución a la Paleontología del Jurásico Sudamericano. *Argent. Dir. Nac. Geol. Min., Publ.*, 4: 1–160. (in part translation from Jaworski, 1915).

Jaworski, E., 1926a. La fauna del Lias y Dogger de la Cordillera Argentina en la parte meridional de la Provincia de Mendoza. *Acad. Nac. Cienc. Cba., Actas*, 9: 135–317.

Jaworski, E., 1926b. Lias und Dogger. *Geol. Rundsch., Sonderband (Steinmann Festschrift)*: pp. 373–427.

Jensen, O.L. and Vicente, J.C., 1977. Estudio geológico del area de 'Las Juntas' del Río Copiapó. *Asoc. Geol. Argent. Rev.*, 31 (3): 145–173.

Jensen, O., Vicente, J.C., Davidson J. and Godoy, E., 1976. Etapas de la evolución marina jurásica de la Cuenca Andina externa (mioliminar) entre los paralelos 26° y 29°30' Sur. *Primer. Congr. Geol. Chileno, Actas*, I: A273–A295.

Klohn, C., 1960. Geología de la Cordillera de los Andes de Chile Central, Provincias de Santiago, O'Higgins, Colchagua y Curicó. *Chile, Inst. Invest. Geol., Bol.*, 8: 1–95.

Krantz, F., 1928. La fauna del Titono superior y medio en la parte meridional de la provincia de Mendoza. *Acad. Nac. Cienc. Cba., Actas*, 10: 9–57.

Lahee, F.H., 1927. The petroliferous belt of Central-Western Mendoza Province, Argentina. *Am. Assoc. Petrol. Geol., Bull.*, 11 (3): 261–278.

Lambert, L.R., 1943. Perfil geológico en el valle superior del río de Los Patos Sur (Provincia de San Juan). *Univ. Nac.*

La Plata, Mus., Rev., Geol., 2(11): 1–10.

Lambert, L.R., 1944. Algunas trigonias del Neuquén. Univ. Nac. La Plata, Mus., Rev., Paleontol., 2 (14): 357–397.

Leanza, A.F., 1942. Los Pelecípodos del Lías de Piedra Pintada en el Neuquén. Univ. Nac. La Plata, Mus., Rev., Paleontol., 2 (10): 143–206.

Leanza, A.F., 1945. Ammonites del Jurásico superior y del Cretácico inferior de la Sierra Azul, en la parte meridional de la provincia de Mendoza. An. Mus. La Plata (N.S.), 1: 1–99.

Leanza, A.F., 1947a. Descripción de la Fáunula Kimmeridgiana de Neuquén. Argent. Dir. Nac. Geol. Min., Inf. Prelim. Comunic., 1: 3–15.

Leanza, A.F., 1947b. Ammonites Coralianos en el Jurásico de Chile. Asoc. Geol. Argent. Rev., 2 (4): 285–295.

Leanza, A.F., 1947c. Upper limit of the Jurassic system. Geol. Soc. Am., Bull., 58: 833–842.

Leanza, A.F., 1968. Anotaciones sobre los fósiles Jurásico–Cretácicos de Patagonia Austral (Colección Feruglio) conservados en la Universidad de Bologna. Acta Geol. Lilloana, 9: 121–186, 3 pl. Tucumán.

Leanza, H.A., 1980. The Lower and Middle Tithonian ammonite fauna from Cerro Lotena, Province Neuquén, Argentina. Zitteliana, 5: 3–49.

Leanza, H.A. and Hugo, C.A., 1978. Sucesión de ammonites y edad de la Formación Vaca Muerta y sincrónicas entre los Paralelos 35° y 40° l.s., Cuenca Neuquina-Mendocina. Asoc. Geol. Argent. Rev., 32 (4): 248–264.

Lesta, P.J. and Ferello, R., 1972. Región Extraandina de Chubut y Norte de Santa Cruz. In: A.F. Leanza (Editor), Geología Regional Argentina. Acad. Nac. Cienc., Córdoba, pp. 601–653.

Lesta, P.J., Ferello, R. and Chebli, G.A., 1980. Geología del Chubut Extraandino. Segundo Simposio de Geología Regional Argentina. Acad. Nac. Cienc., Córdoba, II, pp. 1307–1387.

Levi, B., 1960. Estratigrafía del Jurásico y Cretáceo inferior de la Cordillera de la Costa entre las latitudes 32°40′ y 33°40′. Univ. Chile, Fac. Cienc. Fis. Mat., Inst. Geol., Publ., 16: 223–269.

Levi, B., Mehech, S. and Munizaga, F., 1963. Edades Radiométricas y Petrografía de Granitos Chilenos. Chile, Inst. Invest. Geol., Bol., 12: 1–42.

Lizuain Fuentes, A., 1980. Las Formaciones Suprapaleozoicas y Jurásicas de la Cordillera Patagónica, Provincia de Río Negro y Chubut. Asoc. Geol. Argent., Rev., XXXV (2): 174–182.

Losert, J., 1973. Genesis of copper mineralizations and associated alterations in the Jurassic volcanic rocks of the Buena Esperanza mining area (Antofagasta Province, Northern Chile). Univ. Chile, Fac. Cienc. Fis. Mat., Inst. Geol., Publ., 40: 1–64.

Malumian, N. and Ploszkiewicz, J.V., 1977. El Liásico fosilífero de Loncopan, Departamento Tehuelches (Provincia de Chubut, República Argentina). Asoc. Geol. Argent., Rev., 31 (4): 279–280.

Manceñido, M.O., 1978. Studies of Early Jurassic Brachiopoda and their Distribution, with Special Reference to Argentina. Ph.D. Thesis, University of Wales, Swansea (unpublished).

Manceñido, M.O., 1981. A revision of Early Jurassic Spiriferinidae (Brachiopoda, Spiriferida) from Argentina. Segundo Congr. Latinoamericano de Paleontol., (Porto Alegre, 1981, Simposio Jurásico–Cretácico, 2: 625–660.

Marchese, H.G., 1971. Litoestratigrafía y variaciones faciales de las sedimentitas Mesozoicas de la Cuenca Neuquina. Provincia de Neuquén, República Argentina. Asoc. Geol. Argent. Rev., 26 (3): 343–410.

McDougall, I. and Rüegg, N.R., 1966. Potassium-argon dates on the Serra Geral Formation of South America. Geochim. Cosmochim. Acta, 30: 191–195.

McGowan, Ch., 1972. Evolutionary trends in longipinnate ichthyosaurs with particular reference to the skull and fore fin. R. Ont. Mus., Life. Sci. Contrib., 83: 1–38.

Menendez, C.A., 1968. Estudio palinológico del Jurásico medio de Picún Leufú (Neuquén). Ameghiniana, 5 (10): 379–405.

Menendez, C.A., 1969. Datos Palinológicos de las Floras Preterciarias de la Argentina. Gondwana Stratigraphy, IUGS Symposium, Buenos Aires, 1967, pp. 55–69.

Mercado, M., 1978. Hojas Chañaral y Potrerillos. Chile, Inst. Invest. Geol., Mapas Geol. Prelim., 2: 1–24.

Minato, M., 1977. Brief Note on the Lower Jurassic Ammonites from the Vichuquen Region, Central Chile. In: T. Ishikawa and L. Aguirre (Editors), Comparative Studies on the Geology of the Circum-Pacific Orogenic Belt in Japan and Chile. 1st. Rep. Japan Soc. Promotion Sci. Tokyo, pp. 119–123.

Montecinos, P., 1979. Plutonismo durante el ciclo tectónico Andino en el norte de Chile, entre los 18°–29° Lat. Sur. Segundo Congr. Geol. Chileno, Actas, 3: E89–E108.

Moraga, A., Chong, G., Fortt, M.A. and Henriquez, H., 1974. Estudio Geológico del Salar de Atacama, Provincia de Antofagasta. Chile, Inst. Invest. Geol., Bol., 29: 1–56.

Möricke, W., 1894. Versteinerungen des Lias und Unteroolith von Chile. Neues Jahrb. Geol. Palaeontol., Abh., 9: 1–100, 6 pl.

Mpodozis, A.C. and Rivano, S., 1976. Evidencias de Tectogénesis en el límite Jurásico–Cretácico en la Alta Cordillera de Ovalle (Provincia de Coquimbo). Primer Congr. Geol. Chileno, Actas, I: B57–B68.

Mpodozis, A., Rivano, S. and Vicente, J.C., 1973. Resultados preliminares del Estudio Geológico de la Alta Cordillera de Ovalle entre los ríos Grande y Los Molles (Prov. de Coquimbo, Chile). Quinto Congr. Geol. Argent., Actas, 4: 117–132.

Muñoz Cristi, J., 1956. Chile. In: W.F. Jenks (Editor), Handbook of South American Geology. Geol. Soc. Am. Mem., 65: 131–165.

Musacchio, E.A., 1975. Sobre algunas consideraciones estratigráficas acerca del Jurásico en Pampa de Agnia, Chubut. Asoc. Geol. Argent., Rev., 30 (1): 115.

Musacchio, E.A., 1978. Microfauna del Jurásico y del Cretácico inferior. Relatorio Geología y Recursos Naturales del Neuquén. Septimo Congr. Geol. Argent., Buenos Aires, pp. 147–161.

Musacchio, E.A., 1980. Algunos microfósiles calcáreos, marinos y continentales, del Jurásico y el Cretácico inferior de la República Argentina. Segundo Congr. Geol. Argent. Paleontol. Bioestr., Actas, V: 67–76.

Musacchio, E.A. and Riccardi, A.C., 1971. Estratigrafía, principalmente del Jurásico, en la Sierra de Agnia, Chubut, República Argentina. *Asoc. Geol. Argent., Rev.*, 26 (2): 272–273.

Nakayama, C., 1973. Sedimentitas pre-bayocianas en el extremo austral de la Sierra de Taquetrén, Chubut. *Quinto Congr. Geol. Argent., Actas*, 3: 269–277.

Naranjo, J.A. and Covacevich, V., 1979. Nuevos antecedentes sobre la geología de la Cordillera de Domeyko en el área de Sierra Vaquillas Altas, Región de Antofagasta. *Segundo Congr. Geol. Chileno, Actas*, 1: A45–A64.

Natland, M.L., Gonzalez, E., Cañon, A. and Ernst, M., 1974. A system of stages for correlation of Magallanes Basin sediments. *Geol. Soc. Am. Mem.*, 139: 126 pp.

Nullo, F., 1974. Reubicación estratigráfica de la Formación El Córdoba, Pampa de Agnia, Provincia de Chubut, República Argentina. *Asoc. Geol. Argent. Rev.*, 29 (3): 377–8.

Nullo, F. and Proserpio, C., 1975. La Formación Taquetrén en Cañadón del Zaino (Chubut) y sus relaciones estratigráficas en el ámbito de la Patagonia, de acuerdo a la flora, República Argentina. *Asoc. Geol. Argent. Rev.*, 30 (2): 133–150.

Nullo, F.E. Proserpio, C., and Ramos, V.A., 1979. Estratigrafía y tectónica de la vertiente Este del Hielo Patagónico, Argentina–Chile. *Septimo Congr. Geol. Argent., Actas*, 1: 455–470.

Padula, E.L., 1972. Subsuelo de la Mesopotamia y regiones adyacentes. In: A.F. Leanza (Editor), *Geología Regional Argentina. Acad. Nac. Cienc. Cba., Córdoba*, pp. 213–235.

Padula, E. and Mingramm, A., 1969. Sub-surface Mesozoic Red-Beds of the Chaco–Mesopotamian Region, Argentina, and their relatives in Uruguay and Brazil. *Gondwana Stratigraphy, IUGS Symposium, Buenos Aires*, 1967, pp. 1053–1071.

Palacios, C., 1974. Geología y metalogénesis de la Formación Volcánica La Negra y las Rocas graníticas en el área de Tocopilla, Provincia de Antofagasta. *Univ. Chile, Fac. Cienc. Fis. Mat., Inst. Geol., Publ.*, 43: 1–47.

Pascual, R., Bondesio, P., Scillato Yane, G.J., Vucetich, M.G. and Gasparini, Z.B. de, 1978. Vertebrados. *Relatorio Geología y Recursos Naturales del Neuquén, Septimo Congr. Geol. Argent.*, pp. 177–185.

Pascual, R., Odreman, O.E. and Tonni, E., 1969. Las unidades estratigráficas del Jurásico de la Argentina portadoras de Vertebrados. Correlaciones y Edades. *Cuartas Jorn. Geol. Argent. Actas*, 1: 469–483.

Perez, E. and Reyes, R., 1977. Las Trigonias Jurásicas de Chile y su valor cronoestratigráfico. *Chile, Inst. Invest. Geol., Bol.*, 30: 5–58.

Philippi, R.A., 1899. Los Fósiles Secundarios de Chile. Santiago, 104 pp.

Piatnitzky, A., 1932. Rético y Liásico en los valles de los ríos Genua y Tecka y sedimentos continentales de la sierra de San Bernardo. *Bol. Inf. Petrol.*, 103: 151–182.

Piatnitzky, A., 1936. Estudio Geológico de la región del río Chubut y del río Genua. *Bol. Inf. Petrol.*, 137: 83–115.

Piraces, R., 1976. Estratigrafía de la Cordillera de la Costa entre la Cuesta El Melón y Limache, Provincia de Valpa-
raíso, Chile. *Primer Congr. Geol. Chileno, Actas*, I: A65–A82.

Ploszkiewicz, J.V. and Ramos, V.A., 1978. Estratigrafía y tectónica de la Sierra de Payaniyeu. *Asoc. Geol. Argent. Rev.*, 32 (3): 209–226.

Proserpio, C.A., 1976. Sedimentitas Jurásicas Continentales en el Norte de la Provincia del Chubut (Departamento de Gastre), República Argentina. *Sexto Congr. Geol. Argent. Actas*, 1: 423–432.

Putzer, H., 1962. *Geologie von Paraguay*. Borntraeger, Berlin, 180 pp.

Ramos, V.A., 1976. Estratigrafía de los lagos La Plata y Fontana, Provincia de Chubut, República Argentina. *Primer Congr. Geol. Chileno, Actas*, I: A43–A64.

Ramos, V.A., 1978. Estructura. *Relatorio Geología y Recursos Naturales del Neuquen. Septimo Congr. Geol. Argent.*, pp. 99–118.

Reig, O., 1961. Noticias sobre un nuevo anuro fósil del Jurásico de Santa Cruz (Patagonia). *Ameghiniana*, 2 (5): 73–78.

Reutter, K.J., 1974. Entwicklung und Bauplan der chilenischen Hochkordillere im Bereich 29° südlicher Breite. *Neues Jahrb. Geol. Palaeontol., Abh.*, 146 (2): 153–178.

Reyes, R. and Perez, E., 1979. Estado Actual del conocimiento de la Familia Trigoniidae (Mollusca, Bivalvia) en Chile. *Rev. Geol. Chile*, 8: 13–64.

Reyes, R. and Perez, E., 1980. *Quadratojaworskiella* nov., a Liassic subgenus of Trigoniidae from Chile. *Pac. Geol.*, 14: 87–93.

Riccardi, A.C., 1970. Favrella R. Douvillé, 1909 (Ammonitina, Cretácico inferior): Edad y distribución. *Ameghiniana*, 7 (2): 119–138.

Riccardi, A.C., 1971. Estratigrafía en el Oriente de la Bahía de la Lancha, lago San Martín, Santa Cruz, Argentina. *Univ. Nac. La Plata, Mus., Rev., Geol.*, 7 (61): 245-318.

Riccardi, A.C., 1976. Paleontología y Edad de la Formación Springhill. *Primer Congr. Geol. Chileno, Actas*, I: C41–C56.

Riccardi, A.C., 1977. Berriasian invertebrate Fauna from the Springhill Formation of Southern Patagonia. *Neues Jahrb. Geol. Palaeontol., Abh.*, 155 (2): 216–252.

Riccardi, A.C. and Rolleri, E.O., 1980. Cordillera Patagónica Austral. In: *Segundo Simposio de Geología Regional Argentina*. Acad. Nac. Cienc., Córdoba, II: 1173–1306.

Rigal, R., 1930. El Liásico en la Cordillera del Espinacito (Provincia de San Juan). *Argent. Dir. Nac. Geol. Min., Publ.*, 74: 5–9.

Robbiano, J.A., 1971. Contribución al conocimiento estratigráfico de la Sierra del Cerro Negro, Pampa de Agnia, Provincia de Chubut, República Argentina. *Asoc. Geol. Argent. Rev.*, 26 (1): 41–56.

Ruiz, C., Aguirre, L., Corvalan, J., Klohn, C., Klohn, E. and Levi, B., 1965. *Geología y Yacimientos Metalíferos de Chile*. Inst. Invest. Geol. Chile, Santiago, 305 pp.

Ruiz, C., Segerström, K., Aguirre, L., Corvalan, J., Rose, H.J. and Stern, T.W., 1960. Edades Plomo-alfa y marco estratigráfico de granitos chilenos. *Chile, Inst. Invest. Geol., Bol.*, 7: 1–26.

Rusconi, C., 1940. Nueva especie de ictiosaurio del Jurásico de Mendoza. *Bol. Paleontol.*, 11: 1–4.

Rusconi, C., 1942. Nuevo género de ictiosaurio argentino. *Bol. Paleontol.*, 13: 1–2.

Rusconi, C., 1943. Presencia de un plesiosaurio en Mendoza. *Bol. Paleontol.*, 15: 1–4.

Rusconi, C., 1948a. Nuevo plesiosaurio, pez y langosta de mar Jurásico de Mendoza. *Rev. Mus. Hist. Nat.*, 2: 3–12.

Rusconi, C., 1948b. Ictiosaurios del Jurásico de Mendoza (Argentina). *Rev. Mus. Hist. Nat.*, 2 (1–2): 17–160.

Rusconi, C., 1949. I. Presencia de ictiosaurios en el liásico de San Juan, II. Otra especie de laberintodonte del Triásico de Mendoza. *Rev. Mus. Hist. Nat.*, 3: 89–94.

Rusconi, C., 1967. *Animales Extinguidos de Mendoza y de la Argentina*. Ofic. Gob. Mendoza, Mendoza, 489 pp.

Sáez, M.D. de, 1939. Noticias sobre peces fósiles argentinos. *Mus. La Plata, Notas*, 4 (19): 425–432.

Sáez, M.D. de, 1940a. Noticias sobre peces fósiles argentinos. Leptolépidos del Titoniense de Plaza Huincul. *Mus. La Plata, Notas*, 5 (26): 299–305.

Sáez, M.D. de, 1940b. Noticias sobre peces fósiles argentinos. Celacántidos titonienses de Plaza Huincul. *Mus. La Plata, Notas*, 5 (25): 295–298.

Sáez, M.D. de, 1949. Noticias sobre peces fósiles argentinos. *Mus. La Plata, Notas*, 14 (96): 443–461.

Saint-Seine, P., 1949. Les poissons des calcaires lithographiques de Cerin (Ain). *Nouv. Arch. Mus. Hist. Nat. Lyon*, 2: 1–357.

Salas, R., Kast, R.F., Montecinos, F. and Salas, I., 1966. Geología y recursos minerales del Departamento de Arica, Provincia de Tarapacá. *Chile, Inst. Invest. Geol., Bol.*, 21: 7–114.

Schiller, W., 1912. La Alta Cordillera de San Juan y Mendoza y partes de la Provincia de San Juan. *An. Min. Agric., Sec. Geol.*, VII (5): 1–68.

Segerström, K., 1959. Cuadrángulo Los Loros, Provincia de Atacama. *Carta Geol. Chile*, 1 (1): 5–33.

Segerström, K., 1968. Geología de las Hojas Copiapó y Ojos del Salado, Provincia de Atacama. *Chile, Inst. Invest. Geol., Bol.*, 24: 5–58.

Sigal, J., Grekoff, N., Singh, N.P., Cañón, A. and Ernst, M., 1970. Sur l'âge et les affinités 'gondwaniennes' de microfaunes (Foraminifères et Ostracodes) malgaches, indiennes, et chiliennes au sommet du Jurassique et à la base du Crétacé. *C.R. Acad. Sci. Paris, Ser. D*, 271: 24–27.

Silva, I., 1976. Antecedentes estratigráficos del Jurásico y estructurales de la Cordillera de la Costa en el Norte Grande de Chile. *Primer Congr. Geol. Chileno, Actas*, I: A83–A95.

Stehn, E., 1923. Beiträge zur Kenntniss des Bathonian und Callovian in Südamerika. *Neues Jahrb. Geol. Palaeontol., Abh.*, 49: 52–158.

Steinmann, G., 1881. Zur Kenntniss der Jura und Kreideformation von Caracoles (Bolivia). *Neues Jahrb. Geol. Palaeontol., Abh.*, 1: 239–301.

Stelzner, A., 1873. Über die argentinische Cordillere zw. 31° und 33° S.Br. *Neues Jahrb. Min. Geol. Paläontol.*, pp. 726–744.

Steuer, A. 1897. Beiträge zur Kenntniss der Geologie und Paläontologie der Argentinischen Anden. *Palaeont. Abh., N.F.*, 3: 127–222.

Stipanicic, P.N., 1951. Sobre la presencia del Oxfordense superior en el arroyo de la Manga (Provincia de Mendoza). *Asoc. Geol. Argent. Rev.*, 6 (4): 213–239.

Stipanicic, P.N., 1966. El Jurásico en Vega de la Veranada (Neuquén), el Oxfordense y el diastrofismo Divesiano (Agassiz-Yaila) en Argentina. *Asoc. Geol. Argent. Rev.*, 20: 403–478.

Stipanicic, P.N., 1967. Consideraciones sobre las edades de algunas fases magmáticas del Neopaleozoico y Mesozoico. *Asoc. Geol. Argent. Rev.*, 22 (2): 101–133.

Stipanicic, P.N., 1969. El avance en los conocimientos del Jurásico Argentino a partir del esquema de Groeber. *Asoc. Geol. Argent. Rev.*, 24: 367–388.

Stipanicic, P.N. and Bonetti, M.I.R., 1970a. Posiciones estratigráficas de las principales floras Jurásicas Argentinas, I. Floras Liásicas. *Ameghiniana*, 7 (1): 57–78.

Stipanicic, P.N. and Bonetti, M.I.R., 1970b. Posiciones estratigráficas y edades de las principales floras Jurásicas Argentinas, II. Floras Doggerianas y Málmicas. *Ameghiniana*, 7 (2): 101–118.

Stipanicic, P.N. and Methol, E.J., 1972. Macizo de Somuncurá. In: A.F. Leanza (Editor), *Geología Regional Argentina*. Acad. Nac. Cienc. Cba., Córdoba, pp. 581–599.

Stipanicic, P.N. and Reig, A.O., 1955. Breve noticia sobre el hallazgo de Anuros en el denominado Complejo Porfírico de la Patagonia Extraandina, con consideraciones acerca de la composición geológica del mismo. *Asoc. Geol. Argent. Rev.*, 10 (4): 215–233.

Stipanicic, P.N. and Reig, E.O., 1957. El Complejo Porfírico de la Patagonia Extraandina y su fauna de anuros. *Acta Geol. Lilloana*, 1: 185–297.

Stipanicic, P.N. and Rodrigo, F., 1970a. El diastrofismo Eo-y Mesocretácico en Argentina y Chile, con referencia a los movimientos Jurásicos de la Patagonia. *Cuartas Jorn. Geol. Argent. Actas*, II: 337–352.

Stipanicic, P.N. and Rodrigo, F., 1970b. El diastrofismo Jurásico en Argentina y Chile. *Cuartas Jorn. Geol. Argent. Actas*, II: 353–368.

Stipanicic, P.N., Rodrigo, F., Baulíes, O.L. and Martínez, C.G., 1968. Las Formaciones presenonianas en el denominado Macizo Nordpatagónico y regiones adyacentes. *Asoc. Geol. Argent. Rev.*, 23 (2): 67–98.

Stipanicic, P.N., Westermann, G.E.G. and Riccardi, A.C., 1975. The Indo-Pacific Ammonite Mayaites in the Oxfordian of the Southern Andes. *Ameghiniana*, 12 (4): 281–305.

Suarez, M. and Pettigrew, T.H., 1976. An Upper Mesozoic island-arc–back-arc system in the southern Andes and South Georgia. *Geol. Mag.*, 113 (4): 305–328.

Tasch, P. and Volkheimer, W., 1970. Jurassic conchostracans from Patagonia. *Kans. Univ., Paleontol. Contrib., Pap.*, 50: 1–23.

Thiele, R., 1964. Reconocimiento geológico de la Alta Cordillera de Elqui. *Univ. Chile, Fac. Cienc. Fis. Mat., Inst. Geol., Publ.*, 27: 135–197.

Thiele, R., 1965. El Triásico y Jurásico del Departamento de Curepto en la Provincia de Talca. *Univ. Chile, Fac. Cienc.*

Fis. Mat., Inst. Geol., Publ., 28: 29–46.

Thiele, R., Castillo, J.C., Hein, R., Romero, G. and Ulloa, M., 1979. Geología del sector fronterizo de Chile Continental entre los 43°00′–43°45′ latitud sur (Comunas de Futaleufú y de Palena). *Septimo Congr. Geol. Argent., Actas*, I: 577–591.

Thomas, A., 1967, Cuadrángulo Mamiña, Provincia de Tarapacá. *Carta Geol. Chile*, 17: 5–49.

Thomas, A., 1970. Cuadrángulos Iquique y Caleta Molle, Provincia de Tarapacá. *Chile, Inst. Invest. Geol., Carta Geol. Chile*, 21–22: 5–52.

Thomas, H., 1958. Geología de la Cordillera de la Costa entre el Valle de La Ligua y la Cuesta de Barriga. *Chile, Inst. Invest. Geol., Bol.*, 2: 5–86.

Thomas, H., 1967. Geología de la Hoja Ovalle, Provincia de Coquimbo. *Chile, Inst. Invest. Geol., Bol.*, 23: 5–58.

Tobar, A., Salas, I. and Kast, R.F., 1968. Cuadrángulos Camaraca y Azapa. *Carta Geol. Chile*, 19–20: 5–20.

Tornquist, A., 1898. Der Dogger am Espinazito Pass. *Palaeontol. Abh., N.F.*, 8: 3–69.

Toubes, R.O. and Spikermann, P., 1974. Algunas edades K/Ar y Rb/Sr de Plutonitas de la Cordillera Patagónica entre los paralelos 40° y 44° de latitud sur. *Asoc. Geol. Argent. Rev.*, 28 (4): 382–396.

Vicente, J.C., 1970. Tectónica. In: *Conferencia sobre Problemas de la Tierra Sólida (Proyecto Internacional del Manto Superior), I. Seminario sobre el Programa Geofísico Andino y Problemas Geológicos y Geofísicos relacionados, Buenos Aires*, pp. 162–188.

Vicente, J.C., 1972. Reflexiones sobre la porción meridional del sistema peripacífico oriental. *Symp. Upper Mantle Invest., Buenos Aires, 1970.*

Vicente, J.C., 1974. Exemple de volcanisme initial euliminaire: les complexes albitophyriques néo-triasiques et méso-jurassiques du secteur côtier des Andes Méridionales centrales (32 à 33° L. Sud). *Proc. Symp. 'Andean and Antarctic Volcanology Problems', Santiago de Chile*, pp. 267–329.

Vila, T., 1976. Secuencia estratigráfica del Morro de Arica, Provincia de Tarapacá. *Primer Congr. Geol. Chileno, Actas*, I: A1–A10.

Volkheimer, W., 1968. Esporas y granos de polen del Jurásico de Neuquén (República Argentina), I. Descripciones Sistemáticas. *Ameghiniana*, 5 (9): 330–370.

Volkheimer, W., 1970a. Neuere Ergebnisse der Anden Stratigraphie von Süd-Mendoza (Argentinien) und benachbarter Gebiete und Bemerkungen zur Klimageschichte des Südlichen Andenraums. *Geol. Rundsch.*, 59 (3): 1088–1124.

Volkheimer, W., 1970b. Jurassic microfloras and paleoclimates in Argentina. *Second Gondwana Symp., Proc. Pap., South Africa*, pp. 543–549.

Volkheimer, W., 1972. Estudio palinológico de un carbón caloviano de Neuquén y consideraciones sobre paleoclimas Jurásicos en la Argentina. *Univ. Nac. La Plata, Mus., Rev., Paleontol.* 4 (40): 101–157.

Volkheimer, W., 1974. Palinología estratigráfica del Jurásico de la Sierra de Chacai-Co y adyacencias (Cuenca Neuquina), II. Descripción de los palinomorfos del Jurásico inferior y Aaleniano (Formaciones Chacai Co y Los Molles). *Ameghiniana*, 11 (2): 135–172.

Volkheimer, W., 1978. Microfloras fósiles. *Relatorio Geología y Recursos Naturales del Neuquén, Septimo Congr. Geol. Argent.*, pp. 193–207.

Volkheimer, W. and Baldis, D.P. de, 1975. Significado estratigráfico de microfloras Paleozoicas y Mesozoicas de la Argentina y países vecinos. *Segundo Congr. Ibero-Americano Geol. Econ., Buenos Aires*, IV: 403–424.

Volkheimer, W. and Quatrocchio, M., 1975a. Palinología estratigráfica del Titoniano (Formación Vaca Muerta) en el área de Caichigüe (Cuenca Neuquina), A. Especies terrestres. *Ameghiniana*, 12 (3): 193–241.

Volkheimer, W. and Quatrocchio, M., 1975b. Sobre el hallazgo de microfloras en el Jurásico superior del Borde Austral de la Cuenca Neuquina (República Argentina). *Primer Congr. Argent. Paleontol. Bioestr., Actas*, I: 589–615.

Volkheimer, W., Manceñido, M.O. and Damborenea, S., 1978a. La Formación Los Patos (nov. form.) Jurásico inferior de la alta cordillera de la provincia de San Juan (República Argentina), en su localidad tipo (río de los Patos sur). *Asoc. Geol. Argent. Rev.*, 32 (4): 300-311.

Volkheimer, W., Manceñido, M.O. and Damborenea, S., 1978b. Zur Biostratigraphie des Lias in der Hochkordillere von San Juan, Argentinien. *Münster Forsch. Geol. Palaeontol.*, 44/45: 205–235.

Wahnish, E., 1942. Observaciones geológicas en el Oeste de Chubut. Estratigrafía y fauna del Liásico en los alrededores del río Genua. *Arg., Dir. Nac. Geol. Min., Bol.*, 51: 1–73.

Weaver, C., 1931. Paleontology of the Jurassic and Cretaceous of West Central Argentina. *Univ. Wash. Mem.*, 1: 1–469.

Weaver, C., 1942. A general summary of the Mesozoic of South America and Central America. *Proc. Am. Sci. Congr., 8th, Washington, 1940. IV Geol.*, 149–193.

Weeks, L.G., 1947. Paleogeography of South America. *Am. Assoc. Petrol. Geol., Bull.*, 31 (7): 1194–1241.

Westermann, G.E.G., 1967. Sucesión de ammonites del Jurásico medio en Antofagasta, Atacama, Mendoza y Neuquén. *Asoc. Geol. Argent. Rev.*, 22 (1): 65–73.

Westermann, G.E.G. and Riccardi, A.C., 1972a. Amonites y estratigrafía del Aaleniano–Bayociano en los Andes Argentino–chilenos. *Ameghiniana*, 9: 357–389.

Westermann, G.E.G. and Riccardi, A.C., 1972b. Middle Jurassic ammonoid fauna and biochronology of the Argentine–Chilean Andes, I. Hildocerataceae. *Palaeontographica*, A140: 1–116.

Westermann, G.E.G. and Riccardi, A.C., 1975. Edad y taxonomía del género *Podagrosiceras* Maubeuge et Lambert (Ammonitina, Jurásico Medio). *Ameghiniana*, 12 (3): 242–252.

Westermann, G.E.G. and Riccardi, A.C., 1976. Middle Jurassic ammonite distributions and the affinities of the Andean faunas. *Primer Congr. Geol. Chileno Actas*, I: C23–C39.

Westermann, G.E.G. and Riccardi, A.C., 1979. Middle Jurassic ammonoid fauna and biochronology of the Argentine–Chilean Andes, II. Bajocian Stephanocerataceae. *Palaeontographica*, 164A: 85–188.

Westermann, G.E.G. and Riccardi, A.C., 1980. The Upper Bajocian ammonite *Strenoceras* in Chile: first circum-

Pacific record of the Subfurcatum Zone. *Newsl. Stratigr.*, 9 (1): 19–29.

Wiedmann, J., 1980a. El límite Jurásico–Cretácico: Problemas y soluciones. *Segundo Congr. Argent. Paleontol. Bioestr., Actas*, V: 103–120.

Wiedmann, J., 1980b. Paläogeographie und Stratigraphie im Grenzbereich Jura/Kreide Südamerikas. *Münster Forsch. Geol. Palaeontol.*, 51: 27–61.

Wood, R. and Freiberg, M., 1977. Redescription of *Notoemys laticentralis* the oldest fossil turtle from South America. *Acta Geol. Lilloana*, 13 (6): 187–204.

Yrigoyen, M., 1976. Observaciones geológicas alrededor del Aconcagua. *Primer Congr. Geol. Chileno, Actas*, I: A169–A190.

Yrigoyen, M., 1979. Cordillera Principal. *Segundo Simposio Geología Regional Argentina*. Acad. Nac. Cienc., Córdoba, I, pp. 651–694.

Zambrano, J.J., 1980. Cuencas sedimentarias de la parte austral del Continente Sudamericano: Esquema preliminar de su evolución a fines del Jurásico y comienzos del Cretácico. *Segundo Congr. Argent. Paleontol. Bioestr., Actas*, V: 15–39.

Zeil, W., 1960. Zur Geologie der nordchilenischen Kordilleren. *Geol. Rundsch.*, 50: 639–673.

Zeil, W., 1964. *Geologie von Chile*. Borntraeger, Berlin, 233 pp.

Zeil, W., 1979. *The Andes, A Geological Review*. Borntraeger, Berlin, 260 pp.

Zeil, W., 1980. Los Plutones de los Andes. *Acad. Nac. Cienc. Córdoba, Bol.,* 53 (1–2): 45–58.

Zeil, W., Damm, K.W. and Pichowiak, S., 1980. Los plutones de la Cordillera de la Costa al norte de Chile. In: *Nuevos Resultados de la Investigación Científica Alemana en Latinoamérica*. Deutsche Forschungsgemeinschaft, Bonn, pp. 112–122.

Chapter 9

THE CRETACEOUS OF ARGENTINA, CHILE, PARAGUAY AND URUGUAY

N. MALUMIÁN, F.E. NULLO and V.A. RAMOS

INTRODUCTION

Geological investigation of the Cretaceous of southern South America, and in the Andean Cordillera and Neuquén Embayment in particular, dates back to the late 19th century with the work of Bodenbender (1889), Steinmann in 1902 and 1903, Beherendsen (1891–1892) and Haupt (1907). The modern phase began with the establishment by Groeber (1929 and subsequent papers) of a series of Jurassic–Cretaceous sedimentary cycles in the Central Andes (Table I). He showed that the classic European boundaries cannot be readily determined since neither the upper nor lower Cretaceous boundary coincides with a diastrophic episode, although through the work of Weaver (1927, 1931) equivalence could be established with the European standards. It is unfortunate that the Magallanes or Austral Basin with its nearly complete Cretaceous sequence is not better known. Systematic work in this complex basin and its relationship with the Antarctic Cordillera began relatively late (Cecioni, 1955; Katz, 1963; Leanza, 1967a; and others) after the pioneering work of Hatcher (1897–1900), Bonarelli and Nágera (1921) and Feruglio (1931).

Subsequently the development of local stage names (Natland et al., 1974) and zonation (Malumián, 1968; Malumián and Masiuk, 1975; and others) based upon foraminifers has permitted stratigraphic correlation between the basins of southern South America as well as tentative correlation with European stages. To this must be added the initial palynological works of Archangelsky and Gamerro (1965–1967) and micropaleontological studies of Martinez Pardo (1965) and Bertels (1969). The latter defined the Cretaceous–Tertiary boundary in northern Patagonia.

Orogenic and epeirogenic Cretaceous phases

In general during the Mesozoic, as in the Cenozoic, it is possible to generalize by saying that the western margin behaved as an orogenic belt, while the eastern was affected by the opening of the Southern Atlantic with the differentiation of a number of discrete basins lying between relatively positive "mesocratonic" areas; their distribution is shown in Fig. 1. There is a remarkable synchronism between coastal fluctuation in the east and tectogenic phases in the Chilean–Argentinian Cordillera during the Andean orogenic cycle, with epeirogenic phases coinciding with periods of compression separated by periods of extension (Charrier and Malumián, 1975). The principal tectonic work was carried out by Stipanicic and Rodrigo (1970), Aubouin and Borrello (1966, 1970), Borello (1969), Vicente (1970) and Charrier and Vicente (1972); the relation of tectonic events to faunal change and events in the South Atlantic was pointed out by Malumián and Báez (1976).

TABLE I

Jurassic–Cretaceous sedimentary cycles (after Groeber, 1929)

Tertiary	Riograndico cycle	Malalhueyan (marine) subcycle
		Neuquenian (continental) subcycle
Cretaceous	Andean cycle	Huitrinian–Diamantian (continental) subcycle
Jurassic		Mendocian (marine) subcycle

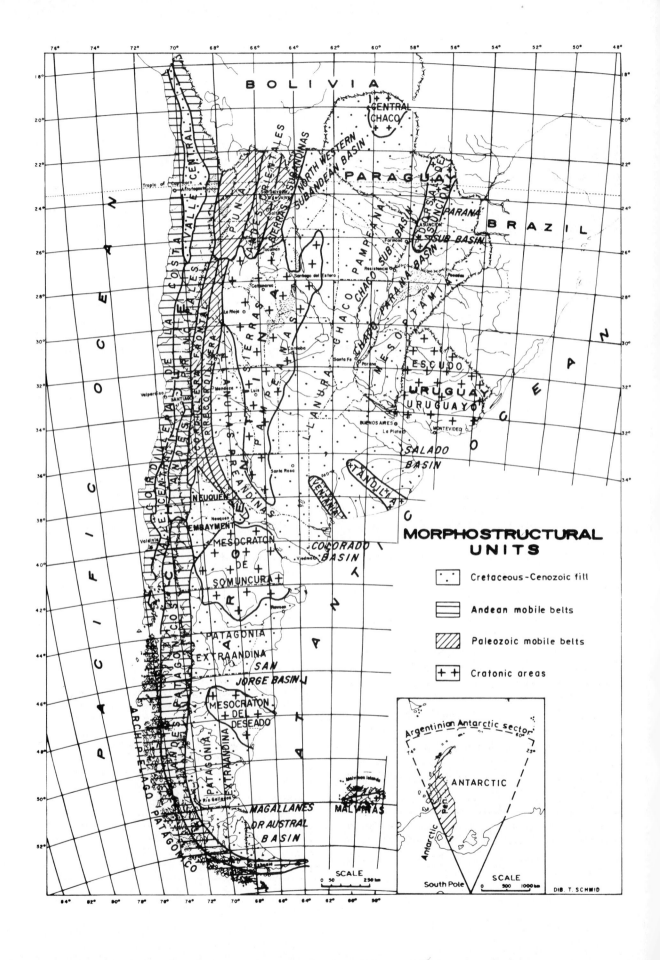

Fig. 1. Morphostructural units. In southern South America a series of morphostructural units each with a distinctive geologic evolution and a tectonic framework can be distinguished. Some of these units acted as the foreland of the Cretaceous Andean transgression from the Pacific Ocean. Others have been activated during the formation of the Atlantic Ocean as continental basins. These differences in evolution and in the distribution of the pre-Mesozoic cratonic areas are responsible for the morphostructural pattern of the region. From the Pacific coast to the east there are the following units.

(1) Cordillera de la Costa (Coastal Range). An Upper Paleozoic positive axis, upon the eastern slopes of which a thick marine and volcanic Mesozoic series of the Andean cycle accumulated.

(2) Valle Central (Central Valley). A central N–S depression which separates the Andean Cordillera from the Coastal Range mainly filled with clastic Cenozoic sediments.

(3) Andes Principales (Main Andean Cordillera). The main Cenozoic cordilleran belt which contains the Mesozoic geosynclinal deposits uplifted during the Late Tertiary Andean orogeny.

(4) Andes Patagonicos (Patagonian Andean Cordillera). South of 39°S the evolution of the Andean belt is strikingly different having a greater oceanic participation particularly marked in the Cretaceous evolution of the Andes.

The austral geosynclinal sector contrasts with the rectilinear structures of the central sector. Three main morphostructural units are distinguishable:

(a) the Patagonian Archipelago, which includes Paleozoic geosynclinal deposits;

(b) Patagonian Range, constituted by (i) an occidental strip with eugeosynclinal sediments (Upper Jurassic–Mid-Cretaceous) graywackes and siltstones with radiolarite intercalations and some flysch-like sediments, and (ii) an oriental strip with Paleozoic metamorphic rocks.

(5) Puna. The southern extension of the 'Altiplano', a high plain up to 4,000 m above sea level, which acted as a mobile belt during Early Paleozoic time and was epeirogenically depressed at the end of the Cretaceous, then elevated, uplifted at the end of the Tertiary.

(6) Andes Orientales (Eastern Andes). A portion of the Paleozoic mobile belt uplifted and thrust to the east by the rigid Puna blocks during the Andean orogeny. Complex imbricate structures characterize the tectonic style.

(7) Sierras Subandinas (Subandean belt). A folded belt of Upper Paleozoic and Mesozoic sediments covered by thick Tertiary sequences which form several continental basins.

(8) Sierras Pampeanas (Pampean Ranges). This is a rigid Precambrian basement, with a block structure, which constituted the foreland of the Paleozoic geosynclinal evolution.

(9) Llanuras Preandinas (Pre-Andean Plains). These plains cover a complex set of continental basins and positive blocks which have been the shore line of the marine Cretaceous ingressions from the Pacific Ocean.

(10) Engolfamiento Neuquino (Neuquén Embayment). An important embayment of the Cretaceous sea over the stable shelf, where conspicuous Neocomian deposits are preserved under Upper Cretaceous continental sequences. It is also known as Neuquén Basin.

(11) "Macizo" de Somuncura (Somuncura "Massif"). Widespread Precambrian basement complex remobilized during the Triassic–Jurassic times covering this area with acid volcanic sequence. It acted as a stable rigid block during the Andean orogeny.

(12) "Macizo" del Deseado (Deseado "Massif"). A similar but slightly older reactivation during the Variscan orogeny characterizes this "Massif", which had cratonic behavior during the Andean evolution.

(13) Patagonia Extra-Andina (Extra-Andean Patagonia). The basaltic plains of this region are structurally interrupted by the Deseado "Massif". The northern area underwent active continental subsidence during the Middle Cretaceous (San Jorge Basin); while the southern area was invaded by the Cretaceous seas (Austral or Magallanes Basin).

(14) Llanura Chacopampeana (Chacopampean plains). A cratonic Late Cenozoic plain which covers the isolated continental Salado and Colorado basins.

(15) Macizo del Paraguay (Paraguayan Massif). This is an Early Paleozoic area which has been fractured during Late Mesozoic times. It is bounded by the Late Paleozoic Paraná Basin.

(16) Mesopotamia. This cratonic region is tectonically bounded by a set of Andean fractures which partially uplifted a continental clastic volcanic sequences of Early Cretaceous age.

(17) Escudo Uruguayo (Uruguayan Shield). This Precambrian basement (1,800 to 2,000 m.y.) was tectonically active during the Atlantic opening. It is bounded by several Cretaceous continental basins.

(18) Sistema de Tandilia (Tandilia System). A set of ranges which emerge from the Pampean Plains as a faulted part of the Precambrian basement of the Uruguayan Shield.

(19) Sistema de Ventania (Ventania System). This is a typical aulacogen controlled by a triple-junction pattern whose western branch was highly mobilized during Paleozoic times. A set of northwestern ranges are interrupted by the Atlantic coast.

The main orogenic phases providing the framework for the sedimentary cycles are the Araucanian, Peruvian, and Laramide phases.

Araucanian phase

Following the Late Oxfordian regression there was a general emergence during Kimmeridgian times and this controlled the pre-Tithonian–Neocomian paleogeography. As a result in the western Chilean coastal ranges there is a para-unconformity and in some places Tithonian–Neocomian sediments transgress over Middle Jurassic beds. In eastern and central Patagonia there was intensive block faulting of Jurassic volcanic rocks followed by tensional release which led to restricted vulcanism. In the Neuquén Embayment and Magallanes Basin beds of the Mendocian subcycle rest unconformably upon older rocks (Digregorio, 1978; Ramos, 1979). This was also the time of the first granodioritic intrusions in the Patagonian Cordillera.

Peruvian phase (= Patagonidica, Intrasenonica, Subhercynian, or Mirano of different authors)

This diastrophic event interrupted a period of subsidence, folded and faulted Lower Cretaceous rocks and was responsible for the unconformity below the basal Upper Cretaceous conglomerates. The conglomerates grade in thickness and in particle size eastwards until they are represented by a few hundred meters of red continental sediments on the eastern flank. This phase was related to the emplacement of the main tectonic Andean granodiorite plutonic cycle (98 ± 4 m.y., Ramos and Ramos, 1979).

Prior to the main event two precursory phases have been identified, the earliest pre-Peruvian (= Infraneocomian, Catanlil, or early Patagonidica; see Stipanicic and Rodrigo, 1970, and Ploszkiewicz and Ramos, 1978) and the second precursory phase (= early Mirano or Patagonidica). The first of these phases is recognized in the Neuquén Embayment and the Main Cordillera by the appearance of a continental facies in the western part of the Mulichinco Formation in Berriasian times, and by continental conglomerates in the Fontana and Pueyrredon lakes during the Neocomian of the Magallanes Basin (Hatcher, 1900; Ramos, 1976). The younger precursory phase was responsible for regression, emergence and vulcanism of the Cordillera Patagonica in the northern part of the Magallanes Basin (Ploszkiewicz and Ramos, 1978); in the Main Cordillera and Neuquén Embayment it is re-

sponsible for an Upper Neocomian regression (Digregorio, 1972). Some basaltic vulcanism in the Chacoparana, central Argentinian and Northwest Subandean basins indicates a phase of tensional release.

The Peruvian phase is considered to be one of the major orogenic phases of the Andean cycle. According to Vicente (1970), during this phase the internal series were thrusted over the external.

Laramide or Late Cretaceous–Early Tertiary phase

This phase is recognized in the Main Cordillera and along the eastern slopes of the Coastal Range. The andesitic–basaltic extrusions and Early Tertiary formation (Lo Valle–Farellones) are clearly unconformable upon older rocks, in particular upon the Cretaceous andesitic sequence of the Abanico Formation and upon the continental deposits of Las Chilcas Formation, which has been dated from its content of dinosaur bones (Casamiquela et al., 1969).

In the eastern side of the continent a marine ingression followed the Huantraico phase during Maastrichtian times.

The north–south trending third granodiorite cycle is related to this phase (Ruiz, F., 1965; Levi, 1968).

Cretaceous sedimentary cycles

Outside the geosynclinal area the effects of the three orogenic phases show up in terms of transgression and regression of the epicontinental seas or by breaks in the sedimentary sequences (Charrier and Malumián, 1975). This variable history and its effects upon sedimentation forms the basis of Groeber's Cretaceous sedimentary cycles mentioned earlier. The paleogeography was controlled by the morphostructural elements depicted in Fig. 1, and an attempt has been made to present the paleogeography for four time intervals (Figs. 1–3). A more notable change took place at the Berriasian–Valanginian boundary than at the Tithonian–Berriasian one. It should be more practicable in this region to establish the base of the Cretaceous System at the base of the Valanginian stage (Leanza, 1981).

The Lower Cretaceous (Mendocian) subcycle

Groeber (1953) recognized a group of Tithonian–Neocomian rocks which form a single transgressive sedimentary cycle in the Neuquén Embayment (Fig. 1) to which he gave the name 'Mendoza Group'. Distinc-

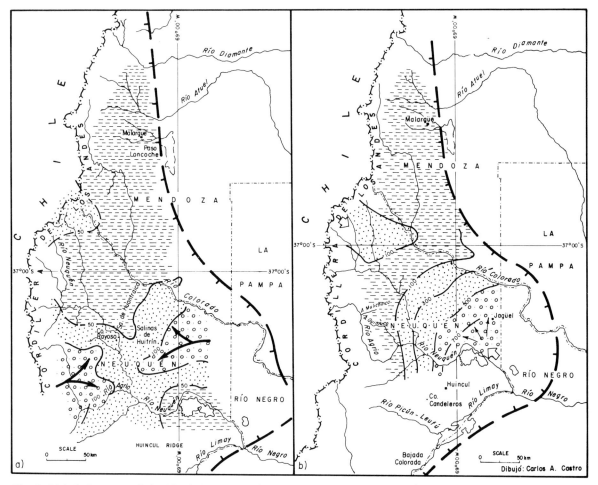

Fig. 2. Lithofacies maps of the Mendocian cycle of the Main Cordillera and Neuquén Embayment. (a) Isolithic map of the Mulichinco and Chachao formations. (b) Idem for the Agrio Formation. Thickness in meters. (After Uliana et al., 1975b.)

tion between Tithonian and Neocomian rocks is only possible paleontologically. A zone of subsidence extended to the eastern margin of the continent where red beds, deposited in the Chacoparaná Basin, are widely distributed (Fig. 2). Detrital clastic sediments form the dominant lithological type and continue to be so throughout the Cretaceous. However, there were some subordinate limestones laid down during this cycle.

The cycle is terminated by a marine regression, the Huitrinian–Diamantian subcycle in the Neuquén Embayment associated with movements of the early Mirano Subhercynian tectonic phase. The basaltic effusions in the Salado, Colorado and Northwestern Subandean basins are contemporary with the opening of the South Atlantic Ocean.

The Middle Cretaceous (Huitrinian–Diamantian) continental subcycle

The Middle Cretaceous (Aptian–Albian) cycle is characterized by the continental regime established in the Northwestern Subandean, Salado, Colorado and San Jorge basins as well as in the Neuquén Embayment (Fig. 4). The locus of subsidence appears to have been displaced westwards, with respect to the previous cycle, with the inception of the Alemania Sub-basin (Fig. 3).

In Patagonia red beds of the Chubut Group reach their maximum development. Particularly in the Austral Basin this cycle is represented by marine sediments, initially formed of pelitic deposits which subsequently become micritic.

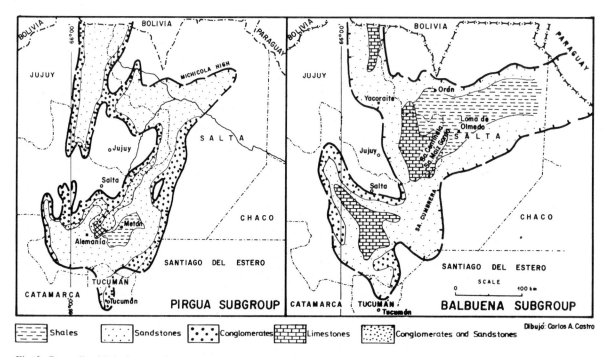

Fig. 3. Generalized lithofacies maps of the Northwestern Subandean Basin. After Moreno, 1970, and Reyes and Salfity, 1973.

The Upper Cretaceous (Neuquenian) subcycle

In the Neuquén Embayment the sediments of this Cenomanian–Campanian cycle are essentially red beds. In northern and central Chile, however, some volcanic and pyroclastic sediments do occur. In Patagonia sedimentation of the Chubut Group continued and correlative beds are found in the Salado and Colorado basins.

Thus continental red-bed sedimentation covers the greater part of the Northwestern Argentine Basin, western Paraguay and southern Bolivia. Only in the Colorado Basin is there some evidence of the first (? Turonian) transgression from the Atlantic.

The terminal Cretaceous (Malalhueyan in part) marine-cycle

The sediments of this group form a continuous sequence into the Cenozoic, and only a small part of the total sequence is Cretaceous. The cycle differs from the preceding by the widespread occurrence of marine horizons due to the Huantraico and equivalent phases.

The transgression began at least as early as lower Maastrichtian; at this time the sea covered northern Patagonia, the Colorado and Salado basins, and continued up to Danian when marine waters flooded the San Jorge and Peninsula de Valdes basins and covered, at least partially, the Áustral Basin. Reyment (1977) suggested that the causes of this singular Atlantic transgression may be sought in the deformation of the geoid. A sudden and general regression set in in late Danian times as a consequence of the Laramide phase.

There are also classical marine sequences in localities in Chile on the Pacific continental margin while in the Northwestern Subandean Basin marine and nonmarine horizons are intercalated.

GENERAL STRATIGRAPHY

The geological history of the region is best summarised by considering the basins (see Fig. 4) successively.

The Andean Basins

The Western Andean Basin (Cordillera de la Costa)

The volcanogenic sequence found in this basin is typical of a fore-arc basin, and is quite distinct from that found in the Cordillera Principal. The succession opened with the Tithonian–Neocomian marine

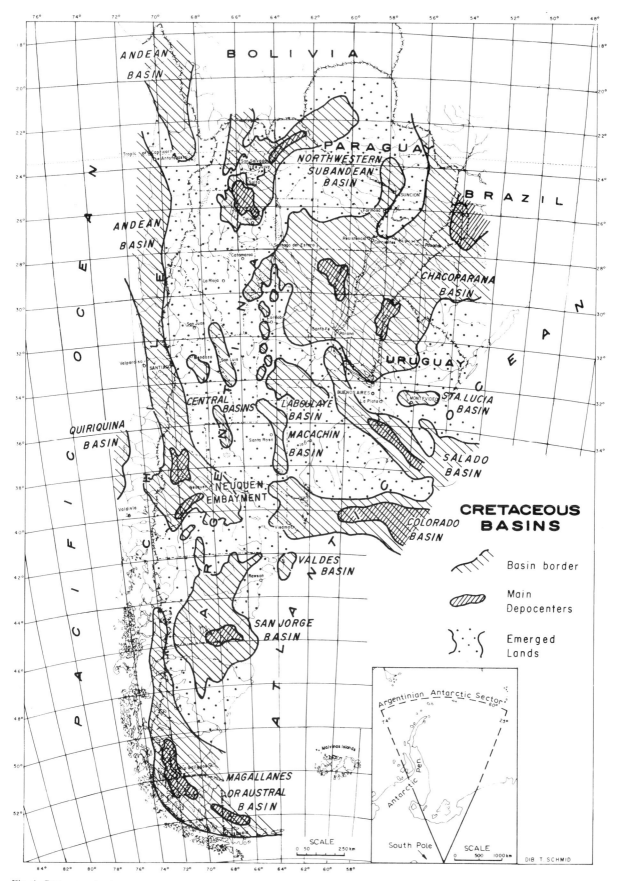

Fig. 4. Cretaceous basins.

TABLE II

Main formations of the Main Andean Cordillera and Coastal Range

AREA / AGE	C. de la COSTA (Coastal Range)	CORDILLERA PRINCIPAL (Main Andean Cordillera)	
		CHILEAN SLOPE	ARGENTINIAN SLOPE
MAASTRICHTIAN	Quiriquina (M)		Loncoche-Roca (M)
CAMPANIAN SANTONIAN CONIACIAN CENOMANIAN	Las Chilcas (C-M)	Abanico (C)	Neuquén Group (C)
ALBIAN	Veta Negra (M)	Colimapu (C)	Diamante (C)
APTIAN			Huitrín (T)
BARREMIAN HAUTERIVIAN	Lo Prado (M)		Agrio (M)
VALANGINIAN	Pachacana (M)	Baños del Flaco (M)	Mulichinco (C-M)
BERRIASIAN			Quintuco (M)
TITHONIAN	Patahua (M)	Leñas Espinoza (M)	Vaca Muerta (M)

C = continental; M = marine; T = transitional. After Fuenzalida, 1964; Zeil, 1964; Ruiz Fuller, 1965; Charrier and Vicente, 1972; Yrigoyen, 1972; Aubouin et al., 1973.

TABLE III

Main features of the most representative formations of the Coastal Range
(After Zeil, 1964, and Aubouin et al., 1973)

Formations	Lithology	Main features
Quiriquina (75 m)	Sandstones and conglomerates	Maastrichtian; marine; fishes, birds, reptiles; cephalopods, gastropods, pelecypods; neritic
Las Chilcas	Volcani-sedimentary conglomerates and tuffs	Maastrichtian; continental
Veta Negra (9,000 m)	Andesites with subordinate red beds	Aptian–Albian; continental
Lo Prado (1,000 m)	Bioclastic limestones and siliceous volcanic rocks	Upper Valanginian–?Barremian; marine; *Thurmannites, Holcoptychites; Isoarca, Trigonia*
Pachacama (1,200 m)	Breccias, tuffs, graywackes and volcanic rocks	Valanginian–Berriasian; marine?
Patagua	Sandstones, limestones, breccias and conglomerates	Tithonian–Berriasian; marine; *Olcostephanus, Argentiniceras, Thurmanniceras*

transgression with littoral marine conglomerates, graywackes and limestone interbedded with andesitic and rhyolitic tuffs and volcanics. These rest unconformably upon Dogger or Liassic tuffs and keratophyric flows. One of the most complete sections in the Cordillera de la Costa, on the western flank of La Campana hill (60 km NW of Santiago de Chile) has been described by Thomas (1958), Aliste et al. (1960) and Corvalán and Davila (1964), and a brief lithological description of the units is given in Table III. The appearance of red beds of the Veta Negra Fm. during Barremian–Aptian time marks emergence of the region. The succession is terminated by the deposition of in excess of 4 km of andesitic porphyrics and tuff which ended in Cenomanian times.

In the latest Cretaceous a fore-arc basin developed in the region of Isla Quiriquina, and a marine sequence, the Quiriquina Formation, preserved between Algarrobo and Puerto Mont in the south, transgressed over the Las Chilcas Formation on the western flank of the Cordillera de la Costa to rest upon metamorphic basement. This transgressive Maastrichtian sequence is made up of conglomerates and littoral sandstones and is dated by the occurrence of *Baculites* and associated faunas.

The Eastern Andean Basin (Cordillera Principal)

The Lower Cretaceous sequence is similar to that found in the Neuquén Embayment (see below). The Tithonian–Neocomian transgression between 18°S and 40°S was extensive and over 600 m of sediment overlie the Kimmeridgian volcaniclastic Rio Damas

Formation on the Chilean slopes of the Cordillera Principal. These sediments are locally divided into two marine formations (Leñas Espinoza and Baños del Flaco formations) which together constitute a mioliminar sequence of the Mendocian cycle (Charrier and Vicente, 1972; Aubouin et al., 1973). A brief lithological description is given in Table IV. Their maximum northerly occurrence is at Quebrada El Way where a typically Lower Cretaceous fauna was recorded by Farias and Vergara (1964).

During Barremian times there were distinctive differences in the depositional environments on either side of the Andes. On the western side in Chile volcanic tuffs and breccias, together with red beds make up the 1,800–2,000 of the Colimapu Formation. On the eastern side volcanism is absent in the 600 m thickness of the Huitrin Formation (Table II) but dolomites occur and gypsum horizons are found within the red argillites which together with sandstone constitute the principal lithologies.

The Cretaceous of the Cordillera Principal closed with a thick volcanic sequence the Abanico Formation, up to 5 km thick in the axial region. It rests unconformably upon Lower to Middle Cretaceous rocks, and is dated by reptilian remains as Maastrichtian. On the Argentinian slope (Table II) the Upper Cretaceous is represented by a 1,000 m red-bed sequence named the Neuquén Group.

The Neuquén Embayment

The marine-continental Cretaceous sequence in the Neuquén Embayment constitutes part of the fill

TABLE IV

Main features of the most representative formations of the Cordillera Principal, Chilean slope
(After Ruiz Fuller, 1965, and Aubouin et al., 1973)

Formations	Lithology	Main features
Abanico (more than 3,000 m)	Gray and reddish andesites; trachites, tuffaceous breccias	Maastrichtian; continental; reptiles
Colimapu (3,000 m)	Red beds, volcanic breccias and tuffs; microconglomerates and lacustrine limestones	Barremian–Coniacian; lacustrine; continental
Baños del Flaco (800 m)	Marls and calcareous sandstones	Valanginian–Barremian?; marine; *Neocomites, Crioceras, Berriasella, Olcostephanus*
Leñas Espinoza (1,500 m)	Calcarenites and sandstones	Tithonian–Berriasian?; marine; *Berriasella, Cuyaniceras*

(The Baños del Flaco and Leñas Espinoza formations are grouped together under RIO DAMAS FM.)

TABLE V

Main formations of the Neuquén Embayment

AREA / AGE	SOUTHERN (Río Limay, Neuquén Prov.)	CENTRAL (Bajada del Agrio, Neuquén Prov.)	NORTHERN (Malargüe, Mendoza Prov.)	
MAASTRICH.	Jagüel ⓜ	Roca (pars) ⓜ		Malargüe Group
CAMPANIAN	Allen ©c	Loncoche ©c		Malargüe Group
to	Río Colorado ©c			Neuquén Group
up	Río Neuquén ©c			Neuquén Group
CENOMANIAN	Río Limay ©c			Neuquén Group
ALBIAN	Bajada Colorada ©c	Rayoso	Diamante	
APTIAN	Bajada Colorada ©c	Huitrín ⓣ		
BARREMIAN	Limay member ⎱ Ortiz member ⎰ Agrio	Upper member ⎱ Avilé member ⎰ Agrio	Agrio ⓜ	Mendoza Group
HAUTERIVIAN	Pichi Picún Leufú m. ⎰	Lower member ⎰	Agrio ⓜ	Mendoza Group
VALANGINIAN	Mulichinco ⓜ		Chachao ⓜ	Mendoza Group
BERRIASIAN	/////////////	Quintuco ⓜ	Chachao ⓜ	Mendoza Group
TITHONIAN	/////////////	Vaca Muerta ⓜ		Mendoza Group

C = continental; M = marine; T = transitional. After Digregorio, 1972, 1978; Cazau and Uliana, 1973; Digregorio and Uliana, 1980.

of a back-arc basin. The Tithonian—Neocomian marine sediments buried a "horst and graben" topography, with the margins formed by step-faulted basement. Some parts of this topography apparently remained active, for example the Huincul Ridge which emerged during part of the Cretaceous. The sediments are mioliminar with no volcanic activity.

With such a topography, facies changes are relatively common and resulted in numerous local formational names being assigned (Table V), and differences of opinion arising concerning many of the stratigraphical relationships. Brief lithological descriptions of the principal units in the Neuquén Embayment, with some paleontological information can be found in Table VI.

The earliest known sediments, dark pelites and limestones of the Vaca Muerta Formation (lower Tithonian—Berriasian) are found in the west. These are followed by a sequence of late Berriasian to late

Valanginian sandstones and conglomeratic sandstones carrying an abundant *Olcostephanus* fauna. In the centre of the basin these are represented by the Quintuco and Mulichinco formations which together are the equivalent of the Chachao Formation found in the north (Table V). Again in the western part of the basin the Mulichinco Formation is overlain by limestones, sandy limestones and coquinas of the lower Hauterivian—lower Barremian Agrio Formation.

The Agrio Formation changes in lithology southwards and becomes essentially terrigenous, and here three members (Pichi Picún Leufú, Ortiz and Limay) were defined which were once assigned to the Early Cretaceous La Amarga Formation (see Weaver, 1931; Groeber, 1946, 1952; Digregorio, 1972, 1978; and Leanza et al., 1978).

Regional uplift and regression terminated deposition of the Agrio Formation and the succeeding lacu-

TABLE VI

Main features of the formations of the Neuquén Embayment*[1]

Formations	Lithology	Main features
Jagüel (150 m)	Yellowish claystones	Maastrichtian; marine; *Ostrea*, *Gryphaea*, *Eubaculites argentinicus*
Loncoche (300 m)	Green-yellowish claystones	Maastrichtian; continental to marine
Allen (100 m)	Gray-yellowish green claystones and mudstones; conglomeratic sandstones	Continental; fresh water pelecypods: *Corbicula*, *Hydrobia*, *Viviparus*
Rio Colorado (300 m)	Red mudstones with levels of calcareous concretions and geodes of calcite; 'Ondulites'; cross-bedding; variegated sandstones	Continental; alluvial to flood plain; dinosaur remains
Río Neuquén (600 m)	Red mudstones with alternating light sandstones; polymictic conglomerates	
Río Limay (600 m)	Quartzose medium coarse yellow sandstones; conglomeratic sandstones; Mn concretions; acid to mesosilicic vulcanites, conglomerates	
Bajada Colorada (300 m)	Red claystones with subordinate conglomeratic layers	Aptian–Albian; continental; flood plain; dinosaur remains
Rayoso (1,000 m)	Red claystones; fine to medium reddish sandstones; variegated claystones	Aptian–Albian; continental; pseudodeltaic, lacunar; bones remains; fresh water pelecypods
Huitrín (500 m)	Gypsiferous sandstones with layers of gypsum; argillaceous sandstones with layers of salt	Aptian–Albian; continental to marine; *Melania, Corbicula, Diplodon*
Agrio (1,500 m)	Argillaceous sandstones, green limestones; gypsum, dolomites; dark gray claystones, shales and marls	Hauterivian–Barremian; neritic, sublittoral; *Pseudofavrella*, *Crioceras*; Ostracods, foraminifers
Chachao (150 m)	Gray, greenish thick bedded bioclastic limestones with *Exogyra*	Valanginian–Berriasian; marine; in southern Mendoza Province include Quintuco and Mulichinco formations
Mulichinco (300 m)	Red sandstones, conglomerates in its southern part; limestones and coquinas, light sandstones, in the central and northern parts	Valanginian; marine; littoral; deltaic; *Neocomites, Olcostephanus, Lissonia*
Quintuco (300 m)	Limestones, oolitic limestones; biostromes; calcareous and marly sandstones	Berriasian; neritic; sublittoral; *Cuyaniceras*, *Spiticeras*, pelecypods
Vaca Muerta (1,200 m)	Greenish shales, claystones with intercalated sandstones, argillaceous tuffs	Tithonian–Berriasian; neritic; littoral–sublittoral; *Virgatosphinctes*, *Argentiniceras*, *Substeueroceras*

*[1] After Groeber, 1952; Marchese, 1971; Digregorio, 1972; Cazau and Uliana, 1973; Musacchio, 1978; Volkheimer, 1978.

strine limestones and evaporites followed by red-beds of the regressive cycle were assigned to the Huitrin Formation. They are overlain by the Rayoso and Bajada Colorada formations, deposits of a fluviatile/lacustrine environment which included also pseudo-deltaic and flood plain phases. The only fossils, bone fragments, carbonaceous plant material and occasional freshwater pelecypods and the ostracods from the Rayoso Formation do not permit the assignment of a precise date. There is a suggestion from the microflora (Volkheimer and Salas, 1975a, b) that the base may be extended back into Albian time.

Unconformably overlying the Rayoso Formation is a thick continental series referred to as the Neuquén Group. The essential lithology of every formation within the Neuquén Group is red mudstone, sandstone and polymictic conglomerate. Although

sauropods have long been known to be present, there are no recent studies and such knowledge of these "Dinosaur beds" dates back to Lydekker (1893) and Von Huene (1929). On stratigraphical grounds the Neuquén Group is probably of Coniacian–Campanian age.

The uppermost Cretaceous is represented by sediments of the Loncoche Formation (or Lower Huantrai-co Formation) and its equivalent in the Rio Negro Province, the Jagüel Formation. The beds record a transition from a continental to a marine environment. In the lower part of the Huantrai-co Formation gray, variegated pelitic sediments with ostracods and characeae give way to argillites and foraminiferal limestones of Maastrichtian age in the upper part (Bertels, 1978). The Roca Formation which concordantly overlies the Loncoche Formation has a Danian microfauna in the upper part.

TABLE VII

Main formations of the Austral and Magallanes basins (all marine)[1]

EUROPEAN STANDARD	CHILEAN STANDARD	Santa Cruz Province		Última Esperanza Departament	Region of I. Riesco and P. Brunswick	ISLA GRANDE DE TIERRA DEL FUEGO				
		North to Shehuen River	Between Shehuen and Gallegos Rivers			Cordillera	Springhill District	NNE	Southern Cordillera	
		ARGENTINA		C	H	I	L	E	ARGENTINA	
MAASTRICHT.	RIESCONIAN		Man Aike	Dorotea	Rocallosa	Cerro Cuchilla	Z. Glauconítica	"Z. Glauconítica"		
			Chorrillo	Tres Pasos	Fuentes Rosa Barcarcel		"Lutitas arenosas"			
CAMPANIAN										
SANTONIAN	LAZIAN	Cardiel	Mata Amarilla	Cerro Toro	Nodales	Cerro Matrero	"Lutitas gris verdosas"	Cabeza de León	Río Claro group	
CONIACIAN			Piedra Clavada							
TURONIAN		Piedra Clavada								
CENOMANIAN	PENINSULIAN			Punta Barrosa		Pizarras de La Paciencia		Arroyo Alfa		
ALBIAN	TENERIFIAN	Río Mayer	Palermo Aike	Zapata	Pizarras de Froward		"Margas"	Nueva Argentina	"Capas Hito XIX"
APTIAN							Lutitas con Ftanitas		
BARREMIAN	PRATIAN								Beauvoir/ Monte Olivia/
HAUTERIVIAN	ESPERANZIAN				Pizarras Río Jakson	Estratos con Favrella	Pampa Rincón	Yaghan/ Hardy
VALANGINIAN								
BERRIASIAN	RINCONIAN	Springhill		Complejo Arenoso Basal	Sylva Palma		Springhill	
TITHONIAN								

[1] After: Feruglio, 1949; Katz, 1963; Leanza, 1963, 1967a, 1969a, 1970, 1972; Herm, 1966; Charrier and Lahsen, 1968; Borrello, 1969; Russo and Flores, 1972; Natland et al., 1974; Malumián and Masiuk, 1975, 1976, 1978.

The Austral or Magallanes Basin

The Austral basin, in contrast to those previously described has a marine sequence in which most of the Cretaceous is represented (Table VII). It is divided into a system of stages on foraminiferal grounds defined in Chile by Natland et al. (1974) and subsequently subdivided in zones in Argentina by Malumián and Masiuk (1975, 1976, 1978). The composition of the foraminiferal assemblages is distinct from the European. Studies of the macrofauna have been made by Leanza (1967a, 1969a, 1970, 1972) and Riccardi (1971, 1976, 1977). The former author suggested that the pattern of sedimentary deposition was out of phase with that in the Cordillera Principal and Neuquén Embayment.

The age of the basal deposit in the Austral Basin, the Springhill Formation sandstones which mark marine ingression into the basin, has been a matter of controversy. Formerly regarded as Kimmeridgian (Natland et al., 1974) or Aptian (Leanza, 1972) it is now assigned to the Tithonian–Valanginian interval (Riccardi, 1976, 1977). The basin progressively deepened to an Aptian–early-Cenomanian maximum (Tenerifian–Peninsulian in local terminology) and then shallowed because of general uplift culminating in the uppermost Cretaceous (Riesconian).

According to age and sedimentological environ-

TABLE VIII

Main features of the most representative formations of the Austral Basin

Formations	Lithology	Main features
(a) *Region north of Shehuen River, Santa Cruz Prov.* [1]		
Cardiel (300 m)	Reddish and violetish claystones and tuffs; reddish and yellowish subordinate sandstones	
Mata Amarilla/Chorrillo (300 m)	Grayish, yellowish, dark grayish mudstones and claystones; subordinate sandstones	M. Amarilla Fm.: continental to marine; *Peroniceras, Placenticeras, Corbula, Exogyra, Potamides;* arenaceous foraminifera Chorrillo Fm.: continental; charophytes; Coniacian
Piedra Clavada (400 m)	Yellowish and gray-yellowish medium to coarse grained sandstones; cross-bedding	Coniacian?; marine, deltaic, mixed; *Acteonella, Corbula, Exogyra, Potamides*
Río Mayer (700 m)	Dark shaly gray pelites; rhythmic alternation with thin-bedded gray and gray-yellowish psammites	Valanginian–Cenomanian; marine; flysch; *Acanthoceras, Mantelliceras, Sanmartinoceras. Parasilesites, Lechites; Favrella, 'Tropaeum', Calycoceras, Belemnopsis*; arenaceous foraminifera
Springhill (100 m)	Sandstones and claystones	Titonian–Hauterivian; marine–continental
(b) *Department of Ultima Esperanza region* [2]		
Dorotea (1,200 m)	Gray to green, yellowish and brown to reddish sandstones, commonly with lenticular conglomerates and some clay intercalations	Maastrichtian; marine shallow water; molasse; *Pholadomya, Inoceramus, Pinna*; shark teeth; plant fragments; *Hoplitoplacenticeras, Holcodiscus, Maorites*
Tres Pasos (2,500 m)	Thick-bedded sandstones, quick lateral change of facies between shales and siltstones; massive rust-colored, moderately indurate sandstones; subordinate shales; mud-chip conglomerates	Upper Campanian; marine shallow water; molasse; *Hoplitoplacenticeras, Pseudokossmaticeras; Inoceramus*

TABLE VIII (continued)

Formations	Lithology	Main features
Cerro Toro (2,000 m)	Shales with indurate sandstones interbedded; include a thick conglomerate in the upper part (Lago Sophia Conglomerate)	Coniacian up to Lower Campanian (?); marine; flysch; *Parabinneyites, Anapachydiscus, Parapuzosia; Inoceramus; Chondrites, Helminthoidea*
Punta Barrosa (400 m)	Dark shaly gray pelites; rhythmic alternations with fine-grained psammites	Albian–Cenomanian; marine; flysch; *Turrilites*; planktonic foraminifera, arenaceous autochthonous forms
Zapata (1,200 m)	Fine-grained black, generally argillaceous sediments, partly silicified	Hauterivian–Aptian?; marine euxinic conditions; *Aucellina, Belemnopsis; Favrella, Olcostephanus*; Radiolaria; Foraminifera (Nodosariacea)

Complejo Arenoso Basal (Springhill equivalent)

(c) *NNE of Tierra del Fuego*[3]

Cabeza de León (350 m)	Gray and greenish silty claystones and argillaceous siltstones with *Inoceramus* prisms	Lower Maastrichtian–Turonian Upper part: agglutinate foraminifers; brackish water Lower part: inner shelf; *Notoplanulina rakauroana, Pseudospiroplectinata ona*
Arroyo Alfa (160 m)	Light gray claystones with *Inoceramus* prisms	Upper Albian–Cenomanian; outer shelf; *Lingulogavelinella, Spiroplectinata, Dorothia, Valvulineria, Tritaxia*
Nueva Argentina (250 m)	Light gray biomicrites and biomicritic claystones ――――――――――――――― Light gray claystones with white specks	Aptian–Albian; outer shelf to upper bathyal; *'Globigerinelloides' gyroidinaeformis*, Radiolaria
Pampa Rincón (120 m)	Dark gray claystones	Hauterivian–Valanginian; shallow neritic; *Lenticulina nodosa, Astacolus gibber, Pseudopolymorphina martinezi*
Springhill (150 m)	Fine- to coarse-grained angular light gray quartzitic sandstone, silty shales and siltstones	Tithonian–Hauterivian; shallow marine to continental

(d) Main features of the formations of the Cordilleran island arc[4]

Yaghan (3,000 m)	Quartz-poor volcaniclastic turbidites and shales; chert levels; (prehnite-pumpellyite facies of metamorphism)	Upper Jurassic–Lower Cretaceous; island-arc-derived flysch sequence; *Favrella, Inoceramus*; Radiolaria; trace fossils
Hardy (1,300 m)	Volcaniclastic rocks, lava flows, rhyodacite to basalts; andesites dominant	Upper Jurassic–Lower Cretaceous; marine; islandic-arc-assemblage; *Favrella, Belemnopsis*; Radiolaria

[1] After: Feruglio, 1949; Leanza, 1969a, 1972; Russo and Flores, 1972.
[2] After: Wilckens, 1904; Paulcke, 1906; Brandmayr, 1945; Feruglio, 1949; Cecioni, 1957; Zeil, 1958; Katz, 1963; Leanza, 1963; Herm, 1966; Scott, 1966; Riccardi, 1979; Blasco et al., 1980b.
[3] After: Flores et al., 1973; Malumián and Masiuk, 1975, 1976, 1978; Malumián and Báez, 1976.
[4] After: Kranck, 1932; Katz and Watters, 1966; Dalziel et al., 1974; Suarez, 1976; Suarez and Pettigrew, 1976; Dalziel and Palmer, 1979.

ment several general divisions can be recognized, a Lower and Middle Cretaceous flysch series in the Cordilleran area, coeval stable shelf sediments in the extra-Andean region and an Upper Cretaceous molasse in both areas. Local stratigraphical details are briefly indicated below, lithological and some paleontological details are given in Table VIII.

Departamento Ultima Esperanza (Chile)

Euxinic marine beds of the Zapata Formation (= Erezcano Fm.) transgressed over the Upper Jurassic quartz porphyries of the Quemado Formation. The former are partially time equivalents of the rocks of the Springhill Formation. The succeeding thick Punta Barrosa and Cerro Toro formations represent a regimen of sedimentation of flysch type, and the presence of planktonic Foraminifera in the lower part of the Cerro Toro Formation indicates further deepening and connection with the open sea. However, a typical autochthonous flysch benthic foraminiferal fauna have also been recovered. Near the top a thick conglomerate, the Lago Sofia Conglomerate also occurs.

The uppermost Cretaceous molasse, assigned to two formations, the Tres Pasos and Dorotea formations, show rapid lithological variations, and this is reflected in the development of local formational names. The upper limit of the Tres Pasos Formations is the rich faunal horizon "f" of Hauthal which contains *Hoplitoplacenticeras* and *Lahillia luisa*. (Wilckens, 1904; Paulcke, 1906; Katz, 1963; Herm, 1966).

Region of Isla Riesco and the Brunswick Peninsula (Chile)

Although there is a lack of detailed information two, rather more altered formations, the Nodales and Pizarras de Froward formations, are equated with the Zapata, Punta Barrosa, Cerro Toro formations. The Pizarras de Froward Formation is a striped dark slate-shale estimated as 2,500 m thick. The Nodales Formation (Complejo de Estero la Pera) consists of an estimated 2,000 m of gray shale with sandy and conglomerate horizons. The Barcarcel Formation consists of 1,000 m of shale with subordinate sandstones (Cecioni, 1960b; Gonzales et al., 1965).

The uppermost Cretaceous rocks consist of a sequence of shales (the Natales and Fuentes formations) overlain by argillaceous rocks which become coarse-grained deltaic deposits assigned to the Rocal-losa and Tres Morros formations. Although the fauna is poor, the Fuentes Formation can be recognized as Maastrichtian from the sporadic occurrence of some *Bolivinoides* (Herm, 1966). They are covered by the arenaceous Chorrillo Chico Formation which has a poor fauna, and may include the Cretaceous—Tertiary boundary time interval (Charrier and Lahsen, 1968).

The Springhill district (Chile) and the NNE of Tierra del Fuego (Argentina)

The sandstones of the Springhill Formation considered by Borrello (1969) as a proto-molasse rest unconformably upon the "Serie Tobifera". They have been subdivided in Chile into a lower, continental Manantiales Formation and the marine, upper Sombrero Formation.

They are followed by the "Estratos con Favrella" in Chile and the equivalent Pampa Rincón Formation in Argentina. The name in Chile derives from the common ammonite genus *Favrella*, but in subsurface it may be identified from the occurrence of the Foraminifera group, *Lenticulina nodosa—Astacolus gibber*. The succession continues with the Nueva Argentina Formation which is equivalent to the radiolarite bearing Lutitas con Ftanitas and the Margas formations in Chile. At least in the southeastern part of the area the contact with the underlying beds is an unconformable one. The upper part of the Nueva Argentina Formation is micritic and near the Chilean—Argentinian border, where it crops out it is referred to as the "Capas del Hito XIX" in Argentina, or as the Vicuña Formation in Chile. It is followed by the Arroyo Alfa Formation (Argentina) (the lower part of the Lutitas Gris Verdosas Formation in Chile) and this marks the end of the transgression in this basin for the facies of the Lutitas Arenosas Formation in Chile and its Argentinian principal equivalent the Cabeza de Leon Formation, mark the development of the Cretaceous regression. The terminal unit, the Zona Glauconítica, is a sedimentary complex which contains sediments ranging in age from Maastrichtian through Danian and into Lower Eocene (Herm, 1966; Flores et al., 1973; Natland et al., 1974; Malumián and Masiuk, 1975, 1976, 1978; Malumián and Báez, 1976).

The interfluve area Shehuen—Gallegos rivers, Santa Cruz Province

In the Cordillera there appears to be a poor development of the Springhill Formation. The succession often begins with black shales of the Rio Mayer Formation the equivalent of the Pampa Rincón, Nueva Argentina and Arroyo Alfa formations of Tierra del Fuego. It is recognized as a succession of dark pelites with subordinate psammites in the region of San Martin, Viedma and Argentino lakes, where it is overlain with presumed unconformity by the sandstones of the Piedra Clavada Formation. In the Cordilleran area it is followed by the Cerro Toro Formation, a flysch type sediment with typical agglutinated Foraminifera. In the southern sector the Piedra Clavada Formation and the overlying Mata Amarilla Formation change to pelitic facies of Río Guanaco Formation (Blasco et al., 1980b). The latter formation contains a microfossil similar to that of the Cerro Toro Formation, including in its lower part calcareous Foraminifera known in the middle Cabeza de León Formation (Malumián and Proserpio, 1979; Malumián, 1978) and Lazian stage. Its main dinoflagelate contents are *Chattangiella* sp., ? *Tenua* sp. and *Oligospharidium* cf. *complex* (D. Pöthe de Baldis, pers. comm., 1981).

The Mata Amarilla Formation changes in lithological character and northwards interfingers with continental pyroclastic sediments similar to those found in the lower part of the Chubut Group.

The terminal Cretaceous beds assigned to late Maastrichtian are the yellowish cross-bedded, conglomeratic sandstones in the Man-Aike Formation (and its equivalent the Calafate Formation).

In the subsurface at the deeper part of the basin the pelitic Palermo Aike Formation would be the equivalent of all the above mentioned units with the exception of the Springhill Formation (Feruglio, 1949; Leanza, 1969a, 1970, 1972; Russo and Flores, 1972; Riccardi and Rolleri, 1980; Nullo et al., 1981).

Region north to the Shehuen River (Argentina)

In addition to the formations found in the preceding section the 300 m of reddish, greyish and yellowish claystones and tuffs of the Cardiel Formation crops out between the Shehuen River and Cardiel Lake. It contains dinosaur remains. Stratigraphically it interfingers with, or covers the Mata Amarilla Formation (Russo and Flores, 1972).

Recently, in the uppermost Río Mayer and lowermost Belgrano beds *Hatchericeras santacrucense*, *Sanmartinoceras patagonicum*, *Colchidites* cf. *colchicus* and *Emericiceras* sp. were found. This ammonitiferous assemblage and the formational boundary were assigned to late Barremian (Blasco et al., 1980a). Besides, it was found the first association between

macrofossils and microfossils in the Río Mayer Formation with ammonites: *Hatchericeras patagonense*, *Cryptocrioceras yrigoyeni* etc., pterosaurus remains and crustacean decapods associated with a foraminiferal assemblage, of very low diversity, dominated by *Epistominella caracolla*, and very rare *Lenticulina nodosa*, *Astacolus gibber*, *Lenticulina reyesi*, *Planularia crepidularis*, *Saracenaria tsaramandrosoensis* (Aguirre Urreta and Ramos, 1981). This microfauna indicates a Pratian age.

Region between Pueyrredon and Fontana lakes, Chubut (Arg.) and Aisen (Chile) provinces

This part of the Austral Basin is marginal to the extra-Andean Rio Chico High, the positive element composed of Jurassic volcanics which forms the northeastern limit to the basin. On the western side of the basin, north of Pueyrredon—Cochrane Lake a Lower Cretaceous volcanic sequence is widely developed in the Cordillera. Marine black shales of the Coyhaique and Katterfeld formations in which *Lenticulina nodosa* (foraminifer) and *Favrella americana* (ammonite) have been found suggest an age (Valanginian—Hauterivian) equivalence with the Pampa Rincón Formation. These beds are separated by a conglomeratic episode from older marine rocks in the Fontana and Ghio Lakes.

Occasional exposures of limestones, sandstones and shales, assigned to the Tres Lagunas Formation provide the sole indication of the Tithonian to Valanginian marine transgression (Ploszkiewicz and Ramos, 1978; Ramos, 1979; Masiuk and Nakayama, 1979).

The succession ends in Argentina with a sandstone sequence passing into continental facies, the Apeleg Formation. This unit represents the deltaic fill of the basin prior to intense volcanicity (Ramos, 1976; Ploszkiewicz and Ramos, 1978; Masiuk and Nakayama, 1979). The volcanic rocks are characteristically calc-alkaline and seem to have developed from a complex arc system (Ramos, 1979).

Region between Futaleufu and Fontana Lake (Argentina and Chile)

At the northern end of the Austral Basin the volcanics of the Lower Cretaceous form a relatively narrow belt. Isolated remains of the sediments deposited during the Early Cretaceous marine transgression have been reported from Palena (Chile), Carrenleufu and Temenhuao (Argentina) (Fuenzalida, 1968; Thiele et al., 1978; Pesce, 1979). They are

TABLE IX

Main formations of the northern sector of the Austral Basin[1]

AGE \ AREA		LAGO PUEYRREDÓN (Arg.)	COYHAIQUE (Chile)	LAGO FONTANA (Arg.)	SIERRA de PAYANIYEU (Arg.)	FUTALEUFÚ PALENA (Chile)	CARRENLEUFÚ (Arg.)
MAASTRICHT.	RIESCONIAN	/////	/////	/////	/////	/////	/////
CAMPANIAN		/////	/////	Muzzio Gabro	/////	/////	Morro Serrano Gabros
SANTONIAN	LAZIAN	/////	/////	/////	/////	/////	/////
CONIACIAN		/////	Granites	La Plata Chico Granite	La Magdalena Granite	Granite	Corcovado Granite
TURONIAN		/////	/////	/////	/////	/////	/////
CENOMANIAN	PENINSULIAN	"Chubut Gr. equivalents" (C)	Nirehuao α	El Gato γ	Cordón de las Tobas α		
ALBIAN	TENERIFIAN			Carrenleufú α	Nirehuao α		Carrenleufú α
APTIAN		Upper Conglomerates			Payaniyeu γ		
BARREMIAN	PRATIAN		Divisadero γ	Apeleg (M)			
HAUTERIVIAN	ESPERANZIAN	Belgrano Beds (M)			Apeleg	Alto	Cerro
VALANGINIAN		Lower Conglomerates		Katterfeld (M)	(C)		Campamento (M)
BERRIASIAN	RINCONIAN	Ghio Beds (M)	Coyhaique (M)	Tres Lagunas (M)	Tres Lagunas (M)	Palena	Arroyo Cajón α
TITHONIAN							

[1] M = marine; C = continental; α = andesites; γ = rhyolites. After: Hatcher, 1897, 1900; Feruglio, 1931, 1950; Heim, 1940; Skarmeta, 1976; Skarmeta and Charrier, 1976; Ramos, 1976; Ploszkiewicz and Ramos, 1978; Thiele et al., 1978; Pesce, 1979.

regarded as equivalent to part of the Katterfeld and Tres Lagunas Formations. Volcanism, however, was nearly continuous from Upper Jurassic to Lower Cretaceous. In the pre-Andean region Valanginian to Hauterivian marine beds are interbedded in a thick andesitic–dacitic volcanic sequence.

North of Futaleufu town no Cretaceous rocks have been reported, and although the Chubut (Juan Fernandez Land) Ridge was still an important positive element separating the Neuquén and Austral basins, it is not impossible that Cretaceous outcrops may yet be found in this difficult and inaccessible region.

Cordillera Fueguina and associated arc systems

Following Dalziel et al. (1974), Suarez (1976), and others, three tectono-sedimentary belts, paralleling the Pacific margin can be recognized in this southernmost region of the Cordillera de los Andes. From east to west these are:

(1) A continental shelf assemblage, an area external to that previously described in the Springhill District—NNE Tierra del Fuego where no magmatic activity has occurred.

(2) An island arc-derived volcani-clastic flysch sequence deposited in a back-arc marine basin (Suarez, 1976) corresponding to the Yaghan Formation which is floored by an ophiolitic sequence (Dalziel, 1974; Dalziel et al., 1974). These rocks were interpreted as having been intruded into continental crust as the result of back-arc extension which created the marginal basin (Katz, 1963; Suarez and Pettigrew, 1976).

(3) An ensialic island-arc assemblage related to an east-dipping subduction zone which includes gently folded andesitic, dacitic and basaltic volcani-clastic rocks and lavas of the Hardy Formation (Suarez and Pettigrew, 1976). This belt extends to the southernmost extremity of the Andes in the Upper Jurassic and Lower Cretaceous reaching the South Georgia Islands.

The Yaghan Formation (Table VII) is the coeval equivalent of the Zapata Formation of Seno Ultima Esperanza and its 3,000 m of volcani-clastic flysch represents the eugeosynclinal facies of the Southern Andes (Katz, 1963; Borrello, 1969). Further east, in Isla de los Estados, the Beauvoir Formation is the equivalent Lower Cretaceous unit (Dalziel et al., 1974; Caminos and Nullo, 1979).

The volcani-clastic Lower Cretaceous sequence with its ophiolitic magmatic activity, has also a regional metamorphism partially obliterated by the post-orogenic intrusion of the Patagonia batholith.

The Northwestern Subandean Basin

This Cretaceous basin is made up of a number of tectonic troughs, in which the evolution of the lower members is somewhat variable. The initial fill which consists of conglomerate and red-beds is assigned to the Pirgua Subgroup and may total in excess of 3,500 m. Several sub-basins can be recognized, e.g., the Alemania, Metan, Lomas de Olmedo and Tres Cruces (Table X). The occurrence of several basalt flows in the first two-named sub-basins permits radiometric dating showing that the rocks are contemporary, in

TABLE X

Main formations of the Northwestern Andean Basin*1

AREA / AGE	EL CADILLAL (Tucumán Prov.)	ALEMANIA (Southern Salta Prov.)		LOMAS DE OLMEDO (Southern Salta Prov.)
PALEOCENE		Santa Bárbara Subgroup Pars / Balbuena Subgroup	Mealla (C) / Olmedo (C)	
MAASTRICH.			Yacoraite (M)	
			Lecho (C)	
CAMPANIAN	Río Loro (C)		Los Blanquitos	
SANTONIAN	?		76 (C) / 78 (β)	
CONIACIAN		Pirgua Subgroup (C)	Las Curtiembres	Pirgua Subgroup (C)
TURONIAN				
CENOMANIAN	97 (β)		99 (β)	
ALBIAN	103 (β)		La Yesera (C)	
APTIAN	112 (β)			
BARREMIAN	112 (β)			
HAUTERIVIAN	El Cadillal (C)		114 (β)	
VALANGINIAN				
BERRIASIAN	128 (β)			

*1 C = continental; M = marine; β = basalts.
After: Bonaparte and Bossi, 1967; Moreno, 1970; Reyes and Salfity, 1973; Reyes et al., 1976; Valencio et al., 1976.

part, with the Serra Geral Basalts (Reyes and Salfity, 1973; Reyes et al., 1976; Valencio et al., 1976).

The entire basin was invaded from the northwest (Bolivia) in Campanian–Maastrichtian time, so that littoral and lagoonal deposits onlap onto rocks of continental type (for descriptions see Table II). These conditions extended eastwards to the Paraguayan Chaco Plains.

The Las Curtiembres Formation is interesting because it contains horizons where pipid frogs have been found. This frog fauna has suggested faunal links with Africa (Reig, 1959; Baez, 1975). The remaining fauna, fish and plant remains are poorly represented. Although there is an abundant fauna preserved in the Balbuena Subgroup — melanid gastropods (Bonarelli, 1921; Danielli and Porto, 1968), dinosaur, carnosaur teeth, and crocodiles (Bonaparte and Bossi, 1967) have all been recorded — to the present time they have proven to be of small chronostratigraphic value. Stromatolites are the most widespread and best preserved lithofacies in most of the basin. The microfauna of charophytes, ostracods and rare benthonic Foraminifera suggest a late Senonian age although certain forms are considered no younger than Turonian (Malumián and Báez, 1976).

The presence of evaporites in the Olmedo Formation may indicate more severe climatic conditions at the end of the Cretaceous, or a general regression.

The Subandean Basin extended towards the Chaco Plains and into Paraguay, where it is locally referred to as the Pirity Basin, terminating against the Central Chaco High (see Fig. 2). The high, formed of Ordovician to Devonian rocks, separates the continental-marine rocks of the Salta Group from the Cretaceous of the Curupaity Sub-basin. The latter, a wedge of continental red beds, reaches important dimensions in Bolivia (Banks and Diaz de Vivar, 1975).

The Chacoparaná Basin

The Chacoparaná Basin is one of the largest cratonic basins in South America. It covers the southern half of the Brazilian Paraná Basin and extends into northwestern Argentina, eastern Paraguay and northwestern Uruguay. The tectonic style and sedimentary fill indicate a typically passive marginal basin of Atlantic type with conspicuous gravity faulting. The basin was repeatedly active during the Upper Paleozoic and in Cretaceous time.

Although the basin subsided fairly uniformly dur-

TABLE XI

Main features of the formation of the Northwestern Andean Basin[*1]

Formations	Lithology	Main features
Yacoraite (800 m)	Oolitic and dolomitic limestones; yellowish and greenish clay; calcareous sandstones with intercalated black shales; stromatolithes	Maastrichtian; marine to transitional; Ostrea, gastropods; foraminifers, ostracods; Charophytes; *Coelodus*
Lecho (300 m)	Yellowish or whitish medium or coarse sandstones, somewhat calcareous	
Los Blanquitos (1,500 m)	Conglomeratic layers; whitish pink, medium to coarse sandstones with conglomerates of Q and granite	Senonian; continental; titanosaurids
Las Curtiembres (2,000 m)	Basaltic flows, dykes and sills; reddish brown mudstones, claystones and shales with layers of fine conglomerates; the pelitic content with intercalated sandstones, increases upward	Middle Cretaceous–Senonian; continental; anurans; plant remains
La Yesera (600 m)	Reddish-brown medium to conglomeratic sandstones; polymictic conglomerates with clast of phyllite, quartzite; basaltic flows	Middle Cretaceous; continental; piedmont deposits

[*1] After: Bonaparte and Bossi, 1967; Leanza, 1969b; Reyes and Salfity, 1973; Valencio et al., 1976.

ing the Cretaceous, it has been subdivided by Putzer (1962) into an eastern Paraná Sub-basin, bounded on the one side by the Assuncion Arch (Banks and Diaz de Vivar, 1975), with the Chaco Sub-basin on the other, west of Paraguay and the middle Paraná River. In this sub-basin no Cretaceous sediments are exposed but its history seems to be similar to that of the Paraná Sub-basin. In the Assuncion Arch rocks ranging in age from Precambrian to Carboniferous are found.

The Misiones Sandstone crops out on both sides of the arch with the same characteristics. Yet Padula (1972) supposed that the southern subsurface continuation of the arch could have been incipiently active during the Cretaceous. The outcrop of the Misiones Sandstone is shown on the maps of Harrington (1956), Eckel (1959) and Putzer (1962), although these authors assigned the formation to the Triassic. Yet they correlate the Misiones Sandstone with the Botucatu Sandstone Formation in Brazil which they recognize as interfingering its upper levels with flows of the Serra Geral for which a Lower Cretaceous radiometric date (125–110 m.y.) has been now determined. Thus there seems little doubt

that the Misiones Sandstone is latest Jurassic to Early Cretaceous in age, in close relationship with the early Gondwana break-up and the opening of the South Atlantic Ocean.

In northeastern Argentina in the Paraná Sub-basin, the San Cristobal Formation, thin bedded sandstones overlain by basalt flows, is partially equivalent to the Misiones Formation. It extends further east into Uruguay, crops out near Tucuarembo and La Paloma and continues into Brazil. Small exposures occur near Concepcion, Rosario and at locations near Asuncion in Paraguay.

In the southern part of the Chacoparana Basin there is no evidence in favor of two separate sub-basins. To the northwest lies the Charata–Presidente Hayes Arch a positive area which stopped the Maastrichtian marine transgression from entering the Chacoparana Basin. This subsurface feature is the northerly extension of the Pampean Ranges. The Chaco Sub-basin extends as far as the Sierras de Cordoba for in its southwestern section small remnants of Cretaceous sandstone and basalts are preserved. However, it is cut off in the south by the Martin Garcia High

(Zambrano and Urien, 1970), a northwest trending structural high along which Precambrian basement has been uplifted with possible significant transcurrent displacement (Yrigoyen, 1975a).

The tholeiitic character of the basalts contrasts strongly with the alkalic character of contemporary basalts further west in the northwestern Argentina region and in the Central basins of San Juan, Mendoza and San Luis.

The Lower Cretaceous rocks near Gaspar on the Uruguay River reach a maximum thickness of 1,400 m, at a further depocenter west of the Paraná River the thickness reaches 800 m, but elsewhere it more commonly varies between 50 m and 250 m.

A regional unconformity separates the relatively thin Upper Cretaceous rocks from the Serra Geral Basalt. Along the Pilcomayo megashear numerous alkalic plugs and stocks have been intruded in southeastern Paraguay and Brazil and possible bodies of alkalic rocks have recently been reported in the Misiones Province of Argentina. The age of these intrusions is in the 80–75 m.y. range.

In Paraguay, Putzer (1962) reported Upper Cretaceous sediments in the Cordillera de Amambay and in the vicinity of the Acaray River which were correlated with the Bauru Formation of southern Brazil. Other isolated outcrops have been reported from Mercedes (Corrientes, Argentina) and in the Guichon area (Uruguay), locally called the Yerua and Guichon formations (Herbst, 1971; Padula, 1972). The age assignment is based upon the recovery of dinosaur bones belonging to *Laplatosaurus araucaniens* Huene, *Antarctosaurus wichmannianus* Huene and *Argyrosaurus superbus* Lydekkker. These Upper Cretaceous beds are also said to occur in the Paraná Sub-basin. However, in the subsurface of the Chaco Sub-basin a different Maastrichtian–Paleocene succession, the Mariano Boedo Formation is found. The rocks of this formation, light to dark sandy clays with some calcareous nodules and gypsum horizons indicate a lacustrine-swampy environment.

Central Argentina basins

These are a series of small Early Cretaceous tectonic basins in the provinces of La Rioja, San Juan, San Luis, Cordoba and Mendoza with a fill of clastic continental sediments. These are pull apart basins related to the Bermejo megashear. The basins are the Ischigualasto–Cerro Bola, Marayes–Guayaguas, northern Mendoza, southern Mendoza and San Luis.

The sediments unconformably overlie Upper Triassic continental rocks, and consist of conglomerates, red-beds and basaltic flows akin to those found in the Paraná Sub-basin. The dating of the flows provides a means of correlation between the basins (Yrigoyen, 1975a). The basins appear to have become inactive during Late Cretaceous.

Atlantic basins

These are a number of basins of passive margin-type characteristic of the Atlantic margin (Table IX).

TABLE XII

Main formations of the central Argentina basins[1]

BASINS / AGE	ISCHIGUALASTO CERRO BOLA	GUAYAGUAS MARAYES	SAN LUIS	NORTHERN MENDOZA	SOUTHERN MENDOZA	SIERRAS DE CÓRDOBA	PARANÁ
CENOMANIAN	/////	/////	/////	/////	/////	/////	/////
ALBIAN		Lagarcito		98 — 98		Conglomerado β	β
APTIAN		La Cruz	La Cruz 107 β / 109	129 β	β	Libertad β	
BARREMIAN	Cerro	Conglomerate	El Congl. Toscal	Barrancas	Pozo	Embalse	San
HAUTERIVIAN		El Toscal / La Cantera	La Cantera		Chimango	Río Tercero	Cristóbal
VALANGINIAN	Rajado	Los Riscos	El Jume				
BERRIASIAN			Los Riscos				

[1] All continental; β = basalts.

After: Flores, 1969; Bossi, 1971; Gordillo, 1972; Stipanicic and Bonaparte, 1972; Yrigoyen, 1975a.

Modern opinion regards them as aulacogenic basins closely related to the opening of the South Atlantic.

The Salado Basin

The Salado Basin is bounded to the north by the Martin Garcia High and to the south by a series of faults parallel to the Tandil High (Zambrano and Urien, 1970) (see Fig. 1). The basin extends northwestwards towards a main depocenter where it is over 3,000 m deep known as the Laboulaye Sub-basin. In the coastal region subsidence may attain 5,000 m and a lithological description of the units which have been defined is given in Table XIII. Although undated, a basalt which floors the succession is regarded as coeval with the Serra Geral. Over this lie continental deposits of the Rio Salado Formation of which only the upper 1,000 m have been drilled although seismic

records indicate it has a thickness of 3,500 m. It is overlain, possibly unconformably, by a predominantly red-bed sequence, the General Belgrano Formation. No marine horizons have been found. This in turn is capped by the Las Chilcas Formation which consists of sandy and silty gypsiferous sediments which become sandier and enriched in gypsum towards the top. They are assigned a Maastrichtian–Paleocene age. They are a neritic-marine facies, partially transitional reflecting an Atlantic ingression which seems to be connected with the conspicuous marine Maastrichtian–Paleocene development in the Neuquén Embayment.

The Colorado Basin

Most of the Colorado Basin lies offshore, and it is there that the succession reaches its maximum thick-

TABLE XIII

Main features of the formations of the Salado and Colorado basins

Formations	Lithology	Main features
(a) *Salado Basin*[1]		
Las Chilcas (1,200 m)	Gray and reddish claystones with anhydrite and gypsum	Marine, neritic, littoral; upper Maastrichtian–Danian
General Belgrano (900 m)	Red, green or variegated fine to coarse sandstones alternating with layers of purple, greenish gray variegated claystones; conglomerates	Upper Cretaceous; equivalent to Guichón Fm. (Uruguay) and Abramo Fm. (Laboulaye Basin)
Río Salado (2,500 m)	Reddish, brown and purple sandstones and claystones partly with light greenish decoloration zones, generally micaceous	Middle Cretaceous; alluvial plain; equivalent to Migues Fm. in Uruguay
	Basaltic rocks	Lower Cretaceous; Arapey lavas in Uruguay
(b) *Colorado Basin*[2]		
Pedro Luro (300 m)	Green, gray or dark gray claystones, generally calcareous	Marine; upper part: Danian, *Globoconusa daubjergensis*; Lower part: Maastrichtian, *Rugoglobigerina* and frequent calcareous Nannoplankton
Colorado (1,700 m)	Red, brown, green or variegated claystones and siltstones, with sandstones generally cross-bedded	Continental; Upper Cretaceous
Fortin (more than 3,500 m)	Dark brown, purplish, red siltstones, claystones and shales; subordinate sandstones	Continental; lacustrine

[1] After: Zambrano and Urien, 1970; Zambrano, 1971, 1974; Yrigoyen, 1975b.
[2] After: Kaasschieter, 1963; Malumián, 1970; Zambrano, 1971, 1974; Yrigoyen, 1975b.

ness. The on-shore limit to the basin is a conspicuous positive zone, the Tandil High or Positivo de las Sierras Bonaerenses of Yrigoyen (1975b). The southern limit is defined by a complex east–west series of faults which constitute a typical aulacogenic basin as proposed by De Wit (1977).

The rocks forming the base of the basin are unknown; upon them lie basalts as in the Salado Basin followed by a red-bed sequence of grey to red clays, and siltstones with sandstones and a few conglomeratic horizons interpreted as a continental alluvial plain environment interspersed with lacustrine intervals. These are followed, with an assumed unconformity indicated by coarser lithologies, by the Colorado Formation (Table XIII). The lithology is variable with variegated sandstones, partly conglomeratic, pyroclastic material, and grey-green shales. The microflora from some grey shales indicate a marine horizon of possible Turonian age, which if true would represent the earliest transgression from the Atlantic.

Calcareous siltstones and shales of the Pedro Luro Formation onlap older rocks and mark the onset of the Maastrichtian–Paleocene transgression which affected most of central and southern Argentina.

The Peninsula de Valdes Basin

A 1,000 m thickness of red purple and green series of sandstones, tuff and tuffaceous sediments terminating in pink argillites resting upon Devonian quartzites and overlain by upper-Danian marine pelites has been found in the subsurface of Peninsula de Valdes (Masiuk et al., 1976). The argillites contain charophytes and ostracods, which to the present time are of no stratigraphical value.

The San Jorge Basin

The San Jorge Basin (Table XIV) lies in central Chubut Province and continues offshore through San Jorge Bay. It is an asymmetrical closed basin, with a gentle north slope and a steeper southern flank (Zambrano and Urien, 1970). Gravity faulting transverse to the Atlantic margin permit one to regard it as an intracratonic basin (Dickinson, 1976).

The continental sandstones, conglomerates, siltstones and pyroclastic rocks are assigned to the Chubut Group and it can be found westwards close to the Cordillera Patagonica; to the south it includes sediments of the Baquero locality. Northward the limits are indefinite. The type locality is in the Sierras de San Bernardo where the sequence is thickest

(Lesta, 1968). There is commonly a basal conglomerate resting unconformably on the Lonco Trapial Formation (Lesta and Ferello, 1972) in central Chubut. Near the Atlantic coast the conglomerate rests on the Chon Aike Formation (Stipanicic and Reig, 1957), and in southern Chubut and northern Santa Cruz provinces upon the Bahia Laura Granite (Feruglio, 1938).

The Pozo D-129 Formation is an Early Cretaceous continental sequence found in subsurface underlying the Chubut Group; it may be equated with the Cañadón Calcáreo Formation which crops out in the middle Chubut River (Proserpio, 1976).

In the north the Chubut Group rests unconformably upon either the Middle Jurassic Cañadón Asfalto Formation (Stipanicic et al., 1968) or the Upper Jurassic–Lower Cretaceous volcanic and clastic sequence of the Taquetrén Formation (Nullo and Proserpio, 1975).

The clastics of the Chubut Group show torrential and cross-bedding with frequent paleochannels. The lower part is essentially conglomerate, the upper part is more fine grained passing at the top to variegated tuffitic sandstone. They may sometimes pass laterally into lagunar grey siltstones.

PALEOGEOGRAPHICAL EVOLUTION

As it has been shown previously, thick clastic continental sediments predominate amongst the Cretace-

TABLE XIV

Main formations of the San Jorge Basin, Chubut and northern Santa Cruz provinces[1]

AREA / AGE	CENTRAL CHUBUT PROV.	MIDDLE CHUBUT RIVER	COMODORO RIVADAVIA	NORTHERN SANTA CRUZ PROV.
Danian		Bororó	Salamanca	
MAASTRICH.		Paso del Sapo	Trébol	
CAMPANIAN to	Talquino		Comodoro Rivadavia	
up	Bajo Barreal	Cerro Barcino		Baqueró ?
	Castillo			
BARREMIAN	Matasiete	Los Adobes	Mina del Carmen	Bajo Grande
HAUTERIVIAN			Pozo D 129	
BERRIASIAN		Taquetrén		
Upper Jurassic		Cañadón Asfalto		

(Central Chubut Prov., Middle Chubut River, and Comodoro Rivadavia columns labelled Chubut Group)

[1] All continental except Bororó and Salamanca formations. After: Lesta, 1968; Stipanicic et al., 1968; Lesta and Ferello, 1972; Nullo and Proserpio, 1975; Codignotto et al., 1978; Lapido and Page, 1979.

ous rocks in the area studied which included Argentina, Chile, Paraguay and Uruguay. In these countries Cretaceous rocks cover nearly half of the surface area.

The paleogeographical evolution of the southern South America during Cretaceous times is governed by two distinct permanent features: (a) to the west by the development of the Andean orogenic belt, more or less related with the active Pacific continental margin; (b) in the central and eastern parts, by the development of intracratonic basins, such as the Paraguayan, Northern and Central Argentinian basins. Only south of the Santa Lucia Basin is this region clearly linked to the evolution of a passive continental margin, i.e., to the early stages of opening of the South Atlantic Ocean.

Four main periods can be recognized in the paleogeographic history of these two different sectors, the Neocomian, Aptian—Albian, Cenomanian—Campanian, and Maastrichtian periods, as illustrated by Figs. 5 to 8.

Neocomian

In the western sector, the paleogeography during the uppermost Jurassic—earliest Cretaceous times is controlled by a subduction process along the Pacific continental margin. As a result the Andean region is characterized by a volcano-plutonic arc related to a series of fore-arc and back-arc basins (Fig. 5). The fore-arc basins were cut by transverse tectonic features and filled with marine platform deposits closely related to the magmatic arc. The back-arc basins were developed on a thinned sialic crust and filled by marine platform deposits assigned to the Mendocian subcycle. Occasionally, the back-arc basins may have been connected with the eastern intracratonic areas, as in the Neuquén Embayment.

The Magallanes Basin shows a more complex structural and depositional setting due to the back-arc spreading which introduced oceanic crust into the floor of the marginal basin. This produced a complex facies pattern where bathyal pelagic deposits are associated with outer to near shore platform sediments, particularly characteristic of the Neocomian sedimentary sequences of the Patagonian and Fuegian Cordilleras. These sediments were frequently deposited under anoxic conditions which prevailed in the South Atlantic Ocean up to the Albian times.

The intracratonic sector is characterized by a series of isolated basins filled with thick clastic deposits. The dominant facies is represented by continental red-bed sequences, with alkali-basic rocks closely associated. For example, tholeiitic basic rocks interfinger with clastic continental deposits in the Paraná Basin. In Southern Patagonia the first continental deposits which fill the San Jorge Basin are associated with pyroclastic material derived from the Andean arc.

According to some authors, subsidence and subsequent clastic continental sedimentation may have been also initiated in pre-Aptian times in the Atlantic basins, such as the Colorado and Salado basins.

Aptian—Albian

Both in the Andean and Atlantic sectors, the succession of geological events during the Aptian—Albian seems to be closely related to an acceleration of the convergence rates between the plate carrying the South American continent and the adjacent oceanic plate.

At that time a general regression occurred in most of the Andean basins. Transitional facies are common in the northern Neuquén Embayment (Fig. 6). Evaporitic deposits interfinger with marine and continental red-bed sequences. At the end of the Albian most of the Andean basins were filled up with continental deposits. Volcanism remained active from the Upper Jurassic, with maximum activity at various times at different parts of the Cordillera.

This high convergence rate is concomitant with a widespread extension of the passive margin in southern South America, resulting in an opening of the South Atlantic Ocean.

The Atlantic basins such as Santa Lucia, Salado, Colorado and San Jorge basins, may have begun their subsidence at this time, in the form of aulacogenic basins developed transverse to the Atlantic margin. The paleogeographic configuration is controlled by the complex fracture pattern of the basement in the extra-Andean region of Patagonia.

The Magallanes Basin is unique because here marine conditions persist until and after the latest Albian. Near shore and outer platform facies are found in the southern part of the basin with widespread pelagic organisms and anoxic bottom conditions, while continental deposits of alluvial and deltaic facies were developed in the extreme northern part of the basin during the same period.

Basic alkaline volcanics are poorly exposed in cen-

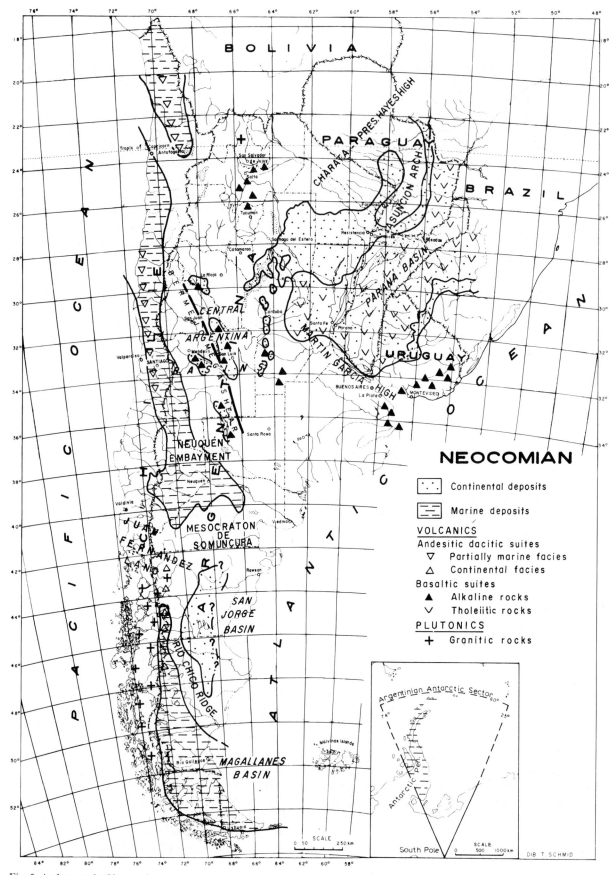

Fig. 5. Andean cycle (Neocomian).

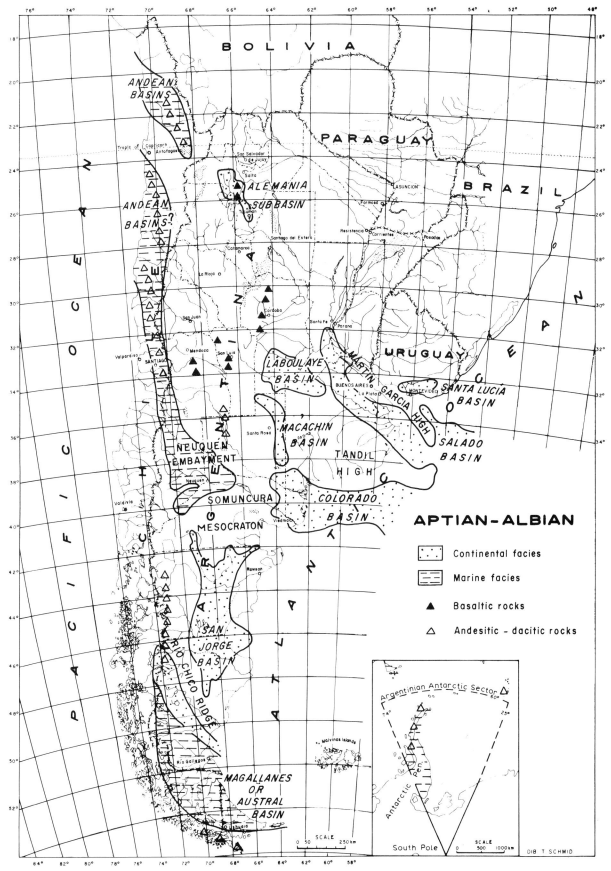

Fig. 6. Huitrinian–Diamantian cycles (Aptian–Albian).

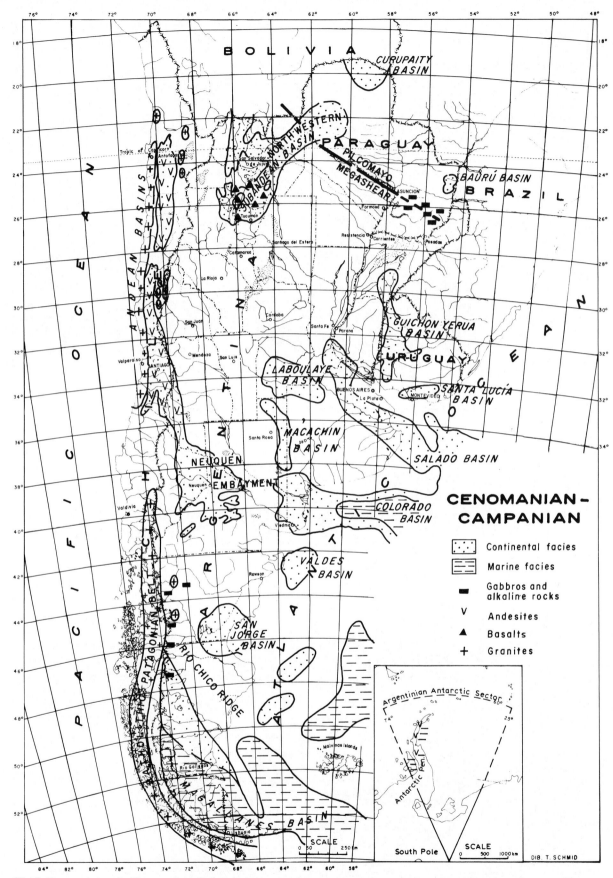

Fig. 7. Neuquenian cycle (Cenomanian–Campanian).

TABLE XV

Main features of the formations of the San Jorge Basin (cf. Table XIV)

Formations	Lithology	Main features
(a) *Central Chubut Province*[1]		
Talquino (250 m)	Red to reddish tuffs, sandstones and conglomerates	Continental-fluviatile; angiosperm remains
Bajo Barreal (250 m)	Gray tuffs and cinerites, greenish tuffaceous claystones. Conglomeratic sandstones	
Castillo (450 m)	Green tuffs and yellow reddish conglomeratic sandstones	Continental-fluviatile
Matasiete (970 m)	Red to reddish tuffs, sandstones and conglomerates	
(b) *Middle Chubut River*[2]		
Cerro Bororo (60 m)	Upper part: greenish claystones; lower part: sandstones and conglomerates	Upper part, marine with Danian foraminifers; lower part, continental, with plant remains
Paso del Sapo (100 m)	Sandstones, conglomerate sandstones, siltstones	Upper Cretaceous; continental to marine
Cerro Barcino (220 m)	Green tuffs; variegate sandy tuffs	Upper Cretaceous (Cenomanian?); *Clavatipollenites, Microcachrydites, Cythidites*; fresh water ostracods
Los Adobes (190 m)	Brown yellowish conglomerates; reddish sandy siltstones; purplish claystones	Continental
Taquetren (1,500 m)	Andesites, basic andesites; tuffs; conglomerates	Upper Jurassic–Lower Cretaceous; continental; *Elatocladus, Cladophlebis*
(c) *Comodoro Rivadavia area*[3]		
Trebol (700 m)	Light gray, blueish, greenish, reddish clays and sandy lens	Upper Cretaceous; continental; lacustrine
Comodoro Rivadavia (800 m)	Sandy layers with clay and tuffaceous intercalations	Upper Cretaceous; continental; deltaic
Mina del Carmen (300 m)	Gray or green tuffs, claystones; locally with reddish or variegated intervales	Lower Cretaceous; continental
Pozo D 129 (700 m)	Brown calcareous and oolitic beds. Light to dark gray argillaceous tuffs and siltstones. Brown finely layered siltstones	Upper Jurassic–Lower Cretaceous; continental; lacustrine; fresh water ostracods
(d) *Northern Santa Cruz Province*[4]		
Baquero (80 m)	Light tuffs and claystones; dark tuffs and tuffaceous sandstones	Continental; abundant palynomorphes
Bajo Grande (60 m)	Tuffs, tuffaceous claystones and sandstones	Continental

CHUBUT GROUP *(for a, b, c)*

[1] After: Lesta and Ferello, 1972.
[2] After: Petersen, 1946; Nullo and Proserpio, 1975.
[3] After: Lesta, 1968; Zambrano, 1971; Lesta and Ferello, 1972.
[4] After: Archangelsky and Gamerro, 1966.

tral Argentina, while in the Andean region several pulses of calc-alkaline magmatic activity are recognized in the Patagonian batholith.

Cenomanian–Campanian

This period is marked by an abrupt change in the paleogeography. As a consequence of the emplacement of the Patagonian batholith and other Middle Cretaceous plutons of the Central Andes, most of the Cordilleran area was uplifted (Fig. 7). Along the entire Andean belt continental deposits were intercalated with volcani-clastic sequences during the Upper Cretaceous.

The intracratonic basins underwent rapid subsidence, especially those presently located in the pre-Andean foothills. Clastic continental sequences are dominant from Bolivia to northern Patagonia with the only exception of the first Atlantic ingression recognized in the present off-shore part of the Colorado Basin.

The Magallanes Basin contains the only marine deposits of this period and also contains the first molassic sediments. Marine shelf facies are dominant in the eastern sector of the basin, the depocenter of which lies near to the uplifted area. A conspicuous intertonguing of coarse clastic deposits (derived from the Proto-Cordillera) with the marine facies is observed in the western part of this sector. No more generalized euxinic conditions occur. A regressive trend is noted from north to south from Cenomanian to the Upper Campanian, as most of the northern Magallanes Basin becomes continental.

Maastrichtian

Concomitant with an important eustatic or epeirogenic change, large parts of the continent were covered by a thin sequence of mainly shallow marine deposits (Fig. 8).

Marine transgression has been detected at three points along the Atlantic margin. The most northerly is indicated by the marine deposits of the Salado Basin and the transitional sediments of the Chacoparana region. The central branch occupied the Neuquén Embayment and the adjacent Patagonian area. Between these two branches a hypothetic Somuncura High was already postulated by some authors. The southern branch covered a restricted area of the previous Magallanes Basin. These marine ingressions are regarded as the first important South Atlantic transgressions.

In summary, it can be said that during the Cretaceous in southern South America the continuous interaction between an active and a passive margin formed a peculiar set of basins in which were deposited continental red beds and marine platform sediments, related with magmatics at different episodes.

MAGMATIC ACTIVITY

There are two different magmatic assemblages present in the Cretaceous of southern South America, each with a characteristic development and tectonic environment. Associated with the Andean orogenic belt there developed at different times and places volcanism of a typical continental andesitic–dacitic suite. West of the Main Cordillera was a marine volcanic chain which preceded the emplacement of the Andean batholith. In the extra-Andean region there existed the typical Atlantic border type activity with alkaline basic plugs, basaltic flows etc. The geographic and chronological setting is illustrated in Figs. 5–7.

Andean magmatic activity

The most striking feature of the Cordilleran volcanic sequence is their lack of longitudinal continuity. Segments may be identified which were activated during the Cretaceous and repeatedly reactivated in Cenozoic time. Although there are insufficient geological data to determine the relationship of the Cenozoic volcanics to the older activity there is a suggestion that some at least are related to the present segmentation of the Nasca Plate.

The earliest igneous activity consisted of andesitic tuffs, rhyolites and andesites associated with keratophyres and interbedded with marine clastics of Neocomian age in the Chilean Coastal Range. They extend from the northeastern limit of Chile to south of the town of Cauquenes (ca. 36°S). These deposits which unconformably overlie Middle Jurassic volcani-clastic deposits may reach thickness varying from 4,500 m to 13,000 m (Aguirre et al., 1974).

The principal phase of activity is dated as Hauterivian by use of the fauna of the included marine beds (Moscoso, 1976). In the Lo Prado Formation a chemical variation from acid to mesosilicic is observed.

Further, Lower Neocomian volcanic activity ex-

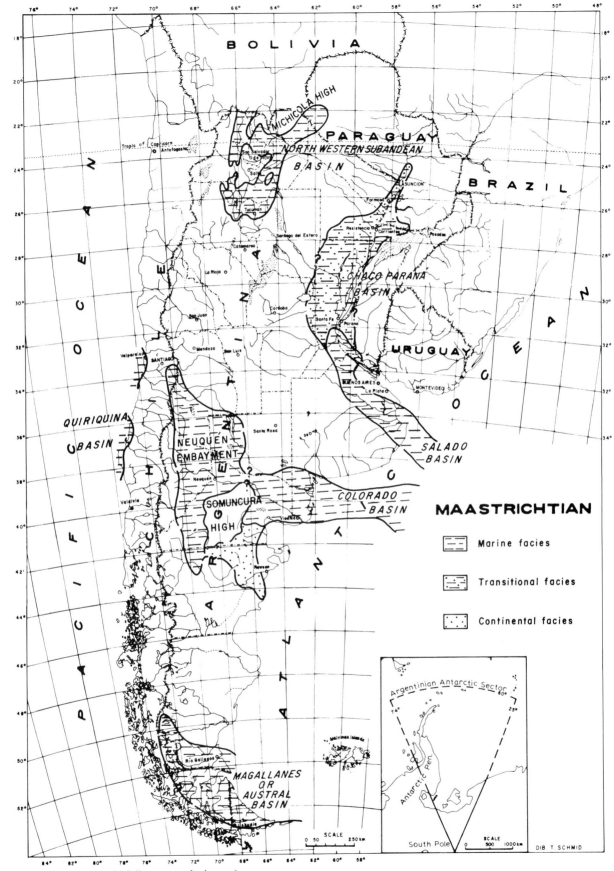

Fig. 8. Maastrichtian (Malalhueyan cycle, in pars).

TABLE XVI

Generalized correlations of the more representative formations and movements

Basins / Europ. standard	Neuquén Embayment	Austral or Magallanes		Atlantic Basins			Norwestern Andean	Coastal Range	Movements
		Chilean stages	Arg. form	Salado	Colorado	San Jorge			
MAASTRICH.	Malalhueyano Group	Riesconian		Las Chilcas	Pedro Luro	Trébol	Balbuena subgroup	Quiriquina	Laramian
CAMPANIAN			Cabeza de León	General Belgrano	Colorado	Comodoro Rivadavia			
SANTONIAN	Neuquén Group	Lazian					β		
CONIACIAN									Peruvian
TURONIAN							Chubut Group	Pirgua subgroup	
CENOMANIAN		Peninsulian	Arroyo Alfa		?		β		
ALBIAN	Diamante	Tenerifian	Nueva	Río	Fortín	Mina del Carmen		Veta Negra	
APTIAN	Huitrín		Argentina	Salado					Mirano inicial
BARREMIAN	Agrio	Pratian		β			β	Lo Prado	
HAUTERIVIAN	Mulichinco	Esperanzian	Pampa Rincón			Pozo D 129			
VALANGINIAN	Quintuco							Pachamama	
BERRIASIAN		Rinconian	Spring hill	?				Patahua	
TITHONIAN	Vaca Muerta								Araucanian

tends from the latitude of the Corcovado Gulf (appt. 44°S) to the Pena Gulf (47°S). Here activity extended from Upper Jurassic (145 m.y.) and continued until Upper Neocomian with the ejection of andesites and some subordinate dacites. The interbedded black shales carry an ammonitic fauna indicating Berriasian to Valanginian ages (Skarmeta, 1976; Skarmeta and Charrier, 1976). The southern termination of these volcanics is abrupt and coincides with the end of the belt of recent volcanicity and seismic activity; presumably it was related to the southern limit of the Nazca Plate.

In the western Coastal Ranges a further 4,000 m of andesites and tuffs (Vicente, 1970) were added in Central Chile in post-Hauterivean–pre-Cenomanian time (the Veta Negra Formation). On the Chilean side of the Main Cordillera a thinner volcanic sequence of andesites, breccias and tuffs interbedded with red-

beds unconformably overlies Neocomian. In the Sierra de Candeleros (25°23'S) Chong Diaz (1976) reported a thickness of 500 m which he tentatively assigned to Aptian–Albian. Further to the south at the latitude of Curico a sequence of 1,800–2,000 m of andesitic breccias and tuffs interbedded with red sandstones and lacustrine limestones, the Colimapu Formation, overlies marine Lower Barremian (Davidson, 1971). These volcanic rocks like the Neocomian volcanics also only extend as far as 36°S.

In the Patagonian Cordillera three distinctive episodes of volcanic activity are recorded during late Barremian, early Aptian and sporadic activity until Albian time (Ramos, 1977). The initial phase, placed in the Divisadero Formation, consists of rhyolitic ignimbrites with dacitic agglomerates and acid tuffs and may reach a thickness of 780 m in the Sierra de Payaniyeu (Ploszkiewicz and Ramos, 1977).

The second, andesitic phase (Nirihuao Formation) 300–350 m in thickness from Cerros de Apeleg to Nirihuao is dated at 115 m.y. thus assigning it to lower Aptian. At its most southerly extent it is locally known as the Chile Chico Formation (Niemeyer, 1975). The final dacitic phase (El Gato Formation) is less widely spread, but appears to have occurred in Aptian time continuing perhaps into early Albian.

These rocks crop out on the latitude of Esquel (43°S) on the Argentinian side of the Cordillera and continue through Corcovado, Palena, Fontana Lake, Coyhaique, Carrera–Buenos Aires lake to the latitude of the Penas Gulf (47°S). All these Lower Cretaceous volcanics were tectonically disturbed during the orogenic activity of the Peruvian phase at some time between late Albian and early Turonian (see p. 268). As a result of these movements they are split up into several blocks and tilted through various angles.

The principal event which followed the main diastrophic phase throughout the Patagonian Andes south of 39°S was the post-orogenic emplacement of Andean batholith. This continuous belt of granitic rocks intrudes all the earlier volcanics resting on the low-grade metamorphosed Paleozoic sediments. Although some activity may date back to the Upper Jurassic (Halpern, 1973), the main intrusive phase according to Ramos and Ramos (1979) was Cenomanian, 98 ± 4 m.y. North of 39°S the batholith has not been emplaced in the Main Andean Cordillera, but is exposed in the Coastal Range of Chile north of 37°S. To the east a number of stocks intrude Neocomian marine sediments and Aptian–Albian volcanics.

Following the batholithic emplacement was another significant volcanic phase, and in the central Western Cordillera the products of this phase may total 4,000–6,000 m. These andesitic volcanic rocks, which however, may include some rhyolitic tuffs in the lower part are assigned to the Abanico Formation (Aguirre et al., 1974).

Extra-Andean magmatic activity

The magmatic activity east of the Main Cordillera belt is much more sporadic, and appears to be more extensive close to the Atlantic margin.

Volcanic activity in eastern Paraguay is represented by the Sapucai Lavas, a series of intrusions of Late Jurassic–Early Cretaceous age consisting of tholeiitic basaltic flows yielding ages comparable to those of the Paraná Basin (Creer et al., 1965; Amaral et al., 1966). In Argentina the equivalent flows are the basalts of the Serra Geral Formation, which although extruded from Late Jurassic time reach their climax at the end of Barremian time (113 m.y.).

Contemporaneous volcanic activity is found in several areas of the Pampean Ranges and Subandean belt, where alkaline basalts were extruded from isolated volcanic centres. Most are included in the Pirgua Formation or its equivalents further south. The basalts are well exposed in the Córdoba and San Luis provinces. In the central Argentine basins basalt flows separate Lower Cretaceous from Middle to Upper Cretaceous deposits (see p. 284)

Upper Cretaceous magmatic activity in the extra-Andean region seems confined to several alkaline plugs intruded in northwestern Paraguay. They appear to be related to complexes of successively injected nepheline basalts, phonolites and other alkaline rocks better developed in Brazil. This activity appears to be centred at about 80 m.y.

At about the same time close to some parts of the Patagonian Andes some gabbro-diorites and other basic intrusives were emplaced in the pre-Andean region as minor stocks. They occur several kilometers east of the Andean batholith near Payaniyeu, Fontana, Corcovado, Aysén and Pueyrredón Lake.

PALEONTOLOGICAL ASPECTS

Paleontological research is most advanced in Argentina and Chile; the Cretaceous faunas and floras of Paraguay and Uruguay on the other hand are poorly known.

Probably best known is the marine invertebrate fauna of the Neuquén Embayment. In the Andean cycle, the lower Tithonian is characterized by ammonites of the genus *Virgatosphinctes*, the Middle Tithonian by species of the genera *Aulacosphinctes* and *Pseudolissoceras*, and the Upper Tithonian by *Windhauseniceras*, *Lytohoplites*, *Substeueroceras*, *Pseudohimalayites*, *Himalayites*, etc. The Berriasian is identified through *Argentiniceras*, *Berriasella*, *Frengueliceras*, *Groebericeras* and *Spiticeras*. The Valanginian–Hauterivian fauna include mainly *Lissonia riveroi* (Lisson), *Acantholissonia gerthi* (Weaver) *Olcostephanus* and *Leopoldia*; the Hauterivian by *Holcoptychites neuquensis* Douvillé and *Pseudofavrella* (Camacho and Riccardi, 1978).

Of the pelecypod fauna, the trigonids have an

austral character and those found in the Agrio and Mulichinco formations show a clear relation with those from the South African Uitenhage Formation (Uhlig, 1911; Weaver, 1931; Hallam, 1967). The ostracods from the Agrio Formation also show affinities with South African forms (Musacchio, 1978). The Aptian–Albian pelecypods of the genera *Melania, Corbicula, Diplodon* and *Modiola* all indicate the presence of brackish water environments.

In the continental beds of the Upper Cretaceous titanosaur sauropods show a remarkable development, they are recorded from Chile (Vinitas Formation), Northwestern Argentine Basin (Bonaparte and Bossi, 1967), Uruguay and southern Brazil (Arid and Vizotto, 1971).

The biostratigraphy of the Austral Basin has recently been revised through the use of ammonites. The most representative ammonite families in the Cretaceous system of the Austral Basin are composed of: Phylloceratidae, Heteroceratidae, Ancyloceratidae, Hamitidae, Baculitidae, Turrilitidae, Scaphitidae, Labeceratidae, Aconeceratidae, Berriasellidae, Olcostephanidae, Neocomitidae, Silesitidae, Kossmaticeratidae, Pachydiscidae, Placenticeratidae, Hoplitidae, Acanthoceratidae, Texanitidae, etc. (Leanza, 1963, 1967b, 1969b, 1970; Riccardi, 1970, 1971, 1977, 1979; Riccardi and Rolleri, 1980; Blasco et al., 1980a, b; Nullo et al., 1981; Aguirre Urreta and Ramos, 1981). The older studies are listed by Feruglio (1949).

The foraminiferal assemblages have been discussed by Herm (1966), and Malumián and Masiuk (1973). There are some clear relations with other microfaunas from the Southern Hemisphere, but also some distinctive features. The fauna of the Esperanzian stage (see Table VII) is similar to that of the Majunga Basin (Madagascar) and South Africa (Sigal et al., 1970; Malumián and Masiuk, 1973, 1975; McLachlan et al., 1976a, b; Malumián and Baez, 1976), but differs by the presence and frequency of occurrence of *Lenticulina nodosa–Astacolus gibber*, and, up to this date, of an endemic species *Pseudopolymorphina martinezi*. The low diversity has been interpreted as a result of a small sea and the instability of the inhabited environment (Malumián and Masiuk, 1975). In the younger assemblages the similarity with South Africa or other parts of the Southern Hemisphere is not so apparent, presumably due to the opening of the South Atlantic.

The Tenerifian stage has the oldest planktonic foraminifers of the basin, with '*Globigerinelloides*' *gyroidinaeformis* Moullade as the characteristic form of the upper part. Typical of the Peninsulian stage are the agglutinated forms *Tritaxia gaultina, Spiroplectinata annectens, S. complanata* and species of the genus *Lingulogavelinella*. These microfauna show some affinities with South Africa and Australia. From Esperanzian up to Tenerifian general anaerobic conditions are registered (Malumián, 1978).

The benthic Foraminifera used to zone the Lazian stage show some specific identity with the fauna of Australia and New Zealand. At this stage there are restricted horizons with double keeled forms (Natland et al., 1974; Malumián and Masiuk, 1975, 1976a, b, 1978; Malumián and Báez, 1976). The Riesconian agglutinated Foraminifera are associated with the molassic sedimentary regime and the withdrawal of the Cretaceous sea from the basin.

The foraminiferal data from the Austral Province during the Cretaceous suggest the presence of temperate to cold water as was proposed by Scheibnerova (1972a, b, 1973) and others. The clear faunal differences in the South American Province can be attributed to the particular sedimentary environment existing as for example in the high proportion of agglutinated forms of *Tritaxia gaultina, Uvigerinammina jankoi* etc., associated with flysch (Malumián and Proserpio, 1979; Malumián, 1978).

The Cretaceous–Tertiary boundary is well established in northern Patagonia and the Colorado Basin through the microfauna. The Maastrichtian has the same *Globotruncana* and calcareous nannoplankton, followed by the Danian with cosmopolitan foraminifers (Bertels, 1964, 1969, 1970; Kaasschieter, 1963; Malumián, 1970). It is not so apparent in the Austral Basin because of the dominance of regressive facies (Charrier and Lahsen, 1968; Malumián, 1968; Malummián et al., 1973; Natland et al., 1974). However, the stages of the Austral Basin, which provide a nearly complete marine Cretaceous sequence constitute a useful standard for the southern temperate-cold regions. Much remains to be done to relate the megafossils studies from outcrop with the microfossils obtained from the subsurface.

The Jurassic and Early Cretaceous microflora of the Neuquén Embayment are characterized by the common genus *Classopollis*. Several subdivisions of the associated morphospecies have been proposed (Volkheimer, in Volkheimer and Pöthe de Baldis, 1976). Of these, the *Callialosporites dampieri–C. seg-*

mentatus is recognized from Toarcian to Albian, the sub-assemblage *Equisetosporites–Trisaccites* is found in the Vaca Muerta Formation (Tithonian), while the *Cyclusphaera* sub-assemblage is found in the Mulichino and Agrio formations and the lower part of the Huitrín Formation (Volkheimer and Sepúlveda, 1976). These assemblages, which also include some cosmopolitan elements define in southern South America a Gondwana sub-province from Valanginian to Barremian times (Volkheimer, 1975). They are also known to occur at the base of the Rio Mayer Formation at St. Martin Lake (Pöthe de Baldis and Ramos, 1978) and in subsurface in Chubut Province (Archangelsky and Seiler, 1978).

The *Huitrinpollinites–Stephanocolpites* sub-assemblage, probably of Albian age (Volkheimer and Salas, 1975a) is well represented in the Upper Huitrín Formation where the first angiosperms also occur (Volkheimer and Salas, 1975b).

The first Cretaceous microfloral study was made with material from the Baquero Formation in southern Patagonia by Archangelsky and Gamerro (1965). They found *Cicatricosisporites hughesii, Trisaccites microsaccutus, Microcachryidites antarcticus* and *Inaperturopollenites limbatus*, forms common to Australia and Antarctica together with elements such as *Trilobosporites apiverrucatus, T. purvenulentus, T. trioreticulosus* and *Clavatiopollenites hughessii* known in extra-Gondwana areas. The flora suggests a Barremian–Aptian age for the Baqueró Formation (Archangelsky and Gamaro, 1966a–d). Lower Cretaceous palynomorphs such as *Taurocusporites cureatus, Ephedripites bilateralis, Monosulcites cuyoensis* and *Classopollis hyalinus* were also found in the La-Cantera Formation in the San Luis Basin, one of the central basins of Argentina, by Stover (in Yrigoyen, 1974).

In Tithonian–Neocomian time limestone deposition was important in the Mendoza and Neuquén provinces, and the occurrence of some *Trocholina* in the Upper Tithonian of the Neuquén Embayment suggests temperate waters (Malumián and Masiuk, 1973). Volkheimer (1969) interpreted the Barremian–Aptian evaporites which are associated with limestones and dolomites of the Huitrín Formation as evidence of aridity. He also considered the Upper Cretaceous oolites of the El Molino and Yacoraite formations as evidence of shallow water temperatures of $20°–30°$ C.

Throughout most of the Cretaceous between modern latitudes $30°–40°$ S red-beds accumulated, and in the Upper Cretaceous in particular, titanosaurs were widely distributed (Viñitas Formation in Chile, Las Blanquitos Formation in the Northwestern Subandean Basin, Neuquén Group and Neuquén Embayment, Bajo Barreal and Castillo formations, and San Jorge Basin) suggesting at least temperate conditions. Since the area was covered by arboreal dicotyledons and large leaves with drip points are common, moist warm temperate conditions were inferred by Volkheimer (1969).

The low diversity and scarcity of keeled forms in the planktonic foraminiferal assemblage from the uppermost Cretaceous sediments of northern Patagonian, Salado and Colorado basins suggests cooler temperate conditions. In the Austral Basin the lack of limestone and the temperate to cold water forms in the microfauna, the low diversity and the restriction of keeled forms to a few species shows analogies with the Boreal Province (Malumián and Masiuk, 1978).

There is thus in southern South America a gradation in the inferred climate from warm temperate in the north to cool to cold temperate in the south, probably very little different from that now existing.

PALEOCLIMATE

The absence of rudists, larger benthic foraminifers, and biohermal corals indicates that southern South America was not influenced by warm Tethyan waters. Northwest of the area, in Early Cretaceous lay part of the most extensive paleodesert known (Bigarella and Salamuni, 1961, 1964; Volkheimer, 1969) with aeolian sandstones known from the Chacoparana Basin (San Cristobal Sandstone), Uruguay (Tacuarembó Sandstone) and Brazil (Botucatú Fm.).

REFERENCES

Abad, E., 1976. Las formaciones Carrillos y Hornitos al norte de Vallenar, prov. de Atacama, Chile. *1° Congr. Geol. Chileno*, I (A), pp. 97–114.

Aguirre, L., 1960. Geología de los Andes de Chile Central, Prov. de Aconcagua. *Inst. Inv. Geol. Chile, Bol.*, 9, 70 pp.

Aguirre, L., Charrier, R., Davidson, J., Mpodozis, A., Rivano, S., Thiele, R., Tidy, E., Vergara, M. and Vicente, J.C., 1974. Andean magmatism: its paleogeographic and structural setting in the central part (30°–35° S) of the southern Andes. *Pacific Geol.*, 8:1.

Aguirre Urreta, M.B. and Ramos, V.A., 1981. Estratigrafía y paleontología de la alta cuenca del Río Robles, Cordillera Patagonica, Prov. de Santa Cruz. *VIII Congr. Geol. Arg., Actas*, III. In press.

Aliste, N., Pérez, E. and Carter, W.D., 1960. Definición y edad de la Formación Patagua, Prov. de Aconcagua, Chile. *Rev. Miner.*, 71: 40–50.

Amaral, G., Cordani, U.G., Kawashita, K. and Reynolds, J.H., 1966. Potassium-argon dates of basaltic rocks from Southern Brazil. *Geochim. Cosmochim. Acta*, 30: 159–189.

Amaral, G., Bushee, J., Cordani, U.G., Kawashita, K. and Reynolds, J.H., 1967. Potassium-argon ages of alkaline rocks from southern Brazil. *Geochim. Cosmochim. Acta*, 31: 117–142.

Archangelsky, S., 1976. Vegetales fósiles de la Formación Springhill, Cretácico, en el subsuelo de la Cuenca Magallánica, Chile. *Ameghiniana*, 13 (2): 141–158.

Archangelsky, S. and Gamerro, J.C., 1965. Estudio palinológico de la Formación Baqueró (Cretácico), Prov. de Santa Cruz, I. *Ameghiniana*, 4 (5): 159–169.

Archangelsky, S. and Gamerro, J.C., 1966. Estudio palinológico de la Formación Baqueró (Cretácico), Prov. de Santa Cruz, II, III and IV. *Ameghiniana*, 4 (6): 201–209; 4 (7): 229–236; 4 (10): 363–372.

Archangelsky, S. and Gamerro, J.C., 1967. Pollen grains found in coniferous cones from the Lower Cretaceous of Patagonia (Argentina). *Rev. Paleobot. Palynol.*, 5: 179–182.

Archangelsky, S. and Seiler, S., 1978. Algunos resultados palinológicos de la perforación UN oil Os 1, sudoeste de la Prov. del Chubut, *Rep. Argentina. II Congr. Arg. Paleontol. Bioestrat.; I Congr. Lat. Paleontol., Buenos Aires*. In press.

Arid, F.M. and Vizotto, L.D., 1971. *Antartosaurus brasiliensis*, um novo sauropodo do Cretaceo Superior do Sul do Brasil. *XXV Congr. Bras. Geol., Annaes*, pp. 297–305.

Aubouin, J. and Borrello, A.V., 1966. Chaînes andines et chaînes alpines: regard sur la géologie de la Cordillère des Andes au parallèle de l'Argentine moyenne. *Soc. Géol. Fr., Bull.*, (7) VIII: 1050–1070.

Aubouin, J. and Borrello, A.V., 1970. Regard sur la géologie de la Cordillère des Andes: relais tectoniques et cycles orogéniques superposés. Le Nord Argentin. *Soc. Géol. Fr., Bull.*, (7) XII: 246–260.

Aubouin, J., Borrello, A.V., Cecioni, G., Charrier, R., Chotin, P., Frutos, J., Thiele, R. and Vicente, J.C., 1973. Esquisse paléogeographique et structurale des Andes Méridionales. *Rev. Geogr. Phys. Geol. Dyn.*, XV (1–2): 11–71.

Báez, Ana M., 1975. *Los Pípidos de la Formación Las Curtiembres (Cretácico, Prov. Salta, Rep. Arg.). Evolución de la Familia Pipidae (Amphibia, Anura) en Relación a la Historia Paleogeográfica*. Tesis Doctoral de la Univ. de Buenos Aires. (*M.S.*).

Banks, L.M. and Diaz de Vivar, V., 1975. Exploration in Paraguay reactivated. *Oil Gas J.*, 6: 160–168.

Beherendsen, O., 1891. Zür Geologie des Ostabhanges der argentinischen Cordillere. *Z. Dtsch. Geol. Ges.*, 43 (1891): 369–420; 44(1) (1892): 1–42; 45 (2).

Bertels, A., 1964. Micropaleontología del Paleoceno de General Roca (Prov. de Río Negro). *Rev. Mus. La Plata, Secc. Paleontol.*, 4 (23): 125–185.

Bertels, A., 1969. Estratigrafía del límite Cretácico–Terciario en Patagonia Septentrional. *Asoc. Geol. Arg., Rev.*, 24 (1): 41–54.

Bertels, A., 1970. Los foraminíferos planctónicos de la cuenca Cretácico–Terciaria en Patagonia Septentrional (Argentina), con consideraciones sobre la estratigrafía de Fortín General Roca (Prov. Río Negro). *Ameghiniana*, 7(1): 1–56.

Bertels, A., 1972. Ostrácodos de agua dulce del miembro inferior de la Formación Huantrai-co (Maastrichtiano inferior), Prov. de Neuquén. Rep. Argentina. *Ameghiniana*, 9 (2): 172–182.

Bertels, A., 1974. Upper Cretaceous (Lower Maastrichtian?) ostracodes from Argentina. *Micropaleontology*, 20 (4): 385–397.

Bertels, A., 1975. Upper Cretaceous (Middle Maastrichtian) ostracodes of Argentina. *Micropaleontology*, 21 (1): 97–190.

Bertels, A., 1978. Microfauna del Cretácico Superior y del Terciario. *VII Congr. Geol. Arg., Relatorio: Geología y recursos naturales de Neuquén*. pp. 163–175.

Bigarella, J.J. and Salamuni, R., 1961. Early Mesozoic wind patterns as suggested by dune bedding in Botucatú Sandstone of Brazil and Uruguay. *Geol. Soc. Am. Bull.*, 72: 1089.

Bigarella, J.J. and Salamuni, R., 1964. Paleowind patterns in the Botucatú Sandstone (Triassic–Jurassic) of Brazil and Uruguay. In: A.E.M. Nairn (Editor). *Problems in Palaeoclimatology*, Wiley, London.

Blasco de Nullo, G., Nullo, F. and Ploszkiewicz, J.V., 1980a. El género Colchidites Djánelidze, 1926, y la posición estratigrafica del género Hatchericeras Stanton, 1901, en la estancia Tucu-Tucu, Prov. de Santa Cruz. *Asoc. Geol. Arg., Rev.*, 35 (1): 41–58.

Blasco de Nullo, G., Nullo, F. and Proserpio, C., 1980b. Santoniano–Campaniano: Estratigrafía y contenido ammonifífero, Cuenca Austral. *Asoc. Geol. Arg., Rev.*, 35 (4): 467–493.

Bodenbender, G., 1889. Expedición al Neuquén. *Inst. Geogr. Arg. Bol.*, 10, Cuad. 10.

Bonaparte, J.F. and Bossi, C., 1967. Sobre la presencia de dinosaurios en la Formación Pirgua del Grupo Salta y su significado cronológico. *Acta Geol. Lilloana*, 9: 25–44.

Bonarelli, G., 1921. Tercera contribución al conocimiento geológico de las regiones petrolíferas subandinas del norte (Prov. de Salta y Jujuy). *Min. Agric., Secc. Geol., An.*, XV-1.

Bonarelli, G. and Nágera, J.J., 1921. Observaciones geológicas en las inmediaciones del Lago San Martin (territorio de Santa Cruz). *Dir. Gral. Min. Geol. Hidrol., Bol., 27 Sec. B (Geol.)*.

Borrello, A.V., 1969. Geosinclinales de la Argentina. *Dir. Nac. Min. Geol. An.*, XIV.

Bossi, G., 1971. Análisis de la cuenca de Ischigualasto-Ischichuca. *I. Congr. Hisp. Amer. Geol. Econ. Sec.*, 1, 2: 611–626.

Bracaccini, I.O., 1968. Panorama General de Geología Patagónica, III. *J. Geol. Arg., Actas*, 1: 17–42.

Brandmayr, J., 1945. Contribución al conocimiento geológico del extremo sud-sudoeste del Territorio de Santa Cruz (region Cerro Cazador-alto Río Turbio) *Bol. Inform. Petroleras*, XXII (256): 415–443.

Brüggen, J., 1950. *Fundamentos de la Geologia de Chile.* Inst. Geogr. Mil., Santiago de Chile, 374 pp.

Camacho, H.H., 1967. Las transgresiones del Cretácico Superior y Terciarcio en la Argentina. *Asoc. Geol. Arg., Rev.*, 22 (4): 253–280.

Camacho, H.H., 1972. The Cretaceous–Tertiary boundary in Argentina. *Int. Geol. Congr., 24th, Sect. 7*, pp. 490–495.

Camacho, H.H. and Riccardi, A., 1978. Invertebrados-Megafauna. *VII Congr. Geol. Arg., Relatorio: Geología y Recursos Naturales del Neuquén.* pp. 137–146.

Caminos, R. and Nullo, F., 1979. Descripción geológica de la Hoja 67 e-f Isla de los Estados (Tierra del Fuego). *Serv. Geol. Nac., Bol.*, 175.

Carter, W.D., 1962. Unconformity marking the Jurassic–Cretaceous boundary in the La Ligua area, Aconcagua Prov., Chile. *U.S. Geol. Surv., Prof. Pap.*, 450–E: 61–63.

Casamiquela, R.M., Corvalán, J. and Franquesa, F., 1969. Hallazgo de Dinosaurios en el Cretácico Superior de Chile. *Inst. Geol. Chile, Bol.*, 25.

Cazau, L. and Uliana, M., 1973. El Cretácico Superior continental de la Cuenca Neuquina. *V Congr. Geol. Arg., Actas*, 3: 151–163.

Cecioni, G., 1955. Distribuzione verticale di alcune Kossmaticeratidae nella Patagonia Chilena. *Bol. Soc. Geol. Ital.*, 74: 141–149.

Cecioni, G., 1957. Cretaceous flysch and molasse in Departamento Ultima Esperanza, Magallanes Province, Chile. *Am. Assoc. Petrol. Geol. Bull.*, 76: 1–18.

Cecioni, G., 1958. Preuves in faceur d'une glaciation Neo-Jurassique en Patagonie. *Soc. Géol. Fr., Bull., Sér. 6*, 8 (5): 413–436.

Cecioni, G., 1960a. Orogénesis subhercínica en el estrecho de Magallanes. *Univ. Chile, Fac. Cienc. Fis. Mat., Inst. Geol. Publ.*, 17: 279–289.

Cecioni, G., 1960b. Perfil geológico entre cabo Froward y cabo San Isidro, estrecho de Magallanes. *Univ. Chile, Fac. Cienc. Fis. Mat., Inst. Geol. Publ.*, 17: 293–310.

Cecioni, G. and Charrier, R., 1974. Relaciones entre la Cuenca Patagónica, la Cuenca Andina y el Canal de Mozambique. *Ameghiniana*, 11 (1): 1–38.

Cecioni, G. and García, F., 1960. Observaciones geológicas en la Cordillera de la Costa, de Tarapacá. *Inst. Inv. Geol. Chile, Bol.*, 6, 28. pp.

Charrier, R., 1973. Interruption of spreading and the compressive tectonic phases of the meridonal Andes. *Earth Planet. Sci. Lett.* 20: 242–249.

Charrier, R. and Lahsen, A., 1968. Contribution à l'étude de la limite Crétacé–Tertiaire de la Province de Magellan, extrème-sud du Chili. *Rev. Micropaleontol.*, 2 (11): 111–120.

Charrier, R. and Malumián, N., 1975. Orogénesis y epeirogénesis en la región austral de América del Sur durante el Mesozoico y el Cenozoico. *Asoc. Geol. Arg., Rev.*, 30 (2): 193–207.

Charrier, R. and Vicente, J.C., 1972. Liminary and geosynclinal Andes: major orogenic phases and synchronical evolutions of Central and Magellean Sectors of the Argentine–Chilean Andes. *Solid Earth Probl. Conf., Upper Mantle Proj., Buenos Aires*, 2: 451–470.

Chong Diaz, G., 1976. Las relaciones de los Sistemas Jurásico y Cretácico en la Zona Preandina del norte de Chile. *I Congr. Geol. Chileno*, 1 (A): 21–42.

Codignotto, J., Nullo, F., Panza, J. and Proserpio, C., 1978. Estratigrafía del Grupo Chubut entre Paso de Indios y Las Plumas, Prov. del Chubut, Rep. Argentina. *VII Congr. Geol. Arg., Actas*, 1: 471–480.

Corvalán, J., 1959. El Titoniano de Río Leñas, Prov. de O'Higgins con una revisión del Titoniano y Neocomiano de la parte chilena del Geosinclinal Andino. *Inst. Inv. Geol. Chile, Bol.*, 3: 65 pp.

Corvalán, J., 1974. Estratigrafía del Neocomiano marino de la Región al Sur de Copiapó, Provincia de Atacama. *Rev. Geol. Chile*, 1: 13–36.

Corvalán, J. and Dávila, A., 1964. Observaciones geológicas en la Cordillera de costa entre los ríos Aconcagua y Mataquito. *Soc. Geol. Chile, Resúmenes*, 9: 1–4.

Corvalán, D.J. and Perez, D.E., 1958. Fósiles guías chilenos (Titoniano Neocomiano). *Inst. Inv. Geol. Chile, Mem.*, 1: 48 pp.

Creer, K.M., Mitchell, J.G. and Abou Deeb, J., 1972. Paleomagnetism and radiometric age of the Jurassic Chon Aike Formation from Santa Cruz Province, Argentina. Implications for the opening of the South Atlantic. *Earth Planet. Sci. Lett.*, 14: 131–140.

Dalziel, I.W.D., 1974. Evolution of the margins of the Scotia Sea. In: C.A. Buret and C.L. Drake (Editors). *The Geology of Continental Margins*, Springer Verlag, New York, N.Y., pp. 567–579.

Dalziel, I.W.D. and Palmer, K.F., 1979. Progressive deformation and orogenic uplift at the southern extremity of the Andes. *Geol. Soc. Am. Bull.*, 90 (3): 259–280.

Dalziel, I.W.D., Caminos, R., Palmer, K.F., Nullo, F. and Casanova, R., 1974. South extremity of the Andes: Geology of Isla de los Estados, Argentina, Tierra del Fuego. *Am. Assoc. Petrol. Geol., Bull.*, 58 (12): 2502–2511.

Danieli, C.A. and Porto, C., 1968. Sobre la extensión austral de las Formaciones Mesozoico–Terciarias de la Provincia de Salta, limítrofe con Tucumán. *III Jornadas Geol. Arg., Actas*, 1: 77–90.

Davidson, J., 1971. *Geología del área de las Nacientes del Terro, Prov. de Curicó.* Tesis Univ. de Chile, Santiago de Chile.

de Ferrariis, C., 1968. El Cretacico del norte de la Patagonia, III. *Jorn Geol. Arg., Actas*, 1: 121–144.

De Wit, M.J., 1977. The evolution of the Scotia Arc as a key to the reconstruction of southwestern Gondwanaland. *Tectonophysics*, 37 (1–3): 53–82.

Dickinson, W., 1976. Plate tectonics evolution of sedimentary basins. *Am. Assoc. Petrol. Geol., Course Note Ser.*, pp. 1–62.

Digregorio, J.H., 1972. Neuquén. In: A.F. Leanza (Editor), *Geología Regional Argentina.* Acad. Nac. Cienc., Córdoba, pp. 439–506.

Digregorio, J.H., 1978. Estratigrafía de las acumulaciones Mesozoicas. *VII Congr. Geol. Arg., Relatorio Geología y Recursos Naturales del Neuquén*, pp. 37–66.

Digregorio, J.H. and Uliana, A., 1980. Cuenca Neuquina. In: J.C.M. Turner (Editor) *II Simp. Geol. Reg. Arg.* Acad. Nac. Cienc., Córdoba.

Eckel, E.B., 1959. Geology and mineral resources of Paraguay – A reconnaissance. *U.S. Geol. Surv., Prof. Pap.*, 327: 1–110.

Farias, A.B. and Vergara, M., 1964. Nuevos antecedentes sobre la geología de la Quebrada El Way. *An. Fac. Cienc. Fis. Nat.*, 20–21: 101–128.

Feruglio, E., 1931. Nuevas observaciones geologicas en la Patagonia Central. *Contrib. Dirección General Y.P.F., Seminario Geografia, Buenos Aires.*

Feruglio, E., 1938. El Cretácico Superior del Lago San Martín (Patagonia) y de las regiones adyacentes. *Physis*, XII: 293–342.

Feruglio, E., 1949–1950. *Descripción Geológica de la Patagonia.* 1949, tomo I; 1950, tomo II. Yacimientos Petrolíferos Fiscales, Min. Industria y Comercio Buenos Aires.

Flores, M.A., 1969. El Bolsón de las Salinas en la Prov. de San Luis. *Actas IV Jorn. Geol. Arg.*, I: 311–327.

Flores, M.A., Malumián, N., Masiuk, V. and Riggi, J.C., 1973. Estratigrafía Cretácica del subsuelo de Tierra del Fuego. *Asoc. Geol. Arg., Rev.*, 28 (4): 407–437.

Fuenzalida, H., 1964. El geosinclinal andino y el geosinclinal de Magallanes. *Com. Esc. Geol.*, 5: 1–27.

Fuenzalida, R., 1968. Reconocimiento geológico de Alto Palena, Chiloé Continental. *Univ. Chile, Dept. Geología, Publ.*, 31.

Gonzales, E. et al., 1965. La cuenca petrolífera de Magallanes. *Rev. Miner.*, XX (91): 43–57.

Gonzalez Bonorino, F., 1950. Geologic cross-section of the Cordillera de los Andes at about parallel 33°S. (Argentina–Chile). *Geol. Soc. Am., Bull.*, 61.

Gordillo, C.E., 1972. Petrografía y composicion química de los basaltos de la Sierra de las Quijadas, San Luis, y sus relaciones con los basaltos cretácicos de Córdoba. *Asoc. Geol. Córdoba, Bol.*, 1 (3–4).

Gordillo, C.E. and Lencinas, A., 1967. Geología y petrología del extremo norte de la sierra de Los Cóndores, Córdoba. *Acad. Nac. Cienc., Bol.*, 46 (1).

Groeber, P., 1929. Lineas fundamentales de la geología del Neuquén, sur de Mendoza y regiones adyacentes. *Dir. Minas Geol. Hidrol. Publ.*, 58.

Groeber, P., 1946. Observaciones geológicas a lo largo del meridiano 70-I. Hoja Chos Malal. *Asoc. Geol. Arg., Rev.*, 1 (2): 177–208.

Groeber, P., 1952. Mesozoico. In: *Geografía de la República Argentina.* Soc. Arg. Geogr. GAEA, 2, 541 pp.

Groeber, P., 1959. Supracretácico. In: *Geografía de la República Argentina.* Soc. Arg. Est. Geogr. GAEA, 2, 165 pp.

Grossling, B., 1954. Geología de petróleo de la formación Springhill en el distrito de Springhill, Magallanes. *An. Inst. Ing. Min.*, 1:55.

Halpern, M., 1973. Regional geochronology of Chile south of the 50° latitude. *Geol. Soc. Am., Bull.*, 89: 2407–2422.

Hallam, A., 1967. The bearing of certain paleozoogeographic data on continental drift. *Palaeogeogr. Palaeoecol. Palaeoclimatol.*, 3 (2): 201–241.

Haller, M. and Lapido, O., 1981(?). El Mesozoico de la Cordillera Patagónica Central. *Asoc. Geol. Arg., Rev.*, 35 (2): 230–247.

Harrington, H.J., 1950. Geología del Paraguay oriental. *Fac. Cienc. Exac. Fis. Nat., Serie E, Geol.*, 1: 82 pp.

Harrington, H.J., 1956. Paraguay. In: Jenks (Editor), *Handbook of South American Geology. Geol. Soc. Am. Mem.*, 65.

Hatcher, J.B., 1897. On the geology of Southern Patagonia. *Am. J. Sci., 4th Ses.*, IV (23): 327–354.

Hatcher, J.B., 1900. Sedimentary rocks of southern Patagonia. *Am. J. Sci., 4th Ses.*, IX (50): 85–108.

Haupt, O., 1907. Beiträge zur Jauna des oberen Malm und der unteren Kreide in der Argentinischen Kordillere. In: G. Steinmann, *Beiträge zur Geologie und Paleontologie von Südamerika. Neues Jahrb. Mineral., Geol. Palaeontol.*, 23: 187–236.

Heim, A., 1940. Geological observations in the Patagonian Cordillera. *Eclogae Geol. Helv.*, 33 (1): 25–57.

Hemmer, A., 1935. Geología de los terrenos petroliferos de Magallanes y las exploraciones realizadas. *Soc. Nac. Min., Bol. Miner.*, 47: 139–149; 181–188.

Herbst, R., 1971. Esquema estratigráfico de la provincia de Corrientes. *Asoc. Geol. Arg., Rev.*, 26 (2): 221–243.

Herm, D., 1966. Micropaleontological aspects of the Magallanes Geosyncline, southern-most Chile, South America. *Proc. 2nd West African Micropaleontol. Coll., Ibadan, June 18th–July 1st*, pp. 72–86.

Herrero Ducloux, A., 1946. Contribución al conocimiento geológico del Neuquén Extraandino. *Bol. Inf. Petrol.*, 13 (266): 245–281.

Hoffstetter, R., Fuenzalida, H. and Cecioni, G., 1957. Chile. *Lexique Stratigraphique International*, 5 (7).

Holmberg, E., 1964. Descripción geológica de la Hoja 33d. Auca Mahuida, Provincia del Neuquén. *Dir. Nac. Geol. Min., Bol.*, 94: 88 pp.

Kaasschieter, J.P.H., 1963. Geology of the Colorado Basin. *Tulsa Geol. Soc. Digest*, 31: 177–187.

Katz, R., 1963. Revision of Cretaceous stratigraphy in Patagonian Cordillera of Ultima Esperanza, Magallanes, Chile. *Am. Assoc. Petrol. Geol., Bull.*, 47 (3): 506–524.

Katz, R. and Watters, W., 1966. Geological investigation of the Yaghan Formation (Upper Mesozoic) and associated igneous rocks of Navarino Island, Southern Chile. *N. Z. J. Geol. Geophys., Wellington*, 9: 323–359.

Klohn, C., 1960. Geología de la Cordillera de los Andes de Chile Central, provincias de Santiago, O'Higgins, Colchagua y Curicó. *Inst. Inv. Geol. Chile, Bol.*, 8: 95 pp.

Kranck, E.H., 1932. Geological investigation in the Cordillera of Tierra del Fuego. *Acta Geograph., Helsinki*, 4 (2): 1–231.

Lahsen, A. and Charrier, R., 1972. Late Cretaceous ammonites from Seno Skyring-Strait of Magellan Area, Magallanes Province, Chile. *J. Paleontol.*, 46 (4): 520–532.

Lapido, O. and Page, R., 1979. Relaciones estratigráficas y estructura del Bajo de la Tierra Colorada (prov. del Chubut). *VII Congr. Geol. Arg., Actas*, I: 299–313.

Leanza, A.F., 1963. *Patagoniceras* gen. nov. (Binneyitidae) y otros amonites del Cretácico Superior de Chile meridional, con notas acerca de su posición estratigráfica. *Acad. Nac. Cienc., Córdoba, Bol.*, 63: 203–225.

Leanza, A.F., 1967a. Descripción de la fauna de Placenticeras del Cretácico Superior de Patagonia Austral con consideraciones acerca de su posición estratigráfica. *Acad. Nac. Cienc., Córdoba, Bol.*, 46 (1): 5–47.

Leanza, A.F., 1967b. Los Baculites de la provincia de La Pampa con notas acerca de la edad del Piso Rocanense. *Acad. Nac. Cienc., Córdoba, Bol.*, 46 (1): 49–59.

Leanza, A.F., 1969a. Sobre el descubrimiento de depósitos del Piso Coniaciano en Patagonia Austral y descripción de una nueva especie de ammonites (*Peroniceras santacrucense* n. sp.). *Acad. Nac. Cienc., Córdoba, Bol.*, 47 (1): 5–20.

Leanza, A.F., 1969b. Sistema de Salta, su edad, sus peces voladores, su asincronismo con el Horizonte Calcáreo–Dolomítico y con las calizas de Miraflores y la hibridez del Sistema Subandino. *Asoc. Geol. Arg., Rev.*, 24 (4): 393–407.

Leanza, A.F., 1970. Ammonites nuevos o poco conocidos del Aptiano, Albiano y Cenomaniano de los Andes Australes con notas acerca de su posición estratigráfica. *Asoc. Geol. Arg., Rev.*, 25 (2): 197–261.

Leanza, A.F., 1972. Andes Patagónicos Australes. In: A.F. Leanza (Editor), *Geología Regional Argentina*. Acad. Nac. Ciencias, Córdoba, pp. 689–706.

Leanza, H.A., 1981. The Jurassic–Cretaceous boundary beds in west-central Argentina and their ammonite zones. *Neues Jahrb. Geol. Paläontol. Abh.*, 161 (1): 62–92.

Leanza, A.F. and Leanza, H., 1973. *Pseudofavrella* gen. nov. (Ammonitina) del Hauteriviano de Neuquén, sus diferencias con *Favrella* R. Douvillé, 1909, del Aptiano de Patagonia Austral y una comparación entre el geosinclinal andino y el geosinclinal Magallánico. *Acad. Nac. Cienc., Córdoba, Bol.*, 50 (1–4): 127–145.

Leanza, H., Marchese, H. and Riggi, J.C., 1978. Estratigrafía del Grupo Mendoza, con especial referencia a la Formación Vaca Muerta entre los paralelos 25° y 40°S – Cuenca neuquina mendocina. *Asoc. Geol. Arg., Rev.*, 32 (3): 190–208.

Lesta, P.J., 1968. Estratigrafía de la Cuenca del Golfo de San Jorge. *III Jorn. Geol. Arg., Actas*, 1: 251–289.

Lesta, P.J. and Ferello, R., 1972. Región extraandina de Chubut y norte de Santa Cruz. In: A.F. Leanza (Editor), *Geología Regional Argentina*. Acad. Nac. Cienc., Córdoba, pp. 601–653.

Lesta, P.J., Turic, M.A. and Mainardi, E., 1979. Actualización de la información estratigráfica en la cuenca del Colorado. *VII Congr. Geol. Arg., Actas*, 1: 701–713.

Levi, B., 1960. Estratigrafía del Jurasico y del Cretácico. Inferior de la Cordillera de la Costa, entre las latitudes 32°40′ y 35°40′. *Univ. Chile, Inst. Geol., Publ.*, 16: 221–269.

Levi, B., 1968. *Cretaceous Volcanic rocks from a Part of the Coast Range West from Santiago, Chile*. Ph. D. Thesis, Univ. California, Berkeley, Calif.

Levi, B. and Corvalán, J., 1968. Espesor y distribución de los depósitos del Geosinclinal andino en Chile Central. *Rev. Miner.*, 13 (101): 13–15.

Ludwig, W.J., Ewing, I.J. and Ewing, M., 1965. Seismic-refraction measurements in the Magellan Straits. *J. Geophys. Res.*, 70: 1855–1876.

Ludwig, W.J., Ewing, I.J. and Ewing, M., 1968. Structure of the Argentine continental margin. *Am. Assoc. Petrol. Geol., Bull.*, 52 (12): 2337–2368.

Lydekker, R.N., 1893. The Dinosaurs of Patagonia. *Mus. La Plata, An. Paleontol.*, 11: 1–14.

Malumián, N., 1968. Foraminiferos del Cretácico Superior y Terciario del subsuelo de la Provincia de Santa Cruz, Argentina. *Ameghiniana*, 5 (6): 191–272.

Malumián, N., 1970. Bioestratigrafía del Terciario marino del subsuelo de la Provincia de Buenos Aires. *Ameghiniana*, 7 (2): 173–204.

Malumián, N., 1978. Aspectos paleoecológicos de los foraminíferos del Cretácico de la cuenca Austral. *Ameghiniana*, 15 (1–2): 149–160.

Malumián, N. and Báez, A.M., 1976. Outline of Cretaceous stratigraphy of Argentina. *Ann. Mus. Hist. Nat. Nice*, 4: 1–10.

Malumián, N. and Masiuk, V., 1973. Asociaciones foraminiferológicas fósiles de la República Argentina. *V Congr. Geol. Arg., Actas*, 3: 433–453.

Malumián, N. and Masiuk, V., 1975. Foraminíferos de la Formación Pampa Rincón (Cretácico Inferior) Tierra del Fuego. *Rev. Esp. Micropaleontol.*, 7 (3): 579–600.

Malumián, N. and Masiuk, V., 1976. Foraminíferos característicos de las Formaciones Nueva Argentina y Arroyo Alfa, Cretácico inferior, Tierra del Fuego, Argentina. *VI Congr. Geol. Arg., Actas*, 1: 393–411.

Malumián, N. and Masiuk, V., 1978. Foraminíferos planctónicos del Cretácico de Tierra del Fuego. República Argentina. *Asoc. Geol. Arg., Rev.*, 33 (1): 36–51.

Malumián, N. and Proserpio, C., 1979. Foraminíferos aglutinados del Cretácico de Cuenca Austral. Su significado geológico ambiental. *VII Congr. Geol. Arg., Actas*, 2: 431–437.

Malumián, N., Masiuk, V. and Riggi, J.C., 1971. Micropaleontología y sedimentología de la perforación SC-1, prov. Santa Cruz, Rep. Argentina, Su importancia y correlaciones. *Asoc. Geol. Arg., Rev.*, 24 (2): 175–208.

Marchese, H.G., 1971. Litoestratigrafía y variaciones faciales de las sedimentitas mesozoicas de la Cuenca Neuquina, Prov. Neuquén. Rep. Argentina. *Asoc. Geol. Arg., Rev.*, 26 (3): 343–410.

Martinez, R. and Osorio, R., 1963. Consideraciones preliminares sobre la presencia de carófitas fósiles en la Formación Colimapu. *Rev. Miner.*, 82: 28–43.

Martinez Pardo, R., 1965. *Bolivinoides draco dorreni* Finlay from the Magellan Basin, Chile. *Micropaleontology*, 11 (3): 360–364.

Masiuk, V. and Nakayama, C., 1979. Sedimentitas marinas mesozoicas del Lago Fontana, su importancia. *VII Congr. Geol. Arg., Actas*, 2: 361–378.

Masiuk, V., Becker, D. and García Espiasse, A., 1976. Micropaleontología y sedimentología del Pozo YPF Ch. Pves-1 (Península de Valdez), Prov. del Chubut, Rep. Argentina, Importancia y correlaciones. *ARPEL XXIV, Yacimientos Petrolíferos Fiscales, Buenos Aires*.

McLachlan, I.R., McMillan, I.K. and Brenner, P.W., 1976a. Micropaleontological study of the Cretaceous beds at Mbotyi and Mngazana, Transkei, South Africa. *Trans. Geol. Soc. S. Afr.*, 79: 321–340.

McLachlan, I.R., Brenner, P.W. and McMillan, I.K., 1976b. The stratigraphy and micropaleontology of the Cretaceous Brenton Formation and the PBA/1 well, near Knysna, Cape Province. *Trans. Geol. Soc. S. Afr.*, 79: 341–370.

Moreno, J.A., 1970. Estratigrafía y paleogeografía del Cretácico superior de la cuenca del noroeste argentino, con especial mención de los subgrupos Balbuena y Santa Bárbara. *Asoc. Geol. Arg., Rev.*, 25 (1): 2–44.

Moscoso, R., 1976. Antecedentes sobre un engranaje volcanico-sedimentario marino del neocomiano en el área de Tres Cruces, IV. Región Chile. *I Congr. Geol. Chileno*, 1 (A): 156–167.

Muñoz, C.J., 1960. Contribución al conocimiento geológico de la cordillera de la Costa de la zona central. *Rev. Miner.*, 69: 28–47.

Musacchio, E., 1978. Microfauna del Jurásico y del Cretácico Inferior. *VII Congr. Geol. Arg., Relatorio: Geología y Recursos Naturales del Neuquén*, pp. 147–161.

Natland, M.L., González, E.P., Cañon, A. and Ernst, M., 1974. A system of stages for correlation of Magallanes Basin sediments. *Geol. Soc. Am., Mem.*, 139: 1–125.

Niemeyer, H., 1975. *Geología de la Región entre lago Gral. Carrera y el Rio Chacabuco (Prov. de Aisen) Chile*. Tesis, Univ. de Chile, Santiago de Chile. Unpublished.

Nullo, F. and Proserpio, C., 1975. La Formación Taquetrén en Cañadón del Zaino (Chubut) y sus relaciones estratigráficas en el ámbito de la Patagonia, de acuerdo a su flora. *Asoc. Geol. Arg., Rev.*, 30 (2): 133–150.

Nullo, F., Proserpio, C. and Blasco, G., 1981. El Cretácico de la Cuenca Austral entre Lago San Martin y Río Turbio. In: W. Volkheimer and E.A. Musachio (Editors), *Contribución Comite Sudamericano del Jurásico y Cretácico, II Congr. Latinoam. Paleontol.*

Padula, E.L., 1972. Subsuelo de la Mesopotamia y regiones adyacentes. In: A.F. Leanza (Editor), *Geología Regional Argentina*. Acad. Nac. Ciencias, Córdoba, pp. 213–236.

Padula, E.L. and Mingramm, A., 1976. Subsurface Mesozoic Red-Beds of the Chaco Mesopotamian Region and their Relatives in Uruguay and Brasil. *IUGS Simp., Buenos Aires, Gond. Estrat.*, pp. 1053–1071.

Paulcke, W., 1906. Die Cephalopoden der oberen Kreide Südpatagoniens. *Ber. Naturforsch. Ges., Freiburg*, 15: 167–248.

Pesce, A.H., 1979. Estratigrafía de la Cordillera Patagónica, entre los paralelos 43°30′ y 44° de l.s. y sus áreas mineralizadas. *VII Congr. Geol. Arg., Actas*, 1: 315–333.

Petersen, C.S., 1946. Estudios geológicos en la región del río Chubut medio. *Dir. Gral. Miner. Geol., Bol.*, 59.

Pisaces, L.R., 1976. Estratigrafía de la Cordillera de la Costa entre la Cuesta El Melón y Limache, Provincia de Valparaiso, Chile. *I Congr. Geol. Chileno*, 1 (A): 65–82.

Ploszkiewicz, J.V. and Ramos, V.A., 1978. Estratigrafía y Tectónica de la Sierra de Payaniyen (Prov. del Chubut). *Asoc. Geol. Arg., Rev.*, 32 (3): 209–226.

Pöthe de Baldis, D. and Ramos, V.A., 1978. Las microfloras de la Formación Río Mayer y su significado estratigráfico. Prov. de Santa Cruz, República Argentina. *II Congr. Arg. Paleontol. Bioestratigr.; I Congr. Lat. Paleontol., Resúmenes, Buenos Aires*.

Proserpio, C., 1976. Descripción Geológica de la Hoja 44e Cañadón Racedo (prov. del Chubut). *Ser. Geol. Nac., Buenos Aires*.

Putzer, H., 1962. *Die Geologie von Paraguay. Beiträge zur regionalen Geologie der Erde*. Borntraeger, Berlin, 182 pp.

Ramos, V.A., 1976. Estratigrafía de los Lagos la Plata y Fontana, prov. del Chubut, Rep. Argentina. *I Congr. Geol. Chileno*, I (A): 43–64.

Ramos, V.A., 1977. Descripción geológica económica de la Hoja 47 a-b, Lago Fontana, prov. del Chubut. *Ser. Geol. Nac., Buenos Aires*.

Ramos, V.A., 1979. El vulcanismo del Cretácico Inferior de la Cordillera Patagónica de Argentina y Chile. *VII Congr. Geol. Arg., Actas*, I: 423–435.

Ramos, E.D. and Ramos, V.A., 1979. Los ciclos magmáticos de la República Argentina. *VII Congr. Geol. Arg., Actas*, I: 771–786.

Reig, O.A., 1959. Primeros datos descriptivos sobre los anuros del Eocretácico de la Provincia de Salta (Rep. Argentina). *Ameghiniana*, 1 (4): 3–8.

Reyes, F.C. and Salfity, J.A., 1973. Consideraciones sobre la estratigrafía del Cretácico (subgrupo Pirgua) del NO argentino. *V Congr. Geol. Arg., Actas*, 3: 355–385.

Reyes, F.C., Salfity, J.A., Viramonte, J.G. and Gutierrez, W., 1976. Consideraciones sobre el vulcanismo del subgrupo Pirgua (Cretácico) en el norte argentino. *VI Congr. Geol. Arg., Actas*, 1: 206–223.

Reyes, R.B., 1970. La Fauna de Trigonias de Aisen. *Inst. Inv. Geol. Chile, Bol.*, 26: 4–39.

Reyes, R.B. and Perez d'A., E., 1978. Las Trigonias del Titoniano y Cretácico inferior de la Cuenca Andina de Chile y su valor cronoestratigráfico. *Inst. Inv. Geol. Chile, Bol.*, 32: 105 pp.

Reyment, R.A., 1972. *Cretaceous history of the South Atlantic Ocean. Implications of Continental Drift for the Earth Science*. Academic Press, 2: 805–814.

Reyment, R., 1977. Las transgresiones del Cretácico medio en el Atlántico Sur. *Asoc. Geol. Arg., Rev.*, 32 (4): 291–299.

Riccardi, A.C., 1970. Favrella R. Douvillé, 1909 (Ammonitina) Cretácico Inferior: edad y distribución. *Ameghiniana*, 7 (2): 119–138.

Riccardi, A.C., 1971. Estratigrafía en el oriente de la Bahía de la Lancha, Lago San Martín, Santa Cruz, Argentina. *Museo La Plata, Rev., VII, N.S., Geol.*, 61: 245–318.

Riccardi, A.C., 1976. Paleontología y edad de la Formación Springhill. *I Congr. Geol. Chileno*, 1: C, pp. 41–54.

Riccardi, A.C., 1977. Berriasian invertebrate fauna from the Springhill Formation of Southern Patagonia. *Neues Jahrb. Geol. Palaeontol. Abh.*, 155: 216–252.

Riccardi, A., 1979. El género Calycoceras Hyatt (Ammonitina, Cretácico Superior) en Patagonia Austral. *Ob. Cent. Mus. La Plata (Paleontol)*, pp. 63–72.

Riccardi, A. and Rolleri, E.O., 1980. Cordillera Patagonica Austral. *II Simp. Geol. Reg. Arg., Acad. Nac. Ciencias, Córdoba*, II: 1163–1306.

Riccardi, A.C., Westermann, G.E. and Levy, R., 1971. The Lower Cretaceous, Leopoldia and Favrella from West Central Argentina. *Palaeontographica*, 136–1: 83–121.

Richter, M., 1925. Beiträge zur Kenntnis der Kreide in Feverland. *Neues Jahrb. Mineral., Geol. Palaeontol., Beil. Bs. Abt. B*, 52: 524–568.

Ruiz Fuller, C., 1965. *Geología y Yacimientos Metalíferos de Chile*. Inst. Inv. Geol. de Chile, Santiago de Chile.

Russo, A. and Flores, M., 1972. Patagonia Austral extraandina. In: A.F. Leanza (Editor), *Geología Regional Argentina*. Acad. Nac. Ciencias, Córdoba, pp. 707–726.

Rutland, R.W.R., Guest, J.E. and Grasty, R.L., 1965. Isotopic ages and Andean uplift. *Nature*, 208: 677–678.

Scheibnerova, Viera, 1971. Paleoecology and paleogeography of Cretaceous deposits of the Great Artesian Basin (Australia). *Geol. Surv. N.S. Wales, Rec.*, 13 (1): 5–48.

Scheibnerova, Viera, 1972a. Foraminifera and their Mesozoic biogeoprovinces. *24th IGC Session*, 7: 331–338.

Scheibnerova, Viera, 1972b. Some interesting Foraminifera from the Cretaceous of the Great Artesian Basin, Australia. *Micropaleontology*, 18 (2): 212–222.

Scheibnerova, Viera, 1973. Non-tropical Cretaceous Foraminifera in Atlantic deep sea cores and their implications for continental drift and palaeoceanography of the South Atlantic Ocean. *Geol. Surv. N.S. Wales, Rec.*, 15 (1): 19–46.

Scott, K.M., 1966. Sedimentology and dispersal pattern of a Cretaceous flysch sequence, Patagonian Andes, Southern Chile. *Am. Assoc. Petrol. Geol., Bull.*, 50: 72–107.

Segerström, K., 1959. Cuadrángulo Los Loros Provincia de Atacama. *Inst. Inv. Geol., Carta Geol. Chile*, 1 (1).

Segerström, K., 1963. Engranaje de sedimentos calcareos con rocas volcánicas y clásticas en el Neocomiano del Geosinclinal Andino. *Soc. Geol. Chile, Simp. Geosinc. Andino*, 1, 1962.

Segerström, K., 1967. Geology and ore deposits of Central Atacama Province, Chile. *Geol. Soc. Am., Bull.*, 78 (3): 305–318.

Sigal, J., Grekoff, N., Singh, N.P., Cañón, A. and Ernst, M., 1970. Sur l'âge et les affinités 'gondwaniennes' de microfaunes (Foramanifères et Ostracodes) malgaches, indiennes et chiliennes au sommet du Jurassique et à la base du Crétacé. *C.R. Acad. Sci. Paris*, 271 : 24–27.

Skarmeta, J., 1976. Evolución tectónica y paleogeografía de los Andes Patagónicos de Aisen (Chile) durante el Neocomiano. *I Congr. Geol. Chile*, 1 (B): 1–15.

Skarmeta, J. and Charrier, R., 1976. Geología del sector fronterizo de Aisén entre los 45° y 46° de l.s. (Chile). *VI Congr. Geol. Arg., Actas*, 1: 267–286.

Stipanicic, P.N. and Bonaparte, J.F., 1972. Cuenca triásica de Ichigualasto. Villa Unión. In: A.F. Leanza (Editor), *Geología Regional Argentina*. Acad. Nac. Ciencias, Córdoba, pp. 507–536.

Stipanicic, P.N. and Reig, A.O., 1957. Breve noticia sobre el hallazgo de anuros en el denominado Complejo Porfírico de la Patagonia Extra-andina con consideraciones acerca de la composición geológica del mismo. *Asoc. Geol. Arg., Rev.* 10 (4): 215–233.

Stipanicic, P.N. and Rodrigo, F., 1970. El distrofismo Jurásico y mesocretácico en Argentina y Chile, con referencia a los movimientos jurásicos de Patagonia. *IV Jorn. Geol. Arg., Actas*, 2: 337–352.

Stipanicic, P.N., Rodrigo, P.F., Boulies, O.I. and Martinez, C.G., 1968. Las formaciones presenonianas en el denominado macizo Nordpatagónico y regiones adyacentes. *Asoc. Geol. Arg., Rev.*, 23 (2): 67–98.

Suarez, M., 1976. Plate tectonic model for southern Antarctic Peninsula and its relation to southern Andes. *Geology*, 4: 211–214.

Suarez, M. and Pettigrew, T.H., 1976. An upper Mesozoic island-arc bank arc system in the southern Andes and Southern Georgia. *Geol. Mag.*, 13 (4): 305–400.

Tavera, J., 1956. Fauna del Cretaceo Inferior de Coiapá. *An. Fac. Cienc. Fis. Mat.*, 13: 205–216.

Thiele, R., Castillo, R., Heim, R., Romero, G. and Ulloa, M., 1978. Geología del sector fronterizo de Chile continental 43°00'–43°45' l.s., Chile. (Columnas de Futaleufú y de Palena). *VII Congr. Geol. Arg., Actas*, 1: 577–591.

Thomas, C.R., 1949. Geology and petroleum exploitation in Magallanes Province, Chile. *Am. Assoc. Petrol. Geol., Bull.*, 33 (9): 1553–1578.

Thomas, C.R., 1958. Geología de la Costa entre el Valle de La Ligua y la Cuesta de Barriga. *Inst. Invest. Geol., Bol.*, 2: 86 pp.

Uhlig, V., 1911. Die marinen Reiche des Juras und der Unter Kreide. *Mitt. Geol. Ges. Wien*, 3: 329–448.

Uliana, M.A., Dellapé, D.A. and Pando, G.A., 1975a. Distribución y génesis de las sedimentitas Rayocianas. Cretácico Inferior de las Provincias de Neuquén y Mendoza. *II Congr. Iberoam. Geol. Econ.*, 1: 151–176.

Uliana, M.A., Dellapé, D.A. and Pando, G.A., 1975b. Estratigrafía de las sedimentitas Rayosianas, Cretácico Inferior. Provincia de Neuquén y Mendoza. *II Congr. Iberoam. Geol. Econ.*, 1: 177–196.

Valencio, D.A., Guidici, A., Mendía, J.E. and Gascón, J.O., 1976. Paleomagnetismo y edades K-Ar del subgrupo Pirgua, Prov. de Salta, Argentina. *VII Congr. Geol. Arg., Actas*, 1: 527–542.

Vicente, J.C., Charrier, K., Davidson, J., Mpodozis, A. and Rivano, S., 1976. La orogénesis subhercínica: Fase mayor de la evolución paleogeográfica y estructural de los Andes Argentino-Chilenos centrales. *V. Congr. Geol. Arg., Actas*, pp. 81–98.

Vicente, J.C., 1970. Reflexiones sobre la porción meridional (del sistema pacifico oriental). In: *Proyecto Int. Manto Superior*. Buenos Aires, pp. 1–37.

Volkheimer, W., 1969. Palaeoclimatic evolution in Argentina and relations with other regions of Gondwana. *Earth Sci., Gondwana Stratigr., IUGS Symp., Paris*, pp. 551–587.

Volkheimer, W., 1978. Microfloras del Jurásico superior y Cretácico inferior de America Latina. *II. Congr. Arg. Paleontol. Bioestratigr., I. Congr. Lat. Paleontol., Besúmenes, Buenos Aires*.

Volkheimer, W. and Pöthe de Baldis, D., 1976. Significado estratigráfico de microfloras paleozoicas y mesozoicas de la Argentina y países vecinos. *XI Congr. Iberoam. Geol. Econ.*, 4: 403–424.

Volkheimer, W. and Salas, A., 1975a. Estudio palinológico de la Formación Huitrín. Cretácico de la Cuenca Neuquina en su localidad tipo. *VI Congr. Geol. Arg., Actas*, 1: 431–453.

Volkheimer, W. and Salas, A., 1975b. Die älteste Angiospermen-PalynoFlor Argentiniens von der Typus und Lokalität des unter-Kretazischen Huitrin-Folge des Neuquén-Beckens. Ihre mikrofloristische Assoziation und biostratigraphische Bedeutung. *Neues Jahrb. Geol. Palaeontol. Monatsh.*, 47: 424–436.

Volkheimer, W. and Sepúlveda, E., 1976. Biostratigraphische Bedeutung und Mikrofloristische Assoziation und Biostratigraphische Bedeutung. *Neues Jahrb. Geol. Palaeontol. Monatsh.*, H. 7: 424–436.

Von Huene, F., 1929. Los saurisquios y ornistisquios del Cretácico Argentino. *Mus. La Plata, An., 2 Ser.*, 3: 1–96.

Waterhouse, J.B. and Riccardi, A.C., 1970. The Lower Cretaceous bivalve *Maccoyella* in Patagonia and its paleogeographic significance for continental drift. *Ameghiniana*, 7 (3): 281–296.

Weaver, Ch., 1931. Paleontology of the Jurassic and Cretaceous of West Central Argentina. *Univ. Wash., Mem.*, 1: 469 pp.

Wilckens, O., 1904. Über Fossilien der oberen Kreide Südpatagoniens. *Centralbl. Min. Geol. Palaeontol.*, pp. 597–599.

Yrigoyen, M., 1962. Evolución de la exploración patrolera en Tierra del Fuego. *Petrotecnia*, (4): 28–38.

Yrigoyen, M.R., 1972. Cordillera Principal. In: D.F. Leanza (Editor), *Geología Regional Argentina*. Acad. Nac. Cienc., Córdoba, pp. 345–364.

Yrigoyen, M., 1975a. La edad cretácica del Grupo Gigante (San Luis) y su relación con cuencas circunvecinas. *I Congr. Arg. Paleontol. Bioestratigr., Actas*, 11: 29–56.

Yrigoyen, M., 1975b. Geología del subsuelo y plataforma continental. *VI Congr. Geol. Arg. Relatorio, Geología de la Provincia de Buenos Aires*, pp. 139–168.

Yrigoyen, M.R., 1976. Observaciones geológicas alrededor del Aconcagua. *I Congr. Geol. Chileno*, 1: 169–190.

Zambrano, J.J., 1971. Las cuencas sedimentaria en la plataforma continental argentina. *Petrotecnia*, 4: 126–137.

Zambrano, J.J., 1974. Cuencas sedimentarias en el subsuelo de la Provincia de Buenos Aires y zonas adyacentes. *Asoc. Geol. Arg., Rev.*, 29 (4): 443–469.

Zambrano, J.J. and Urien, C.M., 1970. Geological outline of the basins in southern Argentina and their continuation off the Atlantic shore. *J. Geophys. Res.*, 75 (8): 1363–1396.

Zeil, W., 1958. Sedimentation in der Magallanes Geosynklinale mit besonderen Berücksichtigung des Flysch. *Geol. Rundsch.*, 47 (1): 425–443.

Zeil, W., 1964. *Geologie von Chile*. Borntraeger, Berlin, 234 pp.

Chapter 10

INDIA

S.N. BHALLA

INTRODUCTION

In Asia a major redistribution of land and sea began towards the end of the Paleozoic. The rigid Indian Shield remained practically unaffected but the
marginal areas and regions lying to the north came
under the influence of this reshuffling. The Mesozoic
was then an era of rapidly changing paleogeography,
the effects of which can be seen in the widely distributed rocks. North of the Indian Shield, Tethys came
into existence. Stretching east—west it received a
thick pile of marine sediments. Over the Shield area,
a thick Gondwana fresh-water sequence accumulated. Along the eastern and western coastal margins,
a pattern of transgression and regression reflect sea
level fluctuations. There are three physiographic provinces, Peninsular and Extra-Peninsular India and the
intervening Ganga Plain (Fig. 1).

The triangular plateau, nearly a peneplain, composed mainly of pre-Cambrian metamorphics, south
of the Vindhyan Range, forms the Indian Shield. It has
been a rigid landmass since the Cambrian Period and,
except for occasional block-faulting, has not experienced major earth movements. This explains the absence of post-Cambrian marine sediments except
along its fringes. It has been exposed to weathering
since the beginning of the Paleozoic Era with residual
mountains and shallow and slow moving rivers. The
strata are, generally, low dipping or horizontal and
show no evidence of orogenic movements.

The Himalayan Mountain Chain extending from
Afghanistan to Burma through Pakistan, Kashmir,
Nepal, and Assam constitutes Extra-Peninsular India.
This region, once occupied by the Tethys, contains a
great thickness of marine strata, metamorphic rocks
and igneous rocks. It was severely affected by orogenic movements and mainly consists of mountains

formed due to folding, faulting, and overthrusting.
The rivers are fast flowing and still in a young stage.

The region lying between Peninsular and Extra-
Peninsular India stretching from the Brahmaputra
River in the east to the Ganga River in the west and
further west to Rajasthan, is mainly drained by the
Ganga and its tributaries and constitutes the Ganga
Plain. It is quite wide and is almost entirely made up
of a considerable thickness of alluvium. Originally a
deep depression between the Peninsula and the Extra-
Peninsula formed by the rise of the Himalayas, it has
since been filled up by alluvium mainly brought down
from the Himalayas. There is considerable interest in
what lies below the alluvium and recently Rao (1973)
gave a lucid account of the sub-surface geology.

Mesozoic rocks occur in Peninsular as well as in
Extra-Peninsular India. They are varied in nature, for
not only do they include marine and non-marine
facies but also volcanics and intrusives. Marine Mesozoic strata are well-developed in Extra-Peninsular India and in the coastal regions of the Peninsula but
non-marine Mesozoic rocks, part of the Gondwana
sequence, are developed in the Peninsula only. In order to facilitate presentation, the marine Mesozoic,
the non-marine Mesozoic and Mesozoic igneous activity have been dealt with separately.

STRATIGRAPHY

Marine Mesozoic rocks

An almost continuous record of the marine Mesozoic rocks have been studied in detail in the Kashmir,
Spiti, and Kumaun regions. Strachey (1851) was the
first to work on the stratigraphy of the Himalayas
and described a fossiliferous sequence of marine Pale-

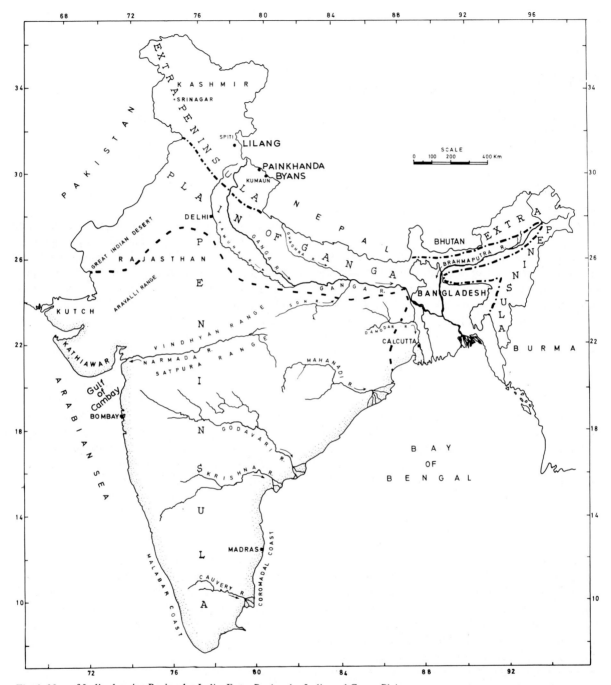

Fig. 1. Map of India showing Peninsular India, Extra-Peninsular India and Ganga Plain.

ozoic and Mesozoic rocks. This was followed by Stoliczka (1866) on the classic area of Spiti. At about
the same time Godwin-Austen (1864) and Ver Chere
and Ver Neiuie (1867) were working on the northwestern Himalayas, including Kashmir. They were fol-

lowed by Lydekker (1883) in Kashmir and Griesbach
(1891) in the Kumaun and Spiti regions. In the recent
past, Pascoe (1959, 1964) and Gansser (1964) have
given a good summary of the geology of the Himalayas.

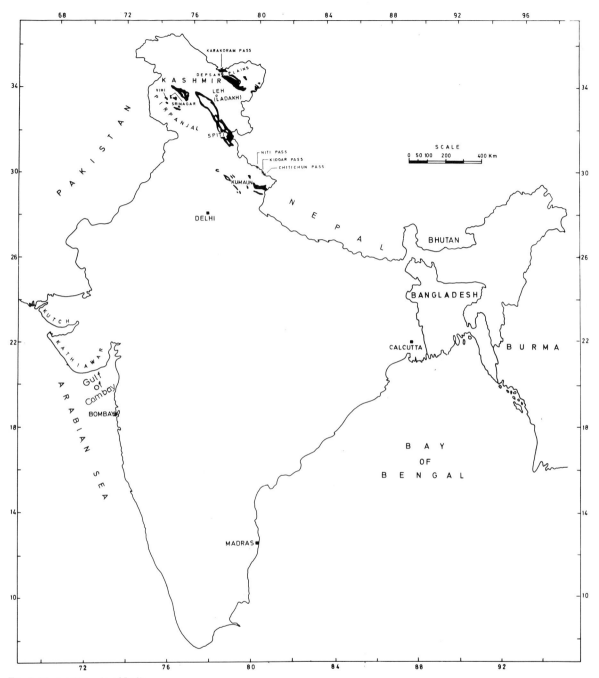

Fig. 2. Marine Triassic of India.

In Peninsular India, the marine Mesozoic rocks are not as extensively developed as in the Extra-Peninsula and are mostly found in the coastal regions.

Triassic

In India, the marine as well as the non-marine facies of Triassic rocks are well-represented. The present-day coast-line of the Peninsula had not yet developed and, consequently, Triassic marine deposits are not found in the present coastal regions. In the interior, arid and semi-arid conditions prevailed and fresh-water deposits forming a part of the Gondwana

TABLE I

Marine Mesozoic sequence in Extra-Peninsular India (After Sastry and Mamgain, 1971)

Period	KASHMIR	SPITI	KUMAUN — PAINKHANDA	KUMAUN — BYANS	European equivalents
C	Karakoram Pass beds (with fauna allied to Ariyalur and Niniyur forms)	"Flysch" — Top limestone with *Globotruncana stuarti* and *G. linneana*		(d) Green and red siliceous sandstone and dense radiolarian hornstone with *Chondrites intricatus* (300–400 m)	Maastrichtian
R					Senonian
E					
T	Chisil Pass beds with *Inoceramus*, *Exogyra*, etc. (with faunal affinity to mostly Uttattur and partly Trichinopoly fauna of east coast); *Chikkim limestone and shales* (33 m)	Middle limestone with *Globotruncana linneana*		(c) Predominantly black shales (slates) and limestone flags with fucoids (500–600 m) — 'Flysch' (Upper Cret.)	Turonian
A	Lingzi Thang beds with *Radiolites* sp., *Gyropleura* sp., *Orbitolina* sp., *Chofatella* sp. and *Pervinquieria inflata*	Basal limestone with *Rotalipora appenninica*		(b) Reddish purple marly shales with Globigerinidae and Rotaliidae (100 m)	Cenomanian
C				(a) Greenish shales (50 m)	Albian
E	*Gieumal sandstone*	*Giumal (Gieumal) sandstones*, loose grits and coarse conglomerates with *Holcostephanus*, *Hoplites*, *Dufrenoyia* (?), *Gryphaea*, *Pseudomonotis*, *Opis* and small Globigerinae, etc.		*Giumal (Gieumal)* sandstones with rare pelecypods	Aptian
O	Dras and Burzil volcanics with intercalated limestones containing *Orbitolina* sp. and *Orbiquia* sp.				Neocomian
U					
S					
J	*Spiti shales* of Pir Panjal and around the Syntaxis and Remo valley	Lochambel stage: with *Spiticeras*, *Blanfordiceras*, *Neocomites*, etc.	Lochambel stage: with *Spiticeras*, *Blanfordiceras*, *Neocomites*, etc.	Lochambel stage: with *Spiticeras*, *Blanfordiceras*, *Neocomites*, etc.	Purbeckian · Portlandian · Kimmeridgian
U	*Spiti shales* partly of Ladakh and Zanskar ranges	Chidamu stage: with *Phylloceras*, *Virgatosphinctes*, *Pseudomonotis*, etc.	Chidamu stage: with *Phylloceras*, *Virgatosphinctes*, *Pseudomonotis*, etc.	Chidamu stage: with *Phylloceras*, *Virgatosphinctes*, *Pseudomonotis*, etc.	
R					
A		Belemnite shale: with *Belemnopsis gerardi*, *Mayaites*	Belemnite shale: with *Belemnopsis gerardi*, *Mayaites*	Belemnite shale: with *Belemnopsis gerardi*, *Mayaites*	Oxfordian
S					
S	Karakoram limestone with *Reineckeia*, *Perisphinctes*, *Hecticoceras*, etc. (south of the Pass)	*Sulcacutus beds*: Black ferruginous oolite with *Belemnites sulcacutus*, *Macrocephalites*, *Dolikophalites*, etc.	Black ferruginous oolite with *Belemnites sulcacutus*, etc. (*Sulcacutus beds*)	Black ferruginous oolite with *Belemnites sulcacutus*, etc. (*Sulcacutus beds*)	Callovian
I					
C	*Megalodon* (Kioto) limestone with *Megalodon ladakhensis* at Matayan, Dras, Sonamarg, Zanskar range and also in Vihi distr.	*Megalodon (Kioto) limestone* (380 m approx.) with *Spiriferina* cf. *obtusa* and some microcephalitid ammonites	Laptal 'Series' lumachelle with oyster viz. *Trigonia*, *Ostrea*, *Cardium*, *Pecten* etc. and *Belemnites* sp.	Laptal 'Series' (80–90 m) lumachelle (agglomerate of shells) with small oysters viz. *Pecten*, *Cardium*, *Trigonia*, etc. and *Belemnites* sp.	Lias
T	Massive unfossiliferous dolomitic limestone (41 m)	Massive limestone and dolomite (with *Spirifera noetlingi*, *Megalodon ladakhensis* and *Dicerocardium himalayense* at 91–121 m above the base); *Kioto (Megalodon) limestone*	*Megalodon* limestone (? 450 m)	*Megalodon* limestone (? 120 m)	
		White and brown *Quartzite series* (91.4 m) with subordinate limestone and black shales with *Spirigera maniensis*	Quartzite series with *Spirigera maniensis* (76 m)		
			Sagenite or *Anodontophora* beds: brown limestone with *Anodontophora griesbachi* (48.7 m)		
R	Unfossiliferous limestones and shales alternating grey or drab limestones and shales (25.6 m)	Sandy and shaly limestone with brown weathering shales and sandstones with *Monotis salinaria* and *Spiriferina griesbachi* (*Monotis shales*, 91.4 m)	Earthy limestones with *Spiriferina griesbachi* passing down into calcareous shales (97.4 m)	Greenish black shales with sandy bands black shales with *Arcestes* (300 m)	Noric

I			Coral limestone (30.4 m)	Limestone with *Spiriferina griesbachi*, coral and crinoid remains	*Halorites* beds: massive grey limestone with numerous cephalopods especially *Halorites procyon* etc. (60.9 m)			
			Juvavites beds (152.3 m)	Brown weathering shales, limestones and sandstones with *Juvavites angulatus*	Nodular and slaty limestone with *Proclydonautilus griesbachi* (30.4 m)		*Juvavites* beds with *Parajuvavites*, *Halorites*, etc.	
A			*Tropites* beds (91.4 m)	Dolomitic limestone with *Dielasma julicum*			*Tropites* limestone with *Tropites subbullatus*	Carnic
			(182 m)	Shales and dark limestone with ammonite bed 122 m above base (*Tropites subbullatus*)				
S	*Myophoria* beds (3.6 m)	*Spiriferina stracheyi* zone, dark limestone with Myophoria (? = Halobis beds)	Grey beds (152 m)	Grey shales and shaly limestone with a bivalve bed 91.4 m above the base and an ammonite bed 15.2 m above the base (*Spiriferina shalshalensis*, and *Joannites cymbiformis*)	*Halobia* beds: black-flaggy limestones, shales, massive earth grey limestones and dolomites passing up into micaceous shales with *Halobia* cf. *comata*			
S			*Halobia* beds (42.7 m)	Dark splintery limestone with *Halobia* cf. *comata*. beds with *Joannites thanamensis* near base	*Traumatocrinus* beds: black flaggy limestone with shale partings with *Traumatocrinus* and *Daonella indica* (3 m)			
	? *Daonella* Limestone (10.6 m)	Hard compact rough weathering limestone with sandy shales	*Daonella* limestone (45.7 m)	Hard dark limestone with *Daonella indica*	Passage beds (shalshal); thin-bedded concretionary limestone with *Daonella indica*, *Spirigera hunica* (6 m)	Grey massive limestone (131 m) (Kuti shales)	Zone of *Paraceratites thuillieri* and *Daonella indica*	Ladinic
I	*Daonella* shales (30 m)	Yellowish sandy shales and sandy limestone with *Daonella*	*Daonella* shales (48.7 m)	Black limestone, shaly limestone and shale with *Daonella lommeli* and *Ptychites gerardi*				
	Upper Muschelkalk (14 m)	Thinly bedded black shales and sandy limestones with *Ceratites*, *Ptychites*, etc.	Upper Muschelkalk (6 m)	Concretionary limestone with shale bands with *Ptychites rugifer*	Upper Muschelkalk limestone with *Ptychites rugifer* (6 m)			Muschelkalk
C	Lower Muschelkalk (2.4 m)	Dark red earthy limestone and hard compact limestone with *Ceratites thuillieri*, *C. ravana*, *Gumnites jollyanus*, etc.	Lower Muschelkalk (1.8 m)	Dark shales and grey limestone with *Keyserlingites dieneri*, *Sibirites prahlada*, *Spiriferina stracheyi*	*Spiriferina stracheyi* beds with *Keyserlingites dieneri* (0.9 m)	(Kalapani limestone	Zone of *Spiriferina stracheyi*	
	Nodular limestone (5.4 m)	Hard nodular limestone unfossiliferous (? = Niti limestone)	Nodular limestone (20 m)	Hard nodular limestone with few fossils	Niti limestone: hard nodular limestone (18.2 m)			
			Basal Muschelkalk (1 m)	Shaly limestone with *Rhynchonella griesbachi*	Shaly limestone with *Rhynchonella griesbachi* and *Sibirites prahlada* (0.9 m)		Zone of *Rhynchonella griesbachi*	
	Nodular limestone *Hedenstroemia* beds (19 m)	Grey nodular limestone with *Danubites* and *Hungarites* beds with *Hungarites middlemissi*, *Flemingites rohilla*, etc.	*Hedenstroemia* beds (10 m)	Limestone with *Pseudomonotis himaica* unfossiliferous shaly limestone and shales alternating; thin-bedded limestone and shales with *Hedenstroemia mojsisovicsi*, *Flemingites rohilla*, etc.	*Hedenstroemia* beds: thin-bedded grey limestone with shale partings with *Flemingites rohilla* and *Pseudomonotis himaica* near the top (7.62 m) Grey limestone with no determinable fossils (1.52 m)	Chocolate limestones and shales (45.7 m)	*Sibirites Spiniger* zone near top	Bunter
	Meekoceras beds (8.2 m)	Limestones and shales with *Meekoceras*, etc.	*Meekoceras* zone (1.0 m)	Thin-bedded limestones and shales with *Meekoceras varaha* and *M. lilangense*	*Meekoceras* bed: dark concretionary limestone with *Meekoceras varaha* and *M. markhami* (0.15 m) Dark blue shales, unfossiliferous (5.4 m)		*Meekoceras* and *Otoceras* fauna near bottom with *Pseudosageceras* and *Ophiceras*	
	Ophiceras beds (47 m)	Grey limestone with *Ophiceras sakuntala*, etc.	*Ophiceras* zone (0.3 m)	Grey limestone with *Ophiceras sakuntala* and *Pseudomonotis griesbachi*	Dark limestone with *Otoceras woodwardi* and *Ophiceras tibeticum* (0.15 m) Dark hard (clay) shale with concretions containing *Episageceras dalailamiae* and *Ptychites scheibleri* (0.45 m)			
	Otoceras beds	Dark grey limestone with *Otoceras*, etc.	*Otoceras* zone (0.6 m)	Brown limestone with *Otoceras woodwardi*	Dark blue limestone with *Otoceras woodwardi* and *Ophiceras sakuntala* (0.3 m)			
PERMIAN		Zewan beds	Productus shales		Productus shales		Productus shales	

sequence, were laid down. In Extra-Peninsular India, however, widespread deposits of marine Triassic sediments were accumulated in the Tethys.

Marine Triassic rocks are enormously developed and widely distributed in Extra-Peninsular India, extending from Kashmir to Nepal, where they succeed the Permian beds without any marked break (Fig. 2). On an average, they are about 1300 m thick and comprise mainly limestones, dolomites, and shales with abundant cephalopods useful in comparisons with the standard subdivisions of the European Triassic.

The Triassic sequence of the entire Extra-Peninsula belt exhibits striking uniformity in the lithology of various subdivisions without observable unconformities. However, some faunal gaps have been reported in the sequence. This may be due to destruction or non-detection of fossils in the areas difficult to reach.

The Triassic sequence in the Spiti region of Extra-Peninsular India, examined by Stoliczka (1866) and Griesbach (1891), was worked out in detail by Hayden (1904) and subsequently modified by Diener (1912). In the Kashmir region, Middlemiss (1910) revised the Triassic sequence followed by Wadia (1934) and others while several additions to the Triassic stratigraphy of the Kumaun region were made by Heim and Gansser (1939).

The principal paleontological studies of the Himalayan Triassic made by Diener (1897–1913), Diener and Kraft (1909), Bittner (1899), Mojsisovics (1899), and Jeannet (1959) brought to light an exceptionally rich marine faunal assemblage, particularly in Spiti and Kumaun. In Kashmir, the Triassic is less well-developed and fossiliferous. The fauna shows a close resemblance to the Alpine Triassic assemblage with which it has been correlated and suggests free and uninterrupted communication between the Indian and European fauna through the Tethys during this period.

Spiti-Kumaun region. A complete section of the Triassic rocks is exposed at Lilang in Spiti and, therefore, the Triassic of Spiti is sometimes termed as the Lilang System in Indian geological literature. The strata conformably overlie the Productus Shale (Permian) and consist mainly of dark limestones and dolomites with subordinate blue-colored shales. Although the depositional environment of these rocks has not yet been studied, the constant characters of rock types over a wide area indicate uniform condi-

tions of deposition, possibly in a clear and deep-water environment without any significant addition of terrigenous material.

The sequence is approximately 1250 m thick and contains abundant ammonites by which it is subdivided into Bunter, Muschelkalk, and Keuper, equivalent to the standard Triassic subdivisions of Europe (Table I). The subdivisions are of vastly different thicknesses: the Lower is 12 m, the Middle 120 m, and the Upper is 1100 m thick.

Lower Triassic. The Lower Triassic beds conformably overlie the Productus Shale (Permian) and chiefly comprise dark-colored limestones with shale intercalations. In Spiti, the sequence is only 12 m thick but eastwards, in the Byans and Painkhanda sectors of the Kumaun region, it attains a thickness of 50 m. There is a prolific ammonite assemblage which persists throughout the entire Spiti–Kumaun belt except in the eastern part, near the Nepal border, where poorly fossiliferous limestone predominates.

Near Lilang, the Lower Triassic can be divided into a lower and an upper stage separated by a few metres of rather unfossiliferous shales and shaly limestones representing, perhaps, a minor fluctuation in the environmental conditions. On the basis of ammonites, the lower stage is further subdivided into *Otoceras, Ophiceras*, and *Meekoceras* zones in an ascending order.

In the Painkhanda sector, northeastern Kumaun, the best section of the lower stage occurs in the Shalshal Cliff where between the *Otoceras* and *Ophiceras* zones, an *Episageceras* zone intervenes. A few metres of dark-blue shales are sandwiched between the *Ophiceras* and *Meekoceras* zones which, in turn, are overlain by a small thickness of grey limestones with indeterminate fossils. The upper stage contains *Hedenstroemia* beds in the basal part, overlain by a small thickness of thin-bedded limestones with shale partings and having abundant *Pseudomonotis himaica* Bittner near the top.

In the Byans sector, on the southeastern side of the Spiti–Kumaun belt near Nepal, the Triassic sequence is much disturbed and comparatively poorly developed, more than in the northwest although the Lower Triassic is thicker than in Spiti. Here, it consists of nearly 50 m of compact, chocolate-colored limestones with some shale intercalations and overlies the Productus Shale (Permian). Near the base of the chocolate limestone, a fauna representing *Otoceras, Ophiceras* and *Meekoceras* zones of Spiti has been recognised, while towards the top a *Sibirites spiniger* zone represents the *Hedenstroemia* beds of the western sector. In northern Kumaun, near the Tibetan border, the Lower Triassic is represented by the Chocolate series. It is a 30—50 m predominantly limestone sequence of Scythian age (Heim and Gansser, 1939).

Middle Triassic. The Middle Triassic is thicker than the Lower Triassic and chiefly consists of concretionary limestone with a very rich Muschelkalk fauna, especially cephalopods. Like the Lower Triassic, the uniformity in lithology and faunal assemblage of the Middle Triassic persists over most of the Spiti—Kumaun belt. The Middle Triassic subdivisions of the Alpine region are recognised and well-developed in the Himalayas.

In the Spiti and Painkhanda sectors the basal Muschelkalk marking the beginning of the Middle Triassic is represented by 1 m thick shaly limestone having *Rhynchonella griesbachi* Bittner, resting conformably on the Lower Triassic. It is succeeded by 20 m of poorly fossiliferous, hard, nodular limestone equivalent to the Niti limestone of the Painkhanda sector. The overlying lower Muschelkalk, corresponding to the lower Anisian of the Alpine region, has abundant cephalopods in the lower part and prolific brachiopods in the upper part. It is followed by the comparatively thick limestone of the upper Muschelkalk containing abundant and widely distributed cephalopods along with a few brachiopods, pelecypods, and gastropods and represents the upper Anisian.

As in Lower Triassic, the Muschelkalk shows a change in lithology and fossil assemblage in the Byans sector. Here, it is represented by poorly fossiliferous limestones having a lower *Rhynchonella griesbachi* zone, a middle *Spiriferina stracheyi* zone, and an upper *Buddhaites rama* zone — the last having fairly abundant cephalopods of upper Anisian age. The latter fossils are generally distorted and the assemblage is more or less endemic.

The upper Muschelkalk gradually passes into the overlying Ladinian stage in Spiti without any break in lithology or fossil content. The lower portion of the Ladinian sequence — the *Daonella* shales — mainly consists of limestone and shale having *Daonella lommeli* Wissman and some Ladinian cephalopods. It is succeeded by a uniform mass of dark and hard limestone, the *Daonella* limestone having *Daonella indica* Bittner along with other fossils, and constitutes the

upper part of the Ladinian succession. However, in the upper portion of the *Daonella* limestone, the lowest Carnian fauna together with the uppermost Ladinian assemblage has been found and, therefore, this part of *Daonella* limestone represents a passage between the Middle and the Upper Triassic.

The Ladinian is poorly developed in the Painkhanda sector and is represented by a few metres of poorly fossiliferous, thin-bedded, concretionary limestone. *Daonella lommeli*, the characteristic Ladinian pelecypod, is not found here but the presence of *Joannites* cf. *proavus* Diener suggests a Ladinian age for these beds.

In northern Kumaun, Heim and Gansser (1939) described 20—60 m of limestone overlying the chocolate limestone and termed it as Kalapani limestone. According to these authors, it ranges from Anisian to Carnian. However, subsequently, Ladinian has been discovered by Sastry (1960) in eastern Byans and Sastry and Mamgain (1971) are of the view that the Kalapani limestone represents the Muschelkalk.

Upper Triassic. The Upper Triassic, beginning with the Carnian, is very well-developed and is thickest of all the Triassic subdivisions of the Himalayas. In the lower part of the Upper Triassic sequence, dark shales and marls predominate but are replaced by limestone and dolomite in the upper part.

The Carnian sequence attains a thickness of 470 m in Spiti. It starts with a bed having *Joannites thanamensis* Diener in the upper portion of *Daonella* limestone and is soon followed by a limestone bed having a fossil characteristic of the Julian Alps in Europe — *Halobia* cf. *comata* Bittner. These two comprise *Halobia* beds which are followed by thick grey shales with a few cephalopods in the lower part and a prolific brachiopod-pelecypod assemblage in the upper part. Overlying this is a thick sequence of *Tropites* beds consisting of shales with limestone intercalations. The lower part has yielded a rich assemblage of poorly preserved cephalopods showing Alpine affinities and having *Tropites* cf. *subbullatus* Hauer belonging to the Tuvalic sub-stage of the Alps. The upper part of the *Tropites* beds consists of dolomitic limestones having a poorly preserved fauna, e.g., *Dielasma julicum* Bittner, etc., and also shows Tuvalic affinities.

The Carnian decreases rapidly in thickness towards the southeastern part of the Spiti—Kumaun belt and is reduced to only half of the Spiti sector. In Painkhanda, the Julian sub-stage of the Himalayas is represented by richly fossiliferous *Traumatocrinus* beds in the lower part and an upper part having a thick succession of dark shales and shaly limestone which constitute the *Halobia* beds, containing *Halobia* cf. *comata* and other Julian fossils.

The Upper Triassic sequence is very much reduced and highly disturbed in the Byans sector. At the top of the limestone sequence containing Muschelkalk fossils, a 1 m thick limestone band has yielded an exceedingly rich faunal assemblage having an admixture of Carnian and Norian fossils. The bed containing the famous *Tropites subbullatus* and forming the *Tropites* limestone, corresponds to the *Tropites* beds of Spiti and marks the top of the Carnian stage in Byans.

The Norian stage in the Spiti sector commences with the *Juvavites* beds, a thick sequence of shales, limestones, and sandstones containing *Juvavites angulatus* Diener. These conformably overlie the *Tropites* beds and, in turn, are overlain by coral limestones which have *Spiriferina griesbachi* Bittner, corals and crinoids. The sequence is capped by *Monotis* shales with pelecypod *Monotis salinaria* Schlotheim, an overlying quartzite with some limestone and shale containing *Spirigera maniensis* von Krafft and finally the thick, massive, poorly fossiliferous, Kioto (*Megalodon*) limestone. About 15 m above the base of the Kioto (*Megalodon*) limestone is a richly fossiliferous horizon with *Megalodon ladakhensis* Bittner and *Dicerocardium himalayense* Stoliczka, termed by Stoliczka (see Diener, 1912) the "Para Limestone" and assigned to the Rhaetic. However, Rhaetic elements are not present and the assemblage is more correctly assigned to the Norian. The upper part of the Kioto limestone contains *Spiriferina* cf. *obtusa* Oppel and belongs to the Lias.

In the Painkhanda sector, the beds of the Norian stage dwindle in thickness. The succession commences with nodular limestone beds having *Proclydonautilus griesbachi* Mojsisovics and a few poorly preserved cephalopods. It passes up to *Halorites* beds with *Halorites procyon* Mojsisovics and other ammonites in abundance. The following limestone beds contain *Spiriferina griesbachi*, brachiopods, and pelecypods and are overlain by *Sagenite* or *Anodontophora* beds. The Quartzite series which follows them passes into the Kioto (*Megalodon*) limestone, marking the top of the Norian.

In the Byans sector, near the Nepal border, the Norian is greatly reduced. It is mainly composed of greenish black shales with *Arcestes* overlain by Kioto (*Megalodon*) limestone. Fossils are rare and their preservation is not good.

Kiogar Chitichun. In the Kiogar Chitichun sector, northwestern Kumaun, small to huge, predominantly limestone, exotic blocks are found. They range in age from Permian to Cretaceous and are supposed to be derived from Tibet to the north. Triassic is fairly well-represented in these blocks.

The Lower Triassic found in blocks near the Kiogar–Chitichun Pass has a fauna comparable to the Lower Triassic of the Spiti–Kumaun tract. The Middle Triassic observed in the blocks near Chitichun has a fauna which exhibits affinities with *Spiriferina stracheyi* zone (Muschelkalk) of the Spiti–Kumaun belt. Some blocks at Malla Johar are rich in Carnian forms showing close affinity with the Alpine assemblage. The uppermost Norian is thought to be represented by unfossiliferous grey dolomitic limestone blocks.

Kashmir. The Triassic sequence in Kashmir is the westward extension of the Spiti–Kumaun belt and attains almost the same thickness. It consists mainly of limestone, dolomite, and shale, and is less fossiliferous and less wellexposed than in the Spiti–Kumaun region. Although the Triassic rocks are present in the Lidar Valley, Ladakh, Pir Panjal and other parts of Kashmir, the best sections are seen in the Vihi district, Guryul ravine and Khrew.

The Zewan beds (Permian) are conformably overlain by poorly developed *Otoceras* beds of Lower Triassic age (Table I). The overlying, profusely fossiliferous *Ophiceras* and *Meekoceras* beds are almost the same as in the Spiti sector. The upper part of the Lower Triassic is marked by a sequence of *Hungarites*-bearing limestone beds.

The Middle Triassic begins with a few metres of unfossiliferous limestone followed by compact limestone of Lower Muschelkalk and shales and limestones of Upper Muschelkalk. The Muschelkalk is richly fossiliferous and the fauna is the same as that of Spiti–Kumaun tract. The overlying Ladinian is represented by *Daonella* shales and ? *Daonella* limestones (Sastry and Mamgain, 1971).

The Upper Triassic in Kashmir is very thick and consists mainly of massive limestone and dolomite forming high cliffs practically devoid of fossils. At the base are fossiliferous, *Spiriferina stracheyi*-bearing, Carnian *Myophoria* beds overlain by a thick, massive, unfossiliferous, limestone and dolomite of Norian age. The sequence is conformably capped by the Liassic Kioto (*Megalodon*) limestone.

Jurassic

The Jurassic was a period of world-wide marine transgression and rocks belonging to this period are well-represented in India. Fresh-water facies of the Jurassic are developed in Peninsular India where they form part of the Gondwana sequence. Marine facies occur in Extra-Peninsular India as well as in the coastal regions of the Peninsula. The prolific development of ammonites with their habitat and short time range, makes them excellent index fossils for the Jurassic.

The Jurassic rocks of Extra-Peninsular India show more uniformity in character than the underlying Triassic but are thinner and comparatively less widespread. Free communication of the Himalayan and Mediterranean fauna through the Tethys which was well-established during the Triassic, continued into the Jurassic and several Jurassic species are common to both the regions.

The present coast-line of India began to develop during the Jurassic. The, then, low lying coastal regions of the Peninsula, mainly Kutch and Rajasthan, were flooded by a shallow Jurassic sea. This sea was connected to the north with the Tethys and extended in the south to Madagascar. This is the first major record of the marine fossiliferous rocks in the Peninsula.

Extra-Peninsular India. The Jurassic rocks are fairly well-developed, with a few gaps, in the Spiti–Kumaun belt, including Garhwal and Kashmir (Fig. 3). In comparison with the underlying Triassic rocks, they are poorly exposed and are less fossiliferous than their counterparts in Kutch.

Spiti–Kumaun. The Jurassic sequence in the Spiti–Kumaun belt conformably overlies the Triassic rocks and attains a thickness of 690 m (Table I).

The Lias in Spiti is represented by the upper 420 m of Kioto (*Megalodon*) limestone with *Spiriferina* cf. *obtusa* Oppel and ammonites. It is continuous with the underlying Triassic portion of this limestone. The Kioto (*Megalodon*) limestone thins eastwards and in Byans, it is reduced to only 360 m as against 690 m in Spiti.

In Kumaun, from Kungribingri to Laptal, a 60–80 m thick sequence of shallow marine strata, the Laptal series (Heim and Gansser, 1939), which contains numerous shell conglomerates with poorly preserved specimens of *Ostrea, Arca, Pecten, Lima, Belemnites*, etc., lies between the Kioto (*Megalodon*) limestone and the overlying Spiti shales. It has a Liassic age. The Laptal series has also been observed at Niti Pass but is said to be absent in Spiti (see Sastry and Mamgain, 1971). According to Arkell (1956), the Laptal series probably represents the lateral variation of the upper part of the Kioto (*Megalodon*) limestone of the Spiti region.

The Laptal series in Kumaun is followed by 3–4 m of black, ferruginous oölites, the *Sulcacutus* beds, named after the predominant fossil *Belemnites sulcacutus* (Suess). These beds also contain ammonites and serve as a good stratigraphic marker extending from Spiti to Byans. Although only a few metres thick, they represent the entire Callovian and form a continuous passage between the Laptal series and the overlying Spiti shales. Detailed examination shows, however, that there are sharp discontinuities at the lower and the upper boundaries, suggesting that Dogger is not represented. Wherever *Sulcacutus* beds are absent, as in the Kuti Valley, there is a well-marked discontinuity between the Kioto limestone or the Laptal series and the Spiti shales, indicating the absence of the Callovian.

The Spiti shales, named after their type-section in Spiti, are widely distributed from Kashmir to Byans and extend further east into Nepal and Sikkim. They range from Upper Oxfordian to Early Cretaceous in age and form a very prominent set of beds in the region. These shales are about 100 m thick but towards the north their thickness increases to a few hundred metres. The Spiti shales are composed mainly of black to dark grey, soft, micaceous and splintery shales with thin sandstone intercalations. They have yielded abundant ammonites, a few pelecypods, and some unidentifiable gastropod moulds. The lithology of the shales is remarkably uniform throughout the entire region and they form an important stratigraphic marker in the otherwise complicated structure of the Himalayas. The fossils have been monographed by Uhlig (1910a, b) and Holdhaus (1913). The megafauna shows a close resemblance to that of East Africa, Kutch, and Central Europe, and is Tethyan in character.

Numerous fossils, especially ammonites, are found in calcareous concretions in the Spiti shales. These concretions are known as "Saligrams" and have been

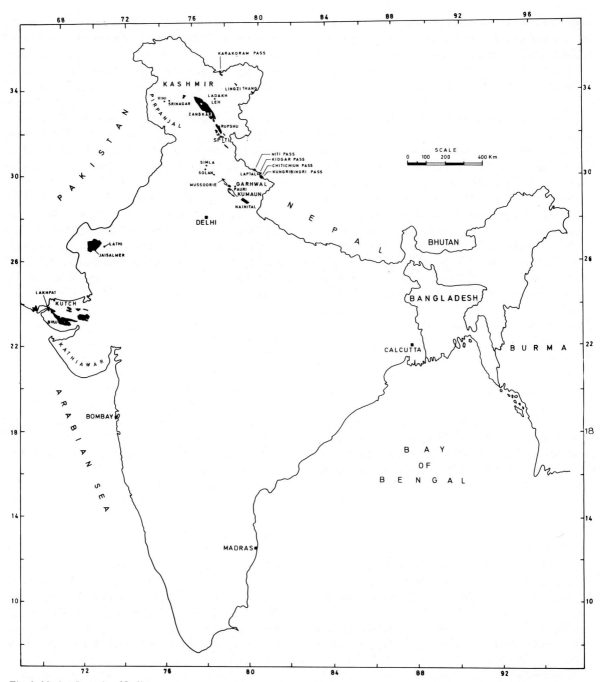

Fig. 3. Marine Jurassic of India.

recently subjected to a detailed investigation by Hagn (1977). The Spiti shales comprise three subdivisions, viz., the Belemnite shale, the Chidamu stage, and the Lochambal stage in an ascending order (Diener, 1895).

The Belemnite shale is mainly composed of black

to brown, friable shales with abundant *Belemnopsis gerardi* Oppel (which extends into the overlying Chidamu stage). Ammonites are poorly represented but the presence of *Mayaites* suggests an Upper Oxfordian age. The Chidamu stage is similar in lithology to the Belemnite shale but is very rich in fossils, with a rich ammonite fauna dominated by Perisphinctids, in shale and limestone concretions. According to Arkell (1956), the fauna of the Chidamu stage is upper Kimmeridgian to lower Tithonian in age, thus the junction between the Belemnite shale and the Chidamu stage marks a time interval covering the lower and middle Kimmeridgian. Sastry and Mamgain (1971) (Table I), however, consider that the Chidamu stage ranges from Kimmeridgian to Purbeckian. The Lochambal stage, the topmost subdivision of the Spiti shales, is similar in lithology to the Chidamu stage. Concretionary ammonites occur in great abundance. *Spiticeras, Blanfordiceras, Neocomites, Holcostephanus*, and several other forms suggest an age range from uppermost Tithonian to Valanginian.

Uhlig (*fide* Pascoe, 1959) observed that the Spiti shales were formed from the muds deposited by rivers from the south in a northward receding Tethys basin.

Garhwal. In the Garhwal Himalaya, the 1500 m Tal Formation (Medlicott, 1864), a sequence of sedimentary strata, occurs between Krol (? Jurassic) and Nummulitic limestone (Paleocene—Eocene). It has been studied by Middlemiss (1887) and others. The lower and the upper boundaries of the Tal Formation are unconformable although the lower may on occasion be sometimes transitional to disconformable. The Tals are important for they yield the first record of pre-Tertiary fossiliferous beds in the Lesser Himalayan zone, although the fossils are fragmentary and poorly preserved.

The well-developed Tal Formation is found in four synclines – Nigalidhar, Korgai, Mussoorie, and Garhwal – extending 180 km from west to east in Himachal Pradesh and Uttar Pradesh. It can be divided into three members, Lower, Middle, and Upper, recently classified lithostratigraphically by Ganesan (1975). The Lower Tal comprises a leached black subarkose with a few plant remains; the Middle Tal contains mainly unfossiliferous arenaceous and argillaceous sediments; and the Upper Tal consists of quartzites, limestone, sandstone with shale and oösparite and has yielded marine fossils.

The age of the Tal Formation is not yet settled. Tewari and Kumar (1966), Bhargava (see Sastry and Mamgain, 1971), Maithani (see Sastry and Mamgain, 1971) and others have obtained algae, bryozoans, foraminifers, and ostracods from the Upper Tal and consider that this part of the Tal Formation belongs to the Lower Cretaceous. However, Ghosh and Srivastava (1962), on palynological evidence, consider the

Tals to be of Jurassic age while Arkell (1956) assigned them to the Upper Jurassic. The consensus is that the major part of the Tal sequence is of Jurassic age, probably Upper Jurassic, but the upper portion passes into the Cretaceous.

The depositional environment of the Tal Formation has not been studied in detail. Arkell (1956) regarded it as the shore facies close to the southern shoreline of the Tethys. Bassi and Vatsa (1971) studied oölitic limestones belonging to the Upper Tal and observed that they accumulated in a shallow, agitated water environment, near to the shoreline. Ganesan (1975) suggested that the Tal sequence of the Mussoorie sector accumulated in a shallow marine basin. He considered that the entire Tal Formation occurring in isolated synclines from Garhwal in the east to Nigalidhar in northwest was deposited in one continuous basin instead of isolated basins as suggested by Auden (1934).

Simla—Garhwal. In the Simla—Garhwal region of the Himalayas, a prominent, 1100 m sequence of sandstone, shale, and limestone occurs and this constitutes the Krol Group. It extends from near Simla in the west to Nainital in the east, a distance of about 300 km. The Krol Group overlies the Blaini Formation and Infra-Krol (? Upper Carboniferous) and is succeeded by the Tals (Jurassic—Cretaceous). It can be divided into different formations from bottom to top and is practically devoid of fossils.

Despite a considerable amount of work, the age of the Krol Group still remains one of the unsolved problems of Indian stratigraphy. This is mainly because until recently, fossils had not been found. The age of this Group was generally considered as probably late Paleozoic to early Mesozoic on stratigraphic, structural or lithological considerations. In recent years, both animal and plant fossils from the Krol and the overlying Tals have been found and, on this basis, Tewari and Kumar (1968) and Bhattacharya and Niyogi (1971) favour a Jurassic age for the upper limit of the Krol Group.

The environment of deposition of the Krol Group has not been studied in detail. Recently, Bhattacharya and Niyogi (1971) studied the geological evolution of the Krol belt in the Simla Hills indicating that the lower part of the sequence (Krol Sandstone Formation) was deposited in a near-shore, high-energy environment with occasional marine and sub-aerial interventions while the upper part (Krol A, B, and C Formations) accumulated in a shallow and stable marine basin.

Kiogar–Chitichun. Some of the exotic blocks of the Kiogar region are composed of earthy-red, thin-bedded, occasionally nodular, limestone with shale and marl intercalations containing cephalopods. Various species of *Phylloceras* dominate the assemblage which also contains *Rhachophyllites, Analytoceras, Ectocentrites, Schlotheimia, Aegoceras, Arietites*, etc. The fauna suggests a Liassic age. According to Arkell (1956), the white, unfossiliferous, Kiogar limestone is probably Upper Jurassic although Heim and Gansser (1939) correlated it with the Triassic Dachsteinkalk of the Alps on lithological grounds.

Kashmir. The Jurassic succession in Kashmir is not as well-developed and exposed as in the Spiti–Kumaun belt. Outcrops are found in the Pir Panjal, Vihi, Ladakh, and the Zanskar and Karakoram ranges which are a direct continuation of the Spiti region.

As in Spiti, the Lias in Kashmir is represented by the upper part of the Kioto (*Megalodon*) limestone (Table I). In the Karakoram area, south of the Karakoram Pass, impure and compact Karakoram limestone occurs. It contains several species of *Reineckeia, Perisphinctes, Ludwigia, Hecticoceras*, etc., with some brachiopods, pelecypods, and foraminifers. The fossils are mostly deformed or fragmentary but clearly indicate a Callovian age.

The fossiliferous Spiti shales showing the same lithology and characters as in Spiti are known to occur in the Remo Valley, around the Syntaxis, Ladakh, and in the Zanskar and Pir Panjal ranges and overlie the Karakoram limestones. As in Spiti, the rich faunal assemblage is dominated by ammonites found in concretions.

Peninsular India. Marine Jurassic rocks are well-developed and exposed in the Kutch and Rajasthan regions of Peninsular India (Fig. 3). They are also found in drill cores on the east coast of India. Unlike Extra-Peninsular India, the Jurassic rocks of the Peninsula accumulated in shallow marine coastal basins established as a result of the transgression of the Jurassic sea over these regions.

Kutch. Kutch is well-known for its rich and varied Tethyan fauna, particularly Upper Jurassic ammonites. The Jurassic rocks are widely distributed, well-developed, and exposed. They are of considerable economic significance as a potential oil producing region.

The Mesozoic sequence of Kutch has been studied by Wynne (1872), Waagen (1876), Oldham (1893),

Spath (1933), Rajnath (1942), Agrawal (1956), Poddar (1964), Biswas (1971), and others. The prolific megafauna has been monographed by Waagen (1876), Gregory (1893, 1900), Kitchin (1900, 1903) and Cox (1940, 1952). The cephalopod fauna was thoroughly revised by Spath (1933) who emended the age of the different subdivisions as proposed by Waagen (1876). In recent years, the Jurassic microfossils of Kutch as well as Rajasthan have also received the attention of micropaleontologists, viz., Subbotina et al. (1960), Lubimova et al. (1960), and Bhalla and Abbas (1978).

In Kutch, the Jurassic rocks crop out principally in three parallel east–west anticlinal ridges (Fig. 4). The northern chain is 161 km in length and forms a series of islands, namely, Patcham, Khadir, Bela and Chorar, while the middle chain, the most prominent, extends for 193 km from Lakhpat in the northwest to Habo in the east. The most southerly of the anticlinal ridges is 64 km long, extending through the Charwar and Katrol Hills south of Bhuj. The anticlines in these three chains are doubly plunging, and due to the quaquaversal nature of the dips they stand out as isolated domes with an east–west alignment. There are four major faults in an east–west direction, corresponding to the three anticlinal belts.

The Kutch Jurassic overlies a Precambrian crystalline basement and has been intruded by numerous sills and dykes, part of the igneous activity of the Deccan Trap. The Jurassic sequence reaches a thickness of about 1950 m and has been divided into four groups, viz., Patcham, Chari, Katrol and Umia in an ascending order, which range in age from upper Bathonian to Tithonian (?Neocomian) (Table II). The youngest group, i.e., the Bhuj Group of Kutch, belongs to the Aptian (Sastry and Mamgain, 1971).

The Patcham Group, the oldest group of Kutch Jurassic sequence, was named after Patcham Island by Waagen (1876). The basal part of the Patcham Group, the Kuar Bet beds, consists of sandstones and limestones with some shale bands. There is a prolific pelecypod fauna with *Corbula lyrata* Sowerby in abundance. A solitary ammonite occurring in these beds is a Stephanoceratid. The beds are followed by Patcham shelly limestone and Patcham coral beds composed of limestones, shales, and marls with rich assemblage of corals, brachiopods, pelecypods, and ammonites, including the characteristic fossil *Macrocephalites triangularis* Spath. The total thickness of the Patcham Group is estimated to be 300 m.

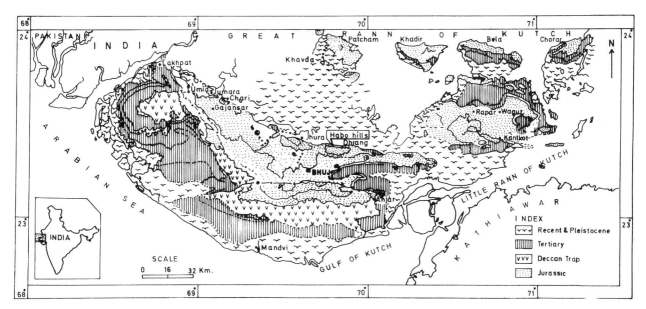

Fig. 4. Geological map of Kutch.

Arkell (1956), on ammonite evidence, considered that the Kuar Bet beds are of late Bathonian age whereas the rest of the Patcham and part of the overlying Chari Group, belong to the Callovian. Spath (1933) assigned an early Bathonian age to the Patcham Group.

The Chari Group (Waagen, 1876) succeeds the Patcham Group. The Chari Group, although generally shaly in comparison with other groups of the Kutch Jurassic, contains hard limestone and calcareous sandstone bands. It is about 366 m thick and has yielded a rich and varied megafaunal assemblage of cephalopods, especially ammonites, with some brachiopods, pelecypods, etc. It has been divided into five subdivisions.

The Chari Group commences with the *Macrocephalus* beds (characteristic fossil *Macrocephalites macrocephalus*) which have been further subdivided by use of ammonites. The uppermost part of the middle subdivision has bands of limestones containing iron-coated, calcareous, oölitic grains with golden color, the Golden Oölite, and these, when present, provide a good stratigraphic marker. The fauna consists of a large number of cephalopods, including the characteristic fossil *Indocephalites diadematus* (Waagen). The *Rehmanni* beds (abundance of *Reineckeia rehmanni* (Oppel)) overlie the *Macrocephalus* beds and are mostly composed of yellow limestone. They pass up into *Anceps* beds consisting of shales in the

lower and limestones in the upper part with abundant *Terebratula* and cephalopods. These three subdivisions range from lower to middle Callovian in age.

The succeeding *Athleta* beds (characteristic fossil "*Ammonites (Peltoceras) athleta*") are mainly composed of light grey shales with bands of white, yellow, or brown limestone. In addition to ammonites, numerous species of pelecypods and some fish remains have also been observed. The *Athleta* beds are of upper Callovian age and their stratigraphic position is easily recognised because they underlie green, reddish or brown, oölitic limestones named Dhosa Oölite — an important stratigraphic marker and the topmost subdivision of the Chari Group. They contain an abundant cephalopod fauna with *Mayaites, Epimayites, Peltoceratoides, Euaspidoceras, Dhosaites*, etc., pelecypods, and terebratulids. This subdivision most probably belongs to the Lower Oxfordian.

The Katrol Group which overlies the Chari Group, attains a thickness of 300 m and is chiefly composed of shales although in the upper part, sandstones predominate. It contains abundant cephalopods and pelecypods. The Katrol Group has been subdivided into Kantkote sandstones, Belemnite marls of Jurun, Lower Katrol beds, Middle Katrol, Upper Katrol sandstone, and Upper Katrol shales in an ascending order.

The Kantkote sandstones comprise grey or pinkish

TABLE II

Mesozoic sequence of the west coast (After Sastry and Mamgain, 1971)

Group	Kutch – Subdivisions with characteristic fossils	Kathiawar	European stratigraphic equivalents
Bhuj	Umia plant beds (= Zamia shales) with *Ptilophyllum* flora	Wadhwan sandstone with limestone and cherty bands containing corals, bryozoa and some ill-preserved ammonites	Aptian
Umia	Ukra beds: calcareous shales with *Australiceras* sp. Unfossiliferous shales and sandstones Trigonia beds Umia ammonite bed with large *Virgatosphinctes* sp.		?Neocomian Tithonian
Katrol	Upper Katrol shales (= Narha and Gajansar beds) with *Hildoglochiceras* Upper Katrol sandstone mostly barren with a doubtful record of *Aulacosphinctoides* sp. Middle Katrol red sandstone with *Torquatisphinctes* sp. and *Katroliceras* sp., etc. Lower Katrol beds: shales with ammonites at base *Belemnites* marls of Jurun Kantkote sandstone with *Euapidoceras*, *Taramelliceras* sp.		Portlandian Upper Kimmeridgian Middle Kimmeridgian Lower Kimmeridgian Up. Oxfordian
	Dhosa (= Mebha) oolite with *Mayaites*, *Epimayaites*, etc. Athleta beds with *Metapeltoceras*, *Peltoceras* and *Reineckeites*		?Lr. Oxfordian Up. Callovian
Chari	Anceps beds {upper: yellow limestone with *Kinkeliniceras* and *Hubertoceras* sp. lower: sandy calcareous shales with *Indosphinctes* sp.}		
	Rehmanni beds {upper: yellow limestone with *Reineckeia tyranniformis* and *Idiocycloceras*, etc. lower: yellow limestone with *Reineckeia rehmanni*}		Middle and Lower Callovian
	Macrocephalus beds {upper: limestones middle: shales with ferruginous nodules lower: white limestone, intercalated shales} with *Macrocephalites*, *Dolikephalites*, etc.		
Patcham	Patcham coral bed with *Macrocephalites*, *Sivajiceras*, *Proceraties*, etc. Patcham shelly limestone with *Macrocephalites* spp. Kuar Bet beds near Khera with *Corbula lyrata*, *Protocardia* and *Pseudotrapezium*, etc.		Up. Bathonian

———————————————————— Archaean granites and gneisses ————————————————————

shales in the lower part and fine to coarse sandstones in the upper part. The rich ammonite assemblage consists of species of *Dichotomosphinctes, Epimayites, Discosphinctes, Torquatisphinctes, Neaspidoceras,* etc., and suggests assignment to the Bimammatum zone. The assemblage includes *Trigonia smeei* Sowerby and indicates an Upper Oxfordian age. The succeeding Belemnite marls of Jurun are composed of marls and Belemnites, e.g., *Hibolites* spp., and some cephalopods which suggest an early Kimmeridgian age. The overlying Lower Katrol beds consist of shales with some sandstones and marls and contain abundant cephalopods at the base. For the first time, Phylloceratids are found in abundance and the assemblage also contains a large number of Oppeliids. They are followed by brown and red iron-stones of the Middle Katrol having a prolific *Pachysphinctes* and *Katroliceras* fauna. The faunal assemblage of the Middle and Lower Katrol indicates a middle Kimmeridgian age. Unfossiliferous sandstones with a doubtful record of *Aulacosphinctoides meridionalis* Spath, constitute the Upper Katrol sandstone and are of a late Kimmeridgian age. The overlying Upper Katrol shales include the Narha and Gajansar beds and have several species of *Hildoglochiceras* and *Subdichotomoceras* with *Phylloceras, Ptychophylloceras,* etc., but *Haploceras elimatum* (Oppel) is the most common element. These shales are of Portlandian age.

The Katrol Group is followed by the 900 m thick Umia Group, the youngest of the Jurassic sequence in Kutch. The group mainly comprises white, pale-brown, sometimes variegated, sandstones with subordinate ferruginous, hard, black or brown grit and a few thin bands of shale. Lithologically, the beds belonging to the Umia Group are similar to the Gondwana rocks of the Peninsula. Cephalopods are not as abundant as in other groups and *Trigonia* and other pelecypods became stratigraphically important.

The Umia Group commences with the Umia ammonite beds consisting of sandstones and shales with conglomerates. Ammonites are abundant, the fauna being dominated by species of *Virgatosphinctes,* particularly *V. denseplicatus* (Waagen). Marine fossils are rare in the rest of the Umia Group. The overlying *Trigonia* beds consist of sandstones having *Trigonia crassa* Kitchin and *T. ventricosa* Krauss and a few other species of *Trigonia.* They are followed, in turn, by unfossiliferous shales and sandstones and then by marine calcareous shales with *Australiceras* (the Ukra beds). According to Sastry and Mamgain (1971), the

Umia Group, including the Ukra beds, is Tithonian–(?)Neocomian in age but Spath (1933), Arkell (1956), Pascoe (1959), Krishnan (1968), and others regard the Ukra beds as Aptian.

The depositional environment of the Jurassic of Kutch has not been studied in detail. Poddar (1964) pointed out that the Patcham, Chari, and the lower portion of the Katrol Group constitute the marine part of the succession while the remaining sequence is mainly paralic to continental in nature. Hardas and Merh (1972) observed that the Chari Group was accumulated in a circalittoral environment, Katrol in infralittoral to paralic conditions, Umia in littoral to lagoonal conditions, and the major part of the Bhuj Group in a fluvial and deltaic environment. According to these authors, the basin gradually became shallow from Chari times onwards. Balagopal (1969) and Bhalla and Abbas (ms.) show that the deposition of the Kutch Jurassic took place in a near-shore, tectonically unstable marine basin with migrating shoreline. Observations made by Biswas (1971) reveal that the Mesozoic sedimentation in Kutch commenced during the Bathonian, that the sea attained its maximum depth during the Callovian–Oxfordian and then regressed from post-Oxfordian to Neocomian times. The sedimentation terminated due to the regional uplift during the Upper Cretaceous diastrophism.

Recently, Mathur et al. (1970), on micropaleontological grounds, pointed out that some of the strata considered earlier as Middle Jurassic are actually Upper Cretaceous to Paleocene in age. However, Bhalla and Abbas (1975) consider that the post-Jurassic microfossils are not in-situ in the Jurassic strata of Kutch but these elements have been brought from the nearby post-Jurassic exposures due to wind transport in Recent times and their subsequent infiltration has placed them in their present position.

Rajasthan. The sea which flooded Kutch during the Jurassic, also covered large parts of Rajasthan, lying about 300 km to the north. Here, the Jurassic rocks although well-developed, are poorly exposed due to vast cover of sand and alluvium of the Rajasthan desert except in the Jaisalmer area (Fig. 3). La Touche (1902) gave a comprehensive account of the stratigraphy of the Jaisalmer area while the fossil collections and their correlation were revised by Spath (1933).

The base of the Jurassic sequence is nowhere exposed in Rajasthan but lies stratigraphically with unconformity above the Badhaura Formation (Permo-

TABLE III

Mesozoic sequence in Rajasthan

Formation	Lithology	Age
Abur Formation	Limestone, sandstone and shale; *Hoplites macconnelli*, *Deshayesites (?)aburensis*, etc.; (108 m)	Aptian
	――――――unconformity――――――	
Parihar Formation	Grit, gritty sandstone, quartzose sandstone; unfossiliferous; (832 m)	Valanginian to Barremian
Bedesir Formation	Grit, sandstone and shale; *Pachysphinctes* aff. *bathyplocus*, *Virgatosphinctes* aff. *oppeli*, etc.; (332 m)	Upper Jurassic
	――――――unconformity――――――	
Baisakhi Formation	Sandstone and gypseous shale (274 m)	Oxfordian to Kimmeridgian
Jaisalmer Formation	Oolitic and shell limestone with layers of calcareous sandstone; *Corbula lyrata*, *Idiocycloceras singulare*, *Sindeites sindensis*, etc.; (355 m)	Callovian to ?Oxfordian
Lathi Formation	Sandstones with plant fossils in the lower part, become marine in upper part having a few limestone layers; (61 m)	Lower to Middle Jurassic
	――――――unconformity――――――	
Badhaura Formation		Permo-Carboniferous

Carboniferous). The entire Mesozoic succession in Rajasthan has been divided into six formations ranging from Bathonian to Aptian in age (see Table III). As in Kutch, the upper part of the sequence extends into the Cretaceous.

The Jurassic succession in Rajasthan commences with the Lathi Formation – a sequence of plant-bearing, current-bedded, sandstones of fresh-water origin which gradually becomes marine towards the top. The higher horizons contain some hard, buff, limestone layers. It has yielded a prolific floral assemblage suggesting a Lower to Middle Jurassic age (Poddar, 1964). However, its fresh-water nature and plant assemblage makes it a part of the Gondwana sequence. There is a gradational contact with the overlying Jaisalmer Formation.

The Jaisalmer Formation is composed of buff, compact, oölitic and shell limestones with layers of calcareous sandstones deposited in a shallow, rather unstable, marine basin near to the shoreline. It has a rich fauna with *Terebratula biplicata* Sowerby, *Corbula lyrata* Sowerby, *Idiocycloceras singulare* Spath, and *Sindeites sindensis* Spath. The fauna suggests a Callovian age but in the oölitic limestones, in the upper part of this formation, Upper Callovian to Upper Oxfordian ammonites and foraminifers have been recorded.

There is a gradual faunal and lithological change from the Jaisalmer Formation into the overlying Baisakhi Formation. The formational name was assigned by Swaminath et al. (1959) who regarded the lower part of the Bedesir Formation as a distinct formation. It consists of sandstones and gypseous shales indicating deposition in a shallow, receding sea. It is considered to range in age from Oxfordian to Kimmeridgian. In contrast to Kutch, the Callovian and Oxfordian are very much condensed in Rajasthan.

The succeeding Bedesir Formation is separated from the Baisakhi Formation by an unconformity and is composed of ferruginous, calcareous grit, current-bedded sandstones, and some shales which, combined with the fossil assemblage, indicate a shallow, open marine environment. The lower part has yielded some Tithonian ammonites, e.g., *Pachysphinctes* aff. *bathyplocus* (Waagen), *Virgatosphinctes* aff. *oppeli* Spath, etc. The formation has been assigned an Upper Jurassic age (Krishnan, 1968).

The overall deposition of the Jurassic sediments of Rajasthan took place mainly in paralic and epineritic environmnments.

Cretaceous

The marine rocks of the Cretaceous are widely distributed in India. At the dawn of this Period, the tectonic conditions were generally stable, but the widespread Cenomanian transgression brought about significant paleogeographic changes in India. Marine conditions were established in several coastal regions of the Peninsula with a particularly rich and varied fauna and sequence in the Tamil Nadu (Madras) region of South India. Alpine–Himalayan orogenic movements during the Upper Cretaceous resulted in the nearly complete cessation of marine sedimentation in Extra-Peninsular India virtually ending the long and eventful role of the Tethys. In the coastal regions of the Peninsula, the accumulation of freshwater sediments of the Gondwana Group continued in the Lower Cretaceous.

At the end of the Cretaceous Period the widespread outpouring began of immense volumes of lava in the Peninsula now forming the magnificient Deccan Trap. The Extra-Peninsula also witnessed some igneous activity during the Cretaceous but on a limited scale.

Extra-Peninsular India. Marine Cretaceous rocks are developed in the Spiti, Byans, and Kashmir regions (Table I) but are not as widespread or fossiliferous as the underlying Triassic or Jurassic strata (Fig. 5). Moreover, the richness and diversity of the South Indian Cretaceous fauna is not mirrored in the Extra-Peninsular Cretaceous.

Spiti. As pointed out earlier, the upper part of the Lochambal stage belonging to the Spiti shale sequence contains an uppermost Tithonian to Valanginian fauna. This stage conformably passes up into a 100 m thick sequence of hard, yellow, brown, and siliceous sandstones and quartzites constituting the Giumal sandstones (Hayden, 1904). The Giumal sandstones accumulated in a northward receding Tethys. They contain a shallow-water fauna but locally, some deeper-water fossils have been recorded. Spitz (1914) described the Giumal fauna and the cephalopod assemblage was subsequently revised by Spath (1939). The fauna suggests an uppermost Valanginian or Hauterivian to Albian age and is markedly different from the Lochambal fauna.

The Giumal sandstones are overlain by more than 33 m of Chikkim limestones and shales after a brief pause in sedimentation. The limestones are light grey to whitish, fossiliferous, and widely distributed. On the basis of fragmentary *Rudistes* and on stratigraphic grounds, the Chikkim limestone was assigned an Upper Cretaceous age. Kohli and Sastri (1956) obtained species of *Globotruncana* from the Chikkim limestone and divided the sequence into a basal portion having *Rotalipora appenninica* (Renz), a middle part containing *Globotruncana linneana* (d'Orbigny), and an upper portion with *G. stuarti* (de Lapparent) and *G. linneana*. On foraminiferal evidence, they considered the basal part Cenomanian, the middle Turonian, and the upper part as Senonian (Campanian).

The Chikkim limestone passes upwards into sandy, calcareous, grey-green Chikkim shales which have yielded a rich microfaunal assemblage (Jain and Gupta, 1973). The presence of *Globotruncana gansseri dicarinata* (Pessagno) suggests a lower Maastrichtian age while other species of *Globotruncana*, e.g., *G. stuarti stuartiformis* (Dalbiez), *G. subcircumnodifer* Gandolfi, *G. linneiana* (d'Orbigny), and *G. tricarinata* (Qureau) indicate a range from Campanian to Maastrichtian (Jain and Gupta, 1973).

Towards Tibet, the Chikkims are overlain by unfossiliferous sandstones and sandy shales of younger Cretaceous age belonging to *flysch* facies. According to Bhargava (in Sastry and Mamgain, 1971), the so-called flysch (Cretaceous) in Spiti may not be flysch s. str. These strata were, however, deposited in a still shallowing and vanishing Tethys and with them, the marine phase of Extra-Peninsular India was almost over.

Kumaun. In the Kumaun Himalaya, the Spiti shales gradually pass into the overlying Giumal sandstones (Table I). These sandstones are also developed

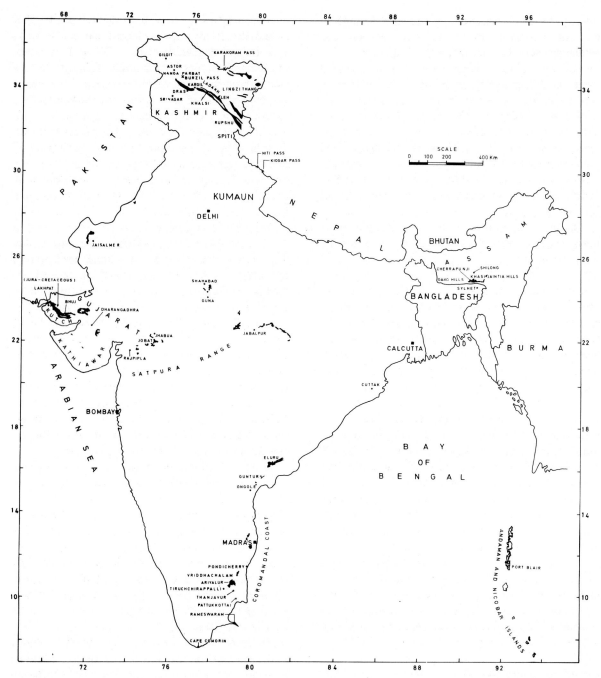

Fig. 5. Marine Cretaceous of India.

from Niti Pass to Hundes, near the Tibetan border, and good sections are exposed at Sangcha Malla in Byans. Here, the Giumal sandstones are glauconitic and contain thin layers of siliceous, greenish-grey shales. Heim and Gansser (1939) estimated their total thickness as 500–700 m but this seems to be exag-

gerated in view of tectonic disturbances and lack of fossils and a suitable map. In the Kiogar area, the Giumal sandstones overlie the Spiti shales. Fossils are rare and show poor preservation.

At Sangcha Malla, the Giumals are overlain by about 1000 m of shales and some sandstones constituting the flysch. Heim and Gansser (1939) divided this sequence into four subdivisions and observed radiolarian chert in the topmost part, indicating the deepening of the basin. According to Sastry and Mamgain (1969; Geological Survey of India Report, see Sastry and Mamgain, 1971)), the so-called Cretaceous "flysch" of Sangcha Malla and Belcha Dhura do not belong to flysch facies s. str. These authors (Sastry and Mamgain, 1971) considered the sequence to be Cenomanian to Maastrichtian in age. The flysch is overlain by sheets of basic igneous rocks having exotic blocks. The blocks are generally considered as the remnants of sheets thrust over the Cretaceous "flysch".

Kashmir. The Cretaceous sequence in Kashmir is less well-developed than the Jurassic (Table I). The outcrops are few and occur in mountainous and inaccessible terrain. In Kashmir, the Indus flows in an approximately southeast–northwest direction following a prominent tectonic zone — the Indus suture (Gansser, 1964) — in which occur Cretaceous flysch-like deposits: the Indus Flysch. Outcrops of the Indus Flysch and associated rocks are seen throughout the Indus Valley, extending from near Nanga Parbat in the northwest to Rupshu and beyond in the southeast. The Indus Flysch consists of limestones, shales, sandstones, conglomerates and volcanics (Dras volcanics) (see Dainelli, 1933).

On the southeastern side of the Nanga Parbat region, from south of Astor and passing through Burzil Pass, Dras, and beyond, a thick sequence of submarine volcanics, intercalated with *Orbitolina*-bearing sediments, forms a continuous belt up to 16 km wide — the Dras volcanics of De Terra (1935). The volcanics consist of basaltic and andesitic, chloritised and epidotised, lava flows with interbedded, finely laminated, purple and green, tuffs and ash beds with shales and cherts or jasper with slates and agglomerates. The sequence is intruded by granites and porphyries with a complicated network of sills and dykes, commonly showing disharmonic folding. The sequence is about 2000 m thick and belongs to volcanic Indus Flysch facies (Gansser, 1964).

The Dras volcanics, according to Wadia (1937), are similar to the Panjal volcanics but De Terra (1935) considers that they are quite different from the older Panjal volcanics. Their Cretaceous age has been confirmed by Sahni and Sastri (1957) and Mamgain and Rao (1965) on the basis of *Orbitolina* species they contain.

The Lower Cretaceous Indus Flysch facies near the Burzil Pass comprises dark, calcareous shales containing *Orbitolina bulgarica* and *O. discoidea* which suggest an upper Barremian age (Douville, 1925). In the nearby Gilgit area, Douville (*op. cit.*) identified *O. bulgarica*, *O.* cf. *discoides*, *Isastraea regularis*, etc., suggesting an upper Barremian or Aptian age.

In the Ladakh region of Kashmir, the Indus Flysch is well-developed (Lydekker, 1883). Several species of *Orbitolina*, including *O. bulgarica*, *O. pileus*, *O. parva*, etc., have been recorded (Douville, 1925; Sahni and Sastri, 1957). The Indus Flysch of the Rupshu area, southeastern Ladakh, has been investigated by Berthelsen (1953).

Northeast of Leh, the principal town of Ladakh, in the Lingzi Thang plains, oölitic limestones belonging to the Lingzi Thang beds have yielded *Sphaerocodium*, *Radiolites* spp., *Gyropleura cenomanensis* (d'Orbigny), etc., suggesting a Turonian age. However, in the eastern sector of the plains, *Orbitolina conulus*, *Choffatella* sp., *Praeradiolites hedini* Douville, *Pervinquieria inflata* (Sowerby), etc., have been recorded which indicate an Albo-Cenomanian age for these beds. Recently, Tewari et al. (1970a) described *Orbitolina* spp. from a limestone band belonging to the Indus Flysch exposed near Khalsi, northwest of Leh, and assigned a Lower Cretaceous (Aptian) age to the limestone.

From the limestones of the Chisil Pass beds, exposed at the Chisil Pass, a Cenomanian fauna showing very close affinity to the Uttattur and partly to the Trichinopoly fauna of the South Indian Cretaceous has been recorded. The Upper Cretaceous Indus Flysch at the Karakoram Pass is represented by Karakoram Pass beds having a Senonian fauna. The assemblage shows strong faunal affinity with the Ariyalur and Niniyur (Campanian–Paleocene) fauna of South India and several species are common to both regions.

A Santonian–Campanian Indus Flysch foraminiferal assemblage has been described by Tewari et al. (1970b) from near Gya in Ladakh. The microfauna includes *Gaudryina bronni* (Reuss), *Dorothia biformis* Finlay, *Citharina geisendorferi* Frank, *Neoflabellina* cf. *rugosa* (d'Orbigny), *Pullenia cretacea* Cushman, *Anomalinoides eriksdalensis* (Brotzen), and some Upper Cretaceous planktonic species, e.g., *Globotruncana fornicata* Plummer, *Globigerinella aspera* (Ehrenberg), *Heterohelix globosa* (Ehrenberg), and *H. pulchera* (Brotzen). From the Kargil area, De Terra

(1935) recorded a gastropod assemblage indicating a Senonian or Maastrichtian age for the Upper Cretaceous Flysch.

The Indus Flysch is thus diachronous ranging in age from Lower to Upper Cretaceous.

Assam. The Cenomanian transgression which covered the eastern part of South India, flooded a large area in Assam during Campanian times. This northward advancing Cretaceous sea deposited the oldest fossiliferous strata in this region.

The Cretaceous rocks of Assam were surveyed by Medlicott (1869) and Palmer (1923). Some alterations and additions have been proposed by Nagappa (1959) and Biswas (1962) while the fossils were chiefly described by Spengler (1923).

In Assam, marine Cretaceous rocks crop out in the Garo, Khasi, and Jaintia Hills, bordering the southern edge of the Shillong plateau and several isolated patches are seen on the Assam plateau. The strata are almost horizontal on the Shillong plateau but dip steeply over its southern edge forming monoclines which, in the Garo Hills, are overfolded and thrust-faulted. The rocks have been grouped into the Mahadeo Formation and the overlying Langpar Formation (Table IV). The sequence is about 320 m thick and is best exposed near Therria Ghat. It was accumulated over gneisses or the Sylhet Trap. Around Cherrapunji,

in the Khasi and Jaintia Hills, there is locally a basal 30 m of conglomerates which merges with the beds of the Mahadeo Formation towards the north.

The Mahadeo Formation is composed chiefly of hard, coarse, and massive sandstones with glauconite. Towards the north, outcrops are restricted. Fossils have been observed in a few thin calcareous beds near the top. The overlying Langpar Formation is more persistent and widely distributed. It consists chiefly of fine-grained, sandy limestone and calcareous shales with some sandstones and sandy shale.

Both the Langpar and Mahadeo Formations have a fairly rich fauna, including foraminifers, with affinities with the Ariyalur assamblage of the South Indian Cretaceous. The important species common to both the regions are *Stigmatopygus elatus* (Forbes), *Alectryonia ungulata* Schlotheim, *Nerita (Otostoma) divaricata* d'Obigny, *Lyria crassicostata* Stoliczka, and *Eubaculites vagina* (Forbes). A total of 50 megafossil species out of 82 are characteristic of the Upper Senonian. However, the presence of *Globotruncana stuarti* (de Lapparent), *Guembelina plummerae* Loetterle, *Orbitoides, Siderolites calcitrapoides* Lamarck in the sequence extends its age to Maastrichtian, and the upper part having *Globigerina pseudobulloides* Plummer and *G. triloculinoides* Plummer, etc., possibly belongs to the Danian.

TABLE IV

Marine Mesozoic sequence in Assam and the Bengal Basin (After Sastry and Mamgain, 1971)

	European stratigraphic equivalents	Assam Shelf		Bengal Basin
U P P E R	Danian (Lower Paleocene)	Therria stage Cherra Formation	Coarse felspathic sandstones and mottled clays	Sylhet limestone
C R E T A C E O U S	Maastrichtian	Langpar Formation	*Globigerina pseudobulloides, G. triloculinoides,* etc., towards the top. Fine grained sandy limestone and calcareous shales with *Globotruncana stuarti, Orbitoides, Siderolites,* etc.	Jalangi Formation
	Campanian	Mahadeo Formation	Glauconitic sandstone with *Siderolites calcitrapoides, Stygmatopygus elatus, Eubaculites vagina, Inoceramus* sp. etc.	Ghatal Formation
			Gneisses/Sylhet Trap	Bolpur Formation

The Cretaceous fauna of Assam is more closely allied to the Cretaceous of South India, Madagascar, and South Africa than that of Baluchistan or the Mediterranean, and belongs to the Indo-Pacific realm.

Peninsular India. Several areas of Peninsular India were invaded by Cretaceous seas (Fig. 5). Those parts of Rajasthan and Kutch which were transgressed during the Jurassic, continued to receive sediments in the Cretaceous. Regression had begun, however, so that marine Cretaceous strata are neither as thick nor as widespread as the Jurassic beds. The site of active and extensive marine sedimentation during the Cretaceous shifted from the Rajasthan–Kutch region to South India. There is a fairly extensive development of marine Cretaceous in Kutch, Kathiawar, and along the Narmada Valley and some Cretaceous beds also occur on the east coast of India.

Rajasthan. The Cretaceous in Rajasthan conformably overlies the Upper Jurassic Bedesir Formation. It is represented by the Parihar Formation and the unconformably overlying Abur Formation (Table III).

The Parihar Formation consists of thick, unfossiliferous, grit, gritty sandstones and quartzose sandstones. Poddar (1964) and Krishnan (1968) consider it to be Valanginian to Barremian in age while Sastry and Mamgain (1971) assigned it an upper Tithonian to Neocomian age. However, its correct age is unknown except from its stratigraphic position.

The Abur Formation is composed of fossiliferous, buff limestones, gritty sandstones and shales. The lower part was deposited in an open marine shelf condition while the upper part was accumulated during the regressive phase. Aptian ammonites, e.g., *Hoplites macconnelli* Whiteaves, *Deshayesites* (?) *aburensis* Spath, etc., have been recorded by Spath (1933).

Kutch. As already noted, the Ukra beds belong to the Aptian. They attain a thickness of 23 m and are exposed in a small area near Lakhpat in northwestern Kutch. A meagre ammonite fauna consisting of *Colombiceras waageni* Spath, *Cheloniceras* aff. *martini* (d'Orbigny), and *Crioceras (Tropaeum)* aff. *australis* (Moore) belonging to the upper Aptian has been recorded by Spath (1933).

Kathiawar. In Kathiawar, southeast of Kutch, a 41 m thick sequence of ferruginous sandstones with thin bands of impure limestone and chert near the top form the Wadhwan sandstone. It conformably overlies the fresh-water Dhrangdhra Formation of Lower Cretaceous age and is unconformably overlain by the

Deccan Trap. The limestone has yielded bryozoa, corals, and poorly preserved ammonites which have not been studied in detail. The faunal assemblage indicates a shallow marine envrionment and suggests an Albian or Cenomanian age (Fedden, 1884). However, Sastry and Mamgain (1971) consider it to be of Aptian age (Table II).

Narmada Valley. The rectilinear Narmada Valley is defined by a group of parallel faults aligned east-northeast–west-southwest. This valley, in the western part of the Indian Shield, probably developed in Jurassic–Cretaceous times, and was flooded during the Cenomanian transgression.

Exposures of marine Cretaceous rocks occur in a chain of widely spaced, isolated patches arranged in a narrow belt following the almost straight east–west Narmada Valley (Fig. 5). The belt extends from near the Gulf of Cambay in the west to Barwaha in the east, covering a distance of about 350 km. The Cretaceous sequence is well-developed near Bagh, Madhya Pradesh, and is commonly referred to as Bagh beds. However, some good exposures are also found at Chirakhan and in the Alirajpur–Jhabua areas of Gujarat State. It was mapped by Blanford (1869) followed by Bose (1884), Roy Chowdhury and Sastri (1962), and others. The fauna and the age of the Bagh beds have been mainly studied by Duncan (1887), Vredenburg (1907), Fourtau (1918), Chiplonkar (1937, 1939a, 1939b, 1941, 1942), Verma (1968), and Jain (1969). Murthy et al. (1963) and Poddar (1964) coined new names of local importance for different lithotopes.

In the Narmada Valley, the Cretaceous sequence rests unconformably over Archaean granites and gneisses. It begins with thick fresh-water Nimar sandstones which have yielded Upper Jurassic to Hauterivian plant fossils (Murthy et al., 1963). In the upper part, marine bivalves and shark teeth have been found and this part is considered as the basal Bagh beds (see Sastry and Mamgain, 1971).

The marine Bagh beds s.str. overlie the Nimar sandstones. The various subdivisions of the Bagh beds in the eastern and western sectors of the Narmada Valley have been summarised in Table V.

The lowest subdivision is the 12 m thick Nodular limestone in the eastern sector of the Narmada Valley. It is richly fossiliferous, whitish, argillaceous, compact, mostly nodular limestone and is widely distributed. Marine fossils, e.g., *Hemiaster fourteavi* Chiplonker, *Parastantoceras mintoi* (Vredenburg),

TABLE V

Marine Mesozoic sequence of the Narmada Valley
(After Sastry and Mamgain, 1971)

European stratigraphic equivalents		Western Narmada Valley	Eastern Narmada Valley
		Deccan Trap/Lameta beds	Deccan Trap/Lameta beds
Senonian	B	Rajpipla limestone unfossiliferous (100 m)	
	A		
Coniacian	G	Oyster beds with *Coilopoceras* and *Proplacenticeras* (2–3 m)	
Turonian	H		Coralline limestone (10 m) with *Proplacenticeras* sp.
	B		Nodular limestone (12 m) with *Placenticeras* sp., *Parastantonoceras* sp.
	E		
Cenomanian	D	Calcareous sandstone (upper Nimars with oysters and shark teeth)	Calcareous sandstone (upper Nimars, 1–2 m, with shark teeth and oysters)
	S		
Lower Cretaceous		Nimar sandstone with plant fossils	Nimar sandstone with plant fossils
		Archaean granites and gneisses	

Proplacenticeras stantoni var. *bolli* (Hyatt), *Coilopoceras scindiae* (Vredenburg), *C. bosei* (Vredenburg), etc., have been recorded from it.

The overlying Coralline limestone is red or yellow in color and mainly consists of abundant small fragments of bryozoa, e.g., *Cerinopora dispar* Stoliczka, and other shells. A thin band (3 m) of marl – Deola and Chirakhan marls – sometimes intervenes between the Nodular limestone and Coralline limestone but recent investigations (Roy Chowdhury and Sastri, 1962; Sahni and Jain, 1968) show that it is not a distinct horizon but has resulted from the weathering of these limestones. Wherever present, it contains abundant fossils also present in the underlying and the overlying limestones.

The other subdivisions, higher in the sequence, are found in the western sector of the Narmada Valley. Here, a 2–3 m thick Oyster bed, a bioclastic calcarinite or rudite, with abundant shells of oysters, e.g., *Ostrea leymeriei* d'Orbigny, disconformably overlies the Nimar sandstone. It passes into the 100 m thick, compact, greyish-white, unfossiliferous Raj-

pipla limestone without any break and, in turn, is overlain by the Deccan Trap or Lameta Group. The Rajpipla limestone is very well-developed around Rajpipla and is the topmost subdivision of the Bagh beds (Sastry and Mamgain, 1971).

The fauna of the Bagh beds is interesting for it illuminates the controversial age of these beds and helps in visualising the Cretaceous paleogeography of western India.

The different groups of fossils occurring in the Bagh beds have provided conflicting evidences for the precise age of these beds. This problem has been reviewed by Verma (1968) and Jain (1969). It is generally considered, however, that Bagh beds extend from Cenomanian to Senonian (Krishnan, 1968; Sastry and Mamgain, 1971).

The Bagh fauna shows close affinity with the Cretaceous fauna of Europe and Arabia and with some elements in the Cretaceous of South India also. This led workers to assume that there was practically no connection between the Gulf of Tethys which penetrated into the Narmada Valley and the Indo-

Pacific sea which flooded the east coast of South India and the Assam region. Verma (1968) recorded fourteen species of sharks from the Bagh beds of which eight are either the same or closely allied to those of the South Indian Cretaceous. Verma (*op. cit.*) discussed the paleogeography and observed that as a result of the break-up of Gondwanaland, the two regions were linked by a strait via Cape Comorin and Ceylon during Turonian times which allowed the intermingling of the faunas. He regarded the comparatively thin succession of the Bagh beds as a condensed sequence homotaxial with the thick development of the Upper Cretaceous in South India.

Central India. A sequence of 6–35 m of cherty limestones with some earthy sandstones and sandy clays underlie the Deccan Trap in Central India and constitute the Lameta Group (Matley, 1921). It crops out as a narrow border around the traps and is of shallow marine origin (Chanda, 1968). It may unconformably overlie the Archaean gneisses, the Upper Gondwana, or sometimes the Bagh beds. It occupies the same or a slightly higher stratigraphic position than the Bagh beds.

The Lameta Group is generally unfossiliferous but has yielded some angiosperm plant fossils, algae, molluscs, and fish remains. However, it is the dinosaurs which constitute the chief element of the fossil assemblage. These have been found in the vicinity of Pisdura in the Chanda district and Jabalpur. Von Huene and Matley (1933) described the vertebrate assemblage which includes *Titanosaurus indicus* Von Huene, *Antarctosaurus septentrionales* Von Huene, *Indosuchus raptorius* Von Huene, *I. matleyi* Von Huene, *Compsosuchus solus* Von Huene, *Laevisuchus indicus* Von Huene, *Laplatasaurus madagascariensis* (Deperet), etc. The Lameta dinosaurian fauna of Jabalpur and Pisdura is closely allied to the Cretaceous of Madagascar, Patagonia and Brazil and suggests a Turonian age (see also De Lapparent, 1957). However, Poddar (1964) and others consider the Lameta Group to be of Senonian age.

East coast. There are marine intercalations in the otherwise fresh-water Upper Gondwana sequence of the east coast of India which occurs in patches from near Cuttack in the north to Tiruchchirappalli (formerly spelled Trichinopoly) in the south along the eastern coast line of the Peninsula. This sequence has been divided into three stages. In the middle stage — Raghavapuram in Eluru, Vammevaram in Ongole, and Sriperumbudur in Madras — is a marine fauna in

addition to a terrestrial flora (Table VI). This stage consists of a moderately thick sequence of shales. It has yielded poorly preserved ammonite assemblage comprising "*Stephanoceras opis* (Sowerby)", *Hocodiscus* cf. *perezianus* (d'Orbigny), *H.* cf. *H. calliaudianus* (d'Orbigny), *Pascoeites budavadaensis* Spath, *P. crassa* Spath, *Gymnoplites simplex* Spath, etc., of Lower Cretaceous age (Spath, 1933). A rich foraminiferal assemblage mainly comprising species of *Ammobaculites* and *Haplophragmoides* has been described by Bhalla (1969a). The fauna compares well with that of the Lower Cretaceous of the Great Artesian Basin, Australia.

The depositional history of the middle stage, though interesting due to the presence of land flora and marine fauna in the same bed, sometimes in a single hand specimen, has not yet been studied in detail. However, Bhalla (1968a) deduced the paleoecology of the Raghavapuram shales and observed that the lower part of the sequence was deposited in an open marine environment while the upper part was accumulated in brackish, shallow-water, near-shore, marshy conditions.

The middle stage of the east-coast Gondwana contains a Jurassic land flora and a Lower Cretaceous marine fauna and as a result its age was much debated. Bhalla (1972) observed that the plant fossils are probably reworked whereas the animal fossils are in-situ, favouring an Aptian or Albian as the upper age limit.

South India. The South Indian Cretaceous was first discovered by Kaye in 1840 (in Blanford, 1862), studied by Blanford (*op. cit.*), and the fossils examined by Stoliczka (1973). The classification of these rocks was modified by Kossmat (1897, 1898), and Rao (1956) reviewed the findings of the earlier investigators. A large volume of literature exists on the South Indian Cretaceous fauna and stratigraphy, but is beyond the scope of the present treatment.

The Cretaceous rocks of Tiruchchirappalli (= Trichinopoly) and adjoining areas in South India, constitute one of the finest developments of the marine fossiliferous Cretaceous sequence in the world. The well-known Cenomanian transgression invaded a large tract of the Coromandel coast, the marine deposits are almost continuous from Upper Albian to Maastrichtian and continue into the Tertiary. This sedimentary sequence, extending from Pondicherry in the north to Rameswaram in the south, accumulated in a pericratonic basin known as Cauvery Basin. In

TABLE VI

Marine Mesozoic sequence in South India and the east coast
(After Sastry and Mamgain, 1971)

European stratigraphic equivalents	Tiruchchirappalli area	Thanjavur and Pattukkottai	Vriddhachalam area		Pondicherry area		Marine intercalations with the east coast Gondwanas		
							Rajahmundry area	Ongole area	Madras area
	Niniyur Group (Paleocene)	Cuddalore sandstones (Mio-Pliocene)	Cuddalore sandstone (Mio-Pliocene)		Paleocene	*Globotruncana gansseri* zone	Infratrappeans, Cuddalore and Rajahmundry sandstones Pleistocene, sub-Recent and Recent Alluvia		
	Kallamedu Formation — Unfossiliferous horizon with fragmentary dinosaurian bone remains		Palakottai Formation — unfossiliferous	*Nerinea* beds	(Pondicherry Formation)				
Maastrichtian (ARIYALUR Group)	Ottakkovil Formation — *Pachydiscus otacodensis* zone		F o r a m i n i f e r a l z o n a t i o n	*Trigonarca* beds	Mettuveli Formation				
	Kallankurichchi Formation — *Hauericeras rembda* zone	Friable sandstone with *Inoceramus* and bryozoans at Budalur. *Globotruncana* cf. *linneiana* in Pattukkottai	*Globotruncana formicata, G. tricarinata* assemblage zone						
				Antisoceras (Valudavur) beds	Valudavur Formation	*Globotruncana tricarinata* zone			
Campanian (TRICHINOPOLY Group)	Sillakkudi Formation — *Karapadites karapadense* zone		*Fronticularia pattiensis* zone						
Santonian — Upper	*Placenticeras tamulicum* zone								
							Tirupati sandstone	Pavalur sandstone	Satyavedu beds
Coniacian — Middle	*Kossmaticeras theobaldeanum* zone						Break	Break	Break
Turonian — Lower (UTTATTUR)	*Lewesiceras vaju* zone								
Turonian — Upper	*Mammites conciliatum* zone	Break	Break		Break				
Cenomanian — Middle	*Calycoceras newboldi* zone								
Albian — Lower	*Schloenbachia inflata* zone						Raghavapuram shales	Vammevaram shales	Sriperumbudur beds
Aptian	Break / Uttattur plant beds								
Neocomian							Golapilli sandstone	Budavada sandstone	
	Archaean	Archaean	Archaean		Archaean		Archaean	Archaean	Archaean

recent years, the Cauvery Basin has been extensively explored for oil and gas possibilities and this has brought to light new information on this region. Important contributions to the structure and tectonics of the Cauvery Basin, mostly from geophysical data, have been made by Kailasam and Simha (1963), Kailasam (1968), Ramanathan (1968), Satpathy and Kanungo (1973), and others.

The Cauvery Basin developed due to block subsidence during different periods of its history although some areas remained tectonically positive throughout. Sub-surface Precambrian crystalline ridges divided the basin into sub-basins and these have been cut by numerous basement faults (Varadarajan and Jagtap, 1968). The maximum depth of the basin is estimated to be 3–3.5 km. A thick sequence of marine Cretaceous is developed in its western part while the Tertiary succession is developed in its eastern part. The sediment source area was nearly peneplained at the beginning of the Cretaceous but was subsequently elevated and material was brought in by easterly flowing streams.

The stratigraphy of the Cauvery Basin has been discussed by Ramanathan (1968), while its sedimentation history and evolution have received the attention of Ramanathan (1968), Datta and Bedi (1968), Venkataraman and Rangaraju (1968), and others. Conditions varied in the different sub-basins at different times due to structural controls.

The Cretaceous rocks of South India crop out from near the coast to more than 100 km west of the present coast-line, and form a nearly continuous belt stretching for about 160 km in a northeast–southwest direction. The outcrops are generally poor due to the vast spread of Cuddalore sandstones (Mio–Pliocene) and alluvium. They occur in four main sectors, viz., Tiruchchirappalli, Pondicherry, Vriddhachalam, and Thanjavur (Fig. 6). The beds overlie mostly the Archaean but occasionally rest on Upper Gondwana rocks. The rocks contain a rich and varied invertebrate fauna of Indo-Pacific nature, also showing some affinities with the Mediterranean and West European fauna. Thus, it illuminates the paleogeography of the region and provides a clue to the communication route followed by the ancient marine fauna. The development of Middle Cretaceous–Lower Tertiary marine sequence is also useful in delineating the much debated Cretaceous–Tertiary boundary.

The best developed sequence of the Cretaceous is found in the Tiruchchirappalli sector.

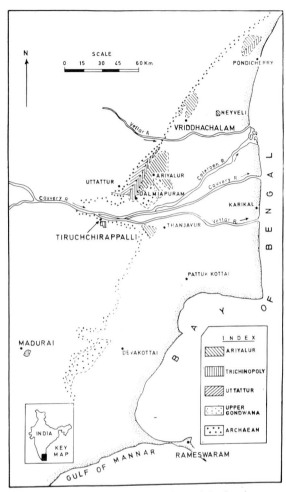

Fig. 6. Geological map of South India, mainly showing occurrences of Cretaceous rocks.

(1) Tiruchchirappalli. The Cretaceous rocks in the Tiruchchirappalli sector occur in a 60 km belt. The sequence has been divided into the Uttattur, Trichinopoly, and Ariyalur groups in an ascending order (Table VI). The overlying Niniyur Group belongs to the Paleocene (Nagappa, 1960; Mamgain et al., 1968) and not to the uppermost Cretaceous as considered by earlier workers.

The Uttattur Group is exposed in the western part where it overlies the Upper Gondwana plant beds or rests directly upon Archaean rocks. The younger groups are exposed successively towards the east coast. On the basis of ammonites, Sastry et al. (1968) proposed nine biostratigraphic zones ranging from Upper Albian to Maastrichtian for the Cretaceous sequence of the Tiruchchirappalli sector.

(a) The Uttattur Group is mainly composed of

fine silts, calcareous shales, and sandy clays with phosphatic and calcareous concretions and gypsum, and attains a thickness of 700 m. At several places, pale, pure, and compact coral limestone is present at the base of the Uttattur sequence. The upper part of the group is arenaceous and exhibits current bedding. On the whole, the group shows characteristics of coastal deposits.

A 27–55 m sequence of grey shales occurring unconformably below the basal coral reef limestone of the Uttattur Group, have been named as Dalmiapuram Formation by Bhatia and Jain (1969). Bhatia and Jain (1972) obtained a rich foraminiferal assemblage, including *Tristix tricarinatum acutangulum* (Reuss,), *Gavelinella rudis* (Reuss), *Eoguttulina anglica* Cushman and Ozawa, *Gavelinopsis umbonella* (Reuss), etc., from the Dalmiapuram Formation and suggested an Aptian–Albian (in part) age. The presence of this formation indicates the existence of pre-Uttattur marine sediments in the Tiruchchirappalli sector.

The Uttattur has a rich fauna and has yielded a variety of foraminifers, cephalopods, pelecypods, gastropods, corals and brachiopods, which show an Indo-Pacific affinity but some cephalopods are allied to the Cenomanian of Europe. The majority of cephalopods are not found in the higher groups. Sastry et al. (1968) divided the Uttattur Group into *Schloenbachia inflata, Calycoceras newboldi,* and *Mammites conciliatum* zones in an ascending order ranging from upper Albian to lower Turonian in age.

(b) Trichinopoly Group unconformably overlies the Uttattur Group. The outcrops form a northeast-southwest belt which thins out towards the north, gradually overlapping the Uttattur and, in turn, almost wholly overlapped by the Ariyalur. It is composed of calcareous grit, sandstone, and some shale and sandy clay with bands of shell limestone, with gastropods and pelecypods, near the base. Granite pebbles commonly occur in the gravels and conglomerates. The beds are about 600 m thick with abundant fossil wood, including huge tree trunks, false bedding, and other features suggesting accumulation in a shallow water, coastal environment.

A large number of invertebrates, especially pelecypods and gastropods with some cephalopods, brachiopods, corals, etc., occur in the Trichinopoly Group but the assemblage is not as rich as that of Uttattur and is somewhat different from it. Sastry et al. (1968) divided the group into *Lewesiceras vaju, Koss-*

maticeras theobaldianum and *Placenticeras tamulicum* zones from bottom to top with ages ranging from middle Turonian to Santonian.

(c) The Ariyalur Group is, in turn, unconformable upon the Trichinopoly Group and is more widely distributed than the other two groups. Exposures are, however, poor. It consists of sandstones with some marly clays, calcareous shales, and limestones and shows uniform bedding and very low dips. It is about 1200 m thick (Ramanathan, 1968) and contains a very rich and well-preserved faunal assemblage which indicates a shallow, rather calm, marine environment. However, towards the upper part, it is mostly unfossiliferous and the strata were deposited probably in a lacustrine environment.

Pelecypods and gastropods are the most abundant fossils in the Ariyalur Group but cephalopods, echinoderms, brachiopods, bryozoans, foraminifers, ostracodes, fishes, and reptiles are also present. Sastry et al. (1968) divided this group into a *Karapadites karapadense* zone (Sillakkudi Formation), a *Hauericeras rembda* zone (Kallankurichchi Formation), and a *Pachydiscus otocodensis* (Ottakkovil Formation), overlain by unfossiliferous beds (Kallamedu Formation) which contain fragments of dinosaurian bones. The entire group ranges from Campanian to Maastrichtian and is unconformably overlain by the Niniyur Group of Paleocene age.

A rich and varied microfaunal assemblage has been described from the Ariyalur Group of Tiruchchirappalli sector. Rao et al. (1968) pointed out that the fossiliferous beds of the Ariyalur Group can be divided into two bio-stratigraphic assemblage zones on the basis of *Globotruncana* species. The lower zone is characterised by *G. lapparenti lapparenti* Brotzen, *G. linneiana* (d'Orbigny), *G. tricarinata* (Qureau), an assemblage suggesting a Campanian age, while the upper zone contains *G. contusa* (Cushman), *G. gansseri* Bolli and *G. stuarti stuartiformis* Dalbiez along with Orbitoides and belongs to the lower Maastrichtian. The overlying beds contain a scanty dinosaurian fauna, viz., *Megalosaurus* sp., *Titanosaurus indicus* Lydekker, which probably belongs to the upper Maastrichtian (Prasad, 1968).

(2) Pondicherry. The Pondicherry sector marks the northern limit of the Cretaceous exposures of South India (Fig. 6). Only the Ariyalur Group is represented here which attains a thickness of more than 250 m and occurs in a poorly exposed 15 km belt, resting directly upon the Archaean. The sequence

consists of massive arenaceous limestones and marls in the lower part, constituting the Valudavur Formation (= *Anisoceras* or Valudavur beds) and calcareous sandstones, shell limestones, and shales in the upper part, belonging to the Mettuveli Formation (= *Trigonarca* beds). The Mettuveli Formation is conformably overlain by *Nerinea* beds (= Pondicherry Formation) (Rajagopalan, 1965, see McGowran, 1968; Murthy, 1968), which are equivalent to the basal Niniyur Group and of Paleocene age.

The Pondicherry Cretaceous has yielded a rich and varied micro- and megafaunal assemblage which is dominated by pelecypods and gastropods but also contains cephalopods and other invertebrates. The fossils are most abundant in the Valudavur Formation and show best preservation. *Anisoceras* occurs frequently and some other ammonites are *Gaudryceras kayei* (Forbes), *Pseudophyllites indra* (Forbes), *Baculites vagina* Forbes, *Brahmaites brahma* (Forbes), etc. The overlying Mettuveli Formation is less fossiliferous and ammonites are few. *Trigonarca gladrina* (d'Orbigny), *Turritella breantiana* d'Orbigny, etc., are, however, numerous. The combined fauna of both these formations suggests a Campanian to Maastrichtian age (Sastry and Mamgain, 1971). Two microfaunal zones – a lower *Globotruncana tricarinata* and an upper *G. gansseri* – have been recognised which also indicate a Campanian to Maastrichtian age for the Cretaceous sequence (Rajagopalan, 1965; McGowran, 1968). There is considerable difference of opinion regarding the precise age of the uppermost part of the Mettuveli Formation and the lowest part of *Nerinea* beds. However, the presence of lower Danian in the Pondicherry sector is not certain (McGowran, 1968) and as such the nature of the contact between Mettuveli Formation (Maastrichtian) and the conformably overlying *Nerinea* beds (Paleocene) remains unsolved.

The depositional environment of the Cretaceous sequence of the Pondicherry sector has been studied by Banerji (1968) who observed that the accumulation of sediments commenced in a moderately deep marine basin which fluctuated between inner and outer shelf conditions and then gradually became shallow towards the close of the Cretaceous.

(3) Vriddhachalam. Only the Ariyalur Group occurs in a 24 km belt stretching northeast–southwest north of Vriddhachalam (Fig. 6). The exposures are poor and in comparison to the Tiruchchirappalli and Pondicherry sectors, the sequence is thin and fossils are sparse. The beds overlie the Archaean and, in turn, are succeeded by Cuddalore sandstones towards the east. They consist of sands and sandy clays with some limestone and sandstone near the base. There is a rich microfaunal assemblage. Stratigraphy of this sector has been worked out by Banerji (1966), mainly on the basis of foraminifers and the age has been discussed by Sastry et al. (1969). The lower, fossiliferous, part of the Cretaceous sequence has been named the Patti Formation and the upper, unfossiliferous part the Palakottai Formation. The Patti Formation has been divided into a lower *Frondicularia pattiensis* zone and an upper *Globotruncana tricarinata–G. fornicata* assemblage zone. The combined age of the Patti and Palakottai Formations was considered as late Santonian to late Maastrichtian.

(4) Thanjavur. Southeast of Tiruchchirappalli, the Cretaceous exposures are seen a few kilometres west of Thanjavur (= Tanjore) and occur in subcrops near Pattukkottai, about 45 km southeast of Thanjavur (Fig. 6). Only the Ariyalur Group is represented in this sector and is composed of friable, fine-grained, argillaceous sandstones directly overlying the Archaean near Budalur. *Cellepora, Inoceramus*, etc., are found and a rich microfaunal assemblage, including *Globotruncana* cf. *linneiana*, has been recorded from a structural well near Pattukkottai.

The zonal distribution of foraminifers in the Cauvery Basin has recently been discussed by Raju (1970).

Other areas – Bengal Basin. Due to thick and widespread development of Ganga alluvium, little is known about the Bengal Basin and the only source of information is the well data obtained by the Indo-Stanvac Company for the exploration of petroleum in this region. Biswas (1959, 1963) pointed out that in the Bengal Basin there is almost a complete sequence from Upper Cretaceous to Lower Tertiary. The Cretaceous rocks have yielded diagnostic Upper Cretaceous foraminifers but they are not richly fossiliferous. The correlation of the Cretaceous sequence of the Bengal Basin and Assam region is given in Table IV.

Andaman Islands. The Andaman group of islands in the Bay of Bengal is the southward extension of the Arakan Yoma mountains of Burma and this tectonic belt passes further down to Sumatra and Java.

In the Andamans, the Cretaceous is represented by basic and ultrabasic intrusives, the Serpentine series

TABLE VII

Mesozoic of the Andaman Islands
(After Sastry and Mamgain, 1971)

European stratigraphic equivalents	Formation with lithology and fossil content		
Eocene Paleocene	Shales and limestone containing *Miscellanea miscella* and *Distichoplax biserialis*	Muthakhari Group (700 m)	Hope Town conglomerate Lipa Black shales
Maastrichtian (Upper Cretaceous)	Ophiolite suite with basic and ultra-basic intrusives intercalated with radiolarian cherts and jaspers; towards the top of the volcanic suite the jaspers contain *Globotruncana arca*, *G. stuarti*, *G. contusa*		
Older Mesozoic	Older sedimentaries with crystalline limestones, quartzites, jaspers, etc.; unfossiliferous		
	Basement not exposed		

which also contains some marble, silicified shales, quartzites, porcellanic limestone, etc., intercalated with radiolarian chert and jasper (Table VII). The porcellanic limestone was considered by Tipper (1911) as similar to Parh limestone (Upper Cretaceous) of Baluchistan and his view was shared by Nagappa (1959). The occurrence of *Globotruncana arca* Cushman, *G. stuarti* (de Lapparent), and *G. contusa* in jasper towards the top of the volcanic suite, indicates a Maastrichtian age (Guha and Mohan, 1965).

Non-marine Mesozoic (Gondwana in part)

Introduction

The Gondwana rocks ranging from Permian to Lower Cretaceous, occupy an important place in Indian stratigraphy. They are mainly composed of sandstones and shales of fresh-water origin, with a few marine intercalations on the east coast, and attain a thickness of about 6100 m. These rocks are widely distributed in Peninsular India with a few occurrences reported from the Extra-Peninsula, especially Kashmir. Except for some faulting and igneous activity, e.g., the Rajmahal Trap and basic intrusives, they are practically undisturbed in the Peninsula but in the Extra-Peninsula they show signs of structural disturbances.

The term "Gondwana" was coined in 1872 by Medlicott in an unpublished report of the Geological Survey of India for a set of fresh-water strata in the erstwhile state of *Gond* in Central India but it was Feistmantel (1876) who brought it into print. As the geological investigations of the different parts of India progressed, more Gondwana outcrops were discovered and they proved to be the major coal-bearing strata in the country. In addition to coal, these sediments contain a prolific floral assemblage too, the Gondwana flora, which is not restricted to India alone but occurs in other continents of the Southern Hemisphere as well.

The Gondwana rocks and the associated coalfields of the Peninsula are best developed in three long and narrow tracts — the Damodar-Son, Mahanadi, and Godavari valleys. These tracts are the main tectonic troughs in which Gondwana sediments and coal have been preserved. On geophysical evidence Bhatia and Rao (1974) estimated the maximum thickness of the Gondwana sediments in the Satpura tract as 2.7 km and in the Godavari tract 4.8 km.

There is a good deal of controversy as to whether the enormous thickness of the Gondwana sediments accumulated in these gradually subsiding troughs due to contemporaneous faulting (Fox, 1931) or whether deposition took place over a vast area but that it was only preserved in these down-faulted tracts (Jowett, 1925; Robinson, 1967; Elliot, 1973). The age of the trough faulting is still uncertain but the faults are considered to be mainly post-Lower Gondwana (Krishnan, 1968).

In the Peninsula, the Gondwana sequence rests unconformably upon the Archaean metamorphics or the Vindhyan or equivalent formations (Cambrian or pre-Cambrian) and are unconformably overlain by the Deccan Trap and associated sedimentary rocks.

Although the Gondwana sequence ranges from Permian to Lower Cretaceous, the coal occurs mainly in the Permian part of the sequence and due to its economic significance, it is this part which has received maximum attention in this country whereas the Mesozoic portion of the sequence is mainly of academic interest.

Distribution

Gondwana rocks are widely distributed in India and a complete sequence is found in the Peninsula. In the Extra-Peninsula, only the Paleozoic part of the sequence is present and, therefore, this region has been excluded.

Gondwana exposures in the Peninsula are found in the Damodar-Son, Mahanadi, Wardha-Godavari, and Narmada Valleys, in the Rajmahal Hills, Chattisgarh, Rajasthan, Kutch, Kathiawar, and in a chain of isolated outcrops along the east coast of the country in the Eluru, Ongole, Madras, and Tiruchchirappalli areas (Fig. 7). In the Chindwara and Nagpur regions of Central India, the Gondwana rocks probably lie concealed below the vast expanse of the Deccan Trap.

Fossil assemblages

The Mesozoic part of the Gondwana sequence in India contains a rich floral assemblage and also a faunal assemblage which includes both vertebrates and invertebrates. The vertebrate fauna predominates in the Triassic part but in Jurassic and Lower Cretaceous portions it is replaced by the prolific development of flora.

Flora. The genus *Dicroidium* makes its first appearance in the Scythian and attains the acme of its development in the Rhaetic. The other plants which are associated with it include *Neocalamites, Danaeopsis, Parsorophyllum, Araucarites,* etc. In the Jurassic and younger strata, an altogether different but still rich floral assemblage comprising *Dictyozamites, Nilsso-*

nia, *Pagiophyllum, Gleichenites,* etc., with the index fossil *Ptilophyllum* is found. In addition, pollens and spores, e.g., *Nidipollenites, Satsangisaccites, Classopollis, Cicatricosisporites, Polycingulatisporites,* etc., are also known in the Mesozoic part of the Indian Gondwana.

Fauna. The Gondwana sediments are chiefly of fresh-water origin and contain a meagre fresh-water invertebrate fauna and a comparatively rich vertebrate fauna (Table VIII). Marine intercalations occurring within the otherwise fresh-water Mesozoic Gondwana on the east coast, contain a prolific foraminiferal assemblage and some ammonites.

In the Scythian and Jurassic, *Unio* and *Estheria* are found; *Tikkia* occurs in the late Upper Triassic; and *Eryon* has been recorded from the Upper Jurassic sediments. Marine invertebrates in the east-coast Gondwana include Lower Cretaceous ammonites, e.g., *Holcodiscus, Lytoceras, Pascoeites, Gymnoplites,* etc. (Spath, 1933) and a rich foraminiferal assemblage dominated by *Haplophragmoides* and *Ammobaculites* and showing Lower Cretaceous affinities (Bhalla, 1969a, b).

Vertebrates are well-represented in the Triassic and play a significant role in the stratigraphic correlation and contribute towards paleogeographic reconstructions. Amongst pisces, *Amblypterus* occurs in the pre-early upper Triassic and *Ceratodus, Depedius, Lepidotus, Jhingrania,* etc., are found in the upper Triassic. The amphibian remains in the Scythian are *Gonioglyptus, Glyptognathus, Indobrachyops, Indobenthosuchus* etc., and in Carnian to Norian *Metaposaur, Mastodonsaurus, Pachygonia,* etc., occur. The reptiles, e.g., *Belodon, Parasuchus, Brachysuchus, Paradepedon,* etc., have been recorded from the post-Permian to pre-early Upper Triassic sediments and dinosaurs occur in early Upper Triassic and younger sediments (Tables VIII, IX). Recently, Chatterjee and Roy-Chowdhury (1974) have given a good review of the Triassic Gondwana vertebrates from India and compared them with those from other parts of the world.

Classification

The classification of the Gondwana sequence has remained one of the ticklish problems of Indian stratigraphy ever since it was discovered by Medlicott more than a century ago. The enormous thickness of the Gondwana rocks distributed widely in space and

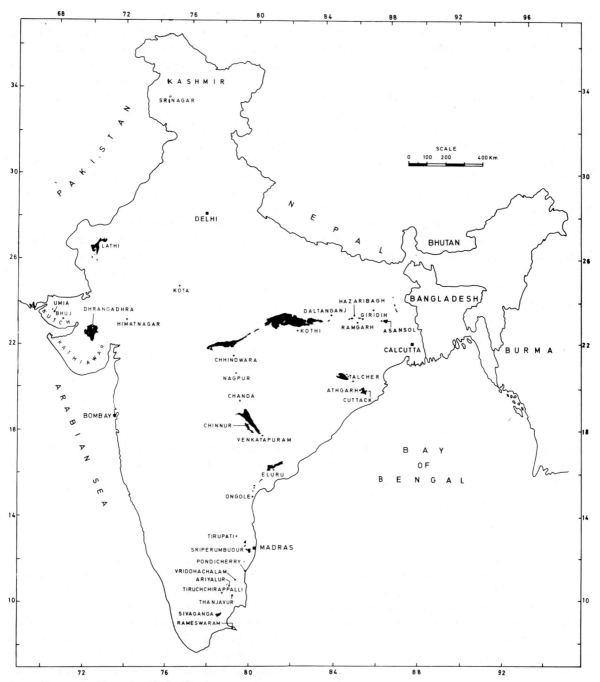

Fig. 7. Non-marine Mesozoic of India (Gondwana, *part*).

ranging from Permian to Lower Cretaceous conveys the idea of a supergroup. As stated elsewhere, the Gondwana sediments contain a prolific floral assemblage and some animal remains which provide a fairly accurate basis for the classification of the sequence.

The main controversy, however, is whether the entire Gondwana sequence can be divided into two or three subdivisions.

The Gondwana floral assemblage can be divided into two major floras — the lower, *Glossopteris* flora and the upper, *Ptilophyllum* flora (Feistmantel, 1876). On this basis, the Gondwana sequence was divided into Lower and Upper Gondwana (Medlicott and Blanford, 1879) and was accepted by Oldham (1893), Cotter (1917), Fox (1931), Pascoe (1959) and others. The different floral assemblages of the Lower and Upper Gondwana are usually separate but may sometimes overlap.

The occurrence of a transitional Triassic flora marked by the presence of *Dicroidium* (Feistmantel, 1882) led Vredenburg (1910) to propose a three-fold classification for the Indian Gondwana—the Lower with a *Glossopteris* flora, the Middle with a *Dicroidium* flora, and the Upper characterised by a *Ptilophyllum* flora, corresponding to the European Permian, Triassic, and Jurassic periods. This found support from Wadia (1949), Lele (1964) and others. According to supporters of the three-fold classification, such a scheme is based not only on floristic limits but takes into consideration the lithological, paleoclimatological, and faunal history of the subdivisions as well.

The Gondwana sequence has been classified mainly on the basis of lithostratigraphy, and different names for the same formation have been given in different areas. In several cases, chronostratigraphic rank terms have been misused for lithostratigraphic units. Fossil contents also have been taken into consideration. On the basis of fossil contents of the different lithostratigraphic units, Shah et al. (1971) have recently proposed a biostratigraphic classification of the Indian Gondwana. The floral assemblage does not contain index fossils to precisely fix the age and, therefore, the authors have based their classification on the occurrence of assemblages and abundance of different forms (Table IX). A correlation of biostratigraphic and lithostratigraphic units of the Indian Gondwana has also been attempted by these authors (Table X). A total of five assemblages zones comprising eleven subzones ranging from Sakmarian to Wealden were recognised for the entire Gondwana sequence by them. Of these, three assemblages zones and six subzones belong to the Mesozoic. Space here does not permit a detailed treatment of the different assemblage zones and subzones which have been discussed by Shah et al. (1971).

Problem of the upper age limit

The upper age limit of the Indian Gondwana is not properly defined and remains a problem. This is main-

ly because in a few places, viz., the East Coast, Kutch, etc., the uppermost part of the Gondwana sequence contains an admixture of terrestrial plants and marine animals and a discrepancy exists in their chronological assessment.

The Upper Gondwana rocks of Eluru, Ongole, and Madras are well-developed and have been divided into three stages, lower, middle, and upper, which have been given local names in the three areas. The middle stage, Raghavapuram in Eluru, Vammevaram in Ongole, and Sriperumbudur in Madras, has yielded an Upper Jurassic Gondwana flora along with Lower Cretaceous ammonites and foraminifers. Bhalla (1972) observed that the fauna of these beds is indigenous while it is yet to be established that the plant fossils are not reworked and on this basis, favoured an Aptian or Albian upper age limit for the East Coast Gondwana.

In the Kutch region, the Umia Group contains Gondwana plant fossils of Upper Jurassic age interstratified with rocks having a Tithonian to Lower Cretaceous marine fauna, including ammonites. The age indicated by flora and fauna exhibits almost the same discrepancy as shown by the assemblages of the East Coast Gondwana. This attracted the attention of workers on Gondwana rocks and some of them (Shah et al., 1971; Rao and Venkatachala, 1971) consider that the Lower Cretaceous "marine intercalations in the Gondwana" belong to marine Cretaceous sequence rather than being a part of the Gondwana succession. However, Bose and Sukh Dev (1956), Singh (1966), and Roy (1968) have confirmed that the *Ptilophyllum* flora ranges up to Wealden and Shah et al. (1971, p. 308) observed, "... the palaeobotanical, palaeogeographical and tectonic considerations tend to fix the upper age limit of the Gondwanas at the top of Wealden."

Climate

The accumulation of the Triassic part of the Gondwana sequence took place in a warm, arid or semi-arid climate with widespread land conditions. The characteristic *Dicroidium* flora of the Triassic developed and reptiles and amphibians were common.

In the Jurassic Period, a milder and humid climate prevailed with the luxuriant growth of ferns, cycadophytes and conifers from which a few coal seams developed. This climate continued into the Lower Cretaceous when the Gondwana sedimentation came to an end.

TABLE VIII

Distribution of fresh-water faunas in the assemblage subzones of the Indian Gondwana (After Shah et al., 1971)

Fauna	Vertebrates				Invertebrates	
Assemblage zones and sub-zones	Pisces	Amphibia	Reptilia	Dinosauria	Lamellibranchs	Crustaceans and others
Weichselia–Onychiopsis sub-zone						
Pagiophyllum–Brachyphyllum sub-zone	? Fish remains	–	–	–	Unio	? Eryon cf. barrovensis
Dictyozamites–Pterophyllum sub-zone	Pholidophorous sp. Lepidotus pachylepis L. deccanensis L. longiceps L. breviceps L. calcaratus Dapedius egertoni Tetragonolepis analis T. rugosus T. oldhami Jhingrenia roonwali		? Rhamphorhynchus sp.	Dinosaurian remains	Unio	Estheria kotaensis Candona kotaensis Insects Blattoids, Hemiptera, Orthoptera
Dicroidium–Neocalamites zone				Remains of plateosaurid and thecodontosaurid prosauropod	Unio	
Ceratodus–Metaposaurus sub-zone	Ceratodus virapa C. hunterianus C. hislopianus C. nageswari	Metaposaurus Mastodonsaurus indicus Pachigonia incurvata Phytosaurs	Paradepedon huxleyi P. (?) indicus Parasuchus Brachysuchus maleriensis Belodon	–	Tikkia corrugata T. navi T. compressa T. subangulata	

Dicynodont sub-zone	—	Remains of neorachitomous amphibians	Dicynodont Theriodont and Theriodont resembling *Erythrosuchus*			*Estheria mangliensis* ? Brachyurous crab ? Insect
Lystrosaurus sub-zone	*Amblypterus* sp.	*Gonioglyptus* sp. *Longirostris huxleyi* *Glyptognathus fragilis* *Indobrachyops panchetensis* *Brachyops laticeps* *Indolyrocephalus* *Indobenthosuchus*	*Lystrosaurus murrayi* *L. rajrukari* *Epicampodon indicus* *Patyeognathus orientale*	*Chasmatosaurus*	*Unio*	
Glossopteris conspicua– *G. retifera* sub-zone	—	*Gondwanosaurus bijoriensis* *Rhinesuchus wadiai*	—	—	*Anthraconauta*	
Cyclodendron sub-zone						? Leaf-like insects
Barakaria dichotoma– *Walkomiella indica* sub-zone						
Gondwanidium– *Buriadia* sub-zone	*Amblypterus kashmirensis* *A. symmetricus*	*Archegosaurus ornatus* *A. wadiai* *Actinodon risinensis* *Chelydosaurus marahomensis*				Insects *Gondwanoblatta kasmiroblatta*
Noeggerathiopsis– *Paranocladus* sub-zone						Neuropterous insects

TABLE IX

Biostratigraphic classification of the Indian Gondwana (After Shah et al., 1971)

Standard scale	Assemblage zones	Assemblage sub-zones	Guide fossils				Type area
			plant fossils		animal fossils		
			mega fossils	micro fossils	vertebrate	invertebrate	
Wealden	Ptilophyllum assemblage zone	Weichselia–Onychiopsis assemblage sub-zone	Cladophlebis salicifolia, Weichselia reticulata, Onychiopsis paradoxus, Gleichenites rewahensis, Brachyphyllum feistmanteli, B. expansum, Araucarites striatus, A. chandaensis, Ginkgoites feistmanteli				Near village Bansa, Shahdol dist, Madhya Pradesh
Upper Jurassic		Pagiophyllum–Brachyphyllum assemblage sub-zone	Cladophlebis indicus, Onychiopsis psilotoides, Nilssonia fissa, Taeniopteris vittata, Ptilophyllum jabalpurense, Pagiophyllum divaricatum	Matoniosporites, Gleichenidites, Boseiosporites, Osmundacidites, Foveotriletes		Unio	Sehoraghat, Narsinghpur dist., Madhya Pradesh
Middle Jurassic		Dictyozamites–Pterophyllum assemblage sub-zone	Ptilophyllum acutifolium, Bucklandia indica, Dictyozamites falcatus, Dictyozamites sp., Nilssonia princeps, Pterophyllum jabalpurense, P. cf. distans	Conbaculatisporites, Aequitriradites, Murospora, Cicatricosisporites			Chaugan Reserve Forest, Narsinghpur dist., Madhya Pradesh
Lower Jurassic							
Rhaetic	Dicroidium–Neocalamites assemblage zone		Parsorophyllum indicum, Dicroidium hughesi, D. odontopteroides, Glossopteris sp., Neocalamites foxtii, Pterophyllum sahnii, Araucarites parsorensis				Near village Parsora, Shahdol dist, Madhya Pradesh
Carnic to Noric	Labyrinthodont assemblage zone	Ceratodus–Metaposaurus assemblage sub-zone			Ceratodus hunterianus, C. hislopianus, C. virapa, C. nageswari, Metaposaurus maleriensis, Paradepedon huxleyi, P. ? indicus, Parasuchus hislopi, Brachysuchus maleriensis, Belodona, Pachygonia incurvata	Tikkia corrugata	Near village Maleri, Adilabad dist., Andhra Pradesh
?		Dicyno-dont assemblage sub-zone			Remains of neorachitomous amphibians, Dicynodon, Theriodon, ? Erythrosuchus		Near Yerrapalli, Adilabad dist., Andhra Pradesh

Age	Assemblage zone	Assemblage sub-zone	Megaflora	Microflora	Fauna	Locality
Scythian	Glossopteris assemblage zone	Lystrosaurus assemblage sub-zone	Schizoneura gondwanensis, Glossopteris communis, G. indica, G. ampla, Cyclopteris pachyrhachis, ?Dicroidium, Podozamites ? lanceolatus, Taeniopteris stenoneura	Epicampodon indicus, Ptycognathus orientalis, Gonioglyptus longirostris, G. huxleyi, Glyptognathus fragilis, Chasmatosaurus, Lystrosaurus murrai, L. rajurkari, Indobrachyops panchetensis	Unio sp., Estheria mangalensis	Panchet Hill, Burdwan dist., Bangladesh
Tatarian		Glossopteris retifera – G. conspicua assemblage sub-zone	(1) Common occurrence of Glossopteris retifera, G. conspicua along with Merianopteris major, Belemnopteris wood-masoniana, Palaeovittaria kurzi, Aletheopteris roylei, Raniganjia bengalensis (2) Maximum richness of Glossopteridae, Equisetales and Filicales	Concavisporites, Eupunctisporites, Ricaspora, Calamospora, Anapiculatisporites, Reticulatisporites, Lycopodiumsporites, Gravisporites, Gondisporites, Punctatosporites, Distriomonosaccites, Kosankeisporites, Tumoripollenites	Anthraconauta sp.	Raniganj Coalfield, Bangladesh
Kazanian		Cyclodendron assemblage sub-zone	(1) Cyclodendron leslii along with Glossopteris conspicua and G. retifera (2) Poor assemblage of megaflora as compared to others	(1) Poor assemblage of microflora (2) Extreme rarity of Aulisporites, Marsupipollenites		Jharia Coalfield, Bihar
		Barakaria dichotoma – Walkomiella indica assemblage sub-zone	(1) Abundance of Glossopteris along with Barakaria dichotoma, Walkomiella indica, Pseudoctensis balli (2) Appearance of Sphenophyllum speciosum	Didecitriletes, Lacinitriletes, Indotriradites, Dentatispora, Barakarites, Korbapollenites, Rhizomaspora, Primuspollenites, Direticuloidispora, Striapollenites, Maculatisporites		Barakar river section, Bihar
Artinskian	Gangamopteris assemblage zone	Gondwanidium – Buriadia assemblage sub-zone	Abundance of Gangamopteris and Noeggerathiopsis along with Buriadia sewardi and Gondwanidium validum	Abundance of monosaccate spore genera with Crucisaccites, Alisporites, Succinctisporites, Crustaesporites, Tetrasaccus		Near village Karharbari in Giridih Coalfield, Bihar
Sakmarian		Noeggerathiopsis – Paranocladus assemblage sub-zone	(1) Abundance of Gangamopteris and Noeggerathiopsis (2) Rarity of Glossopteris and Filicales	(1) Abundance of monosaccati: Plicatipollenites, Virkipollenites, Potoniesporites (2) Rarity of bisaccate and trilete ferns		Near village Rikba, Karanpura Coalfield, Bihar

TABLE X

Correlation of biostratigraphic and lithostratigraphic units of the Indian Gondwana (After Shah et al., 1971)

Standard scale	Biostratigraphic assemblage zone	Assemblage sub-zone	Damodar Valley	Upper Narmada Valley — Satpura	Upper Narmada Valley — Bansa-Parsora	Lower Narmada Valley	Son Valley	Rajmahal region	Wardha-Godavari Valley	Mahanadi Valley	Rajasthan region	West coast — Kutch region	West coast — Kathiawar	East coast — Ongole	East coast — Eluru	East coast — Madras	East coast — Tiruchchirappalli and Far South	Kashmir
LOWER CRETACEOUS — Wealden	Ptilophyllum assemblage zone	Weichselia–Onychiopsis sub-zone		Bansa Fm.	Bansa Fm.								Himatnagar Fm./Up. Dhrangadhra Fm.					
JURASSIC — Upper	Ptilophyllum assemblage zone	Pagiophyllum–Brachyphyllum sub-zone		Jabalpur Fm.		? Lower Nimar Fm.		Nipania beds				Bhuj Fm.	Lr. Dhrangadhra Fm.	? Pavalur Fm. / ? Vammevaram Fm.	? Tirupati Fm. / ? Raghavapuram Fm.	? Satyavedu Fm. / ? Sriperumbudur Fm.	Uttatur plant beds/Sivaganga Fm.	
JURASSIC — Middle / Lower	Ptilophyllum assemblage zone	Dictyozamites–Pterophyllum sub-zone		Chaugan Fm.				Rajmahal plant Fm.	Kota Limestone		Lathi Fm.			? Budavada Fm.	Golapilli Fm.			
TRIASSIC — Upper	Dicroidium–Neocalamites assemblage zone				Parsora Fm.			Dubrajpur Fm.	Dharmaram Fm.									
TRIASSIC — Upper	Labyrinthodon assemblage zone	Ceratodus–Metaposaurus sub-zone		Denwa Fm.	Tiki Fm.				Maleri Fm.									
TRIASSIC — Middle	Labyrinthodon assemblage zone	Dicynodon sub-zone	Supra-Panchet beds	Pachmarhi Fm.					Bhimaram Fm. / Yerrapalli Fm.									
TRIASSIC — Lower	Labyrinthodon assemblage zone	Lystrosaurus sub-zone	Panchet Fm.	Almod beds	Daigaon Fm.		Panchet Fm.		Mangli beds									
PERMIAN — Tatarian	Glossopteris assemblage zone	Glossopteris conspicua–G. retifera sub-zone	Raniganj Fm.	Bijori Fm.	Pali Fm.				Kamthi Fm.	Himgir Fm.								
PERMIAN — Kazanian	Glossopteris assemblage zone	Cyclodendron sub-zone	Barren Measure Fm. or Ironstone Shale	Motur Fm.	Barren Measure Fm.				Ironstone Shale	Barren Measure Fm.								
PERMIAN — Artinskian (Up.)	Glossopteris assemblage zone	Barakaria dichotoma–Walkomiella indica sub-zone	Barakar Fm.	Barakar Fm.	Barakar Fm.		Barakar Fm.	Barakar Fm.	Barakar Fm.	Barakar Fm.								
PERMIAN — Artinskian (Lr.)	Gangamopteris assemblage zone	Gondwanidium–Buriadia sub-zone	Karharbari Fm.	Karharbari Fm.	Karharbari Fm.		Karharbari Fm.	Karharbari Fm.	Karharbari Fm.	Karharbari Fm.	Badhaura Fm.							Gangamopteris beds
PERMIAN — Sakmarian	Gangamopteris assemblage zone	Noeggerathiopsis–Paranocladus sub-zone	Talchir Fm.	Talchir Fm.	Talchir Fm.		Talchir Fm.	Talchir Fm.	Talchir Fm.	Talchir Fm.	Bap Fm.							Gangamopteris beds

Paleomagnetic studies

Paleomagnetic studies of Parsora, Tirupati, and Satyavedu sandstones have been made. The paleomagnetic data obtained from Parsora sandstones (Triassic) by Bhalla and Verma (1969) indicates that during the Upper Triassic times, the Indian landmass was in the Southern Hemisphere. Pullaiah and Verma (1970) have shown that Tirupati sandstones (? Albian) are normally magnetised in the lower part which is just the opposite of what is observed in the Deccan Trap (Upper Cretaceous to Eocene or Oligocene). The paleomagnetic results obtained by Mital et al. (1970) from Satyavedu sandstones (? Albian) lends credence to their correlation with the Tirupati sandstones based on stratigraphic and paleontological considerations. The paleomagnetic studies strongly support northward drifting of the Indian landmass during Cretaceous times while it was still in the Southern Hemisphere.

MESOZOIC IGNEOUS ACTIVITY

The major episodes of the Mesozoic igneous activity, both extrusive and intrusive, are confined to Peninsular India (Fig. 8). The extrusive rocks are chiefly basaltic lavas of the Deccan Trap and Rajmahal Trap in the Peninsula and the Sylhet Trap of Assam and the Dras volcanics of Kashmir in Extra-Peninsular India. The Sylhet Trap is considered as equivalent to the Rajmahal Trap while the Dras volcanics, discussed earlier, are the counterpart of the Deccan Trap of the Peninsula. The Rajahmundry Trap, on the East Coast, hitherto considered as Cretaceous in age, belongs to Early Eocene (Bhalla, 1968b) and is therefore not discussed here.

The intrusive igneous activity of the Mesozoic Era in the Peninsula occurred in the region occupied by the Gondwana rocks lying between the Deccan Trap and the Rajmahal Trap while in Extra-Peninsular India it is probably represented by the basic and ultrabasic intrusives of Simla (Pilgrim and West, 1928), Garhwal (Jain, 1972), and other areas of the Himalayas, some of which, on stratigraphic considerations, may be doubtfully equivalent to the Dras volcanics (Cretaceous).

The disruption of Gondwanaland in the Mesozoic Era brought about significant changes in the continents of the Southern Hemisphere and generated intense diastrophism. Fissures opened in the Indian landmass allowing an immense volume of lava to well out. They are called *traps* (Swedish word meaning "steps") due to their step-like outcrops. Due to their horizontal nature and consequent wide extension, practically nothing is known of the underlying rocks.

Deccan Trap

The Deccan Trap is extensively developed in Peninsular India, extending from 69.5° to 82°E and 16° to 23°N and occupying an area of about 512,000 km² (Fig. 8). It has a volume of about 700,000 km³, attains a thickness of over 2000 m near the Bombay coast and thins out towards the east where 30–60 m thick traps occur near Rajahmundry. Bore hole data collected by "Glomar Challenger" in 1972 from the Arabian Sea reveal the presence of the Deccan Trap there and Pepper and Everhart (1963) believe that these traps extend below the Ganga Plain. It is evident that the Deccan Trap must have originally covered a much wider area than seen to-day.

The characteristic flat-topped hills and step-like appearance of their outcrops constitute some of the picturesque scenery of the Deccan Trap country. Flows vary in thickness from a few metres to 35 m and individual flows have been traced for 100 km near Nagpur. They are horizontal but sometimes show low dips and were subjected to gentle folding and faulting.

In addition to surface flows, dykes and sills have also been found in the Deccan Trap. Sills are generally doleritic and have been found in the Upper Gondwana strata of the Satpura and Rewa areas and also in the Jurassic rocks of Kutch. Dykes are more numerous than sills, range from a few metres to more than 60 m in thickness, and are seen in the Gujarat, Satpura, and Hyderabad areas and also in the Narmada Valley. Dykes may represent a post-trap hypabyssal phase (Auden, 1949) or some of them may have acted as feeders to the Deccan Trap flows (Krishnan, 1953).

The Deccan Trap flows have been divided into lower, middle, and upper traps. In places, their monotony is disrupted by the presence of volcanic ash beds, especially in the upper part, or by intercalated "inter-trappean" sedimentary strata. The inter-trappean sediments have been assigned a Paleocene or Lower Eocene age on the basis of their fossil assemblage.

The trap rock is generally a non-porphyritic basalt

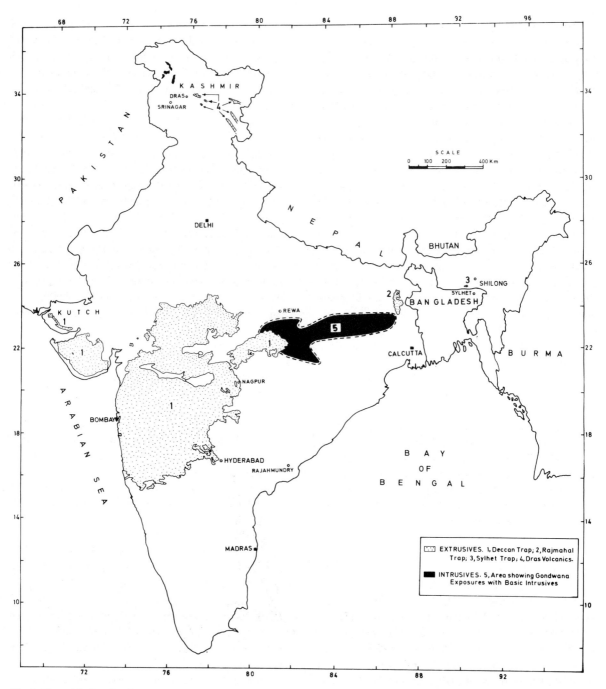

Fig. 8. Map of India, showing occurrences of Mesozoic igneous rocks.

or dolerite in dark green to almost black shades, amygdaloidal and vasicular, very hard, compact and exhibiting characteristic spheroidal weathering. It may vary from finely crystalline to coarse grained. The traps are uniform in mineral and chemical composition and include some acid varities also. Their average specific gravity is 2.9. Balakrishna (1967) observed that the traps are elastically isotropic and noted compressional wave velocity to range from 5.1 to 6.2 km/sec which is related to the porosity in the trap rock, 18–4.5%, respectively.

The common type of Deccan Trap rock is mostly

composed of abundant labradorite and clinopyroxene and generally shows an ophitic texture. Phenocrysts of felspar in dolerite and interstitial glass in basalts frequently occur. Magnetite and ilmenite are commonly found as minor constituents. If present, olivine is rare. Zeolites, chalcedony, calcite, etc., generally occur as fillings in amygdules. West (1958) gave a lucid account of the petrography and petrogenesis of the Deccan Trap flows of western India. Chemically, it is almost uniform in composition (Washington, 1922; Sinha and Karkare, 1964; Chaterjee and Ghosh, 1970). The studies of West (1958), Rao (1964) and others have shown that the parent magma of the Deccan Trap was tholeiitic, somewhat more basic than normal.

The paleomagnetic study of the Deccan Trap by Verma and Mital (1972) and others show that there was only one major geomagnetic field reversal in the Deccan Trap from reverse polarity in the lower part to normal polarity in the upper part — the change occurring at an elevation of about 650 m from the mean sea level near the Bombay coast (Deutsch et al., 1959), although Bhimasankaram and Pal (1970) opined that there was probably more than one field reversal in the Deccan Trap. Deutsch et al. (1958) estimated a drift of about 50° in latitude for the Indian landmass in the past nearly 70 m.y. From the magnetic data, McElhinny (1968) suggested a few degrees drift during the Deccan Trap activity but this has been refuted by Verma et al. (1971). However, these studies confirm that during the Deccan Trap eruptions, India was in the Southern Hemisphere.

The mode of Deccan Trap eruption is a controversial topic. The majority of workers believe that they are mainly fissure eruptions, the fissures now being represented by the position of dykes (West, 1959). West (1971) observed that the northward drifting India at the time of the breaking up of Gondwanaland in Late Cretaceous generated tension in the rear allowing the development of numerous fissures through which the Deccan Trap magma welled out. On geophysical evidence, Krishna Brahmam and Negi (1973) reported the presence of rifts below the Deccan Trap in western India while the Narmada rift in the north and the Godavari rift in the east are already known. According to the authors (Krishna Brahmam and Negi, 1973), fractures associated with these rifts served as fissures through which magma came out. The central type of igneous activity is also known (Rao, 1964; Agashe and Gupte, 1968). How-

ever, Biswas and Deshpande (1973) consider that the majority of the Deccan Trap eruptions are from "shield volcanoes" of Hawaiian type.

Pascoe (1964) summarised various evidences for the age of the Deccan Trap and concluded that it ranges from Upper Cretaceous to Lower Eocene while Krishnan (1968) considered it to be Upper Cretaceous to Oligocene. It is, however, generally regarded that the effusion of the Deccan Trap lava commenced in the Upper Cretaceous and continued well into the Tertiary. Only a few radiometric dates are available for the Deccan Trap. In recent years, Rama (1964) obtained a range of 43 ± 2–65 ± 5 m.y. for the Deccan Trap while Wellman and McElhinny (1970) have shown a range from 59.0 ± 1.1 m.y. to 64.1 ± 1.9 m.y., i.e. around uppermost-Cretaceous–lowermost-Tertiary time, for different flows from various parts. However, more data are required before radiometric dates for the Deccan Trap can be relied upon.

Rajmahal Trap

Named after the Rajmahal Hills in Bihar where it attains a thickness of over 600 m, the Rajmahal Trap occupies an area of about 110,000 km^2 (Fig. 8). It is also found in the Bengal Basin below a thick blanket of younger formations and alluvium (Biswas, 1963) and may be in continuity with the Sylhet Trap further east. The geology of the Rajmahal area was described by Ball (1877).

The Rajmahal Trap consists mainly of flows of basaltic lava with almost uniform composition but, at places, they may be associated with dolerite and dacite. Dacite forms the earliest flows followed by a number of basaltic flows, and andesite which is the youngest of all, occurs in the form of dykes. The thickness of individual flows varies from 20 to 75 m. Some dolerite dykes have penetrated Rajmahal Trap which shows local faults, particularly at its western boundary. The sequence incorporates a few interstratifications of clay and sandstones containing plant fossils and also those having tuffs and ash beds.

The petrography, chemical composition, etc., of the Rajmahal Trap have been dealt with by a few workers (Deshmukh, 1964; Sarbadhikari, 1968). Although the Rajmahal Trap closely resembles the Deccan Trap in lithology, mineral and chemical composition, their exact relationship has not yet been established.

The principal mode of eruption of the Rajmahal

Trap is the same as that of the Deccan Trap, i.e., fissure eruption. Sarbadhikari (1968) observed that the western border of the Rajmahal Trap marks the main vent and, in general, lava flowed from there and covered the Rajmahal region in the east. There was one primitive basic magma of tholeiite composition from very early times and this served as the parent magma for the Deccan Trap as wel as for the Rajmahal Trap (West, 1958).

The precise age of the Rajmahal Trap is controversial. On floral evidence from the associated Gondwana sediments, the Rajmahal Trap was considered as not younger than Upper Jurassic (Krishnan, 1968). However, Biswas (1959) and Dettmann (1963) noted the Early Cretaceous affinity of the Rajmahal flora making *inter alia* the Rajmahal Trap also to be of Early Cretaceous age. Sah and Jain (1965) recognised certain Early Cretaceous spores and pollens from the Rajmahal intertraps but considered the assemblage as transitional between Early Jurassic and Early Cretaceous. Krishnan (1968) believes that these traps may be of Upper Jurassic to Early Cretaceous age (see also Pascoe, 1964). The Early Cretaceous age of the Rajmahal Trap gets further credence from radiometric-age determinations by Radhakrishnamurty (1963) and McDougall and McElhinny (1970). The chronological testimony of different data, in recent years, is strongly in favour of Early Cretaceous age for the Rajmahal Trap.

Paleomagnetic studies of the Rajmahal Trap by Clegg et al. (1958), Athavale et al. (1963), and Radhakrishnamurty (1963) indicate that the traps are of normal polarity and at the time of their effusion, the Indian landmass was in the Southern Hemisphere.

Sylhet Trap

Deriving its name from the Sylhet district in Bangladesh, near the Indo-Bangladesh border, where it is well-developed, the Sylhet Trap occurs along a 80 km long, 4 km wide tract in the Shillong plateau in Assam, India (Fig. 8). Separated from the nearest outcrop of the Rajmahal Trap to the west by a distance of about 300 km, the Sylhet Trap is over 600 m thick and is probably equivalent to the Rajmahal Trap and of the same age (Krishnan, 1968). While these traps may be in continuity below the vast spread of the Ganga Plain (Pascoe, 1959; Biswas, 1963), they do not resemble each other.

The trap rock is mainly dark blue, bedded ande-

site. In places, it is basaltic in nature and shows amygdules which are generally filled with calcite. Ash beds, especially in the upper part, are present. The Sylhet Trap has been pierced by dykes at some places.

The paleomagnetic study of the Sylhet Trap by Athavale et al. (1963) indicates that it has normal polarity, contemporary to the Rajmahal Trap, and of Upper Jurassic to Lower Cretaceous age (see also Athavale and Verma, 1970, table 1).

Basic intrusives

A 500 km long tract between the northeastern margin of the Deccan Trap and the western margin of the Rajmahal Trap (Fig. 8) contains a series of Gondwana exposures in which lie important coalfields, viz., Korea-Chirimiri, Karanpura, Bokaro, Jharia, and Raniganj belonging to the Son-Damodar belt. These Gondwana rocks are pierced by numerous dykes and sills of two main types: (1) mica peridotite or lamprophyre, and (2) dolerite or basalt.

The peridotite intrusives occur in an anastomosing manner and have affected beds up to Panchets only (Lower Gondwana) whereas doleritic and basaltic dykes occur in a regular pattern, cutting across coal seams, fault planes, and peridotite intrusives in the Lower Gondwana strata and have penetrated the overlying Mesozoic part of the Gondwana sequence. These dykes are also found in the Satpura and Assam regions but are uncommon.

The doleritic and basaltic intrusives occur mainly in the form of regularly disposed dykes, are sometimes over 122 m thick and are unaffected by fault planes or coal seams. They occur along faults and are considered younger than the faulting. Petrologically, they are similar to those in the Deccan Trap. The age of the dolerite dykes is a matter of controversy and so is their relationship with the Deccan Trap and Rajmahal Trap. Generally, they are considered as the hypabyssal phase of the Rajmahal igneous activity but may well belong to the Deccan Trap or both.

Athavale and Verma (1970) made a paleomagnetic study of the dolerite dykes and observed that at the time of their intrusion, India was in the Southern Hemisphere. These authors strongly supported Pascoe (1964) that the Mesozoic igneous activity in India began with an extrusive phase in the east giving rise to the Sylhet Trap and Rajmahal Trap, followed by an intrusive phase represented by dolerite dykes in the adjoining country in the west, and then passed further west in the form of the Deccan Trap eruptions. This contention gets further credence from the paleomagnetic studies, including radiometric age determinations of these rocks (Pascoe, 1964). It is therefore evident that the Sylhet Trap, the Rajmahal Trap, dolerite intrusives, and the Deccan Trap are different manifestations of a continuous igneous activity which commenced during the Lower Cretaceous in the east and vanished in the west during Eocene or Oligocene times.

MESOZOIC GEOLOGICAL EVOLUTION

The Mesozoic history of India was varied and eventful. However, in view of the limited information available, only broad generalisations can be made in respect of the Mesozoic paleogeography of India which are tentative and subject to modifications.

The Mesozoic Era dawned with the Tethys occupying Extra-Peninsular India and large parts of Peninsular India. During the Triassic a thick pile of fossiliferous marine strata were accumulated which are magnificently exposed in a nearly continuous belt, stretching from Kashmir in the northwest through Spiti to Kumaun in the southeast. The remarkable uniformity in lithology of different Triassic subdivisions, without significant breaks, suggests a practically uniform depositional environment throughout the Extra-Peninsula. The Himalayan Triassic fauna exhibits close resemblance with the European counterpart and several species are common to both the regions, thereby indicating a free communication between Extra-Peninsular India and Europe through the Tethys. In Peninsular India, there are sporadic records of Triassic continental deposits, mainly restricted to the Damodar Valley, Mahadeva, and Pranhita-Godavari basins. A *Dicroidium* flora developed and reptiles and amphibians became frequent. The fossil and sedimentological evidences indicate warm, arid or semi-arid, climatic conditions in India during the Triassic Period. Although Gondwana sedimentation continued into the Mesozoic, the intensive coal-forming phase of the late Paleozoic was over. The overall deposition of the Triassic Gondwana sediments took place in a fluviatile environment which locally became lacustrine. The red bed assemblage of the Lower and Middle Triassic (Panchet Group) accumulated most probably in flood plains of meandering rivers, while those of the Upper Triassic

(Mahadeva Group), occurring in Central India, were deposited largely in channels of braided streams. The regional paleoslope was dominant towards the northwest and north up to the close of the Middle Triassic (Pachmarhi Formation). During Late Triassic and Middle–Upper Jurassic times, it shifted towards the west and southwest in Central India, and in Late Jurassic to Lower Cretaceous, it was completely reversed towards the southeast in the east-coast region (Casshyap, 1977). These reversals of the paleoslopes were tectonically induced.

The Tethys in Extra-Peninsular India continued to receive sediments during the Jurassic, and the marine rocks of this period occur in the entire Kashmir–Kumaun belt. The free intermingling of the Himalayan and Mediterranean faunas through the Tethys was maintained. As elsewhere in the world, the Jurassic witnessed a widespread transgression of the sea in several coastal regions of India. During Middle and Upper Jurassic times, almost the entire Kutch–Rajasthan region of western India was covered by a gulf which was evidently in direct communication with the Tethys in the north. Indeed, the Jurassic rocks of Kutch and Rajasthan form the first significant record of the fossiliferous marine deposits in the Indian Peninsula. The close affinity of the Jurassic fauna of Kutch with those of Africa and Madagascar suggests that the Tethyan Gulf which covered the Kutch–Rajasthan region extended to Madagascar after skirting the east coast of Africa. The occurrence of Upper Jurassic rocks, encountered in bore-holes along the eastern coast of South India, indicates that the present-day eastern coast-line of India started to develop during the closing phase of the Jurassic Period. The fluviatile sedimentation, constituting a part of the Gondwana sequence, continued to progress in Peninsular India. The semi-arid climate of the Triassic was progressively replaced by milder and humid conditions and a new flora, *Ptilophyllum* flora, emerged which continued into the Lower Cretaceous. The Gondwana rocks of Middle–Upper Jurassic age (Jabalpur Group) in Central India indicate a rapid rate of sedimentation and are comparable to flood plain deposits (Casshyap, 1977).

India witnessed significant paleogeographical changes during the Cretaceous Period. Some marine rocks of this period occur in the Extra-Peninsula but they are not as widespread and richly fossiliferous as their counterparts in Peninsular India. The Tethys receded northward, coinciding perhaps with the initial pulses of the Alpine—Himalayan orogeny due to the northward moving Indian Shield. Although the Tethyan Gulf occupied much of the Kutch—Rajasthan region during the Jurassic, the absence of marine post-Aptian strata indicates its regression from this region before the onset of the Cenomanian transgression over the craton. However, during the Cenomanian transgression an arm of this gulf invaded the linear Narmada Valley in which sediments of Bagh beds and also perhaps of the Lameta Group were accumulated. The eastern coast-line of India which started to develop in Late Jurassic, continued to cleave further due to rupturing of the Indian and East Antartica plates. This gave way to the Indo-Pacific sea which transgressed on the Coromandal coast during Lower Cretaceous times. This sea also flooded some coastal areas of Andhra Pradesh and later on, it engulfed the southern part of Assam.

The Bagh beds fauna shows close resemblance with the Cretaceous faunas of Arabia and Europe and also exhibits affinity with the South Indian Cretaceous assemblage. The similarity of the Chisil Pass fauna (Cenomanian) and Karakoram Pass fauna (Senonian) of Kashmir with the Cretaceous assemblage of South India, is also well-marked. This widespread faunal similarity may indicate that the arm of the sea which ingressed into the Narmada Valley during Cenomanian times was in communication with the Tethys in the north and the Cretaceous sea in the southern part of India. Although the Campanian fauna of Assam shows strong affinity with the Cretaceous assemblage of South India, it does not resemble either to the Bagh or Kashmir Cretaceous assemblage. This would then suggest that the South Indian Cretaceous sea perhaps served as a connecting link between the widely separated regions of Assam and Bagh—Kashmir, probably belonging to different paleobiogeographical realms.

The fresh-water Gondwana rocks continued to be deposited on the margins of the Peninsular Shield through the Lower Cretaceous. The mild and humid climate which was established during the Jurassic, persisted till the closing phase of the Gondwana sedimentation. The Gondwana rocks occurring in the coastal areas of Peninsular India, are generally fluvio-deltaic but locally paralic and marine due to proximity of the basins to the Lower Cretaceous sea.

During the Late Cretaceous, intense diastrophism was generated due to fragmentation of Gondwanaland which developed fissures in Peninsular India allowing an immense volume of lava to come out, now in the form of the Deccan Trap. The Dras volcanics of Kashmir which are equivalent to the Deccan Trap, represent the Cretaceous igneous activity in Extra-Peninsular India.

ACKNOWLEDGEMENTS

I am indeed grateful to Professor S.H. Rasul, Head of the Geology Department, Aligarh Muslim University, Aligarh, for providing facilities to complete the work. I thank the Director General of the Geological Survey of India for giving me permission to reproduce certain tables published by the officers of that organisation; Mr. M.V.A. Sastry, Director of the Paleontology and Stratigraphy Division, Geological Survey of India, Calcutta, for helpful suggestions and also for allowing me to reproduce tables concerning the Indian Gondwana; and Dr. Hari Narain, Director of the National Geophysical Research Institute, Hyderabad, for making available for me a complete set of publications on the Mesozoic of India by the scientists of the Institute.

Dr. S.M. Abbas and Messrs. Rajiv Nigam, Abu Talib, and Rama K. Saini, Research Scholars in the Geology Department of the Aligarh Muslim University, Aligarh, assisted me in the preparation of the manuscript and their help is thankfully acknowledged. Cartography is by Mr. Syed Mohammad and I am also thankful to him for the accomplishment of this painstaking work.

Finally, I express my sincere gratitude to my wife, Mrs. Raj Bhalla, for her ungrudging co-operation in the completion of the manuscript by patiently bearing long hours of solitude.

REFERENCES

Agashe, L.V. and Gupte, R.B., 1968. Some significant features of the Deccan Trap. *Geol Soc. India, Mem.*, 2: 309–319.

Agrawal, S.K., 1956. Contribution à l'étude stratigraphique et paléontologique du Jurassique du Kutch (Inde). *Ann. Centre Etud. Document. Paléontol., Paris*, 19: 1–188.

Agrawal, S.K., 1957. A study of the Jurassic rocks of Kutch with special reference to Jhura dome. *J. Paleontol. Soc. India*, 2: 119–129.

Arkell, W.J., 1956. *Jurassic Geology of the World.* Oliver and Boyd Ltd., London, 806 pp.

Athavale, R.N., Radhakrishnamurty, C. and Sahasrabudhe, P.W., 1963. Paleomagnetism of some Indian rocks. *Geophys. J. R. Astron. Soc.*, 7 (3): 304–313.

Athavale, R.N. and Verma, R.K., 1970. Paleomagnetic results on Gondwana dykes from Damodar Valley coalfields and their bearing on the sequence of Mesozoic igneous activity in India. *Geophys. J. R. Astron. Soc.*, 20: 303–316.

Auden, J.B., 1934. The geology of the Krol Belt. *Rec. Geol. Surv. India*, 67 (4): 357–454.

Auden, J.B., 1949. Dykes in Western India. A discussion of their relationship with Deccan Traps. *Trans. Nat. Inst. Sci. India*, 3: 123–157.

Balagopal, A.T., 1969. *Coarse Clastic and Carbonate Sedimentation in Jurassic Rocks of Central Kutch*. Ph. D. Thesis, Aligarh Muslim University, Aligarh, 201 pp. (unpublished).

Balakrishna, S., 1967. Physical properties of Deccan Traps. *Bull. Volcanol.*, 30: 5–27.

Ball, V., 1877. Geology of the Rajmehal Hills. *Geol. Surv. India, Mem.*, 13 (2): 155–248.

Banerji, R.K., 1966. The genus *Globotruncana* and biostratigraphy of the Lower Ariyalur Stage (Upper Cretaceous) of Vridhachalam, South India. *J. Geol. Soc. India*, 7: 51–69.

Banerji, R.K., 1968. Late Cretaceous foraminiferal biostratigraphy of Pondicherry area, South India. *Geol. Soc. India, Mem.*, 2: 30–49.

Bassi, Udai Kumar and Vatsa, Upendra S., 1971. A study of the Tal oölites from Rishikesh, Garhwal Himalaya. In: *Himalayan Geology*, 1. Hindustan Publ. Corp. (India), Delhi, pp. 244–250.

Berthelsen, A., 1953. On the geology of the Rupshu District, N.W. Himalaya. *Medd. Dan. Geol. Foren.*, 12: 350–414.

Bhalla, M.S. and Verma, R.K., 1969. Paleomagnetism of Triassic Parsora sandstones from India. *Phys. Earth Planet, Inter.*, 2: 138–146.

Bhalla, S.N., 1968a. Paleoecology of the Raghavapuram shales (Early Cretaceous), East Coast Gondwanas, India. *Palaeogeogr., Palaeoclimatol., Palaeoecol.*, 5: 345–357.

Bhalla, S.N., 1968b. Cretaceous–Tertiary boundary in the Pangadi area, West Godavari District, Andhra Pradesh. *Geol. Soc. India, Mem.*, 2: 226–233.

Bhalla, S.N., 1969a. Foraminifera from the type Raghavapuram shales, East Coast Gondwanas, India. *Micropaleontology*, 15 (1): 61–84.

Bhalla, S.N., 1969b. Occurrence of Foraminifera in the Budavada beds of the East Coast Gondwanas, India. *Bull. Geol. Soc. India*, 6 (3): 103–104.

Bhalla, S.N., 1972. Upper age limit of the East Coast Gondwanas, India. *Lethaia*, 5: 271–280.

Bhalla, S.N. and Abbas, S.M., 1975. Post-Jurassic elements in the Jurassic foraminiferal assemblage from Kutch. *J. Geol. Soc. India*, 16 (3): 379–381.

Bhalla, S.N. and Abbas, S.M., ms. Jurassic Foraminifera from Kutch, India. *Micropaleontology*, 24 (2): 160–209.

Bhalla, S.N. and Abbas, S.M., 1979. Depositional environment of the Jurassic rocks of Kutch, India. *J. Geol. Soc. India*.

Bhatia, M.R. and Jain, S.P., 1972. Smaller Foraminifera from the Dalmiapuram Formation (Lower Cretaceous), South India. *Bull. Indian Geol. Assoc.*, 5: 45–46.

Bhatia, S.B. and Jain, S.P., 1969. Dalmiapuram Formation: A new Lower Cretaceous horizon in South India. *Bull. Indian Geol. Assoc.*, 2: 105–108.

Bhatia, S.C. and Rao, D.V.S., 1974. Gravity anomalies over Gondwanas of India. *J. Geol. Soc. India*, 15 (1): 93–97.

Bhattacharya, S.C. and Niyogi, D., 1971. Geological evolution of the Krol Belt in Simla Hills, H.P. In: *Himalayan Geology, 1*. Hindustan Publ. Corp. (India), Delhi, pp. 178–212.

Bhimasankaram, V.L.S. and Pal, P.C., 1970. Paleomagnetic studies in India – A Review. In: *Proc. Symp. Upper Mantle Project, 2nd, Nat. Geophys. Res. Inst., Hyderabad*, pp. 223–249.

Biswas, B., 1959. Subsurface geology of West Bengal, India. *Econ. Comm. Asia Far East, Miner. Resour. Dev., Bankok, Ser.*, 10: 159–161.

Biswas, B., 1962. Stratigraphy of the Mahadeo, Langpur, Cherra and Tura formations, Assam, India. *Bull. Geol. Min. Metall. Soc. India*, 25: 1–48.

Biswas, B., 1963. Exploration for petroleum in western part of the Bengal Basin. *Econ. Comm. Asia Far East, Miner. Resour. Dev., Bankok, Ser.*, 18 (1): 241–244.

Biswas, S.K., 1971. Note on the geology of Kutch. *Q. J. Geol. Min. Metall. Soc. India*, 43 (4): 223–235.

Biswas, S.K. and Deshpande, S.V., 1973. A note on the mode of eruption of the Deccan Trap lavas with special reference to Kutch. *J. Geol. Soc. India*, 14 (2): 134–141.

Bittner, A., 1899. Triassic Brachiopoda and Lamellibranchiata. *Geol. Surv. India, Palaeontol. Indica, Ser.*, 15, 3 (2): 1–76.

Blanford, H.F., 1862. On the Cretaceous and other rocks of the South Arcot, and Trichinopoly districts, Madras. *Geol. Surv. India, Mem.*, 4 (1): 1–217.

Blanford, W.T., 1869. On the geology of the Taptee and Lower Nerbudda Valleys, and some adjoining districts. *Geol. Surv. India, Mem.*, 6 (3): 1–222.

Bose, M.N. and Sukh Dev, 1956. Occurrence of two characteristic Wealden ferns in the Jabalpur series. *Nature*, 183 (4654): 130–181.

Bose, P.N., 1884. Geology of the Lower Narbada Valley between Nimawar and Kawant. *Geol. Surv. India, Mem.*, 21 (1): 1–72.

Burrard, S.G. and Hayden, H.H., 1934. *A Sketch of the Geography and Geology of the Himalaya Mountains and Tibet*. Revised and edited by S. Burrard and A.M. Heron. Govt. of India Publ., Delhi.

Casshyap, S.M., 1977. Patterns of sedimentation in Gondwana basins. *4th Int. Gondwana Symp., Calcutta, 1977*, Vol. II: 525–551. Hindustan Publ. Corp. Delhi, 1979.

Chanda, S.K., 1968. Juras–Cretaceous stratigraphy and sedimentation around Jabalpur, M.P., and their paleogeographic implications. *J. Geol. Soc. India*, 9 (1): 21–31.

Chatterjee, A.C. and Ghosh, Sukomal, 1970. Studies of the Deccan Trap magmas on the basis of their chemical characters. In: *Proc. Symp. Upper Mantle Project, 2nd, Nat. Geophys. Res. Inst., Hyderabad*, pp. 493–506.

Chatterjee, Sankar and Roy-Chowdhury, Tapan, 1974. Triassic Gondwana vertebrates from India. *Ind. J. Earth Sci.*, 1 (1): 96–112.

Chiplonkar, G.W., 1937. Echinoids from the Bagh Beds, I. *Proc. Indian Acad. Sci., Sect. B*, 6 (1): 60–71.

Chiplonkar, G.W., 1939a. Echinoids from the Bagh Beds, II. *Proc. Indian Acad. Sci., Sect. B*, 9 (5): 236–246.

Chiplonkar, G.W., 1939b. Lamellibranchs from the Bagh Beds. *Proc. Indian Acad. Sci., Sect. B*, 10 (4): 254–274.

Chiplonkar, G.W., 1941. Ammonites from the Bagh Beds. *Proc. Indian Acad. Sci., Sect. B*, 14 (3): 271–276.

Chiplonkar, G.W., 1942. Age and affinities of the Bagh fauna. *Proc. Indian Acad. Sci.*, 15: 148–152.

Clegg, J.A., Radhakrishnamurty, C. and Sahasrabudhe, P.W., 1958. Remanent magnetism of the Rajmahal Traps of northeastern India. *Nature*, 181: 830–831.

Cotter, G. de P., 1917. A revised classification of the Gondwana System. *Rec. Geol. Surv. India*, 48 (1): 23–33.

Cox, L.R., 1940. The Jurassic Lamellibranch fauna of Kachh (Cutch). *Geol. Surv. India, Palaeontol. Indica, Ser.*, 9, 3: 1–157.

Cox, L.R., 1952. The Jurassic Lamellibranch fauna of Cutch (Kachh). *Geol. Surv. India, Palaeontol. Indica, Ser.*, 9, 3 (4): 1–128.

Dainelli, G., 1933. Spedizione italiana de Filippi nell'Himàlaia, Caracorùm e Turchestàn cinese (1913–1914): *Ser. 2. Risultati geologici e geografici, Vol. 2. La serie dei Terreni.* Zanichelli, Bolonga, 1: 1–458; 2: 459–1105.

Datta, A.K. and Bedi, T.S., 1968. Faunal aspects and the evolution of the Cauvery Basin. *Geol. Soc. India, Mem.*, 2: 168–177.

De Lapparent, A.F., 1957. The Cretaceous dinosaurs of Africa and India. *J. Paleontol. Soc. India*, 2: 109–112.

Deshmukh, S.S., 1964. Geology of the area around Tiljhari and Berhait, Rajmahal Hills, Santhal Paraganas, Bihar, with a discussion on the differentiation trends in the Rajmahal Traps. *Proc. Int. Geol. Congr., 22nd, New Delhi, 1964*, 7: 61–84.

De Terra, H., 1935. Geological studies in the North-west Himalaya between the Kashmir and Indus Valleys. *Conn. Acad. Arts Sci., Mem.*, 8: 18–76.

Dettmann, Mary E., 1963. Upper Mesozoic microfloras from South-Eastern Australia. *Proc. R. Soc. Victoria, N. Ser.*, 77: 1–148.

Deutsch, E.R., Radhakrishnamurty, C. and Sahasrabudhe, P.W., 1958. The remanent magnetism of some lavas in the Deccan Traps. *Phil. Mag.*, 3: 170–184.

Deutsch, E.R., Radhakrishnamurty, C. and Sahasrabudhe, P.W., 1959. Paleomagnetism of the Deccan Traps. *Ann. Géophys. (Paris)*, 15: 39–59.

Diener, C., 1895. Ergebnisse einer geologischen Expedition in die Zentral-Himalaya von Johar, Hundes und Painkhanda. *Denkschr. Akad. Wiss. Wien, Math.-naturwiss. Kl.*, 62: 533–603.

Diener, C., 1897. The Cephalopoda of the Lower Trias. *Geol. Surv. India, Palaeontol. Indica, Ser.*, 15, 2 (1): 1–181.

Diener, C., 1906. The fauna of the Tropites-limestone of Byans. *Geol. Surv. India, Palaeontol. Indica, Ser.*, 15, 5 (1): 1–201.

Diener, C., 1907. The fauna of the Himalayan Muschelkalk. *Geol. Surv. India, Palaeontol. Indica, Ser.*, 15, 5 (2): 1–140.

Diener, C., 1908a. Upper Triassic and Liassic fauna of the exotic blocks of Malla Johar in the Bhot Mahals of Kumaon. *Geol. Surv. India, Palaeontol. Indica, Ser.*, 15, 1 (1): 1–100.

Diener, C., 1908b. Ladinic, Carnic and Noric fauna of Spiti. *Geol. Surv. India, Palaeontol. Indica, Ser.*, 15, 5 (3): 1–157.

Diener, C., 1910. Fauna of the Traumatocrinus-limestone of Painkhanda. *Geol. Surv. India, Palaeontol. Indica, Ser.*, 15 6 (2): 1–39.

Diener, C., 1912. The Trias of the Himalayas. *Geol. Surv. India, Mem.*, 36 (3): 202–360.

Diener, C., 1913. Triassic faunae of Kashmir. *Geol. Surv. India, Palaeontol. Indica, N. Ser.*, 5 (1): 1–133.

Diener, C. and von Kraft, A., 1909. Lower Triassic Cephalopoda from Spiti, Malla Johar and Byans. *Geol. Surv. India, Palaeontol. Indica, Ser.*, 15, 6 (1): 1–186.

Douville, H., 1925. Fossiles du Kashmir et les Pamirs. *Rec. Geol. Surv. India*, 58 (4): 349–357.

Duncan, P.M., 1887. Echinoidea of Cretaceous Series of Lower Narbada Valley. *Rec. Geol. Surv. India*, 20 (2): 81–92.

Elliot, D.H., 1973. Gondwana Basins of Antarctica. *Proc. Symp. Gondwana, 3rd, Canberra*, pp. 493–536.

Fedden, F., 1884. The geology of Kathiawar Peninsula in Guzerat. *Geol. Surv. India, Mem.*, 21 (2): 73–135.

Feistmantel, O., 1876. Note on the age of some fossil floras of India. *Rec. Geol. Surv. India*, 9 (2): 28–42.

Feistmantel, O., 1882. Fossil flora of the South Rewah Gondwana Basin. *Geol. Surv. India, Palaeontol. Indica, Ser.*, 12, 4 (1): 1–52.

Fourtau, R., 1918. Les Echinides des "Bagh Beds". *Rec. Geol. Surv. India*, 49 (1): 34–53.

Fox, C.S., 1931. Coal in India, II. The Gondwana System and related formations. *Geol. Surv. India, Mem.*, 58: 1–241.

Ganesan, T.M., 1975. Paleocurrent pattern in the Upper Tal rocks of Nigali, Korgai synclines (H.P.) and Mussoorie syncline (U.P.). *J. Geol. Soc. India*, 16 (4): 503–507.

Gansser, A., 1964. *Geology of the Himalayas.* Interscience, New York, N.Y., 289 pp.

Ghosh, A.K. and Srivastava, S.K., 1962. Microfloristic evidence on the age of Krol beds and associated formations. *Proc. Ind. Sci.*, 28 (5): 710–717.

Godwin-Austen, H.H., 1864. Geological notes on part of the Northwestern Himalayas. *Q. J. Geol. Soc. Lond.*, 20: 383–387.

Gregory, J.W., 1893. Jurassic fauna of Kutch: The Echonoida of Cutch. *Geol. Surv. India, Palaeontol. Indica, Ser.*, 9, 2 (1): 1–11.

Gregory, J.W., 1900. Jurassic fauna of Kutch: The corals. *Geol. Surv. India, Palaeontol. Indica, Ser.*, 9, 2 (2): 12–196.

Griesbach, C.L., 1891. Geology of the Central Himalayas. *Geol. Surv. India, Mem.*, 23: 1–232.

Guha, D.K. and Mohan, Madan, 1965. A note on Upper Cretaceous microfauna from the Middle Andaman Island. *Bull. Geol. Min. Metall. Soc. India*, 33: 1–4.

Hagn, H., 1977. Saligrame-Gerölle von Malm-Kalken mit Ammoniten als Kultgegenstände Indiens. *Mitt. Bayer. Staatssamml. Palaeontol. Hist. Geol.*, 17: 71–102.

Hardas, M.G. and Merh, S.S., 1972. Significance of CM diagrams of some Jurassic and Cretaceous sediments of Kutch (Gujarat). *J. Geol. Soc. India*, 13 (3): 292–297.

Hayden, H.H., 1904. The geology of Spiti, with parts of Bashahr and Rupshu. *Geol. Surv. India, Mem.*, 36 (1): 1–121

Heim, A. and Gansser, A., 1939. Central Himalaya: Geological observations of the Swiss Expedition (1936). *Soc. Helv. Sci. Nat., Mem.*, 73 (1): 1–245.

Holdhaus, H., 1913. The Fauna of the Spiti Shales: Lamellibranchiata and Gastropoda. *Geol. Surv. India, Palaeontol. Indica, Ser.*, 15, 4 (2) Fasc. 4: 397–456.

Jain, A.K., 1972. Overthrusting and emplacement of basic rocks in Lesser Himalaya, Garhwal, U.P. *J. Geol. Soc. India*, 13 (3): 226–237.

Jain, S.P., 1969. Taxonomic comments on the ammonites from the Bagh Beds, Madhya Pradesh, with remarks on the age of these beds. *Bull. Indian Geol. Assoc.*, 2: 45–50.

Jain, S.P. and Gupta, V.J., 1973. Smaller foraminifera and ostracoda from Chikkim Shales (Upper Cretaceous) of Spiti. *Sci. Cult.*, 39: 53.

Jeannet, A., 1959. Ammonites Permiennes et Faunaes Triasiques de L'Himalaya Central. *Geol. Surv. India, Palaeontol. Indica, N. Ser.*, 34 (1): 1–168.

Jowett, A., 1925. On the geological structure of the Karanpura Coalfields, Bihar and Orissa. *Geol. Surv. India, Mem.*, 52 (1): 1–144.

Kailasam, L.N., 1968. Some results of geophysical exploration over the Cretaceous–Tertiary formations of the Madras coast. *Geol. Soc. India, Mem.*, 2: 178–195.

Kailasam, L.N. and Simha, K.R.M., 1963. A reflection seismic traverse across the coastal sedimentary belt of South Arcot District, Madras State. *Bull. Nat. Geophys. Res. Inst.*, 1 (3): 151–167.

Kitchin, F.L., 1900. Jurassic fauna of Kutch. The Brachiopoda. *Geol. Surv. India, Palaeontol. Indica, Ser.*, 9, 3 (1): 1–87.

Kitchin, F.L., 1903. Jurassic fauna of Kutch: Lamellibranchiata, genus Trigonia. *Geol. Surv. India, Palaeontol. Indica, Ser.*, 9, 3 (2): 1–122.

Kohli, G. and Sastri, V.V., 1956. On the age of Chikkim Series. *J. Paleontol. Soc. India*, 1 (1): 199–201.

Kossmat, F., 1897. The Cretaceous deposits of Pondicherry. *Rec. Geol. Surv. India*, 30 (2): 51–110.

Kossmat, F., 1898 (1895–1898). Untersuchungen über die Sub-Indische Kreide Formationen. *Beitr. Palaeontol. Geol. Osterungarns*, 9: 97–203; 11: 1–46; 89–152.

Krishna Brahmam, N. and Negi, Janardan, G., 1973. Rift valleys beneath Deccan Traps (India). *Bull. Geophys. Res.*, 11 (3): 207–237.

Krishnan, M.S., 1953. The structural and tectonic history of India. *Geol. Surv. India, Mem.*, 81: 1–109.

Krishnan, M.S., 1968. *Geology of India and Burma*. Higginbothams, Madras, 536 pp.

La Touche, T.H.D., 1902. Geology of Western Rajputana. *Geol. Surv. India, Mem.*, 35 (1): 1–116.

Lele, K.M., 1964. The problem of Middle Gondwana in India. *Proc. Int. Geol. Congr., 22nd, New Delhi, 1964*, 9: 181–202.

Lubimova, P.S., Guha, D.K. and Mohan, M., 1960. Ostracoda of Jurassic and Tertiary deposits from Kutch and Rajasthan (Jaisalmer), India. *Bull. Geol. Min. Metall. Soc. India*, 22: 1–61.

Lydekker, R., 1883. The geology of Kashmir and Chamba territories, and the British district of Khagan. *Geol. Surv. India, Mem.*, 22: 1–344.

Mamgain, V.D. and Rao, B.R.J., 1965. Orbitolines from the limestone intercalations of Dras Volcanics, J and K State. *J. Geol. Soc. India*, 6: 122–129.

Mamgain, V.D., Rao, B.R.J. and Sastry, M.V.A., 1968. The Niniyur group of Trichinopoly, South India. *Geol. Soc. India, Mem.*, 2: 85–91.

Mathur, Y.K., Soodan, K.S., Mathur, Kawal, Bhatia, M.L., Juyal, N.P. and Pant, J., 1970. Microfossil evidences on the presence of Upper Cretaceous and Paleocene sediments in Kutch. *Bull. Oil Nat. Gas Comm.*, 7 (2): 109–114.

Matley, C.A., 1921. On the stratigraphy, fossils and geological relationships of the Lameta Beds of Jabbalpore. *Rec. Geol. Surv. India*, 53 (2): 142–164.

McDougall, I. and McElhinny, M.W., 1970. The Rajmahal Traps of India – K-Ar ages and paleomagnetism. *Earth Planet. Sci. Lett.*, 9 (4): 371–378.

McElhinny, M.W., 1968. Northward drift of India – Examination of recent paleomagnetic results. *Nature*, 217: 342–344.

McGowran, Brian, 1968. Late Cretaceous and Early Tertiary correlations in the Indo-Pacific region. *Geol. Soc. India, Mem.*, 2: 335–360.

Medlicott, H.B., 1864. On the geological structure and relations of the southern portion of the Himalayan Range between the rivers Ganges and Ravee. *Geol. Surv. India, Mem.*, 3 (2): 1–206.

Medlicott, H.B., 1869. Geological sketch of the Shillong Plateau in Northeastern Bengal. *Geol. Surv. India, Mem.*, 7 (1): 151–207.

Medlicott, H.B. and Blanford, W.T., 1879. *A Manual of the Geology of India*, 1. Govt. of India Publ., Delhi, 1st ed., 444 pp.

Middlemiss, C.S., 1887. Physical geology of West British Garhwal. *Rec. Geol. Surv. India*, 20 (1): 26–40.

Middlemiss, C.S., 1910. A revision of the Silurian Trias sequence in Kashmir. *Rec. Geol. Surv. India*, 40 (3): 206–260.

Mital, G.S., Verma, R.K. and Pullaiah, G., 1970. Paleomagnetic study of Satyavedu sandstones of Cretaceous age from Andhra Pradesh, India. *Rev. Pure Appl. Geophys.*, 81: 177–191.

Mojsisovics, E. von, 1899. Upper Triassic Cephalopoda faunae of the Himalaya. *Geol. Surv. India, Palaeontol. Indica, Ser.*, 15, 3 (1): 1–157.

Murthy, N.G.K., 1968. The Upper Cretaceous and Tertiary Formations of Pondicherry. *Geol. Soc. India, Mem.*, 2: 113–119.

Murthy, K.N., Rao, R.P., Dhokarikar, B.G. and Verma, C.P., 1963. On the occurrence of plant fossils in the Nimar Sandstones near Umrali, District Jhabua, M.P. *Curr. Sci.*, 32 (1): 21.

Nagappa, Y., 1959. Foraminiferal biostratigraphy of the

Cretaceous–Eocene succession in India, Pakistan, Burma region. *Micropaleontology*, 5 (2): 141–192.

Nagappa, Y., 1960. The Cretaceous–Tertiary boundary in the India–Pakistan sub-continent. *Rept. Int. Geol. Congr., 21st, Copenhagen, 1960*, 5: 41–49.

Oldham, R.D., 1893. *A Manual of the Geology of India*. Govt. of India Press, Calcutta, 2nd ed., 543 pp.

Palmer, R.W., 1923. Geology of a part of the Khasi and Jaintia Hills, Assam. *Rec. Geol. Surv. India*, 55 (2): 95–168.

Pascoe, E.H., 1959. *A Manual of the Geology of India and Burma*. Govt. of India Publ., Delhi, 2: 485–1343.

Pascoe, E.H., 1964. *A Manual of the Geology of India and Burma*. Govt. of India Publ., Delhi, 3: 1345–2130.

Pepper, J.F. and Everhart, G.M., 1963. The Indian Ocean – the geology of its bordering lands and the configuration of its floor. *U.S. Geol. Surv., Misc. Geol. Invest.*, Map I–380.

Pilgrim, G.E. and West, W.D., 1928. The structure and correlation of the Simla Rocks. *Geol. Surv. India, Mem.*, 53: 1–140.

Poddar, M.C., 1964. Mesozoics of Western India – their geology and oil possibilities. *Proc. Int. Geol. Congr., 22nd, New Delhi, 1964*, 1: 126–143.

Prasad, K.N., 1968. Some observations on the Cretaceous dinosaurs of India. *Geol. Soc. India, Mem.*, 2: 248–255.

Pullaiah, G. and Verma, R.K., 1970. Geomagnetic field reversal in Cretaceous Tirupati sandstone formation from India. *Phys. Earth Planet Int.*, 2: 158–162.

Radhakrishnamurty, C., 1963. *Permanent Magnetism of the Igneous Rocks in the Gondwana Formation of India*. D. Sci. thesis, Andhra University, Waltair, India.

Rajagopalan, N., 1965. Late Cretaceous and Early Tertiary stratigraphy of Pondicherry, South India. *J. Geol. Soc. India*, 6: 108–121.

Rajnath, 1942. The Jurassic rocks of Cutch – their bearing on some problems of Indian geology. *Proc. Indian Sci. Congr., 29th, Baroda*, pp. 93–106.

Raju, D.S.N., 1970. Zonal distribution of selected foraminifera in the Cretaceous and Cenozoic sediments of Cauvery Basin and some problems of Indian biostratigraphic classification. *Publ. Cent. Adv. Stud. Geol. Panjab Univ., Chandigarh*, 7: 85–110.

Rama, 1964. Potassium-Argon dates of some samples from Deccan Traps. *Proc. Int. Geol. Congr., 22 nd, New Delhi*, 1964, 7: 139–140.

Ramanathan, S., 1968. Stratigraphy of Cauvery Basin with reference to its oil prospects. *Geol. Soc. India, Mem.*, 2: 153–167.

Rao, B.R.J., Mamgain, V.D. and Sastry, M.V.A., 1968. *Globotruncana* in Ariyalur Group of Trichinopoly Cretaceous, South India. *Geol. Soc. India, Mem.*, 2: 18–29.

Rao, L.R., 1956. Recent contributions to our knowledge of the Cretaceous rocks of South India. *Proc. Indian Acad. Sci., Sect. B*, 44 (4): 185–245.

Rao, M.B.R., 1973. The subsurface geology of the Indo-Gangetic plains. *J. Geol. Soc. India*, 14 (3): 217–242.

Rao, S. Subba, 1964. The geology of the igneous complex of the Girnar hills, Gujarat State, India. *Proc. Int. Geol. Congr., 22nd, New Delhi*, 1964, 7: 42–60.

Rao, V.R. and Venkatachala, B.S., 1971. Upper Gondwana marine intercalations in Peninsular India. *Proc. Symp. Gondwana System, Aligarh Muslim Univ., Aligarh*, pp. 353–389.

Robinson, P.L., 1967. The Indian Gondwana Formation – a review. *Symp. Gondwana Stratigraphy, 1st, India*, pp. 202–268.

Roy Chowdhury, M.K. and Sastri, V.V., 1962. On the revised classification of the Cretaceous and associated rocks of the Man River Section of the Lower Narbada Valley. *Rec. Geol. Surv. India*, 91 (2): 283–304.

Roy, S.K., 1968. Pteridophytic remains from Kutch and Kathiawar, India. *Palaeobotanist*, 16 (2): 108–114.

Sah, S.C.D. and Jain, K.D., 1965. Jurassic spores and pollen grains from the Rajmahal Hills, Bihar, India: with a discussion on the age of the Rajmahal intertrappean beds. *Palaeobotanist*, 13: 264–290.

Sahni, M.R. and Jain, S.P., 1968. Note on a revised classification of the Bagh beds, Madhya Pradesh. *J. Paleontol. Soc. India*, 11: 24–25.

Sahni, M.R. and Sastri, V.V., 1957. A monograph of the Orbitolines found in the Indian continent (Chitral, Gilgit and Kashmir), Tibet and Burma, with observations on the age of the associated volcanic series. *Geol. Surv. India, Palaeontol. Indica, N. Ser.*, 33 (3): 1–44.

Sarbadhikari, T.R., 1968. Petrology of a northeastern portion of the Rajmahal Traps. *Q. J. Geol. Min. Metall. Soc. India*, 40 (3): 151–171.

Sastry, M.V.A., 1960. Some Triassic fossils from the Eastern Byans, Central Himalayas. *Rec. Geol. Surv. India*, 89 (2): 383–398.

Sastry, M.V.A. and Mamgain, V.D., 1971. The marine Mesozoic formations of India: a review. *Rec. Geol. Surv. India*, 101 (2): 162–177.

Sastry, M.V.A., Rao, B.R.J. and Mamgain, V.D., 1968. Biostratigraphic zonation of the Upper Cretaceous formations of Trichinopoly district, South India. *Geol. Soc. India, Mem.*, 2: 10–17.

Sastry, M.V.A., Rao, B.R.J. and Mamgain, V.D., 1969. On the lower age limit of the Ariyalur Group (Upper Cretaceous). *Rec. Geol. Surv. India*, 98 (2): 136–143.

Satpathy, B.N. and Kanungo, D.N., 1973. Basement structure as inferred from geophysical data around Karaikudi, Tamil Nadu. *J. Geol. Soc. India*, 14 (1): 12–22.

Shah, S.C., Singh, Gopal and Sastry, M.V.A., 1971. Biostratigraphic classification of Indian Gondwanas. *Proc. Symp. Gondwana System, Aligarh Muslim Univ., Aligarh*, pp. 306–326.

Singh, H.P., 1966. Re-appraisal of the microflora from the Jabalpur series of India with remarks on the age of the beds. *Palaeobotanist*, 15 (1): 87–92.

Sinha, R.C. and Karkare, S.G., 1964. Geochemistry of Deccan Basalts: A study of the behaviour of major and trace elements in the basaltic flows of India. *Proc. Int. Geol. Congr., 22nd, New Delhi*, 1964, 7: 85–103.

Spath, L.F., 1933 (1927–1933). Revision of the Jurassic Cephalopod Fauna of Kachh (Cutch). *Geol. Surv. India, Palaeontol. Indica, N. Ser.*, 9 (2): 1–6; 1–945.

Spath, L.F., 1939. The Cephalopoda of the Neocomian

Belemnite beds of the Salt Range. *Geol. Surv. India, Palaeontol. Indica, N. Ser.*, 25 (1): 1–154.

Spengler, E., 1923. Contributions to the palaeontology of Assam. *Geol. Surv. India, Palaeontol. Indica, N. Ser.*, 8 (1): 1–74.

Spitz, A., 1914. A Lower Cretaceous fauna from the Himalayan Gieumal Sandstone, together with a description of a few fossils from the Chikkim Series (translated by E. Vredenburg). *Rec. Geol. Surv. India*, 44 (3): 197–224.

Stoliczka, F., 1866. Geological sections across the Himalayan Mountains, from Wangtu bridge on the river Sutlej to Sungdo on the Indus: with an account of the formations in Spiti, accompanied by a revision of all known fossils from that district. *Geol. Surv. India, Mem.*, 5 (1): 1–153.

Stoliczka, F., 1873 (1866–1873). Cretaceous fauna of Southern India. *Geol. Surv. India, Palaeontol. Indica*, 3 (1): 1–216; 5 (2): 1–498; 6 (3): 1–537; 8 (4): 1–202.

Strachey, R., 1851. On the Geology of part of the Himalaya Mountains and Tibet. *Q. J. Geol. Soc. Lond.*, 7: 292–310.

Subbotina, N.N., Datta, A.K. and Srivastava, B.N., 1960. Foraminifera from the Upper Jurassic deposits of Rajasthan (Jaisalmer) and Kutch, India. *Bull. Geol. Min. Metall. Soc. India*, 23: 1–48.

Swaminath, J., Krishnamurthy, J.G., Verma, K.K. and Chandak, G.J., 1959. General geology of the Jaisalmer area, Rajasthan, Western India. *Proc. Symp. Dev. Petrol. Resour., Asia Far East, Miner. Resour. Dev., Bankok, Ser.*, 10: 154–155.

Tewari, B.S. and Kumar, Ratesh, 1966. Foraminifera from Nummulitic beds of Nilkanth and organic remains from Tal limestone, Garhwal Himalayas. *Publ. Cent. Adv. Stud. Geol., Panjab Univ., Chandigarh*, 3: 33–42.

Tewari, B.S. and Kumar, Ratesh, 1968. On the Upper age limit of the Krols – especially on observations in Nilkanth area, Garhwal Himalayas. *Publ. Cent. Adv. Stud. Geol., Panjab Univ., Chandigarh*, 5: 121–130.

Tewari, B.S., Pande, I.C. and Kumar, R., 1970a. Lower Cretaceous fossiliferous limestone from Khalsi, Ladakh. *Publ. Cent. Adv. Stud. Geol., Panjab Univ., Chandigarh*, 7: 197–200.

Tewari, B.S., Gupta, V.J., Mahajan, G., Kumar, S., Chadha, D.K., Bisaria, P.C., Virdi, N.S., Kochhar, N. and Kashyap, S.R., 1970b. Some Foraminifera from Indus Flysch, Ladakh. *Publ. Cent. Adv. Stud. Geol., Panjab Univ., Chandigarh*, 7: 191–196.

Tipper, G.H., 1911. The geology of the Andaman Islands, with references to the Nicobars. *Geol. Surv. India, Mem.*, 35 (4): 195–216.

Uhlig, V., 1910a (1903–1910). The fauna of the Spiti shales. *Geol. Surv. India, Palaeontol. Indica, Ser.*, 15, 4 (1): 1–396.

Uhlig, V., 1910b. Die fauna der Spiti-Schiefer des Himalaya, ihr geol. Alter und ihre Weltstellung. *Denkschr. Akad. Wiss. Wien., Math.-Naturwiss. Kl.*, 85: 1 p.

Varadarajan, K. and Jagtap, P.N., 1968. Photogeological char-

acters of the Cretaceous–Tertiary sediments of Tiruchirapalli district, Madras State. *Geol. Soc. India, Mem.*, 2: 196–200.

Venkataraman, S. and Rangaraju, M.K., 1968. Oscillatory movements in Cretaceous–Tertiary Basin of South India. *Geol. Soc. India, Mem.*, 2: 201–207.

Ver Chere, A.M. and Ver Neiuie, M.E. de, 1867. Kashmir, the Western Himalaya and Afghan Mountains. *J. Asiat. Soc. Bengal*, 35 (2).

Verma, K.K., 1968. Bagh Beds – their fauna and affinities with the South India Cretaceous Formations. *Geol. Soc. India, Mem.*, 2: 239–247.

Verma, R.K., Bhalla, M.S., Pullaiah, G., Athavale, R.N. and Mital, G.S., 1971. Paleomagnetic evidence in support of continental drift of the Indian landmass. *Proc. Symp. Gondwana System, Aligarh Muslim Univ., Aligarh*, pp. 88–100.

Verma, R.K. and Mital, G.S., 1972. Palaeomagnetism of a vertical sequence of Traps from Mount Girnar, Gujrat, India. *Geophys J. R. Astron. Soc.*, 29: 275–287.

Von Huene, F.B. and Matley, C.A., 1933. The Cretaceous Saurischia and Ornithischia of the Central Provinces of India. *Geol. Surv. India, Palaeontol. Indica, N. Ser.*, 21 (1): 1–74.

Vredenburg, E.W., 1907. Ammonites of Bagh Beds. *Rec. Geol. Surv. India*, 36 (2): 102–125.

Vredenburg, E.W., 1910. *A Summary of the Geology of India*. 2nd. Ed.

Waagen, W., 1876 (1873–1876). Jurassic Fauna of Kutch: The Cephalopoda, Belemnitidae and Nautilidae. *Geol. Surv. India, Palaeontol. Indica, Ser.*, 9, 1: 1–247.

Wadia, D.N., 1934. The Cambrian–Trias sequence of northwestern Kashmir (parts of Muzaffarabad and Baramula districts). *Rec. Geol. Surv. India*, 68 (2): 121–176.

Wadia, D.N., 1937. The Cretaceous volcanic series of Astor-Deosai, Kashmir and its intrusions. *Rec. Geol. Surv. India*, 72 (2): 151–161.

Wadia, D.N., 1949. *Geology of India*. McMillan, London, 460 pp.

Washington, H.S., 1922. Deccan Traps and other plateau basalts. *Bull. Geol. Soc. Am.*, 33: 765–804.

Wellman, P. and McElhinny, M.W., 1970. K-Ar age of the Deccan Traps, India. *Nature*, 227: 595–596.

West, W.D., 1958. The petrology and petrogenesis of fortyeight flows of Deccan Traps penetrated by borings in western India. *Trans. Nat. Inst. Sci. India*, 14 (1): 1–56.

West, W.D., 1959. The source of the Deccan Trap flows. *Trans. Geol. Soc. India*, 1: 44–52.

West, W.D., 1971. Presidential Address to the "International Symposium on Deccan Trap and other flood Eruptions". *Bull. Volcanol.*, 35: 515–518.

Wynne, A.B., 1872. Memoir on the geology of Kutch, to accompany a map compiled by A.B. Wynne and F. Fedden, during the seasons 1867–68 and 1868–69. *Geol. Surv. India, Mem.*, 9 (1): 1–289.

Chapter 11

PAKISTAN

A.A. KURESHY

INTRODUCTION

Pakistan lies at the western end of the Himalayan chain and contains within its boundaries rocks of all geological ages. Paleozoic sediments have a restricted distribution but Mesozoic rocks are more widely distributed and total a few thousands of metres in thickness. They represent for the most part deposits in a pelagic, Tethyan environment. Based upon the stratigraphic and depositional history three distinct sedimentary basins (Fig. 1) can be recognized: the Upper and Lower Indus basins and the Baluchistan Basin (Kureshy, 1972). The Mesozoic rocks are best exposed in the Lower and Upper Indus basins, while in the Baluchistan Basin exposure and time-range of the Mesozoic rocks are limited. Although each basin has its own distinctive character, there are sufficient common features to permit correlation.

In general both Paleozoic and Mesozoic rocks crop out in a folded belt in the Indus Basin, from west of Karachi on the coast to near Islamabad to the north-northeast. These Mesozoic outcrops in the folded belt, i.e. at the western margin of the Upper and Lower Indus basins and to the east and southeast, disappear below the alluvial plains of the Indus. In the extreme north, south of Islamabad, there is a change in trend associated with the Salt Range. Only rocks of Cretaceous age crop out in Baluchistan, where they lie in an east—west belt south of the Afghanistan border.

The general geological history of Pakistan during the Mesozoic is relatively simple. It is one of progressive subsidence resulting in a pelagic marine sequence in most parts of Pakistan until near the end of the Cretaceous. At that time an upper Maastrichtian uplift in the southern part of the Lower Indus Basin resulted first in neritic and subsequently in non-marine sediments being deposited. However, in parts of the Lower Indus Basin there was continuous marine sedimentation through the Maastrichtian and Danian (Kureshy, 1970).

Detailed subdivision of the calcareous and argillaceous deposits of Triassic and Jurassic age is generally difficult for fossils are few and the succession monotonous. During the Cretaceous in particular, the extensive development of foraminifers (Haque, 1959; Latif, 1970; Dorreen 1974; and Kureshy 1970, 1971, 1972a, 1973, 1975, 1976 and 1977) does permit correlation with both austral and boreal faunal assemblages, in particular with the Caribbean and Gulf Coast region of the United States which are related with the Tethys Sea. The foraminifers of the Triassic and Jurassic have been described by Zaninnetti and Bronnimann (1975) and Kureshy (1981). They are mostly benthonic forms, and their use in correlation is still provisional.

Igneous activity is known both in the Triassic, when a number of minor lava flows occurred in the Upper Indus Basin, and in the Cretaceous. Late Cretaceous submarine volcanism occurs near Chagai in the northern part of the Baluchistan Basin, while west of Fort Sandman and Hindubagh in the Lower Indus Basin there are ultrabasic intrusions in some of which chromite is presently mined. Igneous intrusion and submarine volcanism of Upper Cretaceous is also reported from Sind province near Bela. These Upper Cretaceous igneous activities may be taken as the precursors of the orogenic storm of the Cenozoic during which the Himalayas were formed.

STRATIGRAPHY

As previously stated the Mesozoic deposits of the Baluchistan Basin are restricted to the Upper Cretaceous. Older horizons may be present in the subsurface

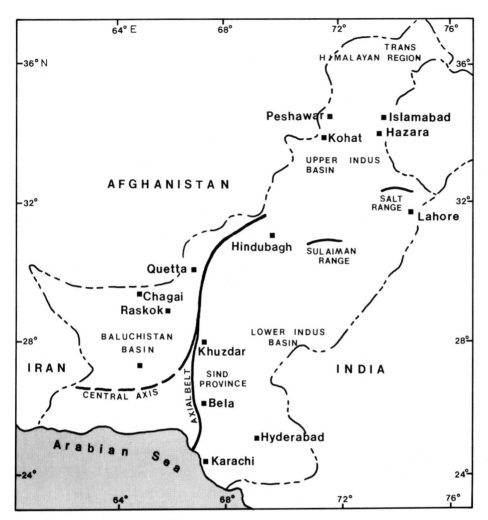

Fig. 1. Major geologic features of Pakistan.

but so far they have not been penetrated by exploratory wells. In the Upper Indus Basin the Mesozoic sequence is much broken up by intense folding and faulting so that the complete section is best studied in the Lower Indus Basin. The distribution of the Mesozoic rocks is shown in Figs. 2–4.

Lower Indus Basin

The Lower Indus Basin lies south of 32°S; it is bounded to the east by the Indian Shield and to the west it is separated from the Baluchistan Basin by a zone of folded strata, known as Axial Belt, the trend of which is approximately north–south. The folded belt includes the Kirthar, Laki, Dunghan and Sulaiman mountains and it is in this region where the

Mesozoic and older rocks crop out. In the eastern part of the basin lie the Indus alluvial plains formed by the deposits of Recent to sub-Recent age.

The maximum Mesozoic thickness in excess of 3,000 m occurs in Sind where Vredenburg (1909) described a complete sequence of Mesozoic rocks. Further descriptions have been given by Kureshy (1964, 1972b) and may also be found in the Hunting Survey Corporation Report (1960). For the purpose of description the stratigraphy of the Lower Indus Basin may be treated as three regions, the southern part (Sind), the northern part (Sulaiman Range), and the Axial Belt (Khuzdar) (see Tables I, II).

Southern part of the Lower Indus Basin (Sind)

A complete succession of Mesozoic is not exposed

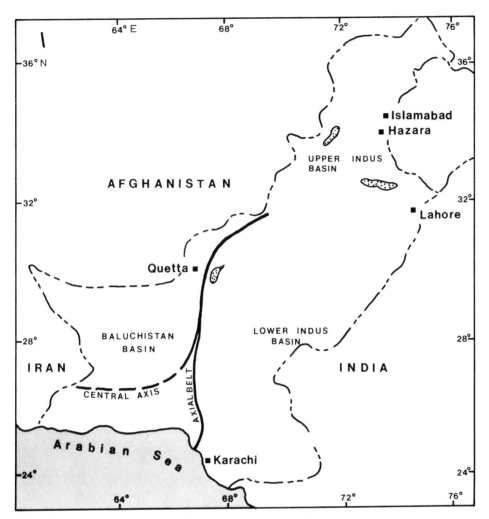

Fig. 2. Geographic distribution of marine Triassic sediments in Pakistan.

in any one part of this vast area. Rocks of Cretaceous age are most widely spread. The generalized succession of Cretaceous outcrops in Sind was given by Vredenburg (1909).

The lowermost unit of the Mesozoic succession in the southern part is of Jurassic age. The lithological units of the Jurassic are the Lorali and Chiltan formations. These formations are approximately equivalent and gradually merge into one another. The upper contact is with the Zidi Formation, predominantly a limestone facies, but the relationship of the Zidi Formation with the Chiltan and Lorali formations is not clear.

The Jurassic rocks are followed unconformably by Cretaceous strata. The lowermost unit of the Cretace-

ous is designated as the Belemnites Shale Formation, which was originally described as the lower member of the Parh Series by Blanford (1879). It is a variegated shale rich in belemnites as the name implies and with diagnostic planktonic foraminiferal assemblages. The shale is glauconitic, usually greenish or brownish in color and has thin interbedded layers of marl and limestone. According to the foraminiferal evidence it has an age range from Neocomian to Albian. The overlying formation, the Parh Limestone Formation, consists primarily of marly limestone and shale in the type locality in hills west of the Kirthar Range. The limestone is typically porcellaneous, white or a light grey to pink or red in color. The shales are generally hard and calcareous with marly

Fig. 3. Geographic distribution of marine Jurassic sediments in Pakistan.

intervals. It is a pelagic deposit and can be dated as Cenomanian to Campanian on the basis of the planktonic foraminifers (Kureshy, 1976).

There was some paleobathymetric change during the later part of the Cretaceous, so that the Parh Formation, a pelagic deposit, may be overlain by neritic limestone which contain larger foraminifers. These deposits, the Hemipneustes Limestone Formation, are Maastrichtian in age based on evidence of the larger foraminifers by Kureshy (1977a). Continued shallowing resulted in local emergence and the formation of the Pab Sandstone, a whitish or brownish sandstone with minor conglomerates of variable thickness found in the Pab Range (Vredenburg, 1909). These deposits are barren of fossils and have been deposited in a littoral environment.

In some areas there appears to be a continuous transitional sequence from Maastrichtian to Danian (Kureshy, 1970). In the Gaj River section and in the Pab Mountains the continuous deposition is represented by a shale sequence to which the name Korara Shale Formation has been assigned.

During Cretaceous time extensive igneous activity took place in the southern part of Lower Indus Basin in the vicinity of Bela, where volcanic rocks are exposed. Vredenburg (1909) referred them to the Deccan Traps, but Hunting Survey Corporation (1960) assigned them to the Bela Volcanic Group, which is associated with the Parh Limestone Formation, and is intruded by the Poralai diorite in Late Cretaceous time. These intrusive and extrusive rocks are discussed in the section on igneous activity.

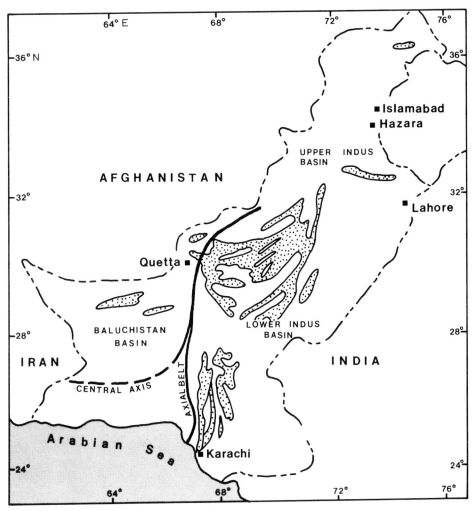

Fig. 4. Geographic distribution of marine Cretaceous sediments in Pakistan.

Axial Belt (Khuzdar)

The Khuzdar area is situated in the Axial Belt where the Las Bela and Central axes join. The Axial Belt is a structural boundary which separates the Lower Indus Basin from the Baluchistan Basin from Cretaceous time onwards. In this area the rocks of the Triassic, Jurassic and Cretaceous are exposed. The Triassic strata crop out in a limited exposure where the rocks named the Shirinab Formation are assigned to the Alazoi Group. The formation consists of a tan to grey, dense and coarse limestone, the limestones are followed by the Jurassic Loralai Formation. The relationship between the Shirinab and Loralai formations is not clear. The Loralai and Chiltan formations are approximately equivalent and gradually merge into one another, and may be considered identical. The

limestone is mostly oolitic, reefoid, usually hard and splintery, black or dark grey in colour. The beds are probably Lower and Middle Jurassic in age, but fossil evidence is lacking. These deposits are followed by the Zidi Formation, probably of the Upper Jurassic age, which is an alternation of limestone and shale. The limestone is dark grey to black, fine grained and argillaceous; the shale is mostly dark grey, calcareous and splintery. The Zidi Formation is overlain by the rocks of the Belemnites Shale Formation, the lower member of the Parh Group.

The Belemnites Shale Formation of Lower Cretaceous age is marked by a change in color and increase in proportion of shale. The Belemnites Shale Formation is followed by the Upper Cretaceous Parh Limestone Formation, a porcellaneous grey to maroon

TABLE I

Mesozoic stratigraphic succession of the lower Indus Basin

Geologic age		Southern part (Sind)	Northern part (Sulaiman Range)	Axial Belt (Khuzdar)
Cretaceous	Upper	Korara shale	Dunghan lst.	
		Pab sandstone	Pab sandstone	Pab sandstone
		Hemipneustes lst.	Orbitoides lst.	
		Parh lst.	Parh lst.	Parh lst.
	Lower	Belemnites sh.	Belemnites sh.	?
Jurassic	Upper	Zidi Fm.	Zidi Fm.	Zidi Fm.
	Middle	Loralai/ Chiltan Fms.		Loralai/Chiltan Fm.
	Lower			
Triassic	Upper	not exposed	Shirinab Fm.	Shirinab Fm.
	Lower			

TABLE II

Generalized Cretaceous succession of Pakistan

Geologic age		European stages		Formations	Lithology and fauna
Cretaceous	Upper	Maastrichtian	Upper	Pab sandstone	Massive, coarse, light brown sandstone with volcanic material
			Lower	Hemipneustes Fm. Orbitoides Fm.	Hemipneustes beds: thick bedded, massive lst, light in color with *Orbitoides* spp. Orbitoides Fm., hard massive, splintry, light brown in color with *Orbitoides* spp.
		Campanian	Upper		
			Lower	Parh Fm.	Parh Fm., marls and marly lst. of maroon white, and gray in color, with intercalation of green shale with abundant planktonic foraminifers
		Santonian			
		Coniacian			
		Turonian			
		Cenomanian			
	Lower	Albian		Belemnites sh.	Belemnites sh., glauconitic khaki and light green shale with intercalation of thin bedded marls with abundant planktonic foraminifers
		Aptian			
		Neocomian			

limestone. The Late Cretaceous (? Maastrichtian) deposits are designated the Pab Sandstone Formation and consist of white to brown, medium- to coarse-grained sandstones, which unconformably overlie the Parh Formation.

Northern part of the Lower Indus Basin (Sulaiman Range)

In this area the Triassic, Jurassic and Cretaceous rocks are exposed. The Triassic and Jurassic strata are better exposed than in the Axial Belt (Khuzdar), however, the Cretaceous strata are less thick than in the southern part (Sind) of the basin.

The Triassic deposits are exposed in the vicinity of Quetta and Zhob districts where they are designated the Shirinab Formation, part of the Alazoi Group. The deposits are fine grained argillaceous shales interbedded with mainly pale grey, greyish grey or dark grey limestone. The rocks of Shirinab Formation of the Alazoi Group are the oldest rocks which crop out in the area, for no lower contact is seen. They are conformably overlain by the Loralai Limestone. The Shirinab Formation can be correlated with the Ceratite Formation of the Salt Range in the Upper Indus Basin.

The Shirinab Formation southwest of Quetta grades upwards into a dark reefoidal and oolitic limestone with subordinate black shale known as the Loralai and Chiltan formations of Jurassic age. The Loralai and Chiltan formations are described as distinct lithological units by Hunting Survey Corporation (1960), but they are approximately equivalent and gradually merge into one another and are considered identical. The upper contact is with the Zidi Limestone Formation, but the relationship of the Zidi Formation to the Loralai and Chiltan formations is not clear.

In most places the Loralai Formation lies directly upon the Belemnites Shale Formation without any transition beds, and the boundary is regarded as a tectonic contact. The lowermost unit of the Cretaceous is the Belemnites Shale Formation, which consists of variegated shale with abundant marine fauna, which passes into the Parh Formation, a more argillaceous deposit than in the southern part of the Lower Indus Basin (Sind). The Parh Limestone Formation is followed by the Orbitoides Limestone Formation, which is characterized by its larger foraminifers and in turn followed by the barren Pab Sandstone Formation. The Orbitoides and Pab Sandstone

formations are correlated with the Hemipneustes Limestone and Pab Sandstone formations of the southern part of the Lower Indus Basin (Sind). In another area, in the Marri Bugti Hills in the northern part of the Lower Indus Basin, the transition of Maastrichtian to Danian is represented by the Dunghan Limestone Formation of Oldham (1890), although Pinfold (1939) reported that in some places the limestone was somewhat younger.

Upper Indus Basin

The Upper Indus Basin is that part of the Indus region north of 32°N. It projects northwards at one end of the Himalayan arc in Kashmir, analagous to the northeasterly projection of the Assam plateau at the other extremity of the Himalayas. Most of the Mesozoic and older rocks are covered by Cenozoic or younger deposits. The Mesozoic sequence itself is considerably disturbed by intense folding and faulting. The best exposures are found in the Salt Range in Punjab, the Samana Range in Kohat, and in Hazara. Limited deposits of Upper Cretaceous rocks are exposed in the Transhimalayan region of Gilgit and Baltistan. The geographic distribution of Mesozoic strata is shown in Figs. 2–4, and the relationship of the various lithostratigraphic units is shown in Table III.

Salt Range (Punjab)

The Triassic deposits of the Salt Range are known as the Ceratite Formation. The Triassic deposits of the Salt Range are about 750 m thick and extend through the Lower, Middle and Upper Triassic. The name Ceratite is derived from the abundance of the genus *Ceratites*. The Ceratite Formation conformably overlies the Permian Productus Limestone (Gee, 1944; Kummel, 1966). The Permo-Triassic contact has been the subject of a study by Kummel and Teichert (1970).

The Jurassic strata unconformably overlie the Triassic deposits. The age of the unconformity is Middle to Upper Jurassic. Lithologically the Jurassic beds are composed of variegated sandstone, yellow limestone and gypseous and pyritic shale, named the Borach Limestone and Variegated Sandstone formations. At different localities the thickness of Jurassic deposits varies appreciably. At Kalabagh the beds are 450 m thick. A few coal and lignite seams occur irregularly distributed in the lower part of Middle Jurassic deposits and are worked near Kalabagh.

TABLE III

Mesozoic stratigraphic succession of the upper Indus Basin

Geologic age		Salt Range	Samana Range (Kohat)	Hazara	Transhimalayan Range
Cretaceous	Upper	Lumshiwal sst.	Lithographic lst.	Chanali lst.	Yasin lst.
	Lower	Belemnite sh.	Belemnite sh.	Giumel sst.	not exposed
Jurassic	Upper	Borach lst.	Samana Suk Fm.	Spiti sh.	
	Middle	Variegated sst.		Sikkar lst.	
	Lower	unexposed		Maira Fm.	
Triassic	Upper	Ceratite Fm.	Lower Samana Suk Fm.	Limestone	
	Lower		Kingerali Fm.	Panjal volc.	

The Cretaceous strata of the Salt Range are best exposed near Kalabagh. The deposits of Lower Cretaceous age consist of white and yellow sandstone and shale with marls designated as Belemnites Shale Formation. These rest on Jurassic deposits with unconformity, and are overlain by the Lumshiwal sandstone of Upper Cretaceous age. The Lumshiwal sandstone is overlain by Paleocene strata with a pronounced unconformity.

Samana Range (Kohat)

The Triassic strata are designated as the Kinggriali Formation, followed by the Lower Samana Suk Limestone Formation. The Samana Suk Formation was regarded as Jurassic in age by Davies (1930), but Zaninnetti and Bronnimann (1975) demonstrated that the lower part of the formation was of Triassic age, and the upper part was Jurassic. It is unconformably followed by the Belemnite Formation.

The Cretaceous strata of the Samana Range consist of the Belemnite Formation of Lower Cretaceous age, which is followed by the Upper Cretaceous Lithographic Limestone Formation. In lithological character, the Lithographic Limestone is comparable to the Parh Limestone Formation of the Lower Indus Basin. So far no microfaunal studies have been undertaken to confirm the lithological correlation.

Hazara

In Hazara district a complete sequence of Mesozoic rocks, from Triassic to Upper Cretaceous, is exposed. The Triassic deposits occupy a fairly large area in the south. These deposits are characterized by 30 m of felsitic lava designated as the Panjal Volcanic Series, succeeded by a poorly fossiliferous limestone characterized by Upper Triassic fossils (Wadia, 1953). The grey limestone is thickly bedded and its thickness varies from 150 to 360 m.

The Jurassic deposits are well developed in Hazara, both north and south of Hazara. These deposits are quite distinct from one another and exhibit different facies. The northern Hazara exposures are characterized by shale and marine limestone and are related to the geosynclinal facies of the northern Himalaya, while the Jurassic deposits of southern Hazara differ sharply from the northern Hazara facies; they are more arenaceous with a greater affinity to the Jurassic strata of the Salt Range. Different lithostratigraphic units were assigned to represent these facies which range in age from Lower to Upper Jurassic according to Latif (1970). These lithostratigraphic units are shown in Table III.

The Cretaceous deposits of Hazara consist of the Giumel Sandstone Formation, which consists of massive sandstone, calcareous shale and shelly limestone, the latter containing ammonites of Albian age. The Giumel Sandstone is unconformably overlain by the Chanali Limestone Formation of Upper Cretaceous age based on the planktonic foraminifers (Latif, 1970). The Chanali Formation is unconformably overlain by the Nummulites Limestone of Eocene age, the contact being marked by a layer of laterite. The Chanali Formation can be correlated with the Lithographic Limestone of the Samana and probably with the Lumshiwal Formation of the Salt Range.

Transhimalyan region

In the Transhimalyan region of Gilgit and Baltistan, Mesozoic deposits of Upper Cretaceous age are exposed overlying the Greenstone Complex in the Yasin Valley, which gives its name to the Yasin Group (Bakr, 1965), a predominantly limestone facies. The lower contact is with Paleozoic strata and upper contact is with Tertiary deposits. This is an isolated exposure which is the result of the great Tethyan transgression over parts of northern Pakistan, Turkestan and Central Asia during Upper Cretaceous time.

Baluchistan Basin

The Baluchistan Basin extends from the Iran border on the west to the Axial Belt of the Lower Indus Basin in the east, and from the Afghanistan border in the north to the Arabian Sea in the south. Within this area the Mesozoic deposits are restricted to Upper Cretaceous (Santonian–Campanian age) and crop out only in the north. There was severe igneous activity during Late Cretaceous time in the northern part of this basin, which is referred to as "Eruptive zone" by the Hunting Survey Corporation (1960). The volcanic rocks of the Eruptive zone are designated as Sinjrani and Kuchakki Volcanic Groups.

The volcanic rocks are associated with a marine sedimentary sequence which is comparable to the Parh Limestone Formation of the Lower Indus Basin. These deposits are intruded by the Late Cretaceous Chagai intrusion. The Sinjrani and Kuchakki Volcanic Groups resulted from a submarine volcanic eruption of Upper Cretaceous age and are similar in character to the Bela Volcanic Group of the Lower Indus Basin. These intrusive and extrusive igneous rocks are discussed in the section on igneous activity (p. 366) and their distribution is shown in Fig. 5.

MICROBIOSTRATIGRAPHY

With few exceptions the greater part of the Mesozoic strata of Pakistan is of marine origin. These deposits are rich in micro- as well as mega-fossils. Among the micro-fossils, with the exception of the foraminifera, little has been published. All three groups, the small benthic, the planktonic and the larger foraminifers are common. Among the megafossils the most common group is the Ammonoidea in the marine deposits of Jurassic and Cretaceous, although there are very few published references available.

Triassic fauna

The Triassic foraminiferal assemblages are poorly represented in the Samana Suk Formation of the Samana Range (Kohat), and Kingriali Formation of the Surghar Range (Kohat), (Zaninnetti and Bronnimann, 1975). The Kingriali Formation was referred to a Lower Triassic age and the Lower Samana Suk Formation to an Upper Triassic age by these authors, although the latter was previously considered to be Jurassic in age by Fatmi and Cheema (1972). Ammonoidea belonging to the genus *Ceratites* are very common in the Ceratite Formation of the Salt Range. The characteristic species of foraminifers are as follows: *Involutina eomesozoica* (Oberhauser), *Involutina impressa* (Kristan-Tollmann), *Meandrospira pustilla* (Ho), *Endothyranella* sp., *Agathammina* sp., *Calcitornella* sp., *Planiinvoluta* sp.

Jurassic fauna

The Jurassic benthic foraminifers from the Chiltan and Zidi formations of Moghal Kot have been described by Kureshy (1981), including: *Pseudocyclammina personata* Tobler, *Lenticulina muensteri* (Roemer), *Haplophragmoides circularis* Said and Barakat, *Ammobaculites glaessneri* Said and Barakat, *Nautiloculina oolithica* Mohler, *Orbitopsella praecursor* Guembel.

In general these assemblages resemble the European and African Jurassic faunas; the zonal value of these forms is uncertain in Pakistan until further detailed biostratigraphic study of the Jurassic outcrop and subsurface data is undertaken.

The megafossils from the Jurassic strata of the Khisor–Marwat and Sheikh Budin Hills of Dera

Islmail Khan were described by Fatmi (1971), and Fatmi and Cheema (1972), as belonging to the Ammonoidea and Nautiloidea groups.

Cretaceous fauna

The microfossils of Cretaceous strata are quite abundant, cosmopolitan and diagnostic in character, with planktonic, small benthic and larger foraminifer groups, ranging in age from Neocomian to Maastrichtian.

The planktonic foraminifers were described from the Belemnites, Parh, and Korara Shale formations of the Lower Indus Basin by Kureshy (1970, 1972, 1976) and Dorreen (1974). Latif (1970) described these forms from the Upper Cretaceous Chanali Formation in Hazara. Kureshy (1976, 1977) divided the Cretaceous pelagic deposits into sixteen planktonic zones, which resemble the zones defined in the Middle East, Mediterranean and Caribbean (Bolli, 1959; Kureshy, 1971; Moullade, 1974). The stratigraphic ranges of the planktonic foraminiferal assemblages and the corresponding lithostratigraphic units are shown in Tables IV and V, and the correlation of these zones with the Caribbean region is shown in Table VI, and the correlation of Lower Cretaceous foraminiferal zones is shown in Table VII.

The benthic foraminifers of Cretaceous deposits are quite abundant and diversified, and closely resemble the Caribbean and Gulf coastal regions of the United States. The Upper Cretaceous benthic foraminifers were described by Haque (1959), and a complete list of benthic foraminifer species of Neocomian to Maastrichtian is given by Kureshy (1972a).

The larger foraminifers of the carbonate facies of the uppermost part of Upper Cretaceous (Orbitoides and Hemipneustes Limestone formations) and of Campanian and Maastrichtian ages were described by Vredenburg (1908), Mark (1962) and Kureshy (1973, 1975, 1977a). Two biozones have been defined in the larger-foraminifers bearing formations, one is the *Orbitoides tissoti* zone of Campanian age and the oth-

TABLE IV

Lower and Middle Cretaceous planktonic foraminiferal zones
(After Kureshy, 1976)

Geological age	Lower Cretaceous					Middle Cretaceous		
European stage	Neocomian	Aptian	Albian			Cenomanian		
Planktonic zones	*H. infra-cretacea*	*G. barri*	*T. bejaou-aensis*	*G. bregg-iensis*	*P. buxtorfi*	*R. apenni-nica*	*R. green-hornensis*	*R. cush-mani*
Formations / Planktonic Species	Belemnite shale					Parh Fm.		
Hedbergella delrioensis								
H. infracretacea								
Globigerinelloides barri								
Hedbergella planispira								
H. trochoidea								
Ticinella bejaouaensis								
Rotalipora ticinensis								
Globigerinelloides breggiensis								
Planomalina buxtorfi								
Rotalipora apenninica								
Praglobotruncana stephani								
Rotalipora greenhornensis								
R. reicheli								
R. cushmani								

TABLE V

Upper Cretaceous planktonic foraminiferal zones of Pakistan
(After Kureshy, 1976)

Geological age	Upper Cretaceous							
European stages	Turonian	Coniacian	Santonian		Campanian		Maastrichtian	
Planktonic zones	G. helvetica	G. renci	G. conca-vata	G. carinata	G. elevata	G. calcarata	G. gansseri	A. mayao-rensis
Formations / Planktonic species	Parh Fm.						Korara sh.	

Planktonic species:

- Globotruncana helvetica
- G. imbricata
- G. sigali
- G. renzi
- G. marginata
- Heterohelix reussi
- Globotruncana concavata
- G. tricarinata
- G. lapparenti
- G. carinata
- G. elevata
- G. ventricosa
- G. fornicata
- G. linneiana
- G. rosetta
- Heterohelix globosa
- H. punctulata
- Globotruncana calcarata
- Rugoglobigerina volutus
- R. naussi
- R. inflata
- Planoglobulina glabrata
- Gluberina robusta
- G. semicostata
- Globotruncana arca
- G. scutilla
- Praeglobotruncana havavensis
- Rugoglobigerina rugosa
- R. rotundata
- R. hexacamerata
- Globotruncanella petaloidea
- Rugoglobigerina pustulosa
- Globotruncana gansseri
- Heterohelix glabrans
- Planoglobulina eggeri
- Racemiguembelina fructicosa
- Pseudotextularia elegans
- Pseudoguembelina excolata
- Globotruncana aegyptiaca
- G. calciformis
- G. conica
- G. stuarti
- Abathomphalus intermedia
- A. mayaorensis
- Globotruncana patelliformis
- Rugoglobigerina scotti

TABLE VI

Correlation and datum planes of Cretaceous planktonic foraminiferal zones of Pakistan with Caribbean zones

Geological age				Planktonic foraminiferal datum planes	Planktonic foraminiferal zones of Pakistan	Planktonic foraminiferal zones of Trinidad (Bolli, 1959)
Upper Cretaceous		Maastrichtian		Single keel *Globotruncana*	*A. mayaroensis*	*A. mayaroensis*
						G. gansseri
					G. gansseri	*G. lapparenti tricarinata*
	Senonian	Campanian			*G. calcarata*	*G. stuarti*
					G. elevata	
		Santonian		Double keel *Globotruncana*	*G. carinata*	*G. fornicata*
					G. concavata	*G. concavata*
		Coniacian			*G. renzi*	*G. renzi*
		Turonian		Single keel *Globotruncana*	*G. helvetica*	*G. inornata*
		Cenomanian		*Rotalipora* extinction	*R. cushmani*	absent
					R. greenhornensis	
					R. apenninica	*R. apenninica* *G. washitensis*
Lower Cretaceous		Albian		*Rotalipora* datum	*P. buxtorfi*	*R. ticinensis*
						absent
				Ticinella datum	*G. breggiensis*	*Praeg. rohri*
					T. bejaouaensis	absent
		Aptian		*H. infracretacea* datum	*G. barri*	
		Neocomian			*H. infracretacea*	

er is the *Orbitoides media* zone of Maastrichtian age. The associations of these zones and their stratigraphic ranges are shown in Table VIII.

The ammonites of the Early Cretaceous of the Sheikh Budin Hills and Dera Islmail Khan were described by Fatmi (1971). No megafossils study has been undertaken in the Lower Indus Basin, although species of Belemnites are quite common in the Belemnite Formation of Lower Cretaceous age.

Paleoecology and paleogeography

The general character of the Mesozoic fauna and its paleoecological interpretation are as follows.

TABLE VII

Correlation of Lower Cretaceous zones

Geologic age	Foraminiferal zones	Van Hinte (1976)	Moullade (1974)	Kureshy (1976)
Albian		Planomalina buxtorfi – Rotalipora apenninica	Schackoina moliniensis	Planomalina buxtorfi
		R. ticinensis – P. buxtorfi	Planomalina buxtorfi	
		Ticinella (Biticinella) breggiensis	Ticinella breggiensis	Globigerinelloides breggiensis
		Ticinella praeticinensis	Hedbergella rischi	Ticinella roberti
		T. bejaouaensis – Gl. gyroidinae-formis – T. primula		
		T. bejaouaensis – Gl. gyroidinae-formis	Pleurostomella subnodosa	
Aptian		G. ferreolensis – T. bejaouaensis	T. bejaouaenensis	Globigerinelloides barri
			H. trocoidea	
		Hedbergella trocoidea – G. ferreolensis	G. algerianus	
		Globigerinelloides algerianus	G. ferreolensis	
		Schackoina cabri	S. cabri	
		Globigerinelloides blowi	Blowiella blowi	
			H. aff. planispira	
Barremian		Hedbergella sigali	Glob. aptiensis	Hedbergella infracretacea
			Gav. barremiana	
			Clavih. simplex	
			Clavih. sigali	
Hauterivian		'Hedbergella' hoterivica	Lenticulina ouachensis	
		Calpionellidae zones	Haplophragmoides vocontianus	
			Dorothia hauteriviana	
Valanginian			Lenticulina busnardoi	
			Lenticulina nodosa	
Berriasian				

TABLE VIII

Range chart of uppermost Cretaceous larger foraminifers of Pakistan
(After Kureshy, 1977a)

Geologic age	Late Cretaceous	
European stages	Campanian	Maastrichtian
Larger foraminiferal zones	*Orbitoides tissoti*	*Orbitoides media*
Formation / Larger foraminifers	Orbitoides Fm.	
Orbitoides tissoti		
O. compressa		
O. minima		
Lepidorbitoides socialis		
Orbitoides apiculata		
O. media		
Lepidorbitoides minor		
Omphalocyclus macropora		
Siderolites calcitrapoides		
Sulcoperculina globosa		

(1) The Samana Suk and Kingriali Formations of Triassic age are characterized by benthic foraminifer species dominated by arenaceous forms.

(2) In the Chiltan and Zidi formations of Jurassic age there are sporadic occurrences of a few arenaceous and calcareous forms, which closely resemble European and African Jurassic forms.

(3) The Belemnite and Parh formations of Cretaceous age are characterized by a much more diversified fauna of calcareous and arenaceous origin. The Parh Limestone Formation is characterized by a rapid expansion and diversification of planktonic foraminifers belonging to the group of the Hedbergellids, Rotaliporids, Globotruncanids, and other related forms of planktonic foraminifers, characteristic of a deep and open marine environment of deposition.

(4) The presence of *Guembelitria* in the Gulf coastal region of the United States and in continental shelf deposits of Nova Scotia indicates a neritic environment of deposition. Thus the absence of this foraminiferal genus in the Parh Formation could be considered as a further evidence of a pelagic environment of deposition.

(5) During the Upper Cretaceous in most parts of the Lower Indus platform the neritic environments of deposition prevailed, in which the Orbitoides and Hemipneustes Limestone Formations were laid down. These deposits are characterized by *Orbitoides* species and other larger foraminiferal assemblages.

(6) The Korara Formation in the Lower Indus Basin was a transitional deposit of Maastrichtian and Danian age in which no break in sedimentation is noted, whereas there is a marked distinction in the planktonic foraminiferal assemblages, one belonging to the Maastrichtian and characterized by *Globotruncana* species and the other belonging to the Danian, which is characterized by *Globigerina* and *Globorotalia* species.

IGNEOUS ACTIVITY

Igneous activity during the Mesozoic is reported at two levels in Pakistan. The earlier consists of lava flows of Triassic age found in the Upper Indus Basin, while the later, Cretaceous, consists of intrusion and submarine flows found in the northern and southern part of the Lower Indus Basin.

None of the Cretaceous intrusions appear to be extensive although they may be well developed locally. Two principal localities are described in the Lower Indus Basin, near Hindubagh where there are numerous small dykes and sills in the sedimentary sequence. The rocks are generally ultrabasic with pyroxenite, dunite and saxonite. They cut the Parh Limestone Formation and therefore are younger than early Maastrichtian.

The second location is in the vicinity of Bela along the Porali River in the southern part of the Lower Indus Basin in Sind. Compositionally the rocks are ultrabasic to basic but range to intermediate in character. They are of the same age as the intrusions in the Hindubagh area.

In the Eruptive zone of the Baluchistan Basin along the border of Afghanistan near Chagai some Upper Cretaceous submarine eruptions of granite and pegmatite took place. Similar eruptions of the same age are also common in the Ras Kok Range. These igneous activities are shown in Fig. 5 and discussed under their respective basins below.

Fig. 5. Mesozoic igneous activity in Pakistan.

Lower Indus Basin

Hindubagh intrusions

The Hindubagh intrusions took place in the marine sedimentary strata of Upper Cretaceous age in the region of the Zhob River Valley near Fort Sandman and Hindubagh. The main concentration of the intrusions is in the south of Hindubagh. These intrusions are small and scattered occurring in the form of dykes and sills in the Parh Limestone Formation of Cretaceous age. Most of the intrusions are serpentinized ultrabasic rocks including saxonite, dunite, pyroxenite, dolerite, gabbro and diorite. The geology of the area was described by Bilgrami (1964) and Ahmad (1974).

Despite of the fact that many of the Hindubagh intrusions bear chromite supporting an important mining industry, little has been published on the geology and petrology of these areas. The main group of ultrabasic, gabbroic and dioritic rocks intruded into the Parh Limestone Formation, suggests that these intrusions are Upper Cretaceous in age. The salient geological features of the Zhob Valley and Hindubagh igneous complex are as follows.

(a) The igneous complex occurs along the "central axis", a zone of intense igneous activity during Cretaceous time.

(b) In the area of the Zhob Valley and Hindubagh the intrusions are emplaced in sedimentary rocks of Cretaceous age.

(c) The intrusive masses are often irregular in shape and vary in area from a few square metres to several square kilometres.

(d) The rock texture is generally hypidiomorphic, granular; poikilitic texture is extremely rare.

(e) Chromite commonly occur in disseminated, banded and nodular forms. The chromite is found as the product of magmatic differentiation in the form of segregated masses and veins.

(f) Chromite concentrations are confined to dunite, peridotite, and serpentine.

(g) Dunite is far more common than the other ultrabasic rocks.

Bela Volcanic Group

The Bela Volcanic Group is confined to the southern part of the Axial Belt in the province of Sind near Bela. Intrusive rocks belonging to the Porali intrusion are common among the volcanic group. In general the Porali intrusive rocks are included in the Bela Volcanic Group (Hunting Survey Corporation, 1960). The Bela Volcanic Group consists of a heterogeneous assemblage of volcanic and sedimentary rocks. The volcanic rocks are mainly basalt and include coarse agglomerate and bedded tuff as well as lava. The sedimentary rocks intercalated with the volcanic beds include shale, marls, limestone, conglomerate and radiolarian chert. All gradations are found, from soft shale to hard maroon and green marls. The limestone is similar to the Parh Limestone Formation.

The Bela Volcanic Group can be correlated with the Sinjrani Volcanic Group of the northern Baluchistan Basin (see below). The Bela Volcanic Group includes submarine lava flows during most of the Cretaceous along the Las Bela geanticline. These igneous rocks were referred to the Deccan Trap by Vredenburg (1909), but they do not correspond to the Deccan Trap of India, for the Traps represent Early Tertiary terrestrial volcanism of the plateau type, whereas the Cretaceous Bela Volcanic Group rocks are of submarine origin and have been recently considered as belonging to some ophiolitic complex (Talent and Mawson, 1979).

Porali intrusions

The rocks of the Porali intrusions are varied in composition and are exposed along the Porali River near Bela. The intrusions are mainly of ultrabasic and intermediate character, granitic rocks are rare. The ultrabasic rocks are mainly altered pyroxenite, and serpentinized peridotite. The rocks of intermediate composition are represented by diorite and gabbro, diorite is common among the Porali intrusives.

The Porali intrusive rocks are broadly similar in its composition to the Hindubagh intrusion. The serpentinized peridotite and pyroxenite are semi-concordant bodies which have intruded Cretaceous marine deposits of the Parh Limestone Formation and even the volcanic rocks of the Bela Group. Thus the Porali intrusions are Late Cretaceous in age and younger than the Bela Volcanic Group.

Upper Indus Basin

Panjal Volcanic Group

The Panjal Volcanic Group consists of volcanic rocks interbedded with marine sedimentary deposits of Triassic age exposed in the Kaghan Valley of Hazara and in the Upper Indus Basin. They can be divided into a lower section of thick pyroclastic material or agglomeratic material called the "Panjal Agglomeratic Slate" and an upper section, "Panjal Trap", a thick series of andesitic and basaltic lavas overlying the agglomeratic material.

The individual flows vary in thickness from a few centimetres to several metres or more. The Panjal Trap is associated with a considerable thickness of "Inter-Trappian" fossiliferous marine limestone of Lower Triassic age. Thickness of the Inter-Trappian limestones varies from 20 to 300 m in parts of the Kaghan Valley, and in Hazara they are followed by thick poorly fossiliferous Upper Triassic limestones, whose thickness varies from 170 to 400 m (Wadia, 1953).

Baluchistan Basin

Sinjrani Volcanic Groups

This group is exposed north of Chagai near and along the Afghanistan border in the eruptive zone of the Baluchistan Basin. It consists of agglomerate, tuff and lava with subordinate shale, sandstone and limestone. Volcanic rocks dominate a sequence which varies in thickness up to a maximum of about 20 m; up to 70 m limestone and calcareous shales occur, commonly interbedded with the volcanic strata.

The Sinjrani Group is the result of submarine volcanism on the evidence of the interbedded limestone and other sedimentary rocks. The limestone is similar to the Parh limestone of the Lower Indus Basin and may be of the same age. The volcanic rocks are also similar to those in the Bela Volcanic Group of the Sind province. Thus it is believed that the Sinjrani Volcanic Group is of Upper Cretaceous age.

Chagai intrusions

The Chagai intrusions are exposed in the Chagai Hills adjacent to the Afghanistan border in the northern part of the Baluchistan Basin. The main rock types are diorite and biotite granite. Along the southern margin of Chagai Hills the range of the composition appears to be wider and the suite includes granite with pegmatite. Most of the intrusions have a medium-grained hypidiomorphic texture.

The main Chagai intrusion is a large batholith intruded into the horizontal strata of the Cretaceous Sinjrani Group. North of the Chagai Hills the dip of the contact is gentle, but in the south the contact is either vertical or steeply inclined, and intrusive masses are more irregular. The age of the Chagai intrusions according to the Hunting Survey Corporation (1960) is Late Cretaceous.

Kuchakki Volcanic Group

The rocks of the Kuchakki Group are confined to Raskok belt in the northern part of the Baluchistan Basin, mainly in a strip along the northern side of the Raskok Range. The group consists of volcanic agglomerate, lava, tuff with limestone, sandstone and shale. The agglomerate forms massive beds up to 15-20 m thick, which consist of coarse red and green fragments of lava and tuff in a matrix of dark green or maroon volcanic material. The sedimentary deposits resemble the Parh Limestone of the Lower Indus Basin. The Kuchakki Group is overlain by shale and sandstone of the Lower Tertiary Rakhshani Formation.

The sedimentary rocks of the Kuchakki Group are barren but from their stratigraphic relation and lithological appearance they are considered to be Cretaceous in age, and are correlated with the Sinjrani Group. The volcanic rocks of both groups are similar and associated with Cretaceous marine sedimentary rocks. The Kuchakki Volcanic Group is intruded by Raskok intrusions accompanied by Paleocene to Eocene Rakshani sedimentary deposits.

THE GEOLOGICAL EVOLUTION OF PAKISTAN

Paleo-oceanography

The majority of Mesozoic deposits in Pakistan are marine, and it is evident that in this area the Mesozoic Era was generally a period of transgression. However, minor regressions occurred at times particularly in the Upper Indus Basin, where the Jurassic and Cretaceous succession is marked by several unconformities as well as non marine sedimentation episodes.

During the Triassic and Jurassic periods the Tethys Sea extended from north to south and even into the Baluchistan Basin. Through the Baluchistan Basin the Tethys Sea was connected with the Middle East. Triassic marine deposits are limited in outcrop, but it is believed that Triassic and Jurassic marine sediments are largely covered by younger sediments, in the Lower Indus and Baluchistan Basins. The Jurassic period in the Lower Indus Basin was characterized by a transgression and is represented by a complete succession of marine deposits. In the Upper Indus Basin the Jurassic Sea retreated in certain areas and nonmarine deposits including coal and lignite were formed. The marine Jurassic fauna of the Lower Indus Basin is identical with the Jurassic benthic foraminiferal fauna of Europe and Africa, indicating that the Jurassic Sea of Pakistan was connected with the European and African epicontinental seas through the Middle East.

An extensive transgression of the Cretaceous Sea took place in the Lower Indus Basin, where the sedimentation is characterized by open marine deposits of the Belemnite Shale, Parh Limestone, and Korara Shale formations of Neocomian to Maastrichtian age. In the Upper Indus Basin the Cretaceous transgression was marked by an embayment of the sea, in which sporadic marine Cretaceous sedimentary deposits were laid down. Nowhere in the Upper Indus Basin area is a complete sequence of the Cretaceous exposed. The foraminiferal fauna of Cretaceous deposits are similar to the assemblages described from the Middle-Eastern, Caribbean and Far East regions. Thus it is evident that during the Cretaceous time Pakistan was in open connection with the western Tethys Sea in the west and the Indo-Pacific seaway in the east.

In the Baluchistan Basin only limited exposures of marine sedimentary rocks of Upper Cretaceous are encountered, but it is believed that complete marine Cretaceous deposits are present under the cover of the Tertiary strata.

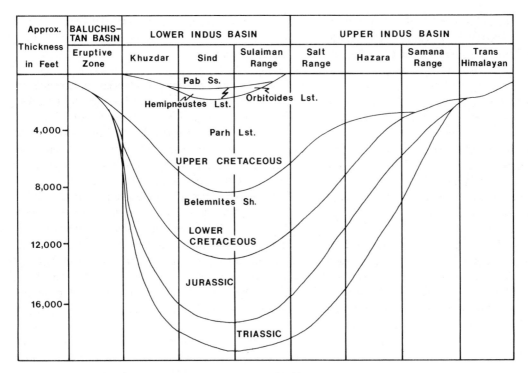

Fig. 6. Tentative correlation in the Mesozoic succession of Pakistan.

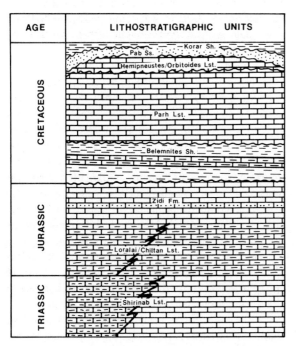

Fig. 7. Facies relationship of the Mesozoic strata of the Lower Indus Basin.

Depositional history (Figs. 6, 7)

During the Mesozoic Pakistan formed part of the downwarping margin of the African and Indian shields, and part of the southern shore of the Tethys Ocean. A thick wedge of more than 3000 m of largely calcareous sediments was deposited under warm, tropical conditions, consistent with a distribution parallel to the equatorial regions. Until the Tertiary no important earth movement affected the region; in most cases the pre-Tertiary movement only interrupted the cycle of sedimentation by causing uplift and resulting in the formation of unconformities and pronounced gaps in the stratigraphic succession. In the Mesozoic these phenomena are encountered particularly in the Upper Indus Basin, whereas in the Lower Indus Basin the greatest thickness of Mesozoic strata was deposited.

Lower Indus Basin

The Triassic deposits of the Lower Indus Basin are mostly clastic with a possible source on the Indian Shield to the east. The Jurassic deposits ranging from

Early Jurassic to Late Jurassic are characterized by carbonate facies. These deposits are rich in ammonites and contain sporadic benthic agglutinating foraminifers, an indication of a deep open marine environment of deposition. The deposits change rapidly towards the Salt Range in the Upper Indus Basin, where the Jurassic deposits are relatively thin and contaminated with carbonaceous material. The Axial Belt started developing as a sub-marine ridge during the Jurassic and continued until the Cretaceous. As previously mentioned, this axis marks the boundary between the Lower Indus Basin and the Baluchistan Basin, and is interpreted as an ophiolite geosuture by Talent and Mawson (1979).

The Cretaceous transgressions were more extensive and outcrops occur over wide areas of the Lower Indus Basin. These deposits are represented by various facies ranging from neritic to abyssal, and even to non-marine environments of deposition. The Cretaceous Period in Pakistan was a time of great tectonic instability witnessed by the significant change in the type of sedimentation, and of widespread igneous activity.

During the Lower Cretaceous the environment of deposition which prevailed in the Lower Indus Basin was that of a shallow sea with an influx of detrital material derived from the Indian Shield. The clastic deposits (Belemnites shale) of Lower Cretaceous age were laid down, followed by a calcareous facies of Parh limestone of Upper Cretaceous age. The Parh Formation is characterized by an abundance of planktonic foraminifers, which indicates that these deposits were formed in a deep and open-marine environment of deposition. In the Upper-Campanian and Maastrichtian a shallow neritic environment of deposition prevailed in which carbonate facies of the Orbitoides and Hemipneustes Limestone formations were deposited. In the late Maastrichtian the area was even emergent and the non-marine Pab sandstone was deposited. The Pab sandstone marked the first great invasion of this area, which can be considered as a geosyncline, by clastic deposits and signalled a progressively greater direct influence of the land margin in the sedimentation.

In the Late Cretaceous Period massive submarine igneous activity (see above) took place in association with the Parh Limestone.

Upper Indus Basin

In the Salt Range marine Triassic deposits, with many species of the genus *Ceratites*, are similar in facies to the Productus Limestone of Permian age, with no evidence of break in sedimentation. In Hazara the Triassic deposits are affected by igneous activity, which is marked by the Pir Panjal volcanic series. The Samana Range Triassic deposits are marine and characterized by benthic foraminifers.

A regional uplift at the end of the Triassic is indicated by the absence of Early Jurassic strata and the marked unconformity followed by Middle Jurassic deposits of Variegated sandstone. A non-marine deposit containing coal seams is found in the Kalabagh area of the Salt Range. In Late Jurassic time the area again subsided and the marine strata of Boarch limestone were formed.

The end of the Jurassic was a period of emergence in all parts of the Upper Indus Basin, followed by a period of subsidence with deposition of the Belemnite shale in Early Cretaceous time, and of Upper Cretaceous Lumshiwal sandstone in the Salt Range.

A calcareous facies rich in planktonic foraminifers of Upper Cretaceous age is exposed in Hazara and known as Chanali Formation. In the Samana Range (Kohat) and in the Transhimalayan region (Gilgit and Baltistan) Upper Cretaceous deposits (Lithographic limestone and Yasin limestone) were reported to be exposed over a limited area, but their environment of deposition is not certain. These deposits may be marine, corresponding to the Parh limestone of the Lower Indus Basin.

Baluchistan Basin

During the Cretaceous the Eruptive zone (see above) was a shallow depression, where volcanic rocks with subordinate marine limestone and shale known as Kuchakki and Sinjarani Volcanic Series were formed.

It is also noteworthy that in Iran, about 100 km west of the Pakistan border at Koh-i-Berg along the extension of the "Central axis", a great thickness of Jurassic and Cretaceous strata of massive and reefoid limestones is exposed. Thus the older Mesozoic sequence may extend at depth under the Baluchistan Basin covered by a huge monotonous sequence of flysch deposits of Tertiary age. So far no exploratory well has penetrated below the Miocene strata in this basin to verify the extension of Mesozoic succession in the sub-surface.

Finally submarine volcanic eruptions associated with marine sedimentary deposits of Cretaceous age occur. These eruptive deposits were subsequently intruded by the Chagai intrusion in Late Cretaceous times.

Geotectonic interpretation

The Mesozoic sediments of Pakistan were deposited in the Tethys Ocean which in the Permian began to divide the Pangea into the Northern Hemisphere (Laurasia) and the Southern Hemisphere (Gondwanaland). The concept of the Tethys was first introduced by Suess (1893) to describe a great equatorial sea of Jurassic age stretching from Mexico to the Himalayas.

In the Late Jurassic there is evidence of a further fragmentation of Gondwanaland, one result of which was Africa, India and Australia ruptured and drifted to form distinct continents. India began moving northwards to Laurasia and collided with the Himalayas. The rotation of Eurasia and the drift of the southern continental blocks began to restrict the eastern end of the Tethys Sea. Lateral movements of the African Plate in relation to Europe created in the Cretaceous an extensive western Tethys Sea which opened into the Atlantic Ocean (Pitman and Talwani, 1972). In this general scheme, Pakistan was part of this oceanic feature which also included the Middle Eastern countries and upon which the Mesozoic sediments were deposited.

The Mesozoic sedimentary rocks of Pakistan are exposed in the foreland (Indus) basin east of the Axial Belt and west of the foreland. These rocks are thickest along the axis of the foreland basin in the Lower Indus Basin, and thin out in the Upper Indus Basin. During the Himalayan orogeny in the Tertiary mountain ranges were formed out of Mesozoic strata of the foreland basin. The eastern limb of the foreland basin was gently depressed to form the alluvial plains, now covered by sub-Recent and Recent river alluvium.

The Central and Las Bela geanticlines became the nuclei of deformation and formed the first mountain ranges. As the deformation continued the folds and faults developed throughout the geosynclinal deposits, and were of great magnitude in the zone of thickest sedimentation, producing considerable topographic irregularities. The Mesozoic earth movements were principally isostatic rather than orogenic, until the commencement of the Himalayan orogeny. The most common of these pre-Himalayan earth movements

which interrupted the cycle of sedimentation caused uplift resulting in depositional unconformities. The other main feature resulting from pre-orogenic movements was the geanticlinal belt which developed longitudinally in the central part of the geosyncline in the Late Mesozoic and divided the region into two sedimentary basins, the Baluchistan and the Lower Indus basins.

These areas become the site of intense igneous activity, both intrusive and extrusive, which was limited to the Triassic deposits of Hazara in the Upper Indus Basin, but is common in the Cretaceous marine sedimentary rocks of the Lower Indus and Baluchistan basins.

The Cretaceous period of Pakistan witnessed great instability and was a time of great tectonic activities, which are evident from the following.

(a) In the Cretaceous Period the full development of the Central and Las Bela geanticlines was accomplished and these structures served as the demarcation line between the Lower Indus Basin and the Baluchistan Basin. This resulted in a complete separation in the types of sedimentation in these basins during the Tertiary.

(b) During Cretaceous times significant changes took place in environment of sedimentary deposition ranging from abyssal to neritic and even to non-marine deposits of Pab sandstone in late Maastrichtian time.

(c) Widespread igneous activity took place particularly during the Upper Cretaceous.

(d) Even in some areas, such as the Pab Mountains and Gaj River, there has been a continuous deposition between the Maastrichtian and Danian, whereas elsewhere the Cretaceous—Tertiary boundary is marked by an abrupt change in lithology and by an unconformity.

The geanticlinal ridges which were responsible for the separation of the Baluchistan Basin and the Lower Indus Basin, extended in fact from the Arabian Sea west of Karachi to the Pakistan—Afghanistan border north of Fort Sandeman. Further north the axis continues eastwards and extends towards Kashmir between the Salt Range and Hazara (Zuberi and Dubois, 1962). The earth movements, as the result of the mobile Afghan craton in the northwest moving towards the Indian Shield in the east, began in Cretaceous time and continued with varying intensity until the present (Krishnan, 1943). The most important tectonic features of the Upper Indus Basin were the irregular ridges or promontories of the Indian Shield that extended northwestwards into the basin. As a result restricted Cretaceous marine sediments are encountered in the Upper Indus Basin.

The Baluchistan Basin is further divided into two structural units, a southern (Makran) and northern (Panjgur), by the central axis at Makran, which merges into the central axis at Khuzdar. The northern part of the Baluchistan Basin is characterized by an igneous belt known as the "Eruptive zone" which has both intrusive and extrusive igneous rocks along with the marine Cretaceous sedimentary deposits and which separated the Baluchistan Basin from the shelf area of the southernmost part of the Afghan craton from Late Cretaceous time.

REFERENCES

Ahmad, Z., 1974. Geology and petrochemistry of a part of the Zhob Valley, Igneous Complex, Baluchistan, Pakistan. *Rec. Geol. Surv. Pakistan*, 24: 1–24.

Bakr, M.A., 1965. Geology of Parts of Trans-Himalayan region in Gilgit and Baltistan, West Pakistan. *Rec. Geol. Surv. Pakistan*, 11 (3): 1–17.

Bilgrami, S.A., 1964. Mineralogy and petrology of the central part of the Hindubagh Igneous Complex, Hindubagh Mining District, Zhob Valley, West Pakistan. *Rec. Geol. Surv. Pakistan*, 10 (2C).

Blanford, W.T., 1879. The geology of Western Sind. *Mem. Geol. Surv. India*, 17: 1–197.

Bolli, H.M., 1959. Planktonic foraminifera from the Cretaceous of Trinidad, B.W.I. *Bull. Am. Paleontol.*, 39 (179): 257–277.

Davies, L.M., 1930. The fossil fauna of Samana Range and some neighbouring areas. *Mem. Geol. Surv. India*, 15: 67–79.

Dorreen, J.M., 1974. The western Gaj River Section in Pakistan and the Cretaceous–Tertiary boundary. *Micropaleontology*, 20: 178–193.

Fatmi, A.N., 1971. Late Jurassic and Early Cretaceous (Berriasian) Ammonites from Shaikh Budin Hills, D.I. Khan, (N.W.F.P.). *Rec. Geol. Surv. Pakistan*, 21 (2).

Fatmi, A.N. and Cheema, M.R., 1972. Early Jurassic Cephalopoda from Khisor-Marwat Range, D.I. Khan, (N.W.F.P.). *Rec. Geol. Surv. Pakistan*, 21 (3).

Gee, E.R., 1944. The age of Saline Series of the Punjab and Kohat. *Nat. Acad. Sci. India, Proc., Sec. B*, 14: 269–311.

Haque, A.F.M.M., 1959. Some Late Cretaceous foraminifera from West Pakistan. *Mem. Geol. Surv. Pakistan*, 2: 1–33.

Hunting Survey Corporation, 1960. *Reconnainance Geology of parts of West Pakistan*. Report, Toronto, Ont., 559 pp.

Krishnan, M.S., 1943. *Geology of India and Burma*. Higginbothams, Madras.

Kummel, B., 1966. The Lower Triassic Formation of the Salt Range and Trans Indus Range, West Pakistan. *Bull. Mus. Comp. Zool.*, 134: 361–429.

Kummel, B. and Teichert, C., 1970. *Stratigraphic Boundary Problems: Permian and Triassic of West Pakistan*. Kansas Univ. Press, Lawrence.

Kureshy, A.A., 1964. Regional stratigraphic studies of West Pakistan. *J. Ind. Nat. Res.*, 3: 14–22.

Kureshy, A.A., 1970. The continuous deposition of Mesozoic and Tertiary boundary of Pab mountain and Gaj River section of Pakistan. *Pakistan J. Sci.*, 22: 30–33.

Kureshy, A.A., 1971. The Upper Cretaceous and Paleocene larger foraminifera of Murree Brewery, West Pakistan. *Proc. Afr. Micropaleontol. Coll., 4th*, pp. 220–227.

Kureshy, A.A., 1971a. Upper Cretaceous and Tertiary Pelagic foraminifera of Iraq. *Proc. Intern. Planktonic Conf., 11th*, pt. 1, pp. 673–676.

Kureshy, A.A., 1972. Sedimentary basins of West Pakistan. *Pakistan J. Sci. Res.*, 24: 132–138.

Kureshy, A.A., 1972a. Stratigraphic Micropaleontology of Pakistan. *Pakistan J. Sci.*, 24: 133–146.

Kureshy, A.A., 1972b. Summary of stratigraphic nomenclature of Pakistan. *Pakistan J. Sci.*, 24: 147–159.

Kureshy, A.A., 1973. The Upper Cretaceous larger foraminifera of Harnai, Pakistan. *Pakistan J. Sci.*, 25: 163–167.

Kureshy, A.A., 1975. The Cretaceous and Tertiary larger foraminifera of West Pakistan. *Rev. Esp. Micropaleontol.*, 7 (2): 553–564.

Kureshy, A.A., 1976. The Cretaceous planktonic foraminiferal zones of Pakistan. *Rev. Esp. Micropaleontol.*, 8: 429–438.

Kureshy, A.A., 1977. The Cretaceous planktonic foraminiferal biostratigraphy of Pakistan. *Paleontol. Soc. Japan, Spec. Pap.*, 21: 223–231.

Kureshy, A.A., 1977a. The Cretaceous larger foraminiferal biostratigraphy of Pakistan. *J. Geol. Soc. India*, 18 (12): 662–667.

Kureshy, A.A., 1981. Biostratigraphy of Chiltan Formation (Jurassic) of Pakistan. *Proc. Afr. Micropaleontol. Coll., 8th*, Paris. In press.

Latif, M.A., 1970. Micropaleontology of the Chanali limestone of the Upper Cretaceous of Hazara, West Pakistan. *Jahrb. Geol. Bundesanst. Sonderb.*, 15: 25–62.

Mark, P., 1962. Variation and evolution in Orbitoides from the Cretaceous of Rakhi nala, West Pakistan. *Geol. Bull. Punjab Univ.*, 2: 15–29.

Moullade, M., 1974. Zones de Foraminifères du Crétacé inférieur mesogeen. *C.R. Acad. Sci. Paris, Ser. D*, 278: 1813–1816.

Oldham, R.D., 1890. Report on the geology and economic resources of the country adjoining the Sind in Pishin Railway between Sharigh Spintangi and of the country between it and Khattan. *Rec. Geol. Surv. India*, 23: 93–110.

Pinfold, E.S., 1939. The Dunghan limestone and the Cretaceous–Eocene unconformity in north west India. *Rec. Geol. Surv. India*, 74: 189–198.

Pitman, W.C. and Talwani, M., 1972. Sea floor spreading in the North Atlantic. *Geol. Soc. Bull.*, 83: 619–646.

Said, R. and Barakat, M.G., 1958. Jurassic microfossils from Gebel Moghara, Sinai, Egypt. *Micropaleontology*, 4: 231-272.

Suess, E., 1893. Are great ocean depths permanent? *Nature*, 2: 180–187.

Talent, J.A. and Mawson, R., 1979. Palaeozoic–Mesozoic biostratigraphy of Pakistan in relation to biogeography and the coalescence of Asia. In: A. Farah and K.A. de Jong (Editors), *Geodynamics of Pakistan*. Geol. Surv. Pakistan, Quetta, pp. 81–102.

Van Hinte, J.E., 1976. A Cretaceous time scale. *Am. Assoc. Petrol. Geol. Bull.*, 60 (4) 498–516.

Vredenburg, E.W., 1908. Cretaceous Orbitoides of India. *Rec. Geol. Surv. India*, 36: 171–213.

Vredenburg, E.W., 1909. Report on the geology of Sarawan Jhalawan Makran and state of Las Bela. *Rec. Geol. Surv. India*, 38: 189–215.

Wadia, D.N., 1953. *Geology of India*. MacMillan, London.

Zaninnetti, I. and Bronnimann, P., 1975. Triassic Foraminifera from Pakistan. *Riv. Ital. Paleontol.*, 81: 257–280.

Zuberi, W. and Dubois, E.P., 1962. Basin architecture, West Pakistan. In: *Symposium on Development of Petroleum Resources of Asia and Far East, Teheran, 1962.*

Chapter 12

PAPUA NEW GUINEA*

S.K. SKWARKO, C.M. BROWN and C.J. PIGRAM

INTRODUCTION

Papua New Guinea, in the west Pacific between 141° and 156°E and north of Australia between 1° and 12°S, comprises the eastern part of the island of New Guinea, groups of small islands to the southeast, the islands surrounding the Bismarck Sea, and Bougainville Island. Its topography is a contrast of high rugged mountain ranges and low widespread alluvial plains, both blanketed by tall grasses and tropical rain forest. Its physiographic regions (Fig. 1) broadly coincide with its major present-day geotectonic provinces (Fig. 2).

The geology of Papua New Guinea reflects its position astride a tectonically active Mobile Belt (Fig. 2), which is the zone of interaction between the northward moving Indo-Australian plate locally represented by the stable Papuan Platform, and the westward moving Pacific plate locally represented by the Oceanic Crust and Island Arcs.

Mesozoic rocks occur on the Papua New Guinea mainland and on the islands to the southeast (Fig. 3), and may underlie at least some of the islands of the Bismarck Sea in the north. Where exposed they are folded, faulted, and partly metamorphosed, so that their original stratigraphic and structural relationships are largely obscured. Subsurface, they have been encountered in petroleum exploration wells.

In this paper we outline the geology, and attempt to reconstruct the geological history of Mesozoic Papua New Guinea by briefly summarising its Mesozoic stratigraphy and by determining the deformational effects of Cainozoic tectonism. Our approach and conclusions reflect personal involvement in the regional mapping of the country, but also unavoidably lean on data written or published prior to 1977 in publications listed below, references to which are made throughout the text.

Systematic regional mapping of Papua New Guinea at 1:250,000 scale is completed. The maps and their accompanying explanatory notes are the main source of the Mesozoic data provided here, and have also been used to compile a set of stratigraphic tables for Papua New Guinea (Skwarko, 1978). A 1:1,000,000 scale geological map was published in 1972 and a more general map at 1:2,500,000 in 1976 (BMR, 1972, 1976).

The geology of Papua New Guinea has been described by Thompson and Fisher (1965), Thompson (1967; 1972), Harrison (1969), Dow (1973; 1977), Bain (1973), Milsom (1974) and Brown et al. (1980); that of the northern New Guinea ranges by Robinson (1973), Robinson et al. (1974), Jaques and Robinson (1977), Norvick and Hutchinson (1980); of the New Guinea Highlands by Rickwood (1955), McMillan and Malone (1960), Dow and Dekker (1964), Dekker and Faulks (1964), Dow et al. (1972), Bain et al. (1975), Pigram (1978); of eastern Papua by Dow and Davies (1964), Davies and Ives (1965), Davies and Smith (1971), Brown et al. (1975); of western Papua by the Australasian Petroleum Co. (1961), Rickwood (1968), Buchan and Robinson (1969), Jenkins et al. (1969), Jenkins and Martin (1969), Hocking et al. (1971), Conybeare and Jessop (1972), Davies and Norvick (1974), Findlay (1974). Geology, geophysics, and drilling operations in the Gulf of Papua have been summarised by Tallis (1975). Compilation of 1:250,000 Sheets which bor-

* Published with the permission of the Director, Bureau of Mineral Resources, Geology and Geophysics, Canberra, A.C.T., Australia.

der the southwest Papuan plains and the Gulf of Papua has relied heavily on data obtained by geologists of the Australasian Petroleum Co., British Petroleum Co., and Phillips Australian Oil Co.

The regional gravity pattern was described by St John (1967, 1970), and the aeromagnetic pattern over the eastern part of the Gulf of Papua and adjacent mainland by Compagnie Générale de Géophysique (1969). Geophysical data on the Gulf of Papua have been presented by Mutter (1972a, b, 1975) and Wilcox (1973).

The Mesozoic microfossils (mostly Foraminifera) have been described by Crespin (1958), and Owen (1973), and microflora by Cookson and Dettman (1958a, b) and Cookson and Eisenack (1958). The larger invertebrates were described by Etheridge (1889), Schluter (1928), Erni (1945), Glaessner (1945, 1949, 1957, 1958, 1960), Skwarko (1967a,b, 1973a,b), Skwarko and Kummel (1974) and Skwarko et al. (1976). Intercontinental correlations using fossils were attempted anonymously (1962) and by Glaessner (1943), and palaeoclimate studies undertaken by Bowen (1961) and Dorman (1968).

PHYSIOGRAPHY

Papua New Guinea is a country of extensive grass and jungle-covered mountains and plains. The rugged relief reflects its youth and position astride a tectonically active zone, and deep erosion due to a hot wet climate. Five physiographic regions have been recognised (Fig. 1).

(a) The northern ranges, made up of Bewani-Torricelli and Adelbert—Finisterre mountains up to 4,160 m high are deeply dissected and largely forested; they are thought to be part of a former Tertiary island arc system.

(b) The swamps and alluvial plains of the Sepik and Ramu rivers extend over 40,000 km² and are vegetated by dense swamp forest and lowland forest.

(c) The extensive and varied central ranges largely formed of deformed geosynclinal Mesozoic and Tertiary sedimentary and igneous rocks are made up of the deeply dissected up to 4,500 m Bismarck Range in the north, less rugged highlands with high-altitude valleys to the south, and the over 4,000 m high Owen Stanley Range in the southeast.

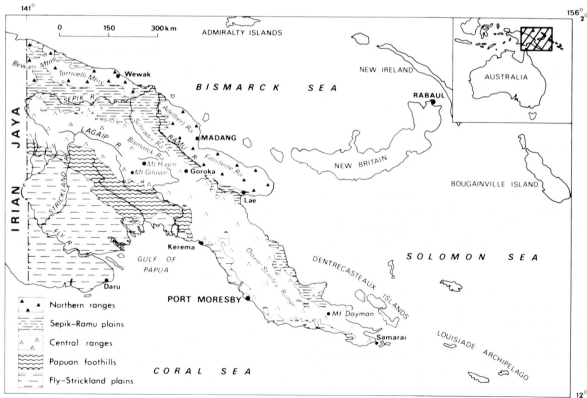

Fig. 1. Physiographic regions.

(d) The Papuan foothills, which mainly consist of over-thrusted and gravity-slumped Cretaceous and Cainozoic sediments, occur on the southern flanks of the central ranges, and attain the maximum elevation of approximately 2,000 m. In the west they consist of limestone scarps weathered to karst, and are partly covered by several large Quaternary strato-volcanoes, the highest of which exceeds 4,000 m.

(e) The low-lying plains, swamps, and deltas of the Fly and Strickland rivers extend over 90,000 km² and support lowland rain forest. The Fly River plains are underlain at shallow depth by generally flat-lying Mesozoic rocks which rest on the continental Platform, while the Strickland river plains possibly rest on Mesozoic former oceanic crust.

TECTONIC AND STRUCTURAL SETTING

In terms of the theory of continental drift and plate tectonics, Papua New Guinea's position is across the junction of the north-moving Indo-Australian plate and the west-moving Pacific plate. The Indo-Australian plate is locally represented by the Papuan Platform (Fig. 2), a rigid and stable extension of Australia. The Pacific plate is represented by sub-plates which comprise the Oceanic Crust and Island Arcs province in the north and east of mainland Papua New Guinea. The zone of Cainozoic interaction between the two plates is the Mobile Belt (Fig. 2), an extensive area of largely disturbed rocks which form the central ranges.

The study of the Mesozoic geology of Papua New Guinea is thus the study of sediments and volcanic rocks which accumulated around the northern and eastern margins of the continental Platform during Mesozoic times, and of their subsequent deformation through plate interaction in the Cainozoic.

The stable Platform is overlain by Mesozoic sediments which are generally sub-horizontal and undeformed. However, fault-bounded basement highs occur at Oriomo, Lake Murray, Komewu, and Pasca, and shallow basins at Morehead, north of Lake Murray and Komewu, and in the Borabi Basin west of the Pasca Ridge (Fig. 2). The deformed northern margin of the Platform forms the Papuan Fold Belt underlain

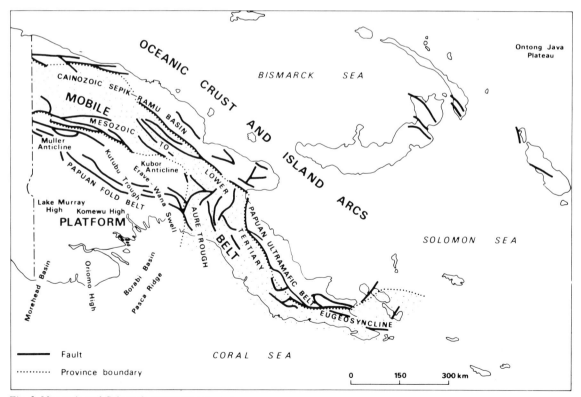

Fig. 2. Mesozoic and Cainozoic structural elements.

by the Kutubu Trough and the Erave-Wana Swell and flanked in the northeast by the Kubor Anticline basement high.

The Pliocene uplift of the central ranges initiated gravity tectonics with southwesterly sliding and thrusting at various levels of detachment within the Mesozoic and Tertiary rocks, giving rise to the Papuan Fold Belt (Jenkins, 1974). Intensity of deformation associated with the thrusting decreases from northeast to southwest. In the northwest the deformation does not extend as far south as in the east, because the rising basement in the core of the Muller Anticline is thought to have acted as barrier to southward thrusting.

The central and western parts of the Mobile Belt are characterised by long, sinuous, vertical or near-vertical bifurcating faults of probable Oligocene to Lower Miocene age. Some of them, particularly in the central part may still be active today. Large vertical movements have previously been postulated for some of the faults, but the style of the faults is typical of transcurrent tectonics, although transcurrent movement is difficult to prove. The system is thought to be left lateral.

Tectonically emplaced ophiolite complexes of the Oceanic Crust and Island Arc province occur at several places along the northern margin of the Mobile Belt (Davies, 1971; Dow et al., 1972; Pieters, 1974; Jaques, 1976), where they appear as relatively undeformed, sheet-like bodies arched and tilted towards the northeast. Commonly they show post-emplacement faulting.

Island arcs, believed to have formed over subduction zones, formed during Eocene times. Several may have been built on a Mesozoic oceanic crust, but the surrounding marginal seas are regarded as Palaeogene to Recent.

STRATIGRAPHY

In Papua New Guinea the Mesozoic rocks occur subsurface on the Platform and are exposed in the Mobile Belt — both on the mainland and on some of the offshore islands in the southeast. They may also be present as Cretaceous oceanic crust and mantle beneath at least some of the islands of the Bismarck Sea.

Papuan Platform (including the *Papuan Fold Belt*)

The isostatically stable Platform consists of a continental block-faulted Palaeozoic basement overlain almost everywhere by up to 2,000 m of mostly flat-lying Mesozoic and Cainozoic fluviatile, shallow marine, and deltaic sediments. In the north and east these are flanked by thicker (7,000 m) miogeosynclinal shallow and deeper marine sediments mixed with volcanic debris deposited on downfaulted Platform margins.

The Mesozoic rocks are concealed by Cainozoic sediments over most of the Platform, but are exposed in its deformed northern margin around the flanks of the Kubor Anticline, and in the core of the Muller Anticline, and as folded and thrust segments in the Papuan Fold Belt. They are also known from petroleum exploration wells as well as from the results of geophysical surveys (Figs. 3, 4, 5, K–P; Tables I, II, columns 1–6).

Triassic. Marine sediments and subaerial intermediate volcanics, and volcanolithic sediments of Middle—Upper Triassic age crop out only locally on the deformed Platform margins (Fig. 3, Table III, columns 9, 10; Bain et al., 1975; Skwarko et al., 1976). Shallow marine to continental sediments of probable Upper Triassic age have been encountered locally on the main body of the Platform in petroleum exploration wells.

Jurassic. Lower Jurassic sediments are absent — even under the surface — over most of the Platform, but may occur over the downfaulted margins (Table II, column 5; Conybeare and Jessop, 1972). The Platform's Palaeozoic basement is overlain by thin transgressive—regressive Middle to Upper Jurassic marginal marine and continental sediments (Table I, column 1, Table II, columns 6—8; Davies and Norvick, 1974).

Cretaceous. Cretaceous sediments are widespread and thick subsurface over the Platform. In many places they conformably overlie the Jurassic sediments (Table I, columns 1, 3, Table II, columns 5—7), but elsewhere sedimentary breaks separate the two (Table II, column 8, Table III, column 9). Locally, as in the Borabi—Pasca area on the Platform's margin they are absent owing to the Upper Cretaceous uplift and erosion.

Mobile Belt

In this zone beyond the Platform margin, shallow to moderately deep-water sediments accumulated

TABLE I

Lithology and correlation of the main Mesozoic sequences, Papuan Platform (for location of numbered sections see Fig. 3)

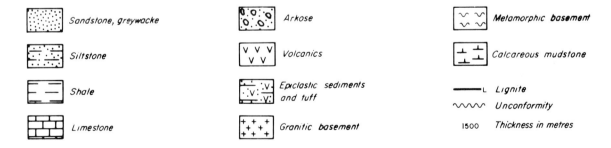

				MOREHEAD BASIN 1	LAKE MURRAY 2	MULLER ANTI 3	SOUTH SEPIK 4
			PALEOCENE TO EOCENE				
MESOZOIC	CRETACEOUS	U	MAASTRICHTIAN		NON-DEPOSITION EROSION	IERU FM 1500 / FEING GROUP 2000	SALUMEI FM 3000+
			CAMPANIAN				
			SANTONIAN				
			CONIACIAN				
			TURONIAN				
			CENOMANIAN	DELTAIC 1500 SHALLOW MARINE ALTERNATING SANDS AND MUDSTONE	IERU FM		
		L	ALBIAN				
			APTIAN	HARD BRITTLE MUDSTONE, MINOR SANDSTONE 500	FEING GP / TORO SST.	TORO SST 400	
			BARREMIAN				
			NEOCOMIAN	SHALLOW MARINE			
	JURASSIC	U	TITHONIAN	BOL ARKOSE 250		IMBURU MDST 600	OM BEDS 3000+
			KIMMERIDGIAN				
			OXFORDIAN			KOI-IANGE SST 550	
		M	CALLOVIAN			ATEMIN SHALE 100	
			BATHONIAN				
			BAJOCIAN			BOL ARKOSE 700	
			AALENIAN				BASE NOT SEEN
		L	TOARCIAN				
			PLIENSBACHIAN				
			SINEMURIAN				
			HETTANGIAN				
	TRIASSIC	U	RHAETIAN				
			NORIAN				
			CARNIAN				
		M	LADINIAN				
			ANISIAN				
		L	SCYTHIAN				
PALAEOZOIC	PERM CARBONIF				DIORITE	STRICKLAND GRANITE	

Legend:
- Sandstone, greywacke
- Siltstone
- Shale
- Limestone
- Arkose
- Volcanics
- Epiclastic sediments and tuff
- Granitic basement
- Metamorphic basement
- Calcareous mudstone
- L Lignite
- Unconformity
- 1500 Thickness in metres

TABLE II

Lithology and correlations of the main Mesozoic sequences, Papuan Platform (for location of numbered sections see Fig. 3; legend see Table I)

				MAGOBU I N°I 5	KOMEWU N°2 6	DARAI PLATEAU 7	KUTUBU 8
	PALEOCENE TO EOCENE						
MESOZOIC	CRETACEOUS	U	MAASTRICHTIAN				CHIM FORMATION SUBSIDING SHALLOW MARINE SHELF 2500 (WAHGI GROUP)
			CAMPANIAN				
			SANTONIAN				
			CONIACIAN				
			TURONIAN				
			CENOMANIAN				
		L	ALBIAN	MUDSTONE WITH RARE SILTSTONE	MUDSTONE SILTSTONE, SST DELTAIC 2000 MARGINAL MARINE	IERU FM 1500 (FEING GROUP)	KERABI FM 2100
			APTIAN	SANDSTONE SHALLOW MARINE			
			BARREMIAN	SHELF-DELTAIC 1300			
			NEOCOMIAN	TORO SST 250	TORO SST 400	TORO SST 400	
	JURASSIC	U	TITHONIAN	SANDSTONE SILTSTONE MUDSTONE SHALLOW MARINE DELTAIC 300	KOI-IANGE SST 600	IMBURU MUDSTONE 600 (KUABGEN GROUP)	MARIL SHALE ?4000
			KIMMERIDGIAN				
			OXFORDIAN			KOI-IANGE SST 700	
		M	CALLOVIAN	? ?	? ?	ATEMIN SH 300	
			BATHONIAN	MARINE SHALE 100 MUDSTONE	ARKOSIC SST CARBONACEOUS SHALE, COAL 100 FLUVIATILE DELTAIC	BOL ARKOSE 1000	
			BAJOCIAN	SILTSTONE SANDSTONE COAL			
			AALENIAN	NON-MARINE FLUVIATILE LACUSTRINE DELTAIC			
		L	TOARCIAN		?	?	?
			PLIENSBACHIAN	SHALLOW MARINE SAND TOWARDS TOP 600			
			SINEMURIAN	LIGNITE COMMON			
			HETTANGIAN	TOWARDS BASE			
	TRIASSIC	U	RHAETIAN		?	COARSE FELDSPATHIC AND VOLCANOLITHIC SST SHALLOW MARINE TO CONTINENTAL 170+	COARSE FELDSPATHIC AND VOLCANOLITHIC SST ? ?700
			NORIAN				
			CARNIAN				
		M	LADINIAN				
			ANISIAN				
		L	SCYTHIAN				
PALAEOZOIC	PERM CARBONIF			VOLCS GRANITE	GRANITE	GRANITE ?	GRANITE GRANODIORITE ?

1300 *Thickness in metres*

probably from the beginning of the Mesozoic until the Cainozoic when they were deformed.

Triassic. The Triassic sediments of the Mobile Belt are the oldest known marine sediments in Papua New Guinea. They consist of Anisian to Rhaetian shallow-water clastic rocks, volcanics, and limestone which overlie the Palaeozoic basement (Table III; Dow and Dekker, 1964; Skwarko, 1967a, 1973a; Dow et al., 1972; Bain et al., 1975; Skwarko et al., 1976).

Jurassic. The Jurassic clastic sediments and the overlying volcanics succeed the Triassic limestone un-

conformably (Dow and Dekker, 1964; Bain et al., 1975; Table III). However, as the oldest known Jurassic sediments are not younger than Sinemurian, the gap in the sequence cannot be a large one (Skwarko, 1967a, 1973b). The Lower Jurassic sediments grade into the widespread Upper Jurassic shale conformably.

Cretaceous. The Cretaceous — particularly the Lower Cretaceous — rocks which are clastic with or without carbonates, and submarine and subaerial volcanics are widespread throughout the Mobile Belt.

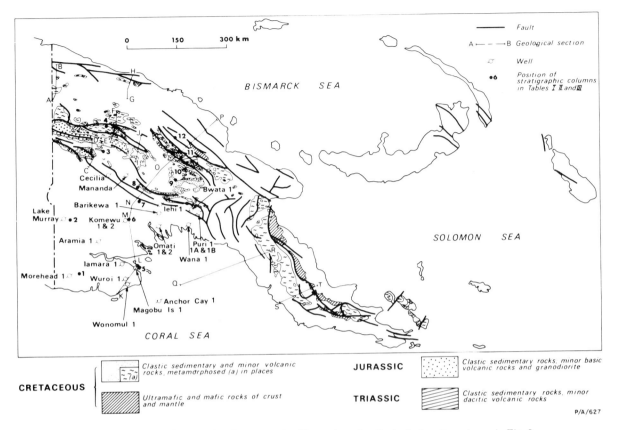

Fig. 3. Mesozoic outcrops, major faults and wells penetrating Mesozoic rocks. Geological sections shown in Fig. 5.

They rest in many places conformably on the Upper Jurassic strata (Tables I–III).

Rocks of probable Cretaceous age in the Mobile Belt

In the Mobile Belt there are metamorphic rocks, igneous rocks, and rocks beneath the Aure Trough whose age is not known, but which are thought to be mainly of Cretaceous age.

Low-grade metamorphic rocks which form a paired belt occur in the northwest Mobile Belt (Ryburn, 1975; Table I, column 4). They also include some amphibolite facies rocks and eclogite in blueschist terrain. Farther east low-grade metamorphics of the Bismarck and Schrader ranges are tentatively correlated with sediments of known Jurassic and Cretaceous age. In the southeast mainland Papua New Guinea, strongly folded metamorphic rocks of low grade greenschist facies (Pieters, 1974) make up most of the Mobile Belt, although amphibolite facies rocks occur farther southeast on some outlying islands. The metamorphic

rocks generally decrease in grade to the south and west, into strongly folded but unmetamorphosed sediments. Rare granodioritic and minor gabbroic intrusive igneous rocks probably of Mesozoic age occur in the central part of the belt.

Beneath the thick Tertiary sediments of the Aure Trough, Mesozoic rocks, if present, would have probably been affected by crustal thinning, rifting, and volcanic activity associated with the opening of the Coral Sea Basin in the Cainozoic. East of the trough, a major unconformity separates strongly folded Mesozoic and associated Lower Cainozoic rocks from the main Tertiary sequence. Immediately west of the trough, open-marine Mesozoic rocks may be about 4,000 m thick; farther west they are relatively undeformed and separated from the succeeding Tertiary sequence by a stratigraphic break and a low angle unconformity. The Mesozoic section seems to thin towards the Pasca Ridge and the eroded uplifted Platform margin.

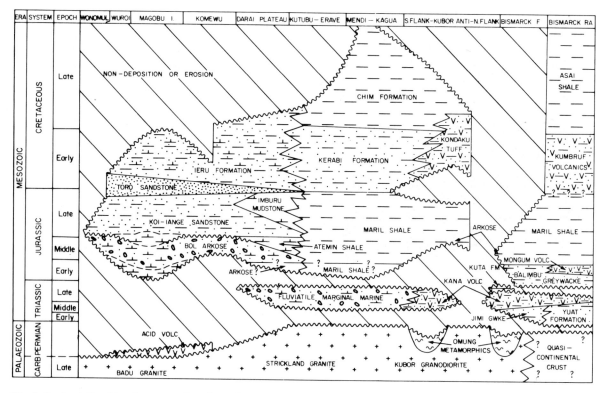

Fig. 4. Rock relationships along part of section K-P (Fig. 3).

Oceanic Crust and Island Arcs

Overlying the Mobile Belt is a series of overthrust slices of Mesozoic oceanic crust and mantle consisting of ultramafics, gabbro, basalt, chert and pelagic limestone, in places strongly tectonised with lawsonite-glaucophaneschist and glaucophane schist, gneiss and associated eclogite.

Basic to acidic Upper Cretaceous to Miocene subvolcanic plutons are present in the central axis of the Bewani and Torricelli Mountains and low-grade greenschist facies metamorphics occur on the southern flanks of these ranges (Hutchinson and Norvick, 1978).

GEOLOGICAL HISTORY

The geological history of Papua New Guinea during the Mesozoic is interpreted as a history of sedimentation on and around a small portion of the north-

moving Indo-Australian plate margin, and the subsequent deformation of the accumulated sediments following collision during the Cainozoic with the island arcs of the westerly moving Pacific plate. The tectonic movements throughout most of the Mesozoic were limited to block-faulting and minor fluctuations of sea level over the Platform, but these were replaced in the Cretaceous by major displacements associated with the formation of a geosyncline.

The Mesozoic sedimentary wedge which overlies the fringes of the continental basement (Fig. 5) is thought to have been deposited in three main tectonic environments: (a) over the major part of the block-faulted Platform in the south and west as a relatively thin sequence of continental, fluvio-deltaic, and marginal marine sediments; (b) at or near the margin of the Platform in rapidly subsiding miogeosynclinal basins and troughs formed by collapse of part of the basement through block-faulting, as thick open-marine fine clastics, while thinner clastics, volcanics, and minor limestone accumulated over and around the basement highs; and (c) over a subsiding

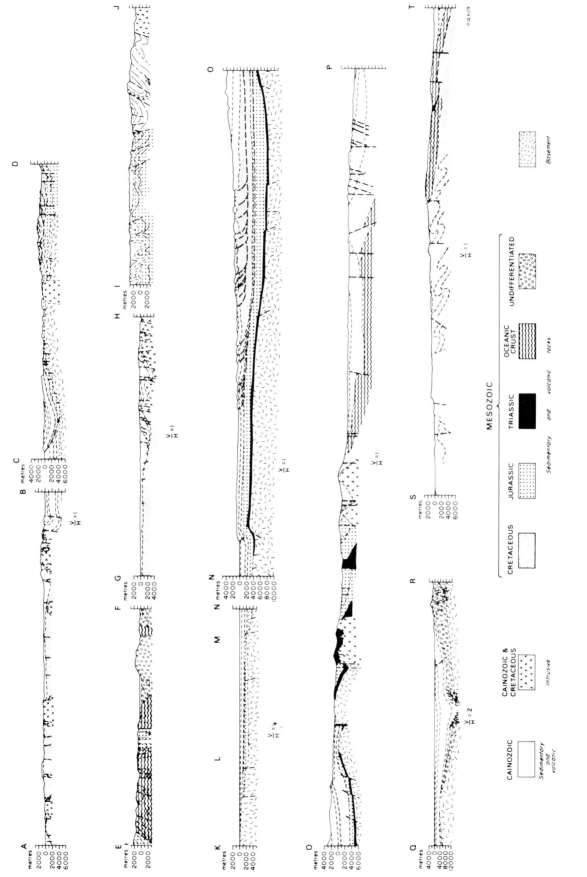

Fig. 5. Cross-sections.

383

TABLE III

Lithology and correlations of the main Mesozoic sequences, Papuan Platform (for location of numbered sections see Fig. 3; legend see Table I)

				STH FLANK KUBOR ANTI 9	NTH FLANK KUBOR ANTI 10	CHIMBU—WAHGI DIVIDE 11	BISMARCK RANGE 12
			PALEOCENE TO EOCENE	MUDSTONE, SST			
MESOZOIC	CRETACEOUS	U	MAASTRICHTIAN				
			CAMPANIAN				
			SANTONIAN	CHIM FM 1500		CHIM FORMATION 1000	ASAI SHALE 4000+
			CONIACIAN				
			TURONIAN		CHIM FM 300		
			CENOMANIAN				
		L	ALBIAN	KONDAKU TUFF 2000	KONDAKU TUFF 1500	KONDAKU TUFF 2000	KUMBRUF VOLCS 1800
			APTIAN				
			BARREMIAN				
			NEOCOMIAN				
	JURASSIC	U	TITHONIAN				
			KIMMERIDGIAN			MARIL SHALE 1500	
			OXFORDIAN				
		M	CALLOVIAN	MARIL SHALE 1500	MARIL SHALE 1200		MARIL SHALE 1000
			BATHONIAN				
			BAJOCIAN				
		L	AALENIAN				MONGUM VOLCS 250
			TOARCIAN			BALIMBU GWKE 2000	BALIMBU GWKE 1500
			PLIENSBACHIAN	?			
			SINEMURIAN				
			HETTANGIAN				
	TRIASSIC	U	RHAETIAN		KUTA FM 250		
			NORIAN	KANA VOLCS 700	KANA VOLCS 700	KANA VOLCANICS 3500	KANA VOLCS 2500
			CARNIAN				
		M	LADINIAN			JIMI GREYWACKE 800+	YUAT FM ?100
			ANISIAN			BASE NOT SEEN	BASE NOT SEEN
		L	SCYTHIAN				
PALAEOZOIC	PERM CARBONIF			KUBOR GRANO OMUNG META	KUBOR GRANO OMUNG META	?	?

1500 *Thickness in metres*

outer continental margin flanking the continental basement, as thicker eugeosynclinal greywacke, sandstone, and shale.

Triassic (Fig. 6a)

In the Lower to Middle Triassic the stable Platform was probably a block-faulted landmass made up of Palaeozoic igneous and metamorphic rocks. Little sedimentation took place, and only minor arkosic fluviatile and marginal marine sediments were deposited over the downfaulted margins. The Kubor basement high was a landmass flanked by arkosic sandstone and reefal limestone. In the north, marine shales passing into greywacke were deposited over the outer continental margin.

In the Upper Triassic continuing erosion removed most of the cover rocks from the main body of the Platform, exposing the igneous basement. A similar sedimentary pattern persisted, except that volcanism occurred in the Kubor area and marine volcanic sediments were deposited over the adjacent downfaulted margins in the south and over the outer continental margin in the north.

Jurassic (Fig. 6, b–g)

By the Lower Jurassic (Fig. 6b) the topography of the Platform was relatively subdued, although remnant fault-bounded topographic highs persisted at Oriomo and Lake Murray. Fluviatile sediments were deposited in intervening topographic lows such as Morehead Basin. Shallow sea probably transgressed in places over the Platform margins. North of the Kubor area feldspathic sandstone and shale accumulated over the continental margin.

During the Middle to Upper Jurassic (Fig. 6c) fluvio-deltaic and marginal marine sediments gradually encroached over the foreland. In the Bajocian, coalesced alluvial fans accumulated around the flanks of the remaining basement highs, and were locally reworked to form braided stream deposits which graded laterally into deltaic-marginal marine deposits with coal beds. Fine clastics continued to accumulate in the fault-bounded collapsed basins of the miogeosynclinal Platform margins and over the subsiding continental shelf. By the Oxfordian (Fig. 6d), these marine clastics overlapped the deltaic sediments of the stable Platform margins. Marginal marine clastics with minor coal beds were being deposited over much of the peneplained Platform. Basement highs such as at Oriomo and Lake Murray probably remained emergent. A minor marine regression of the marginal marine facies occurred in the Upper Oxfordian to lower Kimmeridgian (Fig. 6e), when sandstone was deposited over the margins of the foreland. However, in the upper Kimmeridgian to lower Tithonian (Fig. 6f), the general transgressive trend was resumed, and marine shale was again deposited over the foreland margins. Farther north, deposition of thick marine shale over the subsiding outer continental shelf margins continued uninterrupted until the Upper Jurassic, when fine quartz sandstone was intercalated with the shale.

Cretaceous (Fig. 6, g–i)

The Lower Cretaceous was a period of regional marine regression. In the upper Tithonian to Neocomian (Fig. 6g), reworked sandstone was deposited over the Platform as a regressive sheet of coalesced barrier bar deposits. Uplift and shallowing also oc-

curred over the downfaulted Platform margins. A volcanic arc formed near the present day Schrader and Bismarck ranges; its volcanic activity ceased toward the end of the Lower Cretaceous. Tuff and volcanoclastic sediments accumulated in a downwarping trough between the emergent Kubor basement high and the volcanic arc. Farther west, eugeosynclinal fine clastics accumulated over the rapidly subsiding continental margin from the earliest Neocomian, and sedimentation continued into the Eocene.

In the upper Neocomian (Fig. 6h) paralic and marginal marine fine clastics again were deposited in a sea which transgressed over the foreland. They continued to accumulate until the lower Cenomanian, when a fall in relative sea-level resulted in a further marine regression.

During the Upper Cretaceous (Fig. 6i), stress began to build up between the northerly moving Indo-Australian and the westerly moving Pacific plates, and a Pacific-type convergent margin began to develop. Compression resulting from the plate movements was accompanied by the formation of an arcuate crustal downwarp adjacent to the northern and eastern margins of the Australian continental basement. The foreland was epeirogenically uplifted and the Mesozoic cover partly eroded. Uplift and warping appear to have been greatest over the northeast apex of the Platform, where faulting and subsequent erosion exposed Upper Jurassic rocks (Australasian Petroleum Co., 1961). Farther south and west over the Platform, the erosion surface cut less deeply into the Mesozoic section, and Upper Cretaceous rocks are generally preserved. The miogeosynclinal Platform margins were also elevated, and the fault-bounded basins and troughs became progressively shallower. To the north and east, sediments eroded from the Platform were deposited in the deepening eugeosynclinal trough as thick turbidites. Rapid sedimentation was accompanied by submarine volcanism and possibly by minor contemporaneous folding and burial metamorphism.

In the Eocene, following the separation of the Australian continent from Antarctica, accelerated northward movement of the Indo-Australian plate increased the inter-plate stress. This stress was relieved by the formation of the Coral Sea, by the northeastward rotation of eastern Papua New Guinea, and by fracturing of the oceanic lithosphere to produce northerly and easterly dipping subduction zones over which the volcanic island arc system developed. The Mesozoic oceanic crust and mantle of the ophiolite belt were also tectonically emplaced in the mid-Tertiary. Deformation of Mesozoic rocks and of the Mobile Belt took place in the Mid-Cainozoic when the Bewani–Torricelli and Adelbert–Finisterre island arc system collided with the deformed eugeosynclinal sediments and became accreted to the Australian continent (Jaques and Robinson, 1977). Isostatic uplift

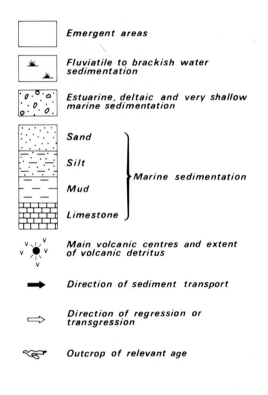

Emergent areas

Fluviatile to brackish water sedimentation

Estuarine, deltaic and very shallow marine sedimentation

Sand
Silt
Mud
Limestone
⎱ Marine sedimentation

Main volcanic centres and extent of volcanic detritus

Direction of sediment transport

Direction of regression or transgression

Outcrop of relevant age

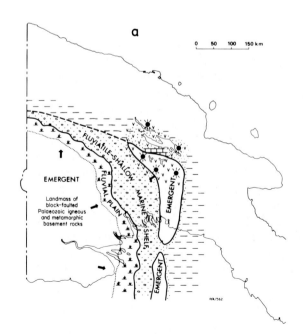

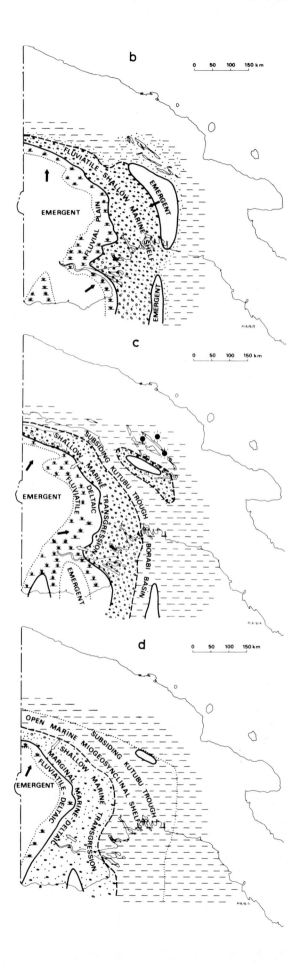

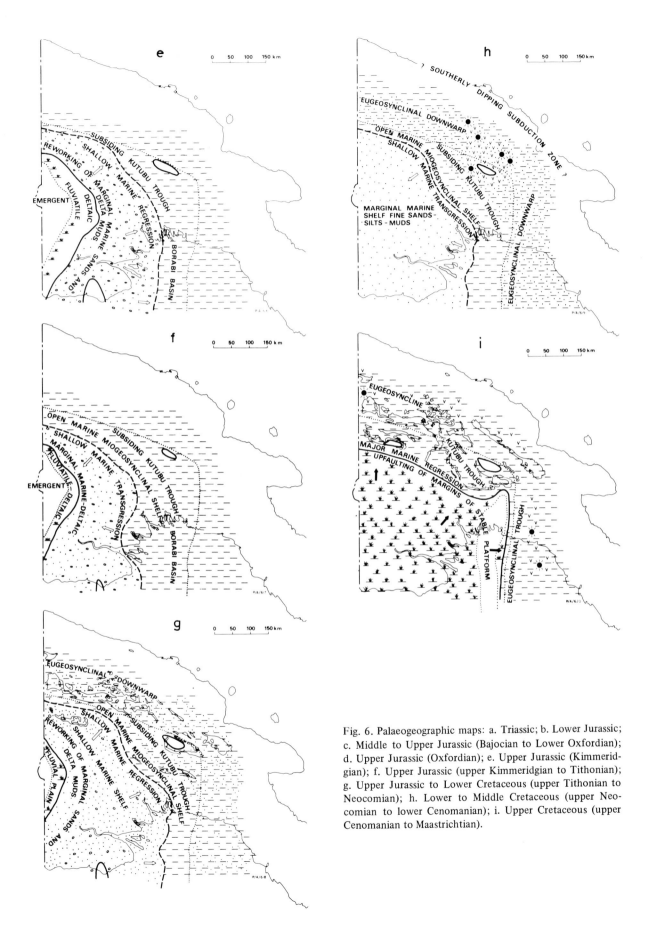

Fig. 6. Palaeogeographic maps: a. Triassic; b. Lower Jurassic; c. Middle to Upper Jurassic (Bajocian to Lower Oxfordian); d. Upper Jurassic (Oxfordian); e. Upper Jurassic (Kimmeridgian); f. Upper Jurassic (upper Kimmeridgian to Tithonian); g. Upper Jurassic to Lower Cretaceous (upper Tithonian to Neocomian); h. Lower to Middle Cretaceous (upper Neocomian to lower Cenomanian); i. Upper Cretaceous (upper Cenomanian to Maastrichtian).

of the Mesozoic rocks of the central ranges rapidly followed in the early Miocene, and was accompanied by igneous activity. In the Pliocene, Miocene limestone and various Cretaceous sediments slid and were thrust southwest over the Platform margin. This gravity sliding continues to the present day.

REFERENCES

Anonymous, 1962. Generalised formation correlation of Australia and Papua. *Oil Gas J.*, 60: 72–75.

Australasian Petroleum Co., 1961. Geological results of petroleum exploration in Western Papua, 1937–1961 by the Australasian Petroleum Co. Pty Ltd. *J. Geol. Soc. Aust.*, 8 (1): 1–133.

Bain, J.H.C., 1973. A summary of the main structural elements of Papua New Guinea. In: P.J. Coleman (Editor), *The Western Pacific Island Arcs, Marginal Seas, Geochemistry. Perth Univ. Press, W.A.*, pp. 14–61.

Bain, J.H.C., Mackenzie, D.E. and Ryburn, R.J., 1975. Geology of the Kubor Anticline, Central Highlands of Papua New Guinea. *Bur. Miner. Resour. Aust. Bull.*, 155.

Bowen, R., 1961. Palaeotemperature analyses of Mesozoic Belemnoidea from Australia and New Guinea. *Bull. Geol. Soc. Am.*, 72: 769–773.

Brown, C.M., Pieters, P.E. and Robinson, G.P., 1975. Stratigraphic and structural development of the Aure Trough and adjacent shelf and slope areas. *APEA J.*, 15 (1): 61–71.

Brown, C.M., Pigram, C.J. and Skwarko, S.K., 1980. Mesozoic stratigraphy and geological history of Papua New Guinea. *Palaeogeogr., Palaeoclimatol. Palaeoecol.*, 29, 301–322.

Buchan, S.H. and Robinson, G.P., 1969. Report on the Kagua-Ere-Lavani detailed geological survey, Permit 46, Papua. *B.P. Petrol. Dev. Aust. Rep.*, 65 (unpubl.).

Bureau of Mineral Resources, 1972. *Geology of Papua New Guinea, 1:1 000 000 geological map.*

Bureau of Mineral Resources, 1976. *Geology of Papua New Guinea, 1:2 500 000 geological map.*

Compagnie Générale de Géophysique, 1969. Papua Basin and basic belt aeromagnetic survey T.P.N.G., 1967. *Bur. Miner. Resour. Aust. Rec.*, 1969/58 (unpubl.).

Conybeare, C.E.B. and Jessop, R.G.C., 1972. Exploration for oil bearing sand trends in the Fly River area, Western Papua. *APEA J.*, 12 (1): 69–73.

Cookson, I.C. and Dettmann, M.E., 1958a. Some trilete spores from Upper Mesozoic deposits in the eastern Australian region. *Proc. R. Soc. Vic.*, 70: 95–128.

Cookson, I.C. and Dettmann, M.E., 1958b. Cretaceous 'megaspores' and a closely associated microspore from the Australian region. *Micropaleontology*, 4: 39–49.

Cookson, I.C. and Eisenack, A., 1958. Microplankton from Australian and New Guinea Upper Mesozoic sediments. *Proc. R. Soc. Vic.*, 70: 19–70.

Crespin, I., 1958. Microfossils in Australian and New Guinea stratigraphy. *J. Proc. R. Soc. N.S.W.*, 92: 113–147.

Davies, H.L., 1971. Peridotite–gabbro–basalt complex in eastern Papua: an overthrust plate of oceanic mantle and crust. *Bur. Miner. Resour. Aust. Bull.*, 128.

Davies, H.L. and Ives, D.J., 1965. The geology of Fergusson and Goodenough Islands, Papua. *Bur. Miner. Resour. Aust. Rep.*, 82.

Davies, H.L. and Norvick, M., 1974. Blucher Range, PNG - 1:250 000 Geological Series. *Bur. Miner. Resour. Aust. Explan. Notes*, SB/54-7.

Davies, H.L. and Smith, I.E., 1971. Geology of eastern Papua. *Bull. Geol. Soc. Am.*, 82: 3299–3312.

Dekker, F.E. and Faulks, I.G., 1964. The geology of the Wabag area, New Guinea. *Bur. Miner. Resour. Aust. Rec.*, 1964/137 (unpubl.).

Dorman, F.M., 1968. Some Australian oxygen isotope temperatures and a 30 million year world temperature cycle. *J. Geol.*, 76, 297–313.

Dow, D.B., 1973. Geology of Papua New Guinea. *Bur. Miner. Resour. Aust. Rec.*, 1973/117 (unpubl.).

Dow, D.B., 1977. A geological synthesis of Papua New Guinea. *Bur. Miner. Resour. Aust. Bull.*, 201.

Dow, D.B. and Davies, H.L., 1964. The geology of the Bowutu Mountains, New Guinea. *Bur. Miner. Resour. Aust. Rep.*, 75.

Dow, D.B. and Dekker, F.E., 1964. Geology of the Bismarck Mountains, New Guinea. *Bur. Miner. Resour. Aust. Rep.*, 76.

Dow, D.B., Smit, J.A.J., Bain, J.H.C. and Ryburn, R.J., 1972. Geology of the South Sepik region, New Guinea. *Bur. Miner. Resour. Aust. Bull.*, 133.

Erni, A., 1945. Ein Cenomanian-Ammonit, *Cummingtoniceras holtkeri* nov. spec. aus Neuguinea, nebst Bemerkungen über einige andere Fossilien von dieser Inseln (mit einem Beitrag von Georg. Holtker). *Eclogae Geol. Helv.*, 37: 486–475.

Etheridge, R. Jr., 1889. Our present knowledge of the palaeontology of New Guinea. *Rec. Surv. N.S.W.*, 1: 172–179.

Findlay, A.L., 1974. The structure of the foothills south of the Kubor Range, Papua New Guinea. *APEA J.*, 14(1): 14–20.

Glaessner, M.F., 1943. Problems of stratigraphic correlation in the Indo-Pacific region. *Proc. R. Soc. Vic.*, 55: 41–80.

Glaessner, M.F., 1945. Mesozoic fossils from the Central Highlands of New Guinea. *Proc. R. Soc. Vic.*, 56: 151–168.

Glaessner, M.F., 1949. Mesozoic fossils from the Snake River, Central New Guinea. *Qld. Mus. Mem.*, 12 (4): 165–180.

Glaessner, M.F., 1957. Cretaceous belemnites from Australia, New Zealand and New Guinea. *Aust. J. Sci.*, 20: 88–89.

Glaessner, M.F., 1958. New Cretaceous fossils from New Guinea. *S. Aust. Mus. Rec.*, 13 (2): 199–226.

Glaessner, M.F., 1960. Upper Cretaceous larger Foraminifera from New Guinea. *Sci. Rep. Tohoko Univ. Hanzawa Memorial Vol.*, 37–44.

Harrison, J., 1969. A review of the sedimentary history of the island of New Guinea. *APEA J.* 9: 41–48.

Hocking, J.B., Jessop, R.G.C. and Cains, L.J., 1971. Magobu Island No. 1 well completion report. *Bur. Miner. Resour. Aust. – Petroleum Search Subsidy Act Rep.* (unpubl.).

Hutchinson, D.J. and Norvick, M., 1978. Wewak, Papua New Guinea – 1:250 000 Geological Series. *Bur. Miner. Resour. Aust. Explan. Notes*, SA/54–16.

Jaques, A.L., 1976. High potash island arc volcanics from the Finisterre and Adelbert Ranges: Papua New Guinea. *Bull. Geol. Soc. Am.*, 87: 861–867.

Jaques, A.L., and Robinson, G.P., 1977. The continent/island arc collision in northern Papua New Guinea. *BMR J. Aust. Geol. Geophys.*, 2: 289–303.

Jenkins, D.A.L., 1974. Detachment tectonics in western Papua New Guinea. *Bull. Geol. Soc. Am.*, 85: 533–548.

Jenkins, D.A.L. and Martin, A.J., 1969. Recent investigation into the geology of the southern highlands, Papua. In: *ECAFE – 4th Symposium on Development of Petroleum Resources in the Far East, Canberra, 1969*. United Nations, New York.

Jenkins, D.A.L., Findlay, A.L. and Robinson, G.P., 1969. Mendi geological survey, Permits 46, 27, Papua. *B.P. Petrol. Dev. Aust. Rep.*, 84 (unpubl.).

McMillan, N.J. and Malone, E.J., 1960. The geology of the eastern central highlands of New Guinea. *Bur. Miner. Resour. Aust. Rep.*, 48.

Milsom, J., 1974. East New Guinea. In: A.M. Spencer (Editor), *Mesozoic–Cenozoic Orogenic Belts, 1974*. Geol. Soc. Lond. Spec. Publ., 4.

Mutter, J.C., 1972a. Marine geophysical survey of the Bismarck Sea and Gulf of Papua, 1970. A structural analysis of the Gulf of Papua and northwest Coral Sea region. *Bur. Miner. Resour. Aust. Rec.*, 1972/134 (unpubl.).

Mutter, J.C., 1972b. A recent geophysical reconnaissance of the Gulf of Papua and northwest Coral Sea. *Bur. Miner. Resour. Aust. Rec.*, 1972/122 (unpubl.).

Mutter, J.C., 1975. A structural analysis of the Gulf of Papua and northwest Coral Sea region. *Bur. Miner. Resour. Aust. Rep.*, 179.

Norvick, M. and Hutchinson, D.J., 1980. Aitape–Vanimo, Papua New Guinea – 1:250 000 Geological Series. *Geol. Surv. P.N.G. Explan. Notes*, SA/54–15; SA/54–11.

Owen, M., 1973. Upper Cretaceous planktonic Foraminifera from Papua New Guinea. *Bur. Miner. Resour. Aust. Bull.*, 140: 47–65.

Pieters, P.E., 1974. Explanatory Notes on the Port Moresby–Kalo–Aroa geological map. *Geol. Surv. PNG Rep.* 74/28 (unpubl.). (In prep. as Port Moresby–Kalo–Aroa, Papua New Guinea – 1:250 000 geological series. *Bur. Miner. Resour. Aust. Explan. Notes*, SC/55-7, 11, 6.)

Pigram, C.J., 1978. Geology of the Schrader Range Papua New Guinea. *Geol. Surv. PNG Rep.*, 76/4.

Rickwood, F.K., 1955. The geology of the western highlands of New Guinea. *J. Geol. Soc. Aust.*, 2: 63–82.

Rickwood, F.K., 1968. The geology of western Papua. *APEA J.*, 8 (2): 51–61.

Robinson, G.P., 1973. Stratigraphy and structure of Huon Peninsula, New Guinea, within the framework of the Outer Melanesian Arc. In: R. Fraser (Editor), *Oceanography of the South Pacific, 1973*. National Commission for UNESCO, Wellington, New Zealand.

Robinson, G.P., Jaques, A.L. and Brown, C.M., 1974. Explanatory notes on Madang Geological Map. *Geol. Surv. PNG Rep.*, 74/13 (unpubl.).

Ryburn, R.J., 1975, South Sepik Blueschist Project, PNG. In: *Geological Branch Summary of Activities 1975*. *Bur. Miner. Resour. Aust. Rec.*, 1975/157.

Schluter, H., 1928. Jura fossilien vom oberen Sepik und New Guinea. *Nova Guinea*, 6 (3): 53–61.

Skwarko, S.K., 1967a. First Triassic and (?) Lower Jurassic marine Mollusca from New Guinea. In: S.K. Skwarko *Mesozoic Mollusca from Australia and New Guinea*. *Bur. Miner. Resour. Aust. Bull.*, 75: 37–84.

Skwarko, S.K., 1967b. Lower Cretaceous Mollusca from the Sampa Beds near Wau, New Guinea. In: S.K. Skwarko, *Mesozoic Mollusca from Australia and New Guinea*. *Bur. Miner. Resour. Aust. Bull.*, 75: 85–101.

Skwarko, S.K., 1973a. Middle and Upper Triassic Mollusca from Yuat River, eastern New Guinea. *Bur. Miner. Resour. Aust. Bull.*, 126: 27–41.

Skwarko, S.K., 1973b. First report of Domerian (Lower Jurassic) marine Mollusca from New Guinea. *Bur. Miner. Resour. Aust. Bull.*, 140: 105–112.

Skwarko, S.K., 1978. Stratigraphic tables, Papua New Guinea. *Bur. Miner. Resour. Aust. Rep.*, 193; BMR Microform MF61.

Skwarko, S.K. and Kummel. B., 1974. Marine Triassic Mollusca of Australia and Papua New Guinea. *Bur. Miner. Resour. Aust. Bull.*, 150: 111–128.

Skwarko, S.K., Nicoll, R.S. and Campbell, K.S.W., 1976. The Late Triassic Mollusca, conodonts and brachiopods of the Kuta Formation and its Triassic palaeogeography, Papua New Guinea. *BMR J. Aust. Geol. Geophys.*, 1 (3): 219–230.

St John, V.P., 1967. *The Gravity Field of New Guinea. Ph.D. Thesis, Univ. Tasmania*, unpublished.

St John, V.P., 1970. The gravity field and structure of Papua and New Guinea. *APEA J.*, 10 (2): 41–55.

Tallis, N.C., 1975. Development of Tertiary offshore Papuan Basin. *APEA J.*, 15: 55–60.

Thompson, J.E., 1967. A geological history of eastern New Guinea. *Bur. Miner. Resour. Aust. Rec.*, 1967/22 (unpubl.).

Thompson, J.E., 1972. Continental drift and the geological history of Papua New Guinea. *APEA J.*, 12: 64–69.

Thompson, J.E. and Fisher, N.R., 1965. Mineral deposits of New Guinea and Papua and their tectonic setting. *Proc. 8th Commonwealth Min. Metall. Congr.*, 6: 115–148.

Wilcox, J.B., 1973. Preview report for the marine geophysical survey of the Gulf of Papua and the Bismarck Sea. *Bur. Miner. Resour. Aust. Rec.*, 1973/38 (unpubl.).

Chapter 13

ANTARCTICA

M.R.A. THOMSON

INTRODUCTION

Antarctica (Fig. 1) is readily divisible on both geographical and geological criteria into two major regions, variously termed Greater, East or eastern Antarctica, and Lesser, West or western Antarctica. Greater Antarctica is a large oval-shaped shield area, with a coast facing the Indian Ocean sector of the Southern Ocean and an 'inland' boundary marked by the Transantarctic Mountains, the eastern margin of the Weddell Sea, and the western margin of the Ross Sea. Lesser Antarctica includes the Antarctic Peninsula, and an archipelago of continental fragments in Ellsworth and Marie Byrd lands, together with the Ellsworth Mountains, and isolated nunataks all joined together by an ice sheet. Since late Palaeozoic times Greater Antarctica has behaved as a stable craton and virtually all the corresponding sedimentary rocks exposed on its surface are continental. By contrast, Lesser Antarctica is largely composed of the rocks formed at an active margin on the Pacific side of Antarctica. Although older rocks are present, the majority are the products of late Palaeozoic–Triassic marine deposition, and of a later volcanic arc and the rocks intruded into it. The Mesozoic history of Lesser Antarctica is best understood in the Antarctic Peninsula area where fossiliferous marine rocks occur in association with the volcanic ones. Small areas of volcanic rocks in the Thurston Island area and Marie Byrd Land are claimed to be Mesozoic (Craddock, 1972) on the basis of lithological similarities to volcanic rocks on the peninsula. However, they are poorly known and some may even be Devonian in age (Grindley and Mildenhall, 1981). The possible continuation of the arc into these regions is perhaps better marked by the presence of Mesozoic intrusions.

The geological relationships between Greater and Lesser Antarctica are far from understood. However, there is a growing body of opinion that the two regions have not always held the same geographical inter-relationships that they do today (e.g., see Gondwana reconstructions by Elliot, 1972a, Barker and Griffiths, 1977, and De Wit, 1977). On geological grounds the Ellsworth Mountains (Schopf, 1969) and nearby Haag Nunataks (Clarkson and Brook, 1977) of Lesser Antarctica appear to be displaced fragments with a stratigraphy more typical of Gondwana than an active margin. These are tenuously linked to Greater Antarctica by a chain of nunataks in the intervening area, and in which Mesozoic intrusions with radiometric ages similar to those of the Jurassic Ferrar tholeiites of the Transantarctic Mountains are prominent (Craddock, 1970a; Kovach and Faure, 1978).

In the present summary, the Mesozoic histories of Greater and Lesser Antarctica are outlined separately.

Geological literature on Antarctica is voluminous and the Mesozoic has already been summarized in a number of papers which are more broad-based than the present one. Among the most important of these are Elliot (1975a, b) for Greater Antarctica, and Adie (1964), Dalziel and Elliot (1973) and Grikurov (1973) for Lesser Antarctica. Many useful references, excluded from this study on grounds of space alone, will be found in these works. Reference should also be made to the geological maps edited by Bushnell and Craddock (1969–70), Craddock (1972), Ravich and Grikurov (1978), Fleming and Thomson (1979) and Thomson and Harris (1981).

GREATER ANTARCTICA

The known sedimentary history of the Mesozoic in Greater Antarctica is entirely terrestrial and this is

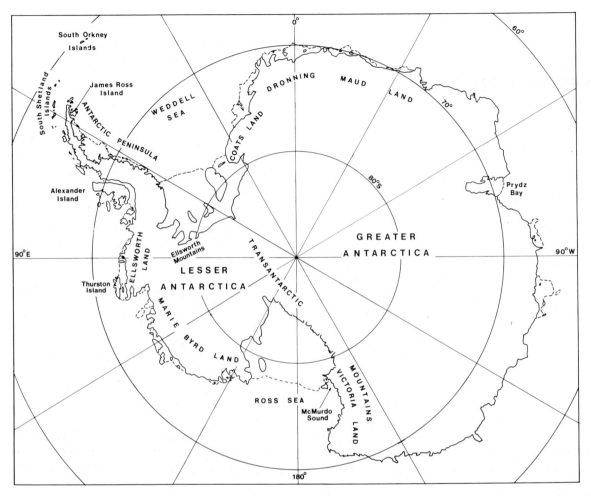

Fig. 1. Sketch map of Antarctica showing the principal areas described in the text. South Georgia, on the northern side of the Scotia Ridge or arc is omitted and is shown in Fig. 7.

consistent with all interpretations of Gondwana, in which Antarctica formed the nucleus of the supercontinent. An extensive Triassic record is preserved as fluviatile sandstone and volcanic sequences in the Transantarctic Mountains, but nowhere else are equivalent strata known to be exposed in Greater Antarctica. Indeed the level of erosion is generally deep and virtually all rock exposures around the Southern Ocean margin are of Precambrian basement. Terrestrial sedimentary sequences of Triassic age are present in the surrounding Gondwana fragments of South Africa, India and Australia, and it is unlikely that the present exposures of Triassic rocks in the Transantarctic Mountains reflect their true original Antarctic distribution. Much may have been eroded away, and some may be preserved in downfaulted

areas beneath the ice cap. Jurassic rocks are more widespread but they are still limited to the Pacific side of the Greater Antarctic shield. They are mostly tholeiitic sills and lavas but thin lacustrine and possibly fluvial intercalations occur at several localities. Cretaceous rocks are unknown in situ but rare erratics, derived fossils and a submarine occurrence suggest that deposits of this age may once have been present.

TRIASSIC

In Greater Antarctica, Triassic strata are limited to the Transantarctic Mountains between northern Victoria Land and the Nilsen Plateau (Fig. 2). The rocks

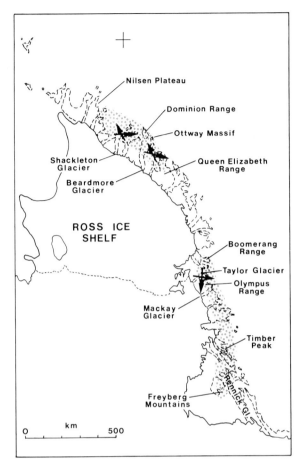

Fig. 2. Sketch map of the Transantarctic Mountains to show the distribution of Triassic strata (open circles). Double-headed arrows indicate Early Triassic sediment dispersal directions and single-headed arrows indicate Late Permian directions. The map is based on Elliot (1975a, fig. 37.6A) and data sources mentioned in the present text.

are nearly all sedimentary and are generally flat-lying. They consist of a variety of sandstones and shales with carbonaceous beds, local developments of coals and minor occurrences of conglomerates. In the central Transantarctic Mountains they pass up into tuffaceous and volcaniclastic strata. All these rocks form the upper part of the Permian–Triassic Victoria Group, itself part of the widespread Beacon Supergroup (? pre-Devonian–Triassic).

From the Nilsen Plateau to western Dronning Maud Land no Triassic strata are known, even though Permian equivalents of the Beacon Supergroup occur in the exposed successions. In the Pensacola Mountains, however, the early Mesozoic is marked by an orogenic phase (Weddell Orogeny of Ford, 1972).

Glossopteris-bearing Permian sandstones in the Ellsworth Mountains of Lesser Antarctica are also deformed and no Triassic strata are present there. Possibly this absence may be attributed to the effects of the same orogeny.

In northern Victoria Land a thin sequence of poorly differentiated Permian and Triassic beds rests directly on an eroded basement of Precambrian to Ordovician metamorphic and intrusive rocks, in southern Victoria Land the Triassic beds are separated from the underlying Permian strata by a conglomeratic sequence (Feather Conglomerate) of Permo–Triassic age, and in the southern part of the area the Permian and Triassic rocks are separated by a disconformity (Elliot, 1975a). Although there are close lithological similarities between the Permian and Triassic successions, important changes in sediment dispersal of the two have been demonstrated in the central Transantarctic Mountains (Fig. 2; Barrett, 1970; Barrett and Kohn, 1975; Elliot, 1975a). However, new data (Collinson et al., 1978) suggest that the reversal observed in the Fremouw Formation (basal Triassic) is not everywhere as marked.

Numerous studies of the Beacon rocks have led to a complicated stratigraphical terminology but much has been done in recent years to simplify and standardize this (cf. Barrett, 1969; Elliot, 1975a), and to effect a correlation between locally established suc-

TABLE I

Stratigraphical correlation of the Upper Victoria Group between the Beardmore Glacier area and south Victoria Land

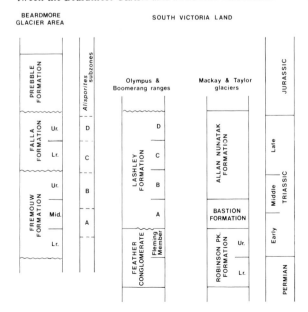

cessions (Table I). Significantly different thickness estimates are often given by various workers in the same areas. These may be due to real variations between individual outcrops, but nevertheless figures quoted in the present summary should be taken as representative rather than definitive.

Evidence for plutonic igneous activity in Greater Antarctica during the Triassic Period is limited to the report of an early Mesozoic syenite from the Gburek Mountains in western Dronning Maud Land (Soloviev, 1972). Situated in a zone of deep regional faulting, the syenite and its related vein and dyke rocks have given K-Ar and Rb-Sr radiometric ages in the range 165–270 Ma. However, the true age of the syenite is thought to be about 200 Ma (Soloviev, 1972). Among the volcanic rocks Furnes and Mitchell (1978) have shown that some basaltic lavas of Jurassic Ferrar Supergroup aspect in the same region are in fact Triassic, and in the central Transantarctic Mountains laharic deposits of largely acidic character occur at the top of the Triassic succession (Barrett and Elliott, 1972).

Northern Victoria Land

Only the upper parts of the Beacon Group are present in this region. Exposed successions are generally thin (< 100 m), although a 270 m thick succession is present in the Freyberg Mountains at Takrouma Bluff (Sturm and Carryer, 1970). A *Glossopteris* flora there indicates the presence of Permian beds but the sequence is believed to be wholly Triassic over much of this region. Permian–Triassic Beacon rocks rest directly on a major erosion surface (Kukri Peneplain) cut in the Precambrian–Ordovician basement, and which further south would be directly overlain by older (? pre-Devonian) parts of the sequence. In the upper Rennick Glacier area the erosion surface is exposed for nearly 160 km and displays very little surface relief (Gair, 1967).

In northern Victoria Land the formations seem to be unnamed. Everywhere they are described as consisting of quartzitic and arkosic sandstones with conglomerate lenses, and minor siltstones and shales. Conglomerate clasts are mainly quartz with minor metamorphic fragments; mud-flake types are common in some areas. Ripple marks and small-scale cross-bedding have been widely reported, and Dow and Neall (1973) have recognized fining upward cycles in the Freyberg Mountains. Carbonaceous

debris is widespread, although little recognizable macro-plant material has been reported. Tree trunks occur at Timber Peak (Skinner and Ricker, 1968).

In this sequence must be included similar sediments 'interbedded' and rafted within the Ferrar dolerite sills, such as were described by Skinner and Ricker (1968) in two sections of 126 and 130 m thickness at Timber Peak and Pudding Bute, respectively.

Southern Victoria Land

Stratigraphical terminology for the Victoria Group in this region is complicated for reasons previously stated, but the most widely used scheme is that initiated by McKelvey et al. (1970) for the Olympus and Boomerang ranges.

The passage from Permian to Triassic is marked by the Feather Conglomerate, which reaches a maximum thickness of 318 m (Barrett et al., 1971) and rests disconformably on the Permian Weller Coal Measures. The formation changes northwards from a quartzose sandstone with rounded quartz pebbles, to a feldspathic sandstone with scattered quartz pebbles. McKelvey et al. (1970) separated off from the Feather Conglomerate an upper division of cross-bedded orthoquartzite and interbedded siltstones as a separate formation (Fleming Formation). However, because the siltstone beds become rare or absent southwards the formation becomes difficult to identify there and Barrett et al. (1971) preferred to accord it member status within the Feather Formation. It has not yet yielded any fossils but Barrett and Kohn (1975) considered that a 90° change in palaeocurrent direction at the base of the Fleming Member corresponds to the complete reversal identified at the Permo–Triassic boundary in the Beardmore Glacier area (Barrett, 1970). The Fleming Member also contains fining-upward cycles reminiscent of the Lower Fremouw Formation (below), and its base is therefore taken as the beginning of the Triassic sedimentary record in the area.

Resting with local disconformity on the Fleming Member is the Lashley Formation (318 m), which is divisible into four members (A, B, C, and D) on lithological criteria. A is a sequence of cyclic sandstones and siltstones, B consists of massive sandstones with wood remains, C is more laminated sandstone and siltstone and D is a return to massive sandstones. Thin coals and the remains of the fern *Dicroidium* occur in members B and C.

In the Mackay and Taylor glacier areas, Matz et al. (1972) erected a different stratigraphical scheme (Table I), but this can be correlated fairly closely with the succession described above. Coals are present in the Allan Nunatak Formation, which is probably equivalent to members B–D of the Lashley Formation (Elliot, 1975a).

Nilsen Plateau to Queen Elizabeth Range

The Victoria Group has been well studied in the Beardmore Glacier area and the stratigraphical scheme established there (Barrett, 1969) is applicable throughout the region. Resting disconformably on the Permian Buckley Formation is the Triassic Fremouw Formation. Its lower 75–125 m consists of cyclic quartzose sandstones and grey-green mudstones with erosional surfaces between cycles. An Early Triassic fauna of reptiles and amphibians, characterized by the reptile genus *Lystrosaurus*, has been obtained near its base in the Beardmore and Shackleton glacier areas (Elliot et al., 1970; Kitching et al., 1972). The Middle Fremouw Formation consists mainly of greenish grey mudstones, whereas the upper part is notable for the dominance of volcanic sandstones. These two parts total more than 530 m and become increasingly carbonaceous upwards; in the upper 100 m thin coals and *Dicroidium* occur.

Although the lower 270 m of the succeeding Falla Formation at the type locality (Barrett, 1969, 1972) consist of cyclic sandstones with some quartz-pebble and mudstone fragments near the base, the upper 110 m consists largely of rhyolitic tuffs in units from 10 cm to 10 m thick. *Dicroidium* is present in thin carbonaceous shales in the lower part.

Grindley's (1963) original use of the term Falla Formation corresponds to Barrett's (1969) Upper Fremouw and redefined Falla Formation. In the Dominion Range, Grindley et al. (1964) recognized a sequence of coal-bearing rocks to which the name Dominion Coal Measures was given. These total approximately 360 m and consist of grey sandstones (frequently current-bedded) and dark shales with silicified logs and seams of anthracite up to 4.6 m thick (McGregor, 1965). According to Elliot (1975a) these appear to be all or part of the Upper Fremouw and Lower Falla Formations of Barrett. Exact correlation, however, is unclear at present as the Dominion Coal Measures rest on beds attributed to some part of

the Falla Formation (Grindley et al., 1964; McGregor, 1965).

The change to a volcanic succession demonstrated in the Upper Falla Formation is continued in the Beardmore Glacier area with the Prebble Formation (Barrett, 1969; Barrett and Elliot, 1972), a mainly laharic succession with pyroclastic breccia, tuff and tuffaceous sediments. This formation reaches a maximum measured thickness of at least 460 m at Mount Pratt, west of the Ottway Massif. Because of its dominantly acidic character like that of the Upper Falla Formation, Elliot (1975a) preferred to include the Prebble Formation within the Beacon Supergroup rather than the Ferrar Supergroup. However, there may be a slight disconformity at its base and basaltic clasts in its upper parts are more akin to the Ferrar Group rocks. Barrett and Elliot (1972) concluded that it probably spanned the Triassic–Jurassic Boundary. The Mawson Formation of southern Victoria Land, has been considered as a possible correlative. However, it is normally considered as being Early to Middle Jurassic in age, and its rocks are basic in composition.

Biostratigraphy

With the exception of a single poorly preserved gastropod from the Lower Fremouw Formation (Barrett, 1969) and frequent reports of trace fossils, palaeontological knowledge of the Triassic part of the Beacon Supergroup is related to plant and land vertebrate remains.

Little varied assemblages of plant macrofossils were described from the Transantarctic Mountains by Plumstead (1962) and Townrow (1967), and were listed by Rigby and Schopf (1969). These floras are characterized by species of the corystosperm *Dicroidium*, whose appearance in the stratigraphical succession is generally taken to denote the passage from the Permian to the Trias. However, *Glossopteris*, the characteristic plant of the Permian, lingers on, and *G. indica* has been identified from beds 'high on Allan Nunatak' in association with *D. odontopteroides* and *Taeniopteris* by Rigby and Schopf (1969). They compared this assemblage to the Panchet (Lower Trias) of India, whereas Townrow (1967) assigned a *Dicroidium* flora from a (?) similar position on the same nunatak to the early Upper Triassic. The *Dicroidium* floras show overall similarities to the floras of Australia, Argentina, South Africa and India, but Townrow

(1967) suggested that their strongest affinities lay with Tasmania.

Detailed work by R.A. Kyle (1977) and Kyle and Schopf (1982) has established four palynological subzones for the greater part of the Triassic succession in Antarctica. They are dominated by non-striate bisaccate pollen of the genus *Alisporites* (which may be related to the megafossil *Dicroidium*), but include a wide range of other forms, all of which suggest a close correlation with the Triassic palynological sequences of Australia. The subzones (Table I), designated *Alisporites* A, B, C and D, are Early to Middle Triassic, Middle Triassic, Upper Triassic, and Rhaetic in age, respectively. They confirm the lithostratigraphical correlations of Barrett (1969, 1972) and Elliot (1975a) between the south Victoria Land and Beardmore Glacier areas. An assemblage from Timber Peak, northern Victoria Land (Norris, 1965) was correlated with *Alisporites* subzones C and D.

The lower part of the Fremouw Formation has a special significance in Antarctic geology for yielding the first known remains of land vertebrates on the continent. Rich collections from the Beardmore and McGregor glaciers area (Colbert, 1974, 1975; Colbert and Cosgriff, 1974; Colbert and Kitching, 1975) produced the following fauna.

Amphibia: *Cryobatrachus kitchingi, Austrobatrachus jenseni.*

Reptilia: *Lystrosaurus murrayi, Lystrosaurus curvatus, Thrinaxodon, Procolophon,* small eosuchians, thecodonts.

New collections from Cumulus Hills in the Queen Maud Mountains have added another two or three labyrynthodont species and two more reptiles to this list (Cosgriff et al., 1978).

The *Lystrosaurus* species are particularly significant because they have direct counterparts in the Early Triassic rocks of South Africa, India and southeastern China (Colbert, 1974), and would seem to offer the ultimate proof of the former existence of Gondwana (Elliot et al., 1970; Colbert, 1973). However, their presence in China, shown as separated from Gondwana by the Tethyan ocean in most reconstructions, is enigmatic. Because of this Colbert (1975) suggested that southeastern China may once have been much nearer to Gondwana and occupied a position close to northern India. The possible significance of *Lystrosaurus* as a typical Gondwana animal is further confused by its discovery in Russia (Kalandadze, 1975).

JURASSIC

The Jurassic was a period of intense igneous activity along the Pacific side of Greater Antarctica (Fig. 3) and it was marked by the extrusion of basaltic lavas (Kirkpatrick Basalt Group) and the intrusion of dolerite mainly as sills (Ferrar Dolerite Group). The two rock units share a common tholeiitic composition and are collectively referred to as the Ferrar Supergroup (Kyle et al., 1981). Their stratigraphical names are based on occurrences in the Transantarctic Mountains and, although petrologically similar rocks of comparable age occur in western Dronning Maud Land and near the head of the Weddell Sea, the names have not been directly applied in those areas. Stephenson (1966) proposed the names Faraway Dolerites and Omega Dolerites for extensive sills in the Theron Mountains and Whichaway Nunataks, respectively.

Kirkpatrick Basalt and equivalents

The Kirkpatrick Basalt and its equivalents are patchily distributed in western Dronning Maud Land and the Transantarctic Mountains (Fig. 3). The basalts have been closely studied in the Transantarctic Mountains (Elliot, 1970; 1972b) where they reach a maximum thickness of 500 m in any one section, with individual flows ranging from 1.5 to 220 m thick. Exposed sequences are punctuated by thin interbeds of plant-bearing acidic tuffs related to the underlying Prebble Formation, and by sedimentary beds which sometimes contain freshwater faunas. In this area the volcanic sequence appears to have been subaerial, with the basalt flows forming large lava ponds. However, in northern Victoria Land pillow lavas (Nathan and Schulte, 1968) suggest that some lavas were erupted into water, presumably as lakes in intermontain depressions. In the upper Rennick Glacier area interbedded sedimentary sequences up to 18 m thick have been reported (Gair, 1967). Consisting of sandstone, siltstone and mudstone with thin coaly lenses, they locally contain silicified tree trunks in growth position. Lacustrine sandstones (Carapace Sandstone) locally precede the lavas in southern Victoria Land (Ballance and Watters, 1971), and palagonitized hyaloclastite breccia is present in the sequence at Carapace Nunatak (Fig. 4).

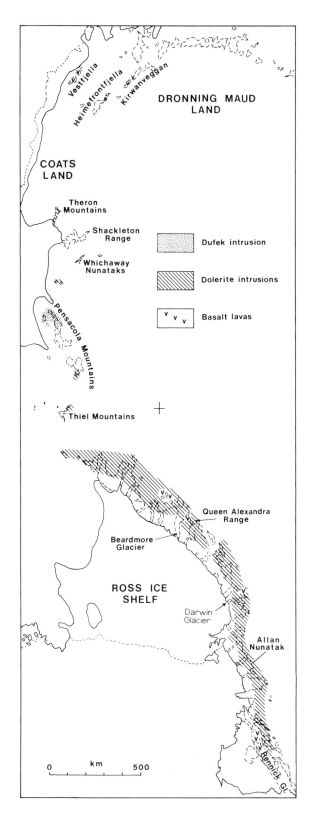

Ferrar Dolerite and equivalents

Although the Ferrar dolerites typically occur as sills, they also form dykes up to 30 m wide, bosses, laccoliths, and 'megadykes' (Gunn and Warren, 1962) 6.5–8 km wide and up to 32 km long. The sills range up to 250 m thick (Barrett, 1969) and five or six may be present in any one cliff section. They represent prodigious amounts of intrusive material, and Hamilton (1964) has calculated that the 'peneplain sill' of southern Victoria Land alone has a volume of more than 4,000 km^3. Layering, interpreted as the result of both differentiation (Gunn and Warren, 1962; Gunn, 1962; Hamilton, 1964; Skinner and Ricker, 1968; Juckes, 1972) and multiple intrusion (Grapes and Reid, 1971) has been observed. Although the dolerites normally occur as intrusions into the sedimentary rocks of the Beacon Supergroup, they also intrude the Kirkpatrick basalts (e.g. Nathan and Schulte, 1968) when the two become extremely difficult to separate in the field. Furthermore it has recently been pointed out that at several localities in southern Victoria Land and the Darwin Glacier area rocks previously mapped as Kirkpatrick Basalt are in fact intrusive dolerite (Kyle, 1979). Some sills are transgressive on a large scale, with the result, that huge slabs of sedimentary rocks are completely surrounded and 'rafted' in the dolerite (Nathan and Schulte, 1968; Skinner and Ricker, 1968).

The Dufek intrusion of the northern Pensacola Mountains deserves special mention. It is a mafic body of unusually large dimensions, at least 50,000 km^2 in areal extent (Behrendt et al., 1981) and perhaps 8–9 km in thickness, although only 3.5 km is exposed (Ford, 1976). It shows pronounced cumulate mineralogical layering which gives the rocks a marked stratiform appearance similar to that of a sedimentary succession. Four major mappable units and several rock members have been recognized (Table II). High concentrations of magnetite are present in the higher levels of the intrusion and, by analogy with similar layered complexes elsewhere in the world, there is speculation as to whether the Dufek intrusion might

Fig. 3. Sketch map of western Dronning Maud Land and the Transantarctic Mountains to illustrate the distribution of Jurassic Ferrar Supergroup tholeiites and their equivalents. The map is based on Elliot (1975a, fig. 37.6B) with reference to original data sources mentioned in the present text.

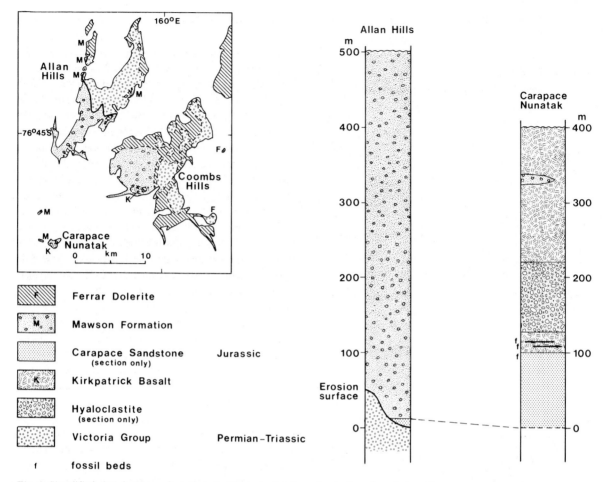

Fig. 4. Simplified sketch map and sections to illustrate the Jurassic stratigraphy of the Allan Hills–Carapace Nunatak area: data taken from Ballance and Watters (1971), Grapes et al. (1974), Hall et al. (1982) and H.W. Ball (personal communication).

contain important reserves of other metallic ores at depth (Wright and Williams, 1974).

Glassy acidic rocks of probable intrusive origin can be traced for 4 km at Butcher Ridge in the Darwin Glacier area of southern Victoria Land and represent the only known extensive area of 'dacitic' or 'rhyolitic' rocks in the Ferrar Supergroup (Kyle, 1979).

Mawson Formation

In the Allan Hills area of south Victoria Land (Fig. 3) the Ferrar Supergroup is represented by upwards of 300 m of massive unstratified diamictite, composed of basaltic and Beacon Supergroup sedimentary debris. Originally interpreted as a tillite (Gunn and Warren, 1962), it was subsequently recognized as a laharic deposit confined in an erosional basin cut in

sedimentary rocks of the Beacon Supergroup. The precise mechanism of formation of the deposit is problematical and it has been described as a mudflow from a volcanic debris source nearby (Ballance and Watters, 1971), or explosion breccias emplaced by

TABLE II

Stratigraphy of the Dufek intrusion

Forrestal Range	Lexington Granophyre	300 m
	Saratoga Gabbro	1400 m
	Ice-covered gap	
	Aughenbaugh Gabbro	1600 m
Dufek Massif		
	Walker Anorthosite	230 m
	5+ km unexposed	

gravity flow after initial extrusion (Hall and Borns, 1972). A detailed study by Grapes et al. (1974) identified explosion breccias, volcanic mudflows, basalt pods and flows, and clastic dykes and sills within the formation, and it was concluded that it formed as the result of shallowly emplaced Ferrar Dolerite intrusions bursting through the sediment cover. On the basis of vesicle formation, Young and Ryburn (1968) independently deduced that the dolerites of Mount Darwin, Beardmore Glacier area were also emplaced shallowly. The presence of a 10+ m thick lens of Mawson-type rock near the top of a 300 m section of Kirkpatrick Basalt lava flows at Carapace Nunatak (Fig. 4; Ballance and Watters, 1971) suggests that the Mawson Formation might be bettter regarded as a facies unit rather than a stratigraphical one. 110 km to the north at Griffin Nunatak a further 200 m of diamictite, thought to be a correlative, of the Mawson Formation, are interbedded within Kirkpatrick Basalt lava (Kyle, 1979).

Carapace Sandstone Formation

In south Victoria Land the Kirkpatrick Basalts of Carapace Nunatak are underlain by about 100 m (120 m: Ballance and Watters, 1971; 85 m: Hall et al., 1982) of volcanic sandstones and fine lithic conglomerates, known as the Carapace Sandstone. Although this sandstone sequence was initially included in the Victoria Group (Gunn and Warren, 1962), Ballance and Watters (1971) argued that its volcanogenic content allied it to the Ferrar Group. Furthermore, at nearby Allan Hills a much reduced but equivalent sequence rests on a marked erosion surface, with 50 m differences in relative elevation, cut in rocks of the Victoria Group (Hall et al., 1982). Shales with conchostrachans occur in the topmost part of the Carapace Sandstone, but plant fragments and a much more diverse lacustrine fauna (mainly of arthropods) are present in two lenses of cherty shale and siltstone in the lower part of the overlying Kirkpatrick Basalt (Fig. 4; H.W. Ball, personal communication, 1975).

Origin of Ferrar Supergroup tholeiites

Determining the origin of the Ferrar Supergroup tholeiites proved problematical. Although the Mesozoic tholeiites of Gondwana have been related in a general way to the latter's break-up, at present it is difficult to offer a precise explanation for their presence in an extensive belt along the Pacific side of Greater Antarctica. The Ferrar tholeiites are continental in origin and presumably formed in an extensional or rift environment (Elliot, 1975b; Kyle et al., 1981). It is possible that those of Dronning Maud Land may have been related to the separation of Africa and Antarctica and/or the opening of the Weddell Sea, whereas those of the Transantarctic Mountains may have been associated with the movement of the Marie Byrd Land area away from Greater Antarctica. In Vestfjella, Dronning Maud Land, Jurassic basaltic dyke activity was preceded by a phase of Triassic lava eruption, perhaps related to an early abortive rifting event (Furnes and Mitchell, 1978).

Despite the time stratigraphical equivalence and general tholeiitic character of the Ferrar Supergroup as a whole, important differences in geochemical composition have been noted throughout the outcrop area. Variations in silica content led Hamilton (1965) to suggest that there was a tendency for the composition of the Antarctic tholeiites to approach that of their nearest neighbours in Gondwana, i.e. the Tasman or Karroo dolerites. Compston et al. (1968) later showed that the $^{87}Sr/^{86}Sr$ initial ratios of the Ferrar dolerites in the Transantarctic Mountains were anomalously high, like those of the Tasman dolerites, whereas those of the Karroo and Serra Geral had the lower values of typical continental tholeiites. The latter are closer to those found in the basalts of Dronning Maud Land (Faure and Elliot, 1971). Although Faure et al. (1972, 1974) ascribed the anomalously high initial ratios and silica contents to crustal contamination of the magma during its ascent from the upper mantle, Kyle (1977) argued that variations in the Ferrar Supergroup rocks along the Transantarctic Mountains can be explained by a differentiation model. Nevertheless the low Rb/Sr ratios of the Dronning Maud Land dolerites (Faure and Elliot, 1971; Furnes and Mitchell, 1978) seem to distinguish them from the Ferrar tholeiites of the Transantarctic Mountains. Published data (Juckes, 1968, 1972; Furnes and Mitchell, 1978; Clarkson, 1981) also indicate that the former have lower MgO/SiO_2 ratios than the latter (cf. Kyle, 1977), except for the values obtained from the Theron Mountains and Whichaway Nunataks (Stephenson, 1966) which partly overlap with the trend for the Ferrar Supergroup of the Transantarctic Mountains. These differences suggest that there are two major tholeiitic subprovinces in Greater Antarctica (Ford and Kistler, 1980).

Radiometric age

With the exception of the Triassic episode recently reported from Vestfjella (Furnes and Mitchell, 1978), K-Ar whole rock and mineral ages for rocks of the Ferrar Group and probable equivalents (Fig. 5) range from 147 Ma in the Queen Elizabeth Range (McDougall, 1963) to 191 at Horn Bluff, north Victoria Land (Starik et al., 1961). However, most ages lie within the range 150–180 Ma (Early to Middle Jurassic). In any one area, ages commonly show a range of 20 Ma or more. Although it is possible that this reflects some local variations in age, the problems of argon loss in such rocks would suggest that the true ages are likely to lie closer to the older ends of the various ranges. A Rb-Sr isochron, based on hypersthene tholeiites from widely scattered samples in the Transantarctic Mountains gave an age of 151 ± 18 Ma (Compston et al., 1968), whereas an isochron from the lower six of fourteen lavas capping Mount Falla, Beardmore Glacier area, gave 173 ± 6 Ma (Faure et al., 1982). New data (Kyle et al., 1981) for the Ferrar Supergroup, suggest that the rocks were erupted over a relatively short period of time (15 Ma) and that the best overall age estimate is 179 ± 7 Ma (Early Jurassic) with some evidence of limited Ferrar dolerite activity around 165 ± 2 Ma ago. Similar results have been reported by Ford and Kistler (1980) for the intrusions of the Pensacola Mountains.

Biostratigraphy

Lacustrine sedimentary beds within the Kirkpatrick Basalt have yielded fresh-water faunas. The most diverse of these is at Carapace Nunatak (Fig. 4) where conchostrachans, ostracods, crustaceans and insects (including the dragon fly *Caraphlebia antarctica* Carpenter, 1969) have been found (Hall and Borns, 1972). Conchostrachan faunas have also been reported from southern Victoria Land (Ricker, 1964) and from the Storm Peak–Blizzard Heights area of Queen Alexandra Range (Elliot and Tasch, 1967). The dominant conchostrachans in both geographical and vertical terms in the Jurassic lake deposits of the Transantarctic Mountains were species of *Cyzicus (Euestheria)* and *C. (Lioestheria)*. Other forms include species of *Paleolimnadia (Paleolimnadia)* and *P. (Grandilimnadia)*. The estheriinids are represented by two species from Storm Peak, and an exotic occurrence of *Cornia* has been noted at Blizzard Heights (Tasch, 1979).

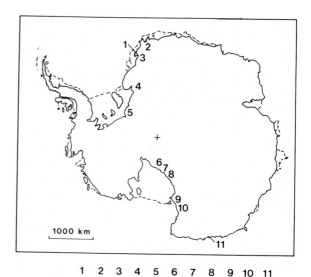

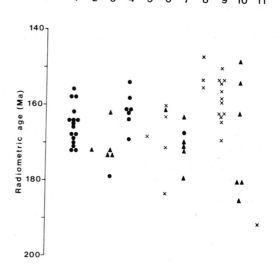

Fig. 5. Graphical representation of K-Ar radiometric age data for the Jurassic tholeiites of the Ferrar Supergroup and related rocks. All dates were calculated using the constants:

$\lambda_\beta = 4.72 \times 10^{-10}$ yr^{-1}, $\lambda_e = 5.84 \times 10^{-11}$ yr^{-1}, Atomic % ^{40}K = 0.011

and should be increased by approximately 2.5% to conform to the revised values given by Steiger and Jäger (1977). Data sources: *1* = Vestfjella (Rex, 1967; Furnes and Mitchell, 1978); *2* = Kirwanveggan (Aucamp et al., 1972); *3* = Milorgfjella (Rex, 1972); *4* = Theron Mountains, Rex 1972); *5* = Dufek Massif (Ford, 1976); *6* = Queen Maud Mountains (McGregor and Wade, 1969); *7* = Queen Alexandra Range (Elliot, 1970); *8* = Queen Elizabeth Range (McDougall, 1963); *9* = South Victoria Land, Escalade Peak–Victoria Valley (McDougall, 1963; Compston et al., 1968); *10* = Allan Hills and Carapace Nunatak (Hall et al., 1982); *11* = Horn Bluff (Starik et al., 1961). Data compiled to July 1979. Solid circles, intrusions (whole rock ages); crosses, intrusions (mineral ages); triangles, lavas (whole rock ages).

Well-preserved fish remains *(Oreochima ellioti)* are associated with conchostrachan remains in Queen Alexandra Range. They belong to the family Archaeomaenidae, previously known only from the Lower Jurassic of Australia (Schaeffer, 1972).

Floral assemblages have been described from only two localities and neither is well-preserved. Cycadophyte and conifer remains from Carapace Nunatak, first thought to be (?) Early Jurassic in age (Plumstead, 1962), were revised by Townrow (1967) who suggested that a Middle Jurassic age was more likely. Spores from a 1.5 m thick mudstone at the base of a 12 m sedimentary intercalation at Section Peak, northern Victoria Land were described by Norris (1965). Although five of the species present also occur in a Triassic assemblage from the nearby Timber Peak, the rest are distinct. Mainly on the basis of similarities to European and Australian forms, Norris suggested an Early Jurassic age, although a Middle Jurassic age could not be ruled out.

CRETACEOUS

No exposures of Cretaceous rocks have been located in Greater Antarctica, although derived fossils in Quaternary deposits suggest that they may have been present once in parts of the region, or are hidden under the ice sheet. Fossiliferous erratics and reworked fossils of Late Cretaceous to Tertiary age have been reported from the McMurdo Sound area (Harrington, 1969; Hertlein, 1969; Webb and Neall, 1972) suggesting that a Late Cretaceous–Tertiary sequence, similar to that of the James Ross Island area of Lesser Antarctica, is or was present nearby. However, these erratics are generally assumed to have come from an unexposed sequence, believed to exist in Marie Byrd Land. Reworked Albian palynomorphs have been recovered from grab samples of Recent marine sediments in Prydz Bay (Kemp, 1972). The absence of marine fossils of comparable age suggests that the palynomorphs were derived, perhaps from a non-marine sequence hidden somewhere beneath the ice covering the adjacent mainland. However, a rich Aptian palynoflora has recently been obtained in situ from a marine core taken close inshore off George V Coast (67°44'S 146°51'E, Domack et al., 1980) and provides the first really firm evidence for the presence of Cretaceous sedimentary rocks in Greater Antarctica.

LESSER ANTARCTICA

In Lesser Antarctica well-documented evidence of Mesozoic sedimentary and volcanic rocks is confined to the Antarctic Peninsula and the Scotia Ridge, a discontinuous loop of islands and submerged continental fragments which links the peninsula to South America. Volcanic rocks of possible Mesozoic age in Ellsworth Land, Thurston Island and Marie Byrd Land (Craddock, 1972) have yet to be fully described and their age properly established. The recognition of Triassic sedimentation in the region is relatively recent, whereas rocks of Jurassic and Cretaceous ages have been documented from the early days of geological exploration in the region. From at least Jurassic times the peninsula was the site of a volcanic arc (Suárez, 1976; Thomson, 1982) into which calk-alkaline intrusions of batholithic proportions were emplaced. West and east of the peninsula spine are thick marine deposits which were derived from the volcanic rocks and their basement. It has been suggested that the westward occurrence of Mesozoic plutons into Marie Byrd Land represents a further extension of this active margin to Gondwana; the Andean Orogen of Craddock (1970b). Together with the processes of erosion and the blanketing effect of the present-day ice sheet, these widespread intrusions have reduced the bedded rocks to a number of local successions which are difficult to correlate stratigraphically, except on a general basis (Table III).

TRIASSIC

Until recently it was generally believed that the Triassic Period in the Antarctic Peninsula was marked by an hiatus in sedimentation, and an orogenic episode which deformed a thick (?) late-Palaeozoic sequence of turbidite greywackes: the so-called Trinity Peninsula Series (Adie, 1957), now Trinity Peninsula Group (Hyden and Tanner, 1981). However, the discovery of marine invertebrate faunas at two localities (Fig. 6; near Cape Legoupil, northern Antarctic Peninsula, and the Lully Foothills of central Alexander Island) proved the presence of Triassic beds within the same sequence (Thomson, 1975c; Edwards, 1982). Unfortunately these faunas are localized and the true extent of Triassic sedimentation is unknown. In the Cape Legoupil area, where the rocks are known as the Legoupil Formation (Halpern, 1965), the fossil-bearing sequence consists of shales, greywacke

TABLE III

Stratigraphical correlation of the Jurassic and Early Cretaceous formations in Lesser Antarctica. Mainly volcanic rocks, v-ornament; mainly marine rocks, horizontal shading; mainly non-marine sediments, open circles

		SOUTH GEORGIA	SOUTH ORKNEY ISLANDS	SOUTH SHETLAND ISLANDS	GRAHAM LAND AREA		ALEXANDER ISLAND	PALMER LAND AREA	
					Peninsula	East coast		West	East
CRETACEOUS	Late			COPPERMINE FORMATION		SNOW HILL ISLAND SERIES			
	Early	ANNENKOV FORMATION / CUMBERLAND BAY FORMATION	conglomerate beds	BYERS FORMATION	ANTARCTIC PENINSULA VOLCANIC GROUP	Longing Gap beds	FOSSIL BLUFF FORMATION	ANTARCTIC PENINSULA VOLCANIC GROUP	Crabeater Pt. beds
JURASSIC	Late				plant beds				LATADY FORMATION
	Middle			Snow Island beds					
	Early								
TRIASSIC	Late								
	Mid			WILLIAMS PT. BEDS					
	Early		GREYWACKE-SHALE FORMATION	MIERS BLUFF FORMATION	TRINITY PENINSULA FORMATION		LEMAY FORMATION		
LATE PALAEOZOIC									

sandstones and granule to pebble conglomerates, with thin interbedded rhyolitic (?) lavas. In Alexander Island the fossiliferous strata consist of tuffaceous sandstones but these are only a small part of the widespread LeMay Formation. This probably forms the 'basement' to the whole of the island and consists of a flysch-type sequence of sandstones, greywackes and argillites with restricted outcrops of chert and pillow lavas (Edwards, 1982).

A chert from an isolated rock in the South Orkney Islands contains Triassic Radiolaria (Dalziel et al., 1981). Its structural attitude suggests that it may be part of the Greywacke-Shale Formation exposed on adjacent islands, a sequence correlated on lithological and structural grounds with the Trinity Peninsula 'Series' (Adie, 1957). However, the chert might alternatively be a tectonically emplaced slice.

Another deformed greywacke sequence equated stratigraphically with the Trinity Peninsula Group is the Miers Bluff Formation of southern Livingston Island (Hobbs, 1968; Dalziel, 1969). Poorly preserved plant material from this formation led Schopf (1973) to conclude that its age was probably post-Carboniferous and that a Mesozoic age could not be excluded.

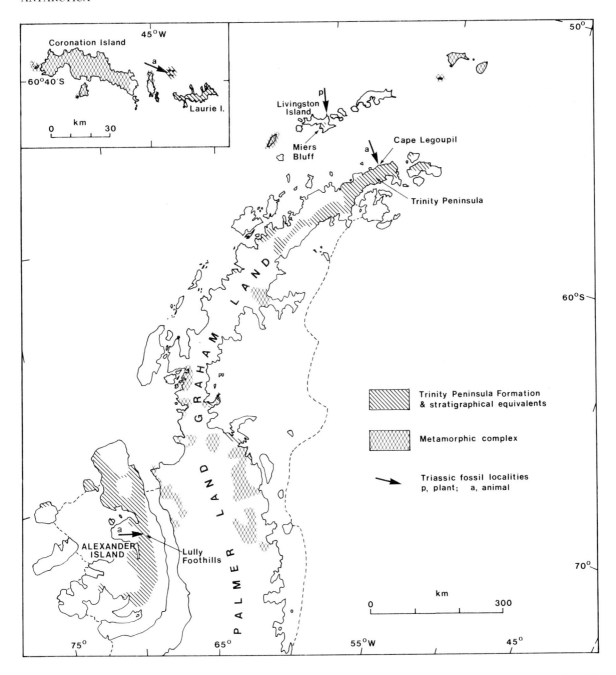

Fig. 6. Sketch map of the Antarctic Peninsula area, showing the distribution of strongly deformed (?) late-Palaeozoic–Triassic strata; the South Orkney Islands are shown in the inset. Previously regarded as late Palaeozoic in age these strata have now yielded Triassic marine fossils at three localities. Because the fossils are rare, the true extent of the Triassic rocks may never be known. Areas of schistose and gneisose rocks (metamorphic complex), once thought to be Precambrian, may be at least in part the metamorphosed equivalents of the (?) late-Palaeozoic–Triassic rocks (cf. Dalziel, 1982). Undeformed strata with a Middle Triassic flora occur at Williams Point on the northern side of Livingston Island.

In contrast to the deformed and indurated marine greywackes, flat-bedded volcaniclastic shales and siltstones at Williams Point, northern Livingston Island have yielded a Triassic land flora (Orlando, 1968). Although these rocks were initially believed to occur as fragments derived from boulders in a nearby conglomerate (Hobbs, 1968), further investigations by the author have shown that the beds occur in situ, and that the fragments which occur widely scattered over the Williams Point headland have been transported to the surface in large squeeze-up structures breaching through a capping sill of younger basalt.

Biostratigraphy

The marine invertebrate faunas of the Legoupil and LeMay formations have no species in common. This may partly be due to facies differences, and partly to a difference in age, the former being Early to Middle Triassic and the latter Middle to Late Triassic. Important bivalve genera include *Bakevelloides* and *Neoschizodus* in the Legoupil Formation, and *Balantioselena* and *Waagenoperna* in the LeMay Formation. These have counterparts in faunas from Japan, New Guinea and New Zealand (Thomson, 1975c; Edwards, 1980a, 1982). Associated with the Alexander Island bivalves are a variety of gastropods with South American affinities, fragments of echinoderms and corals, and a smooth, indeterminate juvenile ammonoid (the only one so far known in the Triassic of Antarctica).

New material from the Williams Point flora has been studied by W.S. Lacey and R.C. Lucas (University College of North Wales, Bangor), who have been able to augment Orlando's (1968) original determinations. The flora consists of possible liverworts, a variety of ferns including *Dicroidium*, and bennetitalean, conifer (*Pagiophyllum*) and ginkgo remains. A Middle Triassic age is suggested for the flora.

JURASSIC AND EARLY CRETACEOUS

Rocks of Jurassic and Early Cretaceous age are discussed together here because they represent the products of a continuous period of volcanic and sedimentary activity, and they are difficult to separate. Even where fossils are present, the Jurassic/Cretaceous boundary is hard to place because of the thick sporadically fossiliferous successions and the presence of a boundary fauna which has both uppermost-Jurassic and lowest Cretaceous affinities.

No bedded rocks of Early Jurassic age are known, and the earliest-Jurassic strata of confidently determined age are the Middle Jurassic marine sandstones of the Behrendt Mountains (Quilty, 1970). A Middle Jurassic age is also claimed for plant beds often found beneath and interbedded with the volcanic rocks. However, numerous attempts to retrieve recognizeable palynological remains from them has so far proved unsuccessful, and the macrofloras of both Jurassic and Early Cretaceous rocks in the peninsula are closely similar.

Virtually all sedimentary rocks in the peninsula area of pre-Middle Jurassic age are strongly deformed. Since they contain rare faunas which indicate that they range as high as Middle or even Late Triassic, it seems that the Early Jurassic in Lesser Antarctica was an orogenic period. This event is generally referred to as the Gondwanian Orogeny (Dalziel and Elliot, 1973; Dalziel, 1982).

South Georgia

The greater part of South Georgia (Fig. 7) consists of approximately 8 km of deformed Late Jurassic–Early Cretaceous turbidites (Cumberland Bay Formation) that filled an old marginal basin. These rocks were derived from a volcanic arc which lay to the south of the present island. In addition, detritus from a mainly continental source is present as the distal turbidites of the Sandebugten Formation (Stone, 1980). The two formations are seen only in faulted contact and stratigraphical relations between the two are uncertain; however, it is possible that they were roughly coeval. Remains of the island arc are no longer exposed above water, but its shelf facies is preserved as the volcaniclastic rocks of the Annenkov Island Formation (Pettigrew, 1981). Metasedimentary rocks and gneisses intruded by plutons (Drygalski Fjord Complex) may represent part of the continental crust which floored the basin (Storey et al., 1977), whereas the pillow lava sequences of the Larsen Harbour Formation (Bell et al., 1977) may be oceanic crust generated during the opening of the marginal basin. Both the Drygalski Fjord Complex and the Larsen Harbour Formation are intruded by sheeted mafic dyke complexes. The timing of the events relating to the formation of the marginal basin, its subsequent infilling and deformation have been discussed by Thomson et al. (1982) and Tanner (1982). In gen-

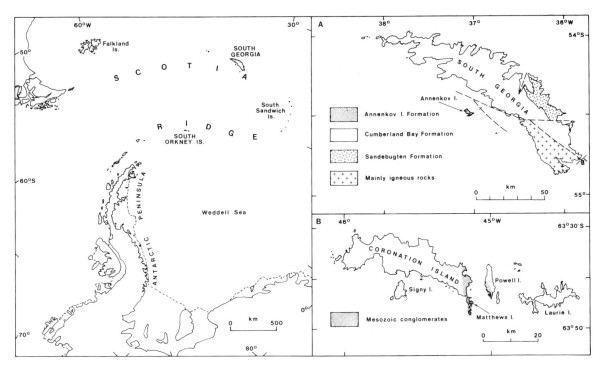

Fig. 7. Map of the Scotia Ridge (or Scotia Arc) to show the chain of islands linking South America to the Antarctic Peninsula. The South Sandwich Islands are recent volcanoes located on a small plate with an active trench to the east. Sketch maps of South Georgia (A) and the South Orkney Islands (B) show the distribution of Late Jurassic–Early Cretaceous rocks in those areas.

eral terms the basin was initiated in the Early Jurassic and received the bulk of its sedimentary infill during the Late Jurassic and Early Cretaceous; deformation and post-orogenic uplift were over by earliest Tertiary times.

Prior to the fragmentation of the Scotia arc, South Georgia probably occupied a position to the east of Navarino Island, along the southern margin of the Burdwood Bank (Dalziel and Elliot, 1971; Dalziel, 1975; Tanner, 1982). In such a location its sedimentary strato-tectonic units can be traced westwards into the equivalent units of a Mesozoic marginal basin in the Southern Andes.

South Orkney Islands

Much of the South Orkney Islands (Figs. 7, 8) consists of a metamorphic complex in the west and a deformed sequence of (?) late-Palaeozoic–Triassic greywackes (Greywacke-Shale Formation) in the east. By contrast, in the central part of the island group Late Jurassic–Early Cretaceous alluvial fan conglomerates which probably accumulated in a north–south

graben are present (Elliot and Wells, 1982). The thickness of these conglomerates is unknown but approximately 518 m are present in a single outcrop on eastern Coronation Island (J.W. Thomson, 1974). The eastern part of the sequence (Powell Island Conglomerate) appears to be entirely terrestrial in origin and contains plants as the only fossils, both as carbonized log moulds, and as fronds in lacustrine siltstones and sandstones. However, the Spence Harbour Conglomerate to the west contains a thick marine intercalation of sandstones with rare poorly preserved fossil invertebrates (Thomson, 1975a, 1981), and it is underlain at two localities by sheared black shale (Gibbon Bay Shale) with the latest-Jurassic–earliest-Cretaceous fauna (Thomson and Willey, 1975).

A puzzling feature of the conglomerates was that, unlike every other Mesozoic sedimentary sequence in the Antarctic Peninsula area, they apparently contained no volcanic debris (Thomson, 1982). However, it has now been shown that the Powell Island Conglomerate contains local concentrations of volcanic clasts (Wells, 1982). Although these clasts are reworked, it is believed that they came from a more or less pene-

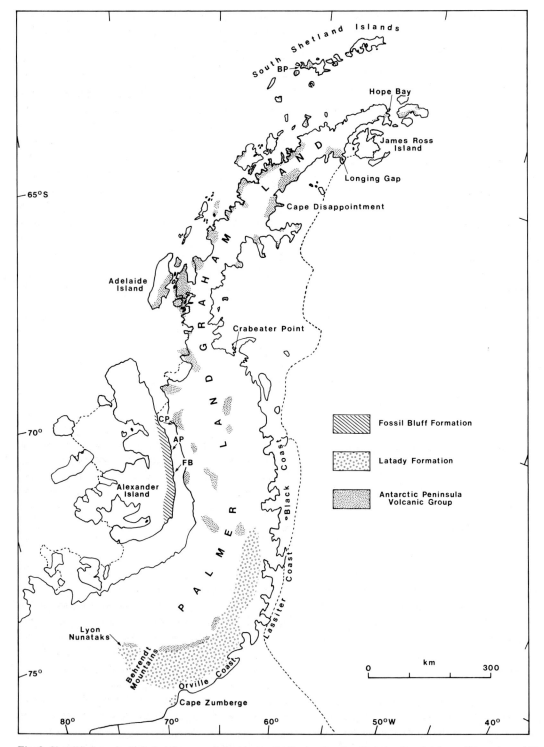

Fig. 8. Simplified geological sketch map of the Antarctic Peninsula, showing the distribution of Jurassic and Early Cretaceous volcanic and sedimentary rocks. *AP* = Ablation Point; *BP* = Byers Peninsula; *CP* = Carse Point; *FB* = Fossil Bluff.

contemporaneous volcanic source, perhaps related to the rifting that produced the graben in which the conglomerates accumulated (Wells, 1982).

Northern Antarctic Peninsula area

In the northern Antarctic Peninsula (or Graham Land) area, most of the preserved Middle Jurassic–Early Cretaceous succession relates to the volcanic arc itself (Fig. 8). Stratigraphical thicknesses of up to 3,000 m have been estimated for sequences in any one area (Dewar, 1970). The constituent calk-alkaline lavas and their pyroclastic equivalents show a general tendency to become more basic in composition from east to west (cf. Adie, 1972a; West, 1974; Weaver et al., 1982). Rhyolitic lavas and acid tuffs predominate on the east coast, whereas andesitic types are more frequent on the west; in the South Shetland Islands basaltic andesites occur (Smellie et al., 1980). All of these rocks have been mapped by British geologists as undifferentiated parts of the Antarctic Peninsula Volcanic Group (previously Upper Jurassic Volcanic Group). However, in the Gerlache Strait area Chilean workers (Alarcón et al., 1976) have recognized two volcanic formations separated by a strong unconformity:

Formación Isla Wienke   Cretaceous
Formación Canal Lautaro   Triassic to Jurassic

As yet the evidence for the ages assigned to these formations is circumstantial.

The base of the Antarctic Peninsula Volcanic Group is exposed at several localities in eastern Graham Land, where there is a strong discordance with the underlying greywackes of the Trinity Peninsula Formation and its equivalents. At some localities, notably Mount Flora, Hope Bay (Fig. 8) and Camp Hill, alluvial fan conglomerate sequences with lacustrine plant beds in their upper parts are interposed between the two (Adie, 1964). Only the fossil flora of Hope Bay has been studied in detail. Its supposed age is controversial (Taylor et al., 1979) and it has been accorded ages ranging from earliest Jurassic to earliest Cretaceous. However, most authors have followed Halle's (1913) view that it is Middle Jurassic in age (cf. Adie, 1962; Bibby, 1966; Fleet, 1968; Aitkenhead, 1975). The suggestion that the conglomerates and plant beds appeared to contain tuffaceous beds has been accorded little significance, but Elliot et al. (1978) have demonstrated recently that such beds are common in the Hope Bay sequence. Thus, in

all probability the plant beds are coeval with much of the volcanic succession exposed in the area, and there is little age difference between the plant beds of Mount Flora and the overlying volcanic rocks. Sedimentary beds are also interbedded with the volcanic rocks on Oscar II Coast (Fleet, 1968), whereas at Cape Disappointment Adie (1958) reported a fossil forest overwhelmed by volcanic eruption.

The westernmost margin of the arc is represented by mixed subaerial volcanic and marine volcaniclastic sequences in the South Shetland Islands. Rapid lateral and vertical changes in facies suggest a complicated palaeogeography in which the South Shetland Islands area may have been a group of small volcanic islands separated from the main arc (Thomson, 1982). The Byers Formation of Livingston Island (Table III) comprises the most extensive sequence. It begins with offshore Tithonian mudstones and thin sandstones, which pass laterally and upwards into volcaniclastic shallow-water marine conglomerates, sandstones and shales, and then into Early Cretaceous subaerial pyroclastic rocks and lavas. The sequence appears to be an intra-arc accumulation, representing the enclosing and filling of an embayment by nearby volcanoes, particularly a major vent in the northwestern part of the area (Smellie et al., 1980). Marine fossils in the volcaniclastic rocks range as high as Valanginian (Covacevich, 1976), but the upper age limit of the pyroclastic succession is uncertain. Radiometric data suggest that it could range as high as Late Cretaceous (Pankhurst et al., 1980). Late Jurassic deep-water mudstones and associated volcanic rocks occur on Low Island (Smellie, 1980), and subaerial volcanic rocks with a (?) Middle Jurassic land flora are present on Snow Island (Fuenzalida et al., 1972).

Farther south on Adelaide Island sedimentary structures suggest a marine environment for the deposition of coarse-grained volcanic successions (Dewar, 1970). Late Jurassic marine Mollusca occur locally in the same rocks (Thomson, 1972a).

The only firm evidence of marine Jurassic deposition on the eastern side of Graham Land is restricted to a few isolated localities of which the best known is at Longing Gap. The preserved succession consists of approximately 100 m of sparsely fossiliferous radiolarian mudstones and thin ash-fall tuffs, whose lack of sedimentary structures, other than bedding, and content of ammonite conchs encrusted with oysters suggest that it accumulated in a quiet shallow-water environment. The absence of terriginous debris in the sediments may indicate further that, at the time of their formation, there was no substantial nearby landmass that could have contributed clastic material to the depositional area. Perisphinctid ammonites (Bibby, 1968) and coarsely ribbed inoceramid bivalves indicate a Late Jurassic (? Kimmeridgian—early-Tithonian age) for these beds.

Southern Antarctic Peninsula area

This area includes not only Palmer Land and Alexander Island but also the adjacent mountains to the southwest as far as the Behrendt Mountains and Lyon Nunataks (Fig. 8). It contains extensive marine deposits which were laid down in both fore- and back-arc positions, thus presenting a near comprehensive section across the volcanic arc. The volcanic rocks themselves are less well studied than those of northern Antarctic Peninsula but they show interbedded relationships with the marine rocks of both sides of the arc, and marine rocks at Carse Point, Palmer Land (Fig. 8) suggest a deep embayment into the volcanic landmass (Thomson, 1975b).

The fore-arc deposits of Alexander Island (Fossil Bluff Formation; Taylor et al., 1979) have a minimum stratigraphical thickness of 5,000 m and they range in age from at least Kimmeridgian to Albian or younger. For the most part they are faulted against the folded greywackes of the (?) late-Palaeozoic–Triassic LeMay Formation but at one locality they rest unconformably on them (Edwards, 1980b). The bulk of the Fossil Bluff Formation consists of marine mudstones with lesser amounts of sandstones and conglomerates. Lavas and coarse pyroclastic rocks are interbedded with marine Tithonian rocks at Ablation Point (Elliot, 1974) and the whole sequence is rich in volcanic debris, much of which was introduced by direct aerial fall-out (cf. Horne and Thomson, 1972). There is a southward increase in sand content and the outcrops at the southeastern corner of Alexander Island are dominated by sandstones which display cyclothemic sequences. They contain highly carbonaceous shales, although no true coal beds are known. Many of the sandstones contain rich leaf floras, and massive petrified tree trunks in growth position indicate a terrestrial environment for parts of the sequence. Horne (1969) suggested that much of the Fossil Bluff Formation accumulated as deltaic, interdeltaic and shelf deposits.

The marine sequences exhibit spectacular indica-

tions of instability during deposition, notably slump zones 100 m or more in thickness which are traceable for several kilometres along the outcrops (Bell, 1975), together with late-stage remobilization features such as clastic dykes (Taylor, 1982). Although much of the Fossil Bluff Formation was probably derived from the Mesozoic arc situated in Palmer Land to the east, there is also some tenuous evidence of detritus with a westerly provenance and for a locally emergent off-shore landmass, perhaps in the form of a chain of non-volcanic islands composed largely of LeMay Formation rocks (Thomson, 1982).

On the eastern side of the area a Jurassic back-arc marine sequence (Latady Formation) has been proved to exist from approximately latitude 72°30'S to Cape Zumberge (Rowley, 1978). The Latady Formation varies from dominantly shaly in the north to dominantly sandy in the south, and its rocks appear to become more arkosic towards the south (cf. Laudon et al., 1969; Rowley and Williams, 1982; Thomson, 1982). Volcanic rocks intertongue in the sequence adjacent to the old arc and they wedge out towards the Weddell Sea. A traverse of the southern part of the depositional area from north to south shows a progressive change from lacustrine or lagoonal deposits within the margin of the arc to open marine shelf and finally deeper water shales (Rowley, 1978). The frequent occurrence in the sandstones of both laterally persistent flat laminations and clay-pellet conglomerates suggest that the shelf area was swept by strong currents. Massive dacitic rocks in southernmost Palmer Land, the Orville Coast and the Behrendt Mountains may represent volcanic centres.

These rocks show a progressive increase in deformation and metamorphism northwards. Together with a decrease in fossil content, this makes them difficult to distinguish from rocks in northeastern Palmer Land assigned to the late-Palaeozoic–Triassic Trinity Peninsula 'Series' (Adie, 1957). Thus it is possible that parts of the latter may eventually be shown to be part of the Latady Formation (Taylor et al., 1979).

Whereas individual stratigraphical sections of up to 830 m have been measured (Thomson et al., 1982), the frequent absence or paucity of fossils, repetitious lithology and complex structure has so far prevented any precise estimate of the total thickness of the Latady Formation. Ammonites in the southern Behrendt Mountains (Quilty, 1970) indicate that beds as old as Middle Jurassic are present, although the base of the formation has not yet been located and is probably not exposed. Ammonites ranging up to Tithonian in age have been reported elsewhere (Thomson, 1980; Rowley and Williams, 1982). The Latady Formation has not yet yielded any fossils younger than Late Jurassic in age. It is intruded by Middle Cretaceous plutons which cut the tectonic structures (Rowley and Williams, 1982), and thus it must have been folded at some time in the Early Cretaceous. Earliest Cretaceous shales at Crabeater Point (Fig. 8; Thomson, 1967; Taylor et al., 1979) are also deformed and may represent the youngest part of the Latady Formation.

Biostratigraphy

Although not everywhere abundant, the Jurassic and Early Cretaceous floras and marine faunas of the Antarctic Peninsula area show great variety (Thomson, 1977).

Floras with assemblages of ferns, cycadophytes, conifers and sometimes ginkgophytes have been reported from the South Orkney Islands, South Shetland Islands, northern Antarctic Peninsula, Adelaide Island, Alexander Island, eastern Palmer Land and the neighbouring mountains to the southwest. Only that from Hope Bay, northern Antarctic Peninsula has been monographed (Halle, 1913), although a Middle Jurassic flora from Snow Island (Fuenzalida et al., 1972) and a Wealden flora from Livingston Island (Hernández and Azcárate, 1971) have also been described. Other floras have been compared in passing to the Hope Bay flora, for which Halle (1913) suggested a Middle Jurassic age. However, this age is open to question (Taylor et al., 1979), and Schopf (in Rowley and Williams, 1982) compared floral remains from Late Jurassic beds in the Latady Formation to those of Hope Bay.

The youngest floras of the period under discussion are probably those of southernmost Alexander Island. These occur in deltaic beds of probable Albian or younger age in the upper part of the Fossil Bluff Formation. Tree trunks up to 6 m high occur in growth position, and at least four occurrences of fossil forest floors with numerous tree stumps and roots in situ have been located (T.H. Jefferson, personal communication, 1979). Associated leaf floras include ferns, cycadophytes, conifers and ginkgophytes.

Marine faunas are dominated by the Mollusca, but they also include Bryozoa, Brachiopoda, Crustacea,

Echinodermata, numerous trace fossils, and fish fragments. The earliest-Jurassic molluscs are the Bajocian and Callovian stephanoceratids and macrocephalitids of the Behrendt Mountains (Quilty, 1970). These are associated with a rich bivalve fauna, as yet undescribed (Quilty, 1977). Possible Oxfordian perisphinctids occur in the Behrendt Mountains (Quilty, 1970). Others of late-Oxfordian–early-Kimmeridgian aspect have been reported from Alexander Island (Howarth, 1958) and as presumed erratics in the James Ross Island area (Spath, 1953). None of this material is very well preserved.

Kimmeridgian faunas can be recognized with more confidence. Inoceramid bivalves, similar to the well-known Indonesian and New Zealand *Retroceramus haasti* and *R. subhaasti*, occur in Alexander Island (Thomson and Willey, 1972). Probably related fragments occur at Longing Gap (new information) and at Adelaide Island (Thomson, 1972a). Faunas with *R. galoi* and *R. subhaasti* from the Behrendt Mountains are likely to be largely Kimmeridgian in age (Quilty, 1977), and a variety of *Retroceramus, Mytiloceramus, Mytiloides*? and *Inoceramus* in the Latady Formation of eastern Palmer Land are 'late' Kimmeridgian to late Tithonian in age (Rowley and Williams, 1982). All known ammonites of possible Kimmeridgian age are poorly preserved. In addition to those of doubtful stratigraphical age mentioned above, there is a possible fragment of *Pachysphinctes* from Ablation Point, Alexander Island (Thomson, 1979) and a variety of perisphinctids from the Orville Coast, which may belong to such genera as *Subdichotomoceras, Torquatisphinctes* and *Pachysphinctes* (Thomson, 1980). However, Stevens (1967) has identified early-Kimmeridgian or older species of the belemnite *Conodicoelites* from Lyon Nunataks, and Willey (1973) has described a Kimmeridgian assemblage of *Belemnopsis* and *Hibolithes* species from Ablation Point.

Tithonian faunas are well-developed in the South Shetland Islands, Alexander Island, and southeastern Palmer Land. The earlier assemblages are characterized by the genera *Virgatosphinctes* and *Aulacosphinctoides*. In the Fossil Bluff Formation of Alexander Island these are associated with phylloceratids, lytoceratids, haploceratids and oppeliids (Thomson, 1979). Overlying faunas in Alexander Island, characterized by *Blanfordiceras*, probably represent the latest part of the Tithonian in the area. 1,000 m above the *Blanfordiceras* beds (Elliot, 1974) is a fauna with *Haplophylloceras, Bochianites* and *Spiti-*

ceras. It has both Tithonian and Berriasian affinities and is associated with species of the belemnites *Belemnopsis* and *Hibolithes* including the characteristic Neocomian form, *H. subfusiformis* (Willey, 1973). It is provisionally taken to mark the Early Berriasian (Thomson, 1979). A similar fauna occurs in the South Shetland Islands and contains *Bochianites, Spiticeras* and *Himalayites* (Smellie et al., 1980).

More diagnostic Berriasian faunas are distinguished by a varied *Belemnopsis/Hibolithes* fauna (Willey, 1973) and densely ribbed inoceramids of the *I. everesti* group (J.A. Crame, personal communication, 1975). Ammonites are represented by a few poorly preserved phylloceratids and neocomitids, and a possible *Himalayites* (Thomson, 1974). The Valanginian has only been recognized in the South Shetland Islands where a *Neocomites* fauna is present (Covacevich, 1976).

Although there is an apparently unbroken sequence from Tithonian to Albian in the Fossil Bluff Formation, the recognition of much of the Neocomian (Valanginian–Barremian) has proved difficult. A report of Barremian faunas (Thomson, 1972b) has been questioned in the light of further collections and local stratigraphical considerations (Thomson, 1974; Taylor et al., 1979). A variety of heteromorphs and silesitids are present which show strong morphological resemblances to late Neocomian species from Europe, yet they occur in association with, or stratigraphically above eotetragonitids and aconeceratids of Aptian or Albian aspect, and dimitobelid belemnites (Willey, 1972). The youngest beds probably occur in the scattered nunataks at the southern end of Alexander Island where a *Hamites* fauna is present.

LATE CRETACEOUS

Although Late Cretaceous marine rocks in the Antarctic Peninsula area have attracted considerable attention because of the superbly preserved faunas they contain (below), comparatively little attention has been paid to the rocks themselves. They are widely, if patchily distributed along the northeastern coast of the peninsula, particularly in the James Ross Island area (Fig. 9; Bibby, 1966; Elliot, 1966; Fleet, 1968). Present outcrops perhaps represent only fragments of a more extensive Cretaceous basin situated below sea-level along the eastern margin of northern Antarctic Peninsula. A sequence of conglomerates, sandstones and shales is believed to total more than 5,000 m on

James Ross Island (Bibby, 1966), although it is difficult to correlate between measured sections and Howarth (1966) queried whether this figure might be too high. No base has been seen, but on James Ross Island the Cretaceous rocks are overlain unconformably by Pliocene volcanic rocks (Bibby, 1966; Rex, 1972), and on Seymour Island they are overlain unconformably or are in fault-contact with Palaeocene marine sands (Elliot and Trautman, 1982). Ammonites in the upper part of the Cretaceous sequence (Snow Hill Island 'Series') are Campanian in age (Spath, 1953; Howarth, 1958, 1966) but lower beds contain less well-studied faunas that point to the presence of significantly older strata (below).

Conglomerates predominate in the western and lower parts of the succession (cf. Elliot, 1966). Most of these are traction types but graded units of probable gravity-flow origin are interbedded with them. Current directions deduced from false bedding and pebble imbrication indicate transport in an easterly direction away from the arc (G.W. Farquharson, personal communication, 1981). The lower parts of the sequence in western James Ross Island comprise a mixed sequence of conglomerates, sandstones and shales, whereas the upper and more easterly exposed parts of the succession (Snow Hill Island 'Series') tend to be finer grained, with poorly consolidated sandstones and sandy clays predominating (Bibby, 1966). Concretions are common in the younger beds and often these contain well-preserved fossils. Bibby (1966) made frequent reference to shallow-water conditions in the rocks of James Ross Island. This interpretation is borne out by the bivalve fauna (Wilckens, 1910; Bibby, 1966) and particularly the number of very thick-shelled forms such as trigoniids.

The clast composition of the conglomerates varies from place to place. Sedimentary rocks of more than one previous depositional cycle are present in the conglomerates of western James Ross Island, together with tuffs and smaller amounts of igneous rocks (Bibby, 1966). Approximately equivalent conglomerates at Sobral Peninsula and Pedersen Nunatak (Elliot, 1966) have a largely volcanic provenance. The volcanic detritus compares lithologically with the rocks of the Antarctic Peninsula Volcanic Group, and it is generally considered that it was derived from the latter. However, the status of the volcanic arc during the Late Cretaceous is uncertain. New information from the South Shetland Islands suggest that volcanic activity there was more or less continuous from Juras-

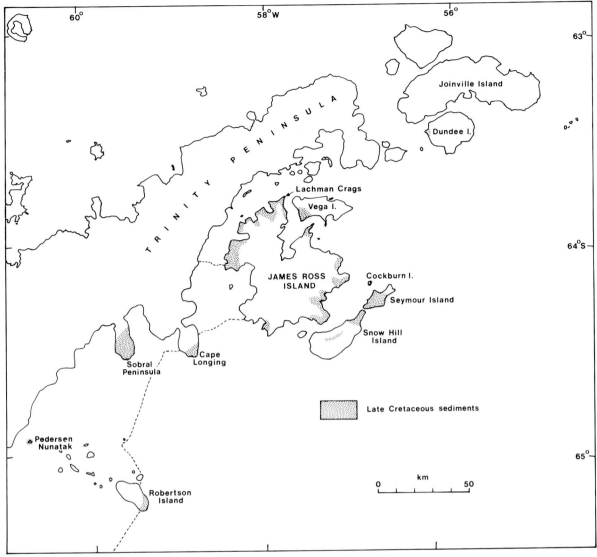

Fig. 9. Geological sketch map of northern Graham Land and the James Ross Island area to illustrate the known extent of Late Cretaceous sedimentary rocks in Lesser Antarctica.

sic to Tertiary times, and there are suggestions of Cretaceous volcanic activity in the Antarctic Peninsula area proper (Halpern, 1965; Adie, 1972a; Alarcón et al., 1976). Thus it is possible that some of the volcanic clasts in the Cretaceous rocks are from contemporaneous as well as older sources. This is supported by the recent discovery of beds of shards up to 2 m thick in the (?) Turonian sandstones of Cape Longing (G.W. Farquharson, personal communication, 1981), clear evidence of nearby volcanic activity in Late Cretaceous times.

Biostratigraphy

Although it is the superbly preserved ammonites of the Snow Hill Island 'Series' which have attracted the most attention from palaeontologists, the Upper Cretaceous beds of the James Ross Island area have yielded a diverse bivalve and gastropod fauna (Wilckens, 1910) and a variety of foraminifers, brachiopods, corals, echinoderms, decapod crustaceans, annelids and a few fish fragments (see Thomson, 1977 for bibliography). Undescribed material, collected by W.N. Croft from Lachman Crags, contains angiosperm leaves in association with Campanian ammonites. The

ammonite faunas are characterized by kossmaticera-
tids but include a range of phylloceratids, lytoceratids, heteromorphs, desmoceratids and pachydiscids.
Nearly all of these come from the Snow Hill Island
Series and consistently indicate Campanian ages
(Spath, 1953; Howarth, 1958, 1966). However, a single specimen of *Tetragonites* cf. *epigonus* from Cape
Longing is stratigraphically less diagnostic as it is
known to range from Turonian to Campanian. Likewise an indeterminate desmoceratid from the earlier
part of the Cretaceous sequence on western James
Ross Island could be any age from Cenomanian to
Campanian (Howarth, 1966). The two specimens
leave room to speculate that pre-Campanian Cretaceous beds might be present and this is confirmed by a
variety of inoceramid bivalves. The presence of *Inoceramus lamarcki* and *I. apicalis* at Cape Longing
strongly suggest a lower to middle Turonian age for
the beds at that locality (Crame, 1981). Other specimens, dominated by members of the *I. pictus* group,
obtained in situ and from moraines in the western
part of James Ross Island also indicate the presence
of pre-Campanian beds. Ammonite fragments from a
calcareous (?) nodule in the Lower Kotick Point beds
of western James Ross Island are similar to late-Aptian–Albian silesitids from Alexander Island (author;
new information). Sandstone eratics on Dundee
Island contain the well-known Albian–Cenomanian
Inoceramus concentricus (Crame, 1980) and an Albian age has been claimed for beds in situ on the same
island (Ramos et al., 1982). Thus the Cretaceous
rocks of the James Ross Island area preserve a far
more extensive stratigraphical record than was first
believed.

Beds without ammonites at the top of the Cretaceous succession on Seymour Island have been assigned
a possible Maastrichtian age on the basis of a comparison with the *Lahillia luisa* (Bivalvia) beds of Santa
Cruz, Argentina (Rinaldi et al., 1978). However, the
diagnostic bivalve and its associated fauna also occur
with Campanian ammonites in the beds below and
further evidence is required to confirm the presence
of Maastrichtian strata on Seymour Island.

MESOZOIC PLUTONISM

In terms of distribution, the most important rocks
of the Mesozoic in Lesser Antarctica are the plutonic
calc-alkaline gabbro–granite intrusions. Although

some Jurassic plutons have been recognized on the
basis of field relations and geochemistry, many calc-
alkaline intrusions in the Antarctic Peninsula have
been loosely referred to the Andean Intrusive Suite,
which is by definition of Late Cretaceous–Early Tertiary age (Adie, 1955). The determination of more
and more radiometric ages (Fig. 10) has shown that
calc-alkaline plutonic activity occurred over a long
period from earliest-Jurassic to Early Tertiary times
(e.g., Halpern, 1971; Rex, 1976), and it is certain that
many rocks referred to the Andean Intrusive Suite
were formed at periods well outside the originally
defined limits of the group. Assuming that the Trias-
sic/Jurassic boundary is approximately 200 Ma B.P.,
the only definite Triassic date so far obtained from a
pluton is one of 218 Ma from the Jones Mountains
(Rutford et al., 1972), although several Triassic ages
have been obtained from metamorphic rocks.

It has been suggested that the intrusive activity
occurred in several distinct phases (Adie, 1972b; Rex,
1976), but compilations of available dates (Fig. 10;
Saunders et al., 1982) demonstrate that the only clear
breaks in activity were for short periods at the Juras-
sic/Cretaceous and Cretaceous/Tertiary boundaries,
and even these may be largely the result of sampling
bias. Frequency plots of radiometric ages (Grikurov,
1973; Saunders et al., 1982), however, suggest that
there was a peak of activity during the mid-Cretaceous 90–110 Ma. Curiously enough no equivalent
peak which might correspond to the Early Mesozoic
Gondwanian Orogeny has yet been identified.

Analysis of the geographical distribution of radiometric ages so far determined on the Mesozoic plutonic rocks of Graham Land suggests that the foci of
plutonic activity migrated trenchwards from west to
east, and from north to south (Saunders et al., 1982).
Age data for Palmer Land are still relatively scarce
and have insufficient coverage to review in a similar
manner. Plutons in southeastern Palmer Land and the
Behrendt Mountains have consistently given Early to
Mid-Cretaceous ages (Fig. 10), whereas the only
Jurassic date so far obtained is from western Palmer
Land. Determinations made on plutonic rocks near
the Black Coast by R.J. Pankhurst also give Early
Cretaceous values (Singleton, 1980), but have been
obtained from rocks which on field evidence appear
to be the youngest.

Both Williams et al. (1972) and Saunders et al.
(1982) commented on the relatively higher potash
levels in the plutons on the eastern side of the Antarc-

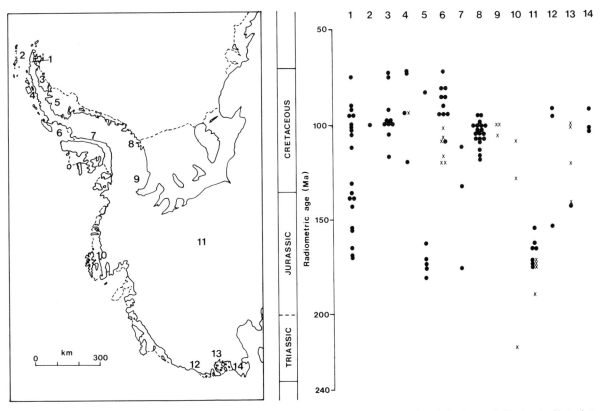

Fig. 10. Graphical representation of radiometric age data for the Mesozoic plutonic rocks of the Antarctic Peninsula. K-Ar dates (solid circles) were calculated using the constants: $\lambda_\beta = 4.72 \times 10^{-10}$ yr^{-1}, $\lambda_e = 5.84 \times 10^{-11}$ yr^{-1}, Atomic % ^{40}K = 0.0119, and should be increased by approximately 2.5% to conform to the revised values given by Steiger and Jäger (1977). Rb-Sr dates (crosses) were calculated using $\lambda = 1.42 \times 10^{-11}$ yr^{-1}. Data sources: *1* = Trinity Peninsula (Grikurov et al., 1970; Halpern, 1971; Rex, 1976); *2* = South Shetland Islands (Grikurov et al., 1970); *3* = Nordenskjöld and Oscar II Coasts (Rex, 1976); *4* = Danco Coast and Argentine Islands (Rex, 1976; Gledhill et al., 1982); *5* = Foyn and Bowman Coasts (Rex, 1976); *6* = Marguerite Bay (Halpern, 1971, 1972; Gledhill et al., 1982); *7* = western Palmer Land (Rex, 1976); *8* = Lassiter Coast (Farrar et al., 1982); *9* = Behrendt Mountains (Halpern, 1971); *10* = Thurston Island and Jones Mountains (Halpern, 1971; Munizaga, 1972); *11* = Pirrit and Nash Hills, and Whitmore Mountains (Craddock, 1972; Halpern, 1971; Kovach and Faure, 1978); *12* = Ruppert Coast (Metcalfe et al., 1978); *13* = Ford Ranges (Halpern, 1971); *14* = Rockefeller Mountains (Wade and Wilbanks, 1972). Data compiled to July 1979.

tic Peninsula. Together with the east—west variations in trace and rare-earth elements, Saunders et al. (1982) distinguished a chemical pattern which parallels that of the age variations in Graham Land, and which can be related to subduction off the western coast of the peninsula and relative differences in level to the Benioff zone.

PALAEOGEOGRAPHY

One of the most perplexing problems in Antarctic geology is understanding the relationship between Greater and Lesser Antarctica. Greater Antarctica formed the nucleus of Gondwana and behaved as a stable craton throughout Mesozoic time. By contrast, the rocks of Lesser Antarctica indicate that the latter was an active margin. The region is now an archipelago of continental fragments (of which the main are the Antarctic Peninsula, Thurston Island—Ellsworth Land, Marie Byrd Land, and Ellsworth Mountains blocks) joined together by ice but separated by deep subglacial troughs. The character of the crust between these blocks is unknown, but there is an increasing body of opinion that each represents a microplate which may have moved independently from an original position in Gondwana that was probably closer to Greater Antarctica. As yet there is no concensus as to the reconstruction of the fragments (e.g., Elliot, 1972a; Barker and Griffiths, 1977; De Wit, 1977) but the microplate concept offers a useful alternative to many Gondwana reconstructions which, by using a rigid present-day outline for Antarctica, result in an overlap of the Antarctic Peninsula onto the Falkland Plateau. Although reconstructions avoiding this problem are also possible using a rigid Antarctica but rotating it so that the peninsula may be brought to lie along the Pacific margin of the southern Andes (Harrison et al., 1979), such a solution introduces a number of geological complications including the doubling-up of arc terrains (Dalziel et al., 1980). The solution of the Antarctic Peninsula overlap problem stands out as one of the most fundamental to be solved in Gondwana geology.

Permian and Triassic sedimentary rocks of the Beacon Supergroup in the Transantarctic Mountains accumulated in an elongate foreland basin along the margin of the Antarctic craton (Elliot, 1975a). The Early to Middle Triassic Fremouw Formation represents a thick wedge of alluvial elastic sediments derived from both sides of the basin (Collinson et al., 1981). A minor contribution of plutonic and metamorphic material was introduced from the craton but the greater part is from a sedimentary and calc-alkaline volcanic terrain (Vavra et al., 1981). The presence of both primary and reworked volcanic detritus in the Fremouw, Falla and Prebble formations suggests that the volcanic source area was close to the basin and much of the Prebble Formation in the Beardmore Glacier area may be locally derived; this suggestion is supported by the presence of a volcanic neck cutting the Falla and Prebble formations (Barrett and Elliot, 1972). However, the principal source area is thought to be the Gondwanian Orogen of Lesser Antarctica (Barrett and Elliot, 1972; Vavra et al., 1981). Current direction data for Triassic rocks in the Transantarctic Mountains (Fig. 2; Elliot, 1975a) indicate that sediment was distributed northwards along the 1800 km basin between Nilsen Plateau and Victoria Land (Nilsen–Mackay Basin) by a river which was probably one of the largest drainage systems in Gondwana (Collinson et al., 1981).

The late-Palaeozoic–Triassic Trinity Peninsula Group and equivalent strata in the Antarctic Peninsula are coeval with at least part of the upper Beacon Supergroup. They were considered by Dalziel (1982) to be parts of an arc–trench gap sequence, and a similar interpretation has been made for the LeMay

Formation of Alexander Island (Smellie, 1981). The turbidites of the Trinity Peninsula Group were derived from a source terrain of mainly acid plutonic, hypabyssal and volcanic rocks with lesser amounts of metamorphic and sedimentary rocks (Hyden and Tanner, 1981). This source area may be a lateral equivalent of the same geanticline (Gondwanian Orogen) that bounded the Pacific-ward margin of the Nilsen–Mackay Basin (above). However, until the problem of the position of the Antarctic Peninsula within Gondwana has been solved the geographical relationships between the Triassic strata of the peninsula and the Transantarctic Mountains will remain uncertain.

Deformation of the trench-accreted assemblages is generally ascribed to the effects of the Gondwanian Orogeny. However, this would have been a more or less continuous process related to subduction, and thus the definition of this as a mainly Early Jurassic event in this paper needs qualification. Elucidating the tectonic history of these rocks is likely to be a lengthy process because of the difficulties of determining the precise ages of the rock units under investigation. Nevertheless, the youngest parts are Middle to Late Triassic in age and the sequence as a whole is overlain with strong unconformity by rocks which appear to be no older than Middle Jurassic. Thus it is possible that during the Early Jurassic the tectonic events reached a climax when the Trinity Peninsula Formation and its equivalents were lifted above sea-level to form much of the Jurassic land surface in the peninsula area.

The Ferrar dolerites are probably related to the initial break-up of Gondwana (Elliot, 1975b; Kyle et al., 1981), and thus suggest an Early to Middle Jurassic timing for the Africa–Antarctica split (cf. Bergh, 1977; Simpson et al., 1979) and perhaps also for some separation of the fragments of Lesser Antarctica from Greater Antarctica. A possible abortive rifting event in the Triassic has been identified by Furnes and Mitchell (1978).

Attempts have been made to assess the Mesozoic palaeogeographical setting of the Antarctic Peninsula (Suárez, 1976; Thomson, 1982), although similar exercises for the other continental fragments of Lesser Antarctica are more difficult because of the almost complete absence of exposures in stratified rocks of this age in those areas. From at least Middle Jurassic to Early Cretaceous times the peninsula was an elongate volcanic landmass or magmatic arc with

area was certainly land by that time. The Fossil Bluff Formation underwent deformation and presumably uplift probably at some time in the Late Cretaceous, although this was not nearly as intense as that suffered by the Latady Formation.

By Late Cretaceous times the fragmentation of Gondwana was well under way and broad seaways were opening up between fragments of the old supercontinent, with the exception of Antarctica—Australia and South America—Antarctic Peninsula which remained attached until Tertiary times. Greater Antarctica was becoming a continent in its own right, being steadily freed from the influences of surrounding lands, but Lesser Antarctica retained its status as an active margin late into Tertiary time. Subduction has now ceased but was active on the northern side of the South Shetland Islands until Late Tertiary (Pliocene) times.

ACKNOWLEDGEMENTS

Any review paper relies heavily on the work of numerous people who have made their individual contributions in the past, and it is influenced not only by those papers referred to in the text but also by others which have gone unrecognized. I wish to express my thanks to the international body of Antarctic geologists who so painstakingly gathered the information so inadequately summarized here. In addition I am indebted to a large number of geologists within the British Antarctic Survey and from many other Antarctic institutes throughout the world who have given me the benefit of their knowledge in fruitful discussions. I am also grateful to Ms. G.M. Tew for typing the manuscript and references without complaint, and to my wife who helped me with the drafting of the diagrams.

This contribution was originally completed in July 1979, and was updated in July 1981.

REFERENCES

Adie, R.J., 1955. The petrology of Graham Land, II. The Andean granite—gabbro intrusive suite. *Sci. Rep. Falkld. Isl. Depend. Surv.*, 12: 39 pp.

Adie, R.J., 1957. The petrology of Graham Land, III. Metamorphic rocks of the Trinity Peninsula Series. *Sci. Rep. Falkld. Isl. Depend. Surv.*, 20: 26 pp.

Adie, R.J., 1958. Geological investigations in the Falkland Islands Dependencies since 1940. *Polar Rec.*, 9: 3—17.

Adie, R.J., 1962. The geology of Antarctica. In: H. Wexler, M.J. Rubin, and J.E. Caskey (Editors), *Antarctic Research: the Matthew Fontaine Maury Memorial Symposium*. Am. Geophys. Union, Washington, D.C., pp. 26—39.

Adie, R.J., 1964. Geological history. In: R. Priestley, R.J. Adie and G. de Q. Robin (Editors), *Antarctic Research*. Butterworth, London, pp. 118—162.

Adie, R.J., 1972a. Evolution of volcanism in the Antarctic Peninsula. In: R.J. Adie (Editor), *Antarctic Geology and Geophysics*. Universitetsforlaget, Oslo, pp. 137—141.

Adie, R.J., 1972b. Recent advances in the geology of the Antarctic Peninsula. In: R.J. Adie (Editor), *Antarctic Geology and Geophysics*. Universitetsforlaget, Oslo, pp. 121—124.

Aitkenhead, N., 1975. The geology of the Duse Bay—Larsen Inlet area, north-east Graham Land (with particular reference to the Trinity Peninsula Series). *Sci. Rep. Br. Antarct. Surv.*, 51: 62 pp.

Alarcón, B., Ambrus, J., Olcay, L. and Vieira, C., 1976. Geología del Estrecho de Gerlache entre los paralelos 64° y 65° lat. sur, Antártica chilena. *Ser. Cient. Inst. Antárt. Chilena*, 4: 4—51.

Aucamp, A.P.H., Wolmarans, L.G. and Neethling, D.C., 1972. The Urfjell Group, a deformed (?) early Palaeozoic sedimentary sequence, western Dronning Maud Land. In: R.J. Adie (Editor), *Antarctic Geology and Geophysics*. Universitetsforlaget, Oslo, pp. 557—562.

Ballance, P.F. and Watters, W.A., 1971. The Mawson Diamictite and the Carapace Sandstone formations of the Ferrar Group at Allan Hills and Carapace Nunatak, Victoria Land, Antarctica. *N.Z. J. Geol. Geophys.*, 14: 512—527.

Barker, P.E. and Griffiths, D.H., 1977. Towards a more certain reconstruction of Gondwanaland. *Phil. Trans. R. Soc. Lond.*, B, 279: 143—159.

Barrett, P.J., 1969. Stratigraphy and petrology of the mainly fluviatile Permian and Triassic Beacon rocks, Beardmore Glacier area, Antarctica. *Rep. Inst. Polar Stud. Ohio State Univ.*, 34: 132 pp.

Barrett, P.J., 1970. Paleocurrent analysis of the mainly fluviatile Permian and Triassic Beacon rocks, Beardmore Glacier area, Antarctica. *J. Sediment. Petrol.*, 40: 395—411.

Barrett, P.J., 1972. Stratigraphy and petrology of the mainly fluviatile Permian and Triassic part of the Beacon Supergroup, Beardmore Glacier area. In: Adie, R.J., (Editor), *Antarctic Geology and Geophysics*. Universitetsforlaget, Oslo, pp. 365—372.

Barrett, P.J. and Elliot, D.H., 1972. The early Mesozoic volcaniclastic Prebble Formation, Beardmore Glacier area. In: R.J. Adie (Editor), *Antarctic Geology and Geophysics*. Universitetsforlaget, Oslo, pp. 403—409.

Barrett, P.J., Kohn, B.P., Askin, R.A. and McPherson, J.G., 1971. Preliminary report on Beacon Supergroup studies between the Hatherton and Mackay Glaciers, Antarctica. *N.Z. J. Geol. Geophys.*, 14: 605—614.

Barrett, P.J. and Kohn, B.P., 1975. Changing sediment transport directions from Devonian to Triassic in the Beacon

Super-Group of south Victoria Land, Antarctica. In: S.K.W. Campbell (Editor), Gondwana Geology. Australian National University Press, Canberra, A.C.T., pp. 15–35.

Behrendt, J.C., Drewry, D.J., Jankowski, E. and Grim, M.S., 1981. Aeromagnetic and radio echo ice-sounding measurements over the Dufek intrusion, Antarctica. J. Geophys. Res., 86 (B4): 3014–3020.

Bell, C.M., 1975. Structural geology of parts of Alexander Island. Bull. Br. Antarct. Surv., 41, 42: 43–58.

Bell, C.M., Mair, B.F. and Storey, B.C., 1977. The geology of part of an island arc–marginal basin system in southern South Georgia. Bull. Br. Antarct. Surv., 46: 109–127.

Bergh, H., 1977. Mesozoic sea floor off Dronning Maud Land, Antarctica. Nature, Lond., 267: 686–687.

Bibby, J.S., 1966. The stratigraphy of part of northeast Graham Land and the James Ross Island group. Bull. Br. Antarct. Surv., 53: 37 pp.

Bushnell, V.C. and Craddock, C. (Editors), 1969–70. Geologic maps of Antarctica. Antarct. Map Folio Ser., 12.

Carpenter, F.M., 1969. Fossil insects from Antarctica. Psyche, 76: 418–425.

Clarkson, P.D., 1981. Geology of the Shackleton Range, IV. The dolerite dykes. Bull. Br. Antarct. Surv., 53: 201–212.

Clarkson, P.D. and Brook, M., 1977. Age and position of the Ellsworth Mountains crustal fragment, Antarctica. Nature, Lond., 265: 615–616.

Colbert, E.H., 1973. Continental drift and the distributions of fossil reptiles. In: D.H. Tarling and S.K. Runcorn (Editors), Implications of Continental Drift to the Earth Sciences. Academic Press, London, New York, pp. 395–412.

Colbert, E.H., 1974. Lystrosaurus from Antarctica. Am. Mus. Novitates, 2535, 44 pp.

Colbert, E.H., 1975. Early Triassic tetrapods and Gondwanaland. Mem. Mus. Natl. Hist. Nat. Paris, N.S., Sér. A, Zoologie, 88: 202–215.

Colbert, E.H., and Cosgriff, J.W., 1974. Labyrinthodont amphibians from Antarctica. Am. Mus. Novitates, 2552: 30 pp.

Colbert, E.H. and Kitching, J.W., 1975. The Triassic reptile Procolophon in Antarctica. Am. Mus. Novitates, 2566: 23 pp.

Collinson, J.W., Stanley, K.O. and Vavra, C.L., 1978. Stratigraphy and sedimentary petrology of the Fremouw Formation (Lower Triassic), Cumulus Hills, central Transantarctic Mountains. Antarct. J. U.S., 13 (4): 21–22.

Collinson, J.W., Stanley, K.O. and Vavra, C.L., 1981. Triassic fluvial depositional systems in the Fremouw Formation, Cumulus Hills, Antarctica. In: M.M. Cresswell and P. Vella (Editors), Gondwana Five. Balkema, Rotterdam, pp. 141–148.

Compston, W., McDougall, I. and Heier, K.S., 1968. Geochemical comparison of the Mesozoic basaltic rocks of Antarctica, South Africa, South America and Tasmania. Geochim. Cosmochim. Acta, 32: 129–149.

Cosgriff, J.W., Hammer, W.R., Zawiskie, J.M. and Kemp, N.R., 1978. New Triassic vertebrates from the Fremouw Formation of Queen Maud Mountains. Antarct. J. U.S., 13 (4): 23–24.

Covacevich, V., 1976. Fauna valanginiana de Peninsula Byers, Isla Livingston, Antártica. Rev. Geol. Chile, 3: 25–56.

Craddock, C., 1970a. Radiometric age map of Antarctica. In: V.C. Bushnell and C. Craddock (Editors), Geologic Maps of Antarctica. Antarct. Map Folio Ser., 12, pl. XIX.

Craddock, C., 1970b. Tectonic map of Antarctica. In: V.C. Bushnell and C. Craddock (Editors), Geologic Maps of Antarctica. Antarct. Map Folio Ser., 12, pl. XXI.

Craddock, C., 1972. Geologic Map of Antarctica, 1:5,000,000. American Geographical Society, New York, N.Y.

Crame, J.A., 1980. The occurrence of the bivalve Inoceramus concentricus on Dundee Island. Bull. Br. Antarct. Surv., 49: 283–286.

Crame, J.A., 1981. Upper Cretaceous inoceramids (Bivalvia) from the James Ross Island group and their stratigraphical significance. Bull. Br. Antarct. Surv., 53: 25–56.

Dalziel, I.W.D., 1969. Structural studies in the Scotia Arc: Livingston Island. Antarct. J. U.S., 4: 137.

Dalziel, I.W.D., 1975. The Scotia Arc tectonics project. 1969–74, summary. Antarct. J. U.S., 10, 79–81.

Dalziel, I.W.D., 1982. The pre-Jurassic history of the Scotia Arc: a review and progress report. In: C. Craddock (Editor), Antarctic Geoscience. University of Wisconsin Press, Madison. In press.

Dalziel, I.W.D. and Elliot, D.H., 1971. Evolution of the Scotia Arc. Nature, Lond., 233: 246–252.

Dalziel, I.W.D. and Elliot, D.H., 1973. The Scotia Arc and Antarctic margin. In: F.G. Stehli and A.E.M. Nairn (Editors), The Ocean Basins and Margins, I. The South Atlantic. Plenum Press, New York, N.Y., pp. 171–245.

Dalziel, I.W.D., and Godoy, E. and Mpodozis, C., 1980. Comments on 'Mesozoic evolution of the Antarctic Peninsula and the southern Andes'. Geology, 8: 260–262.

Dalziel, I.W.D., Elliot, D.H., Jones, D.L., Thomson, J.W., Thomson, M.R.A., Wells, N.A., and Zinsmeister, J., 1981. The geological significance of some Triassic microfossils from the South Orkney Islands, Scotia Ridge. Geol. Mag, 118: 15–25.

Dewar, G.J., 1970. The geology of Adelaide Island. Sci. Rep. Br. Antarct. Surv., 57: 66 pp.

De Wit, M.J., 1977. Evolution of the Scotia Arc: key to the reconstruction of S.W. Gondwanaland. Tectonophysics, 37: 53–81.

Domack, E.W., Fairchild, W.W. and Anderson, J.B., 1980. Lower Cretaceous sediment from the East Antarctic continental shelf. Nature, Lond., 287: 625–626.

Dow, J.A.S. and Neall, V.E., 1973. Geology of the Lower Rennick Glacier, northern Victoria Land, Antarctica. N.Z. J. Geol. Geophys., 17: 659–714.

Edwards, C.W., 1980a. Early Mesozoic marine fossils from central Alexander Island. Bull. Br. Antarct. Surv., 49: 33–58.

Edwards, C.W., 1980b. New evidence of major faulting on Alexander Island. Bull. Br. Antarct. Surv., 49: 15–20.

Edwards, C.W.E., 1982. Further paleontological evidence of Triassic sedimentation in western Antarctica. In: C. Craddock (Editor), Antarctic Geoscience. University of Wisconsin Press, Madison. In press.

Elliot, D.H., 1966. Geology of Nordenskjöld Coast and a comparison with northwest Trinity Peninsula, Graham Land. *Bull. Br. Antarct. Surv.*, 10: 1–43.

Elliot, D.H., 1970. Jurassic tholeiites of the central Transantarctic Mountains. In: E.H. Gilmour and D. Standing (Editors), *Proceedings of the Second Columbia River Basalt Symposium*. Eastern Washington State College Press, Cheyney, Wash., pp. 301–325.

Elliot, D.H., 1972a. Aspects of Antarctic geology and drift reconstructions. In: R.J. Adie (Editor), *Antarctic Geology and Geophysics*. Universitetsforlaget, Oslo, pp. 849–858

Elliot, D.H., 1972b. Major oxide chemistry of the Kirkpatrick Basalt, central Transantarctic Mountains. In: R.J. Adie (Editor), *Antarctic Geology and Geophysics*. Universitetsforlaget, Oslo, pp. 413–418.

Elliot, D.H., 1975a. Gondwana basins of Antarctica. In: S.K.W. Campbell (Editor), *Gondwana Geology*. Australian National University Press, Canberra, A.C.T., pp. 496–536.

Elliot, D.H., 1975b. Tectonics of Antarctica. A review. *Am. J. Sci.*, 275-A: 45–106.

Elliot, D.H. and Tasch, P., 1967. Lioestheriid conchostrachans: a new Jurassic locality and regional and Gondwana correlations. *J. Paleontol.*, 41: 1561–1563.

Elliot, D.H. and Trautman, T.A., 1982. Lower Tertiary strata on Seymour Island, Antarctica Peninsula. In: C. Craddock (Editor), *Antarctic Geoscience*. University of Wisconsin Press, Madison. In press.

Elliot, D.H. and Wells, N.A., 1982. Mesozoic alluvial fans of the South Orkney Islands. In: C. Craddock (Editor), *Antarctic Geoscience*. University of Wisconsin Press, Madison. In press.

Elliot, D.H., Colbert, E.H., Breed, W.J., Jensen, J.A. and Powell, J.S., 1970. Triassic tetrapods from Antarctica: evidence for continental drift. *Science*, 169: 1197–1201.

Elliot, D.H., Watts, D.R., Alley, R.B. and Gracanin, T.M., 1978. Geologic studies in the northern Antarctic Peninsula, R/V *Hero* cruise 78-1B, February 1978. *Antarct. J. U.S.*, 13 (4): 12–13.

Elliott, M.H., 1974. Stratigraphy and sedimentary petrology of the Ablation Point area, Alexander Island. *Bull. Br. Antarct. Surv.*, 39: 87–113.

Farrar, E., McBride, S.L. and Rowley, P.D., 1982. Ages and tectonic implications of Andean plutonism in the southern Antarctic Peninsula. In: C. Craddock (Editor), *Antarctic Geoscience*. University of Wisconsin Press, Madison. In press.

Faure, G. and Elliot, D.H., 1971. Isotope composition of strontium in Mesozoic basalt and dolerite from Dronning Maud Land. *Bull. Br. Antarct. Surv.*, 25: 23–27.

Faure, G., Hill, R.L., Jones, L.M. and Elliot, D.H., 1972. Isotope composition of strontium and silica content of Mesozoic basalt and dolerite from Antarctica. In: R.J. Adie (Editor), *Antarctic Geology and Geophysics*. Universitetsforlaget, Oslo, pp. 617–624.

Faure, G., Bowman, J.R., Elliot, D.H. and Jones, L.M., 1974. Strontium isotope composition and petrogenesis of the Kirkpatrick Basalt, Queen Alexandra Range, Antarctica. *Contrib. Miner. Petrol.*, 48: 153–169.

Faure, G., Pace, K.K. and Elliot, D.H., 1982. Systematic variations of $^{87}Sr/^{86}Sr$ ratios and major element concentrations in the Kirkpatrick Basalt of Mt. Falla, Queen Alexandra Range, Transantarctic Mountains. In: C. Craddock (Editor), *Antarctic Geoscience*, University of Wisconsin Press, Madison. In press.

Fleet, M., 1968. The geology of the Oscar II Coast, Graham Land. *Sci. Rep. Br. Antarct. Surv.*, 59: 46 pp.

Fleming, E.A. and Thomson, J.W. (Compilers), 1979. *British Antarctic Territory Geological Map, 1:500,000. Sheet 2. Northern Graham Land and South Shetland Islands*. British Antarctic Survey, Cambridge.

Ford, A.B., 1972. Weddell Orogeny – Latest Permian to early Mesozoic deformation at the Weddell Sea margin of the Transantarctic Mountains. In: R.J. Adie (Editor), *Antarctic Geology and Geophysics*. Universitetsforlaget, Oslo, pp. 419–425.

Ford, A.B., 1976. Stratigraphy of the layered gabbroic Dufek intrusion, Antarctica. *Bull. U.S. Geol. Surv.*, 1405-D: 36 pp.

Ford, A.B. and Kistler, R.W., 1980. K-Ar age, composition and origin of Mesozoic mafic rocks related to Ferrar Group, Pensacola Mountains, Antarctica. *N.Z. J. Geol. Geophys.*, 23: 371–390.

Fuenzalida, H., Araya, R. and Hervé, F., 1972. Middle Jurassic flora from northeastern Snow Island, South Shetland Islands. In: R.J. Adie (Editor), *Antarctic Geology and Geophysics*. Universitetsforlaget, Oslo, pp. 93–97.

Furnes, H. and Mitchell, J.G., 1978. Age relationships of Mesozoic basalt lava and dykes in Vestfjella, Dronning Maud Land, Antarctica. *Norsk Polarinst. Skr.*, 169: 45–68.

Gair, H.S., 1967. The geology from the Upper Rennick Glacier to the coast, northern Victoria Land, Antarctica. *N.Z. J. Geol. Geophys.*, 10: 309–344.

Gledhill, A., Rex, D.C. and Tanner, P.W.G., 1982. Rb-Sr and K-Ar geochronology of rocks from the Antarctic Peninsula between Anvers Island and Marguerite Bay. In: C. Craddock (Editor), *Antarctic Geoscience*. University of Wisconsin Press, Madison. In press.

Grapes, R.H. and Reid, D.L., 1971. Rhythmic layering and multiple intrusion in the Ferrar Dolerite of south Victoria Land, Antarctica. *N.Z. J. Geol. Geophys.*, 14: 600–604.

Grapes, R.H., Reid, D.L. and McPherson, J.G., 1974. Shallow dolerite intrusion and phreatic eruption in the Allan Hills region, Antarctica. *N.Z. J. Geol. Geophys.*, 17: 563–577.

Grikurov, G.E., 1973. *Geologiya Antarkticheskogo Poluostrova*. Izdatelstvo 'Nauka', Moscow. (English translation: *Geology of the Antarctic Peninsula*. Amerind Publishing Co. Pvt. Ltd., New Delhi, for National Science Foundation, Washington, D.C., 1978.)

Grikurov, G.E., Krylov, A.Ya., Polyakov, M.M. and Tsovbun, Ya.N., 1970. Vozrast porod v severnoi chasti Antarkticheskogo poluostrova i na Yuzhnikh Shetlandskikh ostrovakh (po dannim Kalli-argonovogo metoda). *Byull. Sovetsk. Antarkt. Eksped.*, 80: 30–33.

Grindley, G.W., 1963. The geology of the Queen Alexandra Range, Beardmore Glacier, Ross Dependency, Antarctica; with notes on the correlation of Gondwana sequences. *N.Z. J. Geol. Geophys.*, 6: 307–347.

Grindley, G.W. and Mildenhall, D.C., 1981. Geological background to a Devonian plant fossil discovery, Ruppert Coast, Marie Byrd Land, West Antarctica. In: M.M. Cresswell and P. Vella (Editors), Gondwana Five. Balkema, Rotterdam, pp. 23–30.

Grindley, G.W., McGregor, V.R. and Walcott, R.I., 1964. Outline of the geology of the Nimrod–Beardmore–Axel Heiberg Glaciers region, Ross Dependency. In: R.J. Adie (Editor), Antarctic Geology. North-Holland Publishing Company, Amsterdam, pp. 206–219.

Gunn, B.M., 1962. Differentiation in Ferrar dolerites, Antarctica. N.Z. J. Geol. Geophys., 5: 820–863.

Gunn, B.M. and Warren, G., 1962. Geology 4. Geology of Victoria Land between the Mawson and Mulock Glaciers, Antarctica. Sci. Rep. Trans-Antarct. Exped., 11: 157 pp.

Hall, B.A. and Borns, H.W., 1972. Jurassic geology of the Allan Battlements–Carapace Nunataks area, Victoria Land. (Abstract only). In: R.J. Adie (Editor), Antarctic Geology and Geophysics. Universitetsforlaget, Oslo, p. 411.

Hall, B.A., Sutter, J.F., and Borns, H.W. Jr., 1982. The inception and duration of Mesozoic volcanism in the Allan Hills–Carapace Nunatak area, Victoria Land, Antarctica. In: C. Craddock (Editor), Antarctic Geoscience. University of Wisconsin Press, Madison. In press.

Halle, T.G., 1913. The Mesozoic flora of Graham Land. Wiss. Ergebn. Schwed. Südpolarexped., 3 (14): 123 pp.

Halpern, M., 1965. The geology of the General Bernado O'Higgins area, northwest Antarctic Peninsula. In: J.B. Hadley (Editor), Geology and Paleontology of the Antarctic. Antarct. Res. Ser., 6. American Geophysical Union, Washington, pp. 177–209.

Halpern, M., 1971. Evidence for Gondwanaland from a review of West Antarctic radiometric ages. In: L.O. Quam (Editor), Research in the Antarctic. American Association for the Advancement of Science, Washington, pp. 717–730.

Halpern, M., 1972. Rb-Sr total rock and mineral ages for the Marguerite Bay area, Kohler Range and Fosdick Mountains. In: R.J. Adie (Editor), Antarctic Geology and Geophysics. Universitetsforlaget, Oslo, pp. 197–204.

Hamilton, W., 1964. Diabase sheets differentiated by liquid fractionation, Taylor Glacier region, south Victoria Land. In: R.J. Adie (Editor), Antarctic Geology. North Holland Publishing Company, Amsterdam, pp. 442–454.

Hamilton, W., 1965. Diabase sheets of the Taylor Glacier region, Victoria Land, Antarctica. Prof. Pap. U.S. Geol. Surv., 456-B: 71 pp.

Harrington, H.J., 1969. Fossiliferous rocks in moraine at Minna Bluff, McMurdo Sound. Antarct. J. U.S., 4: 134–135.

Harrison, C.G.A., Barron, E.J. and Hay, W.W., 1979. Mesozoic evolution of the Antarctic Peninsula and the southern Andes. Geology, 7: 374–378.

Hernández, P.J. and Azcárate, V., 1971. Estudio paleobotánico preliminar sobre restos de una tafoflora de la Península Byers (Cerro Negro), Isla Livingston; Isla Shetland del Sur, Antártica. Ser. Cient. Inst. Antárt. Chilena, 2: 15–50.

Hertlein, L.G., 1969. Fossiliferous boulders of early Tertiary age from Ross Island, Antarctica. Antarct. J. U.S., 4: 315–321.

Hobbs, G.J., 1968. The geology of the South Shetland Islands: IV. The geology of Livingston Island. Sci. Rep. Br. Antarct. Surv., 47: 34 pp.

Horne, R.R., 1969. Sedimentology and palaeogeography of the Lower Cretaceous depositional trough of south-eastern Alexander Island. Bull. Br. Antarct. Surv., 22: 61–76.

Horne, R.R. and Thomson, M.R.A., 1972. Airborne and detrital volcanic material in the Lower Cretaceous sediments of south-eastern Alexander Island. Bull. Br. Antarct. Surv., 29: 103–111.

Howarth, M.K., 1958. Upper Jurassic and Cretaceous ammonite faunas of Alexander Land and Graham Land. Sci. Rep. Falkld. Isl. Depend. Surv., 21: 16 pp.

Howarth, M.K., 1966. Ammonites from the Upper Cretaceous of the James Ross Island group. Bull. Br. Antarct. Surv., 10: 55–69.

Hyden, G. and Tanner, P.W.G., 1981. Late Palaeozoic–Early Mesozoic fore-arc basin sedimentary rocks at the Pacific margin in western Antarctica. Geol. Rundsch., 70: 529–541.

Juckes, L.J., 1968. The geology of Mannefallknausane and part of Vestfjella, Dronning Maud Land. Bull. Br. Antarct. Surv., 18: 65–78.

Juckes, L.J., 1972. The geology of northeastern Heimefrontfjella, Dronning Maud Land. Sci. Rep. Br. Antarct. Surv., 65: 44 pp.

Kalandadze, N.N., 1975. Pervaya nakhodka listrozavra na territorii yevropeyskoy chasti SSSR. Paleont. Zh, 1975 (4): 140–142. (English translation: The first discovery of Lystrosaurus in the European regions of the USSR. Paleontol. J., 9 (4): 447–460).

Kemp, E.M., 1972. Reworked palynomorphs from the West Ice Shelf area, East Antarctica, and their possible geological and palaeoclimatological significance, Marine Geol., 13: 145–157.

Kitching, J.W., Collinson, J.W., Elliot, D.H. and Colbert, E.H., 1972. Lystrosaurus zone (Triassic) fauna from Antarctica. Science, 175: 524–527.

Kovach, J. and Faure, G., 1978. Rubidium–strontium geochronology of granitic rocks from Mt. Chapman, Whitmore Mountains, West Antarctica. Antarct. J. U.S., 13 (4): 17–18.

Kyle, P.R., 1977. Petrogenesis of Ferrar Group rocks. Antarct. J. U.S., 12: 108–110.

Kyle, P.R., 1979. Geochemical studies of Ferrar Group rocks from southern Victoria Land. Antarct. J. U.S., 14 (5): 25–26.

Kyle, P.R., Elliot, D.H. and Sutter, J.F., 1981. Jurassic Ferrar Supergroup tholeiites from the Transantarctic Mountains, Antarctica, and their relationship to the initial fragmentation of Gondwana. In: M.M. Cresswell and P. Vella (Editors). Gondwana Five. Balkema, Rotterdam, pp. 283–287.

Kyle, R.A., 1977. Palynostratigraphy of the Victoria Group of South Victoria Land, Antarctica. N.Z. J. Geol. Geophys., 20: 1081–1102.

Kyle, R.A. and Schopf, J.M., 1982. Permian and Triassic

palynostratigraphy of the Victoria Group, Transantarctic Mountains. In: C. Craddock (Editor), *Antarctic Geoscience*. University of Wisconsin Press, Madison. In press.

Laudon, T.S., Lackey, L.L., Quilty, P.G. and Otway, P.M., 1969. Geology of eastern Ellsworth Land. (Sheet 3, eastern Ellsworth Land). In: V.C. Bushnell and C. Craddock (Editors), *Geologic Maps of Antarctica. Antarct. Map Folio Ser.*, 12, pl. III.

McDougall, I., 1963. Potassium-argon age measurements on dolerites from Antarctica and South Africa. *J. Geophys. Res.*, 68: 1535–1545.

McGregor, V.R., 1965. Notes on the geology of the area between the heads of the Beardmore and Shackleton Glaciers, Antarctica. *N.Z. J. Geol. Geophys.*, 8: 278–291.

McGregor, V.R. and Wade, F.A., 1969. Geology of western Queen Maud Mountains (Sheet 16, western Queen Maud Mountains). In: V.C. Bushnell and C. Craddock (Editors), *Geologic maps of Antarctica. Antarct. Map Folio Ser.*, 12, pl. XV.

McKelvey, B.C., Webb, P.N., Gorton, M.P. and Kohn, B.P., 1970. Stratigraphy of the Beacon Supergroup between the Olympus and Boomerang Ranges, Victoria Land, Antarctica. *Nature, Lond.*, 227: 1126–1128.

Matz, D.B., Pinet, P.R. and Hayes, M.O., 1972. Stratigraphy and petrology of the Beacon Supergroup, southern Victoria Land. In: R.J. Adie (Editor), *Antarctic Geology and Geophysics*. Universitetsforlaget, Oslo, pp. 353–358.

Metcalfe, A.P., Spörli, K.B. and Craddock, C., 1978. Plutonic rocks from the Ruppert Coast, West Antarctica. *Antarct. J. U.S.*, 13 (4): 5–7.

Munizaga, F., 1972. Rb-Sr isotopic ages of intrusive rocks from Thurston Island and adjacent islands. In: R.J. Adie (Editor), *Antarctic Geology and Geophysics*. Universitetsforlaget, Oslo, pp. 205–206.

Nathan, S. and Schulte, F.J., 1968. Geology and petrology of the Campbell–Aviator divide, northern Victoria Land, Antarctica, 1. Post-Palaeozoic rocks, *N.Z. J. Geol. Geophys.*. 11: 940–975.

Norris, G., 1965. Triassic and Jurassic miospores and acritarchs from the Beacon and Ferrar Groups, Victoria Land, Antarctica. *N.Z. J. Geol. Geophys.*, 8: 236–277.

Orlando, H.A., 1968. A new Triassic flora from Livingston Island, South Shetland Islands. *Bull. Br. Antarct. Surv.*, 16: 1–13.

Pankhurst, R.J., Weaver, S.D., Brook, M. and Saunders, A.D., 1980. K-Ar chronology of Byers Peninsula, Livingston Island, South Shetland Islands. *Bull. Br. Antarct. Surv.*, 49: 277–282.

Pettigrew, T.H., 1981. The geology of Annenkov Island. *Bull. Br. Antarct. Surv.*, 53: 213–254.

Plumstead, E.P., 1962. Geology 2. Fossil floras of Antarctica (with an appendix on Antarctic fossil wood by R. Krausel). *Sci. Rep. Transantarct. Exped.*, 9: 154 pp.

Quilty, P.G., 1970. Jurassic ammonites from Ellsworth Land, Antarctica. *J. Paleontol.*, 44: 110–116.

Quilty, P.G., 1977. Late Jurassic bivalves from Ellsworth Land, Antarctica: their systematics and paleogeographic implications. *N.Z. J. Geol. Geophys.*, 20: 1033–1080.

Ramos, A.M., Medina, F.A., Macchiavello, J.C.A.M. and del

Valle, R.A., 1982. Informe preliminar sobre las sedimentitas del Cretácico medio de Cabo Welchness, Isla Dundee, Antártica. *Contrnes Inst. Antart. Argent.*, 249. In press.

Ravich, M.G. and Grikurov, G.E. (Editors), 1978. *Geologicheskaya Karta Antarktidi 1:5,000,000*. Research Institute of the geology of the Arctic, Leningrad.

Rex, D.C., 1967. Age of a dolerite from Dronning Maud Land. *Bull. Br. Antarct. Surv.*, 11: 101.

Rex, D.C., 1972. K-Ar determinations on volcanic and associated rocks from the Antarctic Peninsula and Dronning Maud Land. In: R.J. Adie (Editor), *Antarctic Geology and Geophysics*. Universitetsforlaget, Oslo, pp. 113–136.

Rex, D.C., 1976. Geochronology in relation to the stratigraphy of the Antarctic Peninsula. *Bull. Br. Antarct. Surv.*, 43: 49–58.

Ricker, J., 1964. Outline of the geology between Mawson and Priestley glaciers, Victoria Land. In: R.J. Adie (Editor), *Antarctic Geology*. North-Holland Publishing Company, Amsterdam, pp. 265–275.

Rigby, J.F. and Schopf, J.M., 1969. Stratigraphic implications of Antarctic paleobotanical studies. In: A.J. Amos (Editor), *Gondwana Stratigraphy*. UNESCO, Paris, pp. 91–106.

Rinaldi, C.A., Massabie, A., Morelli, J., Rosenman, H.L. and del Valle, R., 1978. Geología de la Isla Vicecomodoro Marambio. *Contrnes Inst. Antárt. Argent.*, 217: 44 pp.

Rowley, P.D., 1978. Geologic studies in Orville Coast and eastern Ellsworth Land, Antarctic Peninsula. *Antarct. J. U.S.*, 13 (4): 7–9.

Rowley, P.D. and Williams, P.L., 1982. Geology of the northern Lassiter Coast and southern Black Coast, Antarctic Peninsula. In: C. Craddock (Editor), *Antarctic Geoscience*. University of Wisconsin Press, Madison. In press.

Rutford, R.H., Craddock, C., White, C.M. and Armstrong, R.L., 1972. Tertiary glaciation in the Jones Mountains. In: R.J. Adie (Editor), *Antarctic Geology and Geophysics*. Universitetsforlaget, Oslo, pp. 239–243.

Saunders, A.D., Weaver, S.D. and Tarney, J., 1982. The pattern of Antarctic Peninsula plutonism. In: C. Craddock (Editor), *Antarctic Geoscience*. University of Wisconsin Press, Madison. In press.

Schaeffer, B., 1972. A Jurassic fish from Antarctica. *Am. Mus. Novitates*, 2495: 1–17.

Schopf, J.M., 1969. Ellsworth Mountains: position in West Antarctica due to sea floor spreading. *Science*, 164: 63–66.

Schopf, J.M., 1973. Plant material from the Miers Bluff Formation of the South Shetland Islands. *Rep. Inst. Polar Stud. Ohio State Univ.*, 45: 45 pp.

Simpson, E.W.S., Sclater, J.G., Parsons, B., Norton, I. and Meinke, L., 1979. Mesozoic magnetic lineations in the Mozambique Basin. *Earth Planet. Sci. Lett.*, 43: 260–264.

Singleton, D.G., 1980. The geology of the central Black Coast, Palmer Land. *Sci. Rep. Br. Antarct. Surv.*, 102: 50 pp.

Skinner, D.N.B. and Ricker, J., 1968. The geology of the region between the Mawson and Priestley Glaciers, north Victoria Land, Antarctica, II. Upper Paleozoic to Quaternary geology. *N.Z. J. Geol. Geophys.*, 11: 1041–1075.

Smellie, J.L., 1980. The geology of Low Island, South Shetland Islands, and Austin Rocks. *Bull. Br. Antarct. Surv.*, 49: 239–257.

Smellie, J.L., 1981. A complete arc-trench system recognized in Gondwana sequences of the Antarctic Peninsula region. *Geol. Mag.*, 118: 139–159.

Smellie, J.L., Davies, R.E.S. and Thomson, M.R.A., 1980. Geology of a Mesozoic intra-arc sequence on Byers Peninsula, Livingston Island, South Shetland Islands. *Bull. Br. Antarct. Surv.*, 50: 55–76.

Soloviev, D.S., 1972. Platform magmatic formations of East Antarctica. In: R.J. Adie (Editor), *Antarctic Geology and Geophysics*. Universitetsforlaget, Oslo, pp. 531–538.

Spath, L.F., 1953. The Upper Cretaceous cephalopod fauna of Graham Land. *Sci. Rep. Falkld. Isl. Depend. Surv.*, 3: 60 pp.

Starik, I.Ye, Krylov, A.Ya, Ravich, M.G. and Silin, Yu I., 1961. The absolute age of East Antarctic rocks. In: J.L. Kulp and F.N. Furness (Editors), *Geochronology of Rock Systems. Ann. N.Y. Acad. Sci.*, 91: 576–582.

Steiger, R.H. and Jäger, E. (Compilers), 1977. Subcommission on geochronology: convention on the use of the decay constants in geo- and cosmochronology. *Earth Planet. Sci. Lett.*, 36: 359–362.

Stephenson, P.J., 1966. Geology 1. Theron Mountains, Shackleton Range and Whichaway Nunataks (with a section on palaeomagnetism of the dolerite intrusions by D.J. Blundell). *Sci. Rep. Transantarct. Exped.*, 1: 79 pp.

Stevens, G.R., 1967. Upper Jurassic fossils from Ellsworth Land, West Antarctica, and notes on the biogeography of the South Pacific region. *N.Z. J. Geol. Geophys.*, 10: 345–393.

Stone, P., 1980. The geology of South Georgia, IV. Barff Peninsula and Royal Bay area. *Sci. Rep. Br. Antarct. Surv.*, 96: 45 pp.

Storey, B.C., Mair, B.F. and Bell, C.M., 1977. The occurrence of Mesozoic oceanic floor and ancient continental crust on South Georgia. *Geol. Mag.*, 114: 203–208.

Sturm, A. and Carryer, S.J., 1970. Geology of the region between the Matusevich and Tucker Glaciers, North Victoria Land, Antarctica. *N.Z. J. Geol. Geophys.*, 13: 408–434.

Suárez, M., 1976. Plate-tectonic model for southern Antarctic Peninsula and its relation to southern Andes. *Geology*, 4: 211–214.

Tanner, P.W.G., 1982. Geological evolution of South Georgia. In: C. Craddock (Editor), *Antarctic Geoscience*. University of Wisconsin Press, Madison. In press.

Tasch, P., 1979. Crustacean branchiopod distribution and speciation in Mesozoic lakes of the southern continents. In: B. Parker (Editor), *Terrestrial Biology, III. Antarct. Res. Ser., 30*. American Geophysical Union, Washington.

Taylor, B.J., 1982. Sedimentary dykes, pipes and related structures in the Mesozoic sediments of southeastern Alexander Island. *Bull. Br. Antarct. Surv.* In press.

Taylor, B.J., Thomson, M.R.A. and Willey, L.E., 1979. The geology of the Ablation Point-Keystone Cliffs area, Alexander Island. *Sci. Rep. Br. Antarct. Surv.*, 82: 65 pp.

Thomson, J.W., 1974. The geology of the South Orkney Islands, III. Coronation Island. *Sci. Rep. Br. Antarct. Surv.*, 86: 39 pp.

Thomson, J.W. and Harris, J. (Compilers), 1981. *British Antarctic Territory Geological Map, 1:500,000. Sheet 3. Southern Graham Land*. British Antarctic Survey, Cambridge.

Thomson, M.R.A., 1967. A probable Cretaceous invertebrate fauna from Crabeater Point, Bowman Coast, Graham Land. *Bull. Br. Antarct. Surv.*, 14: 1–14.

Thomson, M.R.A., 1972a. New discoveries of fossils in the Upper Jurassic Volcanic Group of Adelaide Island. *Bull. Br. Antarct. Surv.*, 30: 95–101.

Thomson, M.R.A., 1972b. Ammonite faunas of south-eastern Alexander Island and their stratigraphical significance. In: R.J. Adie (Editor), *Antarctic Geology and Geophysics*. Universitetsforlaget, Oslo, pp. 155–160.

Thomson, M.R.A., 1974. Ammonite faunas of the Lower Cretaceous of south-eastern Alexander Island. *Sci. Rep. Br. Antarct. Surv.*, 80: 44 pp.

Thomson, M.R.A., 1975a. Fossils from the South Orkney Islands, II. Matthews Island. *Bull. Br. Antarct. Surv.*, 40: 75–79.

Thomson, M.R.A., 1975b. Upper Jurassic Mollusca from Carse Point, Palmer Land. *Bull. Br. Antarct. Surv.*, 41, 42: 31–42.

Thomson, M.R.A., 1975c. New palaeontological and lithological observations on the Legoupil Formation, north-west Antarctic Peninsula. *Bull. Br. Antarct. Surv.*, 41, 42: 169–185.

Thomson, M.R.A., 1977. An annotated bibliography of the palaeontology of Lesser Antarctica and the Scotia Ridge. *N.Z. J. Geol. Geophys.*, 20: 865–904.

Thomson, M.R.A., 1979. Upper Jurassic and Lower Cretaceous ammonite faunas of the Ablation Point area, Alexander Island. *Sci. Rep. Br. Antarct. Surv.*, 97: 37 pp.

Thomson, M.R.A., 1980. Late Jurassic ammonite faunas from the Latady Formation, Orville Coast. *Antarct. J. U.S.*, 15 (5): 28–30.

Thomson, M.R.A., 1981. Late Mesozoic stratigraphy and invertebrate palaeontology of the South Orkney Islands. *Bull. Br. Antarct. Surv.*, 54: 65–83.

Thomson, M.R.A., 1982. Mesozoic paleogeography of western Antarctica. In: C. Craddock (Editor), *Antarctic Geoscience*. University of Wisconsin Press, Madison. In press.

Thomson, M.R.A. and Willey, L.E., 1972. Upper Jurassic and Lower Cretaceous *Inoceramus* (Bivalvia) from south-east Alexander Island. *Bull. Br. Antarct. Surv.*, 29: 1–19.

Thomson, M.R.A. and Willey, L.E., 1975. Fossils from the South Orkney Islands, I. Coronation Island. *Bull. Br. Antarct. Surv.*, 40: 15–21.

Thomson, M.R.A., Tanner, P.W.G. and Rex, D.C., 1982. Fossil and radiometric evidence for ages of deposition and metamorphism of sedimentary sequences on South Georgia. In: C. Craddock (Editor), *Antarctic Geoscience*. University of Wisconsin Press, Madison. In press.

Townrow, J.A., 1967. Fossil plants from Allan and Carapace nunataks, and from the upper Mill and Shackleton glaciers, Antarctica. *N.Z. J. Geol. Geophys.*, 10: 456–473.

Vavra, C.L., Stanley, K.O. and Collinson, J.W., 1981. Provenance and alteration of Triassic Fremouw Formation, central Transantarctic Mountains. In: M.M. Cresswell and P. Vella (Editors), *Gondwana Five*. Balkema, Rotterdam, pp. 149–153.

Wade, F.A. and Wilbanks, J.R., 1972. Geology of Marie Byrd and Ellsworth lands. In: R.J. Adie (Editor), *Antarctic Geology and Geophysics*. Universitetsforlaget, Oslo, pp. 207–214.

Weaver, S.D., Saunders, A.D. and Tarney, J., 1982. Mesozoic–Cainozoic volcanism in the South Shetland Islands and the Antarctic Peninsula: geochemical nature and plate tectonic significance. In: C. Craddock (Editor). *Antarctic Geoscience*. University of Wisconsin Press, Madison. In press.

Webb, P.N. and Neall, V.E., 1972. Cretaceous Foraminifera in Quaternary deposits from Taylor Valley, Victoria Land. In: R.J. Adie (Editor), *Antarctic Geology and Geophysics*. Universtetsforlaget, Oslo, pp. 653-657.

Wells, N.A., 1982. The alluvial fans and other Late Mesozoic sedimentary deposits of the South Orkney Islands, Antarctica. *Rep. Inst. Polar Stud., Ohio State Univ.* In press.

West, S.M., 1974. The geology of the Danco Coast, Graham Land. *Sci. Rep. Br. Antarct. Surv.*, 84: 58 pp.

Wilckens, O., 1910. Die Anneliden, Bivalven und Gastropoden der antarktischen Kreideformation. *Wiss. Ergebn. Schwed. Südpolarexped.*, 3 (12): 132 pp.

Willey, L.E., 1972. Belemnites from southeastern Alexander Island, I. The occurrence of the family Dimitobelidae in the Lower Cretaceous. *Bull. Br. Antarct. Surv.*, 28: 29–42.

Willey, L.E., 1973. Belemnites from southeastern Alexander Island, II. The occurrence of the family Belemnopseidae in the Upper Jurassic and Lower Cretaceous. *Bull. Br. Antarct. Surv.*, 36: 35–59.

Williams, P.L., Schmidt, D.L., Plummer, C.C. and Brown, L.E., 1972. Geology of the Lassiter Coast area, Antarctic Peninsula: preliminary report. In: R.J. Adie (Editor), *Antarctic Geology and Geophysics*. Universitetsforlaget, Oslo, pp. 143–148.

Wright, N.A. and Williams, P.L., 1974. Mineral resources of Antarctica. *Circ. U.S. Geol. Surv.*, 705: 29 pp.

Young, D.J. and Ryburn, R.J., 1968. The geology of Buckley and Darwin nunataks, Beardmore Glacier, Ross Dependency, Antarctica. *N.Z. J. Geol. Geophys.*, 11: 922–939.

Zinsmeister, W.J., 1979. Biogeographic significance of the Late Mesozoic and Early Tertiary molluscan faunas of Seymour Island (Antarctic Peninsula) to the final breakup of Gondwanaland. In: J. Gray and A.J. Boucot (Editors), *Historical Biogeography, Plate Tectonics and the Changing Environment*. Oregon State University Press, Oregon, pp. 349–355.

AUTHOR INDEX

Batten, D.J., 9, 24, *28*
Baulíes, O.L., *261*
Bauman Jr., C.F., 75, *84*
Bayle, E., 221, 254
Bebout, D.G., 71, *84, 85*
Beck, C., *149*
Beck, C.M., *118*, 139, 140, *145*
Becker, D., *301*
Beckmann, J.P., *113, 146*
Bedi, T.S., 329, *348*
Beets, D.J., 101, 103, 110, 110, *113, 116*, 139, 144, *145*
Behrendsen, O., 250, *254*, 265, *298*
Behrendt, J.C., 397, *417*
Belcher, B., 64–66, 82, 83, *84*
Bell, C.M., 404, 408, *417, 421*
Bell, J.S., 110, *113*, 121, 122, 125, 127, 136, 140, *145*
Bellizzia, A., *118*
Bellizzia, G.A., 110, *113*, 127, 130, 137, 139–141, *145*
Bello Barrádas, A., 72, 75, *85*
Bellon, H., *113*
Berg, E., *148*
Berg, E.L., 70, *84*
Bergh, H., 414, *417*
Bermúdez, P.J., 103, *113*, 139, *145*
Berryhill Jr., H.L., *113*
Bertels, A., 265, 276, 296, *298*
Berthelsen, A., 323, *347*
Bertrand, J., *113*
Bhalla, M.S., 341, *347, 351*
Bhalla, S.N., 316, 319, 327, 333, 335, 341, *347*
Bhatia, M.L., *349*
Bhatia, M.R., 330, *347*
Bhatia, S.B., 330, *347*
Bhatia, S.C., 332, *347*
Bhattacharya, S.C., 315, *347*
Bhimasankaram, V.L.S., 343, *347*
Bibby, J.S., 407, 408, 410, 415, *417*
Biese, W., 70, 223, 228, 252, *254*
Bigarella, J.J., 169, *178*, 297, *298*
Bilgrami, S.A., 367, *373*
Birkelund, T., 9, 24, *28, 29*
Biro, L., 221, 250, *254*
Bisaria, P.C., *351*
Bishop, B.A., 75, *84*
Bisol, D.L., 170, *179*
Biswas, B., 324, 331, 343, 344, *347*
Biswas, S.K., 316, 319, 343, *347*
Bittner, A., 310, *347*
Blanchet, R., *118*
Blanford, H.F., 327, *347*
Blanford, W.T., 325, 335, *347, 349*, 355, *373*
Blasco, G., 205, 208, 209, 250, *254, 302*
Blasco de Nullo, G., 278, 280, 296, *298*
Blesch, R.R., *113*
Blondel, F., 124, *145*
Bloomstein, E., 5, *30*
Bocchino, A., 252, *254*
Bock, W., *113*

Bodenbender, G., 201, *254*, 265, *298*
Boelrijk, N.A.I.M., *118*
Bohnenberger, O.H., *114*
Boiteau, A., 92, 95, *114*
Bolli, H.M., *146*, 362, *373*
Bonaparte, J.F., 183, 189–196, *197*, 252, 253, *254*, 282–284, 296, *298, 303*
Bonarelli, G., 265, 282, *298*
Bonet, F., 76, 77, 81, *84*
Bonetti, M.I.R., 190, 191, 193, 195, *197, 199*, 206–208, 219, 246, 248, *261*
Bordas, A., 193, *197*
Borns, H.W., 398, 400, *419*
Borns Jr., H.W., *419*
Borrello, A.V., 191, 192, 195, *197*, 229–231, 234, *254*, 265, 276, 279, 282, *298*
Böse, E., 68, 69, 71, 78, 79, 82, *84*
Bose, M.N., 335, *347*
Bose, P.N., 325, *347*
Bossi, C., 282, 283, 296, *298*
Bossi, G., 196, 284, *298*
Botero, A.G., 125, *145*
Boulies, O.I., *303*
Bourgois, J., *119*
Bouysse, P., 90
Bowen, J.M., 129, 130, 137, *145*
Bowen, R., 253, *254*, 376, *388*
Bowes, W., *197*
Bowin, C.O., 100, 107, *114, 118*
Bowman, J.R., *418*
Brabb, E.E., 18, *28*
Bracaccini, I.O., 192, *197, 298*, 229, 231, *254*
Brandmayr, J., 278, *298*
Breed, W.J., *418*
Brenner, P.W., *301*
Brideaux, W.W., 1, *28*
Briden, J.C., 6, *30*
Bridges II, L.W., 63, 68, *84*
Brineman, J.H., *119*
Bronnimann, O., *114*
Bronniman, P., *114*, 353, 360, 361, *374*
Brook, M., 391, *417, 420*
Brosgé, W.P., 1, 3, 5, *28, 30, 31*
Brown, C.M., 375, *388, 389*
Brown, J.B., *87*
Brown, J.F., *115*
Brown, J.S., *119*
Brown, L.E., *422*
Brown, P.E., 9, 24, *28*
Brown, P.M., 36, *58*
Brüggen, J., *299*
Bryant, W.R., *118*
Buchan, S.H., 375, *388*
Bucher, W.H., 112, *114*
Buffler, R.T., *58*
Bullard, E., 110, *114*
Bunce, E.T., 111, *114*
Burbank, W.S., *119*

GENERAL INDEX